Discovering Mathematics 2e

A Quantitative Reasoning Approach

Richard N. Aufmann

‹›‹ Cengage

Australia • Brazil • Canada • Mexico • Singapore • United Kingdom • United States

Discovering Mathematics: A Quantitative Reasoning Approach, Second Edition

Richard N. Aufmann

Portfolio Product Director: Rita Lombard

Portfolio Product Manager: Taylor Shenberger

Sr. Learning Designer: Powell Vacha

Sr. Product Assistant: Anna Hamilton

Content Manager: Emma Collins, Lynh Pham

Content Acquisition Analyst: Nichole Nalenz

Production Service: MPS Limited

Compositor: MPS Limited

Text and Cover Designer: Tim Biddick

Cover Image: Stephen Kaps/EyeEm/Getty Images

Previous Edition: © 2019

For product information and technology assistance, contact us at
Cengage Customer & Sales Support, 1-800-354-9706 or support.cengage.com.

For permission to use material from this text or product, submit all requests online at **www.copyright.com.**

Library of Congress Control Number: 2027076345

Student Edition:
ISBN: 978-0-357-76003-1

Loose-leaf Edition:
ISBN: 978-0-357-76004-8

Cengage
5191 Natorp Boulevard
Mason, OH 45040
USA

Cengage is a leading provider of customized learning solutions. Our employees reside in nearly 40 different countries and serve digital learners in 165 countries around the world. Find your local representative at: **www.cengage.com.**

To learn more about Cengage platforms and services, register or access your online learning solution, or purchase materials for your course, visit **www.cengage.com.**

Printed at CLDPC, USA, 08-24

Contents

8 Introduction to Statistics   355

Module 5: Selected Topics   407

9 Voting and Apportionment   409

10 Circuits and Networks   457

11 Geometry   509

Appendices:

Preface

Discovering Mathematics: A Quantitative Reasoning Approach is designed for today's Quantitative Reasoning course. Written to enhance students' quantitative skills, *Discovering Mathematics* explores topics relevant in today's data-driven economy. Students will examine wide-ranging topics applicable to real-life work and personal scenarios. Topics include how to identify and analyze arguments and fallacies, assess whether buying or leasing a car is a better option, understand the power of investing early for retirement, divide an estate into equal shares, understand the use of radioactive elements in medical treatments, evaluate student loan options, analyze data, and many other topics that will expand a student's quantitative skills.

To support the development of these skills, a WebAssign® course, Activities Booklet, Guided Notes, and other support materials are provided, offering a comprehensive learning program that is both approachable and engaging.

Emphasizing What's Important

Student success depends on engaging the student in relevant content that acknowledges the diversity of a student's experience and the individual goals of the student. These key principles are an essential part of a Quantitative Reasoning course. *Discovering Mathematics* fosters, embraces, and emphasizes these principles:

- **Connecting mathematical concepts to student experiences:** Real-life applications are a central theme of the text and were chosen to involve the student in a meaningful way. *Discovering Mathematics* maintains a focus on connecting mathematical concepts both to students' present and potential future experiences.
- **Encouraging a growth mindset:** Mathematics is a process. The entire *Discovering Mathematics* program emphasizes the values of productive struggle and persistence not only in the application of mathematics but to life endeavors as well. Our goal is to ensure the student experience in this course is meaningful and promotes critical thinking.
- **Developing skills beyond the course:** For many students, this may be their last college math course. However, our program seeks to foster practical skills that will travel with students beyond this course into their working lives. By emphasizing problem solving and critical thinking skills while also providing hands-on opportunities to work with data using Excel, students will become more informed consumers of information and improve their quantitative literacy.

Flexible and Adaptable

The guiding principle in the creation of this text has been to provide core content around which the scope of the course can be modified to meet the curricular needs of the department. A broad summary of the topics included in the core content are problem-solving strategies; fair division; exploring logical reasoning and fallacies; and ratio, rates, and percent. The options to seamlessly integrate additional topics into your course include financial decision-making, linear and nonlinear functions, probability, statistics, apportionment and voting, graph theory, and geometry.

We understand the needs for a Quantitative Reasoning course are varied. This text and the accompanying program were developed to support your requirements. They take into consideration multiple classroom models, including lecture, hybrid, fully online, and flipped classrooms.

The Text Pedagogy

Our goal is to provide targeted, scaffolded support in the most meaningful way possible to students. To achieve this, careful thought went into streamlining the text design to be uncluttered, based on student research that underscored the need to ensure the pedagogy was intentional and purposeful. Each of the following features was developed with student needs in mind, ensuring it had a place and a purpose.

Chapter Opener: Organized to give students mathematical context for the material they are about to learn, the Chapter Opener provides an at-a-glance preview of the main ideas and topics in a concise, easy-to-navigate format.

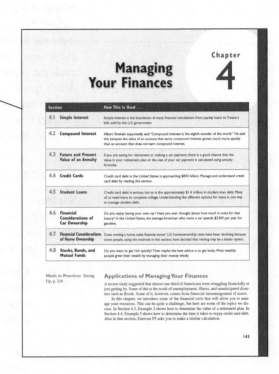

Be Prepared: At the beginning of the first section of every chapter, students are directed to Prep Tests found on the Companion Site, offering students an opportunity to test the algebra skills used in the chapter. Each question is followed by a reference directing the student to a specific resource in an online algebra review appendix so that the student may review the concept used in the question.

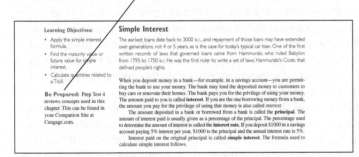

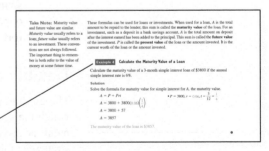

Examples: The real-world examples make up the heart of the text with its emphasis on mathematical context. To connect context to understanding and practice, Preliminary Exercises at the beginning of the exercise set tie directly to the examples from the preceding section. This combination of learning and practice builds students' skills while preparing them for more robust applications of the material found later in the exercise sets.

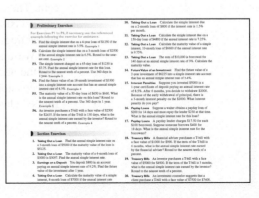

Take Note: Appearing right when they're needed, Take Notes call attention to concepts that require special attention or deeper understanding.

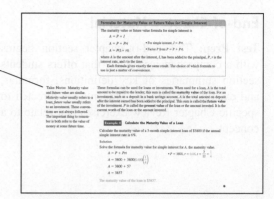

Infographics: Infographics combine a visual break in the narrative with illustrated concepts that are purposeful and informative, presenting data in a familiar style encountered in daily life and the media.

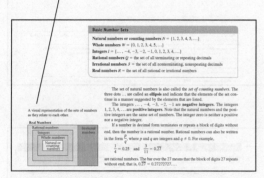

End-of-Section Exercises

Think About It: Beginning with **Think About It** exercises, the end-of-section problems start by assessing a student's understanding of key concepts.

Preliminary Exercises: Preliminary Exercises provide an opportunity to test students' understanding of concepts by working an exercise related to the section example. A complete solution for each Preliminary Exercise can be found in the Solutions to Preliminary Exercises section in the text. This allows students to not only check an answer but to compare their solution to the model solution.

Section Exercises: Following the Preliminary Exercises, **Section Exercises** reimagine traditional skill-building problems as word problems, providing context to the mathematics and allowing students to practice interpreting and extracting data from sentences.

Investigations: Providing opportunities for group work and class discussion, these end-of-section activities also guide students on how to navigate and organize their work. Select projects are coded and assignable in WebAssign.

End-of-Chapter Material

Test Prep: This end-of-chapter section begins with a comprehensive review that offers students section summaries of key definitions, formulas, and example references. Students are directed to specific Chapter Review exercises that relate to a concept.

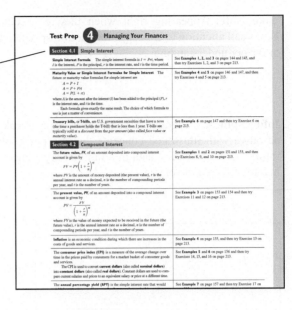

Review Exercises: The Review Exercises provide students an opportunity to practice concepts that cover the breadth and depth of the chapter.

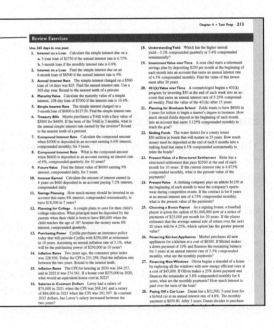

Projects: Similar to the end-of-section Investigations, the end-of-chapter projects provide extended opportunities for group work and class discussion. More in-depth than Investigations, Projects are designed to fit multiple scenarios for use, including in-class collaboration and at-home project work. They were written to provide flexibility. Instructors can assign part of a Project if time is short or have students complete the entire Project if extended work is desired. Select projects are coded and assignable in WebAssign.

Math in Practice: These End of Chapter, multi-part exercises demonstrate "math at work," expanding on chapter learning objectives with real-world, applicable situations students can relate to in their daily lives, or their intended career tracks.

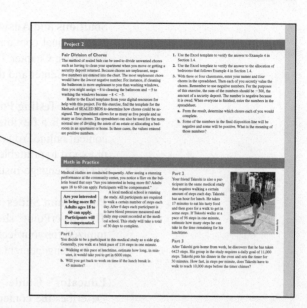

The Development of This Edition

We took several things into account when developing this latest edition. Guiding our work were the Four Core Principles of Learning Design: authenticity, inclusivity, intentionality, and personalization, with a focus on removing barriers to learning to offer a carefully planned and inclusive experience for all students. Our narrative, examples, and exercises were also reviewed extensively throughout the revision process. The DEIB (diversity, equity, inclusion, and belonging) review was an iterative process designed apply our learning design principles and best practices. Cengage is committed to continue this process in future editions. If you or your students have concerns about any of our content, please do not hesitate to reach out to your Cengage Account Executive. We are always open to learning and take every concern seriously.

Lastly, we wanted to ensure the text could be used in a variety of ways. To maintain optimum flexibility, many chapters are sufficiently independent to allow instructors to pick and choose topics that are relevant to their students. Additionally, the accompanying WebAssign course is a critical extension of our approach including new features as coded problem types where possible.

Ancillaries for the Instructor

WebAssign

Cengage WebAssign

Built by educators, WebAssign provides flexible settings at every step to customize your course with online activities and secure testing to meet learners' unique needs. Students get everything in one place, including rich content and study resources designed to fuel deeper understanding, plus access to a dynamic, interactive eTextbook. Proven to help hone problem-solving skills, WebAssign helps you help learners in any course format. For more information, visit https://www.cengage.com/webassign. Additional instructor resources for this product are available online at the Cengage Instructor Center—an all-in-one resource for class preparation, presentation, and testing. Resources available for download from the Cengage Instructor Center include:

Instructor's Manual

Includes activities and assessments correlated by learning objectives, chapter and section outline as well as key formulas and terms with definitions.

Solutions and Answer Guide

This manual contains solutions to all exercises from the text, including Chapter Review Exercises, Chapter Tests, and Cumulative Review Exercises. Located on the Cengage Instructor Center and the WebAssign Resource Tab.

Cengage Testing Powered by Cognero® (978-0-357-76007-9)

Cognero is a flexible online system that allows you to author, edit, and manage test bank content online. You can create multiple tests in an instant and deliver them from your LMS, or export to printable PDF or Word format for in-class assessment. This is available online via the Cengage Instructor Center.

PowerPoint Slides

The PowerPoint® slides are ready to use, visual outlines of each section can be easily customized for your lectures. Presentations include activities, examples, and opportunities for student engagement and interaction.

Educator's Guide

Describes the content and activities available in the accompanying WebAssign course—including videos, prebuilt assignments, and other exercise types—that you can integrate into your course to support your learning outcomes and enhance student engagement and success.

Guided Notes with Answers

New to this edition is a series of chapter-by-chapter guided notes to help your students engage further with the concepts covered in the text. These can be used as is or customized to your specific needs. Available in WebAssign in the Resources tab as well as on the companion website. Sign up or sign in at www.cengage.com to search for and access this textbook and its online resources.

Student Activities Booklet

Developed to prompt meaningful discussion or provide active learning opportunities for a group as well as individual students within or outside of class, the Student Activities Booklet is a key component of the *Discovering Mathematics* program that further builds connections from real-life context to mathematical concepts.

Ancillaries for the Student

WebAssign

Prepare for class with confidence using WebAssign from Cengage. This online learning platform for your course includes an interactive eTextbook that fuels practice, so that you truly absorb what you learn and prepare better for tests. Videos and tutorials walk you through concepts and deliver instant feedback and grading, so you always know where you stand in class. Focus your study time and get extra practice where you need it most. Ask your instructor today how you can get access to WebAssign or learn about self-study options at https://www.cengage.com/webassign.

Student Solutions Manual (978-0-357-76006-2)

Go beyond the answers and see what it takes to understand the concepts and improve your grade! This manual provides worked-out, step-by-step solutions to the odd-numbered problems in the text, giving you the information needed to understand how these problems are solved. Available in WebAssign and on the companion website.

Guided Notes

New to this edition is a series of chapter-by-chapter guided notes to help you further engage with the concepts covered in the text. Available in WebAssign as well as on the companion website.

Additional student resources are available online. Sign up or sign in at www.cengage.com to search for and access this product and its online resources

Acknowledgments

I would like to thank the many instructors who have contributed their time and insights into the development of our quantitative reasoning program, through reviews, interviews, focus groups, and our course redesign events over both editions:

- Susan Addington, *California State University, San Bernardino*
- Froozan Afiat, *College of Southern Nevada*
- John Bennett, *Halifax Community College*
- Michael Bailey, *Brookhaven College*
- Brandon Bartley, *Jefferson Community & Technical College*
- Michele Bilton, *Metropolitan Community College-Blue River*
- Rachel Black, *Central New Mexico Community College*
- Jason Boehm, *Saint Louis Community College, Florissant Valley*
- Martin Bredeck, *Northern Virginia Community College*
- Don Brown, *Middle Georgia State College, Macon*
- Kristein Buddemeyer, *Seminole State College of Florida*
- Helen Burn, *Highline College*
- Israel Castro, *Pasadena City College*
- Suzanne Caulifield, *Cardinal Stritch University*
- Mukesh Chhajer, *Danville Community College*
- Phil Compton, *Arizona Christian University and Glendale CC and High School*
- David Crombecque, *University of Southern California*
- Joyati Debnath, *Winona State University*
- Alok Dhital, *University of New Mexico-Gallup*
- Qiang Dotzel, *University of Missouri-St. Louis*
- John Fay, *Chaffey College*
- Jessica Fowler, *Rowan Cabarrus*
- John Fuhrmann, *Glendale Community College*
- Valdez Gant, *North Lake College*
- Mehrnaz Ghaffarian, *Tarrant County College*
- Carol Gick, *Germanna Community College*
- Carri Hales, *South Dakota State University*
- Kathryn Hernandez, *Lee College*
- Lori Heymans, *Northern Essex Community College*
- Carrie Hibbs, *Germanna Community College*
- Timothy Huber, *University of Texas, Rio Grande Valley*
- Brian Huyvaet, *Rowan Cabarrus*
- Mortaza Jamshidian, *California State University, Fullerton*
- Sue Ann Jones-Dobbyn, *Pellissippi State Technical Community College*
- Mickie Karcher, *Jefferson Community & Technical College*
- Robert Lam, *University of California, Riverside*
- John Landrum, *Tarrant County College*
- Femar Lee, *Honolulu Community*
- Richard Leedy, *Polk State College*
- Shakir Manshad, *Doña Ana Community College*
- Arda Melkonian, *Victor Valley College*
- Jonathan Meshes, *Harper College*
- Charles Moore, *Washington State University*
- Diane Murray, *Glendale Community College*
- Lamies Nazzal, *California State University, San Bernardino*
- Reina Ojiri, *Leeward Community College*
- Diall Ousmane, *Northern Virginia Community College*

- Jenny Paden, *Midland State*
- Suzanne Palermo, *Glendale Community College*
- Leonardo Pinheiro, *Rhode Island College*
- Sarah Reznikoff, *Kansas State University*
- Sabrina Ripp, *Tulsa Community College*
- Chrisine Ritter, *Randolph Community College*
- LeeAnn Roberts, *Georgia Gwinnett College*
- Rebecca Rose, *Northern Essex Community College*
- Elizabeth Russell, *Glendale Community College*
- Paul Runnion, *Missouri University of Science and Technology*
- Imran Shah, *Jefferson College*
- Christopher Shaw, *Columbia College Chicago*
- Alison Sutton, *Austin Community College*
- Constantine Terzopoulos, *Danville Community College*
- Maureen Thayer, *Rowan Cabarrus*
- Theresa Thomas, *Blue Ridge Community College*
- Anil Venkatesh, *Ferris State University*
- John Ward, *Jefferson Community & Technical-Downtown*
- Danny Whited, *Southwest Virginia Community College*
- Judy Williams, *Tidewater Community College*
- Nathan Winkles, *Shelton State Community College*
- Jim Wolper, *Idaho State University*
- Dong Ye, *Middle Tennessee State University*

Additionally, I would like to thank the entire team that has worked on this program: Rita Lombard, Portfolio Product Director; Taylor Shenberger, Portfolio Product Manager Powell Vacha, Sr. Learning Designer; Emma Collins, Content Manager; Danielle Derbenti, freelance Development Editor; Anna Hamilton, Sr. Product Assistant; Ivan Corriher, Technical Content Program Manager, Jennifer Burt, and Erica Pultar, In-house Subject Matter Experts, for their work on the accompanying WebAssign course; and accuracy reviewers and ancillary authors, Jon Booze, Scott Barnett, and Christi Verity.

With thanks—Dick Aufmann

A Letter to Students

Welcome to a new experience in mathematics! *Discovering Mathematics* is a program that provides you with quantitative reasoning skills to enhance your problem-solving skills, interpret data, analyze arguments, and solve relevant, real-world problems. Briefly review the Table of Contents to see all the different ways that quantitative reasoning can enhance skills that are valuable to you whether you start your own business or work for a company. On a personal side, these skills will guide you to be an informed consumer.

Here are some frequently asked questions that we have received while teaching this course. The answers to these may help motivate you to succeed in this course.

Why do I have to take this course? You may have heard that "Math is everywhere." That is probably a slight exaggeration, but math does find its way into many disciplines. There are obvious places like engineering, science, and medicine. There are other disciplines such as business, social science, and political science where math may be less obvious but still essential. Each of these branches of learning relies heavily on statistics, one part of the math curriculum. If you are going to be an artist, writer, or musician, the direct connection to math may be even less obvious. Even so, as art historians who have studied the Mona Lisa have shown, there is a connection to math. Suppose, however, you find these reasons not all that compelling. There is still a reason to have quantitative reasoning skills: You will be a better consumer and able to make better financial choices for you and people important to you. For instance, is it better to buy a car or lease a car? Math can provide an answer.

I find math difficult. Why is that? It is true that some people, even very smart people, find math difficult. Some of this can be traced to previous math experiences. If your basic skills are lacking, it is more difficult to understand the math in a new math course. Some of the difficulty can be attributed to the ideas and concepts in math. They can be quite challenging to learn. Nonetheless, most of us can learn and understand the ideas in a quantitative reasoning course. If you want math to be less difficult, practice. When you have finished practicing, practice some more. Ask an athlete, actor, singer, dancer, artist, doctor, or skateboarder what it takes to become successful and the one common characteristic they all share is that they practiced—a lot. And do not be afraid to ask for help. Your instructor wants you to succeed!

Why is math important? As we mentioned earlier, math is found in many fields of study. There are, however, other reasons to take a math course. Primary among these reasons is to become a better problem solver. Math can help you learn critical-thinking skills. It can help you develop a logical plan to solve a problem. Math can help you see relationships between ideas and identify patterns. Whether you are an entrepreneur or an employee, being a problem solver is primary indicator of success. Studying math can help you be a better problem solver.

What do I need to do to pass this course? The most important thing you must do is to know and understand the requirements outlined by your professor. These requirements are usually given to you in a syllabus. Once you know what is required, you can chart a course of action. Set aside time to study and do homework. If possible, choose your classes so that you have a free hour after your math class. Use this time to review your lecture notes, rework examples given in class, and begin your homework. All of us eventually need help, so know where you can get assistance with this class. This means knowing your professor's office hours. Know the hours of the math help center. If there are online resources available, learn how to access them. And finally, do not get behind. Try to do some math EVERY day, even if it is for only 20 minutes.

Why did you write this book? Simply put, this book was written to help you succeed. Your instructor, the textbook, campus math tutorial centers, online resources: all these resources exist to help you succeed.

Validity Studies

Introduction to Problem Solving

Section	How This Is Used . . .
1.1 Inductive and Deductive Reasoning	Reasoning is a rational way of thinking about something and drawing a conclusion. Drawing a conclusion may be based on observations or it may be based on known facts. These conclusions are then tested to determine whether they are always true and, if not, when they are not.
1.2 Estimation and Graphs	Estimation is a rough calculation used to approximate a true value. If we are driving to work or school, we might estimate the time it will take to get there. A contractor gives an estimate on the cost to build a house and event planners estimate the number of attendees at an event.
1.3 Problem-Solving Strategies	Whether it be the president of a company, an engineer, a physician, or a sales representative, to be successful, that person must have good problem-solving skills. Effective problem solvers use strategies that produce powerful solutions.
1.4 Introduction to Fair Division	A problem frequently encountered in the legal profession is the fair distribution of the assets in an estate. These strategies have applications to other situations as well, such as fairly dividing the rent among several roommates.

Math in Practice: Medical Studies, p. 51

Application of Estimation

To provide a safe environment, city planners need estimates of the number of people who can safely occupy a location where a rally can be held. Exercise 11, page 21, shows one method used to estimate how many people can safely occupy a piece of land.

1.1   Inductive and Deductive Reasoning

Learning Objectives:

- Use inductive reasoning.
- Use a counterexample to show a statement is false.
- Use deductive reasoning.
- Solve a logic problem using an elimination grid.

Be Prepared: Prep Test 1 reviews concepts used in this chapter. This can be found in your Companion Site at Cengage.com

Inductive Reasoning

The type of reasoning that forms a conclusion based on the examination of specific examples is called *inductive reasoning*. The conclusion formed by using inductive reasoning is a **conjecture** or **theory**, an assumption based on the collected evidence.

Famous examples of theories include Einstein's *theory* of general relativity. All evidence appears to verify the theory.

In our legal system, a prosecutor presents collected evidence to a jury, which evaluates it and considers whether a conclusion of guilty is warranted. Unlike the general theory of relativity, there has been no evidence (so far) presented to suggest the theory is wrong. Unfortunately, sometimes the evidence in a trail leads to a wrong conclusion and an innocent person is wrongly convicted.

> **Inductive Reasoning**
>
> **Inductive reasoning** is the process of reaching a general conclusion by examining specific examples.

Example 1   Use Inductive Reasoning to Predict a Number

Use inductive reasoning to predict the next number in each of the following sequences.

a. 3, 6, 9, 12, 15, ?

b. 1, 3, 6, 10, 15, ?

c. 1, 2, 4, 8, 16, ?

Solution

a. Each successive number is 3 larger than the preceding number. Thus, we predict that the next number in the list is 3 larger than 15, which is 18.

b. The first two numbers differ by 2. The second and the third numbers differ by 3. The difference between any two successive numbers always seems to be 1 more than the preceding difference. Because 10 and 15 differ by 5, we predict that the next number in the list will be 6 larger than 15, which is 21.

c. $2 = 1 \cdot 2, 4 = 2 \cdot 2, 8 = 4 \cdot 2, 16 = 8 \cdot 2$. The next number in the list appears to be 2 times the preceding number. Because $16 \cdot 2 = 32$, we predict that the next number in the list is 32.

    Just to emphasize the point about a conjecture from specific examples, consider the list 3, 5, 7, ? Is the next number 9? Could be, because the list appears to be odd numbers. However, the next number could be 11. This would then be a list of the first four odd prime numbers (3, 5, 7, 11). Predicting the next number in a finite list of numbers is not an exact art.

    Inductive reasoning can be applied to figures.

Example 2   Use Inductive Reasoning to Predict a Pattern

Use inductive reasoning to predict the next figure.

a.

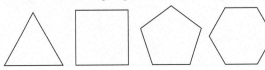

b.

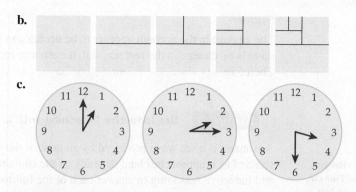

c.

Solution

a. Each figure is a regular polygon (all sides are equal) with the number of sides increasing by 1 for each figure. The next figure would be a regular polygon with seven sides.

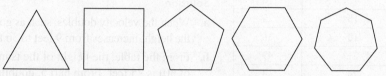

b. Alternating horizontal and vertical lines are used to divide the figure. The next figure would have a horizontal line as shown in the following.

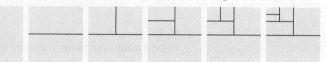

c. Each image shows a clock that is 1 hr and 15 minutes later than the previous image. The next image would be a clock showing a time of 4:45.

Example 3   **Use Inductive Reasoning with a Graph**

The following graph shows the amount of money remaining in a rental budget account at the beginning of each month.

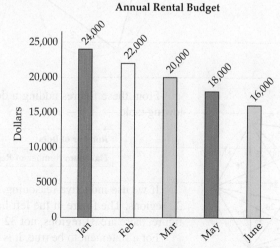

Annual Rental Budget

What amount is remaining at the beginning of August?

Solution

The amount in the account appears to be decreasing by $2000 each month. Assuming there is no change in the rent amount, the amount remaining in July would be $14,000. Therefore, the amount remaining at the beginning of August would be $12,000.

●

Example 4   Use Inductive Reasoning with a Table of Data

A tsunami is a sea wave produced by an underwater earthquake. The height of a tsunami as it approaches land depends on the tsunami's velocity. Use the table at the left and inductive reasoning to answer each of the following questions.

a. What happens to the height of a tsunami when its velocity is doubled?

b. What should be the height of a tsunami if its velocity is 30 feet per second?

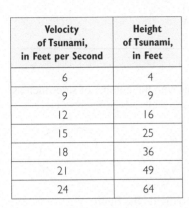

Velocity of Tsunami, in Feet per Second	Height of Tsunami, in Feet
6	4
9	9
12	16
15	25
18	36
21	49
24	64

Solution

a. When the velocity doubles, such as going from 9 feet per second to 18 feet per second, the height increases from 9 feet to 36 feet. That is, the height is quadrupled.

b. From the table, the height of the tsunami traveling at 15 feet per second (one-half of 30) is 25 feet. From part a, doubling the speed (from 15 feet per second to 30 feet per second) quadruples the height. The height of a tsunami travelling 30 feet per second is 100 feet.

●

Conclusions based on inductive reasoning may be incorrect. As an illustration, consider the following circles. For each circle, all possible line segments have been drawn to connect each dot on the circle to every other dot on the circle. The dots are drawn so that no three lines intersect at a point.

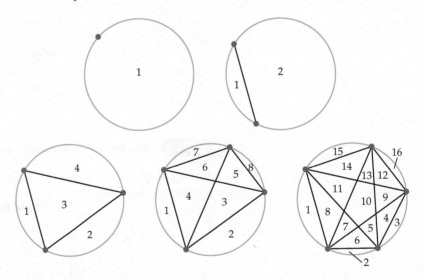

From these figures adding a dot doubles the number of regions as shown in the following table.

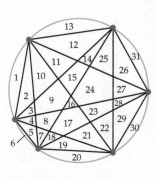

Number of Dots	1	2	3	4	5	6
Maximum Number of Regions	1	2	4	8	16	?

If we use inductive reasoning, we might conjecture that with six dots there would be 32 regions. The figure at the left has six dots and all of the regions numbered. From that figure, there are 31 regions, not 32.

For a statement to be true, it is true in *all* cases. If you can find one example for which a statement is not true, called a *counterexample*, then the statement is a false statement. The fact that there are 31 regions in a circle with six dots is a counterexample to the conjecture that adding a dot doubles the number of regions.

Counterexample

A **counterexample** is an example that disproves a conjecture.

Example 5   Find a Counterexample

Find a counterexample to each of the following statements.

a. My sisters are mammals. My dog is a mammal. My cat is a mammal. The horse in the field is a mammal. By inductive reasoning, all animals are mammals.

b. 12 is divisible by 4. 20 is divisible by 4. 36 is divisible by 4. By inductive reasoning, all even numbers are divisible by 4.

c. $2^2 > 2$, $3^2 > 3$, $4^2 > 4$, and $5^2 > 5$. By inductive reasoning, the square of any number is greater than the number.

d. All roads lead to Rome.

Solution

a. My pet turtle is not a mammal. (There are many other counterexamples.)

b. 6 is an even number that is not divisible by 4.

c. $\left(\dfrac{1}{2}\right)^2 = \dfrac{1}{4} < \dfrac{1}{2}$

d. The road to my house does not lead to Rome.

Deductive Reasoning

Deductive Reasoning

Deductive reasoning is a process by which a conclusion is reached based on premises that are assumed to be true.

Whereas inductive reasoning proceeds from specific instances to a general conclusion, deductive reasoning proceeds from general premises to a specific conclusion. The following figure shows four specific instances of a shape called a *triangle*. By inductive reasoning, the general conclusion is that three-sided closed figures are triangles.

Specific Instances

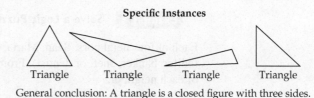

Triangle   Triangle   Triangle   Triangle

General conclusion: A triangle is a closed figure with three sides.

The following figure makes general premises about a certain triangle. Using deductive reasoning, the specific conclusion is that a certain triangle has a 90° angle.

General premise: A right triangle has a 90° angle.
General premise: Triangle *ABC* is a right triangle.
Specific conclusion: Triangle *ABC* has a 90° angle.

Example 6 **Distinguish Between Inductive and Deductive Reasoning**

For the following two scenarios, determine whether inductive or deductive reasoning is being used.

I. You choose two odd numbers, say 7 and 15, and add them. The result is 22, an even number. Now you try two other odd numbers, 13 and 37, and add them. The result is 50, an even number. You try this for another five pairs of odd numbers and note that the sum of the two odd numbers is always even. Based on these observations, you conclude that the sum of two odd numbers is always an even number.

II. Einstein established that no particle with mass can exceed the speed of light. An electron is a particle that has mass. Therefore, an electron cannot exceed the speed of light.

Solution

For scenario I, specific instances (adding two odd numbers) have led to the general conclusion that the sum of two odd numbers is an even number. This is inductive reasoning.

For scenario II, there is a general premise that no particle with mass can exceed the speed of light. An electron is a particle with mass. By deductive reasoning, the speed of an electron cannot exceed the speed of light.

●

Example 7 **Use Deductive Reasoning to Solve a Puzzle**

Of nine total coins, eight have the same weight and one is heavier. There are no other distinguishing characteristics. Determine which is the heavier coin using two weighings with a balance scale similar to the one shown at the left.

Solution

Number the coins from 1 to 9. Place coins 1, 2, and 3 on the left pan of the balance, and place coins 4, 5, and 6 on the right pan of the balance. There are three possibilities. The left side goes down (coin 1, 2, or 3 is the heavy one). The right side goes down (coin 4, 5, or 6 is the heavy one). The scale balances (coin 7, 8, or 9 is the heavy one).

Take the three coins that include the heavy coin and label them A, B, and C. Now weigh A and B. Again, three possibilities exist: the side with coin A goes down (A is the heaviest coin), the side with coin B goes down (B is the heaviest coin), or the scale balances (C is the heaviest coin).

●

Example 8 **Solve a Logic Puzzle**

Each of four neighbors, Sean, Maria, Sarah, and Brian, has a different occupation (editor, banker, chef, or dentist). From the following clues, determine the occupation of each neighbor.

1. Maria gets home from work after the banker but before the dentist.
2. Sarah, who is the last to get home from work, is not the editor.
3. The dentist and Sarah leave for work at the same time.
4. The banker lives next door to Brian.

Solution

From clue 1, Maria is not the banker or the dentist. In the following chart, write X1 (which stands for "ruled out by clue 1") in the Banker and the Dentist columns of Maria's row.

	Editor	Banker	Chef	Dentist
Sean				
Maria		X1		X1
Sarah				
Brian				

From clue 2, Sarah is not the editor. Write X2 (ruled out by clue 2) in the Editor column of Sarah's row. We know from clue 1 that the banker is not the last to get home, and we know from clue 2 that Sarah is the last to get home; therefore, Sarah is not the banker. Write X2 in the Banker column of Sarah's row.

	Editor	Banker	Chef	Dentist
Sean				
Maria		X1		X1
Sarah	X2	X2		
Brian				

From clue 3, Sarah is not the dentist. Write X3 for this condition. There are now X's for three of the four occupations in Sarah's row; therefore, Sarah must be the chef. Place a ✓ in that box. Because Sarah is the chef, none of the other three people can be the chef. Write X3 for these conditions. There are now X's for three of the four occupations in Maria's row; therefore, Maria must be the editor. Insert a ✓ to indicate that Maria is the editor, and write X3 twice to indicate that neither Sean nor Brian is the editor.

	Editor	Banker	Chef	Dentist
Sean	X3		X3	
Maria	✓	X1	X3	X1
Sarah	X2	X2	✓	X3
Brian	X3		X3	

From clue 4, Brian is not the banker. Write X4 for this condition. See the following table. Because there are three X's in the Banker column, Sean must be the banker. Place a ✓ in that box. Thus, Sean cannot be the dentist. Write X4 in that box. Because there are three X's in the dentist column, Brian must be the dentist. Place a ✓ in that box.

	Editor	Banker	Chef	Dentist
Sean	X3	✓	X3	X4
Maria	✓	X1	X3	X1
Sarah	X2	X2	✓	X3
Brian	X3	X4	X3	✓

Sean is the banker, Maria is the editor, Sarah is the chef, and Brian is the dentist.

1.1 Exercise Set

▶ Think About It

For Exercises 1 and 2, fill in the blanks.

1. _____ reasoning proceeds from specific instances to a general conclusion.

2. Deductive reasoning proceeds from general premises to a _____ conclusion.

3. Find a counterexample to the statement "All numbers are even numbers."

▶ Preliminary Exercises

For Exercises P1 to P8, if necessary, use the referenced example following the exercise for assistance.

P1. Use inductive reasoning to predict the next number in the sequence. **Example 1**

 a. 4, 7, 12, 19, 28, . . .

 b. 1, 2, 6, 24, 120, . . .

P2. Use inductive reasoning to predict the next image in the sequence. **Example 2**

P3. The following graph shows the value of an original investment of $5000 at the end of a year for 5 years. **Example 3**

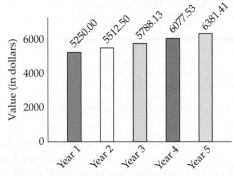

Using inductive reasoning, will the value of the investment at the end of year 6 be less than or more than $6600?

P4. The following table shows the number of grams of sugar that can be dissolved in water for various temperatures, in Celsius.

Temperature	10	20	30	40	50	60
Grams	11.05	12.2	13.48	14.89	16.45	18.17

Using inductive reasoning, will the number of grams of sugar that can be dissolved in water that is 70°C be less than or greater than 19 g? **Example 4**

P5. Find a counterexample to the statement "If x is a positive number, then $x^3 > x^2$." **Example 5**

P6. Determine whether the following is inductive or deductive reasoning. If $x = 4$, then $x + 7 = 11$. **Example 6**

P7. Of nine coins, eight have the same weight. The other one is counterfeit, being either heavier or lighter than the other eight. There are no other distinguishing characteristics. Determine which is the counterfeit coin using three weighings with a balance scale. **Example 7**

P8. Brianna, Ryan, Tyler, and Ashley were recently elected as the new class officers (president, vice president, secretary, treasurer) of the sophomore class at Summit College. From the following clues, determine which position each holds. **Example 8**

1. Ashley is younger than the president but older than the treasurer.

2. Brianna and the secretary are both the same age, and they are the youngest members of the group.

3. Tyler and the secretary are next-door neighbors.

▶ Section Exercises

For Exercises 1 to 6, use inductive reasoning to predict the next number in the given sequence.

1. 1, 4, 9, 16, 25, 36, 49, ?

2. 80, 70, 61, 53, 46, 40, ?

3. $\dfrac{3}{5}, \dfrac{5}{7}, \dfrac{7}{9}, \dfrac{9}{11}, \dfrac{11}{13}$, ?

4. $\dfrac{1}{2}, \dfrac{2}{3}, \dfrac{3}{4}, \dfrac{4}{5}, \dfrac{5}{6}$, ?

5. 2, 7, −3, 2, −8, −3, −13, −8, −18, ?

6. 5, −3, 9, −7, 13, −11, 17, −15, ?

For Exercises 7 to 12, use inductive reasoning to predict the next figure in the sequence.

7.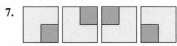

8.

9.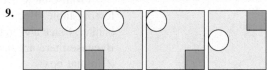

10.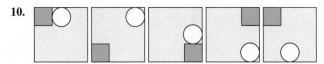

11.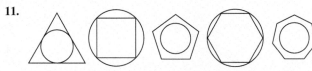

12.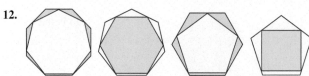

For Exercises 13 to 18, use inductive reasoning to answer a question about a graph.

13. Business Equipment   For income tax purposes, a business owner sets up a depreciation schedule for a machine that is used to help make 3-D printers. The value of the machine over time is shown in the following graph.

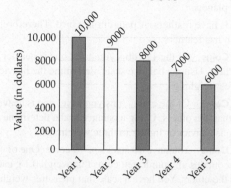

Using inductive reasoning, predict the value of the machine in year 6.

14. Northeast Temperatures   The following graph shows the average daily temperature in a Northeastern U.S. city.

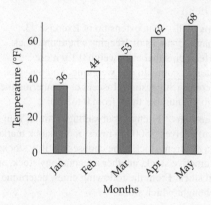

a. Using inductive reasoning, will the temperature in June be greater than or less than the temperature in May?

b. Would it make sense to assume that the pattern in the graph continues for one year?

15. Teacher Salaries   The following graph shows the average annual salary for an elementary school teacher in a certain district.

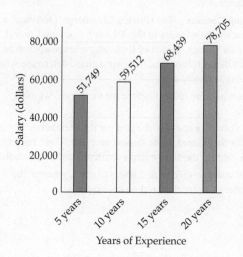

Using inductive reasoning, is the average annual salary of a teacher in this district with 25 years of experience more or less than $88,000?

16. Cell Phone Prices   An economist for a telecommunication company determines that the number of cell phones the company can sell at various prices can be represented in the following graph.

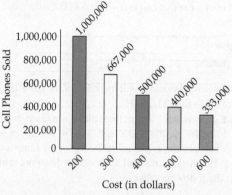

By inductive reasoning, will the company sell fewer or more cell phones if the price of the phone is $700?

17. Runner Distance   The following graph shows the distance an athlete has run over 60 seconds.

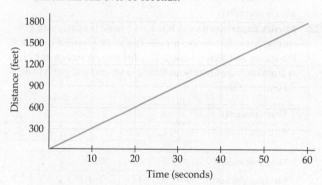

Use inductive reasoning to predict the distance the athlete will run in 70 seconds.

18. Driving Distances   A driver starts on a 500-mile journey. The following graph shows the distance remaining in the journey after driving at a constant speed.

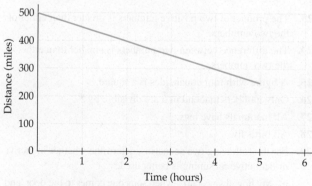

Using inductive reasoning, how many miles will remain in the trip after 6 hours?

For Exercises 19 to 22, use inductive reasoning to answer a question about data in a table.

19. Ocean Temperatures   The following table shows the temperature, in Celsius, of ocean water at various depths.

Depth (meters)	0	5	10	15	20	25	30	35	40
Temperature (°C)	25	23	20	18	17	15	14	12	11

Use inductive reasoning to determine whether the temperature will be more than or less than 11°C when the depth is 45 meters.

20. Car Depreciation The following table shows the depreciated value of a car that originally cost $35,000 after each year for 5 years.

Age (years)	1	2	3	4	5
Value (dollars)	28,078	22,742	17,551	14,621	11,743

Using inductive reasoning, will the value of the car after year 6 be more or less than $11,000?

21. Physics Experiments An inclined plane is a ramp with a groove down the middle of the ramp used in physics experiments. A ball bearing is placed on the top of the ramp and released. The distance the ball has traveled down the ramp is given in the following table.

Time (seconds)	0	1	2	3	4	5	6
Distance (feet)	0	1	4	9	16	25	36

Using inductive reasoning, how far will the ball have traveled after 8 seconds?

22. Physics Experiments A block on a table is attached to a spring that is fastened to a wall. The block is pulled to stretch the spring and then released. The distance, in inches, the block is from the wall for various times, in seconds, is given in the following tables.

Time (seconds)	0	1	2	3	4	5	6
Distance (inches)	10.	17.1	20.	17.1	10.	2.9	0.

Time (seconds)	7	8	9	10	11	12	13
Distance (inches)	2.9	10.	17.1	20.	17.1	10.	2.9

Using inductive reasoning, what distance will the block be from the wall after 14 seconds?

For Exercises 23 to 28, give a counterexample to the given statement.

23. The product of two positive numbers is greater than either of the two numbers.

24. The difference between two numbers is smaller than either of the two numbers.

25. A figure with four equal sides is a square.

26. Only numbers that end in 5 are divisible by 5.

27. All mammals have legs.

28. All birds fly.

29. For the following two scenarios, determine whether inductive or deductive reasoning is being used.

    a. My dog always barks when someone comes to the door, and my dog did not bark. Therefore, no one came to the door.

    b. The product of 3 and 5 is 15, an odd number. The product of 7 and 13 is 91, an odd number. Therefore, the product of two odd numbers is an odd number.

30. For the next two scenarios, determine whether inductive or deductive reasoning is being used.

    a. Our moon revolves around Earth. Titan is a moon of Saturn and revolves around Saturn. Dione is a moon of Saturn and revolves around Saturn. Io is a moon of Jupiter and revolves around Jupiter. Therefore, moons revolve around planets.

    b. All birds have feathers. A penguin is a bird. Therefore, a penguin has feathers.

31. Weighing Coins Of the 12 coins you have, one is heavier than the others. Using a balance scale, determine the heavier coin in three weighings.

32. Weighing Coins Of the eight coins you have, one is heavier or lighter than the others. Using a balance scale, determine which coin is heavier or lighter in three weighings.

33. Weighing Coins You have 10 stacks of 10 coins. One of the stacks contains all counterfeit coins that weigh 0.1 g each more than the others. You have a scale that measures weight accurately to a tenth of a gram. In one weighing, determine which stack contains the counterfeit coins. (Suggestion: Number the stacks from 1 to 10. Now think of a way to choose a different number of coins from each stack, weigh the selected coins, and, based on the weight, conclude which stack is counterfeit.)

34. Weighing Coins This is an extension of Exercise 33. Suppose you have 11 stacks of 10 coins where one stack has all counterfeit coins that each weigh 0.1 g more than non-counterfeit coins. In one weighing, find the stack of counterfeit coins. (Suggestion: Instead of numbering the stacks from 1 to 10, number them from 0 to 10.)

35. Investment Decisions Each of four siblings (Anita, Tony, Maria, and Jose) is given $5000 to invest in the stock market. Each chooses a different stock. One chooses a utility stock, another an automotive stock, another a technology stock, and the other an oil stock. From the following clues, determine which sibling bought which stock.

    1. Anita and the owner of the utility stock purchased their shares through an online brokerage, whereas Tony and the owner of the automotive stock did not.

    2. The gain in value of Maria's stock is twice the gain in value of the automotive stock.

    3. The technology stock is traded on NASDAQ, whereas the stock that Tony bought is traded on the New York Stock Exchange.

36. A Cooking Contest The Changs, Steinbergs, Ontkeans, and Gonzaleses were winners in the All-State Cooking Contest. There was a winner in each of four categories: soup, entrée, salad, and dessert. From the following clues, determine which category each family won.

    1. The soups were judged before the Ontkeans' winning entry.

    2. This year's contest was the first for the Steinbergs and for the winner in the dessert category. The Changs and the winner of the soup category entered last year's contest.

    3. The winning entrée took 2 hours to cook, whereas the Steinbergs' entry required no cooking at all.

37. Collectors' Conventions The cities of Atlanta, Chicago, Philadelphia, and San Diego held conventions this summer for collectors of coins, stamps, comic books, and baseball cards. From the following clues, determine which collectors met in which city.

1. The comic book collectors' convention was in August, as was the convention held in Chicago.

2. The baseball card collectors did not meet in Philadelphia, and the coin collectors did not meet in San Diego or Chicago.

3. The convention in Atlanta was held during the week of July 4, whereas the coin collectors' convention was held the week after that.

4. The convention in Chicago had more collectors attending it than did the stamp collectors' convention.

38. Map Coloring The following map shows eight states in the central time zone of the United States. Four colors have been used to color the states so that no two bordering states are the same color.

a. Can this map be colored, using only three colors, such that no two bordering states are the same color? Explain.

b. Can this map be colored, using only two colors, such that no two bordering states are the same color? Explain.

Investigation

Interview Questions The following problem is attributed to a biotech company and was used as a question to assess job candidates. "There are 17 red and 17 blue balls in a large bucket. Next to the bucket is a pile of red and blue balls. Without being able to see into the bucket, you remove two balls. If the two balls are the same color, take a blue ball from the pile and place it in the bucket. If the two balls are different in color, take a red ball from the pile and place it in the bucket. What color is the last ball removed?" Solve this problem.

Section 1.2 Estimation and Graphs

Learning Objectives:

- Estimate various quantities.
- Use scientific notation.
- Estimate data from a graph.

Estimating Various Quantities

An important skill in mathematics and problem solving is the ability to determine whether or not an answer to a problem is reasonable. One method of determining if an answer is reasonable is to use estimation. An **estimate** is an approximation.

Estimation is especially valuable when using a calculator. Suppose you are adding 1578 and 2318 with a calculator. You enter 1578 correctly but accidentally enter 231 instead of 2318. The calculator display shows 1809. You should ask, "Is this reasonable?"

Using estimation, you can determine that 1809 is not a reasonable answer. Round each number to the highest place value of the number; the first digit of each number will be nonzero and all other digits will be zero. Perform the calculation using the rounded numbers.

$$1578 \rightarrow 2000$$
$$\underline{+ 2318} \rightarrow \underline{+ 2000}$$
$$ 4000$$

The number 4000 is an estimate or approximation of the sum of 1578 and 2318. Because 4000 is not close to the incorrectly calculated sum 1809, an error must have occurred inputting the numbers.

Estimation is a useful skill in everyday experiences. For instance, if someone tells you that one-half of the 84,297 people who voted in an election voted in favor of a new park, you could estimate that about 42,000 $\left(\frac{1}{2} \cdot 84,000 \right)$ people voted for the new park. When estimating, remember we are looking for an approximation—not an exact answer. Estimation is supposed to be an "in your head" kind of calculation, not one that requires a calculator.

Example I   **Estimating a Quantity**

The fuel efficiency rating of your car is 28 miles per gallon. About how many gallons of gas are required for a trip of 620 miles?

Solution

To get an estimate, 620 miles ≈ 600 miles and 28 miles per gallon ≈ 30 miles per gallon. The number of gallons of gas needed is 600 ÷ 30 = 20.
You will need approximately 20 gallons of gas.

Example 2   **Estimating a Cost**

A Little League coach wants to take the team of 17 children to lunch. If the coach estimates that the average order for each child is $6.25, what is the approximate cost of the lunch?

Solution

To get an estimate, 17 ≈ 20 and $6.25 ≈ $6. The approximate cost of the lunch is 20 · $6 = $120. The approximate cost of the lunch will be $120.

There are various ways to estimate the height of a tree, cell tower, or any other tall structure. To estimate the height of a tree, for instance, one method requires a sunny day and a yardstick.

Example 3   **Estimating the Height of a Tree**

The following figure shows a tree, the shadow it casts and its length, a yardstick, and the shadow it casts and its length.

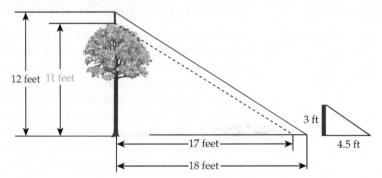

Use this information to estimate the height of the tree.

Solution

The concept behind the estimation of the tree's height is based on the fact that when the height of a known object is multiplied by a certain factor, the length of the shadow is multiplied by the same factor. For instance, if the height of the known object is multiplied by 5, the length of the shadow is multiplied by 5. From the figure,

> A 3-foot yardstick casts a 4.5-foot shadow.
> A 6-foot (2 × 3) stick casts a shadow of 9 (2 × 4.5) feet.
> A 9-foot (3 × 3) stick casts a shadow of 13.5 (3 × 4.5) feet.
> A 12-foot (4 × 3) stick casts a shadow of 18 (4 × 4.5) feet.

Because the shadow cast by the tree is 17 feet and a 12-foot stick would cast a shadow of 18 feet, the tree must be taller than 9 feet but shorter than 12 feet. Because 17 is close to 18, we estimate the height of the tree to be 11 feet.

Example 4   **Visual Estimation**

Estimate the number of blueberries in the top layer of the following photo.

Daniel Hurst Photography/Moment/Getty Images

Solution

Divide the photo into rectangular sections like a grid, all with equal areas. You can divide the photo into as many rectangular regions as you wish, but generally estimates are more accurate with a greater number of sections. We chose a 3-by-5 grid as shown in the following photo.

Daniel Hurst Photography/Moment/Getty Images

Select a sector that appears to be a good representation of all sectors. Then count the number of blueberries in the selected area. We will use the sector in the middle. Our count is 13 blueberries, including pieces. Remember this is an estimation and the count may be an estimate as well.

Multiply 13 by the number of sections, 15. The product is 195. There are approximately 195 blueberries in the photo.

●

Estimating the size of a crowd can be accomplished much as in Example 4 by putting a grid over the crowd. Another method is based on city maps, from which the area of the space a crowd occupies can be determined. Knowing the area of the space and using estimates of crowd density, the size of the crowd can be estimated. Here are some typical crowd density estimates: very dense, 2.5 square feet per person; dense, 5 square feet per person; loose, 10 square feet per person.

For instance, on New Year's Eve, people crowd into Times Square in New York City to celebrate the coming new year. The generally accepted boundaries of Times Square are the five blocks between West 42nd to West 47th, between Broadway and Seventh Avenue. The approximate area of this region is 330,000 square feet. If we assume that people are very densely packed in this region, then the approximate number of people in Times Square on New Year's Eve is $\frac{330,000}{2.5} = 132,000$ people.

This estimate is way short of estimates given by various organizations, which usually range around 1 million or more. The discrepancy may be the result of several factors, one of which is the boundaries of Times Square.

Scientific Notation

Oftentimes when estimating, approximations of numbers can be either extremely large or extremely small. For instance, the U.S. debt is approximately \$21,000,000,000,000; the size of mitochondria (an element of a cell) is approximately 0.0000005 meter; the mass of the sun is approximately 2,000,000,000,000,000,000,000,000,000,000 kilograms. The mass of an electron is approximately 0.000000000000000000000000000000009 kilogram. Writing these numbers as powers of 10 can help us understand the size of the estimation.

Review of Powers of Ten

When multiplying by a positive integer power of 10, move the decimal point to the right.

$$8.44 \times 10^7 = 8.44 \times 10,000,000 = 8.4400000 = 84,400,000$$

7 Places right

When multiplying by a negative integer power of 10, move the decimal point to the left.

$$8.44 \times 10^{-7} = \frac{8.44}{10,000,000} = 0.000000844$$

7 Places left

Using positive exponents, the mass of the sun is approximately 2,000,000,000,000,000,000,000,000,000,000 kg $= 2 \times 10^{30}$ kg.

Using negative exponents, the mass of an electron is approximately

$$0.000000000000000000000000000000009 = \frac{9}{10^{31}} = 9 \times 10^{-31} \text{ kg.}$$

Scientific Notation

In **scientific notation**, a number is expressed as the product of two factors, one a number between 1 and 10 (not including 10), and the other an integer power of 10.

Examples
1. 2×10^{30} kg, the mass of the sun, is a number expressed in scientific notation.
2. 9×10^{-31} kg, the mass of an electron, is a number expressed in scientific notation.

To write a number in scientific notation:

- For numbers 10 or greater, move the decimal point to the right of the first digit.
- The exponent is positive and equal to the number of places the decimal point has been moved.

$$240,000 = 2.4 \times 10^5$$

$$93,000,000 = 9.3 \times 10^7$$

- For numbers less than 1, move the decimal point to the right of the first nonzero digit.

$$0.00030 = 3 \times 10^{-4}$$

- The exponent is negative. The absolute value of the exponent is equal to the number of places the decimal point has been moved.

$$0.0000832 = 8.32 \times 10^{-5}$$

Example 5 **Writing in Scientific Notation**

a. The weight of 31,000,000 orchid seeds is approximately one ounce. Write the number of orchid seeds in one ounce in scientific notation.

b. The length of a femtoplankton (a sea virus) is approximately 0.000000018 meter. Write the length of a femtoplankton in scientific notation.

Solution

a. $31,000,000 = 3.1 \times 10^7$

b. $0.000000018 = 1.8 \times 10^{-8}$

Estimating with Graphs

Graphs are a visual image of data or equations. The purpose of a graph is to give a visual approximation of data. For instance, the following table gives the average rainfall, in centimeters, in a rainforest for each month of a year.

Month	Jan	Feb	Mar	Apr	May	June	July	Aug	Sept	Oct	Nov	Dec
Rain (cm)	230	109	135	167	300	351	284	168	94	112	175	330

The data are precise, but it is difficult to see the "whole" picture at a glance. Here is a bar graph of the data.

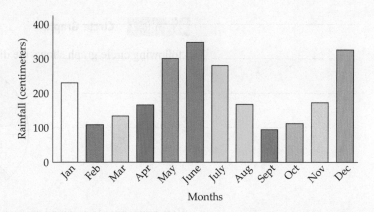

From the graph, we can estimate that June receives about 350 cm of rain and that September receives the least amount of rain.

Example 6 **Solve a Problem Using a Bar Graph**

The following double bar graph shows the average number of walks per game and the average number of runs per game in Major League Baseball for selected years. For instance, in Year 1 there were approximately 3.9 walks per game. In Year 5, there were approximately 4.5 runs scored per game.

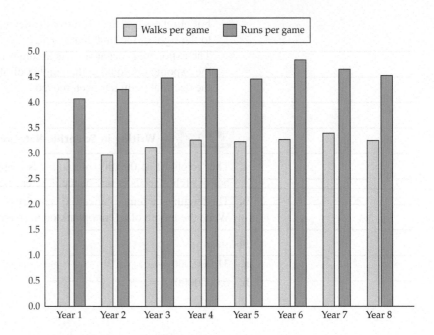

a. Which year had the most walks per game?
b. Which year had the fewest runs scored per game?
c. Between which two years did walks per game decrease the most?
d. Between which two years did runs scored per game increase the most?

Solution

a. The tallest orange bar is for Year 7. In Year 7, there were the most walks per game.
b. The shortest blue bar is for Year 1. In Year 1, there were the least runs scored per game.
c. Walks per game decreased the most between Year 7 and Year 8.
d. Runs scored per game increased the most between Year 5 and Year 6.

Example 7   Circle Graphs

The following circle graph shows the distribution of grades on an exam.

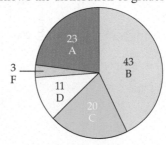

a. How many people received an A on the test?
b. How many people received a B on the test?
c. How many people received below a C on the test?
d. How many people took the test?

Solution

The circle graph shows the number of people who received each letter grade on a test.
a. 23
b. 43
c. 11 + 3 = 14
d. 23 + 43 + 20 + 11 + 3 = 100

Example 8   Line Graphs

The following graph shows the closing price of one ounce of gold at the end of each year for selected years.

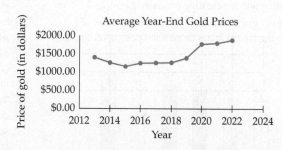

Average Year-End Gold Prices

a. For the years shown, was the price of an ounce of gold ever less than $1000?
b. For the years shown, was the price of an ounce of gold ever more than $1500?
c. Did the price of one ounce of gold increase or decrease between 2014 and 2015?

Solution

The line graph shows the price of gold for eight different years.

a. The graph never goes below $1000. The price of gold for the years shown was never less than $1000.
b. The graph is above $1500 in 2020, 2021, and 2022. The price of gold was above $1500 in 2020, 2021, and 2022.
c. The price of gold decreased between 2014 and 2015.

1.2 Exercise Set

Think About It

1. Which of the following is an estimate and which is an exact number?
    a. It takes me about 25 minutes to drive to work.
    b. The average distance of the sun from Earth is 93 million miles.
    c. There are 60 seconds in 1 minute.
    d. The circumference of Earth is approximately 25,000 miles.
    e. There are 100 centimeters in 1 meter.
    f. There are 2.54 centimeters in 1 inch.

2. Is the number 5.32×10^{-6}
    a. less than 0?
    b. greater than 0 but less than 1?
    c. greater than 1?

3. True or false? $10^9 = 1$ billion

4. True or false? $10^{-3} = \dfrac{1}{100}$

5. If two shadows are the same length, are the objects casting the shadows the same height? Assume the measurements are made at the same time of day and in the same location.

Preliminary Exercises

For Exercises P1 to P8, if necessary, use the referenced example following the exercise for assistance.

P1. The base price of a new car is $24,789. The option for the sports package is $2529. Estimate the price of the car with the sports package. **Example 1**

P2. An engineer decides that the minimum width of a parking space in a parking garage is 9 feet 8 inches. If the proposed length of one row in the parking garage is 587 feet, approximately how many cars can be parked in the row? **Example 2**

P3. Estimate the number of raspberries in the following photo. **Example 3**

Romariolen/Shutterstock.com

P4. A cell tower casts a shadow of 188 feet while a 6-foot pole casts a shadow of 7.5 feet. Estimate the height of the cell tower. **Example 4**

P5. A wavelength of a certain blue light is 0.000000495 meter. Write the wavelength in scientific notation. **Example 5**

P6. The following bar graph shows the approximate number of millionaires in the United States for selected years. **Example 6**

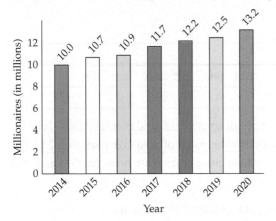

a. Between which two years did the number of millionaires increase the least?

b. Between which two years did the number of millionaires increase the most?

P7. The circle graph shows the distribution of blood types in humans. **Example 7**

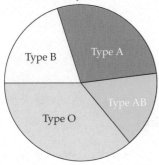

a. Which blood type occurs most frequently?

b. Which blood type occurs least frequently?

P8. The following broken-line graph shows how much an item that cost $10 in 1995 would cost, because of inflation, for the years shown. **Example 8**

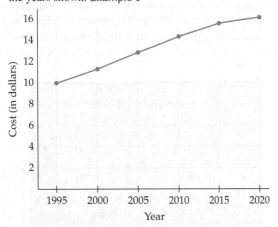

What would be the item's approximate cost in 2020?

Section Exercises

1. **Planning a Road Trip** You are planning a trip to visit a friend. According to Google Maps, the trip is 1487 miles. If you would like to make the trip in three days, approximately how many miles a day must you drive?

2. **Planning a Pizza Party** You are going to order pizza for 11 children. You estimate that each child will eat three slices of pizza. If each pizza is cut into eight pieces, how many pizzas must you order?

3. **Temperature Conversions** There is an exact formula to convert Celsius temperatures to Fahrenheit. However, a simple approximation is twice the Celsius temperature plus 30. What is the approximate Fahrenheit temperature for 25°C?

4. **Estimating Carpet Cleaning** A carpet-cleaning company charges $0.28 per square foot to clean carpets in an apartment. If you have 238 square feet of carpet to be cleaned, approximately what will it cost to clean the carpet?

5. **Event Planning** A convention planner estimates that the size, in square feet, of a room needed for a large assembly is 20 times the number of attendees. If the planner estimates that 400 people will attend, what minimum size room is required?

6. **Painting a Room** A typical bedroom has 400 square feet of wall space, including the windows and doors. If you are going to paint this room and one gallon of paint covers 350 square feet, how many gallons of paint will you need to purchase?

7. **Parking Lots** Use a grid to estimate the number of cars in the parking lot.

8. Use a grid to estimate the number of wildflowers in the field.

9. **Armed Forces**   Use a grid to estimate the number of people in the following photo.

10. **Marching Band**   Use a grid to determine the number of people marching on the field.

11. **Estimating Event Size**   The National Mall is approximately 6,400,000 square feet. If an event organizer estimates the density at 4 square feet per person, approximately how many people are attending the event on the mall?

12. **Population Size**   The population density of Washington, D.C., is approximately 10,000 people per square mile. If the area of Washington, D.C., is 68.3 square miles, approximately how many people live in Washington, D.C.?

13. **Height of a Telephone Pole**   A telephone pole casts a shadow of 72 feet, and a 10-foot pole casts a shadow of 18 feet. Estimate the height of the telephone pole.

14. **Height of Apartment Building**   An apartment building casts a 94-foot shadow, whereas an 8-foot pole casts a shadow of 6 feet. Estimate the height of the apartment building.

15. **Gamma Rays**   A gamma ray has an approximate wavelength of 0.0000000000089 meter. Write this number in scientific notation.

16. **Distance in Parsecs**   One parsec is the distance light can travel in approximately 3.3 years. One parsec is 30,900,000,000,000,000 meters. Write the length of one parsec in meters in scientific notation.

17. **Number of Cells**   A person who weighs approximately 150 pounds has 35,000,000,000,000 cells. Write the number of cells in scientific notation.

18. **Human Fat Cells**   A human fat cell is approximately 0.000000076 meter wide. Write the width of a human fat cell in scientific notation.

19. **Education and Lifetime Earnings**   The following graph shows the expected lifetime earnings, in thousands of dollars, for a person who attains certain academic degrees. For instance, the lifetime earnings of a person who receives an associate degree (AS) is approximately $1,730,000.

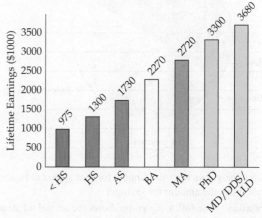

Degree Attained

From the graph, does it appear that as more advanced degrees are earned, there will be higher lifetime earnings?

20. **Golfing**   The following graph shows the distance a golf ball will carry for various club head speeds of a driver. For instance, if the head of the driver is traveling at 110 miles per hour, the approximate distance the ball will travel in the air is 284 yards.

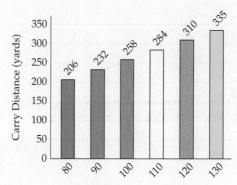

Swing Speed (miles per hour)

From the graph, does it appear that as swing speed increases by 10 miles per hour, carry distance increases by approximately 25 yards?

21. **Pet Owners**   A survey of pet owners who keep their pets in their homes or apartments asked what kind of pet they owned. The percent of owners who responded for each pet type is shown in the following circle graph.

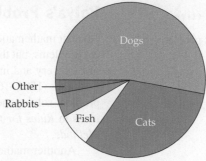

From the graph, is it reasonable to estimate that more than 50% of the respondents had dogs?

22. Family Budgets   The monthly budget for a family who earns $3500 per month is shown in the following circle graph.

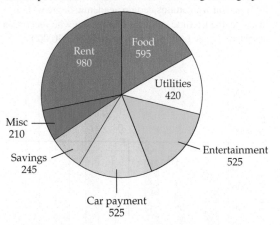

From the graph, does the amount for rent appear to be about four times the amount for savings?

23. Inflation   The following graph shows the annual inflation rate of the United States for selected years.

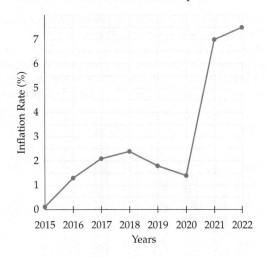

a. Between which years did the inflation rate decrease?

b. Did the inflation rate increase more or less from 2015 to 2018 than it did from 2021 to 2022?

24. Politics   The graph below shows the number of Democrats and the number of Republicans in the U.S. House of Representatives for selected Congress.

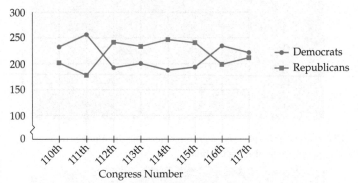

a. For which Congress was the number of Democrats more than the number of Republicans?

b. For which Congress was the difference between the number of Representatives from each party the greatest?

c. For which Congress was the difference between the number of Representatives from each party the least?

▶ **Investigation**

Bad Graphs   Do an Internet search for "bad graphs." You will get a lot of hits. Read through some of the articles and then select two or three articles to share with other students. Be sure to explain why each graph is bad and what should have been done to display the data more accurately.

Section **1.3**   **Problem-Solving Strategies**

Learning Objective:

• Apply Polya's problem-solving strategy.

Polya's Problem-Solving Strategy

Ancient mathematicians such as Euclid and Pappus were interested in solving mathematical problems, but they were also interested in *heuristics*, the study of the methods and rules of discovery and invention. In the 17th century, the mathematician and philosopher René Descartes (1596–1650) contributed to the field of heuristics. He tried to develop a universal problem-solving method. Although he did not achieve this goal, he did publish some of his ideas in *Rules for the Direction of the Mind* and his better-known work *Discourse de la Methode*.

Another mathematician and philosopher, Gottfried Wilhelm Leibnitz (1646–1716), planned to write a book on heuristics titled *Art of Invention*. Of the problem-solving process, Leibnitz wrote, "Nothing is more important than to see the sources of invention which are, in my opinion, more interesting than the inventions themselves."

One of the foremost recent mathematicians to make a study of problem solving was George Polya (1887–1985). He was born in Hungary and moved to the United States in 1940. The basic problem-solving strategy that Polya advocated consisted of the following four steps.

Polya's Four-Step Problem-Solving Strategy

1. Understand the problem.
2. Devise a plan.
3. Carry out the plan.
4. Review the solution.

Polya's four steps are deceptively simple. To become a good problem solver, it helps to examine each step and determine what is involved.

Understand the Problem   This part of Polya's four-step strategy is often overlooked. You must clearly understand the problem. To help you focus on understanding the problem, consider the following questions.

- Can you restate the problem in your own words?
- Can you determine what is known about these types of problems?
- Is there missing information that, if known, would allow you to solve the problem?
- Is there extraneous information that is not needed to solve the problem?
- What is the goal?

Devise a Plan   Successful problem solvers use a variety of techniques when they attempt to solve a problem. Here are some frequently used procedures.

- Make a list of the known information.
- Make a list of information that is needed.
- Draw a diagram.
- Make an organized list that shows all the possibilities.
- Make a table or a chart.
- Work backward.
- Try to solve a similar but simpler problem.
- Look for a pattern.
- Write an equation. If necessary, define what each variable represents.
- Perform an experiment.
- Guess at a solution and then check your result.

Carry Out the Plan   Once you have devised a plan, you must carry it out.

- Work carefully.
- Keep an accurate and neat record of all your attempts.
- Realize that some of your initial plans will not work and that you may have to devise another plan or modify your existing plan.

Review the Solution   Once you have found a solution, check the solution.

- Ensure that the solution is consistent with the facts of the problem.
- Interpret the solution in the context of the problem.
- Ask yourself whether there are generalizations of the solution that could apply to other problems.

In Example 1, we apply Polya's four-step problem-solving strategy to solve a problem involving the number of routes between two points.

Example 1 **Apply Polya's Strategy** (Solve a similar but simpler problem)

Consider the map shown in Figure 1.1. Allison wishes to walk along the streets from point A to point B. How many direct routes can Allison take?

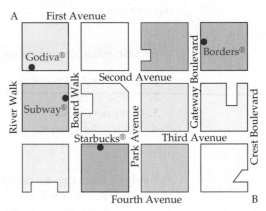

Figure 1.1 City map

A simple diagram of the street map in Figure 1.1

Solution

Understand the Problem We would not be able to answer the question if Allison retraced her path or traveled away from point B. Thus, we assume that on a direct route, she always travels along a street in a direction that gets her closer to point B.

Devise a Plan The map in Figure 1.1 has many extraneous details. Thus, we make a diagram that allows us to concentrate on the essential information.

Because there are many routes, we consider the similar but simpler diagrams to the right. The number at each street intersection represents the number of routes from point A to that particular intersection.

Look for patterns. It appears that the number of routes to an intersection is the *sum* of the number of routes to the adjacent intersection to its left and the number of routes to the intersection directly above. For instance, the number of routes to the intersection labeled 6 is the sum of the number of routes to the intersection to its left, which is 3, and the number of routes to the intersection directly above, which is also 3.

Simple street diagrams

Take Note: The strategy of working a similar but simpler problem is an important problem-solving strategy that can be used to solve many problems.

Carry Out the Plan Using the pattern just noted, we see from the figure at the left that the number of routes from point A to point B is 20 + 15 = 35.

Review the Solution Ask yourself whether a result of 35 seems reasonable. If you were required to draw each route, could you devise a scheme that would enable you to draw each route without missing or duplicating a route?

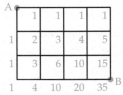

A street diagram with the number of routes to each intersection labeled

Example 2 illustrates the technique of using an organized list.

Example 2 **Apply Polya's Strategy** (Make an organized list)

A baseball team won two out of its last four games. In how many different orders could the team have two wins and two losses in four games?

Solution

Understand the Problem   There are many different orders. The team may have won two straight games and lost the last two (WWLL). Or maybe the team lost the first two games and won the last two (LLWW). Of course there are other possibilities, such as WLWL.

Devise a Plan   We will make an *organized list* of all the possible orders. An organized list is a list that is produced using a system that ensures that each of the different orders will be listed once and only once.

Carry Out the Plan   Each entry in our list must contain two W's and two L's. We will use a strategy that makes sure each order is considered with no duplications. One such strategy is to always write a W unless doing so will produce too many W's or a duplicate of one of the previous orders. If it is not possible to write a W, then and only then do we write an L. This strategy produces the following six different orders.

1. WWLL (start with two wins)
2. WLWL (start with one win)
3. WLLW
4. LWWL (start with one loss)
5. LWLW
6. LLWW (start with two losses)

Review the Solution   We have made an organized list. The list has no duplicates and it considers all possibilities, so we are confident there are six different orders in which a baseball team can win exactly two of four games.

In Example 3, we make use of several problem-solving strategies to solve a problem involving the total number of games to be played.

Example 3   **Apply Polya's Strategy** (Solve a similar but simpler problem)

In a basketball league consisting of 10 teams, each team plays each of the other teams exactly three times. How many league games will be played?

Solution

Understand the Problem   There are 10 teams in the league, and each team plays exactly three games against each of the other teams. The problem is to determine the total number of league games that will be played.

Devise a Plan   Try the strategy of working a similar but simpler problem. Consider a league with only four teams (denoted by A, B, C, and D) in which each team plays each of the other teams only once. The diagram at the left illustrates that the games can be represented by line segments that connect the points A, B, C, and D.

Because each of the four teams will play a game against each of the other three, we might conclude that this would result in $4 \cdot 3 = 12$ games. However, the diagram shows only six line segments. It appears that our procedure has counted each game twice. For instance, when team A plays team B, team B also plays team A. To produce the correct result, we must divide our previous result, 12, by 2. Hence, four teams can play each other once in $\frac{4 \cdot 3}{2} = 6$ games.

Carry Out the Plan   Using the process already developed, we see that 10 teams can play each other *once* in a total of $\frac{10 \cdot 9}{2} = 45$ games. Because each team plays each opponent exactly three times, the total number of games is $45 \cdot 3 = 135$.

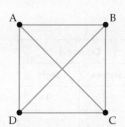

The possible pairings of a league with only four teams

AB AC AD AE AF AG AH AI AJ
BC BD BE BF BG BH BI BJ
CD CE CF CG CH CI CJ
DE DF DG DH DI DJ
EF EG EH EI EJ
FG FH FI FJ
GH GI GJ
HI HJ
IJ

An organized list of all possible games

Review the Solution   We could check our work by making a diagram that includes all 10 teams represented by dots labeled A, B, C, D, E, F, G, H, I, and J. Because this diagram would be somewhat complicated, let us try the method of making an organized list. The figure at the left on the previous page shows an organized list in which the notation BC represents a game between team B and team C. The notation CB is not shown because it also represents a game between team B and team C. This list shows that 45 games are required for each team to play each of the other teams once. Also notice that the first row has nine items, the second row has eight items, the third row has seven items, and so on. Thus, 10 teams require

$$9 + 8 + 7 + 6 + 5 + 4 + 3 + 2 + 1 = 45$$

games if each team plays every other team once, and $45 \cdot 3 = 135$ games if each team plays exactly three games against each opponent.

●

Example 4   **Apply Polya's Strategy** (Make a table and look for a pattern)

Determine the digit 100 places to the right of the decimal point in the decimal representation $\frac{7}{27}$.

Solution

Calculator Note: A TI-84 calculator displays $\frac{7}{27}$ as 0.2592592593. See the screenshot below. However, the last digit 3 is not correct. It is a result of the rounding process. The actual decimal representation of $\frac{7}{27}$ is the decimal 0.259259... or $0.\overline{259}$, in which the digits continue to repeat the 259 pattern forever.

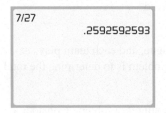

7/27
.2592592593

Understand the Problem   Express the fraction $\frac{7}{27}$ as a decimal and look for a pattern that will enable us to determine the digit 100 places to the right of the decimal point.

Devise a Plan   Dividing 27 into 7 by long division or with a calculator produces the decimal 0.259259259 . . . . Because the decimal representation repeats the digits 259 over and over forever, we know that the digit located 100 places to the right of the decimal point is either a 2, a 5, or a 9. A table may help us to see a pattern and enable us to determine which of these digits is in the 100th decimal digit. Because the decimal digits repeat every three digits, we use a table with three columns.

The First 15 Decimal Digits of $\frac{7}{27}$

Column 1		Column 2		Column 3	
Location	Digit	Location	Digit	Location	Digit
1st	2	2nd	5	3rd	9
4th	2	5th	5	6th	9
7th	2	8th	5	9th	9
10th	2	11th	5	12th	9
13th	2	14th	5	15th	9
. . .		. . .		. . .	

Carry Out the Plan   Only in column 3 is each of the decimal digit *locations* evenly divisible by 3. From this pattern we can tell that the 99th decimal digit (because 99 is evenly divisible by 3) must be a 9. Because a 2 always follows a 9 in the pattern, the 100th decimal digit must be a 2.

Review the Solution   The preceding table illustrates additional patterns. For instance, if each of the location numbers in column 1 is divided by 3, a remainder of 1 is produced. If each of the location numbers in column 2 is divided by 3, a remainder of 2 is produced. Thus, we can find the decimal digit in any location by dividing the location number by 3 and examining the remainder. For instance, to find the digit in the 3200th decimal place of $\frac{7}{27}$ merely divide 3200 by 3 and examine the remainder, which is 2. Thus, the digit 3200 places to the right of the decimal point is a 5.

●

Example 5 illustrates the method of working backward. In problems in which you know a final result, this method may require the least effort.

Example 5   **Apply Polya's Strategy** (Work backward)

In consecutive turns of a Monopoly game, Stacy first paid $800 for a hotel. She then lost half her money when she landed on Boardwalk. Next, she collected $200 for passing GO. She then lost half her remaining money when she landed on Illinois Avenue. Stacy now has $2500. How much did she have just before she purchased the hotel?

Solution

Understand the Problem   We need to determine the number of dollars Stacy had just before her $800 hotel purchase.

Devise a Plan   We could guess and check, but we might need to make several guesses before we found the correct solution. An algebraic method might work, but setting up the necessary equation could be a challenge. Because we know the end result, let us try the method of working backward.

Carry Out the Plan   Stacy must have had $5000 just before she landed on Illinois Avenue, $4800 just before she passed GO, and $9600 just before landing on Boardwalk. This means she had $10,400 just before she purchased the hotel.

Review the Solution   To check our solution we start with $10,400 and proceed through each of the transactions. $10,400 less $800 is $9600. Half of $9600 is $4800. $4800 increased by $200 is $5000. Half of $5000 is $2500.

●

Some problems can be solved by making guesses and checking. Your first few guesses may not produce a solution, but quite often they will provide additional information that will lead to a solution.

Example 6   **Apply Polya's Strategy** (Guess and check)

The product of the ages, in years, of three teenagers is 4590. None of the teens are the same age. What are the ages of the teenagers?

Solution

Understand the Problem   We need to determine three distinct counting numbers, from the list 13, 14, 15, 16, 17, 18, and 19, that have a product of 4590.

Devise a Plan   If we represent the ages by x, y, and z, then $xyz = 4590$. We are unable to solve this equation, but we notice that 4590 ends in a zero. Hence, 4590 has a factor of 2 and a factor of 5, which means that at least one of the numbers we seek must be an even number and at least one number must have 5 as a factor. The only number in our list that has 5 as a factor is 15. Thus, 15 is one of the numbers, and at least one of the other numbers must be an even number. At this point we try to solve by *guessing and checking*.

Carry Out the Plan

$15 \cdot 16 \cdot 18 = 4320$ • No. This product is too small.

$15 \cdot 16 \cdot 19 = 4560$ • No. This product is too small.

$15 \cdot 17 \cdot 18 = 4590$ • Yes. This is the correct product.

The ages of the teenagers are 15, 17, and 18.

Review the Solution Because $15 \cdot 17 \cdot 18 = 4590$ and each of the ages represents the age of a teenager, we know our solution is correct. None of the numbers 13, 14, 16, and 19 is a factor (divisor) of 4590, so there are no other solutions.

●

Some problems are deceptive. After reading one of these problems, you may think that the solution is obvious or impossible. These deceptive problems generally require that you carefully read the problem several times and that you check your solution to make sure it satisfies all the conditions of the problem.

Example 7 **Solve a Deceptive Problem**

A hat and a jacket together cost $100. The jacket costs $90 more than the hat. What are the cost of the hat and the cost of the jacket?

Solution

Understand the Problem After reading the problem for the first time, you may think that the jacket costs $90 and the hat costs $10. The sum of these costs is $100, but the cost of the jacket is only $80 more than the cost of the hat. We need to find two dollar amounts that differ by $90 and whose sum is $100.

Devise a Plan Write an equation using h for the cost of the hat and $h + 90$ for the cost of the jacket.

$$h + h + 90 = 100$$

Carry Out the Plan Solve the above equation for h.

$2h + 90 = 100$ • Collect like terms.

$2h = 10$ • Solve for h.

$h = 5$

The cost of the hat is $5 and the cost of the jacket is $90 + $5 = $95.

Review the Solution The sum of the costs is $5 + $95 = $100, and the cost of the jacket is $90 more than the cost of the hat. This check confirms that the hat costs $5 and the jacket costs $95.

●

1.3 Exercise Set

Think About It

1. What are the four steps in Polya's problem-solving strategy?

2. Consider this problem. "Jennifer enters the 100-yard dash in a track-and-field event. How many seconds did it take Jennifer to finish the race?" Is there enough information to answer this question?

Preliminary Exercises

For Exercises P1 to P7, if necessary, use the referenced example following the exercise for assistance.

P1. Consider the street map in Figure 1.1 on page 24. Allison wishes to walk directly from point A to point B. How many different routes can she take if she wants to go past Starbucks on Third Avenue? **Example 1**

P2. A true–false quiz contains five questions. In how many ways can a student answer the questions if the student answers two of the questions with "false" and the other three with "true"? **Example 2**

P3. If six people greet each other at a meeting by shaking hands with one another, how many handshakes will take place? **Example 3**

P4. Determine the units digit (ones digit) of 4^{200}. **Example 4**

P5. Melody picks a number. She doubles the number, squares the result, divides the square by 3, subtracts 30 from the quotient, and gets 18. What are the possible numbers that Melody could have picked? What operation does Melody perform that prevents us from knowing with 100% certainty which number she picked? **Example 5**

P6. Nothing is known about the personal life of the ancient Greek mathematician Diophantus except for the information in the following epigram. "Diophantus passed $\frac{1}{6}$ of his life in childhood, $\frac{1}{12}$ in youth, and $\frac{1}{7}$ more as a bachelor. Five years after his marriage was born a son who died four years before his father, at $\frac{1}{2}$ his father's (final) age." At what age did Diophantus die? **Example 6**

P7. Two U.S. coins have a total value of 35¢. One of the coins is not a quarter. What are the two coins? **Example 7**

Exercises

Use Polya's four-step problem-solving strategy and the problem-solving procedures presented in this section to solve each of the following exercises.

1. **Number of Girls**   There are 364 first-grade students in Park Elementary School. If there are 26 more girls than boys, how many girls are there?

2. **Heights of Ladders**   If two ladders are placed end to end, their combined height is 31.5 feet. One ladder is 6.5 feet shorter than the other ladder. What are the heights of the two ladders?

3. **Number of Squares**   How many squares are in the following figure?

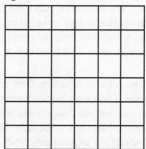

4. **Determine a Digit**   What is the 44th decimal digit in the decimal representation of $\frac{1}{11}$?

$$\frac{1}{11} = 0.09090909\ldots$$

5. **Cost of a Shirt**   A shirt and a tie together cost $50. The shirt costs $30 more than the tie. What is the cost of the shirt?

6. **Number of Games**   In a basketball league consisting of 12 teams, each team plays each of the other teams exactly twice. How many league games will be played?

7. **Number of Routes**   Consider the following map. Tyler wishes to walk along the streets from point A to point B. How many direct routes (no backtracking) can Tyler take?

8. **Number of Routes**   Use the map in Exercise 7 to answer each of the following.

   a. How many direct routes are there from A to B if you want to pass by Starbucks?

   b. How many direct routes are there from A to B if you want to stop at Subway for a sandwich?

   c. How many direct routes are there from A to B if you want to stop at Starbucks and at Subway?

9. **True–False Test**   In how many ways can you answer a 12-question true–false test if you answer each question with either a "true" or a "false"?

10. **A Puzzle**   A frog is at the bottom of a 17-foot well. Each time the frog leaps, it moves up 3 feet. If the frog has not reached the top of the well, then the frog slides back 1 foot before it is ready to make another leap. How many leaps will the frog need to escape the well?

11. **Number of Handshakes** If eight people greet each other at a meeting by shaking hands with one another, how many handshakes take place?

12. **Number of Line Segments** Twenty-four points are placed around a circle. A line segment is drawn between each pair of points. How many line segments are drawn?

13. **Number of Ducks and Pigs** The number of ducks and pigs in a field totals 35. The total number of legs among them is 98. Assuming each duck has exactly two legs and each pig has exactly four legs, determine how many ducks and how many pigs are in the field.

14. **Racing Strategies** Carla and Allison are sisters. They are on their way from school to home. Carla runs half the time and walks half the time. Allison runs half the distance and walks half the distance. If Carla and Allison walk at the same speed and run at the same speed, which one arrives home first? Explain.

15. **Change for a Quarter** How many ways can you make change for 25¢ using combinations of dimes, nickels, and pennies?

16. **Carpet for a Room** A room measures 12 feet by 15 feet. How many 3-foot by 3-foot squares of carpet are needed to cover the floor of this room?

In Exercises 17 to 20, determine the units digit (ones digit) of the counting number represented by the exponential expression.

17. 4^{300} 18. 2^{725} 19. 3^{412} 20. 7^{146}

21. **A Puzzle** Three volumes of the series *Mathematics: Its Content, Methods, and Meaning* are on a shelf with no space between the volumes. Each volume is 1 inch thick without its covers. Each cover is $\frac{1}{8}$ inch thick. See the following figure. A bookworm bores horizontally from the first page of Volume 1 to the last page of Volume III. How far does the bookworm travel?

22. **Connect the Dots** Nine dots are arranged as shown. Is it possible to connect the nine dots with exactly four lines if you are not allowed to retrace any part of a line and you are not allowed to remove your pencil from the paper? If it can be done, demonstrate with a drawing.

23. **Floor Design** A square floor is tiled with congruent square tiles. The tiles on the two diagonals of the floor are blue. The rest of the tiles are green. If 101 blue tiles are used, find the total number of tiles on the floor.

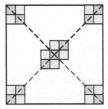

24. **Votes in an Election** In a school election, one candidate for class president received more than 94%, but less than 100%, of the votes cast. What is the least possible number of votes cast?

25. **Brothers and Sisters** I have two more sisters than brothers. Each of my sisters has two more sisters than brothers. How many more sisters than brothers does my youngest brother have?

26. **Number of Children** How many children are there in a family in which each girl has as many brothers as sisters, but each boy has twice as many sisters as brothers?

27. **Bacterial Growth** The bacteria in a petri dish grow in a manner such that each day the number of bacteria doubles. On what day will the number of bacteria be half of the number present on the 12th day?

28. **A Coin Problem** If you take 22 pennies from a pile of 57 pennies, how many pennies do you have?

29. **Examination Scores** On three examinations, Dana received scores of 82, 91, and 76. What score does Dana need on the fourth examination to raise his average to 85?

30. **Number of River Crossings** Four people on one side of a river need to cross the river in a boat that can carry a maximum load of 180 pounds. The weights of the people are 80, 100, 150, and 170 pounds.

    a. Explain how the people can use the boat to get everyone to the opposite side of the river.

    b. What is the minimum number of crossings that must be made by the boat?

31. **Find the Fake Coin** You have eight coins. They all look identical, but one is a fake and is slightly lighter than the others. Explain how you can use a balance scale to determine which coin is the fake in exactly

    a. three weighings.

    b. two weighings.

32. **Puzzle from a Movie** In the movie *Die Hard: With a Vengeance,* Bruce Willis and Samuel L. Jackson are given a 5-gallon jug and a 3-gallon jug and they must put *exactly* 4 gallons of water on a scale to keep a bomb from exploding. Explain how they could accomplish this feat.

Mensa is a society that welcomes people from every walk of life whose IQ is in the top 2% of the population. The multiple-choice Exercises 33 to 36 are from the Mensa Workout, which is posted on the Internet at www.mensa.org

33. Sally likes 225 but not 224; she likes 900 but not 800; she likes 144 but not 145. Which of the following does she like?

    a. 1600   **b.** 1700

34. If it were 2 hours later, it would be half as long until midnight as it would be if it were an hour later. What time is it now?

    a. 18:30   **b.** 20:00   **c.** 21:00
    d. 22:00   **e.** 23:30

35. Following the pattern shown in the number sequence below, what is the missing number?

    $$1 \quad 8 \quad 27 \quad ? \quad 125 \quad 216$$

    a. 36   **b.** 45   **c.** 46   **d.** 64   **e.** 99

36. There are 1200 elephants in a herd. Some have pink and green stripes, some are all pink, and some are all blue. One-third are pure pink. Is it true that 400 elephants are definitely blue?

    a. Yes   **b.** No

Source: Mensa Workout questions © Dr. Abbie Salny. www.mensa.org. Reprinted by permission.

▶ **Investigation**

Cryptarithms   A *cryptarithm* is a mathematical puzzle consisting of an equation with unknown numbers whose digits are expressed as letters. Here is an example.

```
  S O M E          8 7 9 5
+ G O O D    is  + 4 7 7 3
---------        ---------
I D E A S        1 3 5 6 8
```

$S = 8$, $O = 7$, $M = 9$, $E = 5$, $G = 4$, $D = 3$, $I = 1$, and $A = 6$.

Try these cryptarithms.

```
   W A R M
1. + W A T E R
   ---------
   E V E N T S

     S E N D
2. + M O R E
   ---------
   M O N E Y

     F I F T Y
3. + S T A T E S
   -----------
   A M E R I C A

   S U N T A N
4. + L O T I O N
   -----------
   P L A N T S
```

Section 1.4 — Introduction to Fair Division

Introduction to Fair Division

Learning Objectives:

- Identify the types of fair division problems.
- Use the lone divider-chooser method for more than two players.
- Use the method of adjusted winner.
- Use the method of sealed bids.
- Use the Knaster Inheritance procedure.

Introduction to Fair Division

The premise behind fair division is that when something of value is divided among several people, each person is satisfied with the share received. Probably the most widely known method of fair division is one that involves two children who decide to share a cookie. One child cuts the cookie into two pieces, and the other child gets to choose one of the pieces. In the case of a cookie, there are an infinite number of ways the cookie can be cut.

Suppose, however, that two people inherit a house, a car, and a sculpture. Now dividing the assets is a little more complicated. Suppose, for instance, that one person wants the house and will not consider selling it.

A fair division problem consists of a set of goods (houses, paintings, money, cookies, jewelry, land, etc.) that are to be divided into N shares so that each player gets a *fair share*. The solution of a fair division problem is based on the following criteria.

1. A player's **fair share** is one that, *in the opinion* of the player, has a value that is *at least* $\frac{1}{N}$ of the total value of all goods *as assigned by that player*.

2. No player knows the value the other players have assigned to the goods. For instance, the fact that you may value say, a house, at $250,000 cannot be known to anyone else.

3. All players act rationally in their own best interest.

4. There are essentially two kinds of fair division problems. One type involves goods that can be divided into smaller pieces (money, cookies, land, natural gas, etc.). These problems are called **continuous division problems**. The second type of problem involves goods that cannot be divided (houses, cars, sculpture, etc.). These problems are called **discrete division problems**.

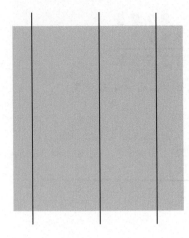

Pay special attention to the italicized phrases under criterion 1. If there are four players, then each player is entitled to *at least* $\frac{1}{4}$ or 25% of the value of the goods as assigned *by the player*. For instance, it is quite possible that someone would divide a plot of land as shown at the left. The areas do not have to be equal. However, the value of each piece *from the divider's value system* must be equal. When dividing the land, one piece may be larger than another piece that has an ocean view.

Suppose a boat, motorcycle, piano, and diamond ring are to be divided among four people. Each person assigns a value (unknown to the others) for each item as in the following table.

	Boat	Motorcycle	Piano	Diamond Ring
Player I Assigned Values	$8000	$4000	$7000	$1000
Player 2 Assigned Values	$5000	$1000	$6000	$5000
Player 3 Assigned Values	$4000	$5000	$3000	$4000
Player 4 Assigned Values	$1000	$1000	$2000	$6000

If each person acts rationally, then player 1 will choose the boat, player 2 will choose the piano, player 3 will choose the motorcycle, and player 4 will choose the diamond ring. In each case, the value received by a player is *at least* (greater than or equal to) 25% of the total value as *assigned by the player*.

$$\text{Player 1:} \frac{8000}{8000 + 4000 + 7000 + 1000} = 45\% \geq 25\%$$

$$\text{Player 2:} \frac{6000}{5000 + 1000 + 6000 + 5000} \approx 35\% \geq 25\%$$

$$\text{Player 3:} \frac{5000}{4000 + 5000 + 3000 + 4000} \approx 31\% \geq 25\%$$

$$\text{Player 4:} \frac{6000}{1000 + 1000 + 2000 + 6000} = 60\% \geq 25\%$$

Note that it does not matter that player 1 assigned a greater value to the piano than did player 2. All that matters is that player 1 chose the item that had the greatest value of goods available, as assigned by player 1. It also does not matter that player 4 received 60% of the perceived value. It is player 4's value, not the value measured by the other players.

Divider–Chooser Method for Two Players

Now we return to the problem of two children sharing a cookie. In the language of fair division, the solution to this problem is called the *divider–chooser method*.

> **Divider–Chooser Method**
>
> Randomly choose (say, by tossing a coin) one of the two players (the **divider**) who divides whatever is to be shared into two pieces of equal value *to the divider*. The other player (the **chooser**) selects one of the two pieces.

The divider–chooser method works well when there are only two players and the item to be shared is continuous. The divider–chooser method, however, does have a bias, which is why the divider should be randomly chosen.

To see the bias, consider Annabel and Bennet who are going to share a cake that has vanilla frosting on one half and chocolate frosting on the other half. Because there are two players, each player is entitled to a share that is at least 50% of the total value as assigned by the player.

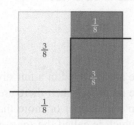

Suppose Annabel likes vanilla and chocolate equally well. However, Bennet hates chocolate and only likes vanilla.

If Annabel is the divider, she can divide the cake into two equal pieces in several ways. Suppose that the cake is 1 square unit in area and is divided as in the figure at the left. Because Annabel values vanilla and chocolate equally, whether she gets the piece, $\frac{1}{2}$ square unit total in area, that is $\frac{3}{8}$ square unit chocolate and $\frac{1}{8}$ square unit vanilla, or whether she gets the piece, also $\frac{1}{2}$ square unit total in area, that is $\frac{3}{8}$ square unit vanilla and $\frac{1}{8}$ square unit chocolate, the values (to her) are the same.

Now Bennet gets to choose. Because he hates chocolate, he will choose the piece that is $\frac{3}{8}$ square unit vanilla and $\frac{1}{8}$ square unit chocolate. Therefore, he values his piece as

$$\frac{\frac{3}{8}}{\frac{3}{8} + \frac{1}{8}} = \frac{3}{4} = 75\% > 50\%$$

This satisfies criterion 1.

Annabel gets the remaining piece, which is the other half of the cake. Because she likes vanilla and chocolate equally well, she values her piece as 50% of the total.

Now let us think about this if Bennet were the divider. Bennet would *not* cut the cake like Annabel has. From his point of view (he does not know Annabel's preferences), Annabel may choose the piece that is three-eighths vanilla and leave him with a large piece of chocolate and a small piece of vanilla.

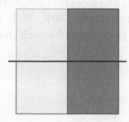

Bennet would cut the cake as shown at the left. No matter which piece Annabel chooses, Bennet will end up with a piece that he values that is 50% of his total share. This shows there is a slight bias in the divider–chooser method to be the divider.

Gambling is not allowed in the divider–chooser method. For instance, Bennet cannot cut the cake so that one piece was all chocolate and one piece was all vanilla. In this case, he either gets all of what he prefers or nothing. If he gets nothing, criterion 1 is not satisfied.

Example 1   Determine Whether a Division Satisfies Criterion 1

Olivia and William divide a $12 cake, half of which is covered in chocolate icing and half of which is covered in vanilla icing. Olivia values the chocolate half as $8 and the vanilla half as $4. William values each half as $6. Olivia is randomly chosen as the divider. In each case, determine whether Olivia has made a cut that satisfies criterion 1.

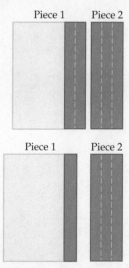

a. Olivia divides the chocolate half into fifths and then cuts the cake so that piece 1 consists of the entire vanilla half and two fifths of the chocolate half. Piece 2 consists of the remaining three fifths of the chocolate half.

b. Olivia divides the chocolate half into fourths and then cuts the cake so that piece 1 consists of the entire vanilla half and one fourth of the chocolate half. Piece 2 consists of the remaining three fourths of the chocolate half.

Solution

a. As Olivia values the cake, piece 2 is worth $\frac{3}{5} \cdot \$8 = \$4.80$. Piece 1 is worth the value of the remaining two fifths of the chocolate half plus the value of the entire vanilla half. Piece 1 is then worth $\frac{2}{5} \cdot \$8 + \$4 = \$7.20$. Because the two pieces do not have the same value, the cut does not satisfy criterion 1.

b. As Olivia values the cake, piece 2 is worth $\frac{3}{4} \cdot \$8 = \$6$. Piece 1 is worth the value of the remaining fourth of the chocolate half plus the value of the entire vanilla half, or $\frac{1}{4} \cdot \$8 + \$4 = \$6$. Because the two pieces have the same value, the cut does satisfy criterion 1.

Example 1 again illustrates that dividing goods depends on the value assigned by the divider. It must be done so that each piece has the same value to the divider.

Lone Divider–Chooser Method for More than Two Players

Take Note: The lone–divider method is discussed here. There is a variation of this method called the lone–chooser method as an Investigation at the end of this section.

We now look at a situation in which more than two people are involved in a fair division. The divider–chooser method discussed previously must be modified when there are more than two players. In these cases, there must be more than one divider or more than one chooser. We will discuss the situation with one divider, which is called the *lone–divider method*. There are two strategies to this method, depending on whether a share is *uncontested* (only one person wants a share) or *contested* (more than one person wants a share).

Uncontested Shares

Suppose investors A, B, C, and D want to divide a piece of land. They begin by randomly selecting one of the investors—say, A—to be the lone divider. Investor A divides the land into four shares, S_1, S_2, S_3, and S_4, so that, to A, all shares are of equal value. To investor A, each share is worth 25% of the whole.

Now B, C, and D look at each piece and decide (without letting the other players know) the value each player places on the four shares. These are called the **bids**. Rather than place a dollar value on each piece, the three choosers assign a percentage value to each share with the sum equaling 100%. For instance, investor B might assign 40%, 30%, 15%, and 15% to each of the shares of the divided land. Record the percentages for each chooser in a table such as the following.

	S_1	S_2	S_3	S_4	Bids
A	25%	25%	25%	25%	S_1, S_2, S_3, S_4
B	40%	30%	15%	15%	S_1, S_2
C	30%	20%	35%	15%	S_1, S_3
D	10%	30%	30%	30%	S_2, S_3, S_4

Looking at the bids, determine whether there is an uncontested share among the *choosers*, investors B, C, and D. In this case, S_4 is uncontested (only consider the choosers). Award that share to investor D. Now look at investors B and C. Between those two investors, S_2 and S_3 are uncontested. Investor B receives S_2, and investor C gets S_3. This leaves the divider (investor A) with S_1. Each investor has received a share, as determined by the investor, that is at least 25% of the total.

Contested Shares

We will use the same model of dividing land among four investors. However, we will change the values that an investor places on a share.

	S_1	S_2	S_3	S_4	Bids
A	25%	25%	25%	25%	S_1, S_2, S_3, S_4
B	45%	20%	15%	20%	S_1
C	20%	40%	25%	15%	S_2, S_3
D	40%	20%	20%	20%	S_1

In this case, B and D view only S_1 as a fair share. Set B and D aside and S_1. Now flip a coin to determine the disposition of S_2 and S_3. Heads player C gets S_2; tails player C gets S_3. Suppose the coin flip results in a tail. Player C receives S_3. Note that the share C receives is random between the shares C values as fair (a percentage greater than or equal to 25%), not the magnitude of the percent. Player A can now be given S_2 or S_4. Another coin flip decides A receives S_2.

A and C each have a fair share. Combine the remaining shares S_1 and S_4. There are now just two players. Choose one of B or D to be the divider and use the divider–chooser method to determine the disposition of the assets.

<u>Example 2</u>   **Use the Lone–Divider Method for Contested Shares**

Four partners who own a real estate company decide they want to dissolve the partnership. To accomplish the dissolution, one partner, Ang, is selected as the divider. Ang divides the assets into four shares that she believes are of equal value. Each remaining partner submits sealed bids on the shares as shown in the following table. Determine who gets each share.

	S_1	S_2	S_3	S_4	Bids
Ang	25%	25%	25%	25%	S_1, S_2, S_3, S_4
Becket	30%	30%	30%	10%	S_1, S_2, S_3
Charlene	20%	40%	25%	15%	S_2, S_3
Damian	35%	30%	15%	20%	S_1, S_2

Solution

S_2 is contested by all partners and S_4 is uncontested by all choosers. Award S_4 to Ang. Ang has her share so eliminate her. Randomly choose Becket, Charlene, or Damian to be the divider of the remaining shares. We will select Damian. Damian divides the remaining shares that, for him, are equal. Becket and Charlene then submit bids on the shares. Suppose the results are as in the following table.

	P_1	P_2	P_3	Bids
Damian	33.3%	33.3%	33.3%	P_1, P_2, P_3
Charlene	40%	40%	20%	P_1, P_2
Becket	30%	40%	30%	P_2

P_3 is uncontested between Charlene and Becket. Give P_3 to Damian. The two remaining shares are combined. The divider–chooser method for two partners is now used. Randomly choose one person—say, Charlene—to be the divider. Charlene divides the remaining shares into Q_1 and Q_2. Then Damian chooses a share. Say Becket chooses Q_1 so Charlene receives Q_2. Ang receives S_4, Damian receives P_3, Charlene receives Q_2, and Becket receives Q_1.

●

Method of Adjusted Winner

Another method of distributing goods between two players is the *method of adjusted winner.* This method may require that one of the items be sold and the proceeds be *proportionally* given to the two players. Each person is given 100 points to distribute among all the items. The person giving an item the larger number of points wins the item.

<u>Example 3</u>   **Use the Method of Adjusted Winner to Split an Inheritance**

Jeffrey and Leticia are to split a car, a diamond ring, a silk tapestry, a gold coin, and a rare violin. The point value for each item is given in the following table.

Item	Jeffrey	Leticia
Car	21	28
Diamond ring	30	21
Silk tapestry	5	11
Gold coin	13	13
Rare violin	31	27

How are the items distributed using the method of adjusted winners?

Solution

1. Initially assign an item to the person who awarded more points to it. In each row, we have highlighted the higher number of points for each corresponding item.

Item	Jeffrey	Leticia	Winner
Car	21	28	Leticia
Diamond ring	30	21	Jeffrey
Silk tapestry	5	11	Leticia
Gold coin	13	13	Initially skip ties
Rare violin	31	27	Jeffrey
Total of items won	61	39	

Because Leticia has fewer points than Jeffrey ($39 < 61$), Leticia is given the tied item. The table now looks like this.

Item	Jeffrey	Leticia	Winner
Car	21	28	Leticia
Diamond ring	30	21	Jeffrey
Silk tapestry	5	11	Leticia
Gold coin	13	13	Leticia
Rare violin	31	27	Jeffrey
Total of items won	61	52	

Because Jeffrey has the larger total points for items won (61 versus 52), Jeffrey is called the **initial winner**; Leticia is called the **initial loser**.

2. Begin the adjusted winner procedure. For each item given to the *initial winner*, Jeffery, (in this case, the diamond ring and rare violin), divide the points awarded by the initial winner by the points awarded by the initial loser.

Item	Jeffrey	Leticia	Winner	Quotients
Car	21	28	Leticia	
Diamond ring	30	21	Jeffrey	$\frac{30}{21} \approx 1.43$
Silk tapestry	5	11	Leticia	
Gold coin	13	13	Leticia	
Rare violin	31	27	Jeffrey	$\frac{31}{27} \approx 1.15$
Total of items won	61	52		

Starting with *smallest* ratio (1.15 in this case), transfer the item from the initial winner to the initial loser *unless* that item would cause the initial winner to have fewer points than the initial loser. In that case, the item is shared between the two players.

3. **The sharing procedure**

   Transferring the rare violin to Leticia would give her $28 + 11 + 13 + 27 = 79$. Jeffery awarded 31 points to the rare violin so his total is now $61 - 31 = 30$. Because this total is less than Leticia's total, 79, a sharing procedure takes effect.

   Let x be the percentage (as a decimal) that needs to be taken from Jeffrey and given to Leticia that will result in each of them having the same number of points. Then $1 - x$ is the percent (as a decimal) Jeffrey retains. After the transfer, the points should be equal. This is expressed by the equation

$$30 + 31(1 - x) = 28 + 11 + 13 + 27x$$

Solve the equation for x.

$$30 + 31(1 - x) = 28 + 11 + 13 + 27x$$

$$61 - 31x = 52 + 27x$$

$$9 = 58x$$

$$0.155 \approx x$$

Write the result as a percent, 15.5%.

Obviously, cutting up a violin is not going to work. In this case, the violin would be sold and 15.5% of the proceeds would be given to Leticia, and 84.5% of the proceeds is given to Jeffrey.

4. Distribute the estate.

Jeffrey gets the diamond ring and 84.5% of the proceeds from the sale of the rare violin. Leticia gets the car, silk tapestry, gold coin, and 15.5% of the proceeds from the sale of the violin.

Method of Sealed Bids

The divider–chooser method works well for goods that can be easily divided such as land and cakes. For goods such as paintings, houses, cars, and boats, the method does not work. Here is an outline of the *method of sealed bids*.

- Each player, independent of the other players, determines the monetary value of each item in the collection and then submits the evaluation as a sealed bid.
- Each item is given to the bidder who gave the item the highest bid.
- Each player then either receives money or pays money to ensure all players have a fair share.
- If there is money left over, it is divided equally among the players.

> **Example 4**   **Use the Method of Sealed Bids**

An estate consists of a sculpture, oil painting, ski boat, and gold watch. These items are to be divided among three family members, Josephine, Walter, and Corella. Find a fair division of the items.

Solution

Each of the players writes down a value for each item.

	Josephine	Walter	Corella
Sculpture	$ 2500	$ 5000	$ 3000
Oil Painting	$11,000	$ 8000	$10,000
Ski Boat	$ 1000	$ 5000	$ 4000
Gold Watch	$12,000	$10,000	$ 9000

For each item, the player with the highest bid is given the item.

	Josephine	Walter	Corella	Who Receives Item
Sculpture	$ 2500	**$ 5000**	$ 3000	Walter
Oil Painting	**$11,000**	$ 8000	$10,000	Josephine
Ski Boat	$ 1000	**$ 5000**	$ 4000	Walter
Gold Watch	**$12,000**	$10,000	$ 9000	Josephine

Because there are three players, a fair share is one-third of the estate as *valued by the player.* Sum the bids of each player (the total valuation by each player) and sum the values of the items won by each player.

	Josephine	Walter	Corella	Who Receives Item
Sculpture	$ 2500	**$ 5000**	$ 3000	Walter
Oil Painting	**$11,000**	$ 8000	$10,000	Josephine
Ski Boat	$ 1000	**$ 5000**	$ 4000	Walter
Gold Watch	**$12,000**	$10,000	$ 9000	Josephine
Value of Items Won	$ 26,500	$28,000	$26,000	
Fair Share	$\frac{1}{3}$($26,500) ≈ $8833	$\frac{1}{3}$($28,000) ≈ $9333	$\frac{1}{3}$($26,000) ≈ $8667	
Fair Share Minus Value of Items Won	−$ 14,167	−$ 667	$ 8667	

A negative number in the table indicates the amount owed to the estate by that person; a positive number indicates the amount received by a person from the estate. Adding the amounts owed to the estate and then subtracting the amount the estate owes gives the estate surplus.

Estate surplus = $14,167 + $667 − $8667 = $6167

The estate surplus is divided equally among the players. Each player receives

$\frac{1}{3}$ · $6167 ≈ $2056

Determine the amounts owed to the estate and owed by the estate.

Josephine: −$14,167 + $2056 = −$12,111

Walter: −$667 + $2056 = $1389

Corella: $8667 + $2056 = $10,723

Disposition of estate:

Josephine gets the oil painting and gold watch and pays $12,111 into the estate.

Walter gets the sculpture and ski boat and receives $1389 from the estate.

Corella gets $10,723 from the estate.

The *actual* value (as judged, for instance, by an appraiser) is not at issue. The value of each item is the value the player decides the item has. *One critical aspect of the method of sealed bids is that each player has the money (in cash) to pay into the estate.*

●

The method of sealed bids can be extended to other kinds of dispositions such as rent. Suppose Drew, Peyton, and Logan move into a three-bedroom apartment for which the monthly rent is $2100. One bedroom is large with an attached bathroom. The second bedroom faces the street and shares a bathroom with the third bedroom. The third bedroom is the smallest. Use the method of sealed bids to determine what portion of the rent each roommate should pay depending on the bedroom chosen.

Each roommate assigns a value to each bedroom with the condition that the sum of the values equals $2100. Suppose the values are distributed as in the following table.

	Drew	Peyton	Logan	Who Receives Room
Bedroom 1	1050	840	840	Drew
Bedroom 2	630	840	630	Peyton
Bedroom 3	420	420	630	Logan
Total Valuation	2100	2100	2100	
Fair Share	700	700	700	
Value of Bedroom Won	1050	840	630	
Owed to Rent	−350	−140		
Receives from Rent			70	Surplus = 420
Share of Surplus	140	140	140	$\frac{1}{3}$ · 420 = 140
Disposition of Money	−210	0	210	

From the last row, Drew owes (the value is negative) $210 more than a fair share or $910; Peyton's share of the rent is $0 more than a fair share or $700; Logan receives $210 (the value is positive) less than a fair share or $490.

Knaster Inheritance Procedure

The **Knaster inheritance procedure** is a variation of the method of sealed bids. The n players submit sealed bids for each item. The player submitting the highest bid receives the object but then places $\dfrac{n-1}{n}$ of the value of the object into a joint escrow account. Each other player receives a credit that equals $\frac{1}{n}$ of the value of the object, *as perceived by the highest bidder*. Here is an example with just one asset.

Suppose an estate has a single asset that is a vintage car and there are three siblings ($n = 3$). Unknown to the others, each sibling makes a bid on the car. Suppose the bids are as in the following table.

	Mario	Joanne	Arnold
Vintage Car	$60,000	$72,000	$66,000

From the bids, Joanne wins the car. She deposits $\dfrac{n-1}{n} = \dfrac{3-1}{3} = \dfrac{2}{3}$ of the value of the car (as she values it) in an escrow account:

$$\frac{2}{3}(72,000) = 48,000$$ • Joanne places $48,000 in escrow account.

The other players are credited in the escrow account with $\frac{1}{n}$ of the amount of their bid.

With $n = 3$, Mario and Arnold receive a credit in escrow of $\frac{1}{3}$ of **their bid**.

Mario: $\dfrac{1}{3}(60,000) = 20,000$ • Mario receives $20,000 from escrow.

Arnold: $\dfrac{1}{3}(66,000) = 22,000$ • Arnold receives $22,000 from escrow.

$$\text{Total in escrow} = \text{Amount paid in} - \text{Amount paid out}$$
$$= \$48,000 - (\$20,000 + \$22,000) = \$6000$$

Because $6000 remains in the escrow account, the $6000, called the surplus, is divided equally ($2000 each) among the three siblings. The *net disposition of the estate* is given by

Net disposition of the estate = item received − amount paid into escrow
+ amount received from escrow + share of the surplus.

Joanne: car − $48,000 + $2000 = car − $46,000

Mario: $20,000 + $2000 = $22,000

Arnold: $22,000 + $2000 = $24,000

We now verify that the division is fair, which means that each person receives at least $\frac{1}{3} \approx 0.33$ the value *that person* placed on the item.

Joanne: $\dfrac{\$72,000(\text{value of car}) - \$46,000}{\$72,000} \approx 0.36 > 0.33$

Mario: $\dfrac{\$22,000}{\$60,000} \approx 0.37 > 0.33$

Arnold: $\dfrac{\$24,000}{\$66,000} \approx 0.36 > 0.33$

Because each of the *perceived* values is greater than one-third, the division is fair.

We cannot emphasize enough that fairness, invoking an old truth, is in the eyes of the beholder. What matters is that the *player* perceives the portion of the estate received to be fair.

> **Example 5**   **Use the Knaster Inheritance Procedure**

Four siblings inherit a house, a condo, and an antique car. Each sibling submits sealed bids for each item as in the following table.

	Megan	Justin	Samuel	Erin
House	$240,000	$276,000	$200,000	$210,000
Condo	$165,000	$210,000	$200,000	$150,000
Car	$ 25,000	$ 20,000	$ 30,000	$ 27,000

Use the Knaster inheritance procedure to distribute the inheritance.

Solution

There are four players, $n = 4$. Consider each item separately and allot the item to the player with the highest bid. If there is a tie, toss a coin to determine which player receives the item.

	Megan	Justin	Samuel	Erin	Winner
House	$240,000	$276,000	$200,000	$210,000	Justin
Condo	$165,000	$210,000	$200,000	$150,000	Justin
Car	$ 25,000	$ 20,000	$ 30,000	$ 27,000	Samuel

Because Justin has won the house, he compensates each other player by placing $\frac{4-1}{4} = \frac{3}{4}$ of the value of the house (as he valued it) into a joint account. Each other player is credited with $\frac{1}{4}$ of the amount **the player bid**.

Justin owes: $276,000 \cdot \dfrac{3}{4} = 207,000$

Megan receives: $240,000 \cdot \dfrac{1}{4} = 60,000$

Samuel receives: $200,000 \cdot \dfrac{1}{4} = 50,000$

Erin receives: $210,000 \cdot \dfrac{1}{4} = 52,500$

Determine the amount of money left in the joint account.

$207,000 - \$60,000 - \$50,000 - \$52,500 = \$44,500$

Distribute that money ($44,500) equally among the players. There are four players, so each receives one-fourth of $44,500.

$\dfrac{1}{4} \cdot 44,500 = 11,125$

Final accounting for the house:

Megan: $60,000 + 11,125 = 71,125$

Justin: house $- 207,000 + 11,125 =$ house $- 195,875$

Samuel: $50,000 + 11,125 = 61,125$

Erin: $52,500 + 11,125 = 63,625$

Restart the procedure for the next item, the condo. Because Justin won the condo, we have:

Justin owes: $210,000 \cdot \dfrac{3}{4} = 157,500$

Megan receives: $165,000 \cdot \dfrac{1}{4} = 41,250$

$$\text{Samuel receives: } 200,000 \cdot \frac{1}{4} = 50,000$$

$$\text{Erin receives: } \quad 150,000 \cdot \frac{1}{4} = 37,500$$

Equally distribute any money left in the joint account.

$$157,500 - 41,250 - 50,000 - 37,500 = 28,750$$

$$\frac{1}{4} \cdot 28,750 = 7187.50$$

Final accounting for the condo:

Megan: $41,250 + 7187.50 = 48,437.50$

Justin: condo $- 157,500 + 7187.50 = $ condo $- 150,312.50$

Samuel: $50,000 + 7187.50 = 57,187.50$

Erin: $37,500 + 7187.50 = 44,687.50$

Restart the procedure for the next item, the car. Because Samuel won the car, we have:

$$\text{Samuel owes: } \quad 30,000 \cdot \frac{3}{4} = 22,500$$

$$\text{Megan receives: } 25,000 \cdot \frac{1}{4} = 6250$$

$$\text{Justin receives: } \quad 20,000 \cdot \frac{1}{4} = 5000$$

$$\text{Erin receives: } \quad 27,000 \cdot \frac{1}{4} = 6750$$

Equally distribute any money left in the joint account.

$$22,500 - 6250 - 5000 - 6750 = 4500$$

$$4500 \cdot \frac{1}{4} = 1125$$

Final accounting for the car:

Megan: $6250 + 1125 = 7375$

Justin: $5000 + 1125 = 6125$

Samuel: car $- 22,500 + 1125 = $ car $- 21,375$

Erin: $6750 + 1125 = 7875$

Disposition of estate and money:

Megan: $71,125 + 48,437.50 + 7375 = 126,937.50$

Justin: house $- 195,875 + $ condo $- 150,312.50 + 6125 = $ house $+$ condo $- 340,062.50$

Samuel: $61,125 + 57,187.50 + $ car $- 21,375 = $ car $+ 96,937.50$

Erin: $63,625 + 44,687.50 + 7875 = 116,187.50$

As with the method of sealed bids, the Knaster inheritance method requires that all players have sufficient money to play. For instance, in the last example Justin must deposit $340,062.50 into the joint account to compensate the other players for the fact he won the house and condo.

We consider one other item before we close this section. Recall that for all fair division problems, a player must receive a share that is least $\frac{1}{n}$ of the value as assigned by the player. We will verify that for Justin in the last example. His value of the items is $276,000 + \$210,000 + \$20,000 = \$506,000$. He wins the house and condo for $486,000 and pays $340,062.50. The result is $145,937.50. The ratio is $\dfrac{145,937.50}{506,000} \approx 0.288 > \dfrac{1}{4} = 0.25$. As Justin values the items, he has received a fair share.

1.4 Exercise Set

▶ Think About It

1. If Jamie is chosen among four people to divide the assets of an estate, then Jamie must divide the estate in such a way that each share, to Jamie, is worth _____ % of the total value of the estate.

2. Using the divider–chooser method for two people, Jeremiah divides a piece of land into two pieces as shown in the following diagram.

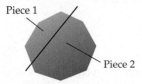

Piece 1

Piece 2

Is this a fair division of the land? Explain.

3. Three people are to divide an estate using the divider–chooser method. The value each has assigned to each share of the estate is shown in the following table.

	P_1	P_2	P_3
Samuel	25%	50%	25%
Craig	$33\frac{1}{3}$%	$33\frac{1}{3}$%	$33\frac{1}{3}$%
Yolinda	40%	50%	10%

Who is the divider?

4. Three partners are dividing the assets of a business into three shares, S_1, S_2, and S_3. Each partner—P_1, P_2, and P_3—secretly decides the value each places on a share as shown in the following table.

	S_1	S_2	S_3
P_1	35%	25%	40%
P_2	40%	25%	35%
P_3	$33\frac{1}{3}$%	$33\frac{1}{3}$%	$33\frac{1}{3}$%

For each partner, list the shares the partner considers a fair share. (You are not trying to find the disposition of the shares, just what each partner considers a fair share.)

▶ Preliminary Exercises

For Exercises P1 to P5, if necessary, use the referenced example following the exercise for assistance.

P1. Suppose Melanie likes strawberry icing and hates chocolate icing. She is given a cake that is one-half strawberry icing and one-half chocolate icing and is tasked with cutting the cake into two pieces. After she cuts the cake, Chris, who likes strawberry icing and chocolate icing equally well, chooses a piece. **Example 1**

a. Referring to the following figure, does cutting the cake as in A satisfy criterion 1 of a fair division problem? Explain.

b. Referring to the following figure, does cutting the cake as in B satisfy criterion 1 of a fair division problem? Explain.

c. If Chris were cutting the cake, can he make a cut that divides the cake in half but does not satisfy criterion 1 of a fair division problem? Explain.

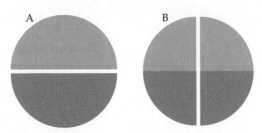

A   B

P2. Kachina, Jameson, and Macon share a parcel of land. Kachina is selected to divide the parcel into three plots she believes are of equal value. Jameson and Macon then assign values to the plots as shown in the following table.

	P_1	P_2	P_3
Kachina	33.3%	33.3%	33.3%
Jameson	40%	40%	20%
Macon	40%	50%	10%

Give an instance where each receives a fair share. **Example 2**

P3. Minh and Abril are getting divorced, and they have decided to use the method of adjusted winner to divide some of their assets. Both have awarded point values to each asset as shown in the following table.

Item	Abril	Minh
Home	48	46
Beachfront condo	25	27
Stamp collection	12	5
Sailboat	5	12
Pollock painting	10	10

Determine the disposition of the assets. **Example 3**

P4. An estate consists of a diamond ring, vintage car, Tiffany lamp, and a coin collection. These items are to be divided among three family members, Carlos, Mina, and Gordon. The family members decide to use the method of sealed bids to divide the estate. Their bids are shown in the following table.

	Carlos	Mina	Gordon
Ring	$ 5000	$ 5000	$ 4000
Car	$50,000	$45,000	$55,000
Lamp	$25,000	$45,000	$40,000
Coin	$12,000	$ 9000	$ 9000

Find a fair division of the items. **Example 4**

P5. An estate consists of a lakeside cottage, a condo, and a sailboat. These items are to be divided among three family members, Amy, Selma, and Pieter. The family members decide to use the Knaster inheritance method to divide the estate. Their bids are shown in the table on the next page.

	Amy	Selma	Pieter
Cottage	$150,000	$225,000	$175,000
Condo	$200,000	$175,000	$150,000
Sailboat	$ 10,000	$ 5000	$ 15,000

Find a fair division of the items. **Example 5**

▶ Section Exercises

Refer to the Excel templates from your digital resources for help. They may be useful to check your work for these exercises.

1. **Sharing a Pizza**   Two friends are to share a pizza that is one-third cheese, one-third pepperoni, and one-third green pepper. To Zelda, the cheese part is worth $9, the pepperoni part is worth $6, and the green pepper part is worth $3.

   a. What is the value of the pizza to Zelda?

   b. If the pizza is cut into such a way that she receives one-third of the cheese part, one-half of the pepperoni part, and none of the green pepper part, does she receive a fair share of the pizza?

2. **Sharing a Pizza**   Abigail and Renaldo divide a $10 pizza that is half cheese and half pepperoni. Abigail values the pepperoni half as $6 and the cheese half as $4. Renaldo values each half as $5. Abigail is randomly chosen as the divider. See the two possible divisions.

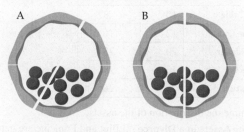

   a. For A, the pizza is cut in half so that one half is one-third cheese and one-sixth pepperoni. The other half is one-third pepperoni and one-sixth cheese. Determine whether Abigail has made a cut that satisfies criterion 1 of a fair division problem.

   b. For B, the pizza is cut in half so that each piece is one-half cheese and one-half pepperoni. Determine whether Abigail has made a cut that satisfies criterion 1 of a fair division problem.

3. **Sharing a Cake**   Martha has a cake that has half chocolate and half strawberry icing as shown in the following figure. The value of the cake is $12. If Martha values the strawberry part at $9 and the chocolate part at $3, how should she cut the cake into two pieces so that each piece is a fair share to her?

4. **Sharing a Piece of Land**   A piece of land is to be divided among three contractors—Charlotte, Ione, and Brandon—on

which they will build homes. They have decided to use the lone–divider method to divide the land into plots A, B, and C. Ione is chosen as the divider. Charlotte values either plot A or B as a fair share. Brandon values plot B or C as a fair share. Determine the fair division of the land.

5. **Sharing a Cake**   Amanda, Marvin, and Angeline want to divide a cake worth $24 that is one-half chocolate icing and one-half strawberry icing. Amanda values chocolate icing at $12 and the strawberry icing at $12. Marvin values the chocolate icing at $16 and the strawberry icing at $8, whereas Angeline values the strawberry icing at $18 and the chocolate icing at $6. If Amanda is selected as the divider and cuts the cake into three pieces as shown in the following figure (each sector is one-sixth of the total), is each person guaranteed a fair share?

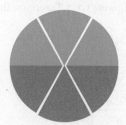

6. **Dividing Business Shares**   Three partners are dividing the assets of a business into three shares, S_1, S_2, and S_3. P_3 is chosen as the divider, and the other two partners value each share as shown in the following table.

	S_1	S_2	S_3
P_1	35%	25%	40%
P_2	40%	30%	30%
P_3	$33\frac{1}{3}\%$	$33\frac{1}{3}\%$	$33\frac{1}{3}\%$

Use the lone–divider method to distribute the shares.

7. **Inheriting a Family Business**   Three sisters are dividing the assets of a business into three shares, S_1, S_2, and S_3. Jean is chosen as the divider; Caroline and Jasmine are the choosers. The value each sister has assigned to a share is shown in the following table.

	S_1	S_2	S_3
Caroline	45%	25%	30%
Jasmine	40%	30%	30%
Jean	$33\frac{1}{3}\%$	$33\frac{1}{3}\%$	$33\frac{1}{3}\%$

Determine a fair division of the assets.

8. **Dividing Business Shares**   Three partners are dividing the assets of a business into three shares, S_1, S_2, and S_3. P_3 is chosen as the divider and the other two partners value each share as shown in the following table.

	S_1	S_2	S_3
P_1	30%	30%	40%
P_2	25%	30%	45%
P_3	$33\frac{1}{3}\%$	$33\frac{1}{3}\%$	$33\frac{1}{3}\%$

Use the lone–divider method to distribute the shares.

9. **Dividing an Estate** Four siblings decide to divide the assets in an estate using the divider–chooser method. Sibling A is chosen as the divider. The bids for each of the siblings are shown in the following table. Determine the disposition of the shares.

	S_1	S_2	S_3	S_4
A	25%	25%	25%	25%
B	30%	10%	30%	30%
C	20%	35%	25%	20%
D	30%	30%	20%	20%

10. **Assets of a Movie Theater** Four partners decide to divide the assets in a movie theater using the divider–chooser method. Jonas is chosen as the divider. The bids for each of the partners are shown in the following table. Determine the disposition of the shares.

	S_1	S_2	S_3	S_4
Jonas	25%	25%	25%	25%
Anne	25%	35%	20%	20%
Gavin	20%	35%	25%	20%
Alisa	20%	10%	30%	40%

11. **Assets of a Medical Office Building** Four business partners decide to divide the assets in a medical office building using the divider–chooser method. Arianna is chosen as the divider. The bids for each partner are shown in the following table. Determine the disposition of the shares.

	S_1	S_2	S_3	S_4
Arianna	25%	25%	25%	25%
Richard	15%	30%	20%	35%
Michel	20%	45%	15%	20%
Carlos	30%	35%	15%	20%

12. **Dividing an Estate** Four siblings decide to divide the assets in an estate using the divider–chooser method. Kyle is chosen as the divider. The bids for each of the siblings are shown in the following table.

	S_1	S_2	S_3	S_4
Kyle	25%	25%	25%	25%
Cody	30%	30%	20%	20%
Colin	15%	30%	20%	35%
Marian	20%	40%	20%	20%

Determine the disposition of the shares.

13. **Skin Care Product Assets** Four investors decide to divide the assets in a skin care product company using the divider–chooser method. Jorge is chosen as the divider. The bids for each of the investors are shown in the following table. Determine the disposition of the shares.

	S_1	S_2	S_3	S_4
Jorge	25%	25%	25%	25%
Yolanda	35%	20%	10%	35%
Maria	15%	40%	35%	10%
Daniel	25%	35%	10%	30%

14. **Real Estate Company Shares** Four partners in a real estate firm decide to dissolve the company and divide the assets using the divider–chooser method. Georgiana is chosen as the divider. The bids for each of the partners are shown in the following table. Determine the disposition of the shares.

	S_1	S_2	S_3	S_4
Georgiana	25%	25%	25%	25%
Macon	15%	35%	30%	20%
Clarisse	40%	30%	25%	25%
Barbara	20%	30%	20%	30%

15. **Dividing an Estate** Chong and Riana decide to use the adjusted winner method to divide the assets of an estate. The assets and the point value each have assigned to an asset are shown in the following table.

	Chong	Riana
Boat	15	5
House	45	65
Condo	40	30

Determine the distribution of the assets.

16. **Antique Furniture Store Assets** Therese and Geraldine decide to use the adjusted winner method to divide the assets in their antique furniture store. The assets and the point value each has assigned to an asset are shown in the following table.

	Therese	Geraldine
Oak Table	10	8
Cedar Chest	12	16
Chandelier	20	25
Tapestry	25	25
Sculpture	33	26

Determine the distribution of the assets.

17. **Dividing Assets in a Divorce** Elliot and Francine are getting divorced and must divide an estate that contains their main residence, a summer home, an airplane, a sailboat, and a desk that originally belonged to Nathaniel Hawthorne. The point values each has assigned to an asset are shown in the following table.

	Elliot	Francine
Residence	20	25
Summer Home	10	10
Airplane	18	25
Sailboat	20	15
Desk	32	25

Using the method of adjusted winner, determine the disposition of the assets.

18. **Dividing an Estate** Three brothers are going to divide the assets of an estate using the method of sealed bids. The bids for the assets by each brother are shown in the following table.

	Office Building	House
Samuel	$255,000	$327,000
Ashton	$276,000	$300,000
Carmelo	$250,000	$275,000

Determine a fair distribution of the assets.

19. **Dividing Business Assets** Three partners are going to divide the assets of their partnership using the method of sealed bids. The bids for the assets by each partner are shown in the following table.

	Medical Suites	Real Estate
Amelia	$510,000	$480,000
Justine	$624,000	$540,000
Colby	$540,000	$545,000

Determine a fair distribution of the partnership.

20. **Dividing an Inheritance** An estate consists of a condo, a car, a yacht, and an emerald ring. The estate will be divided among three sisters, Lois, Juliette, and Melody, using the method of sealed bids. The bids for each of the sisters are given in the following table.

	Lois	Juliette	Melody
Condo	$177,000	$222,000	$195,000
Car	$ 18,000	$ 9000	$ 12,000
Yacht	$ 51,000	$ 66,000	$ 75,000
Emerald Ring	$ 12,000	$ 12,600	$ 9000

Find a fair division of the estate.

21. **Dividing an Inheritance** An estate consists of a home, SUV, boat, and a LeRoy Neiman painting. The estate will be divided among three siblings, Leonard, Taylor, and Madison, using the method of sealed bids. The bids, in thousands of dollars, for each of the siblings are given in the following table.

	Leonard	Taylor	Madison
Home	$600	$520	$475
SUV	$ 12	$ 10	$ 14
Boat	$ 58	$ 70	$ 50
Painting	$110	$120	$121

Find a fair division of the estate.

22. **Sharing an Apartment** Suppose Jacquelyn, Maria, and Petra move into a three-bedroom apartment for which the monthly rent is $2100. The three bedrooms are of various sizes, and two of the bedrooms share a bathroom. The following table shows how the roommates value each bedroom. Use the method of sealed bids to determine what portion of the rent each roommate should pay depending on the bedroom chosen.

	Jacquelyn	Maria	Petra
Bedroom 1	$700	$600	$800
Bedroom 2	$600	$900	$700
Bedroom 3	$800	$600	$600

23. **Sharing an Apartment** Suppose Clara, Manuel, Charles, and Beatrice move into a four-bedroom apartment for which the monthly rent is $2500. The bedrooms have different values to each person, as shown in the following table. Use the method of sealed bids to determine what portion of the rent each roommate should pay depending on the bedroom chosen.

	Clara	Manuel	Charles	Beatrice
Bed 1	$500	$800	$500	$600
Bed 2	$600	$400	$500	$700
Bed 3	$750	$700	$900	$800
Bed 4	$650	$600	$600	$400

24. **Dividing an Inheritance** An estate is left to two children, Annabel and Sebastian, and consists of one item, a Ferrari. The bids by the children are shown in the following table. Use the Knaster inheritance procedure to divide the estate based on the bids of the two children.

	Sebastian	Annabel
Ferrari	$240,000	$250,000

Determine a fair distribution of the Ferrari.

25. **Dividing an Inheritance** An estate is left to two children, Lenore and Raven, and consists of one item, a ski chalet. The bids by the children are shown in the following table. Use the Knaster inheritance procedure to divide the estate based on the bids of the two children.

	Lenore	Raven
Chalet	$500,000	$600,000

26. **Dividing an Inheritance** An estate is left to three children, Megan, Hudson, and Jeremiah, and consists of two items, a diamond pendant and a country house. The bids of the siblings are shown in the following table. Use the Knaster inheritance procedure to divide the estate based on the bids of the children shown in the following table.

	Megan	Hudson	Jeremiah
Pendant	$ 24,000	$ 30,000	$ 21,000
Country House	$360,000	$300,000	$330,000

Find a fair division of the estate.

27. **Dividing an Inheritance** An estate consists of a sailboat, an RV, and a mountain cabin. The estate will be divided among three siblings, Marshall, Jordan, and Jan, using the Knaster inheritance procedure. The bids for each of the siblings are given in the following table.

	Marshall	Jordan	Jan
Sailboat	$75,000	$60,000	$63,000
RV	$66,000	$63,000	$57,000
Cabin	$87,000	$90,000	$75,000

Find a fair division of the estate.

28. **Dividing an Inheritance** Rework Exercise 20 using the Knaster inheritance procedure. Does this result in a different distribution of the estate?

29. **Dividing an Inheritance** An estate consists of a piano, an organ, a recording studio, and a condominium. The estate will be divided among four children, June, Melissa, Lloyd, and Diane, using the Knaster inheritance procedure. The bids are shown in the following table.

	June	Melissa	Lloyd	Diane
Piano	$ 15,000	$ 10,000	$ 12,000	$ 10,000
Organ	$ 9000	$ 12,000	$ 10,000	$ 9000
Studio	$195,000	$225,000	$175,000	$200,000
Condo	$175,000	$250,000	$225,000	$200,000

Find a fair division of the estate.

Investigation

Sharing a Cake The lone–chooser method is a variation of the lone–divider method. Here is an outline of the procedure for three players, Camille, Emily, and Hamilton.

- One person is randomly selected as the chooser. We will assume it is Emily.
- The remaining two players use the divider–chooser method to divide a cake that is one-half vanilla and one-half chocolate. One of them—say, Hamilton—is randomly chosen as the divider and cuts the cake into two pieces in such a way that each piece has the same value to him.
- Camille chooses one of the pieces, and Hamilton receives the other.
- Camille and Hamilton then divide their respective pieces of the cake, according to their value system, into three equal shares.
- Emily chooses one piece from Hamilton's three pieces and one piece from Camille's three pieces. Each person now has a fair share.

Suppose a sheet cake that cost $30 has frosting that is half vanilla and half chocolate. Camille likes vanilla and chocolate equally. Hamilton likes chocolate twice as much as vanilla. Emily likes vanilla twice as much as chocolate. Hamilton cuts the cake as shown here.

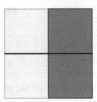

1. Is this a legal cut?
2. Suppose Camille chooses the piece on the bottom. Now Camille and Hamilton must divide their pieces into pieces that are of equal value to them. Give an example of how Camille and Hamilton can make legal slices for their halves of the cake.
3. Have Emily choose one piece from Camille's three pieces and one piece from Hamilton's three pieces. (You will just have to make the choice for her.)
4. Show that the value of the pieces each player has is a fair share.

Test Prep Introduction to Problem Solving

Section 1.1 Inductive and Deductive Reasoning

Inductive reasoning is the process of reaching a general conclusion by examining specific examples. A conclusion based on inductive reasoning is called a *conjecture*. A conjecture may or may not be correct.	See **Examples 1, 2, 3,** and **4** on pages 4 through 6, and then try Exercises 1, 3, 9, 10, and 12 on page 48.
Counterexample A statement is a true statement provided it is true in all cases. If you can find one case in which a statement is not true—called a *counterexample*—then the statement is a false statement.	See **Example 5** on page 7, and then try Exercise 16 on page 48.
Deductive reasoning is the process of reaching a conclusion by applying general assumptions, procedures, or principles.	See **Examples 6** and **7** on page 8, and then try Exercise 19 on page 48.

Section 1.2 Estimation and Graphs

Estimation is used to approximate a number.	See **Examples 1** and **2** on page 14, and then try Exercises 22 and 23 on page 49.
Visual estimation is a technique used to estimate a quantity based on a photo or image.	See **Example 3** on page 14, and then try Exercise 25 on page 49.
Scientific notation is used to express small or very large numbers.	See **Example 4** on page 15, and then try Exercises 27 and 28 on page 49.
Graphs are representations of data or equations.	See **Examples 5, 6,** and **7** on pages 17 and 18, and then try Exercises 29, 30, and 31 on page 49.

Section 1.3 Problem-Solving Strategies

Polya's Four-Step Problem-Solving Strategy 1. Understand the problem. 2. Devise a plan. 3. Carry out the plan. 4. Review the solution.	See **Examples 1, 2, 3,** and **4** on pages 24 through 27, and then try Exercises 32 and 34 on page 50.

Section 1.4 Introduction to Fair Division

A fair division problem consists of a set of goods (houses, paintings, money, cookies, jewelry, land, etc.) that are to be divided into N shares so that each player gets a *fair share*. For two players, the **Divider–Chooser Method for Two Players** is used. One person is randomly selected as the divider who divides the goods into two pieces that are of equal value to the divider. The chooser then selects one of the pieces and the divider gets the remaining piece.	See **Example 1** on page 33, and then try Exercise 35 on page 50.
Lone Divider–Chooser Method for More than Two Players Randomly choose (say, by tossing a coin) one of the two players (the divider) who divides whatever is to be shared into *n* pieces of equal value *to the divider*. The other players (the choosers) then secretly value each of the pieces.	See **Example 2** on page 35, and then try Exercise 36 on page 50.
The **Method of Adjusted Winner** is another method of dividing goods between two players.	See **Example 3** on pages 35 through 37, and then try Exercise 37 on page 50.
The **Method of Sealed Bids** is a way to divide assets that are not easily divided such as paintings, houses, cars, diamond rings, and boats. The method of sealed bids can also be used to fairly divide rent among roommates.	See **Example 4** on pages 37 and 38, and then try Exercises 38 and 39 on page 50.
The **Knaster inheritance procedure** is a variation of the Method of Sealed Bids.	See **Example 5** on pages 40 and 41, and then try Exercise 40 on page 50.

Review Exercises

For Exercises 1 to 6, use inductive reasoning to predict the next number in the list.

1. $1, \dfrac{1}{2}, \dfrac{1}{4}, \dfrac{1}{8}, \dfrac{1}{16},$?

2. $2, -4, 8, -16, 32,$?

3. $4, 1, 3, 0, 2, -1, 1,$?

4. $1, 4, 2, 8, 4, 16, 8, 32, 16,$?

5. $0, 1, 10, 11, 100, 101, 110, 111,$?

6. What is the next row of numbers in the following?

$$
\begin{array}{ccccccccc}
 & & & & 1 & & 1 & & \\
 & & & 1 & & 2 & & 1 & \\
 & & 1 & & 3 & & 3 & & 1 \\
 & 1 & & 4 & & 6 & & 4 & & 1 \\
1 & & 5 & & 10 & & 10 & & 5 & & 1
\end{array}
$$

For Exercises 7 to 10, use inductive reasoning to predict the next image in the list.

7.

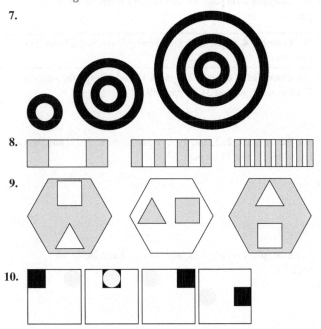

8.

9.

10.

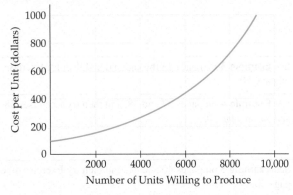

11. **Determining Supply** The graph below, called a *supply graph*, is used by economists to show the number of units of a product a manufacturer is willing to produce at a certain price.

If this manufacturer produces 10,000 units, does the manufacturer think that each unit will sell for more or less than $1000?

12. **Warehouse Inventory** A store orders cases of tomato sauce from a warehouse. The following bar graph shows the number of cases of tomato sauce in the warehouse for the first four months of a year.

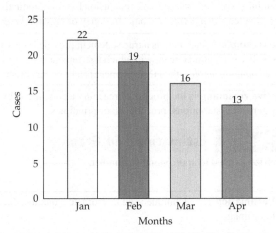

Using inductive reasoning, how many cases of tomato sauce will be in the warehouse in May?

13. **Falling Bodies** The following table shows the distance a rock has fallen after various times.

Time (seconds)	0	0.5	1	1.5	2	2.5	3	3.5	4
Distance (meters)	0	4	16	36	64	100	144	196	256

Using inductive reasoning, will the distance the rock has fallen after 4.5 seconds be less than or more than 316 meters?

14. **Ocean Depths** The following table shows the percent of light that reaches various depths in the ocean.

Depth (meters)	0	20	40	60	80	100	120	140	160
Percent of Light	100	74	55	40	30	22	16	12	9

Using inductive reasoning, will the percentage of light reaching a depth of 180 meters be more or less than 9%?

For Exercises 15 to 18, find a counterexample for the given statement.

15. All positive numbers are greater than 1.

16. Every number has a reciprocal.

17. The product of two numbers is always greater than either of the factors.

18. If two different rectangles have a perimeter of 100 feet, then the rectangles have the same area.

19. **College Majors** Michael, Clarissa, Reggie, and Ellen are attending Florida State University (FSU). One student is a computer science major, one is a chemistry major, one is a business major, and one is a biology major. From the following clues, determine which major each student is pursuing.

1. Michael and the computer science major are next-door neighbors.

2. Clarissa and the chemistry major have attended FSU for 2 years. Reggie has attended FSU for 3 years, and the biology major has attended FSU for 4 years.

3. Ellen has attended FSU for fewer years than Michael.

4. The business major has attended FSU for 2 years.

20. Little League Sponsors Each of the Little League teams in a small rural community is sponsored by a different local business. The names of the teams are the Dodgers, the Pirates, the Tigers, and the Giants. The businesses that sponsor the teams are the bank, the supermarket, the service station, and the drugstore. From the following clues, determine which business sponsors each team.

   1. The Tigers and the team sponsored by the service station have winning records this season.

   2. The Pirates and the team sponsored by the bank are coached by parents of the players, whereas the Giants and the team sponsored by the drugstore are coached by the director of the community center.

   3. Jake is the pitcher for the team sponsored by the supermarket and coached by his father.

   4. The game between the Tigers and the team sponsored by the drugstore was rained out yesterday.

21. Weighing Coins You have 10 coins, one of which is heavier than the others. Using a balance scale, determine the heavier coin in three weighings.

22. Hauling Bricks A standard red brick weighs about 5 pounds. A pallet of bricks contains 500 bricks. Would a truck rated to carry a load of one ton (2000 pounds) be able to haul a pallet of bricks?

23. Estimating Measurements One inch is a little more than 2.5 centimeters. Is 10 inches more or less than 25 centimeters?

24. Reading Speeds The average reading rate of an adult is approximately 300 words per minute. About how long would it take an adult to read a magazine article that is 3000 words?

25. Estimating Berries Estimate the number of acai berries in the picture below.

lazyllama/Shutterstock.com

26. Leonardo da Vinci Leonardo da Vinci made numerous measurements of the human body. For one measurement, he noted that the distance from the bottom of the chin to the top of the head was one-eighth the height of a man. If the height of a man is 72 inches, what is the approximate distance from the bottom of the chin to the top of the head?

27. Astronomy The distance to Alpha Centauri, our nearest star other than the sun, is 41,320,000,000,000 kilometers. Write that distance in scientific notation.

28. Electron Microscopes An electron microscope can see objects as small as 0.00000000012 meter. Write that number in scientific notation.

29. U.S. Pet Spending The following graph shows the total amount spent (in millions of dollars) on pets in the United States for the years shown.

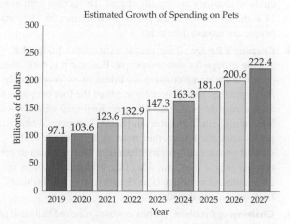

Between which two years was the increase in the amount spent on pets the greatest?

30. Healthcare Spending The following pie chart shows what portion of each $1 spent on health care goes to various segments of the healthcare industry.

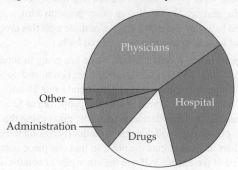

Which segment of the healthcare industry receives the largest portion of each $1 spent?

31. Car Depreciation The following line graph shows the decline in the value of a car that was originally purchased for $35,000 after each year of ownership.

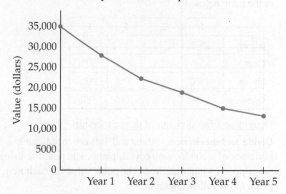

   a. Between which two years did the car decrease in value the most?

b. At the end of 4 years, is the value of the car less than or more than one-half of its original value?

In Exercises 32 to 34, solve each problem using Polya's four-step problem-solving strategy. Label your work so that each of the four steps is clearly identified.

32. Number Games An unknown number of people are each given a number, in sequence beginning with 1, until each person has a number. The people then stand, in order, around a circle so that they are equally spaced. The person with number 11 is directly opposite the person with number 35. How many people are around the circle?

33. Crossing a Bridge Four people must cross a bridge that is wide enough for just two people. Because it is dark, one person in the group crossing the bridge must have the only flashlight they have. The rates at which the four people can walk across the bridge are different: Kevin, 10 minutes; Asher, 5 minutes; Seth, 2 minutes; and Sarah, 1 minute. When two people cross the bridge, they must walk at the rate of the slowest person and have the flashlight. Devise a plan to get the four people from one side of the bridge to the other in 17 minutes. (The bridge is too long to throw the flashlight from one side to the other.)

34. Challenging Problem A box contains nine red balls and nine blue balls along with a bag containing extra red and blue balls. Without looking into the box, two balls are removed. After you look at the balls, if the two balls are the same color, a blue ball is selected from the bag and placed in the box. If the balls are different colors, a red ball is selected from the bag and placed in the box. What is the color of the last ball removed from the box? (Suggestion: Try to solve an easier problem with, say, three red balls and three blue balls. Continue with this procedure and look at what happens to the red balls.)

35. Sharing a Pizza Raymond and Connie are going to share a pizza that costs $12.00 that is one-half pepperoni and one-half cheese. Raymond values the pepperoni portion at $8 and the cheese portion at $4. Connie values the pepperoni at $3 and the cheese at $9. They decide to share the pizza using the divider–chooser method. Connie is randomly chosen to be the divider. If Connie cuts the pizza so that one piece contains one-third of the cheese half and the other piece contains one-sixth of the pepperoni half, did she make a fair cut according to the rules of fair division?

36. Dividing a Partnership Jeffrey, Clara, Alyssa, and Cameron decide to sell the assets in their partnership using the lone–divider method with Jeffrey as the divider. The bids are shown in the table below.

	S_1	S_2	S_3	S_4
Jeffrey	25%	25%	25%	25%
Clara	15%	35%	20%	30%
Alyssa	15%	45%	20%	20%
Cameron	20%	35%	10%	35%

Determine a fair division of the partnership.

37. Divide an Inheritance Alana and Taft are to split an inheritance of a condo, motorcycle, piano, and painting using the adjusted winner method. The point value for each item is given in the following table.

	Alana	Taft
Condo	40	50
Motorcycle	10	5
Piano	25	20
Painting	25	25

Determine a fair division of the estate.

38. Divide an Estate Three brothers are going to divide the assets of an estate using the method of sealed bids. The bids for the assets by each brother are shown in the table below.

	House	Condo
Miguel	$550,000	$300,000
Cannon	$450,000	$350,000
Ralph	$350,000	$400,000

Determine a fair distribution of the assets.

39. Share an Apartment Suppose Thomas, Michael, and Claudia move into a three-bedroom apartment for which the monthly rent is $2700. The three bedrooms are of various sizes, with one bedroom having a garden view, another bedroom having city skyline view, and the third bedroom no view. The table below shows how the roommates value each bedroom. Use the method of sealed bids to determine what portion of the rent each roommate should pay depending on the bedroom chosen.

	Thomas	Michael	Claudia
Garden View	$1000	$1100	$1000
Skyline View	$ 900	$ 900	$1000
No View	$ 800	$ 700	$ 700

40. Divide an Estate An estate consists of a Ming vase, Stradivarius violin, and Manet painting. The estate will be divided among three siblings, Jocelyn, Harrison, and Colby, using the Knaster inheritance procedure. The bids, in thousands, for each of the siblings are given in the table below.

	Jocelyn	Harrison	Colby
Vase	$ 810	$ 900	$ 750
Violin	$1200	$1500	$1200
Painting	$1500	$1650	$1800

Find a fair division of the estate. Round to nearest dollar.

41. Divide an Estate An estate consists of a Bentley, a beachfront condo, a harp, and a Rodin sculpture. The estate will be divided among four siblings, Santoro, Benjamin, Bianca, and Vivian, using the Knaster inheritance procedure. The bids, in thousands of dollars, for each item are given in the table below.

	Santoro	Benjamin	Bianca	Vivian
Bentley	$ 150	$ 125	$ 120	$ 130
Condo	$1000	$1200	$1500	$1300
Harp	$ 12	$ 12	$ 8	$ 16
Sculpture	$1500	$2000	$1800	$1600

Find a fair division of the estate.

Project 1

A Famous Argument

Read the U.S. Declaration of Independence. Copies can be found on the Internet.

1. Make a list of some of the premises. You should have at least six.

2. What is the conclusion?

3. Was Thomas Jefferson trying to make an inductive or deductive argument?

Project 2

Fair Division of Chores

The method of sealed bids can be used to divide unwanted chores such as having to clean your apartment when you move or getting a security deposit returned. Because chores are unpleasant, negative numbers are entered into the chart. The most unpleasant chore would have the *lowest* negative number. For instance, if cleaning the bathroom is more unpleasant to you than washing windows, then you might assign -8 to cleaning the bathroom and -5 to washing the windows because $-8 < -5$.

Refer to the Excel templates from your digital resources for help with this project. For this exercise, find the template for the Method of SEALED BIDS to determine how chores could be assigned. The spreadsheet allows for as many as five people and as many as four chores. The spreadsheet can also be used for the more normal use of dividing the assets of an estate or allocating a bedroom in an apartment or home. In these cases, the values entered are positive numbers.

1. Use the Excel template to verify the answer to Example 4 in Section 1.4.

2. Use the Excel template to verify the answer to the allocation of bedrooms that follows Example 4 in Section 1.4.

3. With three or four classmates, enter your names and four chores in the spreadsheet. Then each of you secretly value the chores. Remember to use negative numbers. For the purposes of this exercise, the sum of the numbers should be -300, the amount of a security deposit. The number is negative because it is owed. When everyone is finished, enter the numbers in the spreadsheet.

   a. From the result, determine which chores each of you would complete.

   b. Some of the numbers in the final disposition line will be negative and some will be positive. What is the meaning of those numbers?

Math in Practice

Medical studies are conducted frequently. After seeing a stunning performance at the community center, you notice a flier on the bulletin board that says "Are you interested in being more fit? Adults ages 18 to 60 can apply. Participants will be compensated."

> **Are you interested in being more fit? Adults ages 18 to 60 can apply. Participants will be compensated.**

A local medical school is running the study. All participants are required to walk a certain number of steps each day. After 6 days each participant is to have blood pressure measured and daily step count recorded at the medical school. This study will take a total of 30 days to complete.

Part 1

You decide to be a participant in this medical study as a side gig. Generally, you walk at a brisk pace of 116 steps in one minute.

a. Walking at this pace at lunchtime, estimate how long, in minutes, it would take you to get in 6000 steps.

b. Will you get back to work on time if the lunch break is 45 minutes?

Part 2

Your friend Takeshi is also a participant in the same medical study that requires walking a certain amount of steps each day. Takeshi has an hour for lunch. He takes 17 minutes to eat his tasty food and then goes for a walk to get in some steps. If Takeshi walks at a pace of 96 steps in one minute, estimate how many steps he can take in the time remaining for his lunchtime.

© Alexis Verity

Part 3

After Takeshi gets home from work, he discovers that he has taken 6423 steps. His group in the study requires a daily goal of 11,000 steps. Takeshi puts his dinner in the oven and sets the timer for 30 minutes. How fast, in steps per minute, does Takeshi have to walk to reach 10,000 steps before the timer chimes?

Part 4

The results from the medical study have been recorded for 30 days. There were four different groups of participants who successfully completed the requirements of the study. Group A had 107 participants who took a total of 35,496,204 steps. Group B had 95 participants who took a total of 27,901,613 steps. Group C had 102 participants who took a total of 30,742,087 steps. Group D had 89 participants who took a total of 24,139,305 steps. Estimate how many steps each participant took. Write your answer in scientific notation.

Hint: First estimate the number of participants and then estimate the total number of steps taken by all the participants, using scientific notation.

Sets and Logic

Section	How This Is Used . . .
2.1 Sets, Set Operations, and Applications	Sets are fundamental to mathematics. They are used to organize similar objects so that the properties of those objects can be studied and their relationships to one another examined. For the various numbers we use—negative numbers, fractions, decimals, and others—can be separated into sets. Set operations are used in analyzing surveys, testing medical treatments, statistics, probability, and other disciplines.
2.2 Categorical Logic	When we make a statement such as "All birds are reptiles," we are making a statement about two categories, birds and reptiles. When these types of statements are used to persuade a person to reach a certain conclusion, we need strategies to determine whether the reasoning of the argument is valid.
2.3 Propositional Logic	Not all statements are categorical. A statement such as "If there is enough snow, then I will go skiing" is a propositional statement. "My next car will be red or blue" is another example of a propositional statement. As a continuation of Section 2.2, in this section, we examine strategies to determine the validity of propositional arguments.
2.4 Investigating Fallacies	A fallacy is an ambiguous, misleading, or unsound argument. We are presented with fallacies on a daily basis, especially in advertisements and political arguments. Detecting these fallacies is an essential part of making informed choices.

Math in Practice: Logic, Sets, and Databases, p. 97

Applications of Sets

Set operations are used when various medical treatments are tested for their effectiveness. For instance, suppose a new flu vaccine is being tested. Several outcomes are possible. A person gets the flu vaccine but still gets the flu. Some people do not get the flu even if they do not get the vaccine. Other people get the vaccine and do not get the flu and still others do not get the vaccine and get the flu. By looking at the numbers in each group, researchers can gauge the effectiveness of the vaccine.

   In Exercise 58, page 65, an example of testing a medical procedure is discussed.

Sets, Set Operations, and Applications

Learning Objectives:

- Identify elements of a set.
- Find the complement of a set.
- Find subsets of a set.
- Find the union and intersection of sets.
- Calculate the number of subsets of a set.
- Solve applications involving sets.

Be Prepared: Prep Test 2 reviews concepts used in this chapter. This can be found in your Companion Site at Cengage.com

Sets

To better understand the relationships among living things, biologists organize living beings into different kingdoms. Humans belong to the kingdom Animalia.

Similarly mathematicians organize objects into *sets*. The objects that belong in a set are the **elements**, or **members** of the set. Braces are used to enclose the elements of a set. For instance, {1, 2, 3, 4, 5} is a set containing the elements 1, 2, 3, 4, and 5.

Listing the elements of a set is called the **roster method** of describing a set.

We designate an element of a set by using the $\in$ symbol.

$$4 \in \{1, 2, 3, 4, 5\} \quad \text{and} \quad 7 \notin \{1, 2, 3, 4, 5\}$$

Sets are general objects and may contain elements other than numbers. For instance, we could have the set of lowercase letters of the English alphabet.

$$A = \{a, b, c, d, e, f, g, h, i, j, k, l, m, n, o, p, q, r, s, t, u, v, w, x, y, z\}$$

As we did previously, a letter (A for the set just given) is frequently used to designate a set. Then we can write $w \in A$, which is more convenient than

$$w \in \{a, b, c, d, e, f, g, h, i, j, k, l, m, n, o, p, q, r, s, t, u, v, w, x, y, z\}$$

The following sets of numbers are used frequently in mathematics.

Basic Number Sets

Natural numbers or counting numbers $N = \{1, 2, 3, 4, 5, \ldots\}$

Whole numbers $W = \{0, 1, 2, 3, 4, 5, \ldots\}$

Integers $I = \{\ldots, -4, -3, -2, -1, 0, 1, 2, 3, 4, \ldots\}$

Rational numbers Q = the set of all terminating or repeating decimals

Irrational numbers $\mathcal{I}$ = the set of all nonterminating, nonrepeating decimals

Real numbers R = the set of all rational or irrational numbers

---

The set of natural numbers is also called the *set of counting numbers*. The three dots … are called an **ellipsis** and indicate that the elements of the set continue in a manner suggested by the elements that are listed.

The integers $\ldots, -4, -3, -2, -1$ are **negative integers**. The integers $1, 2, 3, 4, \ldots$ are **positive integers**. Note that the natural numbers and the positive integers are the same set of numbers. The integer zero is neither a positive nor a negative integer.

If a number in decimal form terminates or repeats a block of digits without end, then the number is a rational number. Rational numbers can also be written in the form $\dfrac{p}{q}$, where p and q are integers and $q \neq 0$. For example,

$$\frac{1}{4} = 0.25 \quad \text{and} \quad \frac{3}{11} = 0.\overline{27}$$

are rational numbers. The bar over the 27 means that the block of digits 27 repeats without end; that is, $0.\overline{27} = 0.27272727\ldots$.

A visual representation of the sets of numbers as they relate to each other.

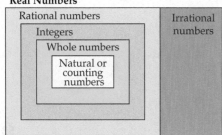

Real Numbers

A decimal number that neither terminates nor repeats a fixed group of digits is an **irrational number**. The number 0.35355355535555... is an irrational number. There is a pattern to the digits, but there is no group of digits that repeats. A famous irrational number is $\pi \approx 3.141592654$. Every real number is either a rational number or an irrational number.

A **prime number** is a natural number greater than 1 that is only divisible by itself and 1. The numbers 2, 5, 11, 23, 31, and 97 are just a few prime numbers. The number 12 is not a prime number because it is divisible by 1, 2, 3, 4, 6, and 12. The number 12 is a **composite number**.

Example 1   Use the Roster Method to Write a Set

a. Write the set of prime numbers less than 20.

b. Write the set of negative integers greater than or equal to -8.

c. Write the set of even whole numbers less than 10.

Solution

a. $\{2, 3, 5, 7, 11, 13, 17, 19\}$

b. $\{-8, -7, -6, -5, -4, -3, -2, -1\}$

c. $\{0, 2, 4, 6, 8\}$

Note: 0 is an even number; it is divisible by 2.

The order in which the elements of a set are listed is not important. For instance, in Example 1, part c, we could have listed the elements as {4, 8, 0, 2, 6}. The sets are the same. Two sets are **equal** if they have exactly the same elements.

The set that has no elements is called the **empty set**. The set of numbers that are both even and odd is the empty set. The empty set is symbolized by $\varnothing$. It is *not correct* to write the empty set as $\{\varnothing\}$. The set $\{\varnothing\}$ has one element, the empty set.

Complement of a Set

In complex problem-solving situations and even in routine daily activities, we need to identify all the elements that are under consideration. For instance, if we are trying to determine all possible area codes, then we should only consider three-digit numbers that do not start with a zero. The set of all elements that are being considered is called a **universal set**. We will use the letter U to denote the universal set.

The Complement of a Set

The **complement** of a set A, denoted by A', is the set of all elements of the universal set U that are not elements of A.

Example 2   Find the Complement of a Set

Let $U = \{1, 2, 3, 4, 5, 6, 7, 8, 9, 10\}$ and $S = \{2, 3, 5, 7\}$. Find S'.

Solution

S' is all the elements of U that are not in S. $S' = \{1, 4, 6, 8, 9, 10\}$.

There are two fundamental results concerning the universal set and the empty set. Because the universal set contains all elements under consideration, the complement of the universal set is the empty set. Conversely, the complement of the empty set is the universal set, because the empty set has no elements and the universal set contains all the elements under consideration. Using mathematical notation, we state these fundamental results as follows:

> **The Complement of the Universal Set and the Complement of the Empty Set**
>
> $$U' = \varnothing \qquad \text{and} \qquad \varnothing' = U$$

Subsets of a Set

Consider the set of natural numbers $\{1, 2, 3, \ldots\}$ and the set of even natural numbers $\{2, 4, 6, \ldots\}$. Every even number is a natural number. The even numbers are said to be a *subset* of the natural numbers.

Take Note: It is always correct to use the symbol $\subseteq$ to denote one set being a subset of another. The symbol $\subset$ is used only when one set is a *proper* subset of the other.

> **A Subset of a Set**
>
> Set A is a subset of set B, denoted by $A \subseteq B$, if every element of A is an element of B.
>
> **Examples**
> 1. $\{a, 2, b, 4, c, 6\} \subseteq \{a, b, c, d, 1, 2, 3, 4, 5, 6\}$
> 2. $\{a, 2, b, 7, c, 8\} \nsubseteq \{a, b, c, d, 1, 2, 3, 4, 5, 6\}$. Not every element of $\{a, 2, b, 7, c, 8\}$ is an element of $\{a, b, c, d, 1, 2, 3, 4, 5, 6\}$.
> 3. $\{2, 3, 5, 7, 11\} \subseteq \{2, 3, 5, 7, 11\}$. Every set is a subset of itself.
> 4. $\varnothing \subseteq \{0\}$. The empty set is a subset of every set.
>
> Set A is a **proper** subset of set B, denoted by $A \subset B$, if every element of A is an element of B, and $A \neq B$. For instance, $\{a, b, c\} \subset \{a, b, c, d\}$. For Example 3, $\{2, 3, 5, 7, 11\}$ is not a proper subset of $\{2, 3, 5, 7, 11\}$. The two sets are equal.
>   The empty set is a proper subset of every set except itself. In Example 4, $\varnothing \subset \{0\}$.

Example 3   **Apply the Definition of a Subset**

Determine whether the statement is true or false.

a. $\{2, 6, 9\} \subseteq \{2, 4, 6, 8, 10\}$

b. $\{1, 3, 5, 7\} \subseteq \{3, 5, 1, 7\}$

c. $\varnothing \subseteq \{0\}$

d. $3 \subset \{1, 3, 5, 7, 9\}$

Solution

a. False. $9 \in \{2, 6, 9\}$ but $9 \notin \{2, 4, 6, 8, 10\}$.

b. True. Every set is a subset of itself.

c. True. The empty set is a subset of every set.

d. False. The correct form is $\{3\} \subset \{1, 3, 5, 7, 9\}$ or $3 \in \{1, 3, 5, 7, 9\}$.

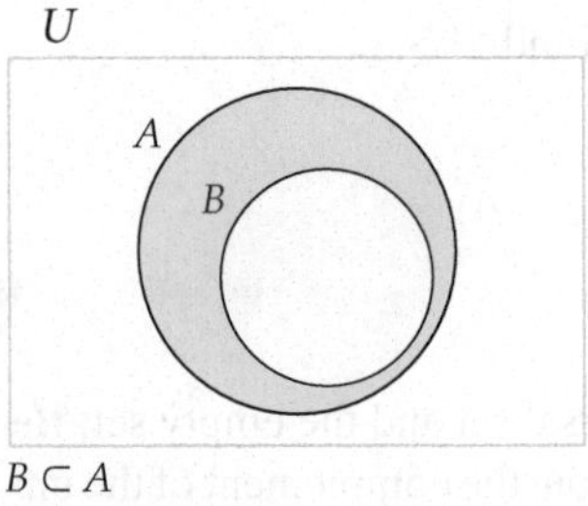

English logician John Venn (1834–1923) developed diagrams that we now refer to as *Venn diagrams* to illustrate sets and relationships among sets. In a **Venn diagram**, a rectangular region represents the universal set, and oval or circular regions drawn inside the rectangle represent subsets of the universal set.

The Venn diagram in Figure 2.1 shows B as a proper subset of A, both of which are in the universal set U.

$B \subset A$

Figure 2.1

Intersection and Union of Sets

Just as we can perform operations such as addition and multiplication of numbers, we can perform operations on sets.

In everyday usage, the word *intersection* refers to the common region where two streets cross. The intersection of two sets is defined in a similar manner.

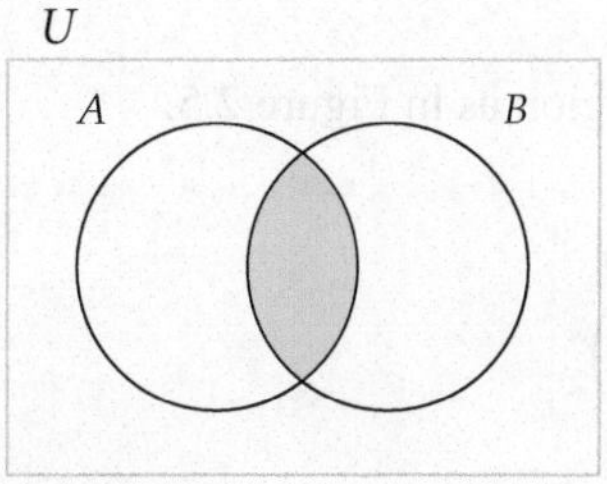

Intersection of Two Sets

The **intersection** of sets A and B, noted by $A \cap B$, is the set of elements that are common to both A and B.

In Figure 2.2, the region shown in blue represents the intersection of sets A and B.

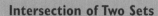

Example 4   Find the Intersection of Sets

Let $A = \{1, 4, 5, 7\}$, $B = \{2, 3, 4, 5, 6\}$, and $C = \{3, 6, 9\}$. Find

a. $A \cap B$ **b.** $A \cap C$

Solution

a. The elements common to A and B are 4 and 5.

$$A \cap B = \{1, 4, 5, 7\} \cap \{2, 3, 4, 5, 6\}$$
$$= \{4, 5\}$$

b. Sets A and C have no common elements. Thus, $A \cap C = \varnothing$.

Disjoint Sets

Two sets are **disjoint** if their intersection is the empty set.

The sets A and C in Example 4b are disjoint. The bottom Venn diagram in Figure 2.3 illustrates two disjoint sets.

In everyday usage, the word *union* refers to the act of uniting or joining together. The union of two sets has a similar meaning.

Union of Two Sets

The **union** of set A and B, denoted by $A \cup B$, is the set of elements that belong to A, belong to B, or belong to A and B.

In Figure 2.4, the region shown in blue represents the union of A and B.

Example 5   Find the Union of Two Sets

Let $A = \{1, 4, 5, 7\}$, $B = \{2, 3, 4, 5, 6\}$, and $C = \{3, 6, 9\}$. Find

a. $A \cup B$ **b.** $A \cup C$

Solution

a. List all the elements of set A, which are 1, 4, 5, and 7. Then add to your list the elements of set B that have not already been listed—in this case 2, 3, and 6.

Enclose all elements with a pair of braces. Thus,

$$A \cup B = \{1, 4, 5, 7\} \cup \{2, 3, 4, 5, 6\}$$
$$= \{1, 2, 3, 4, 5, 6, 7\}$$

b. $A \cup C = \{1, 4, 5, 7\} \cup \{3, 6, 9\}$
$$= \{1, 3, 4, 5, 6, 7, 9\}$$

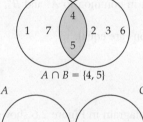

$A \cap B$

Figure 2.2

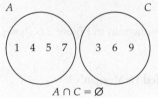

$A \cap B = \{4, 5\}$

$A \cap C = \varnothing$

Figure 2.3

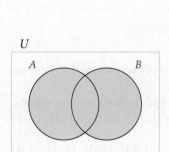

$A \cup B$

Figure 2.4

Venn diagrams can be used to illustrate whether two sets are equal.

Example 6   **Draw a Venn Diagram for a Set Operation**

Let A and B be two subsets of a universal set U. Draw a Venn diagram to determine whether $(A \cup B)'$ and $A' \cap B'$ are equal.

Solution

For each operation, draw a Venn diagram and label each region as in Figure 2.5.

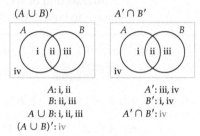

A: i, ii
B: ii, iii
$A \cup B$: i, ii, iii
$(A \cup B)'$: iv

A': iii, iv
B': i, iv
$A' \cap B'$: iv

Figure 2.5

For $(A \cup B)'$, A is regions i and ii; B is regions ii and iii; $A \cup B$ is i, ii, and iii (the union contains all the elements of A and all the elements of B); $(A \cup B)'$ is all the elements not in the union of A and B, which is region iv.

For $A' \cap B'$, A' is regions iii and iv (the regions that are not in A); B' is regions i and iv (the regions that are not in B); $A' \cap B'$ is the region in common to A' and B', which is region iv.

Because the two regions are the same, the two expressions are equal.

$(A \cup B)' = A' \cap B'$.

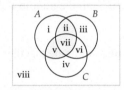

A: i, ii, v, vii
B: ii, iii, vi, vii
C: iv, v, vi, vii
$B \cup C$: ii, iii, iv, v, vi, vii
$A \cap (B \cup C)$: ii, v, vii

Figure 2.6

A Venn diagram with three sets has eight regions. The Venn diagram in Figure 2.6 shows sets defined by $A \cap (B \cup C)$.

The result from Example 6 is part of an important result called DeMorgan's law.

DeMorgan's Law

For any two sets A and B, we have the following:

$$(A \cup B)' = A' \cap B' \qquad \text{and} \qquad (A \cap B)' = A' \cup B'$$

Number of Subsets of a Set

Let $A = \{1\}$. The subsets of A are $\varnothing$ (the empty set is a subset of every set) and A (the entire set is a subset of itself). There are two subsets.

Now let $B = \{1, 2\}$. The subsets of B are $\varnothing$, $\{1\}$, $\{2\}$, and B. There are four subsets.

Consider $C = \{1, 2, 3\}$. The subsets of C are $\varnothing$, $\{1\}$, $\{2\}$, $\{3\}$, $\{1, 2\}$, $\{1, 3\}$, $\{2, 3\}$, and C. There are eight subsets.

If $D = \{1, 2, 3, 4\}$, then the subsets of D are $\varnothing$, $\{1\}$, $\{2\}$, $\{3\}$, $\{4\}$, $\{1, 2\}$, $\{1, 3\}$, $\{1, 4\}$, $\{2, 3\}$, $\{2, 4\}$, $\{3, 4\}$, $\{1, 2, 3\}$, $\{1, 2, 4\}$, $\{1, 3, 4\}$, $\{2, 3, 4\}$, and D. There are 16 subsets.

Table 2.1 summarizes the results so far.

Table 2.1

Number of Elements in a Set: n	1	2	3	4	5	6	...
Number of Subsets of the Set:	2	4	8	16	?	?	...

It appears that if the number of elements in a set is increased by 1, then the number of subsets of the set is doubled. Thus, we suspect that a set with five elements will have $2 \times 16 = 32$ subsets and a set with 6 elements will have $2 \times 32 = 64$ subsets.

This can be described by using powers of 2.

$$2 = 2^1 \quad 4 = 2^2 \quad 8 = 2^3 \quad 16 = 2^4 \quad 32 = 2^5 \quad 64 = 2^6$$

These observations lend support for the following theorem.

Number of Subsets of a Set

A set with n elements has 2^n subsets.

Examples
- $\{1, 2, 3, 4, 5, 6, 7\}$ has seven elements, so it has $2^7 = 128$ subsets.
- $\{4, 5, 6, 7, 8, 9, 10, 11, 12, 13, 14, 15\}$ has 12 elements, so it has $2^{12} = 4096$ subsets.
- The empty set has no elements, so it has $2^0 = 1$ subset.

Consider the set A with n elements. All of the 2^n subsets are proper subsets of A, except for A itself. Thus, the number of proper subsets of A is $2^n - 1$.

Number of Proper Subsets of a Set

A set with n elements has $2^n - 1$ subsets.

Example
- $\{2, 3, 5, 7, 11\}$ has 5 elements, so it has $2^5 - 1 = 31$ proper subsets.

There are practical applications of this result.

Example 7 **Pizza Variations**

A restaurant sells basic cheese pizzas with seven possible toppings.

a. How many variations of pizzas can the restaurant serve?

b. What is the minimum number of toppings the restaurant must provide if it wishes to advertise that it offers more than 1000 variations of its pizzas?

Solution

a. The restaurant can serve a pizza with no topping, one topping, two toppings, three toppings, and so forth, up to all seven toppings. Let T be the set consisting of the seven toppings. The elements in each subset of T describe exactly one of the variations of toppings that the restaurant can serve. Consequently, the number of different variations of pizzas that the restaurant can serve is the same as the number of subsets of T. Thus, the restaurant can serve $2^7 = 128$ different variations of its pizzas.

b. Use the method of guessing and checking to find the smallest natural number n for which $2^n > 1000$.

$$2^8 = 256$$
$$2^9 = 512$$
$$2^{10} = 1024$$

The restaurant must provide a minimum of 10 toppings if it wishes to offer more than 1000 variations of its pizzas.

Applications

The **cardinality** of a set is the number of elements in the set. For instance, if the cardinality of $A = \{a, c, d, g, i, m\}$ is 6, we write $n(A) = 6$. The empty set has no elements, so $n(\varnothing) = 0$.

When counting the number of elements in a union of two sets, we must ensure that we do not count some elements twice. For instance, if $A = \{2, 4, 6, 8\}$ and $B = \{6, 7, 8, 9, 10\}$, then $A \cup B = \{2, 4, 6, 7, 8, 9, 10\}$ and $n(A \cup B) = 7$. However, if we were to add the cardinalities of set A and set B, we have $n(A) + n(B) = 4 + 5 = 9$. The difficulty is that 6 and 8 are common to each set and were counted twice. The correct procedure is given in the following.

The Inclusion-Exclusion Principle

For all finite sets A and B,
$$n(A \cup B) = n(A) + n(B) - n(A \cap B)$$

Using this principle for $A = \{2, 4, 6, 8\}$ and $B = \{6, 7, 8, 9, 10\}$, we have

$$n(A \cup B) = n(A) + n(B) - n(A \cap B)$$
$$= 4 + 5 - 2 = 7$$

This is the same result as we previously found.

Example 8   **Use the Inclusion-Exclusion Principle**

A school finds that 430 of its students are registered in chemistry, 560 in mathematics, and 225 in both chemistry and mathematics. How many students are registered in chemistry or mathematics?

Solution

Let $C = \{$students registered in chemistry$\}$ and let $M = \{$students registered in mathematics$\}$.

$$n(C \cup M) = n(C) + n(M) - n(C \cap M)$$
$$= 430 + 560 - 225$$
$$= 765$$

There are 765 students registered in chemistry or mathematics.

We can extend Venn diagrams to three sets.

Example 9   **Determine Values in a Three-Set Venn Diagram**

A survey asked people their preference for cheese on a grilled hamburger. The results are in the Venn diagram in Figure 2.7 with A representing American cheese, S representing Swiss cheese, and C representing cheddar cheese.

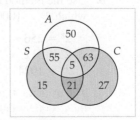

Figure 2.7

a. How many people preferred exactly one cheese?

b. How many people liked all three cheeses?

c. How many people preferred American and cheddar but not Swiss?

d. How many people liked exactly two of the cheeses?

e. If 250 people were surveyed, how many people did not like any of the cheeses?

Solution

a. The people who preferred exactly one cheese are those not in the intersection of any of the sets, which is $50 + 27 + 15 = 92$. There were 92 people who preferred exactly one cheese.

b. The people who like all three cheeses are those in $A \cap C \cap S = 5$. Five people liked all three cheeses.

c. The people who preferred American and cheddar but not Swiss are those in $A \cap C$, excluding those in $A \cap C \cap S$ who liked all three cheeses. From the Venn diagram, this is 63 people. Sixty-three people preferred American and cheddar but not Swiss.

d. The people who preferred exactly two of the three cheeses are those in $A \cap C$, $A \cap S$, and $S \cap C$, excluding those in $A \cap C \cap S$. This is $63 + 55 + 21 = 139$. There were 139 people who preferred exactly two of the cheeses.

e. The number of people who did not like any of the cheeses is 250 minus the number who liked exactly one cheese, 92 (from part a), plus those who like exactly two cheeses, 139 (from part d), plus the number who liked all three cheeses, five (from part c). This gives $250 - (92 + 139 + 5) = 14$. Fourteen people did not like any of the three cheeses.

Analyzing tabular data can be done with set operations.

Example 10 **A Survey in Tabular Form**

A survey of students from the School of Business (S), School of Humanities (H), and the School of Physical Sciences (P) at a university concerning the use of the Internet search engines Google (G), Yahoo! (Y), and Bing (B) yielded the results in Table 2.2.

Table 2.2

	Google (G)	Yahoo! (Y)	Bing (B)
Business (S)	440	310	275
Humanities (H)	390	280	325
Physical Sciences (P)	140	410	40

Use the data in the table to find each of the following
a. $n(H \cap Y)$ b. $n(G \cap P')$ c. $n(S \cup B)$

Solution

a. ($H \cap Y$) represents the students for the School of Humanities who use Yahoo!. From the table, that number is 280.

b. P' represents the set of students from the School of Business or the School of Humanities (students not from the School of Physical Sciences). Then $G \cap P'$ represents the set of Google users who are students from the School of Business or the School of Humanities. $n(G \cap P') = 440 + 390 = 830$.

c. $S \cup B$ represents students from the School of Business users or students who use Bing. The number of students from the School of Business surveyed is $440 + 310 + 275 = 1025$. The number of students surveyed who use Bing is $275 + 325 + 40 = 640$.

$$n(S \cup B) = n(S) + n(B) - n(S \cap B)$$
$$= 1025 + 640 - 275$$
$$= 1390$$
$$n(S \cup B) = 1390.$$

Many news stories concern testing athletes for performance-enhancing drugs. It is possible for an athlete to test positive for a banned substance even without actually taking the banned drug. This is called a **false positive**, a result that incorrectly shows that a drug (or other condition) is present. A **false negative** occurs when a test shows a person does not have a certain drug present but, in fact, does have the drug present.

Table 2.3 shows the possibilities for any kind of test.

Table 2.3

	Test Says Yes	Test Says No
You Have a Condition	Correct outcome	False negative
You Do Not Have a Condition	False positive	Correct outcome

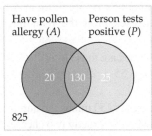

Have pollen allergy (A) Person tests positive (P)

20 130 25

825

Figure 2.8

Example 11 **Determine a False Positive**

A test for a pollen allergy is given to 1000 people. The results are shown in the Venn diagram in Figure 2.8. Using the Venn diagram, complete the following table.

	Test Is Positive	Test Is Negative
Person Has Pollen Allergy		
Person Does Not Have Pollen Allergy		

Solution

The number of people with a pollen allergy and who tested positive is $n(A \cap P) = 130$. The number of people with a pollen allergy who did not test positive but have a pollen allergy is 20. These are the false negatives. The number of people who tested positive but do not have a pollen allergy is 25. These are the false positives. The people without a pollen allergy and tested negative are 825. Enter these numbers in the table.

	Test Is Positive	Test Is Negative
Person Has Pollen Allergy	130	20
Person Does Not Have Pollen Allergy	25	825

2.1 Exercise Set

Think About It

1. True or False
  a. 5 is an integer.
  b. $-\dfrac{5}{3}$ is a rational number.
  c. 0 is a whole number.
  d. 49 is a prime number.
  e. Some rational numbers are also irrational numbers.
  f. All integers are rational numbers.
  g. All rational numbers are real numbers.
  h. π is a real number.

2. Suppose C and D are two nonempty sets.
  a. If $C \subset D$, is it possible that $C = D$?
  b. If $C \subseteq D$, is it possible that $C = D$?

3. If A and B are two finite, nonempty sets and $n(A \cup B) = n(A) + n(B)$, what can be said about A and B?

4. If A is a subset of a universal set U, what is $(A')'$?

5. When a person with chicken pox is given a test for chicken pox, the test comes back negative. This is an example of a

_____ .

6. When a person with chicken pox is given a test for chicken pox, the test comes back positive. This is an example of a

_____ .

Preliminary Exercises

For Exercises P1 through P11, if necessary, use the referenced example following the exercise for assistance.

P1. Use the roster method to list the natural numbers less than 20 that are divisible by 3. **Example 1**

P2. Let $U = \{$natural numbers less than 20$\}$ and $P = \{$prime numbers less than 20$\}$. Find P'. **Example 2**

P3. Let $A = \{4, 6, 8, 9, 10, 12\}$. For each of the following, answer True or False. **Example 3**

a. $\{4, 8, 12\} \subset A$

b. $\{6, 7, 8\} \subset A$

c. $\varnothing \subseteq A$

d. $\{6\} \subseteq A$

P4. Let $A = \{-3, -2, -1, 0, 1, 2, 3\}$, $B = \{-6, -4, 4, 6\}$, and $C = \{-2, 0, 2, 4, 6\}$. Find **a.** $A \cap B$, **b.** $A \cap C$. **Example 4**

P5. Let $A = \{-3, -2, -1, 0, 1, 2, 3\}$, $B = \{-6, -4, 4, 6\}$, and $C = \{-2, 0, 2, 4, 6\}$. Find **a.** $A \cup B$, **b.** $A \cup C$. **Example 5**

P6. Draw a Venn diagram to show the set $A \cup (B \cap C)$. **Example 6**

P7. A car comes with upgrade options that include metallic paint, park assist, blind-spot monitoring, navigation system, sport wheels, and heated seats. How many variations are possible for a person considering this car? **Example 7**

P8. A survey of people who liked chocolate or peanut butter found that 52 people liked chocolate, 38 people like peanut butter, and 28 people liked both. How many people were in the survey? **Example 8**

P9. A survey asked students which team sport they preferred to watch. The results are in the following Venn diagram with H representing hockey, S representing soccer, and F representing football. **Example 9**

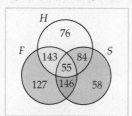

a. How many people preferred exactly one sport?

b. How many people liked all three sports?

c. How many people preferred hockey and football but not soccer?

d. How many people liked exactly two of the sports?

e. If 800 people were surveyed, how many people did not like any of the sports?

P10. A city council surveyed members of a community regarding a proposal for a new discount superstore. The results are shown in the following table. **Example 10**

	Age (I) 18–25	Age (II) 26–50	Age (III) 50+
Yes (Y)	524	409	265
No (N)	321	108	55
Unsure (U)	75	82	63

Find

a. $n(Y \cap I)$ **b.** $n(U \cup III)$ **c.** $n(II \cap U')$

P11. A test for a performance-enhancing drug is given to 400 athletes. The results are shown in the following Venn diagram. **Example 11**

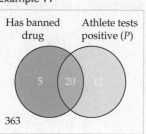

Using the Venn diagram, complete the following table.

	Positive Test	Negative Test
Athlete Has Banned Drug		
Athlete Does Not Have Banned Drug		

Section Exercises

For Exercises 1 to 8, use the roster method to write each given set.

1. The set of U.S. coins with a value of less than $1.

2. The set of months of the year with a name that ends with the letter y.

3. The set of planets in our solar system with a name that starts with the letter M.

4. The set of negative integers greater than -6.

5. The set of whole numbers less than 8.

6. The set of integers x that satisfy $x - 4 = 3$.

7. The set of integers x that satisfy $2x - 1 = -11$.

8. The set of natural numbers x that satisfy $x + 4 < 1$.

For Exercises 9 to 16, determine whether the given statement is true or false.

9. $b \in \{a, b, c\}$

10. $\{0\} \subset \{\text{natural numbers}\}$

11. $\{b\} \subset \{a, b, c\}$

12. $0 \subset \varnothing$

13. $\{3, 9\} \subseteq \{0, 3, 6, 9, 12\}$

14. $\{-5, -4, -3, -2\} \subseteq \{\text{integers}\}$

15. $5 \subset \{1, 2, 3, 4, 5\}$

16. $\{a, b, c\} \subset \{a, b, c\}$

In Exercises 17 to 22, find the complement of the set given that $U = \{0, 1, 2, 3, 4, 5, 6, 7, 8\}$.

17. $\{2, 4, 6, 7\}$

18. $\{3, 6\}$

19. $\varnothing$

20. $\{0, 1, 2, 3, 4, 5, 6, 7, 8\}$

21. $\{\text{prime numbers}\}$

22. $\{\text{composite numbers}\}$

In Exercises 23 to 26, list all of the subsets of the given set.

23. $\{a, z\}$

24. $\{1, 2, 3\}$

25. $\{I, II, III, IV\}$

26. $\varnothing$

In Exercises 27 to 30, find the number of subsets of the given set.

27. $\{a, e, i, o, u\}$

28. $\{\text{Monday, Tuesday, Wednesday, Thursday, Friday, Saturday, Sunday}\}$

29. **Upgrade Options** A truck company makes a pickup truck with 12 upgrade options. How many different versions of this truck can the company produce?

30. Brunch Options A restaurant provides a brunch where the omelets are individually prepared. Each guest is allowed to choose from 10 different ingredients.

   a. How many different types of omelets can the restaurant prepare?

   b. What is the minimum number of ingredients that must be available if the restaurant wants to advertise that it offers more than 4000 different omelets?

In Exercises 31 to 38, let $U = \{1, 2, 3, 4, 5, 6, 7, 8\}$, $A = \{2, 4, 6\}$, $B = \{1, 2, 5, 8\}$, and $C = \{1, 3, 7\}$. Find each of the following.

31. $A \cup B$

32. $A \cap B$

33. $A \cap B'$

34. $B \cap C'$

35. $(A \cup B)'$

36. $A' \cup B'$

37. $A \cap (B \cap C)$

38. $C \cap C'$

For Exercises 39 to 42, answer True or False.

39. For any two sets A and B, $A \cap B \subseteq A$.

40. For any two sets A and B, $A \subseteq A \cup B$.

41. If A is a subset of a universal set U, then $A \cap U = A$.

42. If B is a subset of a universal set U, then $B \cup U = B$.

For Exercises 43 to 46, draw a Venn diagram to represent the result of the set operation.

43. $A' \cup B$

44. $A' \cap B$

45. $A' \cup B'$

46. $(A' \cup B')'$

For Exercises 47 to 50, use Venn diagrams to determine whether the two expressions are equal.

47. $(A \cap B)'$; $A' \cup B'$ (This is the other half of DeMorgan's law.)

48. $A' \cup B'$; $(A \cup B)'$

49. $A \cap (B \cap C)$; $(A \cap B) \cap C$ (This is the associative property for intersection and is similar to the idea that $4 \times (5 \times 6) = (4 \times 5) \times 6$.)

50. $A \cap (B \cup C)$; $(A \cap B) \cup (A \cap C)$. This is a distributive property for sets and is similar to the distributive property for real numbers: $4 \times (5 + 6) = (4 \times 5) + (4 \times 6)$.

51. Investment Options Of 600 investors, 380 invested in stocks, 325 invested in bonds, and 75 did not invest in stocks or bonds.

   a. How many investors had invested in both stocks and bonds?

   b. How many investors had invested only in stocks?

52. Public Transportation Options A survey of 1500 commuters in New York City showed that 1140 take the subway, 680 take the bus, and 120 do not take either the bus or the subway.

   a. How many commuters take both the bus and the subway?

   b. How many commuters take only the subway?

53. Hotel Tipping Practices The management of a hotel conducted a survey of 6000 guests and their tipping habits. The results are in the following Venn diagram, with W representing those who tipped the wait staff, M representing those who tipped room-cleaning staff, and L representing those who tipped the concierge.

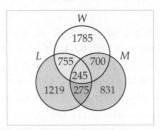

   a. How many people tipped exactly one service?

   b. How many people tipped all three services?

   c. How many people tipped exactly two of the services?

   d. How many people did not tip any of the services?

54. Computer Preferences A computer company surveyed 2000 customers to determine whether they liked a desktop computer, laptop computer, or a tablet computer. The result is shown in the following Venn diagram with D representing the people who liked a desktop computer, L representing those who liked a laptop computer, and T representing those who liked a tablet computer.

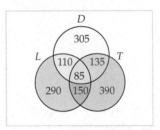

   a. How many people liked only one computer type?

   b. How many people liked all three computer types?

   c. How many people liked exactly two of the computer types?

   d. How many people did not like any of the computer types?

55. Fitness A survey was completed of individuals who were currently practicing Hatha yoga (Z), Ashtanga yoga (S), or Vinyasa yoga (W). All persons surveyed were also asked whether they were regularly swimming (E), walking (P), or running (M). The following table shows the results of the survey.

		Activity Type			
		E	**P**	**M**	**Totals**
	Z	124	82	65	271
Yoga Type	**S**	101	66	51	218
	W	133	41	48	222
	Totals	358	189	164	711

Find the number of surveyed people in each of the following sets.

   a. $S \cap E$

   b. $Z \cup M$

   c. $S' \cap (E \cup P)$

   d. $(Z \cup S) \cap (M')$

   e. $W' \cap (P \cup M')$

   f. $W' \cup P$

56. **Financial Aid** A college study categorized its seniors (*S*), juniors (*J*), and sophomores (*M*) who are currently receiving financial assistance. The types of financial assistance consist of full scholarships (*F*), partial scholarships (*P*), and government loans (*G*). The following table shows the results of the survey.

		Financial Assistance			
		F	**P**	**G**	**Totals**
	S	210	175	190	575
Year	**J**	180	162	110	452
	M	114	126	86	326
	Totals	504	463	386	1353

Find the number of students who are currently receiving financial assistance in each of the following sets.

a. $S \cap P$

b. $J \cup G$

c. $M \cup F'$

d. $S \cap (F \cup P)$

e. $J \cap (F \cup P)'$

f. $(S \cup J) \cap (F \cup P)$

57. **Computer Viruses** A computer is known to have files that are infected with various viruses. A new software product that detects viruses is run on the computer. The results are shown in the following Venn diagram where the green circle represents the infected files on the computer and the blue circle represents the files that the software identified as infected.

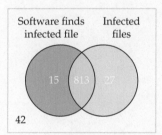

a. Use the Venn diagram to complete the following table.

	File Is Infected	File Is Not Infected
Found an Infected File		
Did Not Find Infected File		

b. How many false positives occurred?

c. Write a sentence that describes a false negative in the context of this test.

58. **Clinical Trials** A new drug to control blood pressure was given to 1000 patients. Another 250 patients were given a placebo. The result showed that 750 of the patients receiving the new drug had a decrease in blood pressure. For the group that received the placebo, 238 had a decrease in blood pressure. The results are shown in the following Venn diagram.

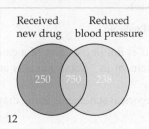

a. Use the Venn diagram to complete the following table.

	Reduced Blood Pressure	No Blood Pressure Reduction
Received New Drug		
Received Placebo		

b. How many false negatives occurred?

c. Write a sentence that describes a false positive in the context of this test.

59. **Genetic Testing** A new test to detect a certain genetic marker was given to 500 children, 207 of whom were known to have the genetic marker. The remaining 293 children did not have the marker. The results of the test are shown in the following table.

	Has Genetic Marker	Does Not Have Genetic Marker
Test Positive	195	5
Test Negative	12	288

a. How many false positives occurred?

b. Write a sentence that describes a false positive in the context of this test.

c. How many false negatives occurred?

d. Write a sentence that describes a false negative in the context of this test.

e. How many true positives occurred?

f. Write a sentence that describes a true positive in the context of this test.

g. How many true negatives occurred?

h. Write a sentence that describes a true negative in the context of this test.

i. Create a Venn diagram for this test.

60. **Rh– Testing** A new procedure to test for the rhesus negative blood protein (Rh–) is tested on 1000 people. Of the people tested, 180 are known to be Rh–. The results of the test are shown in the following table.

	Has Rh–	Does Not Have Rh–
Test Positive	159	4
Test Negative	21	816

a. How many false positives occurred?

b. Write a sentence that describes a false positive in the context of this test.

c. How many false negatives occurred?

d. Write a sentence that describes a false negative in the context of this test.

e. How many true positives occurred?

f. Write a sentence that describes a true positive in the context of this test.

g. How many true negatives occurred?

h. Write a sentence that describes a true negative in the context of this test.

i. Create a Venn diagram for this test.

Investigation

Subtractive Color Mixing Artists who paint with pigments use *subtractive color mixing* to produce different colors. In a subtractive color mixing system, the primary colors are cyan *C*, magenta *M*, and yellow *Y*. The following figure shows that when the three primary colors are mixed in equal amounts, using subtractive color mixing, they form black, *K*. Using set notation, we state this as $C \cap M \cap Y = K$. In subtractive color mixing, the colors red *R*, blue *B*, and green *G* are the secondary colors. As mentioned earlier, a secondary color is produced by mixing equal amounts of exactly two of the primary colors.

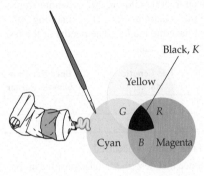

For Exercises 1 to 3, determine the color represented by each of the following.

1. $C \cap M \cap Y'$ 2. $C' \cap M \cap Y$ 3. $C \cap M' \cap Y$

Additive Color Mixing Computers and televisions make use of *additive color mixing*. The following figure shows that when the *primary colors* red *R*, green *G*, and blue *B* are mixed together using additive color mixing, they produce white, *W*. Using set notation, we state this as $R \cap B \cap G = W$. The colors yellow *Y*, cyan *C*, and magenta *M* are called *secondary colors*. A secondary color is produced by mixing exactly two of the primary colors.

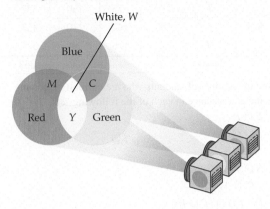

In Exercises 4 to 6, determine the color represented by each of the following.

4. $R \cap G \cap B'$
5. $R \cap G' \cap B$
6. $R' \cap G \cap B$

Section 2.2 <big>**Categorical Logic**</big>

Learning Objectives:

- Identify quantifiers in a categorical statement.
- Form the negation of a categorical statement.
- Determine the validity of an argument using an Euler diagram.

Quantifiers

A **categorical statement** is one that expresses how two sets relate to one another. In a statement, the word *some* and the phrases *there exists* and at *least one* are called **existential quantifiers**. Existential quantifiers are used as prefixes to assert the existence of something. "Some mammals are antelope" and "Some crayons are not blue" are categorical statements.

The words *none*, *no*, *all*, and *every* are called **universal quantifiers**. The universal quantifiers *none* and *no* deny the existence of something, whereas the universal quantifiers *all* and *every* are used to assert that every element of a given set satisfies some condition. "All dogs are mammals" and "No elephant is a bird" are categorical statements.

A categorical statement has a special form as shown here:

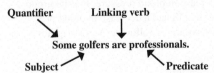

For instance, the statement "Some humans cannot swim" is not in the form of a categorical statement. We can rewrite this as a categorical statement as "Some humans are not swimmers."

If *P* and *Q* are statements, then we can represent categorical statements using diagrams as shown in Figure 2.9.

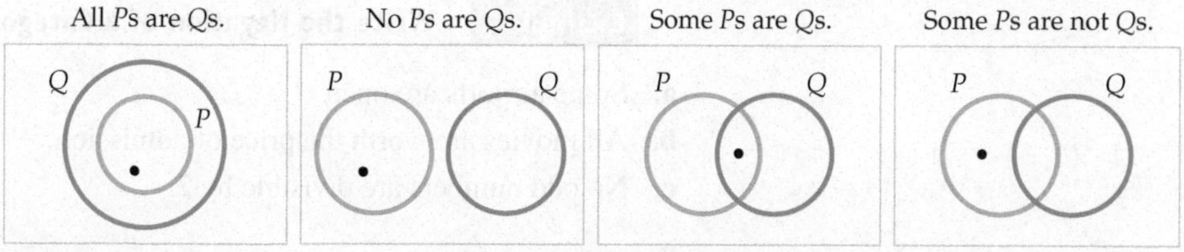

All *P*s are *Q*s.　　　　No *P*s are *Q*s.　　　　Some *P*s are *Q*s.　　　　Some *P*s are not *Q*s.

Figure 2.9

Although these look like Venn diagrams, these diagrams are referred to as **Euler diagrams** as a tribute to mathematician Leonard Euler (1707–1783), who used similar diagrams to determine whether arguments are valid.

| Example 1 | **Draw an Euler Diagram** |

Draw an Euler diagram for each of the following.

a. Some chefs are French.

b. No historians like politics.

Solution

a. Let one circle represent chefs and another represent a person from France. Because some chefs are French, the two categories intersect. The Euler diagram is shown in Figure 2.10.

b. Let one circle represent historians and another represent people who like politics. The statement asserts that no historian likes politics. This means the two categories do not intersect. The Euler diagram is shown in Figure 2.11.

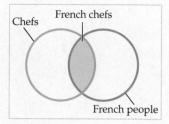

Figure 2.10

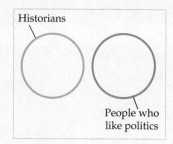

Figure 2.11

Example 1b illustrates an important point. A categorical statement does not have to be true. It is an assertion about the relationship between two categories.

Negation of a Categorical Statement

The **negation** of a true statement is a false statement, and the negation of a false statement is a true statement. Creating the negation of a categorical statement needs some care.

Consider the false statement "All sculptures are expensive." Because the statement is false, its negation must be true. As a categorical statement, this is expressed as "Some sculptures are not expensive." We could also write "At least one sculpture is not expensive."

Now consider the true statement "No dogs are cats." The negation of that statement must be false. As a categorical statement, this is expressed as "Some dogs are cats." Table 2.4 shows categorical statements and their negation.

Table 2.4

Statement	Negation
All *X* are *Y*.	Some *X* are not *Y*.
No *X* are *Y*.	Some *X* are *Y*.
Some *X* are not *Y*.	All *X* are *Y*.
Some *X* are *Y*.	No *X* are *Y*.

Example 2 **Write the Negation of a Categorical Statement**

a. Some airports are open.

b. All movies are worth the price of admission.

c. No odd numbers are divisible by 2.

Solution

a. No airports are open.

b. Some movies are not worth the price of admission.

c. Some odd numbers are divisible by 2.

Arguments

A major component of logic is analyzing arguments to determine their validity.

An Argument and a Valid Argument

An **argument** consists of a set of statements called **premises** and another statement called the **conclusion**. A **valid argument** is one that has a *logical form* that makes it impossible for the premises to be true and the conclusion false. An argument is invalid if it is not a valid argument.

A **syllogism** is a form of logical reasoning in which a conclusion is determined (whether validly or not) from linked premises, a **major premise** and a **minor premise**. Here is an example of a syllogism.

All mammals breathe air (major premise). A dog is a mammal (minor premise). Therefore, a dog breathes air (conclusion).

In this particular case, the argument is valid. In the next few examples, we will explore valid and invalid arguments.

Example 3 **Use an Euler Diagram to Determine the Validity of an Argument**

Use an Euler diagram to determine whether the following argument is valid or invalid.

All apples are fruit.

Granny Smith is a type of apple.

Therefore, Granny Smith is a fruit.

Solution

The premises are "All apples are fruit" and "Granny Smith is a type of apple." The conclusion is "Granny Smith is a fruit." The first premise states that the set of apples is a subset of the set of fruit. The second premise states the Granny Smith is a member of the set of apples. If we let x represent Granny Smith apples, then x must lie inside the set of apples. The Euler diagrams in Figure 2.12 shows the premises. The diagram in Figure 2.13 shows the premises and the conclusion.

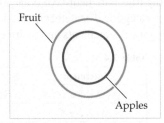

Figure 2.12

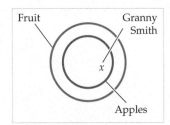

Figure 2.13

Because x is inside the set of fruit, the argument is valid.

If an Euler diagram can be drawn so that the conclusion does not necessarily follow from the premises, then the argument is invalid. This concept is illustrated in the next example.

> **Example 4**   **Use an Euler Diagram to Determine the Validity of an Argument**

Use an Euler diagram to determine whether the following argument is valid or invalid.

Some Impressionist paintings are Renoirs.

Dance at Bougival is an Impressionist painting.

∴ *Dance at Bougival* is a Renoir.

The symbol ∴ is read Therefore.

The premises are "Some Impressionist paintings are Renoirs" and "*Dance at Bougival* is an Impressionist painting." The Euler diagram in Figure 2.14 illustrates the premise that "Some Impressionist paintings are Renoirs." The intersection of the sets are the Impressionist paintings by Renoir. Let *x* represent the painting *Dance at Bougival*. The *x* can be placed in the intersection of the two sets as shown in the middle figure. However, the *x* can be placed in the set of Impressionist paintings *not* in the intersection of the two sets as shown in the figure on the right. Because there are two possible placements for *x*, the argument is invalid.

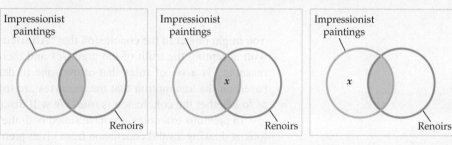

Figure 2.14

An Euler diagram that involves three sets, as shown in Example 5, can represent some arguments.

> **Example 5**   **Use an Euler Diagram to Determine the Validity of an Argument**

Use an Euler diagram to determine whether the following argument is valid or invalid.

No psychologist can juggle.

All clowns can juggle.

∴ No psychologist is a clown.

Solution

The premises are "No psychologist can juggle" and "All clowns can juggle." The Euler diagram in Figure 2.15 illustrates the premise that "No psychologist can juggle." Because there are no psychologists who can juggle, the sets are disjoint.

Figure 2.16 shows that the set of clowns is a subset of the set of jugglers (premise 2). Because the set of psychologists is disjoint with the set of jugglers, it is also disjoint with the set of clowns. This means, "No psychologist is a clown." Thus, the argument is valid.

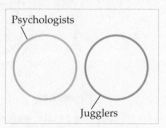

Figure 2.15

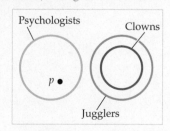

Figure 2.16

Transitive reasoning is drawing a conclusion from a series of interrelated premises. Example 6 uses Euler diagrams to visually illustrate transitive reasoning.

Figure 2.17

Example 6   **Use an Euler Diagram to Determine the Validity of an Argument**

Use an Euler diagram to determine whether the following argument is valid or invalid.

    All fried foods are greasy.

    All greasy foods are delicious.

    <u>All delicious foods are healthy.</u>

∴ All fried foods are healthy.

Solution

Figure 2.17 illustrates that every fried food is an element of the set of healthy foods, so the argument is valid.

You might object to the conclusion that "All fried foods are healthy." If so, the fault lies with accepting the truth of the premises and their relationship to one another. Logical reasoning is a set of rules that allows one to determine whether arguments are valid based on the assumption that the premises are true. Logical reasoning makes no claim as to whether the conclusion is true. We will discuss fallacies later in this chapter.

In previous examples, we have stated both the premises and the conclusion. Now we look at drawing a valid conclusion from given premises.

Example 7   **Use an Euler Diagram to Determine a Conclusion for an Argument**

Use an Euler diagram and all of the premises in the following argument to determine a valid conclusion for the argument.

    Some dogs are terriers.

    <u>All terriers like to take walks.</u>

∴ ?

Solution

The premises are "Some dogs are terriers" and "All terriers like to take walks." The Euler diagram in Figure 2.18 shows the relationship among the sets. The set of terriers is a subset of the set of animals that like to take walks. Therefore, the set of dogs intersects the set of animals that like to take walks. The shaded area represents dogs that like to take walks (which includes terriers).

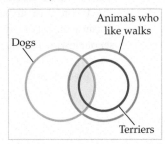

Figure 2.18

The conclusion is some dogs like to take walks.

2.2 Exercise Set

Think About It

1. Give two examples of a universal qualifier.
2. Give two examples of an existential qualifier.
3. In the statement "All Ferraris are red," is the negation of the statement "No Ferraris are red"?
4. Can an argument be valid and yet the conclusion be false? Support your answer.

Preliminary Exercises

For Exercises P1 to P7, if necessary, use the referenced example following the exercise for assistance.

P1. Draw an Euler diagram for each of the following. **Example 1**
   a. No smartphones are expensive.
   b. Some vegetables are not green.

P2. Write the negation of each of the following statements. **Example 2**
   a. Some fruit is not good.
   b. No buildings are tall.

P3. Use an Euler diagram to determine whether the argument is valid or invalid. **Example 3**

   No mathematics professors are good-looking.
   All good-looking people are models.
   ∴ No mathematics professor is a model.

P4. Use an Euler diagram to determine whether the argument is valid or invalid. **Example 4**

   No prime numbers are negative.
   The number 7 is not negative.
   ∴ The number 7 is a prime number.

P5. Use an Euler diagram to determine whether the argument is valid or invalid. **Example 5**

   No football players are race car drivers.
   All race car drivers are risk takers.
   ∴ No football players are risk takers.

P6. Use an Euler diagram to determine whether the argument is valid or invalid. **Example 6**

   All squares are rhombi.
   All rhombi are parallelograms.
   All parallelograms are quadrilaterals.
   ∴ All squares are quadrilaterals.

P7. Use an Euler diagram and all of the premises in the following argument to determine a valid conclusion for the argument. **Example 7**

   All journalists are writers.
   No writers are scientists.
   ∴ ?

Section Exercises

In Exercises 1 to 8, write the negation of the statement. Start each negation with the word "Some," "No," or "All."

1. Some lions are playful.
2. Some dogs are not friendly.
3. All classic movies were first produced in black and white.
4. Everybody enjoyed the dinner.
5. No even numbers are odd numbers.
6. Some actors are not rich.
7. All cars run on gasoline.
8. None of the students took my advice.

In Exercises 9 to 28, use an Euler diagram to determine whether the argument is valid or invalid.

9.   All frogs are poetical.
     Kermit is a frog.
   ∴ Kermit is poetical.

10. All Oreo cookies have a filling.
     All Fig Newtons have a filling.
   ∴ All Fig Newtons are Oreo cookies.

11. Some plants have flowers.
     All things that have flowers are beautiful.
   ∴ Some plants are beautiful.

12. No squares are triangles.
     Some triangles are equilateral.
   ∴ No squares are equilateral.

13. No rocker would do the mariachi.
     All baseball fans do the mariachi.
   ∴ No rocker is a baseball fan.

14. Nuclear energy is not safe.
     Some electric energy is safe.
   ∴ No electric energy is nuclear energy.

15. Some birds bite.
     All things that bite are dangerous.
   ∴ Some birds are dangerous.

16. All fish can swim.
     That barracuda can swim.
   ∴ That barracuda is a fish.

17. All men behave badly.
     Some hockey players behave badly.
   ∴ Some hockey players are men.

18. All grass is green.
     That ground cover is not green.
   ∴ That ground cover is not grass.

19. Most teenagers drink soda.
     No CEOs drink soda.
   ∴ No CEO is a teenager.

20. Some students like history.
     Vern is a student.
   ∴ Vern likes history.

21. No mathematics test is fun.
     All fun things are worth your time.
   ∴ No mathematics test is worth your time.

22. All prudent people shun sharks.
     No accountant is imprudent.
   ∴ No accountant fails to shun sharks.

23. All candidates without a master's degree will not be considered for the position of director.

    All candidates who are not considered for the position of director should apply for the position of assistant.

    ∴ All candidates without a master's degree should apply for the position of assistant.

24. Some whales make good pets.

    Some good pets are cute.

    Some cute pets bite.

    ∴ Some whales bite.

25. All prime numbers are odd.

    2 is a prime number.

    ∴ 2 is an odd number.

26. All Lewis Carroll arguments are valid.

    Some valid arguments are syllogisms.

    ∴ Some Lewis Carroll arguments are syllogisms.

27. All aerobics classes are fun.

    Jan's class is fun.

    ∴ Jan's class is an aerobics class.

28. No tall person takes a math class.

    Some students that take a math class can juggle.

    ∴ No tall person can juggle.

In Exercises 29 to 34, use all of the premises in each argument to determine a valid conclusion for the argument.

29. All Reuben sandwiches are good.

    All good sandwiches have pastrami.

    All sandwiches with pastrami need mustard.

    ∴ ?

30. All cats are strange.

    Boomer is not strange.

    ∴ ?

31. All multiples of 11 end with a 5.

    1001 is a multiple of 11.

    ∴ ?

32. If it isn't broken, then I do not fix it.

    If I do not fix it, then I do not get paid.

    ∴ ?

33. Some horses are frisky.

    All frisky horses are gray.

    ∴ ?

34. If we like to ski, then we will move to Vail.

    If we move to Vail, then we will not buy a house.

    If we do not buy a condo, then we will buy a house.

    ∴ ?

35. Examine the following three premises:

    1. All people who have an Xbox play video games.

    2. All people who play video games enjoy life.

    3. Some mathematics professors enjoy life.

    Now consider each of the following six conclusions. For each conclusion, determine whether the argument formed by the three premises and the conclusion is valid or invalid.

    a. ∴ Some mathematics professors have an Xbox.

    b. ∴ Some mathematics professors play video games.

    c. ∴ Some people who play video games are mathematics professors.

    d. ∴ Mathematics professors never play video games.

    e. ∴ All people who have an Xbox enjoy life.

    f. ∴ Some people who enjoy life are mathematics professors.

36. Examine the following three premises:

    1. All people who drive pickup trucks like Garth Brooks.

    2. All people who like Garth Brooks like country western music.

    3. Some people who like heavy metal music like Garth Brooks.

    Now consider each of the following five conclusions. For each conclusion, determine whether the argument formed by the three premises and the conclusion is valid or invalid.

    a. ∴ Some people who like heavy metal music drive a pickup truck.

    b. ∴ Some people who like heavy metal music like country western music.

    c. ∴ Some people who like Garth Brooks like heavy metal music.

    d. ∴ All people who drive a pickup truck like country western music.

    e. ∴ People who like heavy metal music never drive a pickup truck.

Investigation

The LSAT and Logic Games   A good score on the Law School Admission Test (LSAT) is viewed by many to be the most important part of getting into a top-tier law school. Rather than testing what you have already learned, it is designed to measure and project your ability to excel in law school.

The following problem is similar to some of the questions from the logic games section of the LSAT.

A cell phone provider sells seven different types of smartphones.

- Each phone has either a touch screen keyboard or a push-button keyboard.
- Each phone has a 4-inch, a 4.7-inch, or a 5.5-inch screen.
- Every phone with a 4.7-inch screen is paired with a touch screen keyboard.
- Of the seven different types of phones, most have a touch screen keyboard.
- No phone with a push-button keyboard is paired with a 4-inch screen.

Which one of the following statements *cannot* be true?

a. Five of the types of phones have 4.7-inch screens.

b. Five of the types of phones have 4-inch screens.

c. Four of the types of phones have a push-button keyboard.

d. Four of the types of phones have 5.5-inch screens.

e. Five of the types of phones have a touch screen keyboard.

Section 2.3 Propositional Logic

Learning Objectives:

- Identify propositions.
- Use a truth table.
- Use DeMorgan's Laws.
- Identify conditional statements.
- Rewrite propositions using the conditional, converse, inverse, or contrapositive of a proposition.
- Determine the validity of a propositional argument.

Take Note: A *categorical statement* has a highly specific form: subject–linking verb–predicate. As we will see soon, propositional logic deals with more general statements. For instance, "If it rains today, I will not play tennis" is a proposition.

Simple and Compound Propositions

Every language contains different types of sentences, such as statements, questions, and commands. For instance,

"Is the test today?" is a question.
"Go get the newspaper" is a command.
"This is a nice car" is an opinion.
"Denver is the capital of Colorado" is a statement of fact that all would agree is true.

In the previous section, we discussed the logic rules around *categorical statements*. In this section, we deal with propositions, a more general kind of statement.

Proposition

A **proposition** is a declarative sentence that is either true or false or one that a declarative sentence asserts to be true or false.

Frequently, we will use the word *statement* as a synonym for proposition. We wanted to start with proposition to emphasize the distinction between a proposition and a categorical statement.

Example 1   Identify Propositions

Determine whether each sentence is a proposition.

a. Florida is a state in the United States.

b. How are you?

c. Humans cause climate change.

d. 10,001 is a prime number.

e. Run for cover.

f. This sentence is false.

Solution

a. Florida is one of the 50 states in the United States, so this sentence is true, and it is a proposition.

b. "How are you?" is a question; it is not a declarative proposition. Thus, it is not a proposition.

c. The sentence "Humans cause climate change" is an example of the second part of the definition of a proposition. It is a declarative proposition that asserts something is true or false. You may not agree with the proposition, but it is an assertion that is either true or false. Therefore, it is a proposition.

d. The sentence "10,001 is a prime number" is a proposition. You may not know whether 10,001 is a prime number. However, the declarative proposition is either true or false. Therefore, it is a proposition.

e. "Run for cover" is a command, therefore it is not a proposition.

f. If we assume the sentence "This sentence is false" is true, then it is false. If we assume it is false, then it is true. Therefore, it is not a proposition.

Simple Statements and Compound Statements

A mnemonic for remembering the standard connective symbols:

$\wedge$ND

OR

NOT

IF...THEN

Simple Statements and Compound Statements

A **simple statement** is a statement that conveys a single idea. A **compound statement** is a statement that conveys two or more ideas.

Connecting simple statements with words and phrases such as *and, or, if ... then*, and *if and only if* creates a compound statement. For instance, "I will attend the meeting or I will go to school" is a compound statement. It is composed of the two simple statements, "I will attend the meeting" and "I will go to school." The word *or* is a connective for the two simple statements.

Table 2.5 shows some standard symbols used to represent connectives.

Table 2.5

Statement	Connective	Symbolic Form	Type of Statement
not p	not	$\sim p$	negation
p and q	and	$p \wedge q$	conjunction
p or q	or	$p \vee q$	disjunction
If p, then q	If ... then	$p \rightarrow q$	conditional

Example 2 **Write the Negation of a Proposition**

Write the negation of each proposition.

a. Soccer is a team sport.

b. The dog does not need to be fed.

Solution

a. Soccer is not a team sport.

b. The dog needs to be fed.

We will often find it useful to write compound propositions in symbolic form.

Example 3 **Write Compound Propositions in Symbolic Form**

Consider the following simple statements.

p: Today is Friday.

q: It is raining.

r: I am going to a movie.

s: I am not going to the basketball game.

Write the following compound statements in symbolic form.

a. Today is Friday, and it is raining.

b. It is not raining, and I am going to a movie.

c. I am going to the basketball game or I am going to a movie.

d. If it is raining, then I am not going to the basketball game.

Solution

a. $p \wedge q$ **b.** $\sim q \wedge r$ **c.** $\sim s \vee r$ **d.** $q \rightarrow s$

In the next example, we translate symbolic statements into English sentences.

> **Example 4**   **Translate Symbolic Statements**
>
> Consider the following statements.
>
> > *p*: The game will be played in Atlanta.
> > *q*: The game will be shown on TV.
> > *r*: The game will not be streamed on the Internet.
> > *s*: The Appolos are favored to win.
>
> Write each of the following symbolic statements in words.
>
> **a.** $p \wedge s$ **b.** $\sim r \vee q$ **c.** $p \rightarrow \sim q$
>
> **Solution**
>
> **a.** The game will be played in Atlanta, and the Appolos are favored to win.
> **b.** The game will be streamed on the Internet or the game will be shown on TV.
> **c.** If the game is played in Atlanta, the game will not be shown on TV.

Truth-Value and Truth Tables

The **truth-value** of a simple statement is either true (T) or false (F).

The **truth-value** of a compound statement depends on the truth-values of its simple statements and its connectives.

A **truth table** is a table that shows the truth-value of a compound statement for all possible truth-values of its simple statements.

Table 2.6 Truth Table for $\sim p$

p	$\sim p$
T	F
F	T

As the definition of the truth table as just given states, it shows the truth-values of *all* possible truth-values of its simple statements. For one statement, it is true or false. There are two rows to the truth table. (See Table 2.6 for $\sim p$.)

If a statement *p* is true, its negation, $\sim p$, is false; if a statement *p* is false, its negation, $\sim p$, is true. See Table 2.6. The negation of the negation of a statement is the original statement. Symbolically, we write this as $\sim (\sim p) = p$.

For two statements, we could have both true, first statement true and second statement false, first statement false and second statement true, and both statements false. There are four rows to the truth table. (See Table 2.7 for $p \wedge q$.)

If you order cake *and* ice cream in a restaurant, the waiter will bring *both* cake and ice cream. In general, the conjunction $p \wedge q$ is true if both *p* and *q* are true, and the conjunction is false in all other cases. The truth table in Table 2.7 shows the four possible cases that arise when we form a conjunction of two statements.

Table 2.7 Truth Table for $p \wedge q$

p	*q*	$p \wedge q$
T	T	T
T	F	F
F	T	F
F	F	F

Truth-Value of a Conjunction

The conjunction $p \wedge q$ is true when *p* and *q* are both true and false otherwise.

Table 2.8 Truth Table for $p \vee q$

p	*q*	$p \vee q$
T	T	T
T	F	T
F	T	T
F	F	F

Consider the sentence "Taking a math course or a logic course will make you smarter." The connective is *or*. This sentence could be true if you take a math course, take a logic course, or take both courses. The connective *or* in this case is called **inclusive**, and we symbolize this by $p \vee q$. See Table 2.8.

Truth-Value of a Disjunction

The disjunction $p \vee q$ is true when p is true, when q is true, and when p and q are both true; it is false otherwise.

Equivalent Statements

Two statements are **equivalent** if the truth-values of the two statements are the same. We write two statements are equivalent as $p \equiv q$.

Of the various equivalent statements, two important ones are based on DeMorgan's laws. These laws are restated here for propositions.

DeMorgan's Laws

If p and q are propositions, then

$$\sim(p \vee q) \equiv (\sim p) \wedge (\sim q) \qquad \text{and} \qquad \sim(p \wedge q) \equiv (\sim p) \vee (\sim q)$$

To verify the equivalence of the statements in DeMorgan's law, we compare the truth-values of each compound statement.

Begin by creating truth tables for each statement. When doing this, use the same approach as math expressions containing parentheses: Work from the inside out. For $\sim(p \vee q)$, we begin with $(p \vee q)$ and then its negation. The negation has the opposite truth-values. This is shown in Table 2.9.

Table 2.9

p	q	$(p \vee q)$	$\sim(p \vee q)$
T	T	T	F
T	F	T	F
F	T	T	F
F	F	F	T

Now create a truth table for the expression on the right of the equivalence as shown in Table 2.10.

Table 2.10

p	q	$\sim p$	$\sim q$	$\sim p \wedge \sim q$
T	T	F	F	F
T	F	F	T	F
F	T	T	F	F
F	F	T	T	T

Because the truth-values of the two statements are the same, the statements are equivalent.

Example 5   Use DeMorgan's Law

A rational number is one that has a decimal representation that terminates (as in 0.25) or repeats (as in 0.3333. . .). An irrational number is a number that is not rational. Use DeMorgan's law to write a definition of an irrational number as a conjunction of two statements.

Solution

Let p represent "A decimal number that is represented by a terminating decimal" and let q represent "A decimal number that is represented by a repeating decimal." Then a rational number is $p \vee q$. An irrational number is one that is not rational.

Symbolically, this is $\sim(p \vee q)$. By the first DeMorgan law, this is equivalent to $(\sim p) \wedge (\sim q)$. An irrational number has a decimal representation that does not terminate $(\sim p)$ and does not repeat $(\sim q)$.

Conditional Statements

Conditional statements can be written in "If p, then q" form or in "If p, q" form (the word *then* is not always present). For instance, all of the following are conditional statements.

- If we order pizza, then we can have it delivered.
- If you go to the movie, you will not be able to meet us for dinner.
- If n is a prime number greater than 2, then n is an odd number.

In any conditional statement represented by "If p, then q" or by "If p, q," the p statement is called the **antecedent**, and the q statement is called the **consequent**. Symbolically, the conditional is represented by $p \rightarrow q$.

Besides being stated as "If p, then q," the conditional may also be stated as "p implies q" (sometimes the conditional is called an *implication*), "q is necessary for p," "p is sufficient for q," and "p only if q" (or "only if q, p"). In each of these, p is the antecedent and q is the consequent.

| Example 6 | Identify the Antecedent and Consequent of a Conditional |

Identify the antecedent and consequent in the following statements.

a. If our school was this nice, I would go there more than once a week.
 —*The Basketball Diaries*

b. If you strike me down, I shall become more powerful than you can possibly imagine.
 —Obi-Wan Kenobi, *Star Wars, Episode IV, A New Hope*

c. Getting a role in this sitcom implies I'll be paid well.

d. It is necessary to score above 1200 on the SAT in order to get into the college I would like to attend.

e. Being a dog is sufficient to belong to the class of mammals.

f. Only if a figure is a quadrilateral can it be a square.

Solution

a. Antecedent: Our school was this nice
 Consequent: I would go there more than once a week.

b. Antecedent: You strike me down
 Consequent: I shall become more powerful than you can possibly imagine.

c. Antecedent: Getting a role in this sitcom
 Consequent: I'll be paid well.

d. Antecedent: To attend the college
 Consequent: Score above 1200 on the SAT

e. Antecedent: Being a dog
 Consequent: Belong to the class of mammals

f. Antecedent: A figure is a square
 Consequent: It is a quadrilateral

To determine the truth table for $p \rightarrow q$, consider the advertising slogan for a web-authoring software product that states, "If you can use a word processor, you can create a web page." This slogan is a conditional statement. The antecedent is p, "you can use a word processor," and the consequent is q, "you can create a web page." Now consider the truth-value of $p \rightarrow q$ for each of the following four possibilities.

> *Antecedent T, Consequent T*: You can use a word processor, and you can create a web page. In this case, the truth-value of the advertisement is true.

> *Antecedent T, Consequent F*: You can use a word processor, but you cannot create a web page. In this case, the truth-value of the advertisement is false.

> *Antecedent F, Consequent T*: You cannot use a word processor, and you can create a web page. Because the advertisement does not make any statement about what you might or might not be able to do if you cannot use a word processor, we cannot state that the advertisement is false, and therefore we assume it is true.

> *Antecedent F, Consequent F*: You cannot use a word processor, and you cannot create a web page. Once again we must consider the truth-value in this case to be true because the advertisement does not make any statement about what you might or might not be able to do if you cannot use a word processor. We therefore assume it is true.

The truth table for $p \rightarrow q$ is given in Table 2.11.

Table 2.11 Truth Table for $p \rightarrow q$

p	q	$p \rightarrow q$
T	T	T
T	F	F
F	T	T
F	F	T

Example 7 **Find the Truth-Value for a Conditional**

Determine the truth-value of each of the following conditional statements.

a. If 2 is an integer, then 2 is a rational number.

b. If 3 is a negative number, then $5 > 7$.

c. If $5 > 3$, then $2 + 7 = 4$.

Solution

a. Because the consequent is true, this is a true statement.

b. Because the antecedent is false, this is a true statement.

c. Because the antecedent is true and the consequent is false, this is a false statement.

●

The situation in Example 7b may be a little puzzling. Remember, we are asking whether the *proposition* is true, not whether each part of the proposition is true. The inequality $5 > 7$ is certainly false. However, the proposition is true because the antecedent is false. From the table for implication, a false antecedent yields a true statement whether the consequent is true or false.

Conditional, Converse, Inverse, and Contrapositive

Let p represent the statement "A number is divisible by 4," and let q represent the statement "A number is even." Four different propositions can be formed from these statements and their negations.

> **Conditional** $p \rightarrow q$: If a number is divisible by 4, then the number is an even number.

> **Converse** $q \rightarrow p$: If a number is even, then it is divisible by 4.

> **Inverse** $\sim p \rightarrow \sim q$: If a number is not divisible by 4, then it is not an even number.

> **Contrapositive** $\sim q \rightarrow \sim p$: If a number is not an even number, then it is not divisible by 4.

The conditional "If a number is divisible by 4, then the number is an even number" is a true statement. The converse "If a number is even, then it is divisible by 4" is not true. For instance, 6 is an even number but not divisible by 4.

The inverse "If a number is not divisible by 4, then it is not an even number" is not true. The even number 6 is not divisible by 4.

The contrapositive "If a number is not an even number, then it is not divisible by 4" is true.

The truth table for each of these statements is given in Table 2.12.

Table 2.12

p	q	$\sim p$	$\sim q$	$p \rightarrow q$	$q \rightarrow p$	$\sim p \rightarrow \sim q$	$\sim q \rightarrow \sim p$
T	T	F	F	T	T	T	T
T	F	F	T	F	T	T	F
F	T	T	F	T	F	F	T
F	F	T	T	T	T	T	T

Note that the columns of the same color have the same truth-values. Recalling the definition of equivalence, this means that **the conditional and contrapositive are equivalent statements** and **the converse and inverse are equivalent statements**.

Example 8   **Write the Converse, Inverse, and Contrapositive for a Conditional Proposition**

Given the conditional proposition "If my bicycle has a flat tire, I'll walk to the store," write the converse, inverse, and contrapositive of the proposition.

Solution

Converse: If I walk to the store, then my bicycle has a flat tire.
Inverse: If my bicycle does not have a flat tire, I will not walk to the store.
Contrapositive: If I do not walk to the store, then my bicycle does not have a flat tire.

Analyzing Propositional Arguments

The validity of propositional arguments is based on the following.

Valid Argument

A propositional argument is **valid** if and only if it is impossible for premises, assumed to be true, to result in a false conclusion.

Example 9   **Determine the Validity of a Propositional Argument**

Analyze the following: If I go to a movie theater, I will choose to watch a comedy. I went to a movie theater. Therefore, I watched a comedy.

Solution

Let p represent "I go to a movie theater" and q represent "I will choose to watch a comedy." The argument can be represented symbolically as

$$p \rightarrow q \qquad \text{First premise}$$
$$\underline{p} \qquad\qquad \text{Second premise}$$
$$\therefore q \qquad\qquad \text{Conclusion}$$

The truth table for this argument is shown in Table 2.13.

Table 2.13

p	q	First Premise $p \rightarrow q$	Second Premise p	Conclusion q
T	T	T	T	T
T	F	F	T	F
F	T	T	F	T
F	F	T	F	F

The shaded row (for $p = T$ and $q = T$) is the only row with both premises and the conclusion true. The argument is valid.

The form of reasoning in Example 9 is named *modus ponens* (mode that affirms). The second premise *p affirms* that *p* is true in the first premise $p \rightarrow q$. Therefore, the conclusion is logically valid. *Modus ponens* **is one of the valid forms of reasoning**.

Valid forms of reasoning allow us to form valid conclusions. There is nothing in valid forms of reasoning that addresses whether the statements are true or false. For instance, consider this argument:

> If children are allowed to drive, then the roads will be safe. Children are allowed to drive. Therefore, the roads will be safe. Let *p* represent "children are allowed to drive" and *q* represent "the roads will be safe." The argument is represented by

$$
\begin{array}{ll}
p \rightarrow q & \text{First premise} \\
\underline{p} & \text{Second premise} \\
\therefore q & \text{Conclusion}
\end{array}
$$

> This is *modus ponens*. Therefore, the argument is logically valid. However, the premises are not true.

Example 10 is a slight restatement of Example 9.

Example 10   **Determine the Validity of a Propositional Argument**

Analyze the following: If I go to a movie theater, I will choose to watch a comedy. I choose to watch a comedy today. Therefore, I went to a movie theater.

Solution

Let *p* represent "I go to a movie theater" and *q* represent "I will choose to watch a comedy." The argument can be represented symbolically as

$$
\begin{array}{ll}
p \rightarrow q & \text{First premise} \\
\underline{q} & \text{Second premise} \\
\therefore p & \text{Conclusion}
\end{array}
$$

The truth table for this argument is shown as follows.

Table 2.14

p	*q*	First Premise $p \rightarrow q$	Second Premise q	Conclusion p
T	T	T	T	T
T	F	F	F	T
F	T	T	T	F
F	F	T	F	F

Note that the next-to-last row, highlighted in light red, has true premises but a false conclusion. The argument is invalid.

For Example 10, the problem is that I could have watched a comedy at home by streaming a movie. Just stating that I watched a movie is not enough to guarantee that I went to a theater. Example 10 is called the **fallacy of the converse**. The form of reasoning is invalid.

Here is a slight restatement of Example 10: If I go to a movie theater, I will choose to watch a comedy. I choose not to watch a comedy today. Therefore, I did not go to a movie theater. This argument can be stated symbolically as

$$
\begin{array}{ll}
p \rightarrow q & \text{First premise} \\
\underline{\sim q} & \text{Second premise} \\
\therefore \sim p & \text{Conclusion}
\end{array}
$$

The truth table for this argument is shown in Table 2.15.

Table 2.15

p	q	First Premise $p \rightarrow q$	Second Premise $\sim q$	Conclusion $\sim p$
T	T	T	F	F
T	F	F	T	F
F	T	T	F	T
F	F	T	T	T

The shaded row at the bottom is the only row with both premises true and the conclusion true. The argument is valid.

This form of reasoning is named *modus tollens* (mode that denies). The second premise $\sim q$ *denies* that q is true in the first premise $p \rightarrow q$. Because q is not true, it must follow that p is not true. **Modus tollens is one of the valid forms of reasoning**.

Table 2.16 summarizes the valid forms of reasoning (there are others) and two fallacies. In the next section, we will discuss other types of fallacies.

Table 2.16   Table of Forms of Valid Reasoning and Forms of Fallacies

Modus Ponens	Modus Tollens	Transitive Reasoning	Fallacy of the Converse	Fallacy of the Inverse
$p \rightarrow q$	$p \rightarrow q$	$p \rightarrow q$	$p \rightarrow q$	$p \rightarrow q$
p	$\sim q$	$q \rightarrow r$	q	$\sim p$
$\therefore q$	$\therefore \sim p$	$\therefore p \rightarrow r$	$\therefore p$	$\therefore \sim q$

Example 11   Determine the Validity of a Propositional Argument

Use Table 2.16 to determine the validity of the following argument: If you know the password for your smartphone, you can listen to your playlists. You do not know your password. Therefore, you cannot listen to your playlists.

Solution

Let p represent "you know the password for your smartphone" and q represent "you can listen to your playlists." The argument can be represented symbolically as

$p \rightarrow q$	First premise
$\sim p$	Second premise
$\therefore \sim q$	Conclusion

By Table 2.16, this is the fallacy of the inverse. You could, for instance, listen to your playlists on your computer. The argument is invalid.

There are various other forms of valid reasoning. We will discuss one other form called **transitive reasoning**.

$p \rightarrow q$	First premise
$q \rightarrow r$	Second premise
$\therefore p \rightarrow r$	Conclusion

Example 12   Use Transitive Reasoning to Determine the Validity of an Argument

Determine whether the following argument is valid or invalid.

If I am going to run the marathon, then I will buy new shoes.

If I buy new shoes, then I will not buy a television.

$\therefore$ If I buy a television, I will not run the marathon.

Solution

Label each simple statement.

m: I am going to run the marathon.
s: I will buy new shoes.
t: I will buy a television.

The symbolic form of the argument is

$$\begin{array}{ll} m \rightarrow s & \text{First premise} \\ \underline{s \rightarrow \sim t} & \text{Second premise} \\ \therefore t \rightarrow \sim m & \text{Conclusion} \end{array}$$

At first glance, this does not fit the transitive reasoning model. We would expect the form to be

$$\begin{array}{ll} m \rightarrow s & \text{First premise} \\ \underline{s \rightarrow \sim t} & \text{Second premise} \\ \therefore m \rightarrow \sim t & \text{Conclusion} \end{array}$$

However, we can use the equivalence of the conditional and the contrapositive to rewrite the conclusion $t \rightarrow \sim m$ as $m \rightarrow \sim t$. Because these two statements are equivalent, the reasoning is valid.

2.3 Exercise Set

Think About It

1. What word is used for disjunction?

2. What word is used for conjunction?

3. What phrase is used for implication?

4. If p is a true statement and q is a false statement, what is the truth-value of $p \vee q$?

5. If p is a true statement and q is a false statement, what is the truth-value of $p \wedge q$?

6. If p is a false statement and q is a true statement, what is the truth-value of $p \rightarrow q$?

Preliminary Exercises

For Exercises P1 to P12, if necessary, use the referenced example following the exercise for assistance.

P1. Determine whether each sentence is a statement. **Example 1**

   a. My sister's birthday is next Saturday.

   b. Do you run marathons?

P2. Write the negation of "The color of the wall was not red." **Example 2**

P3. Write the compound statement in symbolic form. "The tea was cold (c) and sweet (s)." **Example 3**

P4. Using p to represent "The stock market went up" and q to represent "The bond market went down," write $p \vee (\sim q)$ in words. **Example 4**

P5. Use DeMorgan's law to restate "It is raining and the sun is not shining" in an equivalent form. **Example 5**

P6. Identify the antecedent and the consequent of the statement "If the snow begins to fall, I will not go skiing." **Example 6**

P7. Determine the truth-value of the conditional "If diamonds are emeralds, then $7 = 8$." **Example 7**

P8. Write the inverse, converse, and contrapositive for the conditional "If it is snowing, I will not go skiing." **Example 8**

P9. Use a truth table to determine the validity of this argument: "If the gemstone is an emerald (e), then we will not sell it ($\sim s$). I sold the gemstone. Therefore, it was not an emerald." **Example 9**

P10. Use a truth table to determine the validity of this argument: "If I don't go to the tennis match, then my daughter is not playing in the match. My daughter is playing in the match. Therefore, I went to the tennis match." **Example 10**

P11. Use Table 2.16 to determine whether this argument is valid: "If Parker can run 100 yards in less than 12 seconds, Parker can qualify for the track team. Parker qualified for the track team. Therefore, Parker can run 100 yards in less than 12 seconds." **Example 11**

P12. I start to fall asleep if I read a math book. I drink soda whenever I start to fall asleep. If I drink a soda, then I must eat a candy bar. Therefore, if I read a math book, then I eat a candy bar. **Example 12**

Section Exercises

In Exercises 1 to 6, determine whether each sentence is a statement.

1. *Star Wars: The Force Awakens* is the greatest movie of all time.

2. Harvey Mudd College is in Oregon.

3. The area code for Storm Lake, Iowa, is 512.

4. January 1, 2024, will be a Sunday.

5. Have a fun trip.

6. Do you like to read?

In Exercises 7 to 10, write the negation of each statement.

7. The Giants lost the game.

8. The lunch was served at noon.

9. The game did not go into overtime.

10. The game was not shown on television.

In Exercises 11 to 16, write each sentence in symbolic form.

11. If today is Wednesday (w), then tomorrow is Thursday (t).

12. I went to the post office (p) and the bookstore (s).

13. If it is a dog (d), it has fleas (f).

14. Polynomials that have exactly three terms (p) are called trinomials (t).

15. I will major in mathematics (m) or computer science (c).

16. All pentagons (p) have exactly five sides (s).

In Exercises 17 to 22, use the given symbols to write each symbolic statement in words.

17. p: I cannot play the piano.
    q: I can play the violin.
    $\sim p \wedge q$

18. p: I ride a bicycle to work.
    q: I take the bus to work.
    $p \vee q$

19. p: The painting is a watercolor.
    q: James will buy the painting.
    $p \rightarrow \sim q$

20. p: You can count your money.
    q: You have less than a million dollars.
    $\sim p \rightarrow \sim q$

21. p: Carmella will major in physics.
    q: Carmella will major in astronomy.
    r: Carmella gets an A in mathematics.
    $r \rightarrow (p \vee q)$

22. p: The car gets less than 25 miles per gallon.
    q: The car does not have a collision avoidance system.
    r: Jordan will buy the car.
    $(\sim p \wedge \sim q) \rightarrow r$

In Exercises 23 to 26, use DeMorgan's laws to restate the given statement in an equivalent form.

23. It is not true that it rained or it snowed.

24. I did not pass the test, and I did not complete the course.

25. Bao did not visit France, and Bao did not visit Italy.

26. It is not true that I bought a new car and I moved to New York.

In Exercises 27 to 32, identify the antecedent and consequent of each conditional statement.

27. If I had the money, I would buy the painting.

28. If Shelly goes on the trip, Shelly will not be able to take part in the graduation ceremony.

29. Emene would go to the movies only if Ariel would go.

30. I will fly to San Francisco only if I can afford the airfare.

31. If Cameron gets a raise, Cameron will buy a new tablet computer.

32. Only if Nuntiya passes biology will Nuntiya get into medical school.

In Exercises 33 to 38, determine the truth-value of the conditional.

33. If Monday is a day of the week, then Saturday is the name of a month.

34. If Saturday is the name of a month, then Monday is the day after Wednesday.

35. If all frogs can dance, then the sun is cold.

36. If all cats are black, then I am a millionaire.

37. If Venus is a planet, then apples are not fruit.

38. If bananas are not a fruit, then broccoli is not a vegetable.

In Exercises 39 to 44, write the converse, inverse, and contrapositive for the given conditional statement.

39. If I were rich, I would quit this job.

40. If we had a car, then we would be able to take the class.

41. If Alex does not return soon, we will not be able to attend the party.

42. If you get the promotion, you will need to move to Denver.

43. Only if we take the train will we be able to take the entire family.

44. Kendall will visit Kauai only if Kendall saves for the vacation.

In Exercises 45 to 48, use a truth table to determine the validity of the argument.

45. If you finish your homework (h), you may attend the reception (r). You did not finish your homework. Therefore, you cannot go to the reception.

46. If I can't buy the house ($\sim b$), then at least I can dream about it (d). I can buy the house or at least I can dream about it. Therefore, I can buy the house.

47. If the winds are from the east (e), then we will not have a big surf ($\sim s$). We do not have a big surf. Therefore, the winds are from the east.

48. If I master college algebra (c), then I will be prepared for trigonometry (t). I am prepared for trigonometry. Therefore, I mastered college algebra.

In Exercises 49 to 58, use Table 2.16 to determine whether the argument is valid.

49. If you take Art 151 in the fall, you will be eligible to take Art 152 in the spring. You were not eligible to take Art 152 in the spring. Therefore, you did not take Art 151 in the fall.

50. If I had a nickel for every logic problem I have solved, then I would be rich. I have not received a nickel for every logic problem I have solved. Therefore, I am not rich.

51. If it is a dog, then it has fleas. It has fleas. Therefore, it is a dog.

52. If Rita buys a new car, then Rita will not go on the cruise. Rita went on the cruise. Therefore, Rita did not buy a new car.

53. If we serve salmon, then Wren will join us for lunch. If Wren joins us for lunch, then Sawyer will not join us for lunch. Therefore, if we serve salmon, Sawyer will not join us for lunch.

54. If my cat is left alone in the apartment, then the cat will claw the sofa. Yesterday I left my cat alone in the apartment. Therefore, my cat clawed the sofa.

55. If we sell the boat (s), then we will not go to the river ($\sim r$). If we don't go to the river, then we will go camping (c). If we do not buy a tent ($\sim t$), then we will not go camping. Therefore, if we sell the boat, then we will buy a tent.

56. If it is an ammonite (*a*), then it is from the Cretaceous period (*c*). If it is not from the Mesozoic era (~*m*), then it is not from the Cretaceous period. If it is from the Mesozoic era, then it is at least 65 million years old (*s*). Therefore, if it is an ammonite, then it is at least 65 million years old.

57. If the computer is not operating (~*o*), then I will not be able to finish my report (~*f*). If the office is closed (*c*), then the computer is not operating. Therefore, if I am able to finish my report, then the office is open.

58. If I get an A on the final exam (*E*), I'll pass the course (*P*). If I pass the course, I'll be eligible for a scholarship (*S*). If I'm not eligible for a scholarship, I can get a student loan (*L*). I got a student loan. Therefore, I did not get an A in the course.

In Exercises 59 to 64, use all the premises to determine whether a valid conclusion can be stated. If so, state the conclusion. If not, answer with "Valid conclusion is not possible."

59. If the moon is made of green cheese, then apples are bananas. Apples are not bananas. Therefore, _____.

60. If Colin joins the Foreign Service, Colin will be stationed in France. Colin joined the Foreign Service. Therefore, _____.

61. Only if I pass the final exam will I get a passing grade in the course. I got a passing score on the final exam. Therefore, _____.

62. If Orla studies hotel management, Orla will get a job at the Historia Hotel. Orla did not study hotel management. Therefore, _____.

63. If it is a theropod, then it is not herbivorous. If it is not herbivorous, then it is not a sauropod. Therefore, if it is a sauropod, then _____.

64. If my friend comes for dinner, then I'll have pizza delivered. If I make spaghetti for dinner, I won't have pizza delivered. Therefore, if my friend does not come for dinner, then _____.

▶ Investigation

Sheffer's Stroke and the NAND Gate In 1913, logician Henry M. Sheffer created a connective that we now refer to as Sheffer's stroke. This connective is often denoted by the symbol |. $p \mid q$ is equivalent to $\sim(p \wedge q)$. Because of this equivalence, the Sheffer stroke is referred to as NAND (Not AND).

Any logic statement can be written using only NAND connectives. This is important in computer science because it is then possible to create only NAND circuits rather than OR, AND, and NOT circuits.

The truth table for $p \mid q$ is given as follows.

Sheffer's Stroke

p	*q*	*p* | *q*
T	T	F
T	F	T
F	T	T
F	F	T

Because every logical connective can be represented in terms of NAND, it must be possible to represent $\sim p$, $p \vee q$, and $p \rightarrow q$ using some combination of NAND. As shown in the following truth table, $\sim p \equiv p \mid p$.

p	*p*	*p* | *p*
T	T	F
T	T	F
F	F	T
F	F	T

1. Use a truth table to show that $p \vee q \equiv (p \mid p) \mid (q \mid q)$.

2. Use a truth table to show that $p \rightarrow q \equiv p \mid (q \mid q)$.

3. Write $p \wedge q$ using only the | connective.

Investigating Fallacies

Learning Objective:

• Investigate fallacies.

Fallacies

"When I got up this morning, I petted the dog in my pajamas." We probably know that what is meant here is that someone awoke in her pajamas and then petted her dog. However, the sentence could be interpreted as the dog wearing the pajamas. A **fallacy** is an ambiguous, misleading, or unsound argument.

There are three main types of fallacies: formal fallacies, informal fallacies, and loaded questions. We discussed formal fallacies in Section 2.3. These occur when the formal rules of logic are distorted.

An **informal fallacy** arises when the content of an argument's premises fail to adequately support the conclusion. The opening sentence in this section is an informal fallacy and is called a *fallacy of amphiboly* (ambiguity from faulty grammatical construction). It occurs from a mistake in grammar or punctuation. If the sentence were written, "When I got up this morning in my pajamas, I petted my dog," there would be no ambiguity. Another example of a fallacy of amphiboly is "The lunch consists of good sandwiches and chips." Are the chips good?

Example 1   Red Herring Fallacy

Analyze the following.

Child: I need a new laptop to do my homework.
Parent: When I was your age, we didn't even have calculators.

Solution

Premise: I didn't have a calculator when I was your age.
Conclusion: Therefore, you don't need a laptop.

A red herring is one of the most commonly used fallacious arguments. The argument attempts to introduce some sort of irrelevancy to distract the audience from the issue at hand. In this case, the fact that the parent did not have a calculator growing up is irrelevant to the child needing a laptop. It is meant to draw the focus from the child's request for a laptop to the parent not having a calculator in school.

Example 2   Straw Man Fallacy

Analyze the following:

My biology teacher told me that all living things evolve. I just can't accept that we evolved from fish.

Solution

Premise: All living things evolve.
Conclusion: Humans evolved from fish.

Straw man fallacies usually contain misrepresentations or oversimplifications. During an argument, one side builds a man made of straw, attacks it, and declares victory while the actual opponent remains unaffected. This visualization helps to understand the nature of such arguments. In this case, the biology teacher did not say anything about humans evolving from fish. The argument has nothing to do with the original premise that all things evolve. Effectively, the arguer has built a new premise to attack—in this case it would be of the form, "Human beings evolved from fish." This premise is the straw man that was built to distract the audience from the real premise, that all living things evolve.

Example 3   Ad Hominem (Personal Attack) Fallacy

Analyze the following:

We all know Mackenzie stole the wallet. Mackenzie only makes $10,000 per year.

Solution

Premise: Mackenzie only makes $10,000 per year.
Conclusion: Therefore, Mackenzie stole the wallet.

Ad hominem fallacies typically attack a person's character or integrity rather than the actual issue. They are frequently used in politics. In this case, Mackenzie financial state alone has nothing to do with the theft of someone's property.

| Example 4 | **Appeal to Popularity Fallacy** |

Analyze the following:

As endorsed by the quarterback of our great team, this cream will cure your athlete's foot.

Solution

Premise: The quarterback of our great team endorses this cream.
Conclusion: Therefore, it will cure your athlete's foot.

This type of fallacy is often used in advertising. A popular figure endorses a product or candidate for political office. However, the popular figure may not have the qualifications to make such an endorsement. The fact that the quarterback says that this cream cures athlete's foot is irrelevant. It is intended to connect a popular figure to a conclusion which, in this case, can more accurately be verified by science.

| Example 5 | **Appeal to Ignorance Fallacy** |

Analyze the following:

Even though we have proven that the moon is not made of cheese, we have not proven that the core cannot be filled with cheese. Therefore, the moon's core is filled with cheese.

Solution

Premise: We have not proved that the moon's core cannot be filled with cheese.
Conclusion: The moon's core is filled with cheese.

This fallacy is an argument that asserts a proposition is true because there is no external evidence to prove it is false. The conclusion cannot be drawn simply because it has not been proven otherwise. An argument from ignorance is a fallacy.

| Example 6 | **Appeal to Authority Fallacy** |

Analyze the following:

The mathematics professor stated that increasing the fluoride in drinking water would prevent tooth decay.

Solution

Premise: The mathematics professor stated that increasing the fluoride in drinking water would prevent tooth decay.
Conclusion: Fluoride in drinking water will prevent tooth decay.

Using expert testimony is almost always a valid form of argumentation. For example, using a doctor's testimony when making a medical argument is valid. However, in this case, the appeal to an authority is fallacious. The mathematics professor is, in fact, an expert. The problem is that expertise in mathematics is not expertise on the effects of fluoride in drinking water. The argument is drawing on the expertise of an irrelevant authority.

Example 7	**Appeal to Emotion Fallacy**

Analyze the following:

You should donate money so animal shelters have enough financial support to save animals from euthanasia.

Solution

Premise: Animal shelters need financial support to avoid euthanasia.
Conclusion: Without your donation, animals will be euthanized.

This argument is an attempt to elicit an emotion to reach a conclusion, bypassing rational thought. These types of fallacies can be directed at a variety of emotions including pity, fear, love, flattery, worry, etc., in order to elicit thoughts of consequence, ridicule, pride, etc. They are also frequently used in advertising. In this case, donations alone are no guarantee of euthanasia prevention.

Example 8	**Appeal to Common Practice Fallacy**

Analyze the following:

Parent: The school called and told us you are suspended for cheating on the test.
Child: Everyone else cheated too! I was just the only one who got caught!

Solution

Premise: Everyone in class cheated on the test.
Conclusion: The child did not do anything that warranted suspension.

Did your mother ever ask you, "If everyone else jumped off a bridge, would you?" This can be thought of as the "Everyone else was doing it" fallacy. In this case, the child's argument is that because everyone else was considered innocent for the same action, the child is therefore innocent as well. Cheating is against the rules, and the fact that no one else got caught is irrelevant.

Example 9	**Post Hoc (False Cause) Fallacy**

Analyze the following:

Parker takes walks on cool evenings and got the flu. Therefore, taking walks on a cool evening causes the flu.

Solution

Premise: Parker takes walks on cool evening and got the flu.
Conclusion: Taking walks on cool evenings causes the flu.

This argument is a post hoc (after the fact) fallacy in which the argument reasons that because one event occurred before the other, the first event *caused* the other. In other words, event y happened after event x. Therefore, x caused y. In this case, the cool evening did not necessarily cause the flu. The flu may have been caused by someone with whom Parker came in contact. The argument merely illustrates a correlation between the two events. Correlation does not always indicate causation.

Example 10 **Black-or-White (False Dilemma) Fallacy**

Analyze the following:

We must go to war or we will lose all of our freedoms.

Solution

Premise: Either we go to war or we lose our freedoms.
Conclusion: Going to war will protect our freedoms.

The black-or-white fallacy is also known as false dilemma or false dichotomy. The error in logic occurs when the argument offers only two choices, either x or y. If not x, then y. In reality, there are more options. In this case, though it is not a formal fallacy, the disjunctive premise is falsely supported. Presenting only two possibilities when there are many makes the argument seem black or white when many "gray" options exist.

Example 11 **Slippery Slope Fallacy**

Analyze the following:

If you keep using gas cars, the emissions will pollute the air, accelerate global climate change, boil the planet, and kill us all.

Solution

Premise: Gas cars pollute the air.
Conclusion: Continued use of gas cars will kill us all.

This is an argument that begins with a relatively insignificant event and then connects purportedly related events to eventually end in an absurd conclusion. Although there may be sound reasons to decrease emissions from cars, the logic between using gas-powered cars and everyone dying is weak.

Example 12 **Circular Reasoning**

Analyze the following:

Our baseball team is the best team in the league because there is no team better.

Solution

Premise: There is no team better than ours.
Conclusion: Our team is the best team.

The conclusion is the same as the premise, just reworded. This is circular reasoning or *circulus in probando* (circle in proving) and a fallacy, a misleading argument. The preceding argument is a simple form of a circular argument. Usually, these types of arguments come in much bigger circles and can be much more difficult to detect. In any case, circular arguments lead to fallacious arguments.

Example 13 **Hasty Generalization Fallacy**

Analyze the following:

Two members of the tennis team improved their play after taking a yoga class. Therefore, we should have all members of the tennis team take a yoga class.

Solution

Premise: Taking a yoga class improves tennis play.
Conclusion: All tennis players should take a yoga class to make the team better.

The hasty generalization fallacy is an argument that is based on a small sample or data set. In this example, the conclusion is based on two tennis players. There is no way to know whether the yoga class improved their play or whether something else was responsible such as more practice. The fallacy is based on concluding something is good for all if it is good for a few.

The third type of fallacy is the **loaded question fallacy**, where a question contains a "loaded" presumption. It is a question with a false or questionable presupposition.

> **Example 14**   **Loaded Question Fallacy**

Analyze the following:

Have you stopped mistreating your pet?

Solution

Because this is a question, it is not technically a fallacious *argument* but an effort to trick someone into an unintended implication. Therefore, a sound premise and conclusion cannot be established. Any direct response implies a deniable statement or one that is downright false. If the responder answers yes, the responder implies mistreating the pet but has quit. If the responder answers no, there is still an implication of mistreating the pet. This type of question is similar to a black-or-white fallacy because it only offers two options. The proper response to such a question is to reject it or refuse to answer.

2.4 Exercise Set

▶ Think About It

1. An advertisement shows a picture of a famous actor next to a new car. What type of fallacy might this be?

2. A television advertisement shows a lively party with everyone drinking a popular beverage. What type of fallacy might this be?

3. An associate asks whether you have stopped making inappropriate comments to coworkers. What type of fallacy might this be?

▶ Preliminary Exercises

For Exercises P1 to P14, if necessary, use the referenced example following the exercise for assistance.

State the logical fallacy of each of the following. Be sure to state the premise and conclusion as well as why the fallacy you chose is correct. If the answer is a loaded question, it will not have a premise or conclusion.

P1. **Child:** It's really hard to make a living on the salary I earn.
Parent: When I was your age, I only made $80 per week.
Example 1

P2. My math teacher told me that we use math every day, sometimes without even realizing it. I can't believe the teacher thinks we are all mindless zombies. **Example 2**

P3. The doctor gave me an exercise plan to follow. However, I'm not going to do it because the doctor is 60 years old. **Example 3**

P4. These shoes will make you run faster because the best running back in the football league endorses them. **Example 4**

P5. Even though we have proven that the jewelry Arya's wearing is yours, we have not proven that Arya stole it from you. Therefore, Arya did not steal it from you. **Example 5**

P6. The police officer told the kids that eating too much red meat is unhealthy. **Example 6**

P7. The advertisement by Health Food Inc. says that eating the company's nutrition bars keeps you from gaining weight. **Example 7**

P8. **Parent:** I received a note from your principal saying that you were making fun of another student. You're grounded!
Child: But everyone else was doing it. **Example 8**

P9. Today, the temperature dropped 10 degrees from yesterday and I got a headache. Therefore, the temperature drop gave me a headache. **Example 9**

P10. You must protest with us or you are against our cause. **Example 10**

P11. If it doesn't rain, the lake will dry up, tourism will decline, the local economy will tank, and we will be forced to move to the city. **Example 11**

P12. This computer is the best because no other computer can compare with its quality. **Example 12**

P13. Three players on our hockey team didn't shave during the playoffs and played much better than usual. Therefore, we should all avoid shaving during the playoffs. **Example 13**

P14. Have you told the teacher that you copied my paper? **Example 14**

Section Exercises

For Exercises 1–42, state the premise and the conclusion. Then write a sentence describing the fallacy. If the answer is a loaded question, there will be no premise or conclusion.

1. Don't listen to Dillon. Dillon doesn't believe in the separation of church and state.

2. Don't let her use the computer, she is mean.

3. You have to buy me a car or I will lose my job.

4. Either you're with me or against me.

5. Shortly after immigration increased, the economy improved. Therefore, immigration improved the economy.

6. I took flu medication yesterday morning and today I feel much better. The flu medication made me feel better.

7. I just landed in this country, and my cab driver from the airport was so rude. Therefore, everyone in this country must be rude.

8. My Ford gets bad gas mileage. Therefore, every Ford must get bad gas mileage.

9. If you don't go to college, you won't get a good job and be able to pay rent. You'll have to move in with your parents and before you know it, you will be on the street begging for change.

10. You shouldn't let your kids play outside because they will want to roam the neighborhood. If they are free to roam the neighborhood, they will get kidnapped by strangers, taken to a safe house, and held for ransom.

11. If he asks you on a date, you have to say yes. He lost his partner a year ago.

12. My parents donate a lot of money to this school. You should consider giving me a better grade on the test.

13. According to the survey, more than 100,000 people believe that astrology affects people's lives. Therefore, there must be some truth to it.

14. At my school, 99% of the students have smartphones. Therefore, they must be the best.

15. I understand that going out of your way is an inconvenience, but surely you know that there are more important things than convenience.

16. **Parent:** You're not allowed to go to your friend's house because your chores are not done.

    Child: You never let me do anything fun with my friends!

17. **Principal:** We need to start a program that makes community service mandatory for teachers.

    Teacher: Teachers already don't get paid enough for what we do.

18. **Political Candidate:** If I am elected, the first thing I will do is cut taxes.

    Opponent: My opponent has consistently misled us as to when a rational economic platform will be announced.

19. Are you still trying to overcome your addiction to cigarettes?

20. Have you stopped using illegal accounting practices?

21. You are a human being because you have human parents.

22. Electrolytes are important, and we need to drink more of them because they are vital.

23. **Police Officer:** I pulled you over because you were driving 75 mph in a 65-mph zone.

    Driver: Yes, but everyone always drives at least 75 mph on this freeway!

24. **Manager:** Did you steal office supplies?

    Employee: I'm just taking a few pens; everyone else takes them too!

25. The actor stated that tight monetary policy is the best way to shorten a recession.

26. My doctor told me that my car isn't starting because the alternator needs to be replaced.

27. How do you know what's best for children? You're not a parent.

28. Why would you let Remy borrow your car? Remy dropped out of high school!

29. If I make an exception for you, Joey will find out and want one too. Then everyone will find out, and I will have to make an exception for everyone.

30. If you eat that cookie, you'll just want to eat more. Soon the whole bag will be gone, and before you know it you will want a glass of milk.

31. The police were unable to prove that a ghost wasn't responsible for the strange activity. Therefore, it was a ghost.

32. You say that you are smarter than most people but you cannot prove it. Therefore, you are not.

33. **Student:** Students need a greater voice when it comes to curriculum changes.

    Teacher: The teachers are the ones who need a greater voice. We don't ever have a say when it comes to budget issues.

34. **Teacher:** You are in violation of dress code again.

    Student: We should be able to wear whatever we want to school.

35. That's soccer. Love it or hate it.

36. You need to get a job or you will end up in the streets.

37. Surely my beautiful and intelligent sister wants to give me a ride to school.

38. If we don't switch to solar power immediately, we are all in trouble.

39. It's okay that the university got caught compensating its athletes. Most universities are doing it these days.

40. It shouldn't matter that we got caught embezzling. This kind of thing goes on in every company.

41. All of my friends have trucks instead of cars. Therefore, trucks are better.

42. Everyone I know watches this TV show. It must be good.

Investigation

Taxonomy of the Logical Fallacies A taxonomy is a classification of something into similar things. For instance, in biology animals are classified in different phyla. There is a taxonomy for fallacies as well. One source for this taxonomy is http://www.fallacyfiles.org/taxonomy.html. Go through Exercises 1 to 42 and try to determine in which class the fallacy belongs. It is possible that you may place a fallacy in a class that is different from one a classmate might use. That is not a problem. Just have a reason for your choice.

Test Prep ② Sets and Logic

Section 2.1   Sets, Set Operations, and Applications

Mathematicians organize objects into *sets*. The objects that belong in a set are the **elements** or **members** of the set. Two sets are **equal** if they have exactly the same elements. The set that has no elements is called the **empty set**. The empty set is symbolized by $\varnothing$. The **roster method** is one way of writing a set.	See **Example 1** on page 55, and then try Exercises 1, 2, and 3 on page 93.
The set of all elements that are being considered is called a **universal set**. The **complement** of a set A, denoted by A', is the set of all elements of the universal set U that are not in A. Two important relationships between a universal set U and $\varnothing$ are $U' = \varnothing$ and $\varnothing' = U$.	See **Example 2** on page 55, and then try Exercises 7 and 9 on page 93.
A set A is a **subset** of set B, denoted by $A \subseteq B$, if every element of A is also a member of B. A is a **proper subset** of B, denoted by $A \subset B$, if every element of A is an element of B, and $A \neq B$. A **Venn diagram** can be used to illustrate sets and relationships among sets.	See **Example 3** on page 56, and then try Exercises 13, 15, and 17 on page 93.
The **intersection** of sets A and B, denoted by $A \cap B$, is the set of elements that are both in A and in B. Two sets are **disjoint** if their intersection is the empty set. The **union** of set A and B, denoted by $A \cup B$, is the set of elements that belong to A, belong to B, or belong to both A and B.	See **Examples 4** and **5** on page 57, and then try Exercises 21, 22, and 27 on page 93.
Venn diagrams can be used to illustrate a set operation and to determine whether two set operations are equal.	See **Example 6** on page 58, and then try Exercise 31 on page 93.
The number of subsets of a set with n elements is 2^n. The number of *proper* subsets is $2^n - 1$.	See **Example 7** on page 59, and then try Exercise 33 on page 93.
The **cardinality** of a set A is the number of elements in the set and is denoted by $n(A)$. The **inclusion-exclusion principle** states that for all finite sets, $n(A \cup B) = n(A) + n(B) - n(A \cap B)$.	See **Example 8** on page 60 and then try Exercises 35 and 36 on page 94.
Applications of sets.	See **Examples 9, 10,** and **11** on pages 60 to 62, and then try Exercises 37, 39, and 40 on page 94.

Section 2.2   Categorical Logic

A **categorical statement** is one that expresses how two sets or categories relate to one another. In a statement, the word *some* and the phrases *there exists* and *at least one* are called **existential quantifiers**. The words *none*, *no*, *all*, and *every* are called **universal quantifiers**. Diagrams similar to Venn diagrams called **Euler diagrams** are used to show the relationship among categories.	See **Example 1** on page 67, and then try Exercise 41 on page 94.
The **negation** of a true statement is a false statement, and the negation of a false statement is a true statement. The following table shows a statement and its negation.	See **Example 2** on page 68, and then try Exercise 42 on page 94.

Statement	Negation
All X are Y.	Some X are not Y.
No X are Y.	Some X are Y.
Some X are not Y.	All X are Y.
Some X are Y.	No X are Y.

An **argument** consists of a set of statements called **premises** and another statement called the **conclusion**. A **valid argument** is one that has a *logical form* that makes it impossible for the premises to be true and the conclusion false. An argument is invalid if it is not a valid argument. Euler diagrams can be used to determine the validity of some arguments.	See **Examples 3, 4, 5,** and **6** on pages 68 to 70, and then try Exercises 43 to 49 on page 94.

Section 2.3   Propositional Logic

A **proposition** is a declarative sentence that is either true or false or one that a declarative sentence asserts to be true or false.	See **Example 1** on page 73, and then try Exercises 50, 51, and 52 on pages 94–95.
The **negation** of a true proposition is false; the negation of a false proposition is true.	See **Example 2** on page 74, and then try Exercises 31 and 32 on page 93.
A **simple statement** is a proposition that conveys a single idea. A **compound statement** is one that conveys two or more ideas. Compound statements are formed using the connective words or phrases *and*; *of*; *if . . . then*; and *if and only if.*	See **Examples 3** and **4** on pages 74 and 75, and then try Exercises 56, 57, 58, and 59 on page 95.
DeMorgan's Laws If p and q are propositions, then $\sim(p \lor q) \equiv (\sim p) \land (\sim q)$ and $\sim(p \land q) \equiv (\sim p) \lor (\sim q)$.	See **Example 5** on pages 76 and 77, and then try Exercises 61 and 62 on page 95.
In any conditional statement represented by "If p, then q" or "If p, q," the p statement is called the **antecedent** and the q statement is called the **consequent**. Symbolically, the conditional is represented by $p \rightarrow q$.	See **Example 6** on page 77, and then try Exercise 63 on page 95.

The truth table for the conditional $p \rightarrow q$ is given as follows. **Truth Table for $p \rightarrow q$**	See **Example 7** on page 78, and then try Exercises 64 and 66 on page 95.

p	q	$p \rightarrow q$
T	T	T
T	F	F
F	T	T
F	F	T

Given the conditional $p \rightarrow q$, there is the **converse** $q \rightarrow p$, the **inverse** $\sim p \rightarrow \sim q$, and the **contrapositive** $\sim q \rightarrow \sim p$.	See **Example 8** on page 79, and then try Exercises 68 and 69 on page 95.

Table of Forms of Valid Reasoning and Forms of Fallacies	See **Examples 9, 10,** and **11** on pages 79 to 81, and then try Exercises 70 to 73 on page 95.

Modus Ponens	Modus Tollens	Transitive Reasoning	Fallacy of the Converse	Fallacy of the Inverse
$p \rightarrow q$	$p \rightarrow q$	$p \rightarrow q$	$p \rightarrow q$	$p \rightarrow q$
p	$\sim q$	$q \rightarrow r$	q	$\sim p$
$\therefore q$	$\therefore \sim p$	$\therefore p \rightarrow r$	$\therefore p$	$\therefore \sim q$

Section 2.4   Investigating Fallacies

An **informal fallacy** arises when the content of an argument's premises fails to adequately support the conclusion. This section discusses some but not all possible fallacies.	See **Examples 1** to **14** on pages 85 to 89, and then try Exercises 74 to 77 on page 95.
Red herring fallacy: An argument that attempts to introduce some sort of irrelevancy in order to distract the audience from the issue at hand.	
Straw man fallacy: Straw man fallacies usually contain misrepresentations or oversimplifications. The idea is that during an argument, one side builds a man made of straw, attacks it, and declares victory while the actual opponent remains unaffected.	

Ad hominem (personal attack) fallacy: The integrity of a person is attacked rather than the issue.

Appeal to popularity fallacy: A popular figure endorses a product or candidate for political office. However, the popular figure may not have the qualifications to make such an endorsement.

Appeal to ignorance fallacy: An argument asserts a proposition is true because there is no evidence to prove that it is false.

Appeal to authority fallacy: An argument that draws on the expertise of an irrelevant authority.

Appeal to emotion fallacy: This argument is an attempt to elicit an emotion to reach a conclusion, bypassing rational thought.

Appeal to common practice fallacy: An argument is assumed valid based on some common practice or that most people do it.

Post hoc (false cause) fallacy: In a post hoc (after the fact) fallacy someone reasons that since one event occurred before the other, the first event *caused* the other.

Black-or-white (false dilemma) fallacy: An argument that offers only two choices when there are actually three or more options available.

Slippery slope fallacy: An argument that begins with a relatively insignificant event and then chains purportedly related events eventually ending in an absurd conclusion.

Circular reasoning: An argument where the conclusion is a restatement, in a little different form, of the premise.

Hasty generalization fallacy: An argument that is based on a small sample or data set.

Loaded question fallacy: This is a question that has no reasonable answer.

Review Exercises

For Exercises 1 to 6, use the roster method to list the elements of the set.

1. The negative integers greater than -6.
2. The composite numbers less than 17.
3. The letters in the name Tennessee.
4. The set of months with exactly 30 days.
5. The set of integers x that satisfy $x - 4 = 3$.
6. The set of whole numbers x that satisfy $x - 1 < 4$.

In Exercises 7 to 10, find the complement of the set given that $U = \{0, 1, 2, 3, 4, 5, 6, 7, 8\}$.

7. $\{2, 4, 6, 7\}$
8. $\{3, 6\}$
9. $\varnothing$
10. $\{0, 1, 2, 3, 4, 5, 6, 7, 8\}$

In Exercises 11 to 20, let $U = \{p, q, r, s, t\}$, $D = \{p, r, s, t\}$, $E = \{q, s\}$, $F = \{p, t\}$, and $G = \{s\}$. Determine whether each statement is true or false.

11. $F \subseteq D$
12. $D \subseteq F$
13. $G \subset E$
14. $F \subset D$
15. $G' \subset D$
16. $E = F'$
17. $\varnothing \subset \varnothing$
18. $G \in E$

19. D has exactly eight subsets and seven proper subsets.
20. $\{0\} = \varnothing$

In Exercises 21 to 30, let $U = \{1, 2, 3, 4, 5, 6, 7, 8\}$, $A = \{2, 4, 6\}$, $B = \{1, 2, 5, 8\}$, and $C = \{1, 3, 7\}$. Find each of the following.

21. $A \cup B$
22. $A \cap B$
23. $A \cap B'$
24. $B \cap C'$
25. $(A \cup B)'$
26. $(A' \cap B)'$
27. $A \cup (B \cap C)$
28. $A \cap (B \cup C)$
29. $A \cup A'$
30. $A \cap A'$

31. Use a Venn diagram to illustrate the region defined by $A' \cap (B \cup C)$.

32. Use a Venn diagram to determine whether $A' \cap B'$ and $(A \cap B)'$ are equal.

33. **Smoothie Options** A store featuring smoothies offers a regular acai smoothie to which blueberries, strawberries, bananas, pineapple, and mango can be added. How many different smoothies can be offered using these ingredients?

34. **Upgrade Options** An automobile company makes a sedan with nine upgrade options. How many different versions of this sedan can the company produce?

35. **College Majors** A college finds that of the 841 students taking math, 525 students are business majors and 202 students are taking math and are business majors. How many students are math or are business majors?

36. **Health Club Survey** In a survey at a health club, 208 members indicated that they enjoy aerobic exercises, 145 indicated they enjoy weight training, 97 indicated they enjoy both aerobics and weight training, and 135 indicated they do not enjoy either of these types of exercise. How many members were surveyed?

37. **Ice Cream Syrup Options** A survey asked 750 people whether they like chocolate (*C*), butterscotch (*B*), or strawberry syrup (*S*) on vanilla ice cream. The results are in the following Venn diagram.

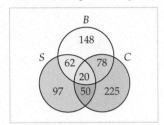

   a. How many people liked exactly one syrup?
   b. How many people liked all three syrups?
   c. How many people liked exactly two of the syrups?
   d. How many people did not like any of the syrups?

38. **Game Preferences** A survey asked 2000 people whether they like Monopoly® (*M*), Scrabble® (*S*), or Randomize® (*R*). The results are in the following Venn diagram.

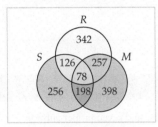

   a. How many people only liked Randomize?
   b. How many people liked all three games?
   c. How many people liked exactly two of the games?
   d. How many people did not like any of the games?

39. **Car Options** A survey asked people of various ages whether they would consider purchasing an American (*A*) made car, a Japanese (*J*) made car, or a German (*G*) made car. The results are shown in the following table.

	A	J	G
18 − 29 (E)	64	48	54
30 − 50 (T)	72	64	56
51 + (F)	58	65	73

Find the number of people in each set.
   a. $J \cup F$
   b. $T \cup G$
   c. $E' \cup A$
   d. $F \cap A'$
   e. $T \cup (A \cup G)$

40. **Clinical Trials** A pharmaceutical company has a new procedure to test whether a person's immune system is reacting properly to the mumps antigen. The test is given to 1000 people, 200 of whom are known to have the mumps antigen. The results are shown in the following Venn diagram.

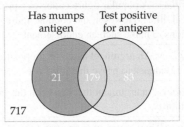

   a. Using the Venn diagram, complete the following table.

	Positive Test	Negative Test
Person Has Mumps Antigen		
Person Does Not Have Mumps Antigen		

   b. How many false negatives occurred?
   c. Write a sentence that describes a false positive in the context of this test.
   d. How many true negatives occurred?

41. Draw an Euler diagram for both of the following.
   a. Some dogs are not poodles.
   b. All cats are not dogs.

42. Write the negation of both of the following.
   a. Some cars are not fuel efficient.
   b. Every parakeet is a bird.

For Exercises 43 to 49, use an Euler diagram to determine whether the argument is valid.

43.   No wizard can yodel.
      All lizards can yodel.
   ∴ No wizard is a lizard.

44.   Some natural numbers are prime numbers.
      8 is a natural number.
   ∴ 8 is a prime number.

45.   Some dogs are brown.
      Some dogs are big.
   ∴ Some big dogs are brown.

46.   All novels are fiction.
      No movies are fiction.
   ∴ No movies are novels.

47.   All sandwiches are good.
      All good sandwiches have pastrami.
      All sandwiches with pastrami need mustard.
   ∴ All sandwiches with mustard are good.

48.   All Italian villas are wonderful. Some wonderful villas are expensive. Therefore, some Italian villas are expensive.

49.   All logicians like to sing, "It's a Small World After All." Some logicians have been presidential candidates. Therefore, some presidential candidates like to sing, "It's a Small World After All."

In Exercises 50 to 52, determine whether each sentence is a proposition.

50.   Saturn is not a planet in our solar system.
51.   Would you like coffee or tea?

52. Prevent wildfires.

53. Write the negation of "The concert music was too loud."

In Exercises 54 to 57, use the following sentences to write the compound proposition in symbolic form.

> *p*: Julian delivers the mail in the snow.
>
> *q*: Julian delivers the mail in the rain.
>
> *r*: Julian delivers the mail in the sleet.

54. Julian delivers the mail in the snow and the rain.

55. Julian delivers the mail in the sleet but not the rain.

56. If Julian delivers the mail in the rain, then Julian does not deliver the mail in the snow.

57. Julian delivers the mail in the rain, sleet, or snow.

In Exercises 58 to 60, use the following sentences to write the symbolic expression as a sentence.

> *p*: The football game will be played on Friday.
>
> *q*: The football game will be played at home.
>
> *r*: The visiting football team is from Boston.

58. $p \land q$

59. $p \rightarrow \sim r$

60. $(q \lor r) \land p$

61. Use DeMorgan's laws to rewrite the negation of "This is the time for action or forever hold your peace."

62. Use DeMorgan's laws to rewrite the negation of "I am committed to this fine state and will not seek the presidential nomination."

63. Identify the antecedent and consequent for the following two propositions.

   a. If I practice my accent, I will sound more like a native speaker.

   b. It is necessary to make a 20% down payment to secure a loan on this house.

For Exercises 64 to 67, determine the truth-value of the conditional propositions.

64. If diamonds are soft, then fire engines are green.

65. If dogs are mammals, then cats are reptiles.

66. If stars burn hydrogen, then Earth has a north pole.

67. If the sun is a planet, then Santa Claus lives at the North Pole.

In Exercises 68 and 69, write the converse, inverse, and contrapositive of the conditional proposition.

68. If I have an electric car, I can ride in the HOV lane on the freeway.

69. If I start exercising, I may lose some weight.

In Exercises 70 to 73, determine whether the argument is valid. Use either truth tables or Table 2.16.

70. If the shoe fits, wear it. I wore the shoe. Therefore, the shoe fits.

71. If I cut my late afternoon class, then I will go to the party. I did not go to the party. Therefore, I did not cut my late afternoon class.

72. If I don't watch television, I'll fall asleep. If I fall asleep, I'll wake up at 3 A.M. Therefore, if I don't wake up at 3 A.M., I watched television.

73. If I win the lottery, I'll buy a Maserati. I won the lottery. Therefore, I bought a Maserati.

In Exercises 74 to 78, write a sentence that explains why the argument is a fallacy.

74. My opponent advocates lowering taxes on the middle class. This from someone who earns a seven-figure salary.

75. This book endorses a gluten-free diet. This is a sound idea because of the author's reasons.

76. This school is extremely selective. However, my child received all A's, so my child will be accepted into the school.

77. By rejecting community involvement, you are rejecting values such as cooperation and entrepreneurship.

Project 1

Switching Circuits and Logic

There is a strong connection between logic and switches in an electric circuit. A switch in a circuit can be represented as in the following diagram.

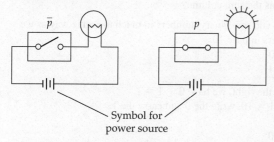

Symbol for power source

When the switch is in the off position, denoted by $\bar{p}$ (for circuits, we use $\bar{p}$ instead of $\sim p$), current cannot reach the light bulb so it remains off. When the switch is in the on position, denoted by p, the circuit is complete and the light bulb turns on.

   Instead of using T and F for true and false in a truth table, we will use 1 for T and 0 for F. A switch has a *closure table* to determine under which conditions a light will be on. We use a 1

to designate that a switch or switching network is closed and a 0 to indicate that it is open. Flipping a switch that is closed opens it, and flipping a switch that is open closes it. This is similar to a truth table for NOT. See Table 2.6.

p	$\bar{p}$
1	0
0	1

   The switches in the following circuit are said to be in *parallel* because the two switches appear to be parallel to one another. The current can pass through p, thereby turning on the light bulb.

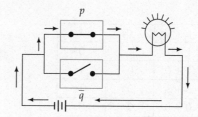

The beginning of the closure table for this circuit is shown as follows.

p	q	On
I	I	
I	0	I
0	I	
0	0	

1. Draw a circuit similar to the preceding for the other three rows of the closure table.
2. Based on the circuits, complete the closure table.
3. What logical operator corresponds to the closure table?

The switches in the following circuit are said to be in *series* because the two switches appear to come one after another as in a line. The current can pass through p but is stopped at $\bar{q}$, so the light bulb is off.

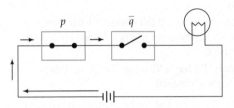

The beginning of the closure table for this circuit is shown as follows.

p	q	On
I	I	
I	0	0
0	I	
0	0	

4. Draw a circuit similar to the preceding for the other three rows of the closure table.
5. Based on the circuits, complete the closure table.
6. What logical operator corresponds to the closure table?

Now consider a fairly common situation. An overhead light illuminates a stairwell. For this light, there is a switch at the top and bottom of the stairwell. The closure table for this circuit is given as follows.

p	q	On
I	I	0
I	0	I
0	I	I
0	0	0

In other words, the light is on when one switch is in the on position and the other is in the off position. This closure table (this applies to truth tables as well) is called exclusive OR and written XOR. The light is on when one switch is on and the other is off. This is given by $p \oplus q = (p \wedge \bar{q}) \vee (\bar{p} \wedge q)$, where the $\oplus$ is symbol for XOR. The XOR circuit (for the second row in the table) looks like the following diagram.

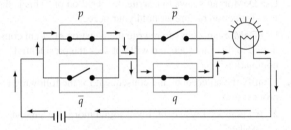

7. Draw the remaining three states of the XOR circuit as indicated in the closure table.

Project 2

Extending Switching Circuits to Binary Arithmetic

There is a relationship between binary arithmetic and the circuits as previously discussed. Before we can establish that relationship, we need a few fundamentals.

A binary number system is one in which the only two symbols are 0 and 1. If we were to count in binary, it would look like 0, 1, 10, 11, 100, 101, 110, 111, 1000, . . . These are the same as decimal (base 10) numbers 0, 1, 2, 3, 4, 5, 6, 7, 8, . . .

1. What are the next four binary numbers after 1000?
   Here is the addition table for binary arithmetic. We are writing this table with an S column for the sum and a C column for the carry when adding.

p	q	p + q	C	S
I	I	I + I	I	0
I	0	I + 0	0	I
0	I	0 + I	0	I
0	0	0 + 0	0	0

2. From the previous project, what circuit has the same closure table as the carry column?
3. From the previous project, what circuit has the same closure table as the sum column?

We can add two binary numbers in much the same way as we add numbers in base 10.

$$\begin{array}{r} 1110 \\ +1011 \end{array}$$

Starting at the right, we have $0 + 1 = 1$. Then $1 + 1 = 10$, write the 0 and carry the 1.

$$\begin{array}{r} 1 \\ 1110 \\ +1011 \\ \hline 01 \end{array}$$

Now $(1 + 1) + 0 = 10 + 0 = 10$. Write the 0 and carry the 1.

$$\begin{array}{r} 11 \\ 1110 \\ +1011 \\ \hline 001 \end{array}$$

$(1 + 1) + 1 = 10 + 1 = 11$. Write 11.

```
   11
 1110
+1011
11001
```

4. Add: 101011 + 110101

The connection between circuits and binary arithmetic is the essence of how computers can perform operations. Rather than use the large diagrams of circuits as we did previously, computer scientists use symbols for each of the circuits. Here are three possible symbols. Other symbols are used.

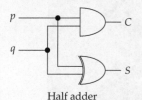

AND   OR   XOR

Using these symbols, we can construct what is called a *half adder* circuit.

Half adder

5. Complete the following closure table for the half adder, thereby verifying the circuit does produce the correct carry and sum of two binary numbers.

p	q	C	S
1	1		
1	0		
0	1		
0	0		

The half adder can add only two numbers. However, by combining half adders and one other circuit type, the OR circuit, a full adder can be constructed.

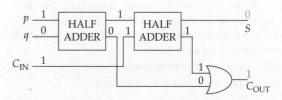

The preceding diagram shows the specific instance of the third step in the preceding addition example. Here it is again.

```
   11
 1110
+1011
  001
```

Note that C_{OUT} is 1, the carry from adding $1 + 1 + 0$.
S is 0, the right digit from adding $1 + 1 + 0$.

6. Draw a fuller adder to show each step is correct for the addition example we previously did. When adding the right-most digits $(0 + 1)$, C_{OUT} is 0. After the initial addition, use the value of C_{OUT} for C_{IN} for each of the next steps.

We will stop here, but by stringing full adders together, we could add binary numbers that have many digits. The main takeaway you should get from these two projects is that electric circuits can be constructed that operate as truth tables for logical operators. From this, we designed a stairwell circuit.

Extending these ideas allowed us to create computer circuits that can add numbers. A computer scientist would continue along this path to find techniques to extend these ideas to the other operators such as subtraction, multiplication, division, square root, and many other math operations.

Math in Practice

There are many jobs that use logic and sets. A database is a large set containing many different subsets. Writing reports for drawing out specific information from a database by writing code with set operations like AND and OR to provide insights into business operations is the job of a Report Developer. A Data Engineer develops, moves, manages, and aggregates raw data into meaningful information housed in a database. A Data Analyst works with business users to identify specific data on sales numbers, market research, logistics, or other behaviors. Then they design processes to help the business identify trends and make better decisions.

Part I

Your company has servers holding millions of records specifying detailed transactions. Every day, thousands of pieces of data are collected at your company and saved to a server. To create a backup, rather than saving the whole database containing millions of pieces of data, you only need to save certain pieces, which are usually in the thousands. It is your job to determine what new data should be uploaded to the backup servers. Your company makes three different products: Widgets (W), Figments (F),

and Doodads (D). Suppose your company has tasked you with saving data about daily sales for each of the products, but due to old software you are only given information about the groups of products sold.

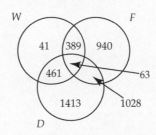

a. Find the total number of Figments (F) sold.

b. Find the total number of Doodads (D) sold.

c. Find the total number of Widgets (W) sold.

d. If the sales goal was to sell more than 2800 Doodads, was that goal achieved?

Part 2

Your company's CEO wants to celebrate your company's success with a burrito bar. The burrito bar consists of tortilla, protein, and add-ons. For a tortilla you can choose either corn or flour. For protein, you can choose from chicken, beef, and tofu. For add-ons, you can choose beans, rice, lettuce, onion, guacamole, salsa, cheese, sour cream, and cilantro.

a. How many variations of proteins can be made?

b. How many variations of add-ons can be made?

Part 3

The Human Resources Department of your mid-sized company (500−1500 employees) requires reporting on benefits. The database gives you the following information.

	Group A <1 year	Group B 1–5 years	Group C > 5 years
Health Insurance	87	902	478
Dental Insurance	35	851	462
Vision Insurance	16	739	381
Health Savings Account	55	899	470
401(k)	75	910	479

a. How many employees participate in Health Insurance benefits?

b. How many employees who have been at the company for more than 1 year have a 401(k)?

c. If there are 480 employees who have been at the company for more than 5 years (Group C), how many of them do not participate in the Vision Insurance benefits?

d. If there are 1479 employees at the company, how many do not participate in the Dental Insurance benefits?

e. If there are 88 employees who have been at the company for less than 1 year (Group A) and 480 employees who have been at the company for more than 5 years (Group C), how many those employees do not participate in the Health Savings Account?

Proportions and Variation

Section	How This Is Used . . .
3.1 Ratios and Rates	A ratio is a comparison of two quantities with the same units. If an adult is 6 feet tall and a child is 3 feet tall, the ratio of the adult's height to the child's height is 6 to 3 or, in simplest form, 2 to 1. A *rate* is a comparison of two quantities with different units. A person who walks 6 miles in 2 hours has a rate of 6 miles to 2 hours or, in simplest form, 3 miles per hour.
3.2 Measurement	A measurement is a quantity that contains a number, its magnitude, and a unit such as feet. For instance, 7 inches is a measurement. The magnitude is 7; the units are inches. The U.S. customary system and the metric system are the two major systems used today.
3.3 Proportions	Proportions are equations that state the equality of two ratios or rates.
3.4 Percent	The word *percent* is derived from Latin *per centum* (100). It is the ratio of one quantity to 100 units of the same quantity. One application of percent is in income tax brackets. The amount of income tax owed is a percent of income earned.
3.5 Variation	Variation considers the change of one quantity as two or more other quantities change. For instance, a person's body mass index changes as the person's height and weight change.

Math in Practice:
Trenchless Sewer Pipe
Lining, p. 140

Applications of Proportions and Variation

One application of a proportion is in estimation. For instance, approximately one person in 10 is left-handed. Knowing this, we can estimate that in a group of 100 people, approximately 10 would be left-handed. Variation applications can be relatively straightforward. For instance, the amount you pay for gasoline varies as the cost per gallon of gasoline and the number of gallons you purchase vary. And variation applications can be especially complicated, such as the force of gravity between two bodies depending on the masses of the bodies and the distance between them.

Exercise 19 on page 133 looks at how the intensity of sound varies as the distance between an airport ramp worker and a jet engine changes.

Section 3.1 Ratios and Rates

Learning Objectives:

- Identify the difference between a ratio and rate.
- Determine a unit rate.
- Determine the more economical purchase.
- Calculate rates in a population.
- Calculate an exchange rate.

Be Prepared: Prep Test 3 reviews concepts used in this chapter. This can be found in your Companion Site at Cengage.com

Ratios

A **ratio** is the comparison of two quantities that have the same units. A ratio can be written in three different ways:

1. As a fraction, $\dfrac{2}{3}$

2. As two numbers separated by a colon (:), and $2:3$

3. As two numbers separated by the word *to*. $2 \text{ to } 3$

Although units such as hours, miles, or dollars are written as part of a rate, units are not written as part of a ratio.

According to the U.S. Bureau of Labor Statistics, there are approximately 123 million married people in the United States, and 82 million of these people are in the labor force. The ratio of the number of married people in the labor force to the total number of married people in the country is calculated as follows. Note that the ratio is written in simplest form.

$$\frac{82,000,000}{123,000,000} = \frac{2}{3} \quad \text{or} \quad 2:3 \quad \text{or} \quad 2 \text{ to } 3$$

The ratio 2 to 3 tells us that two out of every three married people in the United States are part of the labor force.

Given that 82 million of the 123 million married people in the country work in the labor force, we can calculate the number of married people who do not work in the labor force.

$$123 \text{ million} - 82 \text{ million} = 41 \text{ million}$$

The ratio of the number of married people who are not in the labor force to the number of married people who are is:

$$\frac{41,000,000}{82,000,000} = \frac{1}{2} \quad \text{or} \quad 1:2 \quad \text{or} \quad 1 \text{ to } 2$$

The ratio 1 to 2 tells us that for every one married person who is not in the labor force, there are two married people who are in the labor force.

Example 1 **Determine a Ratio in Simplest Form**

A survey revealed that, on average, adults aged 18 to 24 used social media about 3 hours per day. Find the ratio, as a fraction in simplest form, of the number of hours spent using social media to the total number of hours in a week.

Solution

A ratio is the comparison of two quantities with the same units. Use hours as the common unit. Change days and weeks to hours.

There are 7 days in 1 week and social media was used 3 hours per day. The number of hours of social media use in 3 weeks is

$$7 \cdot 3 = 21 \text{ hours per week}$$

There are 24 hours in 1 day and 7 days in 1 week. The number of hours in 1 week is

$$7 \cdot 24 = 168 \text{ hours in 1 week}$$

Write in simplest form the ratio of the number of hours spent using social media to the number of hours in 1 week.

$$\frac{\text{Hours social media use}}{\text{Hours in 1 week}} = \frac{21 \text{ hrs}}{168 \text{ hrs}} = \frac{21}{168} = \frac{1}{8}$$

The ratio is $\frac{1}{8}$.

A **unit ratio** is a ratio in which the number in the denominator is 1. To find a unit ratio, divide the numerator by the denominator. One situation in which a unit ratio is used is student-faculty ratios.

Example 2 **Determine a Unit Ratio**

In a recent year, UCLA had approximately 31,500 undergraduates students, 14,300 graduate students, and 7800 faculty. Find the ratio of students to faculty as a unit ratio. Round to the nearest whole number.

Solution

Find the total number of students.

Total number of students = 31,500 + 14,300 = 45,800

Find the unit ratio. Divide the total number of students by the number of faculty.

$$\frac{\text{Number of students}}{\text{Number of faculty}} = \frac{45,800}{7800} \approx 5.9$$

The ratio of students to faculty is 6 to 1. This ratio means there are about 6 students to each faculty member.

Example 3 **Use a Unit Ratio**

The ratio of the number of high school basketball players to the number of college basketball players is approximately 15 to 1. In one year, there were approximately 28,000 students playing college basketball. In that year, how many students were playing high school basketball?

Solution

The ratio 15 to 1 means there are 15 student high school basketball players to every one student college player. To find the number of high school players, multiply the number of college players by 15.

$$28,000 \cdot 15 = 420,000$$

There were approximately 420,000 students playing high school basketball.

Rates

The word *rate* is used frequently in our everyday lives. It is used in such contexts as unemployment rate, tax rate, interest rate, hourly rate, infant mortality rate, school dropout rate, inflation rate, and postage rate.

A **rate** is a comparison of two quantities and can be written as a fraction. For instance, if a car travels 135 mi on 6 gal of gas, then the miles-to-gallons rate is written

$$\frac{135 \text{ mi}}{6 \text{ gal}}$$

Note that the units (miles and gallons) are written as part of the rate.

A **unit rate** is a rate in which the number in the denominator is 1. To find a unit rate, divide the number in the numerator of the rate by the number in the denominator of the rate. For the preceding example,

$$135 \div 6 = 22.5$$

The unit rate is $\dfrac{22.5 \text{ mi}}{1 \text{ gal}}$.

This rate can be written 22.5 mi/gal or 22.5 miles per gallon, where the word *per* has the meaning "for every."

Unit rates make comparisons easier. For example, if you travel 37 mph and I travel 43 mph, we know that I am traveling faster than you are. It is more difficult to compare speeds if we are told that you are traveling $\dfrac{111 \text{ mi}}{3 \text{ hrs}}$ and I am traveling $\dfrac{172 \text{ mi}}{4 \text{ hrs}}$.

Example 4   Calculate a Unit Rate

A dental hygienist earns $1304 for working a 40-hour week. What is the hygienist's hourly rate of pay?

Solution

The hygienist's rate of pay is $\dfrac{\$1304}{40 \text{ hrs}}$.

To find the hourly rate of pay, divide 1304 by 40.

$$1304 \div 40 = 32.6$$

$$\frac{\$1304}{40 \text{ hrs}} = \frac{\$32.60}{1 \text{ hr}} = \$32.60/\text{hr}$$

The hygienist's hourly rate of pay is $32.60/hr.

Another application of unit rate is the unit price information that grocery stores are required to provide to customers. The **unit price** of a product is its cost per unit of measure.

Suppose that the price of a 2-pound box of spaghetti is $2.79. The unit price of the spaghetti is the cost per pound. To find the unit price, write the rate as a unit rate.

The numerator is the price and the denominator is the quantity. Divide the number in the numerator by the number in the denominator.

$$\frac{\$2.79}{2 \text{ lb}} = \frac{\$1.395}{1 \text{ lb}}$$

The unit price of the spaghetti is $1.395/lb.

Unit pricing is used by consumers to answer the question "Which is the better buy?" The answer is that the product with the lower unit price is the more economical purchase.

Example 5   Determine the More Economical Purchase

Which is the more economical purchase, an 18-ounce jar of peanut butter priced at $2.69 or a 28-ounce jar of peanut butter priced at $3.99?

Solution

Find the unit price for each item.

$$\frac{\$2.69}{18 \text{ oz}} \approx \frac{\$0.149}{1 \text{ oz}} \qquad \frac{\$3.99}{28 \text{ oz}} \approx \frac{\$0.143}{1 \text{ oz}}$$

Compare the two prices per ounce.

$$\$0.149 > \$0.143$$

The item with the lower unit price is the more economical purchase.

The more economical purchase is the 28-ounce jar priced at $3.99.

Table 3.1
Cancer Cases per 100,000 People

Country	Number
USA	354
Canada	334
Germany	305
Japan	250
Finland	265
Israel	233

Rates such as crime statistics or data on fatalities are often written as rates per hundred, per thousand, per hundred thousand, or per million. For example, Table 3.1 shows infant mortality rates per 100,000 live births for selected countries.

The rates in Table 3.1 are easier to read than unit rates. For instance, the cancer rate in the United States would be 0.00354 as a unit rate. It's much easier to understand that 354 cancer cases occurred per 100,000 people.

Presenting data in this form makes it easier to compare things. For instance, Massachusetts has approximately 6000 dentists. Illinois has approximately 9000 dentists. However, Massachusetts has about 83 dentists per 100,000 residents and Illinois has about 68 dentists per 100,000 people.

Example 6   **Calculate a Rate per 100,000 Population**

In a recent year, there were approximately 1,200,000 violent crimes in the United States. Given that the United States population that year was 310,000,000 people, find the rate per 100,000. Round to the nearest whole number.

Solution

The idea behind this and similar calculations is to find the unit rate for the entire population and then multiply by 100,000 (or, for other problems, 100, 1000, 10,000, etc.).

$$\frac{1,200,000 \text{ (crimes)}}{310,000,000 \text{ (population)}} \times 100,000 \approx 0.00387 \times 100,000 = 387$$

The violent crime rate was 387 per 100,000 population.

Given a rate, we can approximate the raw data.

Example 7   **Use a Rate to Find Raw Data**

In the United States, approximately 91 adults per 1000 population were between ages of 18 and 24. If the U.S. population was 330,000,000 when the rate was calculated, how many adults in the United States were between 18 and 24?

Solution

$$\frac{91}{1000} \times 330,000,000 = 30,030,000$$

There were approximately 30,030,000 adults between 18 and 24.

A monthly car or mortgage payment is sometimes given as a rate per $100 or $1000. These rates are given because it makes it easier to calculate the monthly payment instead of using the formula to calculate the monthly payment. A caution to using the rate method is that sometimes the interest rate and the time period for the loan are not as obvious to the borrower.

Example 8   **Calculate a Monthly Car Payment**

A bank offers a 5-year car loan for which the monthly payment is $18.02 per $1000 borrowed. Find the monthly payment for a car loan of $35,351.

Solution

This calculation is similar to Example 7.

$$\frac{\$18.02}{\$1000} \times \$35,351 \approx \$637.03$$

The monthly car payment is $637.03.

Table 3.2

Exchange Rate per U.S. Dollar	
British pound	0.77362
Euro	0.88023
Japanese yen	113.266
Canadian dollar	1.2973

Another application of rates is in the area of international trade. Suppose a company in France purchases a shipment of sneakers from an American company. The French company must exchange euros, which is France's currency, for U.S. dollars in order to pay for the order. The number of euros that are equivalent to one U.S. dollar is called the *exchange rate*. Table 3.2 shows the exchange rates per U.S. dollar for three foreign currencies and the euro.

Example 9 **Solve an Application Using Exchange Rates**

a. Suppose a Japanese company places an order for iron ore that costs $20,000. What is the cost in Japanese yen?

b. An American tourist purchases a vacation package for £3125 (£ is the symbol for the British pound) to travel to England. What is the cost of the vacation in U.S. dollars?

Solution

a. Multiply the unit rate that converts Japanese yen to U.S. dollars by the dollar cost.

$$\frac{113.266 \text{ yen}}{\$1} \cdot \$20{,}000 = 2{,}265{,}320 \text{ yen}$$

The cost is 2,265,320 Japanese yen.

b. The table gives the unit rate that converts British pounds to U.S. dollars. The rate that converts U.S. dollars to British pounds is the reciprocal of the rate in the table.

$$\frac{\$1}{\pounds0.77362} \approx \frac{\$1.292624}{\pounds1} \qquad \text{Use this rate to convert } \pounds \text{ to } \$.$$

$$\frac{\$1.292624}{\pounds1} \cdot \pounds3125 \approx \$4039.45$$

The cost of the vacation is $4039.45.

3.1 **Exercise Set**

Think About It

1. If the ratio of red jelly beans to blue in a can is 1 to 1 and there are 25 red jelly beans in the can, how many blue jelly beans are in the can?

2. Is 25 miles per gallon of gas a unit rate?

3. If the prevalence of a gene is given as 67 per thousand, how many people in a group of 1000 would be expected to have the gene?

4. Suppose you have two choices for a car loan: $22 interest per $1000 borrowed or $25 interest per $1000 borrowed. Which will result in the lower monthly payment?

Preliminary Exercises

For Exercises P1 to P9, if necessary, use the referenced example following the exercise for assistance.

P1. A guideline recommended by some financial advisors is that the ratio of the monthly cost for housing to monthly income should be less than 3 to 10. Does a veterinary technician who earns $2700 per month and pays $900 per month for rent meet the recommended guideline? **Example 1**

P2. In 2022, the University of Texas had an enrollment of 41,309 undergraduate students and 11,075 graduate students. The total number of full-time instructional faculty was 3133. Find the ratio of students to full-time instructional faculty as a unit ratio. Round to the nearest whole number. **Example 2**

P3. The ratio of the total number of draft-eligible NCAA football players to the number of NCAA football players that get drafted into the NFL is approximately 65 to 1. In a year when 251 draft-eligible football players were drafted into the NFL, how many draft-eligible NFL football players were there that year? **Example 3**

P4. You pay $6.75 for 1.5 lb of hamburger. What is the cost per pound? **Example 4**

P5. Which is the more economical purchase: 32 oz of detergent for $6.29 or 48 oz of detergent for $8.29? **Example 5**

P6. A city with a population of 452,100 people has 668 physicians. Find the rate of physicians per 100,000 population for this city. **Example 6**

P7. During a recent year, there were approximately 1405 injuries in a collision per 100,000 licensed drivers in the United States. If there were 231 million licensed drivers during that year, how many injuries in a collision occurred that year? **Example 7**

P8. A bank offers a 4-year car loan for which the monthly payment is $22.801 per $1000 borrowed. Find the monthly payment for a car loan of $28,254. **Example 8**

P9. The exchange rate between Swiss francs and U.S. dollars is 0.930789 franc per $1. **Example 9**

   a. If a Swiss tourist purchased a vacation package for $4500, what is the cost of the vacation in Swiss francs? Round to the nearest franc.

   b. A U.S. business executive entertained clients at a restaurant in Switzerland at a cost of 254.67 Swiss francs. What is the cost of the restaurant bill in U.S. dollars? Round to the nearest cent.

Section Exercises

1. Provide two examples of situations in which unit rates are used.

2. Provide two examples of situations in which ratios are used.

For Exercises 3 to 8, write ratios in simplest form using a fraction.

Family Budget ($)						
Housing	Food	Transportation	Taxes	Utilities	Miscellaneous	Total
1600	800	600	700	300	800	4800

3. **Budgets** Use the table to find the ratio of housing costs to total expenses.

4. **Budgets** Use the table to find the ratio of food costs to total expenses.

5. **Budgets** Use the table to find the ratio of utilities costs to food costs.

6. **Budgets** A financial advisor recommends that the ratio of housing cost to food cost be less than 1.7. Does the family budget in the table satisfy that recommendation?

7. **Real Estate** A house with an original value of $180,000 increased in value to $220,000 in 5 years. What is the ratio of the increase in value to the original value of the house?

8. **Energy Prices** The price of gasoline jumped from $2.70 per gallon to $3.24 per gallon in 1 year. What is the ratio of the increase in price to the original price?

In Exercises 9 to 14, write the expression as a unit rate.

9. 582 mi in 12 hrs

10. 138 mi on 6 gal of gasoline

11. 544 words typed in 8 min

12. 100 m in 8 s

13. $9100 for 350 shares of stock

14. 1000 ft^2 of wall covered with 2.5 gal of paint

15. **Wages** A machinist earns $682.50 for working a 35-hour week. What is the machinist's hourly rate of pay?

16. **Space Vehicles** The space shuttle's solid rocket boosters were pairs of rockets used during the first 2 min of powered flight. Each booster burned 680,400 kg of propellant in

2.5 min. How much propellant did each booster burn in 1 min?

17. **Photography** During filming, an IMAX camera uses 65-mm film at a rate of 5.6 ft/s.

   a. At what rate per minute does the camera go through film?

   b. How much time does it take the camera to use a 500-foot roll of 65-mm film? Round to the nearest second.

18. **Consumerism** Which is the more economical purchase: a 30-ounce jar of mayonnaise for $4.29 or a 48-ounce jar of mayonnaise for $6.29?

19. **Consumerism** Which is the more economical purchase: an 18-ounce box of corn flakes for $3.49 or a 24-ounce box of corn flakes for $3.89?

20. **Wages** You have a choice of receiving a wage of $34,000/year, $2840/month, $650/week, or $16.50/hr. Which pay choice would you take? Assume a 40-hour work week and 52 weeks of work per year.

21. **Baseball** Baseball statisticians calculate a hitter's at bats per home run by dividing the number of times the player has been at bat by the number of home runs the player has hit.

   a. Calculate the at bats per home run for Babe Ruth, who had 60 home runs and 540 at bats. Write the ratio using the word *to*.

   b. Why is this rate used for comparison rather than the number of home runs a player has hit?

22. **Population Density** The following table shows the population and area of three countries. The population density of a country is the number of people per square mile.

   a. Which country has the lowest population density?

   b. How many more people per square mile are there in India than in the United States? Round to the nearest whole number.

Country	Population	Area (square miles)
Australia	25,879,000	2,938,000
India	1,326,093,000	1,146,000
United States	329,543,000	3,535,000

23. **E-mail** An e-mail tracking company compiled the following estimates on consumer use of e-mail worldwide (data for 2023 are estimated).

   a. Complete the last column of the following table by calculating the estimated number of messages per day that each user receives. Round to the nearest tenth.

   b. The predicted number of messages per person per day in 2023 is how many times the estimated number in 2019? Round to the nearest hundredth.

Year	Number of Users (millions)	Messages per Day (billions)	Messages per Person per Day
2019	2850	219.8	
2021	3192	243.9	
2023	3480	275.7	

24. **Real Estate**   One approach to answering the buy-versus-rent question uses the price-to-rent ratio. Find two houses of similar size and quality in comparable neighborhoods, one for sale and the other for rent. Divide the price of the house for sale by the total cost of the rental for 1 year. If the quotient is higher than 20, renting might be the better option. If the quotient is below 15, buying might be the better option.

    a. A house in San Diego, California, is priced at $530,000. The rent on a comparable house is $3100 per month. Find the price-to-rent ratio. Round to the nearest tenth. Does the ratio suggest that you buy or rent a home in San Diego?

    b. A house in Orlando, Florida, is priced at $295,000. The rent on a comparable house is $1150 per month. Find the price-to-rent ratio. Round to the nearest tenth. Does the ratio suggest that you buy or rent a home in Orlando?

25. **Student–Faculty Ratios**   The following table shows the number of full-time students and the number of full-time faculty at several universities in a recent year. Using this table, round ratios to the nearest whole number.

University	Students	Faculty
Swarthmore	1581	200
Rice	6719	1049
Vassar	2435	312
UCLA	41,908	932
Villanova	10,711	893

Source: National Center for Education Statistics, nces.ed.gov

    a. Calculate the student–faculty ratio at Villanova University. Write the ratio using the word *to*. What does this ratio mean?

    b. Which school listed has the lowest student-faculty ratio?

26. **Firefighters**   There are 819 firefighters in a city of 78,500 people. What is the rate of firefighters per 1000 population? Round to the nearest tenth.

27. **Women in the Workforce**   According to the Bureau of Labor Statistics, in a recent year there were 16,369,000 workers who identify as women aged 25 to 34 who were employed and 5,307,000 who were unemployed but looking for work. What is the rate per 1000 of women aged 25 to 34 who were employed? Round to the nearest whole number.

28. **Melanoma Rates**   It is estimated that there will be 93,270 new cases of melanoma of the skin this year in the United States. If the U.S. population is currently 325,377,000, what is the rate of melanoma of the skin per 100,000 in the United States? Round to the nearest whole number.

29. **Dentists**   There are approximately 196,000 dentists in the United States. If the U.S. population is currently 325,377,000, what is the rate of dentist per 100,000 population in the United States? Round to the nearest whole number.

30. **Web Advertising**   One measure of the cost of advertising on the Internet is CPM, cost per 1000 (M, in CPM, is the Roman numeral for 1000) impressions (the number of times an ad is shown). Suppose a company decides to advertise on a website that charges $8 CPM. If the company has a $5000 budget to advertise on this website, how many impressions will the company get?

31. **Home Loan Payments**   A bank charges a monthly payment of $5.06 per $1000 borrowed for a home loan. What is the monthly payment for a home loan of $313,900?

    Exchange Rates   The table below gives the exchange rate per U.S. dollar for selected currencies. Use this table for Exercises 32 to 37.

Exchange Rates per U.S. Dollar	
Mexican peso	18.0774
Australian dollar	1.31545
Israeli shekel	3.54239
Swedish krona	8.42484

32. How many Swedish krona are equivalent to $5000?

33. How many U.S. dollars are equivalent to 5000 Mexican pesos?

34. If an Australian contractor ordered $200,000 U.S. dollars worth of steel from an American company, what is the cost of the steel in Australian dollars?

35. If the United States imports computer chips from Israel that have a value of 21.5 million shekels, what is the value of the computer chips in U.S. dollars?

36. What is the value of 1 Mexican peso in Australian dollars? (Suggestion: Use the fact that $1 is equivalent to 18.0774 pesos and $1 is equivalent to 1.31545 Australian dollars.)

37. What is the value of 1 Israeli shekel in Swedish krona?

Investigation

Moneyball   The movie *Moneyball* is based on statistical methods developed by Bill James to evaluate baseball teams. One such statistic is the win ratio.

$$\text{Win ratio} = \frac{(\text{runs scored})^2}{(\text{runs scored})^2 + (\text{runs allowed})^2}$$

This is sometimes referred to as a *Pythagorean ratio* because of the similarity in the denominator to the Pythagorean theorem.

1. Suppose a team scored 850 runs in one season and allowed 700 runs in that season. What is the team's win ratio? Round to the nearest thousandth.

2. Over time, this formula has been modified so that the exponent is 1.83 rather than 2. Calculate the win ratio using the new exponent. Is the win ratio greater or less than the win ratio from Exercise 1?

3. Suppose a team scored the same number of runs that it allowed. What is the win ratio?

4. Multiplying the win ratio by the number of games in a season gives the number of games a team *should* win. (The formula is not foolproof.) Suppose a team scored 800 runs in one season and allowed 600 runs in that season. How many games should this team win in a regular 162-game season?

5. The win ratio is sometimes written as

$$\text{Win ratio} = \frac{1}{1 + \left(\dfrac{\text{runs allowed}}{\text{runs scored}}\right)^2}.$$ How is this derived from the formula given above?

6. Using the formula in Exercise 5, suppose runs allowed is a fixed number, say 500. How does the win ratio change as the number of runs scored increases?

7. Would a team prefer a small or large win ratio?

Section

3.2 Measurement

Learning Objectives:

- Convert between units of length in the U.S. Customary System.
- Convert between units of weight in the U.S. Customary System.
- Convert between units of capacity in the U.S. Customary System.
- Convert between units of length in the Metric System.
- Convert between units of mass in the Metric System.
- Convert between units of capacity in the Metric System.
- Convert between units in the U.S. Customary System and the Metric System.

U.S. Customary Units of Measure

An important application of rates is converting between various units of measurement. The following are some of the U.S. customary units and their equivalences. **Dimensional analysis** involves using conversion rates to change from one unit of measurement to another.

Equivalences Between Units of Length in the U.S. Customary System
12 inches (in.) = 1 foot (ft)
3 ft = 1 yard (yd)
5280 ft = 1 mile (mi)

Equivalences Between Units of Weight in the U.S. Customary System
16 ounces (oz) = 1 pound (lb)
2000 lb = 1 ton

Equivalences Between Units of Capacity in the U.S. Customary System
8 fluid ounces (fl oz) = 1 cup (c)
2 c = 1 pint (pt)
2 pt = 1 quart (qt)
4 qt = 1 gallon (gal)

These equivalences can be used to change from one unit to another unit by using a **conversion rate**. For instance, because 12 in. = 1 ft, we can write two conversion rates,

$$\frac{12 \text{ in.}}{1 \text{ ft}} \quad \text{and} \quad \frac{1 \text{ ft}}{12 \text{ in.}}$$

Because 12 in. = 1 ft, the magnitude (number without the units) of a conversion rate is 1. Therefore, multiplying by a conversion rate does not change the magnitude of a measurement.

Example 1 **Convert Between Units of Length**

Convert 18 in. to feet.

Solution

Choose a conversion rate so that the unit in the numerator of the conversion rate is the same as the unit needed in the answer ("feet" for this problem). The unit in the denominator of the conversion rate is the same as the unit in the given measurement ("inches" for this problem). The conversion rate is $\frac{1 \text{ ft}}{12 \text{ in.}}$.

The unit in the numerator is the same as the unit needed in the answer.

$$18 \text{ in.} = 18 \text{ in.} \times \frac{1 \text{ ft}}{12 \text{ in.}}$$

The unit in the denominator is the same as the unit in the given measurement.

$$= \frac{18 \text{ ft}}{12} = 1.5 \text{ ft}$$

$$18 \text{ in.} = 1.5 \text{ ft}$$

Example 2   Convert Between Units of Weight

Convert 3.5 tons to pounds.

Solution

Write a conversion rate with pounds in the numerator and tons in the denominator. The conversion rate is $\frac{2000 \text{ lb}}{1 \text{ ton}}$.

$$3.5 \text{ tons} = 3.5 \text{ tons} \times \frac{2000 \text{ lb}}{1 \text{ ton}}$$

$$= 7000 \text{ lb}$$

$$3.5 \text{ tons} = 7000 \text{ lb}$$

In some cases, a direct conversion rate may not be available between two units. In that case, it may be necessary to use more than one conversion rate. For instance, to convert quarts to cups, we can use two conversion rates.

Convert 3 qt to cups.

$$3 \text{ qt} = 3 \text{ qt} \times \frac{2 \text{ pt}}{1 \text{ qt}} \times \frac{2 \text{ c}}{1 \text{ pt}}$$

- The direct equivalence between quarts and cups was not given earlier in the section. Use two conversion rates. First convert quarts to pints, and then convert pints to cups.

$$= \frac{3 \text{ qt}}{1} \times \frac{2 \text{ pt}}{1 \text{ qt}} \times \frac{2 \text{ c}}{1 \text{ pt}}$$

$$= \frac{12 \text{ c}}{1} = 12 \text{ c}$$

$$3 \text{ qt} = 12 \text{ c}$$

Example 3   Convert Between Units of Capacity

Convert 42 cups to quarts.

Solution

First convert cups to pints, and then convert pints to quarts.

$$42 \text{ c} = 42 \text{ c} \times \frac{1 \text{ pt}}{2 \text{ c}} \times \frac{1 \text{ qt}}{2 \text{ pt}}$$

$$= \frac{42 \text{ qt}}{4} = 10.5 \text{ qt}$$

$$42 \text{ c} = 10.5 \text{ qt}$$

The Metric System

We can use conversion rates to change between U.S. customary units and metric units. Before doing so, we will briefly review the metric system.

The basic unit of length in the metric system is the **meter**. One meter is approximately the distance from a doorknob to the floor. All units of length in the metric system are derived from the meter. Prefixes to the basic unit denote the length of each unit. For example, the prefix *centi* means hundredth, so one centimeter is one hundredth of a meter.

Take Note: These are just some of the prefixes in the metric system. There are larger units such as giga (1 billion) and smaller such as pico (1 trillionth).

Prefixes and Units of Length in the Metric System	
kilo- = 1000	1 kilometer (km) = 1000 meters (m)
hecto- = 100	1 hectometer (hm) = 100 m
deca- = 10	1 decameter (dam) = 10 m
	1 meter (m) = 1 m
deci- = 0.1	1 decimeter (dm) = 0.1 m
centi- = 0.01	1 centimeter (cm) = 0.01 m
milli- = 0.001	1 millimeter (mm) = 0.001 m

Conversion between units of length in the metric system involves moving the decimal point to the right or to the left. Listing the units in order from largest to smallest will indicate how many places to move the decimal point and in which direction.

To convert 4200 cm to meters, write the units in order from largest to smallest.

km   hm   dam   m   dm   cm   mm
                    ⎵_____⎵
                    2 positions

• Converting centimeters to meters requires moving two positions to the left.

4200 cm = 42.00 m
              ⎵⎵
           2 places

• Move the decimal point the same number of places and in the same direction.

A metric measurement that involves two units is customarily written in terms of one unit. Convert the smaller unit to the larger unit and then add.

To convert 8 km 32 m to kilometers, first convert 32 m to kilometers.

km   hm   dam   m   dm   cm   mm
     ⎵_____⎵

• Converting meters to kilometers requires moving three positions to the left.

32 m = 0.032 km
          ⎵⎵⎵

• Move the decimal point the same number of places and in the same direction.

8 km 32 m = 8 km + 0.032 km

• Add the result to 8 km.

= 8.032 km

Example 4 **Convert Between Units of Length in the Metric System**

Convert 0.38 m to millimeters.

Solution

0.38 m = 380 mm

Mass and weight are closely related. Weight is a measure of how strongly Earth is pulling on an object. Therefore, an object's weight is less in space than on Earth's surface. However, the amount of material in the object, its **mass**, remains the same.

The **gram** is the unit of mass in the metric system to which prefixes are added. One gram is about the weight of a paper clip.

Units of Mass in the Metric System

1 kilogram (kg) = 1000 grams (g)
1 hectogram (hg) = 100 g
1 decagram (dag) = 10 g
1 gram (g) = 1 g
1 decigram (dg) = 0.1 g
1 centigram (cg) = 0.01 g
1 milligram (mg) = 0.001 g

Conversion between units of mass in the metric system involves moving the decimal point to the right or to the left. Listing the units in order from largest to smallest will indicate how many places to move the decimal point and in which direction.

To convert 324 g to kilograms, write the units in order from largest to smallest.

kg   hg   dag   g   dg   cg   mg
3 positions

324 g = 0.324 kg
3 places

- Converting grams to kilograms requires moving three positions to the left.
- Move the decimal point the same number of places and in the same direction.

Example 5   **Convert Between Units of Mass**

Convert 4.23 g to milligrams.

Solution

$$4.23 \text{ g} = 4230 \text{ mg}$$

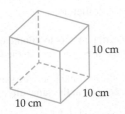

The basic unit of capacity in the metric system is the liter. One **liter** is defined as the capacity of a box that is 10 cm long on each side.

10 cm
10 cm
10 cm

The units of capacity in the metric system have the same prefixes as the units of length.

Units of Capacity in the Metric System

1 kiloliter (kl) = 1000 L
1 hectoliter (hl) = 100 L
1 decaliter (dal) = 10 L
1 liter (L) = 1 L
1 deciliter (dl) = 0.1 L
1 centiliter (cl) = 0.01 L
1 milliliter (ml) = 0.001 L

1 cm
1 cm   1 cm
1 ml = 1 cm³

The milliliter is equal to 1 **cubic centimeter** (cm³). In medicine, cubic centimeter is often abbreviated cc.

Conversion between units of capacity in the metric system involves moving the decimal point to the right or to the left. Listing the units in order from largest to smallest will indicate how many places to move the decimal point and in which direction.

To convert 824 ml to liters, first write the units in order from largest to smallest.

kl hl dal L dl cl ml

3 positions

• Converting milliliters to liters requires moving three positions to the left.

824 ml = 0.824 L

3 places

• Move the decimal point the same number of places and in the same direction.

Example 6   **Convert Between Units of Capacity**

Convert 4327 cl to liters.

Solution

4327 cl = 43.27 L

Visually represent how a meter and foot relate to each other (with meter-stick and foot-long ruler stacked up to show the height of a 6-foot person) or how liters and quarts relate (a liter of milk vs. a quart of milk).

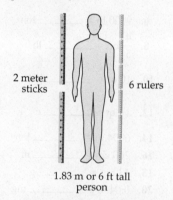

2 meter sticks   6 rulers

1.83 m or 6 ft tall person

1 liter is approximately 6% larger than 1 quart.

Convert Between U.S. Customary Units and Metric Units

To convert between different units, it is necessary to know various conversion rates.

Units of Length	Units of Weight	Units of Capacity
1 in. = 2.54 cm	1 oz ≈ 28.35 g	1 L ≈ 1.06 qt
1 m ≈ 3.28 ft	1 lb ≈ 454 g	1 gal ≈ 3.79 L
1 m ≈ 1.09 yd	1 kg ≈ 2.2 lb	
1 mi ≈ 1.61 km		

These equivalences can be used to form conversion rates to change from one unit of measurement to another. For example, because $1 \text{ mi} \approx 1.61 \text{ km}$, the conversion rates $\frac{1 \text{ mi}}{1.61 \text{ km}}$ and $\frac{1.61 \text{ km}}{1 \text{ mi}}$ are both approximately equal to 1.

Convert 55 mi to kilometers.

$$55 \text{ mi} \approx 55 \text{ mi} \times \boxed{\frac{1.61 \text{ km}}{1 \text{ mi}}}$$

• The conversion rate must contain kilometers in the numerator and miles in the denominator.

$$= \frac{55 \text{ mi}}{1} \times \frac{1.61 \text{ km}}{1 \text{ mi}}$$

$$= \frac{88.55 \text{ km}}{1}$$

$$55 \text{ mi} \approx 88.55 \text{ km}$$

Example 7   **Convert U.S. Customary Units to Metric Units**

The price of gasoline is $3.89/gal. Find the cost per liter. Round to the nearest tenth of a cent.

Solution

$$\frac{\$3.89}{\text{gal}} \approx \frac{\$3.89}{\text{gal}} \times \frac{1 \text{ gal}}{3.79 \text{ L}} = \frac{\$3.89}{3.79 \text{ L}} \approx \frac{\$1.026}{1 \text{ L}}$$

$$\$3.89/\text{gal} \approx \$1.026/\text{L}$$

Example 8   Convert Metric Units to U.S. Customary Units

Convert 200 m to feet.

Solution

$$200 \text{ m} \approx 200 \text{ m} \times \frac{3.28 \text{ ft}}{1 \text{ m}} = \frac{656 \text{ ft}}{1}$$

$$200 \text{ m} \approx 656 \text{ ft}$$

Example 9   Convert Metric Units to U.S. Customary Units

Convert 90 km/hr to miles per hour. Round to the nearest hundredth.

Solution

$$\frac{90 \text{ km}}{\text{hr}} \approx \frac{90 \text{ km}}{\text{hr}} \times \frac{1 \text{ mi}}{1.61 \text{ km}} = \frac{90 \text{ mi}}{1.61 \text{ hr}} \approx \frac{55.90 \text{ mi}}{1 \text{ hr}}$$

$$90 \text{ km/hr} \approx 55.90 \text{ mi/hr}$$

3.2   Exercise Set

Think About It

1. To convert from inches to feet, what unit rate is used?
2. Would you use kilogram, kilometer, or kiloliter to measure the weight of an object?
3. To convert from millimeters to meters, move the decimal point _____ places to the _____.
4. Is 1 meter longer or shorter than 1 yard?
5. Which is larger, 1 quart or 1 liter?

Preliminary Exercises

For Exercises P1 to P9, if necessary, use the referenced example following the exercise for assistance.

P1. Convert 15 ft to yards. **Example 1**
P2. Convert 3 lb to ounces. **Example 2**
P3. Convert 18 pt to gallons. **Example 3**
P4. Convert 0.38 m to millimeters. **Example 4**
P5. Convert 42.3 mg to grams. **Example 5**
P6. Convert 2 kl 167 L to liters. **Example 6**
P7. The price of milk is $3.69/gal. Find the cost per liter. Round to the nearest cent. **Example 7**
P8. Convert 45 cm to inches. Round to the nearest hundredth. **Example 8**
P9. Express 65 mph in kilometers per hour. Round to the nearest hundredth. **Example 9**

Section Exercises

For Exercises 1 to 24, convert between the two measurements.

1. 6 ft = _____ in.
2. 7920 ft = _____ mi
3. 5 yd = _____ in.
4. $\frac{1}{2}$ mi = _____ yd

5. 64 oz = _____ lb
6. 9000 lb = _____ tons
7. $1\frac{1}{2}$ lb = _____ oz
8. $\frac{4}{5}$ ton = _____ lb
9. $2\frac{1}{2}$ c = _____ fl oz
10. 10 qt = _____ gal
11. 7 gal = _____ pt
12. $1\frac{1}{2}$ qt = _____ c
13. 62 cm = _____ mm
14. 6804 m = _____ km
15. 3.21 m = _____ cm
16. 260 cm = _____ m
17. 7421 g = _____ kg
18. 43 mg = _____ g
19. 0.45 g = _____ dg
20. 0.0456 g = _____ mg
21. 7.5 ml = _____ L
22. 0.037 L = _____ ml
23. 0.435 L = _____ cm³
24. 897 L = _____ kl

For Exercises 25 to 34, solve. Round to the nearest hundredth if necessary.

25. Find the weight in kilograms of a 145-pound person.
26. Find the number of liters in 14.3 gal of gasoline.
27. Express 30 mph in kilometers per hour.
28. Seedless watermelon costs $0.59/lb. Find the cost per kilogram.
29. Deck stain costs $32.99/gal. Find the cost per liter.
30. Find the weight, in pounds, of an 86-kilogram person.
31. Find the width, in inches, of 35-mm film.
32. Express 30 m/s in feet per second.
33. A 5-kg ham costs $10/kg. Find the cost per pound.
34. A 2.5-kg bag of grass seed costs $10.99. Find the cost per pound.

Investigation

New Prefixes for Large and Small Units As our scientific and technical knowledge has increased, so has our need for ever-smaller and ever-larger units of measure. The prefixes used to denote some of these units of measure are listed in the following table.

Prefixes for Large Units		10^n	Prefixes for Small Units		10^n
Name	Symbol		Name	Symbol	
yotta	Y	10^{24}	yocto	y	10^{-24}
zetta	Z	10^{21}	zepto	z	10^{-21}
exa	E	10^{18}	atto	a	10^{-18}
peta	P	10^{15}	femto	f	10^{-15}
tera	T	10^{12}	pico	p	10^{-12}
giga	G	10^{9}	nano	n	10^{-9}
mega	M	10^{6}	micro	μ	10^{-6}

These new units of measure are quickly working their way into our everyday lives. For example, it is quite easy to purchase a 1-terabyte hard drive. One terabyte (TB) is equal to 10^{12} bytes. As another example, many computers can do multiple operations in 1 nanosecond (ns). One nanosecond is equal to 10^{-9} s.

For Exercises 1 to 6, convert between the two units.

1. 2.3 T = _____ Y
2. 4.51 n = _____ p
3. 0.65 Z = _____ G
4. 9.46 a = _____ μ
5. 4.01 G = _____ E
6. 7.15 y = _____ f

7. The speed of light is approximately 3×10^8 m/s. What is the speed of light in Zm (zettameters) per second?

8. A light year is approximately 6,000,000,000,000,000 mi. Describe a light year using Ym (yottameters).

9. In the 1980s, it became possible to measure optical events in nanoseconds and picoseconds. Express 1 ps as a decimal.

10. A tau lepton, which is an extremely small elementary particle, has a lifetime of 3×10^{-13} s. What is the lifetime of a tau lepton in femtoseconds?

Section 3.3 Proportions

Learning Objectives:

- Solve proportions.
- Solve multiple ratio problems.

Proportions

A **proportion** is an equation that states the equality of two rates or ratios. The following are examples of proportions.

$$\frac{250 \text{ mi}}{5 \text{ hrs}} = \frac{50 \text{ mi}}{1 \text{ hr}} \qquad \frac{3}{6} = \frac{1}{2}$$

The first example shows the equality of two rates. Note that the units in the numerators (miles) are the same and the units in the denominators (hours) are the same. The second example is the equality of two ratios. Remember that units are not written as part of a ratio.

The definition of a proportion can be stated as follows: If $\frac{a}{b}$ and $\frac{c}{d}$ are equal ratios or rates, then $\frac{a}{b} = \frac{c}{d}$ is a proportion.

Each of the four members in a proportion is called a **term**. Each is numbered as follows.

First term → $\frac{a}{b} = \frac{c}{d}$ ← Third term

Second term → ← Fourth term

The second and third terms of the proportion are called the **means**, and the first and fourth terms are called the **extremes**.

If we multiply both sides of the proportion by the product of the denominators, we obtain the following result.

$$\frac{a}{b} = \frac{c}{d}$$

$$bd\left(\frac{a}{b}\right) = bd\left(\frac{c}{d}\right)$$

$$ad = bc$$

Think About It!

How do you know which are the extremes and which are the means? If we rewrite the following proportion as a set of ratios, notice the "extremes" end up on the outside (or on the *extreme* ends).

Also, if you say the proportion out loud, the extremes are the first and last values you mention.

$$\frac{a}{b} = \frac{c}{d}$$

— extremes —

$a : b = c : d$

—means—

———— extremes ————

a divided by b is equal to c divided by d

—means—

Note that ad is the product of the extremes, and bc is the product of the means. In any proportion, the product of the means equals the product of the extremes. This is sometimes phrased, "the cross products are equal."

In the proportion $\frac{3}{4} = \frac{9}{12}$, the cross products are equal.

$$\frac{3}{4} \quad\diagdown\diagup\quad \frac{9}{12} \quad\rightarrow\quad 4 \cdot 9 = 36 \longleftarrow \text{Product of the means}$$
$$3 \cdot 12 = 36 \longleftarrow \text{Product of the extremes}$$

Sometimes one of the terms in a proportion is unknown. In this case, it is necessary to solve the proportion for the unknown number. The **cross-products method**, which is based on the fact that the product of the means equals the product of the extremes, can be used to solve the proportion. Remember that the cross-products method is just a shortcut for multiplying each side of the equation by the least common multiple of the denominators.

Cross-Products Method of Solving a Proportion

If $\dfrac{a}{b} = \dfrac{c}{d}$, then $ad = bc$.

Example 1 **Solve a Proportion**

Solve: $\dfrac{8}{5} = \dfrac{n}{6}$

Solution

Use the cross-products method of solving a proportion: the product of the means equals the product of the extremes. Then solve the resulting equation for n.

$$\frac{8}{5} = \frac{n}{6}$$
$$8 \cdot 6 = 5 \cdot n$$
$$48 = 5n$$
$$\frac{48}{5} = \frac{5n}{5}$$
$$9.6 = n$$

The solution is 9.6.

Proportions are useful for solving a wide variety of application problems. Remember that when we use the given information to write a proportion involving two rates, the units in the numerators of the rates need to be the same and the units in the denominators of the rates need to be the same. Keep in mind that when we write a proportion we are stating that two rates or ratios are equal.

Example 2 **Solve an Application Using a Proportion**

If you travel 290 mi in your car on 15 gal of gasoline, how far can you travel in your car on 12 gal of gasoline under similar driving conditions?

Solution

Let $x =$ the unknown number of miles.

Write a proportion and then solve the proportion for x.

$$\frac{290 \text{ mi}}{15 \text{ gal}} = \frac{x \text{ mi}}{12 \text{ gal}}$$ • The unit miles is in the numerators.
  The unit gallons is in the denominators.

$$\frac{290}{15} = \frac{x}{12}$$

$$290 \cdot 12 = 15 \cdot x$$ • Use the cross-products method of solving a proportion.

$$3480 = 15x$$

$$232 = x$$

You can travel 232 mi on 12 gal of gasoline.

Example 3   **Determine the Prevalence of a Genetic Trait**

The A-negative blood type occurs in approximately 3 people out of 50. If the blood type of a random sample of 1256 is examined, approximately how many people will have A-negative blood type? Round to the nearest whole number.

Solution

Let x be the number of people in the random sample that have A-negative blood type. Write and solve a proportion.

$$\frac{x}{1256} = \frac{3}{50}$$

$$50x = 3(1256)$$

$$50x = 3768$$

$$x = 75.36$$

There are approximately 75 people with A-negative blood type in the random sample.

Example 4   **Solve an Application Using a Proportion**

The dose for a certain medication is 0.05 milligrams of medication per kilogram of weight per minute. How many milligrams should a patient who weighs 60 kilograms receive in one hour?

Solution

Use a proportion to find the number of milligrams per minute a patient weighing 60 kilograms should receive.

$$\frac{0.05 \text{ mg medication}}{1 \text{ kg of weight}} = \frac{x \text{ mg medication}}{60 \text{ kg of weight}}$$

Solve for x.

$$0.05(60) = x$$

$$3 = x$$

The patient needs 3 mg of medication per minute. To find the number of milligrams in one hour, multiply by 60.

$$(3 \text{ mg/min})(60 \text{ min}) = 180 \text{ mg}$$

The patient requires 180 mg of medication in one hour.

| Example 5 | Solve a Multiple Ratio Problem |

A fertilizer label shows that the ratio of nitrogen, phosphorous, and potassium compounds is in the order of $12:8:5$. How many pounds of each compound are in a 100-pound bag of fertilizer?

Solution

Twenty-five parts ($12 + 8 + 5 = 25$) make up the fertilizer. Each compound is in the same ratio to the total number of parts as the number of pounds of that compound to the total number of pounds (100). Write and solve three proportions.

Let x be the number of pounds of the nitrogen compound.

$$\frac{12}{25} = \frac{x}{100}$$

$$1200 = 25x$$

$$48 = x$$

Let y be the number of pounds of the phosphorous compound.

$$\frac{8}{25} = \frac{y}{100}$$

$$800 = 25y$$

$$32 = y$$

Let z be the number of pounds of the potassium compound.

$$\frac{5}{25} = \frac{z}{100}$$

$$500 = 25z$$

$$20 = z$$

There are 48 pounds of the nitrogen compound, 32 pounds of the phosphorous compound, and 20 pounds of the potassium compound.

3.3   Exercise Set

Think About It

1. For the proportion $\frac{3}{4} = \frac{a}{b}$ which terms are the means and which terms are the extremes?

2. For a proportion, the product of the means _____ the product of the extremes.

3. If $\frac{a}{b} = \frac{3}{4}$ and $\frac{x}{y} = \frac{3}{4}$, is it always true that $\frac{a}{b} = \frac{x}{y}$?

4. If $\frac{a}{b} = \frac{3}{4}$, is it always true that $a = 3$ and $b = 4$?

5. What is wrong with this equation: $\frac{\$12}{2 \text{ feet}} = \frac{24 \text{ feet}}{\$4}$?

Preliminary Exercises

For Exercises P1–P5, if necessary, use the referenced example following the exercise for assistance.

P1. Solve: $\frac{42}{x} = \frac{5}{8}$ Example 1

P2. On a map, a distance of 2 cm represents 15 km. What is the distance between two cities that are 7 cm apart on the map? Example 2

P3. The profits of a firm are shared by its two partners in the ratio $7:5$. If the partner receiving the larger amount of this year's profits receives $84,000, what amount does the other partner receive? Example 3

P4. A certain allergen affects approximately 3 people per 10,000 in the United States. Assuming the U.S. population is approximately 325 million people, how many people suffer from this allergen? Example 4

P5. For the right triangle shown to the right, the sides a, b, and c are in the ratio $3:4:5$, in that order. If side b has a length of 12 inches, what are the lengths of sides a and c? Example 5

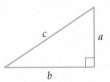

Section Exercises

In Exercises 1 to 12, solve the proportion. Round to the nearest hundredth.

1. $\frac{3}{8} = \frac{x}{12}$

2. $\frac{3}{y} = \frac{7}{40}$

3. $\frac{7}{12} = \frac{25}{d}$

4. $\frac{16}{d} = \frac{25}{40}$

5. $\frac{15}{45} = \frac{72}{c}$

6. $\frac{120}{c} = \frac{144}{25}$

7. $\dfrac{65}{20} = \dfrac{14}{a}$ **8.** $\dfrac{4}{a} = \dfrac{9}{5}$ **9.** $\dfrac{0.5}{2.3} = \dfrac{b}{20}$

10. $\dfrac{1.2}{2.8} = \dfrac{b}{32}$ **11.** $\dfrac{0.7}{1.2} = \dfrac{6.4}{x}$ **12.** $\dfrac{2.5}{0.6} = \dfrac{165}{x}$

13. Gravity The ratio of weight on the moon to weight on Earth is $1:6$. How much would a 174-pound person weigh on the moon?

14. Management A management-consulting firm recommends that the ratio of middle-management salaries to management trainee salaries be $5:4$. Using this recommendation, what is the annual middle-management salary if the annual management trainee salary is $52,000?

15. Medication The dosage of a cold medication is 2 mg for every 80 lb of body weight. How many milligrams of this medication are required for a person who weighs 220 lb?

16. Fuel Consumption If your car can travel 70.5 mi on 3 gal of gasoline, how far can the car travel on 14 gal of gasoline under similar driving conditions?

17. Scale Drawings The scale on the architectural plans for a new house is 1 in. equals 3 ft. Find the length and width of a room that measures 5 in. by 8 in. on the drawing.

18. Scale Drawings The scale on a map is 1.25 in. equals 10 mi. Find the distance between two cities that are 2 in. apart on the map.

19. Art Leonardo da Vinci measured various distances on the human body in order to make accurate drawings. He determined that generally the ratio of the kneeling height of a person to the standing height of that person was $3:4$. Using this ratio, determine the standing height of a person who has a kneeling height of 48 in.

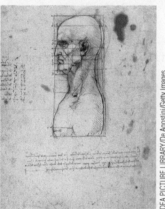

DEA PICTURE LIBRARY/De Agostini/Getty Images

20. Art In one of Leonardo da Vinci's notebooks, he wrote that "from the top to the bottom of the chin is the sixth part of a face, and it is the fifty-fourth part of the man." Suppose the distance from the top to the bottom of the chin of a person is 1.25 in. Using da Vinci's measurements, find the height of the person.

21. Elections A pre-election survey showed that two out of every three eligible voters would cast ballots in the county election. There are 240,000 eligible voters in the county. How many people are expected to vote in the election?

22. Lotteries Three people pool their money to buy lottery tickets. The first person put in $25, the second $30, and the third $35. One of their tickets was a winning ticket. If they won $4.5 million, what was the first person's share of the winnings?

23. Nutrition A pancake 4 in. in diameter contains 5 g of fat. How many grams of fat are in a pancake 6 in. in diameter? Explain how you arrived at your answer.

24. Ratios For the right triangle shown at the right, the sides a, b, and c are in the ratio of $5:12:13$, in that order. If side c has a length of 26 centimeters, what are the lengths of sides a and b?

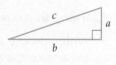

25. Cookie Recipe The ratios of flour, sugar, and water for a basic cookie dough recipe are $32:8:1$, in that order. If a pastry chef is going to make 50 pounds of the basic cookie dough recipe, how many pounds of each does the chef need? Round to the nearest tenth of a pound.

26. Aging According to data from the graph below, the number of people in some age groups is predicted to decline for the next several years.

a. From the graph, what is the population per 1000 of 20 to 24 year olds in 2023?

b. From the graph, what will be the population per 1000 of 25 to 29 year olds in 2030?

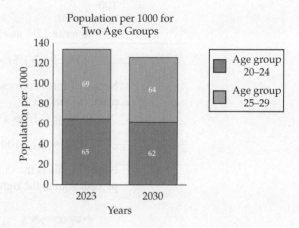

Population per 1000 for Two Age Groups

Investigation

The House of Representatives The U.S. House of Representatives has a total of 435 members. These members represent the 50 states in proportion to each state's population. The population is determined by the Census Bureau and is conducted every 10 years. The last Census was in 2020. As stated in Article XIV, Section 2, of the Constitution of the United States, "Representatives shall be apportioned among the several states according to their respective numbers, counting the whole number of persons in each state."

1. Find the population of each state according to the 2020 U.S. Census. Based on the state populations, determine how many representatives each state should elect to Congress.

2. Compare your list against the actual number of representatives that each state has.

Percent

Learning Objectives:

- Write a decimal as a percent and a percent as a decimal.
- Write a fraction as a percent and a percent as a fraction.
- Solve application problems involving percents.
- Solve percent increase problems.
- Solve percent decrease problems.

Percents

An understanding of percent is vital to comprehending the events that take place in our world today. We are constantly confronted with phrases such as "unemployment of 7%," "annual inflation of 4%," "6% increase in fuel prices," "25% of the daily minimum requirement," and "increase in tuition and fees of 10%."

Percent means "for every 100." Therefore, unemployment of 5% means that 5 out of every 100 people are unemployed. An increase in tuition of 10% means that tuition has gone up $10 for every $100 it cost previously.

A percent is a ratio of a number to 100. Thus $\frac{1}{100} = 1\%$, $\frac{50}{100} = 50\%$, and $\frac{99}{100} = 99\%$. Because $1\% = \frac{1}{100}$ and $\frac{1}{100} = 0.01$, we can also write 1% as 0.01.

$$1\% = \frac{1}{100} = 0.01$$

The equivalence $1\% = 0.01$ is used to write a percent as a decimal. Because $100\% = \frac{100}{100} = 1$, we can write 100% as 1. The equivalence $100\% = 1$ is used to write a decimal as a percent.

To write 17% as a decimal, multiply by 0.01:

$$17\% = 17(1\%) = 17(0.01) = 0.17$$

Note that this is the same as removing the percent sign and moving the decimal point two places to the left.

To write 0.17 as a percent, multiply by 100%:

$$0.17 = 0.17(100\%) = 17\%$$

Note that this is the same as moving the decimal point two places to the right and writing a percent sign at the right of the number.

Example 1 **Write a Percent as a Decimal**

Write the percent as a decimal.
a. 24% **b.** 183% **c.** 6.5% **d.** 0.9%

Solution

To write a percent as a decimal, remove the percent sign and move the decimal point two places to the left.
a. $24\% = 0.24$
b. $183\% = 1.83$
c. $6.5\% = 0.065$
d. $0.9\% = 0.009$

Example 2 **Write a Decimal as a Percent**

Write the decimal as a percent.
a. 0.62 **b.** 1.5 **c.** 0.059 **d.** 0.008

Solution

To write a decimal as a percent, move the decimal point two places to the right and write a percent sign.

a. $0.62 = 62\%$

b. $1.5 = 150\%$

c. $0.059 = 5.9\%$

d. $0.008 = 0.8\%$

The equivalence $1\% = \dfrac{1}{100}$ is used to write a percent as a fraction.

To write 16% as a fraction:

$$16\% = 16(1\%) = 16\left(\frac{1}{100}\right) = \frac{16}{100} = \frac{4}{25}$$

Note that this is the same as removing the percent sign and multiplying by $\dfrac{1}{100}$. The fraction is written in simplest form.

Example 3   **Write a Percent as a Fraction**

Write the percent as a fraction.

a. 25%   b. 120%   c. 7.5%   d. $33\dfrac{1}{3}\%$

Solution

To write a percent as a fraction, remove the percent sign and multiply by $\dfrac{1}{100}$. Then write the fraction in simplest form.

a. $25\% = 25\left(\dfrac{1}{100}\right) = \dfrac{25}{100} = \dfrac{1}{4}$

b. $120\% = 120\left(\dfrac{1}{100}\right) = \dfrac{120}{100} = 1\dfrac{20}{100} = 1\dfrac{1}{5}$

c. $7.5\% = 7.5\left(\dfrac{1}{100}\right) = \dfrac{7.5}{100} = \dfrac{75}{1000} = \dfrac{3}{40}$

d. $33\dfrac{1}{3}\% = \dfrac{100}{3}\% = \dfrac{100}{3}\left(\dfrac{1}{100}\right) = \dfrac{1}{3}$

To write a fraction as a percent, first write the fraction as a decimal. Then write the decimal as a percent.

Example 4   **Write a Fraction as a Percent**

Write the fraction as a percent.

a. $\dfrac{3}{4}$ b. $\dfrac{5}{8}$ c. $\dfrac{7}{400}$ d. $1\dfrac{1}{2}$

Solution

To write a fraction as a percent, write the fraction as a decimal. Then write the decimal as a percent.

a. $\dfrac{3}{4} = 0.75 = 75\%$

b. $\dfrac{5}{8} = 0.625 = 62.5\%$

c. $\dfrac{7}{400} = 0.0175 = 1.75\%$

d. $1\dfrac{1}{2} = 1.5 = 150\%$

The Basic Percent Equation

The Basic Percent Equation

$PB = A$, where P is the percent written as a decimal or fraction, B is the base, and A is the amount.

When solving a percent problem, we have to first identify the percent, the base, and the amount. The base usually follows the phrase "percent of."

Example 5 Solve a Percent Problem for the Amount

A real estate broker receives a commission of 3% of the selling price of a house. What is the amount the broker receives on the sale of a $275,000 home?

Solution

We want to answer the question "3% of $275,000 is what number?" Use the basic percent equation. The percent is $3\% = 0.03$. The base is 275,000. The amount is the amount the broker receives on the sale of the home.

$$PB = A$$
$$0.03(275,000) = A$$
$$8250 = A$$

The real estate broker receives a commission of $8250 on the sale.

Example 6 Solve a Percent Problem for the Base

An investor received a payment of $480, which was 12% of the value of the investment. Find the value of the investment.

Solution

We want to answer the question "12% of what number is 480?" Use the basic percent equation. The percent is $12\% = 0.12$. The amount is 480. The base is the value of the investment.

$$PB = A$$
$$0.12B = 480$$
$$\frac{0.12B}{0.12} = \frac{480}{0.12}$$
$$B = 4000$$

The value of the investment is $4000.

Example 7 **Solve a Percent Problem for the Percent**

If you answer 96 questions correctly on a 120-question exam, what percent of the questions did you answer correctly?

Solution

We want to answer the question "What percent of 120 questions is 96 questions?" Use the basic percent equation. The base is 120. The amount is 96. The percent is unknown.

$$PB = A$$
$$P \cdot 120 = 96$$
$$\frac{P \cdot 120}{120} = \frac{96}{120}$$
$$P = 0.8$$
$$P = 80\%$$

You answered 80% of the questions correctly.

Percent Increase

When a family moves from one part of the country to another, they are concerned about the difference in the cost of living. Will food, housing, and gasoline cost more in that part of the country? Will they need a larger salary in order to make ends meet?

We can use one number to represent the increased cost of living from one city to another so that no matter what salary you make, you can determine how much you will need to earn to maintain the same standard of living. That one number is a percent.

For example, look at the information in Table 3.3.

Table 3.3

If You Live in	and Move to	You Will Need to Earn Your Current Salary Plus This Percent of Your Current Salary
San Diego, CA	New York City, NY	78%
Pittsburg, KS	Philadelphia, PA	39%
Birmingham, AL	Honolulu, HI	134%

A person in Pittsburg earning $67,500 (the approximate median household income in the United States) would need to earn $67,500 plus 39% of $67,500 to maintain the same standard of living in Philadelphia.

$$\$67,500 + \$67,500(39\%) = \$67,500 + \$67,500(0.39) = \$93,825$$

A household income of $93,825 in Philadelphia is the same as a household income of $67,500 in Pittsburg.

Similarly, a person in Birmingham earning $67,500 would need to earn 134% of $67,500 to maintain the same standard of living in Honolulu.

$$\$67,500 + \$67,500(134\%) = \$67,500 + \$67,500(1.34) = \$157,950$$

A household income of $157,950 in Honolulu is the same as a household income of $67,500 in Birmingham.

A person moving from Birmingham to Honolulu will need a 134% *increase* in salary to maintain the same standard of living. **Percent increase** is used to show the relative increase of a quantity to its *original or beginning* value and expressed as a percent. Statements that illustrate the use of percent increase include "sales volume increased by 11% over last year's sales volume," "there was a 5% increase in inflation," and "employees received an 8% pay increase."

For the example of moving from Pittsburg to Philadelphia.

$$\text{Increase in salary required} = \$93,825 - \$67,500 = \$26,325$$

$$\text{Percent increase} = \frac{\text{Increase in salary required}}{\text{Original salary}} = \frac{26,325}{67,500} = 0.39 = 39\%$$

Note that 39% is the same as the value in Table 3.3.

The broken-line graph in Figure 3.1 shows total income tax collected from individuals and corporations by the U.S. government for various years.

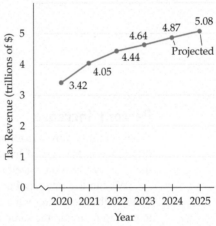

Figure 3.1 Income Tax Revenue

Source: https://www.statista.com/statistics/200405
/receipts-of-the-us-government-since-fiscal-year
-2000/

Example 8 **Solve an Application Involving Percent Increase**

Using the graph in Figure 3.1, find the percent increase in tax revenue from 2021 to 2024. Round to the nearest tenth of a percent.

Solution

From the graph, the income tax revenue increased from $4.05 trillion in 2021 to $4.87 trillion in 2024.

$$\text{Amount of increase} = \$4.87 - \$4.05 = \$0.82.$$

$$\text{Percent increase} = \frac{\text{Increase in tax revenue}}{\text{Tax revenue in 2021}} = \frac{0.82}{4.05} \approx 0.202 = 20.2\%$$

The percent increase in tax revenue from 2021 to 2024 is approximately 20.2%.

Notice in Example 8 that the percent increase is a measure of the amount of increase relative to a base or starting value.

Percent Decrease

The *federal deficit* is the amount by which government spending exceeds its income which includes tax revenue, fees, and investment income. Figure 3.2 shows federal deficits for 2020 through 2025.

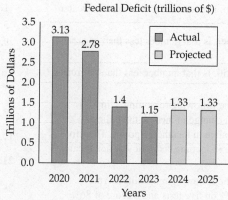

Federal Deficit (trillions of $)

Figure 3.2

Source: https://www.statista.com/statistics/200410/surplus
-or-deficit-of-the-us-governments-budget-since-2000/

Note that the deficit listed for 2022 is less than the deficit listed for 2021. This decrease can be expressed as a percent. First find the amount of decrease in the deficit from 2021 to 2022.

Amount of decrease in the deficit = 2.78 − 1.4 = 1.38

$$\frac{\text{Decrease in the deficit}}{\text{Deficit in 2021}} = \frac{1.38}{2.78} \approx 0.496 = 49.6\%$$

Percent decrease is used to show the relative decrease of a quantity to its *original* or *beginning* value and expressed as a percent. Statements that illustrate the use of percent decrease include "the president's approval rating has decreased 9% over last month" and "there has been a 15% decrease in the number of industrial accidents."

Notice from the deficit example that percent decrease is a measure of the amount of decrease relative to a base or starting value.

Example 9   **Solve an Application Involving Percent Decrease**

The cost for fertilizer, seed, pesticide, and rent to grow corn decreased from $310 per acre to $280 per acre over a 2-year period. Find the percent decrease for these costs over the 2 year period.

Solution

Find the amount of decrease.

Amount of decrease = $310 − $280 = $30

Use the basic percent equation. The base is the original cost per acre ($310); the amount is the decrease over the 2-year period ($30).

$$PB = A$$
$$P(310) = 30$$
$$\frac{P(310)}{310} = \frac{30}{310}$$
$$P \approx 0.097$$

The percent decrease in cost per acre is approximately 9.7%.

3.4 Exercise Set

▶ Think About It

For Exercises 1 through 3, answer the question without doing a calculation.

1. 20% of a number is 16. Is that number less than or greater than 16?

2. 90 is 150% of a number. Is that number less than or greater than 90?

3. 200% of a number is 50. Is that number less than or greater than 50?

4. What is wrong with the following reasoning? "In a survey, 30% favored a change in income tax policy and 40% favored a change in healthcare policy. Therefore, 70% favored a change in income tax policy or healthcare policy."

5. What is wrong with the following reasoning? "You received scores of 90% on five tests and a score of 86% on one test. Therefore, your average score for the six tests is 88%."

▶ Preliminary Exercises

For Exercises P1–P9, if necessary, use the referenced example for assistance.

P1. Write the percent as a decimal. **Example 1**
   a. 74%   **b.** 152%   **c.** 8.3%   **d.** 0.6%

P2. Write the decimal as a percent. **Example 2**
   a. 0.3   **b.** 1.65   **c.** 0.072   **d.** 0.004

P3. Write the percent as a fraction. **Example 3**
   a. 8%   **b.** 180%   **c.** 2.5%   **d.** 66 2/3%

P4. Write the fraction as a percent. **Example 4**
   a. 1/4   **b.** 3/8   **c.** 5/6   **d.** 1 2/3

P5. A General Motors buyer-incentive program offered a 3.5% rebate on the selling price of a new car. What rebate would a customer receive who purchased a $32,500 car under this program? **Example 5**

P6. A real estate broker receives a commission of 3% of the selling price of a house. If the broker receives a commission of $14,370 on the sale of a home, what was the selling price of the home? **Example 6**

P7. A copy editor found 8 misspelled words in a document containing 329 words. What percent of the words in the document were not misspelled? Round to the nearest tenth of a percent. **Example 7**

P8. Using Figure 3.1, find the percent increase in projected income tax revenue from 2022 to 2025. Round to the nearest tenth of a percent. **Example 8**

P9. During strenuous exercise, an athlete may lose 2 liters of water per hour. If the athlete has 55 liters of body water, what percent decrease in water will the athlete experience in 1 hour of strenuous exercise? Round to the nearest tenth of a percent. **Example 9**

▶ Section Exercises

For Exercises 1 to 12, write the percent as a decimal and as a fraction.

1. 40%   **2.** 25%   **3.** 73%   **4.** 150%
5. 2%   **6.** 9%   **7.** 0.5%   **8.** 0.08%
9. 1.5%   **10.** 12.5%   **11.** 56.25%   **12.** 0.75%

For Exercises 13 to 24, write the fraction or decimal as a percent.

13. $\dfrac{3}{5}$ **14.** $\dfrac{11}{20}$ **15.** $\dfrac{3}{8}$ **16.** $\dfrac{9}{50}$
17. 0.35   **18.** 0.29   **19.** 0.6   **20.** 0.75
21. 0.05   **22.** 0.002   **23.** $\dfrac{3}{200}$ **24.** $\dfrac{73}{500}$

25. e-Filed Tax Returns   In a recent year, the IRS reported that it had received 136 million tax returns. Of these, 92% were filed electronically. How many of the returns were filed electronically? Round to the nearest million.

26. Credit Cards   A credit card company offers an annual 2% cash-back rebate on all gasoline purchases. If a family spent $6200 on gasoline purchases over the course of a year, what was the family's rebate at the end of the year?

27. Charitable Contributions   During a recent year, charitable contributions in the United States totaled $390 billion. The following graph shows to whom this money was donated. Determine how much money was donated to educational organizations.

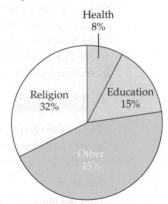

Recipients of charitable contributions in the United States

Source: *Giving USA Foundation*

28. Television   A survey by a university's communications department questioned elementary and middle-school students about television. Sixty-eight students, or 42.5% of those surveyed, said that they had a television in their bedroom at home. How many students were included in the survey?

29. Motorists   A survey of 1236 adults nationwide asked, "What irks you most about the actions of other motorists?" The response "tailgaters" was given by 293 people. What percent of those surveyed were most irked by tailgaters? Round to the nearest tenth of a percent. (*Source:* Reuters/Zogby)

30. **Wind Energy** In a recent year, wind machines in the United States generated 181.7 billion kWh of electricity, enough to serve more than 16 million households. The nation's total electricity production that year was 4094 billion kWh. (*Source:* Energy Information Administration.) What percent of the total energy production was generated by wind machines? Round to the nearest tenth of a percent.

31. **Mining** Approximately 2,240,000 oz of gold went into the manufacturing of electronic equipment in the United States in 1 year. This is 16% of all the gold mined in the United States that year. How many ounces of gold were mined in the United States that year?

32. **Time Management** The two circle graphs show how surveyed employees actually spend their time and how they would prefer to spend their time. Assume that employees have 112 hours a week that are not spent sleeping. Round answers to the nearest tenth of an hour.

    a. What is the actual number of hours per week that employees spend with family and friends?

    b. What is the number of hours that employees would prefer to spend on their jobs or careers?

    c. What is the difference between the number of hours an employee would prefer to spend on him- or herself and the actual amount of time the employee spends on him- or herself?

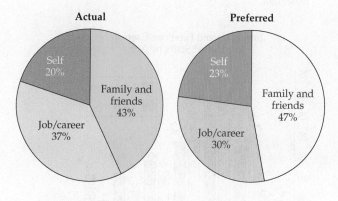

33. **Taxes** A TurboTax online survey asked people how they planned to use their tax refunds. Seven hundred forty people, or 22% of the respondents, said they would save the money. How many people responded to the survey?

34. **Diabetes** Approximately 9.3% of the American population has diabetes. Within this group, 21.0 million are diagnosed, while 8.1 million are undiagnosed. (*Source:* Centers for Disease Control and Prevention.) What percent of Americans with diabetes have not been diagnosed with the disease? Round to the nearest tenth of a percent.

35. **Education** Recent data from the Bureau of Labor Statistics showed that the median weekly earnings of a person with a Bachelor's Degree was approximately $1574 while the median weekly earnings of a person with a high school diploma was $809. What percent of the median weekly earnings of a person with a Bachelor's Degree is the weekly median earnings of a person with a high school diploma? (*Source:* https://www.bls.gov/emp/chart-unemployment-earningseducation.htm)

36. **Telecommunications** The number of Internet users worldwide is projected to increase from 4.5 billion in 2020 to 5.3 billion in 2023. Find the percent increase in the number of Internet users from 2020 to 2023.

37. **Demographics** The following graph shows the projected growth of the number of Americans age 85 and older.

    a. What is the percent increase in the population of this age group from 1995 to 2030?

    b. What is the percent increase in the population of this age group from 2030 to 2050?

    c. What is the percent increase in the population of this age group from 1995 to 2050?

    d. How many times as large is the population in 2050 as that in 1995? How could you determine this number from the answer to part c?

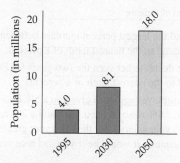

Projected growth (in millions) of the population of Americans aged 85 and older

Source: U.S. Census Bureau

38. **Auto Sales** An automotive dealer trade group estimates that new car sales will decrease next year from 17.6 million cars to 17.1 million cars. Find the percent decrease in auto sales between the two years. Round to the nearest tenth of a percent.

39. **Cable TV** A recent survey showed that 22.2 million adults have cut the cord on cable TV services, a 30% increase from the previous year. If this trend continues, how many adults can be expected to cut the cord next year?

40. **Water Levels** Lake Powell is a water storage facility on the Colorado River in Utah and Arizona. In one year, the elevation of the water in the lake decreased from 3562 feet to 3522 feet. Find the percent decrease in the water level. Round to the nearest tenth of a percent.

41. **Breakfast Cereal Consumption** The table below shows the approximate per capita number of pounds of breakfast cereal consumed in the United States for selected years.

Year	Pounds of Cereal per Person
2020	16.6
2021	16.4
2022	16.1
2023	15.9

    a. What is the percent decrease in the per capita consumption of breakfast cereal from 2022 to 2023?

    b. If the percent decrease between 2022 and 2023 continued for 2023 to 2024, how many pounds of breakfast cereal consumption would there be in 2024?

42. Mobile Apps The graph below shows the number of downloads, in millions, for various social networking apps.

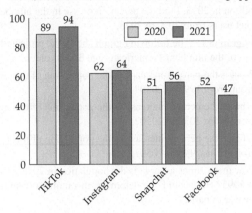

Source: Department of Defense

a. Which app had the largest percent increase between the two years? Round to the nearest tenth of a percent.

b. What was the decrease between the two years for Facebook? Round to the nearest tenth of a percent.

43. Salaries Your employer agrees to give you a 5% raise after 1 year on the job, a 6% raise the next year, and a 7% raise the following year. Is your salary after the third year greater than, less than, or the same as it would be if you had received a 6% raise each year?

44. Work Habits Approximately 74% of undergraduate students at 4-year institutions were enrolled full time. The table below shows the average number of hours per night these student reported studying each night. Approximately what percent of these students spend between 6 and 8 hours per night on homework?

Number of Hours Spent on Homework	Percent
0–2	38
3–5	43
6–8	14
9 or more	5

▶ Investigation

Social Media Facebook, YouTube, WhatsApp, and Instagram were the most popular social networking sites worldwide for a recent year. The table below shows the percent of worldwide users of these social networking sites.

Social Media Platform	Percent
Facebook	36
YouTube	33
WhatsApp	25
Instagram	18

1. Why might the information in this table not be useful to a company that may want to advertise on one of these platforms?

When counting users of social media, a common unit is MAU, monthly active users, the number of unique users who visit a particular website per month. The table below shows the MAU for Facebook and WhatsApp for a recent year and selected countries. All units in millions.

Country	FB MAU	Population
United States	249.8	334.8
Mexico	78.1	131.6
United Kingdom	45.7	68.5
India	416.9	1406.6
Canada	23.2	38.4

2. Which country has the greatest MAU per one hundred population?

3. Suppose an advertisement on Facebook costs $400 per month. What is the cost per MAU for the advertisement?

There is some controversy among social platform companies as to the validity of MAU. Some companies suggest that DAU (Daily Active Users) would be a more appropriate statistic.

4. List some of the pros and cons of MAU and DAU.

The graph below shows the percent of the U.S. age-group population of the United States that uses Facebook and Instagram.

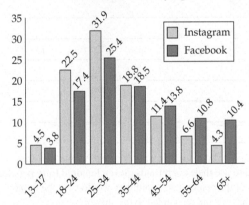

Instagram and Facebook Users in the United States by Age

5. There are approximately 44,000,000 people between 18 and 24 year-olds in the United States. How many more Instagram users are in this group than Facebook users?

6. There are approximately 55,000,000 people of age 65 or older in the United States. How many more Facebook users are in this group than Instagram users?

Analysis of social media users and the platforms they use is a very active area of research in the United States and worldwide.

7. Choose a demographic such as annual income, gender identity, race/ethnicity, political affiliation, or a demographic of your choice and determine how your chosen demographic uses three different social media platforms.

Section 3.5 Variation

Learning Objectives:

- Solve direct variation problems.
- Solve inverse variation problems.
- Solve joint and combined variation problems.

Direct Variation

The analysis of how one quantity varies or changes as another quantity is an important part of mathematics.

> **Direct Variation**
>
> If the relationship between two quantities x and y can be expressed by the equation $y = kx$, then we say that y **varies directly** as the quantity x. The constant k is called the **constant of proportionality** or the **variation constant**. Direct variation can also be expressed as y is **directly proportional** to x.

Direct variation occurs in many daily applications. For example, suppose the price of quinoa is \$4.75 per pound. The price P to purchase n pounds of quinoa is directly proportional to n. That is, $P = 4.75n$. In this example, the variation constant is 4.75.

To solve a problem that involves a variation, we typically write a general equation that relates the variables and then use given information to solve for the variation constant.

Example 1   Solve a Direct Variation Problem

Echolocation is a sensory system in dolphins, and other animals, where a high-pitched sound is emitted and the echo of the sound is used to determine information about the environment. The distance that sound travels in water varies directly to the time it travels. If a dolphin emits a sound that is received by a second dolphin 1000 feet away in 0.21 second, what is the speed of sound in water? Round to the nearest whole number. How far would a sound emitted by a dolphin travel in 2 seconds?

Solution

Write an equation that relates distance d to time t. Because d varies directly as t, $d = kt$. In this case, the constant of proportionality, k, is the speed of sound in water. Use the equation and the known information to find k.

$$d = kt$$
$$1000 = k(0.21) \qquad \bullet\, d = 1000, t = 0.21.$$
$$4762 \approx k$$

The speed of sound in water is approximately 4762 feet per second. The direct variation equation is $d = 4762t$.

Use the variation equation to find the distance a sound emitted by a dolphin would travel in 2 seconds.

$$d = 4762t$$
$$= 4762(2) \qquad \bullet\, t = 2.$$
$$= 9524$$

Under the same conditions, the sound will travel 9524 feet in 2 seconds.

One quantity may vary directly as a power of another quantity. For instance, if air resistance is ignored (say, on the moon), the distance an object falls when released from some point above the ground varies as the *square* of the time it has been falling.

Direct Variation of an *n*th Power

If the relationship between two quantities x and y can be expressed by the equation $y = kx^n$, then we say that y **varies directly** as the nth power of x where k is called the **constant of proportionality** or the **variation constant**. Direct variation of the nth power of a quantity can also be expressed as y is **directly proportional** to the nth power of x.

Example 2   **Solve a Variation of the Form $y = kx^n$**

The distance an object falls on the moon is directly proportional to the square of the time it falls. If a rock falls 47.7 feet in 3 seconds, how far will it fall in 5 seconds?

Solution

Write an equation that relates distance s to time t. Because s varies directly as the square of t, $s = kt^2$. Use the equation and the known information to find k.

$$s = kt^2$$
$$47.7 = k(3)^2 \qquad \bullet\, s = 47.7, t = 3.$$
$$47.7 = 9k$$
$$5.3 = k$$

The direct variation equation is $s = 5.3t^2$.

Use the equation to find the distance an object will fall in 5 seconds.

$$s = 5.3t^2$$
$$= 5.3(5)^2 \qquad \bullet\, t = 5.$$
$$= 5.3 \cdot 25$$
$$= 132.5$$

The rock will fall 132.5 feet in 5 seconds.

Example 3   **Solve a Variation of the Form $y = kx^n$**

The height h, in meters, of a tree varies directly as the two-thirds power of the diameter d, in meters, of the tree when measured 25 centimeters from the ground. If a tree with diameter 0.46 meter is 5 meters tall, find an equation for the height of the tree in terms of its diameter. Use the equation to find the height, to the nearest tenth of a meter, of a tree with a diameter of 0.75 meter. Round the proportionality constant to the nearest tenth.

Solution

Write an equation that relates height h to diameter d. Because h varies directly as the two-thirds power of d, $h = kd^{2/3}$. Use the equation and the known information to find k.

$$h = kd^{2/3}$$
$$5 = k(0.46)^{2/3} \qquad \bullet\, h = 5, d = 0.46.$$
$$5 \approx k(.5959)$$
$$8.4 \approx k$$

The direct variation equation is $h = 8.4d^{2/3}$.

Use the equation to estimate the height of a tree with a diameter of 0.75 meter.

$$h = 8.4d^{2/3}$$
$$= 8.4(0.75)^{2/3} \qquad \bullet\, d = 0.75.$$
$$\approx 8.4(0.8255)$$
$$\approx 6.9$$

The height of the tree is approximately 6.9 meters.

Inverse Variation

One quantity may vary as the *reciprocal* of another quantity.

Inverse Variation

If the relationship between two quantities x and y can be expressed by the equation $y = \dfrac{k}{x}$, then we say that y **varies inversely** as x. The constant k is called the **constant of proportionality** or the **variation constant**.

Example 4   Solve an Inverse Variation

Boyle's law states that the volume V, in milliliters, of a gas (at a constant temperature) varies inversely as the pressure P. If the volume of a gas in a piston chamber is 75 milliliters when the pressure is 1.5 atmospheres, find an equation for the volume of a gas in terms of the pressure. Use your equation to find the volume of the gas when the pressure is increased to 2.5 atmospheres.

Solution

Write an equation that relates volume V to pressure P. Because V varies inversely as the pressure P, $V = \dfrac{k}{P}$. Use the equation and the known information to find k.

$$V = \frac{k}{P}$$

$$75 = \frac{k}{1.5} \qquad \bullet\, V = 75, P = 1.5.$$

$$75(1.5) = k$$

$$112.5 = k$$

The inverse variation equation is $V = \dfrac{112.5}{P}$.

Use the equation to find the volume of the gas when the pressure is increased to 2.5 atmospheres.

$$V = \frac{112.5}{P}.$$

$$= \frac{112.5}{2.5} \qquad \bullet\, P = 2.5.$$

$$= 45$$

The volume of the gas is 45 milliliters.

Some quantities vary inversely to the nth power of another quantity.

Inverse Variation of an nth Power

If the relationship between two quantities x and y can be expressed by the equation $y = \dfrac{k}{x^n}$, then we say that y **varies inversely to the nth power** of x. The constant k is called **constant of proportionality** or the **variation constant**.

> **Example 5** **Solve an Inverse Variation of the Form** $y = \dfrac{k}{x^n}$
>
> The decibel level of a sound is inversely proportional to the square of the distance between the sound source and a listener. If the decibel level of music at a distance of 3 meters from a loudspeaker was 60 decibels, find an equation for the decibel level of music in terms of its distance from the source. Use your equation to find the decibel level of the music for a person dancing 2 meters from the speaker.
>
> **Solution**
>
> Write an equation that relates decibel level (dB) to distance (d). Because dB varies inversely as the square of distance d, $dB = \dfrac{k}{d^2}$. Use the equation and the known information to find k.
>
> $$dB = \frac{k}{d^2}$$
>
> $$60 = \frac{k}{3^2} \qquad \bullet \ dB = 60, d = 3.$$
>
> $$60 \cdot 3^2 = k$$
>
> $$540 = k$$
>
> The inverse variation equation is $dB = \dfrac{540}{d^2}$.
>
> Use the equation to find the decibel level of the music 2 meters from the source.
>
> $$dB = \frac{540}{d^2}$$
>
> $$= \frac{540}{2^2} \qquad \bullet \ d = 2.$$
>
> $$= 135$$
>
> The decibel level 2 meters from the source is 135 dB.

Joint and Combined Variations

The preceding problems involved one quantity changing as a second quantity changed. However, there are important applications where the value of one quantity may depend on two or more other quantities.

> **Joint Variation**
>
> A quantity z **varies jointly** as quantities x and y if $z = kxy$, where k is a constant.

Notes on the Joint Variation Formula:

1. *Jointly* indicates that one quantity varies as the *product* of other quantities.
2. There may be more than two quantities. For instance, the volume of a triangular pyramid varies jointly with A, b, and h: $V = kAbh$, where A is the altitude, b is a base of the triangular bottom, and h is the height of the triangular bottom.
3. Some of the variables may include powers.

Example 6   Solve a Joint Variation Problem

The cost of insulating the ceiling of a house varies jointly as the thickness of the insulation and the area of the ceiling. It costs \$1750 to insulate a 2100-square foot ceiling with 4-in.-thick insulation. Find the cost of insulating a 2400-square foot ceiling with 6-in.-thick insulation.

Solution

Write an equation that relates the cost C to thickness of the insulation T and the area of the ceiling A. Because C varies jointly as the thickness T and the area A, $C = kTA$. Use the equation and the known information to find k.

$$C = kTA$$

$$1750 = k \cdot 4 \cdot 2100 \qquad \bullet\ C = 1750, T = 4, A = 2100.$$

$$1750 = 8400k$$

$$\frac{5}{24} = k$$

The joint variation equation is $C = \dfrac{5}{24}TA$.

Use the equation to find the cost to insulate a 2400-square foot ceiling with insulation 6 inches thick.

$$C = \frac{5}{24}TA$$

$$= \frac{5}{24}(6)(2400) \qquad \bullet\ T = 6, A = 2400.$$

$$= 3000$$

The cost to insulate the ceiling is \$3000.

Combined variation involves direct and inverse variation.

Example 7   Solve a Combined Variation Problem

The weight that a horizontal beam with a rectangular cross section can safely support varies jointly as the width and the square of the depth of the cross section and inversely as the length of the beam. If a 10-foot-long 4-inch by 4-inch beam safely supports a load of 256 pounds, what load L can be safely supported by a beam made of the same material and with a width w of 4 inches, a depth d of 6 inches, and a length l of 16 feet?

Solution

The general variation equation is $L = k \cdot \dfrac{wd^2}{l}$.
Use the given information to find k.

$$L = k \cdot \frac{wd^2}{l}$$

$$256 = k \cdot \frac{4(4)^2}{10} \qquad \bullet\ w = 4, d = 4, l = 10.$$

$$256 = \frac{64k}{10}$$

$$40 = k$$

The formula is $L = 40 \cdot \dfrac{wd^2}{l}$. Use this formula to find the load that can be safely supported by a beam made of the same material with a width of 4 inches, a depth of 6 inches, and a length of 16 feet.

$$L = 40 \cdot \frac{wd^2}{l}$$

$$L = 40 \cdot \frac{4(6)^2}{16} = 360$$

The beam can safely support a weight of 360 pounds.

3.5   Exercise Set

Think About It

1. Fill in the blank: If y varies directly as x, then as x increases, y _____.
2. Write the equation for "y varies directly as the square of x."
3. Fill in the blank: If y varies inversely as x, then as x decreases, y _____.
4. Write the equation for "y varies inversely as the square of x."
5. Suppose you have a job that pays you an hourly rate. Does the amount you earn vary directly or inversely to the number of hours you work?
6. Suppose a single speaker is in a large room. Does the intensity of the sound from the speaker vary directly or inversely as the distance you are standing from the speaker?

Preliminary Exercises

For Exercises P1 through P7, if necessary, use the referenced example for assistance.

P1. The number of gallons of a waterproof sealant needed to cover a wood deck varies directly as the area of the deck. If one gallon of sealant will cover 150 square feet, how many gallons must you purchase to cover a deck that is 630 square feet? **Example 1**

P2. A child's toy soap bubble wand creates approximately spherical bubbles. The surface area of a soap bubble varies directly as the square of the radius of the bubble. If a bubble with a radius of 3 inches has a surface area of 113 square inches, what is the surface area of a bubble that has a radius of 4 inches? **Example 2**

P3. The distance, in miles, a lifeguard can see to the horizon varies directly as the square root of the height, in feet, of the lifeguard's eyes above ground level. Suppose a lifeguard whose eye level is 9 feet above ground can see 4 miles to the horizon. How far could this same lifeguard see when standing on a lifeguard station that is 10 feet above ground? **Example 3**

P4. The current in a simple electrical circuit varies inversely as the resistance. If the current is 40 amps when the resistance is 25 ohms, what is the current when the resistance is 30 ohms? **Example 4**

P5. The percent of x-rays that pass through a lead shield varies inversely as the square of the thickness of the shield. If 75% of the x-rays pass through a shield 125 millimeters thick, what percent will pass through a shield that is 200 millimeters thick? **Example 5**

P6. The potential energy of an object varies jointly as its mass and height above the ground. A 5-kilogram object positioned 10 meters above the ground has a potential energy of 490 joules. Find the mass of an object placed 5 meters above the ground that has a potential energy of 500 joules. **Example 6**

P7. The centripetal force of an object moving in a circle varies jointly with the mass of the object and the square of its velocity and inversely as the radius of the circle. The centripetal force of a 6-g object moving in a circle with a radius of 25 centimeters and at velocity of 20 centimeters per second has a centripetal force of 96 dynes. Find the radius of the circle on which a 14-g object moving in a circle travels with velocity of 40 centimeters per second and the force on the object is 4480 dynes. **Example 7**

Section Exercises

In Exercises 1 to 8, write the equation that expresses the relationship between the variables and then use the given data to solve for the variation constant.

1. y varies directly as x, and $y = 64$ when $x = 48$.
2. m is directly proportional to n, and $m = 92$ when $n = 23$.
3. r is directly proportional to the square of t, and $r = 144$ when $t = 108$.
4. C varies directly as r, and $C = 94.2$ when $r = 15$.
5. T varies jointly as r and the square of s, and $T = 210$ when $r = 30$ and $s = 5$.
6. u varies directly as v and inversely as the square root of w, and $u = 0.04$ when $v = 8$ and $w = 0.04$.
7. V varies jointly as l, w, and h, and $V = 240$ when $l = 8$, $w = 6$, and $h = 5$.
8. t varies directly as the cube of r and inversely as the square root of s, and $t = 10$ when $r = 5$ and $s = 0.09$.
9. **Charles's Law**   Charles's law states that the volume V occupied by a gas (at a constant pressure) is directly proportional to its absolute temperature T. An experiment with a balloon shows that the volume of the balloon is 0.85 liter at 270 K (kelvins). What will the volume of the balloon be when its temperature is 324 K? (Absolute temperature is measured on the Kelvin scale. 0 on the Kelvin scale corresponds to $-273.15°C$.)

Gas expands and
the balloon inflates

Ice water 270 K   Hot water 324 K

10. **Hooke's Law** Hooke's law states that the distance a spring stretches varies directly as the weight on the spring. A weight of 80 pounds stretches a spring 6 inches. How far will a weight of 100 pounds stretch the spring?

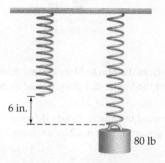

6 in.

80 lb

11. **Semester Hours vs. Quarter Hours** A student plans to transfer from a college that uses the quarter system to a college that uses the semester system. The number of semester hours a student receives credit for is directly proportional to the number of quarter hours the student has earned. A student with 51 quarter hours is given credit for 34 semester hours. How many semester hours credit should a student receive after completing 93 quarter hours?

12. **Pressure and Depth** The pressure a liquid exerts at a given point on a submarine is directly proportional to the depth of the point below the surface of the liquid. If the pressure at a depth of 3 feet is 187.5 pounds per square foot, what is the pressure at a depth of 9 feet?

13. **Amount of Juice Contained in a Grapefruit** The amount of juice in a grapefruit is directly proportional to the cube of its diameter. A grapefruit with a 4-inch diameter contains 6 fluid ounces of juice. How much juice is contained in a grapefruit with a 5-inch diameter? Round to the nearest tenth of a fluid ounce.

14. **Motorcycle Jump** The range of a projectile is directly proportional to the square of its velocity. If a motorcyclist can make a jump of 140 feet by coming off a ramp at 60 mph, find the distance the motorcyclist could expect to jump if the speed coming off the ramp were increased to 65 mph. Round to the nearest tenth of a foot.

15. **Period of a Pendulum** The period T of a pendulum (the time it takes the pendulum to make one complete oscillation) varies directly as the square root of its length L. A pendulum 3 feet long has a period of 1.8 seconds.

   a. Find the period of a pendulum that is 10 feet long. Round to the nearest tenth of a second.

   b. What is the length of a pendulum that *beats seconds* (that is, has a 2-second period)? Round to the nearest tenth of a foot.

16. **Area of a Projected Picture** The area of a projected picture on a movie screen varies directly as the square of the distance from the projector to the screen. If a distance of 20 feet produces a picture with an area of 64 square feet, what distance produces a picture with an area of 100 square feet?

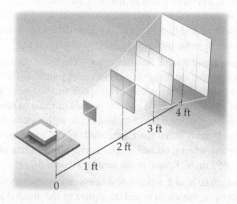

4 ft

3 ft

2 ft

1 ft

0

17. **Speed of a Bicycle Gear** The speed of a bicycle gear, in revolutions per minute, is inversely proportional to the number of teeth on the gear. If a gear with 64 teeth has a speed of 30 revolutions per minute, what will be the speed of a gear with 48 teeth?

18. **Vibration of a Guitar String** The frequency of vibration of a guitar string under constant tension varies inversely as the length of the string. A guitar string with a length of 20 inches has a frequency of 144 vibrations per second. Find the frequency of a guitar string with a length of 18 inches. Assume the tension is the same for both strings.

19. **Jet Engine Noise** The sound intensity of a jet engine, measured in watts per meter squared (W/m^2), is inversely proportional to the square of the distance between the engine and an airport ramp worker. For a certain jet, the sound intensity measures 0.5 W/m^2 at a distance of 7 meters from the ramp worker. What is the sound intensity for a ramp worker 10 meters from the jet?

20. **Illumination** The illumination a light source provides is inversely proportional to the square of the distance from the source. If the illumination at a distance of 10 feet from the source is 50 footcandles, what is the illumination at a distance of 15 feet from the source? Round to the nearest tenth of a footcandle.

21. **Volume Relationships** The volume V of a right circular cone varies jointly as the square of the radius r and the height h. Tell what happens to V when

   a. r is tripled

   b. h is tripled

   c. both r and h are tripled

22. **Safe Load** The load L that a horizontal beam can safely support varies jointly as the width w and the square of the depth d. If a beam with a width of 2 inches and a depth of 6 inches safely supports up to 200 pounds, how many pounds can a beam of the same length that has width 4 inches and depth 5 inches be expected to support? Assume the two beams are made of the same material. Round to the nearest pound.

23. **Ideal Gas Law** The ideal gas law states that the volume V of a gas varies jointly as the number of moles of gas n and the absolute temperature T and inversely as the pressure P. What

happens to V when n is tripled and P is reduced by a factor of one-half?

24. **Maximum Load** The maximum load a cylindrical column of circular cross section can support varies directly as the fourth power of the diameter and inversely as the square of the height. If a column 2 feet in diameter and 10 feet high supports up to 6 tons, how many tons can a column 3 feet in diameter and 14 feet high support? Assume the two columns are made of the same material. Round to the nearest ton.

25. **Force, Speed, and Radius Relationships** The force needed to keep a car from skidding on a curve varies jointly as the weight of the car and the square of its speed and inversely as the radius of the curve. It takes 2800 pounds of force to keep an 1800-pound car from skidding on a curve with a radius of 425 feet at 45 mph. What force is needed to keep the same car from skidding when it takes a similar curve with a radius of 450 feet at 55 mph? Round to the nearest 10 pounds.

26. **Safe Load** The load L a horizontal beam can safely support varies jointly as the width w and the square of the depth d and inversely as the length l. If a 12-foot beam with a width of 4 inches and a depth of 8 inches safely supports 800 pounds, how many pounds can a 16-foot beam that has a width of 3.5 inches and a depth of 6 inches be expected to support? Round to the nearest pound. Assume the two beams are made of the same material.

27. **Stiffness of a Beam** A cylindrical log is to be cut so that it will yield a beam that has a rectangular cross section of depth d and width w. The stiffness of a beam of given length is directly proportional to the width and the cube of the depth. The diameter of the log is 18 inches. What depth will yield the "stiffest" beam: $d = 10$ inches, $d = 12$ inches, $d = 14$ inches, or $d = 16$ inches?

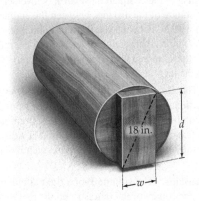

Investigation

Kepler's Third Law Kepler's Third Law states that the square of the time t, in days, needed for a planet to make one complete revolution about the Sun is directly proportional to the cube of the average distance d, in millions of miles, between the planet and the Sun. Earth, which averages 93 million miles from the Sun, completes one revolution in 365 days.

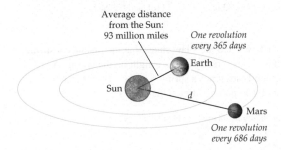

a. Find the average distance from the Sun to Mars if Mars completes one revolution about the Sun in 686 days. Round to the nearest million miles.

b. Find the average distance from the Sun to Jupiter if Jupiter completes one revolution about the Sun in 12 years. Round to the nearest million miles.

Test Prep ③ Proportions and Variation

Section 3.1 Ratios and Rates

A **ratio** is the quotient of two quantities with the same unit. A **unit ratio** is one whose denominator is 1. To find a unit ratio, divide the numerator by the denominator.	See **Examples 1, 2,** and **3** on pages 100 and 101, and then try Exercises 1, 2, and 3 on page 136.
A **rate** is the quotient of two quantities with the different units. A **unit rate** is one whose denominator is 1. To find a unit rate, divide the numerator by the denominator.	See **Examples 4** and **5** on page 102, and then try Exercises 5 and 6 on page 136.
Some rates are expressed in terms of occurrences per 100,000, and so on.	See **Examples 6, 7,** and **8** on page 103 and then try Exercises 8, 9, and 10 on page 137.
When companies or people do business internationally, it is necessary to convert from one currency to another.	See **Example 9** on page 104 and then try Exercise 11 on page 137.

Section 3.2 Measurement

U.S. Customary Units of Measure ● Measures of length: inches, feet, yards, miles ● Measures of weight: ounces, pounds, tons ● Measures of capacity: fluid ounces, cups, pints, quarts, gallons **Dimensional analysis** involves using conversion rates to change from one unit of measurement to another.	See **Examples 1, 2,** and **3** on pages 107 and 108, and then try Exercises 12, 13, and 14 on page 137.
The Metric System The metric system uses a prefix to a base unit to denote the magnitude of a measurement. ● Base unit of length: meter ● Base unit of mass: gram ● Base unit of capacity: liter Some of the prefixes used in the metric system are kilo (1000), hecto (100), deca (10), deci (0.1), centi (0.01), and milli (0.001).	See **Examples 4, 5,** and **6** on pages 109 to 111, and then try Exercises 15, 16, and 17 on page 137.
Convert Between U.S. Customary Units and Metric Units <table><tr><th>Units of Length</th><th>Units of Weight</th><th>Units of Capacity</th></tr><tr><td>1 in. = 2.54 cm 1 m ≈ 3.28 ft 1 m ≈ 1.09 yd 1 mi ≈ 1.61 km</td><td>1 oz ≈ 28.35 g 1 lb ≈ 454 g 1 kg ≈ 2.2 lb</td><td>1 L ≈ 1.06 qt 1 gal ≈ 3.79 L</td></tr></table>	See **Examples 7, 8,** and **9** on pages 111 and 112, and then try Exercises 18 and 19 on page 137.

Section 3.3 Proportions

A **proportion** is an equation that states the equality of two ratios or rates. The **cross-product** method is one technique used to solve a proportion. If $\dfrac{a}{b} = \dfrac{c}{d}$, then $ad = bc$.	See **Example 1** on page 114, and then try Exercises 21 to 24 on page 137.
Application of Proportions To solve an application, write an equation that relates the quantities. One way to ensure the proportion is written correctly is to check the units of the application. The numerator of each fraction should have the same units; the denominator of the fraction should have the same units.	See **Example 2, 3, 4,** and **5** on pages 114 to 116 and then try Exercises 25 to 28 on page 137.

Section 3.4 Percent

Percent means "for every 100." It is a ratio of a number to 100. To write a percent as a decimal, remove the percent symbol. Then multiply by 0.01. This is equivalent to removing the percent symbol and moving the decimal point two places to the left. 　To write a percent as a fraction, remove the percent sign and multiply by $\dfrac{1}{100}$. Then write the fraction in simplest form.	See **Examples 1** and **2** on pages 118 and 119 and then try Exercises 29 to 32 on page 137.
To write a decimal as a percent, multiply by 100%. This is equivalent to moving the decimal point two places to the right and then adding the percent symbol. 　To write a fraction as a percent, first write the fraction as a decimal. Then write the decimal as a percent.	See **Examples 3** and **4** on pages 119 and 120 and then try Exercises 33 to 36 on page 137.
Basic Percent Equation　$PB = A$, where P is the percent written as a fraction or decimal, B is the base, and A is the amount. When solving a percent application, the base usually follows the phrase "percent of."	See **Examples 5, 6,** and **7** on pages 120 and 121 and then try Exercises 37 to 39 on page 137.
Percent increase is the amount of increase divided by the original value, expressed as a percent.	See **Example 8** on page 122 and then try Exercise 40 on page 137.
Percent decrease is the amount of decrease divided by the original value, expressed as a percent.	See **Example 9** on page 123 and then try Exercise 41 on page 137.

Section 3.5 Variation

If a quantity y **varies directly** as a quantity x, or y is **directly proportional** to x, then $y = kx$, where k is a constant called the **constant of proportionality** or the **variation constant**.	See **Example 1** on page 127 and then try Exercise 45 on page 138.
If a quantity y **varies directly as the nth power** of a quantity x, or y is **directly proportional to the nth power** x, then $y = kx^n$, where k is a constant called the constant of proportionality or the variation constant.	See **Examples 2** and **3** on page 128 and then try Exercise 46 on page 138.
If a quantity y **varies inversely** as a quantity x, or y is **inversely proportional** to x, then $y = \dfrac{k}{x}$, where k is a constant called the constant of proportionality or the variation constant.	See **Example 4** on page 129 and then try Exercise 47 on page 138.
If a quantity y **varies inversely as the nth power** of a quantity x, or if y is **inversely proportional as the nth power** of x, then $y = \dfrac{k}{x^n}$, where k is the constant of proportionality or the variation constant.	See **Example 5** on page 130 and then try Exercise 48 on page 138.
A quantity z **varies jointly** as quantities x and y if $z = kxy$, where k is a constant.	See **Example 6** on page 131 and then try Exercise 49 on page 138.
Combined variation involves direct and inverse variation.	See **Example 7** on pages 131 and 132 and then try Exercise 50 on page 138.

Review Exercises

1. **Real Estate**　A house with an original value of $280,000 increased in value to $350,000 in 5 years. Write the ratio of the increase in value to the original value of the house as a fraction in simplest form.

2. **Body Ratios**　A person 6 feet tall has a head size of 10 inches as measured from the bottom of the chin to the top of the head. What is the unit ratio of total height to head height? Round to the nearest tenth.

3. **Student-to-Faculty Ratios**　There are 29,524 students enrolled at a large university that has 2405 faculty members. What is the student-to-faculty unit ratio? Round to the nearest tenth.

4. **Mortgage-to-Income Ratios**　A family has a monthly income of $4395; $1896 of that goes to a mortgage payment. Is the ratio of the mortgage payment to monthly income greater than or less than one-third of the monthly income?

5. **Gas Mileage**　An automobile was driven 326.6 mi on 11.5 gal of gasoline. Find the number of miles driven per gallon of gas.

6. **Finance**　Which is the more economical purchase: a 20-ounce box of cereal for $4.23 or an 18-ounce box for $3.75?

7. **Population Density** The following table shows the population and area of the four most populous cities in the United States.

City	Population	Area (in square miles)
New York	8,400,000	321.8
Los Angeles	3,900,000	467.4
Chicago	2,900,000	228.469
Houston	2,300,000	594.03

The cities are listed from largest to smallest population. Rank the cities according to population density from largest to smallest.

8. **Traffic Safety** There were 270 pedestrians injured in traffic accidents in a city during one year. If the population of the city is 1.2 million people, what is the rate of pedestrian accidents per 100,000 population?

9. **Genetic Traits** A certain genetic trait occurs in 27 people per 100,000. If the population of a state is 1,271,000 people, approximately how many people would have the genetic trait? Round to the nearest whole number.

10. **Home Loans** A bank offers a home-improvement loan at a cost of $22.38 per month per $1000 borrowed. Find the monthly payment for a loan of $125,000.

11. **Exchange Rates** A U.S. food company is paying a company based in India 16,000,000 Indian rupees for tin cans. If the exchange rate is 64.6108 rupees per $1, what is the cost of the contract to the U.S. food company in dollars?

For Exercises 12 to 17, convert one measurement to another.

12. 27 in. = _____ ft

13. 2.5 lb = _____ oz

14. 16 c = _____ qt

15. 37 mm = _____ cm

16. 3.2 kg = _____ mg

17. 5.8 dl = _____ L

18. **Metric Conversion** The price of a beverage is $3.56 per liter. What is the price of this beverage in dollars per quart? Round to the nearest cent.

19. **Metric Conversion** The price of propane is $2.45 per gallon. What is the price per liter of propane? Round to the nearest cent.

20. **Metric Conversion** A car is traveling 45 miles per hour. Find the speed in kilometers per hour. Round to the nearest whole number.

For Exercises 21 to 24, solve the proportion. If necessary, round to the nearest tenth.

21. $\dfrac{n}{4} = \dfrac{5}{12}$

22. $\dfrac{6}{x} = \dfrac{3}{7}$

23. $\dfrac{27}{25} = \dfrac{n}{9}$

24. $\dfrac{9}{2} = \dfrac{6}{p}$

25. **Gardening** Three tablespoons of a liquid plant fertilizer are to be added to 4 gal of water. How many tablespoons of fertilizer would be required for 10 gal of water?

26. **Fire Science** A firefighting truck can hold 3750 gallons of water and deliver 150 gallons of water every 2 minutes. How much water can be delivered in 7 minutes? How long before the truck is empty?

27. **Ecology** To determine the number of elk in a forest, a biologist captures and tags 173 elk and then releases them back into the forest. Later 200 elk are captured; 28 have tags. Estimate the number of elk in the forest. Round to the nearest whole number.

28. **Sharing a Lottery** Tien, Catherine, and Roberta each put $1, $2, and $3, respectively, into a pool to purchase lottery tickets. If one of the tickets is a winner with an award of $240,000, how much should Catherine receive if her payout is in proportion to the amount she paid into the pool?

For Exercises 29 to 32, write the percent as a decimal and as a fraction.

29. 4%

30. 23%

31. 12.5%

32. 150%

For Exercises 33 to 36, write the decimal or fraction as a percent.

33. 0.08

34. 1.2

35. $\dfrac{3}{200}$

36. $\dfrac{5}{8}$

37. **Sports** Sixty-five percent of the pitches thrown by a good major league baseball pitcher are strikes. If, during one season, a starting pitcher throws 35,000 pitches, how many of those pitches are strikes?

38. **Education** In a large history class, 18 students—9% of the class—received a grade of A on a test. How many students were in the class?

39. **Pets** Approximately 47.1 million U.S. households own at least one cat. If there are 125 million households in the United States, what percent own one or more cats? Round to the nearest tenth.

40. **Minimum Wage** The minimum wage in a city increased from $11 per hour to $13 per hour. What is the percent increase in the city's minimum wage? Round to the nearest tenth.

41. **Sports** As a golfer ages, the speed at which the golfer can swing a driver decreases. Alex's driver swing speed decreased from 95 miles per hour to 88 miles per hour. What percent decrease in driver speed did Alex experience?

42. **Retirement Plans** A company's retirement plan allows employees to contribute 11% of their annual salaries to the program. What maximum amount can an employee who earns $64,500 contribute to the plan?

43. **Population Studies** According to the U.S. Bureau of the Census, the population of people under 18 years old and 65 or older in the United States in 2035 and 2060 is projected to be as shown in the following table.

Year	0–17	65+
2035	76,500,000	77,000,000
2060	80,100,000	94,400,000

a. Is the percent increase in the projected population of 0–17 year-olds from 2035 ro 2060 greater than or less than the percent increase of the projected populations of the 65+ age group?

b. In 2060, the population that are 65+ in age will be 23% of the population of the United States What is the projected population of the United States in 2060?

44. **Boston Marathon**   In a recent year, 27,222 people entered the Boston Marathon. Of those, 26,400 finished the race. What percent of the runners who started the race did not finish the marathon? Round to the nearest tenth.

45. **Physics**   Force F is directly proportional to acceleration a. If a force of 10 pounds produces an acceleration of 2 feet per second squared on a certain object, what acceleration will a 15-pound force on the same object produce?

46. **Physics**   The distance an object will fall on the moon is directly proportional to the square of the time it falls. If an object falls 10.6 feet in 2 seconds, how far would an object fall in 3 seconds?

47. **Video Games**   The number of video games a company can sell is inversely proportional to the price of the game. If

5000 games can be sold when the price is $45, how many games could be sold if the price is $30? Round to the nearest whole number.

48. **Magnetism**   The repulsive force between the north poles of two magnets is inversely proportional to the square of the distance between the poles. If the repulsive force is 40 pounds when the distance between the poles is 2 inches, what is the repulsive force when the distance between the two poles is 4 inches?

49. **Wind Resistance**   Wind resistance varies jointly as the object's surface area and its velocity. An object with a surface area of 20 square feet traveling 30 miles per hour experiences 50 pounds of force. What resistance force would a similar object traveling at 45 miles per hour experience?

50. **Chemistry**   The volume of gas varies directly as the temperature and inversely as the pressure. If the volume is 200 cubic centimeters when the temperature is 100 K and the pressure is 20 pounds per square centimeter, what is the volume when the temperature is 120 K and the pressure is 30 pounds per square centimeter? Note: the kelvin (K) is based on absolute zero, which is approximately 459.67 degrees Fahrenheit below zero.

Project 1

Rates and Ratios in Sports

Here are some rates and ratios used in different sports. Answer each question based on the given rate or ratio.

1. Football, touchdowns per interception:

$$\frac{\text{Touchdowns}}{\text{per interception}} = \frac{\text{Number of passing touchdowns}}{\text{Number of interceptions}}.$$

Suppose a quarterback threw 36 touchdown passes and eight interceptions in one season. What is the quarterback's ratio of touchdowns to interceptions?

2. Women's NCAA softball earned run average (ERA): The number of earned runs a pitcher gives up for every seven innings pitched. It is calculated as

$$\frac{\text{Number of earned runs}}{\text{Number of innings pitched}} = \frac{\text{ERA}}{7 \text{ innings}}.$$ Find the ERA of a pitcher who pitches 300 innings and gives up 60 earned runs. (*Note:* The same formula is used in Major League Baseball but with 9 instead of 7.)

3. Basketball: $p_{a,b} = \dfrac{p_a - (p_a \cdot p_b)}{p_a + p_b - 2 \cdot p_a \cdot p_b}$, where $p_{a,b}$ is the percent chance that team A will beat team B, p_a is the percent of games won by team A and p_b is the percent of games won by team B.

a. Suppose team A has won 50% of its games and team B has won 60% of its games. Find $p_{a,b}$.

b. If teams A and B have the same winning percent, find $p_{a,b}$. (*Suggestion:* You may try this algebraically by letting $p = p_a = p_b$. Alternatively, try a couple examples where the teams have equal percent.) Explain why the answer you get should be expected. (*Note:* This formula also applies to other sports as well.)

4. Soccer, usage rate: The definition for *usage rate* is "The number of passes a player attempts per 90 minutes divided by the number of passes a team attempts per game." Before continuing, here is a little background. A soccer game is 90 minutes long. A player normally does not play all 90 minutes. If a player played 40 minutes one game and 50 minutes in another game, then the player would have played 90 minutes in two games.

Suppose player A, playing for team A, averaged 50 passes per 90 minutes played while the team averaged 400 passes per game. Player B, playing for B, averaged 60 passes per 90 minutes played while the team averaged 500 passes per game. Which player had the higher usage rate? Why might usage rate be a better statistic than just passes per 90 minutes?

Project 2

Federal Income Tax

Income taxes are the chief source of revenue for the federal government. If you are employed, your employer probably withholds some money from each of your paychecks for federal income tax. At the end of each year, your employer sends you a **Wage and Tax Statement Form** (**W-2 form**) that states the amount of money you earned that year and how much was withheld for taxes.

Every employee is required by law to prepare an income tax return by April 15 of each year and send it to the Internal Revenue Service (IRS). On the income tax return, you must report your total income, or **gross income**. Then you subtract from the gross income any adjustments (such as deductions for charitable contributions or exemptions for people who are dependent on your income) to determine your **adjusted gross income**. You use your adjusted gross income and either a tax table or a tax rate schedule to determine your **tax liability**, or the amount of income tax you owe to the federal government.

After calculating your tax liability, compare it with the amount withheld for federal income tax, as shown on your W-2 form. If the tax liability is less than the amount withheld, you are entitled to a tax refund. If the tax liability is greater than the amount withheld, you owe the IRS money; you have a **balance due**.

The 2022 tax rate schedules are shown below. To use this table for the exercises that follow, first classify the taxpayer as single, married filing jointly, or married filing separately. Next determine into which range the adjusted gross income falls. Then perform the calculations shown to the right of that range to determine the tax liability.

For example, consider a taxpayer who is unmarried and has an adjusted gross income of $58,720. To find this taxpayer's tax liability, use the portion of the table headed Unmarried.

An income of $58,720 falls in the range $41,775 to $89,075. The tax is $4177.50 + 22% of the amount over $41,775. Find the amount over $41,775.

$$\$58,720 - \$41,775 = \$16,945$$

Calculate the tax liability:

$$\$4177.50 + 22\%(\$16,945) = \$4177.50 + 0.22(\$16,945)$$
$$= \$4177.50 + \$3727.90$$
$$= \$7905.40$$

The taxpayer's liability is $7905.40.

Project Exercises

Use the 2022 Tax Rate Schedules to solve Exercises 1–8.

1. Neko Abruzzio is married, and filing jointly, has an adjusted gross income of $63,850. Find Neko's tax liability.

2. Angela Lopez is unmarried and has an adjusted gross income of $91,680. Find Angela's tax liability.

3. Dee Pinckney is married, and filing jointly. She has an adjusted gross income of $75,120. The W-2 form shows the amount withheld as $8211. Find Dee's tax liability and determine their tax refund or balance due.

4. Jeremy Littlefield is unmarried and has an adjusted gross income of $87,800. His W-2 form lists the amount withheld as $16,420. Find Jeremy's tax liability and determine his tax refund or balance due.

5. Does a taxpayer in a 32% tax bracket pay 32% of his or her earnings in income tax? Explain your answer.

2022 Unmarried Individuals

Taxable income is over—	But not over—	The tax is:	of the amount over—
$0	$10,275	$0 + 10%	$0
$10,275	$41,775	$1027.50 + 12%	$10,275
$41,775	$89,075	$4177.50 + 22%	$41,775
$89,075	$170,050	$6907.50 + 24%	$89,075
$170,050	$215,950	$17,005.00 + 32%	$170,050
$215,950	$539,900	$21,595.00 + 35%	$215,950
$539,900	…	$134,662.50 + 37%	$539,900

2022 Married Individuals Filing Joint Returns

Taxable income is over—	But not over—	The tax is:	of the amount over—
$0	$20,550	$0 + 10%	$0
$20,550	$83,550	$2050.00 + 12%	$20,550
$83,550	$178,150	$8355.00 + 22%	$83,550
$178,150	$340,100	$17,815.00 + 24%	$178,150
$340,100	$431,900	$34,010.00 + 32%	$340,100
$431,900	$647,850	$43,190.00 + 35%	$431,900
$647,850	…	$118,772.50 + 37%	$647,850

Math in Practice

It is recommended to inspect sewer lines if a house is more than 40 years old. After inspection, the plumber tells you that the sewer line from the street main to your house is damaged with tree roots and cracks in the pipe. It would cost thousands of dollars and take many days to dig a trench to remove and replace that piece of the sewer line. There is a less expensive alternative: trenchless sewer pipe lining. The process to fix your sewer line starts with hydro jetting the sewer line, which will cut out tree root build-up and clean the interior of the pipe. Then a hole is dug at the end of the sewer line, near the street. Next, in just a few hours, the sewer pipe liner is installed inside the sewer pipe. Finally, the hole is refilled and your sewer line is ready for another 40 years of use.

Part 1

You find out that the sewer pipe between your house and the street is cracked. Rather than digging up and replacing the old sewer pipe, you decide to use trenchless sewer pipe liners.

a. If the distance from your house to the street is 22 feet, what is that distance in meters? Round to the nearest hundredth.

b. Suppose the installation of trenchless sewer pipe liners requires 500 ml of resin for every meter of the liner. How many liters of resin will be required for the installation at your house?

c. If one standard liner is available in 30-foot lengths, what percent of a standard liner is used at your house? Round to the nearest tenth of a percent.

Part 2

You have a water tank that collects rain water. The hydrostatic pressure, P, in pascals at the valve at the bottom of the tank, varies jointly as p, the density of water, in kg per cubic meter, and h, the height of the tank, in meters. At your house, $p = 1000$ kg/m³. A tank 2 meters tall has a pressure 19,620 pascals.

a. If your tank is 8 feet tall, calculate the pressure at the valve at the bottom of the tank.

b. The conversion rate from pascals to PSI (pounds per square inch) is 1 PSI = 6894.76 pascals. Find the pressure of your tank in PSI.

c. The plumber says the hydro jetting uses 3000 PSI to clean the sewer pipe. Does your 8-foot-tall rain barrel have adequate water pressure to clean the sewer pipe?

Part 3

While the team installs the trenchless sewer pipe liner, you go to your job as a nurse at a hospital. The doctor orders a heparin drip for a patient with a blood clot. The starting rate is 18 units/kg/hr. Your patient is 165 pounds. The stocked bag of heparin is 25,000 units in 500 ml of normal saline.

a. Convert the patient's weight from pounds to kg.

b. What is the rate for the heparin drip in ml/hr?

Vladimir Gjorgiev/Shutterstock.com

Financial Literacy

Chapter 4   Managing Your Finances

Managing Your Finances

Section	How This Is Used . . .
4.1 Simple Interest	Simple interest is the foundation of many financial calculations from payday loans to Treasury bills sold by the U.S. government.
4.2 Compound Interest	Albert Einstein supposedly said "Compound interest is the eighth wonder of the world." He said this because the value of an account that earns compound interest grows much more quickly than an account that does not earn compound interest.
4.3 Future and Present Value of an Annuity	If you are saving for retirement or making a car payment, there is a good chance that the value in your retirement plan or the size of your car payment is calculated using annuity formulas.
4.4 Credit Cards	Credit card debt in the United States is approaching $850 billion. Manage and understand credit card debt by reading this section.
4.5 Student Loans	Credit card debt is serious, but so is the approximately $1.4 trillion in student-loan debt. Many of us need loans to complete college. Understanding the different options for loans is one way to manage student debt.
4.6 Financial Considerations of Car Ownership	Do you enjoy having your own car? Have you ever thought about how much it costs for that luxury? In the United States, the average American who owns a car spends $2300 per year for gasoline.
4.7 Financial Considerations of Home Ownership	Does owning a home make financial sense? U.S. homeownership rates have been declining because some people, using the methods in this section, have decided that renting may be a better option.
4.8 Stocks, Bonds, and Mutual Funds	Do you want to get rich quickly? Then maybe the best advice is to get lucky. Most wealthy people grew their wealth by managing their money wisely.

Math in Practice: Saving Up, p. 216

Applications of Managing Your Finances

A recent study suggested that almost one-third of Americans were struggling financially or just getting by. Some of this is the result of unemployment, illness, and unanticipated disasters such as floods. Some of it, however, comes from financial mismanagement of assets.

In this chapter, we introduce some of the financial tools that will allow you to manage your resources. This can be quite a challenge, but here are some of the topics we discuss. In Section 4.3, Example 2 shows how to determine the value of a retirement plan. In Section 4.4, Example 5 shows how to determine the time it takes to repay credit card debt. Also in that section, Exercise P5 asks you to make a similar calculation.

Section 4.1   Simple Interest

Learning Objectives:

- Apply the simple interest formula.
- Find the maturity value or future value for simple interest.
- Calculate quantities related to a T-bill.

Be Prepared: Prep Test 4 reviews concepts used in this chapter. This can be found in your Companion Site at Cengage.com.

Simple Interest

The earliest loans date back to 3000 B.C., and repayment of those loans may have extended over generations, not 4 or 5 years, as is the case for today's typical car loan. One of the first written records of laws that governed loans came from Hammurabi, who ruled Babylon from 1795 to 1750 B.C. He was the first ruler to write a set of laws, Hammurabi's Code, that defined people's rights.

When you deposit money in a bank—for example, in a savings account—you are permitting the bank to use your money. The bank may lend the deposited money to customers to buy cars or renovate their homes. The bank pays you for the privilege of using your money. The amount paid to you is called **interest**. If you are the one borrowing money from a bank, the amount you pay for the privilege of using that money is also called *interest*.

The amount deposited in a bank or borrowed from a bank is called the **principal**. The amount of interest paid is usually given as a percentage of the principal. The percentage used to determine the amount of interest is called the **interest rate**. If you deposit $1000 in a savings account paying 5% interest per year, $1000 is the principal and the annual interest rate is 5%.

Interest paid on the original principal is called **simple interest**. The formula used to calculate simple interest follows.

Simple Interest Formula

The simple interest formula is

$$I = Prt$$

where I is the interest, P is the principal, r is the interest rate, and t is the time period.

Example 1   Calculate Simple Interest

a. Calculate the simple interest due on a 2-year loan of $2500 if the annual simple interest rate is 5%.

b. Find the simple interest due on a 4-month loan of $1300 if the monthly interest rate is 1.25%.

Solution

a. Use the simple interest rate formula with $P = 2500$, $r = 5\% = 0.05$, and $t = 2$.

$$I = Prt$$
$$I = 2500(0.05)(2)$$
$$I = 250$$

• Note that the interest rate is an *annual* rate and the time of the loan is in *years*. The period for the interest rate (annual) and the time units for the loan (years) must be the same.

The simple interest due on the loan is $250.

b. Use the simple interest rate formula with $P = 1300$, $r = 1.25\% = 0.0125$, and $t = 4$.

$$I = Prt$$
$$I = 1300(0.0125)4$$
$$I = 65$$

• Note that the interest rate is a *monthly* rate and the time of the loan is in *months*. The period for the interest rate (monthly) and the time units for the loan (months) must be the same.

The simple interest due on the loan is $65.

In Example 1, the period for the interest rate and the time units are the same. Example 2 shows how to work with different time units.

Example 2   **Calculate Simple Interest**

a. Find the simple interest due on a 4-month loan of $1175 if the annual interest rate is 3.2%. Round to the nearest cent.

b. Calculate the simple interest due on a 45-day loan of $1500 if the annual interest rate is 5.25%. Round to the nearest cent.

Solution

a. Use the simple interest rate formula with $P = 1175$ and $r = 3.2\% = 0.032$. Because the interest rate is an *annual* rate, change 4 months to years.

$$t = \frac{4 \text{ months}}{1 \text{ year}} = \frac{4 \text{ months}}{12 \text{ months}} = \frac{1}{3}$$

$$I = Prt$$

$$I = 1175(0.032)\frac{1}{3} \qquad \bullet \, P = 1175, r = 0.032, t = \frac{1}{3}$$

$$I \approx 12.5333$$

The simple interest due, rounded to the nearest cent, is $12.53.

b. Use the simple interest rate formula with $P = 1500$ and $r = 5.25\% = 0.0525$. Because the interest rate is an *annual* rate, change 45 days to years.

$$t = \frac{45 \text{ days}}{1 \text{ year}} = \frac{45 \text{ days}}{360 \text{ days}} = \frac{1}{8}$$

• It is standard practice, especially for U.S. government debt, to use 360 days in 1 year. We will use that practice for *some* exercises in this text.

$$I = Prt$$

$$I = 1500(0.0525)\left(\frac{1}{8}\right) \qquad \bullet \, P = 1500, r = 0.0525, t = \frac{1}{8}$$

$$I \approx 9.8438$$

The simple interest due, rounded to the nearest cent, is $9.84.

The simple interest formula can be used to find the interest rate on a loan when the interest, principal, and time period of the loan are known.

Example 3   **Calculate an Annual Simple Interest Rate**

The simple interest charged on a 6-month loan of $3000 is $150. Find the annual simple interest rate for this loan.

Solution

Solve the simple interest equation for r.

$$I = Prt$$

$$150 = 3000r\left(\frac{1}{2}\right) \qquad \bullet \text{ Convert 6 months to years: } \frac{6 \text{ months}}{1 \text{ year}} = \frac{6 \text{ months}}{12 \text{ months}} = \frac{1}{2}$$

$$150 = 1500r \qquad \bullet \, 3000\left(\frac{1}{2}\right) = 1500$$

$$0.10 = r \qquad \bullet \text{ Divide each side of the equation by 1500.}$$

$$r = 10\% \qquad \bullet \text{ Write the decimal as a percentage.}$$

The annual simple interest rate on the loan is 10%.

Maturity Value or Future Value for Simple Interest

When you borrow money, the total amount to be repaid to the lender is the sum of the principal and interest. This sum is calculated using the following maturity value or future value formula for simple interest:

$$A = P + I$$

Formulas for Maturity Value or Future Value for Simple Interest

The maturity value or future value formula for simple interest is

$$A = P + I$$

$$A = P + Prt \qquad \bullet \text{ For simple interest, } I = Prt.$$

$$A = P(1 + rt) \qquad \bullet \text{ Factor } P \text{ from } P = P + Prt.$$

where A is the amount after the interest, I, has been added to the principal, P, r is the interest rate, and t is the time.

Each formula gives exactly the same result. The choice of which formula to use is just a matter of convenience.

Take Note: Maturity value and future value are similar. *Maturity value* usually refers to a loan; *future value* usually refers to an investment. These conventions are not always followed. The important thing to remember is both refer to the value of money at some future time.

These formulas can be used for loans or investments. When used for a loan, A is the total amount to be repaid to the lender; this sum is called the **maturity value** of the loan. For an investment, such as a deposit in a bank savings account, A is the total amount on deposit after the interest earned has been added to the principal. This sum is called the **future value** of the investment. P is called the **present value** of the loan or the amount invested. It is the current worth of the loan or the amount invested.

Example 4   Calculate the Maturity Value of a Loan

Calculate the maturity value of a 3-month simple interest loan of $3800 if the annual simple interest rate is 6%.

Solution

Solve the formula for maturity value for simple interest for A, the maturity value.

$$A = P + Prt \qquad\qquad \bullet P = 3800, r = 0.06, t = \frac{3}{12} = \frac{1}{4}$$

$$A = 3800 + 3800(0.06)\left(\frac{1}{4}\right)$$

$$A = 3800 + 57$$

$$A = 3857$$

The maturity value of the loan is $3857.

For Example 4, we could have used the formula in the form $A = P(1 + rt)$.

$$A = 3800\left[1 + 0.06\left(\frac{1}{4}\right)\right] \qquad\qquad \bullet P = 3800, r = 0.06, t = \frac{3}{12} = \frac{1}{4}$$

$$A = 3800[1 + 0.015] = 3800[1.015]$$

$$A = 3857$$

Note that the answers are the same.

Example 5   **Calculate the Interest Rate on an Investment**

In 2 years, the future value of a $5000 investment will be $5375. What is the annual simple interest rate for this investment?

Solution

Solve the formula for maturity value for simple interest for r, the annual interest rate.

$$A = P + Prt$$

$$5375 = 5000 + 5000r(2) \qquad \bullet\ A = 5375, P = 5000, t = 2$$

$$5375 = 5000 + 10{,}000r \qquad \bullet\ 5000(2) = 10{,}000.$$

$$375 = 10{,}000r \qquad \bullet\ \text{Subtract 5000 from each side of the equation.}$$

$$0.0375 = r \qquad \bullet\ \text{Divide each side of the equation by 10,000.}$$

Write the decimal as a percentage. The annual simple interest rate is 3.75%.

Treasury Bills

Treasury bills, or **T-bills**, are U.S. government securities that have a *term* (the period of time a purchaser holds the T-bill) that is less than 1 year. T-bills are typically sold at a *discount* from the *par amount* (also called *face value* or *maturity value*). For instance, you might pay $990 (the discount) for a $1000 (face value) T-bill. When the T-bill matures (at the end of the term), you would be paid $1000. The difference between the purchase price (the discounted value) and face value is interest. The interest paid on T-bills is simple interest.

The U.S. Department of the Treasury auctions T-bills—that is, they offer, say, a 3-month (the term) T-bill that has a face value of $1000 and ask what an investor is willing to pay today for the $1000. The more an investor is willing to pay for a T-bill, the less the interest rate on the T-bill. Note that the interest rate of the T-bill is not given. By offering to pay less than the face value for the T-bill, the investor, not the government, sets the interest rate for the T-bill.

Example 6   **Calculate the Interest Rate on a T-bill**

An investor purchases a T-bill with a face value (maturity value) of $500 for $492. If the term of the T-bill is 3 months, what is the annual simple interest rate earned by the investor? Round to the nearest tenth of a percent.

Solution

Solve the maturity value formula for r with the face value, $A = 500$; the discounted value, $P = 492$; and $t = \dfrac{3 \text{ months}}{12 \text{ months}} = \dfrac{1}{4}$.

$$A = P + Prt$$

$$500 = 492 + 492r\left(\frac{1}{4}\right)$$

$$500 = 492 + 123r$$

$$8 = 123r$$

$$\frac{8}{123} = r$$

$$0.06504 \approx r$$

The annual simple interest rate, rounded to the nearest tenth, is 6.5%.

4.1 Exercise Set

Think About It

1. If a simple interest rate on a loan of $1000 for 1 year is doubled, is the amount of interest owed on the loan doubled?

2. If Aubrey obtains a simple interest rate loan of $500 and must repay $525, identify the present value and maturity value.

3. A person purchases a T-bill with a face value of $1000 for $980. How much interest does the person receive at maturity?

4. If Raul purchases a 6-month T-bill with a face value of $500 for $490 and Helena purchases a similar T-bill for $495, which person receives the higher interest rate? Try to answer this question without doing a calculation.

Preliminary Exercises

For Exercises P1 to P6, if necessary, use the referenced example following the exercise for assistance.

P1. Find the simple interest due on a 4-year loan of $1250 if the annual simple interest rate is 3.5%. **Example 1**

P2. Calculate the simple interest due on a 3-month loan of $2500 if the annual simple interest rate is 6.5%. Round to the nearest cent. **Example 2**

P3. The simple interest charged on a 65-day loan of $1230 is $7.75. Find the annual simple interest rate for this loan. Round to the nearest tenth of a percent. Use 360 days in 1 year. **Example 3**

P4. Find the future value of an 18-month investment of $3300 into a simple interest rate account that has an annual simple interest rate of 4.3%. **Example 4**

P5. The maturity value of a 50-day loan of $650 is $660. What is the annual simple interest rate on this loan? Round to the nearest tenth of a percent. Use 360 days in 1 year. **Example 5**

P6. An investor purchases a T-bill with a face value of $2500 for $2435. If the term of the T-bill is 130 days, what is the annual simple interest rate earned by the investor? Round to the nearest tenth of a percent. **Example 6**

Section Exercises

1. **Taking Out a Loan** Find the annual simple interest rate on a 3-month loan of $5000 if the maturity value of the loan is $5125.

2. **Taking Out a Loan** The maturity value of a 4-month loan of $3000 is $3097. Find the annual simple interest rate.

3. **Earnings on a Deposit** You deposit $880 in an account paying an annual simple interest rate of 9.2%. Find the future value of the investment after 1 year.

4. **Taking Out a Loan** Calculate the maturity value of a simple interest, 8-month loan of $7000 if the annual interest rate is 8.7%.

5. **Interest on Past Due Bills** Your electric bill is $132. You are charged 9% annual simple interest for late payments.

How much do you owe if you pay the bill 1 month past the due date?

6. **Earnings on a Deposit** You deposit $750 in an account paying an annual simple interest rate of 7.3%. Find the future value of the investment after 1 year.

7. **Property Taxes** Your property tax bill is $1200. The county charges a penalty of 11% annual simple interest rate for late payments. How much do you owe if you pay the bill 2 months past the due date?

8. **Earnings on a Deposit** You deposit $1500 in an account earning an annual interest rate of 5.2%. Calculate the simple interest earned in 6 months.

9. **Taking Out a Loan** Calculate the simple interest due on a 45-day loan of $1600 if the annual interest rate is 9%.

10. **Taking Out a Loan** Calculate the simple interest due on a 2-month loan of $800 if the interest rate is 1.5% per month.

11. **Taking Out a Loan** Calculate the simple interest due on a 150-day loan of $4800 if the annual interest rate is 7.25%.

12. **Taking Out a Loan** Calculate the maturity value of a simple interest, 10-month loan of $6600 if the annual interest rate is 9.75%.

13. **Taking Out a Loan** The sum of $10,000 is borrowed for 140 days at an annual simple interest rate of 9%. Calculate the maturity value.

14. **Future Value of an Investment** Find the future value of a 2-year investment of $6225 into a simple interest rate account that has an annual simple interest rate of 3.4%.

15. **Interest Penalties** Suppose you invested $5000 in a 1-year certificate of deposit paying an annual interest rate of 8.5%. After 6 months, you decide to withdraw $2000. Because of the early withdrawal of principal, there is a 3-month interest penalty on the $2000. What interest penalty do you pay?

16. **Payday Loans** Suppose a waiter obtains a payday loan of $200 for 14 days and must repay the lender $230 at that time. What is the annual simple interest rate for this loan?

17. **Payday Loans** A payday lender charges $17.50 for each $100 borrowed. Suppose someone borrows $400 for 18 days. What is the annual simple interest rate for the borrowers?

18. **Treasury Bills** A financial adviser purchases a T-bill with a face value of $1000 for $990. If the term of the T-bill is 6 months, what is the annual simple interest rate earned by the financial adviser? Round to the nearest tenth of a percent.

19. **Treasury Bills** An investor purchases a T-bill with a face value of $5000 for $4900. If the term of the T-bill is 3 months, what is the annual simple interest rate earned by the investor? Round to the nearest tenth of a percent.

20. **Treasury Bills** An investment counselor suggests that a client purchase a T-bill with a face value of $7500 for $7400. If the term of the T-bill is 80 days, what is the annual simple interest rate earned by the client? Round to the nearest tenth of a percent.

21. **Treasury Bills**   An accountant purchases a T-bill with a face value of $6000 for $5945. If the term of the T-bill is 125 days, what is the annual simple interest rate earned by the client? Round to the nearest tenth of a percent.

22. **Loan Interest Rates**   The simple interest charged on a 2-month loan of $450 is $3.75. Find the annual simple interest rate for this loan.

23. **Loan Interest**   Find the simple interest due on a 35-day loan of $650 if the annual simple interest rate is 3.3%. Round to the nearest cent.

24. **Loan Interest**   The maturity value of a 4-month loan of $6250 is $6362.50. What is the annual simple interest rate on this loan?

25. **Loan Interest**   Find the simple interest due on a 4-month loan of $725 if the monthly simple interest rate is 1.5%.

26. **Treasury Bills**   An investor purchases a T-bill with a face value (maturity value) of $500 for $492. If the term of the T-bill is 4 months, what is the annual simple interest rate earned by the investor? Round to the nearest tenth of a percent.

27. **Treasury Bills**   An investor purchases a T-bill with a face value (maturity value) of $1000 for $992. If the term of the T-bill is 8 months, what is the annual simple interest rate? Round to the nearest tenth of a percent.

▶ **Investigations**

Understanding Payday Loans   The schedule for the cost of a payday loan from an online payday loan company is given in the following table. Payday loans usually range from 14 to 30 days.

Amount Borrowed ($)	Finance Charge ($)	Total Amount Due ($)
100	16	116
150	24	174
200	32	232
250	40	290
300	48	348

1. Is the finance charge per $100 borrowed constant for this company?

2. A website gives a formula for calculating the annual simple interest rate for a payday loan as:

$$\frac{\text{Annual interest rate}} {} = \frac{\text{Finance charge}}{\text{Amount of the loan}} \div \text{Term of loan (in days)} \cdot 365 \text{ days}$$

Use this formula to find the annual interest rate (APR) for a 14-day loan of $250.

3. Use the formula in part 2 to find the APR for a 21-day loan of $250.

4. Using the formula $I = Prt$, show how the lender came up with that formula in part 2.

Payday Loan Fees   In some states, there are two fees shown to obtain a payday loan. One fee is called a *credit services organization* (CSO) fee. That fee is a certain dollar amount for each $100 borrowed. A second fee is a percentage of the amount borrowed. The finance charge is the sum of these two fees. Use the Internet to find a company that uses the CSO model. Give an example of what the annual simple interest rate would be for a 20-day loan of $300 using this company.

Section 4.2   Compound Interest

Learning Objectives:

- Calculate compound interest.
- Find the present value of an investment or loan.
- Calculate the effects of inflation.
- Calculate the consumer price index (CPI).
- Calculate equivalent dollar amounts.
- Calculate the annual percentage yield of an investment.

Compound Interest

Warren Buffett, one of the richest men in the world, is sometimes referred to as the "Oracle of Omaha," a nod to his hometown of Omaha, Nebraska. When asked about the secret to his success, Buffett replied, "My wealth has come from a combination of living in America, some lucky genes, and compound interest." Compound interest is the topic of this section.

Simple interest is generally used for loans of 1 year or less. For loans of more than 1 year, the interest paid on the money borrowed is called *compound interest*. **Compound interest** is interest calculated not only on the original principal but also on any interest that has already been earned.

To illustrate compound interest, you deposit $1000 in a savings account earning 5% interest compounded annually (once a year).

During the first year, the interest earned is calculated as follows:

$$I = Prt$$

$$I = \$1000(0.05)(1) = \$50$$

At the end of the first year, the total amount in the account is

$$A = P + I$$

$$A = \$1000 + \$50 = \$1050$$

During the second year, earned interest is calculated using the amount in the account at the end of the first year:

$$I = Prt$$

$$I = \$1050(0.05)(1) = \$52.50$$

Note that the interest earned during the second year ($52.50) is greater than the interest earned during the first year ($50) because the interest earned during the first year was added to the original principal, and the interest for the second year was calculated using this new principal. If the account earned simple interest rather than compound interest, the interest earned each year would be the same ($50).

At the end of the second year, the total amount in the account is the sum of the amount in the account at the end of the first year and the interest earned during the second year:

$$A = P + I$$

$$A = \$1050 + \$52.50 = \$1102.50$$

The interest earned during the third year is calculated using the amount in the account at the end of the second year ($1102.50):

$$I = Prt$$

$$I = \$1102.50(0.05)(1) = \$55.125 \approx \$55.13$$

The interest earned each year keeps increasing. This is the effect of compound interest.

In this example, the interest is compounded annually. However, compound interest can be compounded semiannually (twice a year), quarterly (four times a year), monthly (12 times a year), or daily (365 times a year). The frequency with which the interest is compounded is called the **compounding period**.

In the preceding example, if interest were compounded quarterly rather than annually, then the first interest payment on the $1000 in the account occurs after 3 months or $\frac{1}{4}$ year.

The following calculations show how the interest and principal increase each quarter (3 months).

End of first quarter:	$I = Prt = 1000(0.05)\left(\frac{1}{4}\right) = 12.50$
	$A = 1000 + 12.50 = 1012.50$
End of second quarter:	$I = Prt = 1012.50(0.05)\left(\frac{1}{4}\right) \approx 12.66$
	$A = 1012.50 + 12.66 = 1025.16$
End of third quarter:	$I = Prt = 1025.16(0.05)\left(\frac{1}{4}\right) \approx 12.81$
	$A = 1025.16 + 12.81 = 1037.97$
End of fourth quarter:	$I = Prt = 1037.97(0.05)\left(\frac{1}{4}\right) \approx 12.97$
	$A = 1037.97 + 12.97 = 1050.94$

The total amount in the account at the end of the first year is $1050.94.

When the interest is compounded quarterly, the account earns $50.94. This is $0.94 more interest than when interest is compounded annually, $50. **In general, an increase in the number of compounding periods per year results in an increase in the interest earned by an account.**

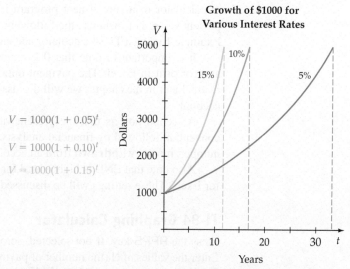

Growth of $1000 for Various Interest Rates

$V = 1000(1 + 0.05)^t$

$V = 1000(1 + 0.10)^t$

$V = 1000(1 + 0.15)^t$

As the interest rate increases, the time required for the investment to reach $5000 decreases. For 5% it takes about 33 years; for 10% it takes about 17 years; for 15% it takes about 11 years.

Future Value of Compound Interest Formula

The future value, FV, of an amount deposited into a compound interest account is given by

$$FV = PV\left(1 + \frac{r}{n}\right)^{nt}$$

where PV is the amount of money deposited (the present value), r is the annual interest rate as a decimal, n is the number of compounding periods per year, and t is the number of years.

For the preceding formula, we used the phrases *future value* and *present value*. These are the phrases commonly used in financial calculations. Future value (FV) is the value of an investment at some future time while present value (PV) is the value of the investment today.

Example 1   Find the Future Value of an Investment

Calculate the future value of $7500 deposited into an account earning an annual interest rate of 5% compounded quarterly after 5 years.

Solution

Use the future value of compound interest formula. The present value is $7500, the amount deposited into the account; $n = 4$, as there are 4 quarters in 1 year; $r = 0.05$, the annual interest rate as a decimal; $t = 5$, the number of years.

$$FV = PV\left(1 + \frac{r}{n}\right)^{nt}$$

$$= 7500\left(1 + \frac{0.05}{4}\right)^{4(5)} \qquad \bullet \; PV = 7500, r = 0.05, n = 4, t = 5.$$

$$= 7500(1 + 0.0125)^{20}$$

$$= 7500(1.0125)^{20}$$

$$\approx 7500(1.282037232)$$

$$\approx 9615.2792$$

The future value is $9615.28.

A calculator or a spreadsheet program is frequently used to calculate future value and present value. For instance, the following are screenshots that show the calculations from Example 1 using a TI-84 calculator and an image from Excel.

It is important to note that 0 is entered for payment (**PMT**) in the TI-84 calculator and for **pmt** in Excel. The payment option is used when there are a series of payments made. Later in the chapter we will discuss the effects of making periodic payments into an account.

A second note is that the present value is entered as a negative number. This is a convention followed by financial analysts. Money *deposited* into an account is a *negative* number; money **withdrawn** from an account is a **positive** number.

Also note that END is highlighted for the TI calculator and that 0 is entered for [**type**] for Excel. These settings will be discussed later in the chapter.

TI-84 Graphing Calculator

Press the **APPS** key. If not selected, scroll to Finance, then press the **ENTER** key twice. Enter the values of **N** (the number of payments, 20 = 4(5)), **I%** (the annual interest rate, 5), **PV** (the present value, −7500), **PMT** (the payment, 0), and **P/Y** = 4 and **C/Y** = 4 (number of compounding periods per year); temporarily skip **FV** (the future value—this is the value to calculate).

```
N=20
I%=5
PV=-7500
PMT=0
■FV=9615.279238
P/Y=4
C/Y=4
PMT: END BEGIN
```

Once all the other values are entered, move the cursor to FV. Press **ALPHA** then **SOLVE** (the **ENTER** key).

Excel

Use the Excel *FV* function. When using an Excel function, precede the function with an equal sign, =.

= FV(rate, nper, pmt, [pv],[type]).

Spreadsheet programs differ from the TI-84 Finance app; for the TI-84, the annual rate is entered. The **rate** for a spreadsheet program is the rate *per compounding period*. This is shown as B1/B2 for the rate in the following Excel spreadsheet. The number or periods, **nper**, is shown as B2*B3, the compounding periods per year times the number of years.

Compounding periods per year (B2) times the number of years (B3).

Annual interst rate (B1) divided by compounding periods per year (B2).

Payment amount
Present value

Type

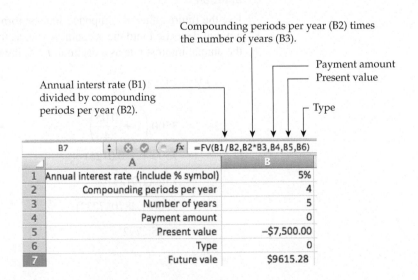

	B7	fx	=FV(B1/B2,B2*B3,B4,B5,B6)
	A		**B**
1	Annual interest rate (include % symbol)		5%
2	Compounding periods per year		4
3	Number of years		5
4	Payment amount		0
5	Present value		−$7,500.00
6	Type		0
7	Future vale		$9615.28

The future value of the investment is **$9615.28**.

Example 2 Calculate the Interest Earned on an Investment

Find the interest earned on an investment of $25,000 deposited into an account earning an annual interest rate of 4.5% compounded monthly after 5 years.

Solution

The interest earned, I, on the investment is the difference between its future value and its present value, $I = FV - PV$. The first step is to calculate the future value. P/Y = 12 and C/Y = 12 because interest is compounded monthly, 12 times a year.

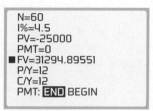

The future value, rounded to the nearest cent, is $31,294.90.

Calculate the interest earned.

$$I = FV - PV$$

$$= 31{,}294.90 - 25{,}000$$

$$= 6294.90$$

The interest earned on the investment was $6294.90.

Present Value

In some instances, we want to calculate the present value of an investment given its future value. The winner of a large lottery prize is given the option of taking, say, a $20 million lottery prize, as a series of payments or $8 million now. $20 million is the future value; $8 million is the present value. To the lottery commission, $20 million paid in the future is the same as paying $8 million today.

Present Value of Compound Interest Formula

Present value, PV, of an investment is the current value of money expected to be received at some future time.

$$PV = \frac{FV}{\left(1 + \dfrac{r}{n}\right)^{nt}}$$

where FV is the value of money expected to be received in the future, r is the annual interest rate as a decimal, n is the number of compounding periods per year, and t is the number of years.

Example 3 Calculate the Present Value of an Investment

How much money should be invested in an account that earns 8% interest, compounded daily, to have $30,000 in 5 years?

Solution

In this case, the future value is given ($30,000), and we need to calculate the present value.

$$PV = \frac{FV}{\left(1 + \dfrac{r}{n}\right)^{nt}}$$

$$PV = \frac{30{,}000}{\left(1 + \dfrac{0.08}{365}\right)^{365(5)}}$$
• $FV = 30{,}000$, $r = 0.08$, $n = 365$, $t = 5$.

$$\approx \frac{30{,}000}{1.491759314}$$

$$\approx 20{,}110.48$$
• Rounded to the nearest cent.

The amount that must be invested today is $20,110.48.

Here is Example 3 calculation using Excel. Note the future value is shown with parentheses around the answer and in red. This is one way Excel shows negative results. The formula for present value is

$$= \text{PV(rate, nper, pmt, [fv], [type])}$$

Compounding periods per year (B2) times
the number of years (B3).

Payment amount
Present value

Annual interst rate (B1)
divided by compounding
periods per year (B2).

Type

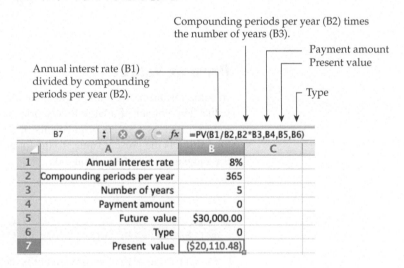

B7		⊗ ⊘ − fx	=PV(B1/B2,B2*B3,B4,B5,B6)	
	A		B	C
1	Annual interest rate		8%	
2	Compounding periods per year		365	
3	Number of years		5	
4	Payment amount		0	
5	Future value		$30,000.00	
6	Type		0	
7	Present value		($20,110.48)	

Inflation

We have discussed compound interest and its effect on the growth of an investment. After your money has been invested for a period of time in an account that pays interest, you will have more money than you originally deposited. But does that mean you will be able to buy more with the compound amount than you were able to buy with the original investment at the time you deposited the money? The answer is, not necessarily. The reason is the effect of inflation.

As an example, consider the cost of tuition at a public college or university. In recent years, the tuition at a public college or university has been increasing at a rate of approximately 3.5% per year. Suppose that tuition this year at a certain school is $5000 and that tuition is expected to rise 3.5% next year. Using the future value of simple interest formula, $A = P(1 + rt)$, we have

$$\text{Tuition next year} = 5000[1 + 0.035(1)] \qquad \text{• Tuition this year} = 5000, i = 0.035, t = 1$$

$$= 5175$$

The tuition next year will be $5175.

If your income this year is, say, $20,000, then tuition, as a percentage of your annual income, is $\frac{5000}{20,000} = 0.25$ or 25% of your income. If your income next year does not increase, then tuition is $\frac{5175}{20,000} \approx 0.259$ or 25.9% of your annual income. This is an example of a phrase you may have heard: "Wages are not keeping up with inflation." You are paying a larger percentage of your annual income for the same product or service (in this case, 1 year of education).

Inflation is an economic condition during which there are increases in the costs of goods and services. Inflation is expressed as a percent; for example, we might speak of an annual inflation rate of 2.3%.

The present value formula can be used to determine the effect of inflation on the future purchasing power of a given amount of money. Substitute the inflation rate for the interest rate in the present value formula. The compounding period is 1 year. Although inflation rates can vary dramatically, we will assume a constant rate of inflation for the examples and exercises.

Example 4 **Calculate the Effect of Inflation on Purchasing Power**

Suppose you purchase an insurance policy today that will provide you with $500,000 when you retire in 30 years. Assuming an annual inflation rate of 2.4%, what will be the purchasing power of $500,000 in 30 years?

Solution

Because you will receive the $500,000 in 30 years, $500,000 is the future value of the policy. To calculate its purchasing power, we need to calculate the present value of the policy, the purchasing power of the policy today.

```
N=30
I%=2.4
■PV=-245454.6733
PMT=0
FV=500000
P/Y=1
C/Y=1
PMT: END BEGIN
```

The purchasing power will be $245,454.67.

Looking back at Example 4, understanding the calculation is important. The result basically states, for the inflation rate, that in 30 years $500,000 will purchase what $245,454.67 will today.

Consumer Price Index (CPI)

According to the Bureau of Labor Statistics (BLS), the **consumer price index (CPI)** is a measure of the average change over time in the prices consumers pay for a market basket of consumer goods and services. Some of the items in this basket of goods and services are milk, coffee, cereal, poultry, shirts, dresses, prescription drugs, physician services, eyeglasses, televisions, tuition, and many other items.

The following graph shows the relative importance of each component of the CPI. For instance, 18% of the value of the CPI is based on the cost of transportation.

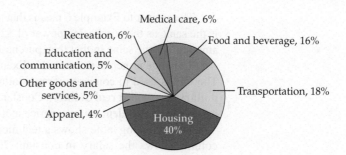

The consumer price index has a base period, the years between 1982 and 1984, from which to make comparisons. The CPI for the base period is 100. The CPI for April 2022 was 289.109. This means that $100 in the period 1982–1984 had the same purchasing power as $289.11 in April 2022.

An index number is a percentage written without a percent sign. The CPI for April 2022 means that the average cost of consumer goods at that time was 289.109% of their cost in the years 1982, 1983, and 1984.

Example 5   Use the CPI to Calculate Inflation Rates

The CPI in 2011 was 224.939; in 2012, it was 229.594. Find the inflation rate between the 2 years. Round to the nearest tenth of a percent.

Solution

The inflation rate between the two years is the percent increase in the CPI between the 2 years.

$$\text{Inflation rate} = 100\left(\frac{\text{CPI in 2012} - \text{CPI in 2011}}{\text{CPI in 2011}}\right)$$

$$= 100\left(\frac{229.594 - 224.939}{224.939}\right)$$

$$\approx 2.0694$$

The inflation rate, rounded to the nearest tenth, between the 2 years was 2.1%.

One use of the CPI is to compare the cost of items at different times.

Example 6   Use the CPI to Calculate Equivalent Dollar Amounts

The consumer price index for food and beverage in December 2021 was 278.802. In December 2022, the CPI for food and beverage was 298.318. If a selection of food and beverage cost $250 in December 2021, what would the same selection cost in December 2022?

Solution

The food and beverage index increased from 278.802 to 298.318 or $\dfrac{298.318}{278.802}$ times. Therefore, the cost of the same selection in December 2022 is

$$250 \cdot \frac{298.318}{278.802} \approx 267.50$$

The same selection of food and beverage would cost $267.50 in December 2022.

The answer to Example 6 means that the purchasing power $267.50 in December 2022 is the same as the purchasing power of $250 in December 2021. In a sense, the two dollar amounts are the same in that they purchase exactly the same amount of goods.

The same principal illustrated in Example 6 can be used to analyze real gain in income. The CPI is used to convert **current dollars** (also called **nominal dollars**) into **constant dollars** (also called **real dollars**). Constant dollars are used to compare current salaries and prices to an equivalent salary or price at a different time.

The following table shows a technician's salary for various time frames. The fourth column shows the salary in constant 2016 dollars. By showing constant dollars, the technician can evaluate whether the current salary in real dollars (purchasing power) is actually increasing. The calculation is similar to the computation in Example 6.

$$\text{Salary in year Y} \cdot \frac{\text{CPI in year X}}{\text{CPI in year Y}} = \text{Salary in constant X year dollars}$$

Year	Salary ($)	CPI	Constant 2016 Dollars	Constant 2022 Dollars
2016	58,000	236.916	$58{,}000 \cdot \dfrac{236.916}{236.916} = 58{,}000$	$58{,}000 \cdot \dfrac{281.148}{236.916} = 68{,}829$
2018	62,000	247.867	$62{,}000 \cdot \dfrac{236.916}{247.867} = 59{,}261$	$62{,}000 \cdot \dfrac{281.148}{247.867} = 70{,}325$
2020	64,000	257.971	$64{,}000 \cdot \dfrac{236.916}{257.971} = 58{,}776$	$64{,}000 \cdot \dfrac{281.148}{257.971} = 69{,}750$
2022	67,000	281.148	$67{,}000 \cdot \dfrac{236.916}{281.148} = 56{,}459$	$67{,}000 \cdot \dfrac{281.148}{281.148} = 67{,}000$

From the table, although the technician's salary has increased by $9000 over 6 years, a 15.5% increase, the purchasing power, in 2016 dollars, has decreased $1541, a 2.7% decrease in purchasing power. If we look at the salary in constant 2022 dollars, it has gone from $68,829 to $67,000, a 2.7% decrease.

Annual Percentage Yield (APY)

When interest is compounded, the annual rate of interest is called the **nominal rate** of interest. The **annual percentage yield (APY)** is the simple interest rate that would yield the same amount of interest after 1 year as the nominal rate. APY is also called **effective annual rate**. When a bank advertises a "7% annual interest rate compounded daily and yielding 7.25%," the nominal interest rate is 7% and the APY is 7.25%.

Formula for Annual Percentage Yield (APY)

$$\text{APY} = \left(1 + \frac{\text{APR}}{n}\right)^n - 1$$

where APR is the annual percentage rate as a decimal, n is the number of compounding periods per year, and APY is the annual percentage yield as a decimal.

Example 7   Calculate an APY

A credit union offers a certificate of deposit (CD) at an annual interest rate of 3% compounded monthly. Find the annual percentage yield. Round to the nearest hundredth of a percent.

Solution

Use the formula for the annual percentage yield where $n = 12$ and APR $= 3\%$ (0.03).

$$\text{APY} = \left(1 + \frac{\text{APR}}{n}\right)^n - 1$$

$$= \left(1 + \frac{0.03}{12}\right)^{12} - 1$$

$$\approx 0.030415957$$

The APY is 3.04%.

4.2 Exercise Set

Think About It

1. If the annual interest rate on two different investments of $1000 is the same but one investment earns interest compounded monthly and the other earns interest compounded daily, which investment will have earned the most interest after 1 year?

2. An investment of $5000 is worth $5100 after 1 year. Identify the future value and present value of the investment.

3. Because of inflation, the purchase of $100 worth of goods today would have cost $94 if bought 5 years ago. Identify the nominal dollars and the real dollars.

4. An investment earns an annual interest rate of 5% compounded daily. The annual simple interest rate that would yield the same amount of interest after 1 year is 5.13%. Identify the nominal interest rate and the annual percentage yield.

Preliminary Exercises

For Exercises P1 to P7, if necessary, use the referenced example following the exercise for assistance. Use 365 for the number of days in a year.

P1. Calculate the future value of $1250 deposited into an account earning an annual simple interest rate of 4% compounded daily after 3 years. **Example 1**

P2. A business analyst deposited $50,000 into an account earning an annual interest rate of 5.1% compounded monthly. If the analyst leaves the money in the account for 3 years, how much interest is earned on the account? **Example 2**

P3. How much money should be invested in an account that earns 7% interest, compounded quarterly, to yield $10,000 in 5 years? **Example 3**

P4. Suppose you purchase an insurance policy today that will provide you with $250,000 when you retire in 25 years. Assuming an annual inflation rate of 4%, what will be the purchasing power of $250,000 in 25 years? **Example 4**

P5. The consumer price index for education in January 2017 was 251.122. In January 2018, the CPI for education was 257.307. Find the percent increase in the education inflation rate between the two years. **Example 5**

P6. Joca was earning $54,500 per year on January 1, 2017, when the CPI was 242.839. On January 1, 2018, she got a raise to $57,000, and the CPI was 251.038. Did she gain in purchasing power? **Example 6**

P7. A bank offers a CD at an annual interest rate of 4% compounded monthly. Find the APY. Round to the nearest hundredth of a percent. **Example 7**

Section Exercises

For these exercises, assume all years have 365 days.

1. **Opening an Interest Earning Account** Jason deposits $5000 into an account that earns 5.1% compounded monthly. How much interest will Jason earn if he leaves the money in the account for 7 years?

2. **A Real Estate Investment** Monika invests in a real estate project that earns 6.8% compounded semiannually. If Monika's initial deposit is $10,000, how much interest will the investment earn in 8 years?

3. **Scholarship Funds** A deposit of $20,000 is placed in a scholarship fund that earns an annual interest rate of 2.75% compounded daily. Find the value of the account after 4 years.

4. **Savings Accounts** What is the future value of $2000 deposited into an account earning an annual simple interest rate of 3.25% compounded quarterly after 5 years?

5. **Planning for Savings** How much money should be invested in an account that earns 7% interest compounded quarterly to yield $10,000 in 5 years?

6. **Planning for Savings** How much money should you invest today into an account that earns an annual interest rate of 8% compounded monthly to have $12,000 in 4 years?

7. **Salaries in Constant Dollars** The consumer price index for selected years is shown in the following table.

Year	CPI	Year	CPI
2015	233.707	2019	251.712
2016	236.916	2020	257.971
2017	242.839	2021	261.582
2018	247.867	2022	281.148

Suppose a video game programmer's salary was $71,000 in 2018, $75,000 in 2020, and $80,000 in 2022. Using the values of the CPI from the table, find the programmer's salary in constant 2018 dollars. Round to the nearest whole dollar.

8. **Purchasing Power** Using the table from Exercise 7, by what percentage did the purchasing power of $1 decline between 2015 and 2022? Round to the nearest tenth of a percent.

9. **Savings Accounts** Buchra Halimi has money in a savings account that earns an annual interest rate of 3%, compounded monthly. What is the APY on Blake's account? Round to the nearest hundredth of a percent.

10. **Savings Accounts** Beth Chawa has money in a savings account that earns an annual interest rate of 3%, compounded quarterly. What is the APY on Beth's account? Round to the nearest hundredth of a percent.

11. **Effect of Inflation on Rents** The average monthly rent for a three-bedroom apartment in Denver, Colorado, is $3115. Using an annual inflation rate of 7%, find the average monthly rent in 15 years.

12. **Fixed Incomes** A retired couple has a fixed income of $3500 per month. Assuming an annual inflation rate of 7%, what is the purchasing power of their monthly income in 5 years?

13. **Inflation Rates** Using the table in Exercise 7, find the inflation rate between 2018 and 2022. Round to the nearest tenth of a percent.

14. **Understanding CPI** Use the table in Exercise 7. If a representative basket of goods as measured by the CPI cost $100 in 2015, what would that same basket of goods cost in 2022?

15. **Interest Earned** Calculate the amount of interest earned in 6 years on $20,000 deposited in an account paying 4% interest compounded monthly.

16. **Interest Earned**   Calculate the amount of interest earned in 8 years on $15,000 deposited in an account paying 10% interest, compounded quarterly.

17. **Future Value**   What is the future value of $4000 earning 6% interest compounded monthly for 6 years?

18. **Future Value**   Calculate the future value of $8000 earning 8% interest compounded quarterly for 10 years.

19. **Interest Earned**   How much interest is earned in 3 years on $2000 deposited in an account paying 6% interest compounded quarterly?

20. **Interest Earned**   How much interest is earned in 5 years on $8500 deposited in an account paying 9% interest compounded semiannually?

21. **Saving for College**   A couple plans to save for their child's college education. What principal must be deposited by the parents when their child is born to have $40,000 when the child reaches the age of 18? Assume the money earns 8% interest compounded quarterly.

22. **CD Investments**   You deposit $7500 in a 2-year certificate of deposit earning 2.4% compounded daily. At the end of the 2 years, you reinvest the compound amount plus an additional $7500 in another 2-year CD. The interest rate on the second CD is 2.9% compounded daily. What is the compound amount when the second CD matures?

23. **Retirement Investments**   You plan to retire in 30 years and would like to have $1,000,000 in investments. How much money would you have to invest today at a 9% interest rate compounded daily to reach your goal in 30 years?

24. **Retirement Investments**   You plan to retire in 40 years and would like to have $1,500,000 in investments. How much money would you have to invest today at a 5% interest rate compounded daily to reach your goal in 40 years?

25. **Savings over Time**   You deposit $5000 in a savings account that earns 1.5% interest compounded daily. After 1 year, you deposit an additional $5000 into the same savings account, but the interest is now 1.75% compounded daily. What is the value of the account after the second year?

26. **Investment Account over Time**   Melanie places $5000 into an account that earns 5% interest compounded daily. Six months later (180 days), she deposits an additional $2500 in the same account. What is the value of the account after 1 year?

27. **Selling an Investment**   An agent is selling an investment that earns 4.875% compounded semiannually. What is the APY of the investment?

28. **Investments**   What is the APY on an investment that earns 5.75% interest compounded quarterly?

▶ **Investigations**

Understanding Compound Interest   We have examined exercises in which interest has been compounded annually, semiannually, quarterly, monthly, and daily. Suppose Elijah invests $1000 into an account that earns 6% interest.

1. Complete the following table to determine the value of his account after 1 year. Use 365 for Daily.

Compounding Periods	Annually	Semiannually	Quarterly	Monthly	Daily
Amount After 1 Year					

2. From the table, does doubling the number of compounding periods per year double the amount of interest earned?

3. Instead of compounding daily, suppose the compounding periods were more frequent. Complete the following table.

Compounding Periods	Every Hour	Every Minute	Every Second
Amount After 1 Year			

4. How much, to the nearest cent, more interest is earned between compounding every minute and every second?

5. From these calculations, it appears there is a limit to the amount of interest that can be earned in 1 year as the number of compounding periods per year increases. From the calculations, it appears that this amount of interest is $61.64. Try to convince yourself this is true by increasing the number of compounding periods to every half-second or every tenth of a second.

6. The formula for compound interest, $FV = PV\left(1 + \dfrac{i}{n}\right)^{nt}$, contains $\left(1 + \dfrac{i}{n}\right)^{nt}$. If the time period is 1 year, then $t = 1$; if we let $i = 1$, the factor becomes $\left(1 + \dfrac{1}{n}\right)^{n}$.

Calculate this expression for the values of n in the following table.

n	10,000	100,000	1,000,000	10,000,000	100,000,000
Value					

From the preceding calculations, it appears that the value of the expression changes little (less than 0.0000012) as n changes from 10,000,000 to 100,000,000. If we continued the process for even larger values of n, we would find that we would be closer and closer to 2.71828182845905. This number is an irrational number (a nonterminating, nonrepeating decimal) and is designated by e. This is one of the most important numbers in mathematics and its applications. Rewriting n as $\dfrac{n}{r} \cdot r$, we have

$$\left(1 + \frac{r}{n}\right)^{nt} = \left(1 + \frac{r}{n}\right)^{\frac{n}{r} \cdot rt}$$

Now let $m = \dfrac{n}{r}$ and then $\dfrac{1}{m} = \dfrac{r}{n}$. Making these substitutions, we have $\left[\left(1 + \dfrac{1}{m}\right)^{m}\right]^{rt}$. For every large value of m, we can replace $\left(1 + \dfrac{1}{m}\right)^{m}$ by e. This gives $FV = PVe^{rt}$. This is known as the **continuous compounding interest formula**. Interest is compounded an infinite number of times each year.

7. Find the value of $1000 after 1 year if interest is compounded continuously. Suggestion: use the e^x key on a calculator.

8. Find the value of $1000 after 10 years if interest is compounded continuously.

Understanding Hyperinflation   Write a few paragraphs on hyperinflation and why it is bad. Give a few examples of countries that have experienced hyperinflation.

Section
4.3

Future and Present Value of an Annuity

Learning Objectives:

• Find the future value of an annuity.

• Find the present value of an annuity.

• Find the monthly payment for an installment loan.

• Determine a loan payoff.

Future Value of an Annuity

An **annuity** is a series of equal payments (sometimes called **cash flows**) paid at regular intervals for a given length of time. A **fixed annuity** is one in which the interest rate is fixed for the term of the annuity; a **variable annuity** is one in which the interest can change over the term of the annuity. Here we will focus only on fixed annuities.

The interest earned on the cash flows is compounded with the same frequency as payments. For instance, if payments are made quarterly, then interest is compounded quarterly; if payments are made monthly, then interest is compounded monthly.

The two basic forms of an annuity are a *due* annuity and an *ordinary* annuity. A **due annuity** is one where the series of payments is made at the *beginning* of each payment period. An example of a due annuity is rent. The payment is due at the beginning of each month.

An **ordinary annuity** is one in which the series of payments is made at the *end* of each payment period. A car payment is an example of an ordinary annuity. The payment is due at the end of each month. Both ordinary and due annuities have the same characteristics: equal payments at equal intervals of time.

The **future value of an annuity** is the sum of the future values of each payment. This could be money you pay or money you receive. Suppose you deposit $1000 on January 1 for the next 5 years into an account that earns 6% annual interest compounded annually. The first deposit will earn interest for 5 years; the second deposit earns interest for 4 years; the third deposit earns interest for 3 years; the fourth deposit earns interest for 2 years; the fifth (last) deposit earns interest for 1 year. The future value of each of the deposits is shown in the following graph.

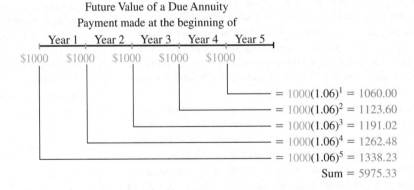

Future Value of a Due Annuity
Payment made at the beginning of

At the end of 5 years, you would have $5975.33, the sum of the future values of each deposit. The deposits were made at the beginning of each year, so this is an example of a *due* annuity.

The formula for calculating the future value of a due annuity (payments made at the beginning of the compounding period) is derived by using a procedure similar to the preceding one.

Formula for the Future Value of a Due Annuity

$$FV = PMT\left[\frac{(1+i)^n - 1}{i}\right](1+i)$$

where PMT is the cash flow amount, n is the number of payments, and i is the interest rate per period, $i = \dfrac{\text{Annual interest rate as a decimal}}{\text{Number of compounding periods per year}}$.

Example 1   **Calculate the Future Value of a Due Annuity**

A flight operations manager for an airline starts a retirement savings plan in which $100 per month is deposited at the beginning of each month into an account that earns an annual interest rate of 5% compounded monthly. Find the value of this investment after 25 years.

Solution

Use the future value of a due annuity formula with $PMT = 100$, $n = 25 \cdot 12 = 300$ (25 years is 300 months), and $i = \dfrac{0.05}{12}$.

$$FV = PMT\left[\frac{(1 + i)^n - 1}{i}\right](1 + i)$$

$$FV = 100\left[\frac{\left(1 + \dfrac{0.05}{12}\right)^{300} - 1}{\dfrac{0.05}{12}}\right]\left(1 + \frac{0.05}{12}\right)$$

$$\approx 59{,}799.10$$

The future value of the annuity is $59,799.10.

Here are the calculations for Example 1 using a calculator and Excel.

```
N=300
I%=5
PV=0
PMT=-100
■FV=59799.09989
P/Y=12
C/Y=12
PMT: END BEGIN
```

TI-84 Calculator

Press **APPS** If necessary, scroll to Finance, Press **ENTER ENTER**. Then input the values of N, I%, PV = 0 (the present value of the savings plan is 0), the monthly payment as a negative number (money is being deposited), and P/Y = C/Y = 12 (payments and compounding periods are 12 times a year). Move the cursor to **BEGIN** (payments at the beginning of the month) and press **ENTER**. Now move the cursor to **FV** and press **ALPHA** then **SOLVE**.

Excel

Use the Excel *FV* function.

= FV(rate, nper, pmt, [pv], [type]).

The rate is the annual rate divided by the number of compounding periods per year (B2/B1), the number of periods is B1*B3, the payment is −$100, and the present value is $0. The last entry [type] is either a 0 or a 1. *Use 1 for a due annuity* (payments made at the beginning of each payment period).

| B7 | ÷ | × ✓ | fx | =FV(B1/B2,B2*B3,B4,B5,B6) |

	A	B	C
1	Annual interest rate	5%	
2	Compounding periods per year	12	
3	Number of years	25	
4	Payment amount	−$100.00	
5	Present value	$0.00	
6	Type	1	
7	Future value	$59,799.10	

For an *ordinary* annuity, deposits are made at the **end** of each compounding period. For instance, some employers offer a 401(k) plan. This plan allows employees to invest part of their paycheck into a retirement plan before income taxes are taken out. The employee has a certain amount deducted from each paycheck at the *end* of the payment period and deposited into the 401(k) account.

Formula for the Future Value of an Ordinary Annuity

$$FV = PMT\left[\frac{(1 + i)^n - 1}{i}\right]$$

where *PMT* is the cash flow amount, *n* is the number of payments, and *i* is the interest rate per period: $i = \dfrac{\text{Annual interest rate as a decimal}}{\text{Number of compounding periods per year}}$.

Look carefully at this formula and the one for a due annuity. The difference between the two formulas is the factor $(1 + i)$. This factor takes into account placing money into an account at the beginning (due annuity) versus the end (ordinary annuity) of a compounding period.

Example 2   Calculate the Future Value of an Ordinary Annuity

A technician for a computer chip company sets up a 401(k) program and invests $25 at the end of each week into an account that earns an annual interest rate of 6% compounded weekly. Find the value of the 401(k) after 10 years.

Solution

Deposits are made at the end of each week, so this is an ordinary annuity. The result using Excel follows. Note that most of the formula for this calculation is exactly the same as for Example 1. The difference is that B6 is 0 instead of 1. *The 0 indicates that this is an ordinary annuity.*

	B7	fx	=FV(B1/B2,B2*B3,B4,B5,B6)	
	A		B	C
1	Annual interest rate		6%	
2	Compounding periods per year		52	
3	Number of years		10	
4	Payment amount		-$25.00	
5	Present value		$0.00	
6	Type		0	
7	Future value		$17,798.92	

The future value of the 401(k) is $17,798.92.

There may be personal reasons for wanting to calculate a payment to reach a financial goal. For instance, the parents of a child might want to start a college savings plan for the child or you may want to save for the down payment on a new car.

Example 3 **Calculate the Payment for the Future Value of a Due Annuity**

Suppose a family wants to have $10,000 in 4 years to take a family vacation to Yellowstone National Park. How much does the family need to invest at the beginning of each month into an account that earns 4.5% interest compounded monthly?

Solution

Here is the solution using both the TI-84 calculator and Excel. Choose software that is easily accessible to you.

TI-84 Calculator

Press APPS If necessary, scroll to Finance. Press ENTER ENTER. Then input the values of N = 12 · 4 = 48, I% = 4.5, PV = 0, temporarily skip PMT, FV = 10000, and P/Y and C/Y = 12. Select BEGIN; payments are made at the beginning of each month. Move the cursor to PMT and press ALPHA then SOLVE.

```
N=48
I%=4.5
PV=0
■PMT=-189.82302....
FV=10000
P/Y=12
C/Y=12
PMT: END BEGIN
```

Excel

Create a spreadsheet similar to the following using the *PMT* function.

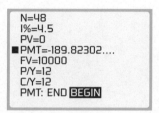

B7	fx	=PMT(B1/B2,B2*B3,B4,B5,B6)	
	A	B	C
1	Annual interest rate	5%	
2	Compounding periods per year	12	
3	Number of years	4	
4	Present value	$0.00	
5	Future value	$10,000.00	
6	Type	1	
7	Payment	($189.82)	

The family must save $189.82 per month to reach its goal.

A **sinking fund** is an annuity established to repay a debt such as bonds a state or company may have sold or to save for future capital expenses such as a new computer system.

Example 4 **Calculate the Payment into a Sinking Fund**

A city issues $25 million in bonds that will mature in 20 years. The city is required to establish a sinking fund so it will have the money to repay that debt. How much money must be deposited at the end of each quarter into an account that earns an annual interest rate of 5.8% compounded quarterly to retire the $25-million bond debt?

Solution

Payments are made at the end of the quarter. This is an ordinary annuity. Solve the future value of an ordinary annuity formula for PMT.

TI-84 Calculator

Enter the values for N = 4 · 20 = 80, I% = 5.8%, PV = 0, skip PMT, FV = 25000000, P/Y = 4, and C/Y = 4. Select END—payments are made at the end of a quarter. Move the cursor to PMT and press ALPHA then SOLVE.

```
N=80
I%=5.8
PV=0
■PMT=-167554.68...
FV=25000000
P/Y=4
C/Y=4
PMT: END BEGIN
```

The city must set aside $167,554.68 each quarter to meet its bond obligation of $25,000,000 in 20 years.

Present Value of an Annuity

The **present value of an annuity** is the sum of the present values of all expected cash flows. As with the future value of an annuity, there is the present value of an ordinary annuity and the present value of a due annuity.

Suppose you purchase an investment that pays $1000 at the end of each year for the next 5 years. The present value of the investment depends on the interest rate that can be earned over the 5 years. For this example, we will assume an interest rate of 6%.

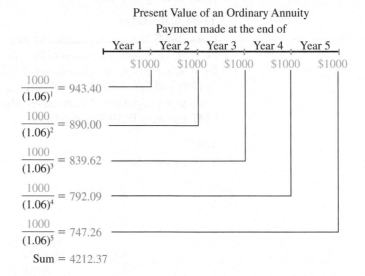

Present Value of an Ordinary Annuity
Payment made at the end of

The present value of the cash flows is $4212.37. This means (assuming a 6% interest rate) that having $4212.37 today (present value) is the same as receiving $1000 a year for the next 5 years. This is the present value of an *ordinary* annuity; the payments are made at the *end* of the compound period.

The formula for calculating the present value of an ordinary annuity is derived by using a procedure similar to the one above.

Formula for the Present Value of an Ordinary Annuity

$$PV = PMT\left[\frac{1 - (1 + i)^{-n}}{i}\right]$$

where PMT is the cash flow amount, n is the number of payments, and i is the interest rate per period, $i = \dfrac{\text{Annual interest rate as a decimal}}{\text{Number of compounding periods per year}}$.

For the example above, $PMT = 1000$, $n = 5$, and

$$i = \frac{\text{Annual interest rate as a decimal}}{\text{Number of compounding periods per year}} = \frac{0.06}{1} = 0.06.$$

$$PV = PMT\left[\frac{1 - (1 + i)^{-n}}{i}\right]$$

$$PV = 1000\left[\frac{1 - (1 + 0.06)^{-5}}{0.06}\right]$$

$$\approx 1000[4.212363786]$$

$$\approx 4212.36$$

The 1 cent difference is due to rounding errors.

Example 5   **Calculate the Present Value of an Ordinary Annuity**

A trust fund pays the benefactor $1200 at the end of each month for 10 years. The current interest rate is 5.2% compounded monthly; find the present value of the annuity.

Solution

Here is the solution using both the TI-84 calculator and Excel. Choose software that is easily accessible to you.

TI-84 Calculator

Press **APPS** If necessary, scroll to Finance. Press **ENTER** **ENTER**. Then input the values of **N** = 10 · 12 = 120, **I%** = 5.2, temporarily skip **PV**, **PMT** = −1200, **FV** = 0, **P/Y** and **C/Y** = 12. Select **END**; payments are made at the end of each month. Now move the cursor to **PV** and press **ALPHA SOLVE**. The present value of the trust fund is $112,101.56.

```
N=120
I%=5.2
■PV=-112101.5573
PMT=-1200
FV=0
P/Y=12
C/Y=12
PMT: END BEGIN
```

Excel

Create a spreadsheet similar to the following using the Excel present value, *PV*, function.

	B7	fx =PV(B1/B2,B2*B3,B4,B5,B6)
	A	B	C
1	Annual interest rate	5%	
2	Compounding periods per year	12	
3	Number of years	10	
4	Payment	-$1,200.00	
5	Future value	$0.00	
6	Type	0	
7	Payment	$1,12,101.56	

The present value of the trust fund is $112,101.56.

The present value of a due annuity is used for things such as rent or a lease agreement in which payments are made at the beginning of a compounding period.

Formula for the Present Value of a Due Annuity

$$PV = PMT\left[\frac{1 - (1 + i)^{-n}}{i}\right](1 + i)$$

where *PMT* is the cash flow amount, *n* is the number of payments, and *i* is the interest rate per period, $i = \dfrac{\text{Annual interest rate as a decimal}}{\text{Number of compounding periods per year}}$.

Example 6   **Calculate the Present Value of a Due Annuity**

The manufacturer of a cell phone leases the rights to the microphone technology from MicroByte Corporation. The lease agreement calls for a payment of $50,000 at the beginning of each month for 4 years. If the current interest rate is 5.25% compounded monthly, find the present value of the payments.

Solution

This spreadsheet image is from Excel, using the present value (*PV*) function.

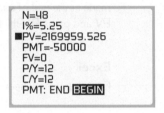

	A	B	C
1	Annual interest rate	5%	
2	Compounding periods per year	12	
3	Number of years	4	
4	Payment	-$50,000.00	
5	Future value	$0.00	
6	Type	1	
7	Payment	$21,69,959.53	

B7 | fx =PV(B1/B2,B2*B3,B4,B5,B6)

TI-84 Calculator

```
N=48
I%=5.25
■PV=2169959.526
PMT=-50000
FV=0
P/Y=12
C/Y=12
PMT: END BEGIN
```

The present value of the payments is $2,169,959.53.

The present value of an annuity can be used in various ways. Probably the one most often encountered is a state lottery. The winner of the lottery is offered two options: a lump sum payment or annual payments of a certain amount over a number of years, typically 30 years. To determine which is the better option, calculate the present value of the payments. The lottery is not the only instance of this calculation. Some insurance settlements are offered as a lump sum payment or monthly payments for several years. Calculating the present value of the payments can help decide which option is best.

Example 7 Calculate the Present Value of a Series of Lottery Payments

The Powerball® lottery offers the winner two options: a lump sum payment of $183 million or $310 million paid in 30 equal payments, with the first payment on the day of winning the lottery and an additional 29 annual payments. If the winner expects the average annual interest rate over the term of the payments to be 3.5%, which is the better option for the winner? Round calculations to the nearest dollar.

Solution

To select the better option, determine the present value of the series of payments. Because payments are made at the beginning of each year, calculate the present value of a due annuity. The calculation using a TI-84 calculator and the Finance app is shown on the right.

```
N=30
I%=3.5
■PV=196702919.3
PMT=-10333333
FV=0
P/Y=1
C/Y=1
PMT: END BEGIN
```

The payment is $\dfrac{\$310 \text{ million}}{30} \approx \$10,333,333$, and the present value is approximately $196,702,919.

Because the present value of the payments is greater than the lump sum offer, the series of payments is the better option.

Take Note: The selection of the payment option assumes an average interest rate of 3.5%. If the average interest rate is greater, the lump sum payment may be a better option.

Monthly Payment for an Installment Loan

The payment calculation for many consumer loans uses the formula for the present value of an ordinary annuity. In this case, the present value is the loan amount. These installment loans are based on the annual percentage rate of the loan.

Example 8   **Calculate a Monthly Payment**

A computer gamer purchases a state-of-the-art gaming computer for $5250 and makes a down payment of 25%. The remainder is financed for 2 years at an annual interest rate of 6.25% compounded monthly. Find the monthly payment for this loan.

Solution

Find the amount financed by subtracting the down payment from the purchase price.

Down payment = 5250.00(0.25) = 1312.50

Amount financed = 5250.00 − 1312.50 = 3937.50

Here is the calculation using a TI-84 calculator.

TI-84 Calculator

Press **APPS** If necessary, scroll to Finance. Press **ENTER ENTER**. Then input the values for N (24 months), annual interest rate (6.25%), amount financed (the present value, 3937.50), the future value (0, no money is owed at the end of the loan), P/Y and C/Y = 12 (monthly payments).

```
N=24
I%=6.25
PV=3937.5
■PMT=-174.95629…
FV=0
P/Y=12
C/Y=12
PMT: END BEGIN
```

The monthly payment is $174.96.

Example 9   **Calculate the Amount of Interest Paid for a Consumer Loan**

For Example 8, calculate the amount of interest paid over the 2-year term of the loan.

Solution

The monthly payment is $174.96 for 2 years or 24 months.

Total amount of monthly payments = 174.96 · 24 = 4199.04

The amount of interest is the difference between the total amount of monthly payments and the amount borrowed.

Amount of interest = 4199.04 − 3937.50 = 261.54

The interest on the loan was $261.54.

Determine a Loan Payoff

Sometimes a consumer wants to pay off a loan before the end of the loan term. For instance, suppose you have a 5-year car loan but would like to purchase a new car after owning your car for 4 years. Because there is still 1 year remaining on the loan, you must pay off the remaining loan amount before purchasing another car.

To calculate a loan payoff for an APR installment loan, use the formula for the present value of an ordinary annuity. The value of n is the number of *remaining* payments to be made.

| Example 10 | **Calculate a Loan Payoff Amount** |

Jordan wants to pay off the loan on a jet ski for which 18 payments have been made. If the monthly payment is $284.67 on a 2-year loan at an annual percentage rate of 8.7%, what is the payoff amount?

Solution

Use the present value of an ordinary annuity formula. Because Jordan has made 18 payments and her loan is 2 years (24 months), she has six payments *remaining*. Therefore, $n = 6$. We will show the calculation using both Excel and a TI-84 calculator.

Excel

Use the *PV* function.

	A	B	C
	B7	fx =PV(B1/B2,B2*B3,B4,B5,B6)	
1	Annual interest rate	8.7%	
2	Compounding periods per year	12	
3	Number of years	0.5	
4	Payment	-$284.67	
5	Future value	$0.00	
6	Type	0	
7	Payment	$1,665.50	

TI-84 Calculator

```
N=6
I%=8.7
■PV=-1665.503446
PMT=284.67
FV=0
P/Y=12
C/Y=12
PMT: END BEGIN
```

The loan payoff is $1665.50.

4.3 Exercise Set

Think About It

1. What is the difference between an ordinary annuity and a due annuity?

2. Give a real-world example of an ordinary annuity and a due annuity.

3. If the interest rate, payment amount, and number of compounding periods per year of an ordinary annuity and due annuity were the same, which would have the greater value after 10 years?

Preliminary Exercises

For Exercises P1 to P10, if necessary, use the referenced example following the exercise for assistance. Assume all years have 365 days.

P1. Margo starts an individual retirement account (IRA) by depositing $450 at the beginning of each month into an account that earns 3.5% interest compounded monthly. If Margo continues this plan for 20 years, what will be the value of the account? **Example 1**

P2. A self-employed Internet security specialist deposits $3000 at the end of every month into a simplified employee pension (SEP) account that earns 4.2% interest compounded monthly. What is the value of the SEP in 25 years? **Example 2**

P3. The manager of a technology store wants to save $20,000 in 3 years to make a down payment on a condominium. How much does the manager need to invest at the beginning of each week into an account that earns 6.25% interest compounded weekly? **Example 3**

P4. A state establishes a sinking fund of $300 million for bonds that will mature in 25 years. How much money must be deposited at the end of each month into an account that earns an annual interest rate of 4% compounded monthly to retire the $300-million bond debt? **Example 4**

P5. The winner of a lottery chooses to receive annual payments of $150,000 at the end of each year for 25 years. If the current interest rate is 5.2%, find the present value of the payments. **Example 5**

P6. A provider of cell phone service leases the rights to use 25 cell towers and makes payments of $100,000 at the beginning of each month for 5 years at an interest rate of 3.75% compounded monthly. Find the present value of the payments. **Example 6**

P7. A logo design company purchases four new computers for $16,500. The company finances the cost of the computers for 3 years at an annual interest rate of 5.125% compounded monthly. Find the monthly payment for this loan. **Example 7**

P8. Tomasz purchased a new heating and air-conditioning system for a home and financed $6100 at an annual interest rate of 3.9% compounded monthly for 3 years. What are Tomasz's monthly payments? **Example 8**

P9. For the loan in P8, how much interest will Tomasz pay over the term of the loan? **Example 9**

P10. Aidan wants to pay off the loan on his refrigerated trailer on which he has made 15 payments. If Aidan's monthly payment is $367.79 on a 3-year loan at an annual percentage rate of 6.5%, what is the payoff amount? **Example 10**

Section Exercises

Assume all years have 365 days.

1. **Compensation Annuity** A professional soccer player has a deferred compensation annuity that pays $2500 at the end of each month for 8 years. If the annual interest rate is 5% compounded monthly, find the present value of the annuity.

2. **Lottery Prize** Paxton won a lottery prize that pays $3000 at the end of each month for 20 years. If the annual interest rate is 4.5% compounded monthly, what is the present value of the annuity?

3. **Purchasing a Barbecue** Chiara purchased a new natural gas barbecue for $2250 and made a down payment that was 30% of the purchase price. She financed the remaining balance for 9 months at an annual interest rate of 5.5%.
   a. What is Chiara's monthly payment?
   b. How much interest will Chiara pay over the term of the loan?

4. **Purchasing a Dining Set** Lear purchased a dining table and chairs for his townhouse at a cost of $4500 plus sales tax of 7.5% of the purchase price and a delivery fee of $100. Lear makes a down payment of $2000 and finances the remainder at an annual interest rate of 7.125% compounded monthly for 2 years.
   a. What are Lear's monthly payments?
   b. How much interest will Lear pay over the term of the loan?

5. **Financing Rain Gutters** Logan had seamless rain gutters installed around a home at a cost of $1500. Logan financed the total amount for 18 months at an annual interest rate of 3.5% compounded monthly. After making 12 payments, Logan decided to repay the loan in full. What is Logan's payoff?

6. **Financing a Wedding** A captain for a regional airline took out a 4-year loan of $7500 for a wedding at an annual interest rate of 5% compounded monthly. After making payments for 28 months, the captain decided to repay the loan in full. What is the captain's payoff?

7. **Retirement Savings** A newspaper editor starts a retirement savings plan in which $250 per month is deposited at the beginning of each month into an account that earns an annual interest rate of 6.4% compounded monthly. Find the value of this investment after 20 years.

8. **Savings Plans** The assistant to the president of an executive search firm starts a saving plan by depositing $75 at the beginning of each week into an account that earns an annual interest rate of 4.75% compounded weekly. Find the value of this investment after 5 years.

9. **Roth IRAs** The timpanist for an orchestra deposits $450 at the beginning of each month in a Roth IRA. The IRA has an annual interest rate of 5.5% compounded monthly. Find the value of the Roth IRA after 12 years.

10. **529 Savings Plans** On June 1, Regina decides to begin a 529 college savings plan for an 8-year-old child. Regina plans on making deposits of $300 at the beginning of each month for 10 years. If the plan pays an interest rate of 5.125% compounded monthly, what will be the value of the account after 10 years?

11. **Purchasing a Cement Mixer** The owner of a construction company purchased a cement mixer truck for $170,000 and made a down payment of $50,000. The remaining balance was financed for 10 years at an annual interest rate of 7.25% compounded monthly. Find the monthly payment.

12. **Financing a Gazebo** Jamie decides to have a gazebo built in the backyard of a home at a cost of $8500. If Jamie finances the total cost for 5 years at an annual interest rate of 6.875%, what are the monthly payments?

13. **Leasing a Warehouse** The maker of cardboard boxes leases a warehouse and pays $5000 at the beginning of each month for 5 years. If interest rates are 4.75% compounded monthly, what is the present value of the payments?

14. **Leasing a Car** A sales representative leases a car and makes payments of $450 at the beginning of each month for 3 years. If interest rates are 3% compounded monthly, what is the present value of the payments?

15. **Financing a New Laboratory** A biotechnology start-up issues $10 million in bonds to build a new laboratory. The company establishes a sinking fund so it will have the money to repay that debt. How much money must be deposited at the end of every 6 months into an account that earns an annual interest rate of 6% compounded semiannually to retire the $10 million bond debt in 15 years?

16. **Financing Bridge Repairs** A state agency issues $500 million in bonds to repair bridges in the state. The state establishes a sinking fund by depositing $3 million at the end of each quarter into an account that earns 4.5% compounded quarterly for 20 years to repay the bonds. Will the state have enough money to repay the debt?

17. **Refurbishing a Studio** A professional photographer wants to have $15,000 in 3 years to refurbish a studio. How much money must the photographer deposit at the beginning of each month into an account that earns 5% interest compounded monthly to reach the goal?

18. **Financing a Ski Lift** The owner of a ski resort estimates that a new 4-person ski lift will cost $5,000,000. The owner begins making payments on March 1 and every 3 months thereafter for 5 years into an account that earns 4% interest compounded quarterly. What payment amount must the owner make to have the cost of the ski lift in 5 years?

Investigation

Understanding Annuities The annuity formulas in this section require that the compounding period and the payment period are equal. For instance, if you have a savings plan in which you deposit $100 at the beginning of each month into an account that earns 6%, then the formula for the future value gives the value of your payments, assuming that interest is compounded monthly, and the compounding period and the payment period (monthly) are the same. However, the bank into which you are depositing the money is probably compounding your money daily. Now the compounding period (daily) and the payment period (monthly) are not the same. To calculate the future value (or present value), we must determine an interest rate, i^*, that corresponds to the payment period. The formula for this interest rate is

$$i^* = \left(1 + \frac{i}{k}\right)^{k/m} - 1$$

where i is the annual interest rate, k is the number of compounding periods per year, and m is the number of payments per year. The formula for the future value of a due annuity in which payment intervals and compounding intervals are different is

$$FV = PMT\left[\frac{(1 + i^*)^n - 1}{i^*}\right](1 + i^*)$$

1. A flight operations manager for an airline starts a retirement savings plan. At the beginning of each month, $100 per month

is deposited into an account that earns an annual interest rate of 5% compounded daily. Find the value of this investment after 25 years.

2. What is the difference between the answer to part 1 and the answer to Example 1?

3. A technician for a computer chip company sets up a 401(k) program and invests $25 at the end of each week into an account that earns an annual interest rate of 6% compounded daily. Find the value of the 401(k) after 10 years.

4. What is the difference between the answer to part 3 and the answer to Example 2?

Section 4.4   Credit Cards

Learning Objectives:

- Find the average daily balance on a credit card.
- Calculate the finance charge for a credit card bill.
- Calculate the payoff amount for credit card debt.

Average Daily Balance

When a customer uses a credit card to make a purchase, the customer is actually receiving a loan, so there is frequently an added cost to the consumer who purchases on credit. This added cost may be in the form of an annual fee or interest charges on purchases. A **finance charge** is an amount paid in excess of the cash price; it is the cost to the customer for the use of credit.

Most credit card companies issue monthly bills. The due date on the bill is usually 1 month after the billing date (the date the bill is prepared and sent to the customer). If the bill is paid in full by the due date, the customer pays no finance charge. If the bill is not paid in full by the due date, a finance charge is added to the next bill.

Suppose a credit card billing date is the 10th day of each month. If a credit card purchase is made on April 15, then May 10 is the billing date (the 10th day of the month following April). The due date is June 10 (1 month from the billing date). If the bill is paid in full before June 10, no finance charge is added. However, if the bill is not paid in full, interest charges on the outstanding balance will start to accrue (be added) on June 10, and any purchase made after June 10 will immediately start accruing interest.

The most common method of determining finance charges is the *average daily balance* method. The **average daily balance** of credit card charges and payments is calculated by dividing the sum of the total amounts owed each day of the billing cycle by the number of days in the billing cycle.

> ### Average Daily Balance
>
> $$\text{Average daily balance} = \frac{\text{Sum of the total amounts owed for each day of the month}}{\text{Number of days in the billing cycle}}$$

Example 1   Calculate an Average Daily Balance

An unpaid bill for $620 had a due date of March 10. A charge of $214 was made on March 15, and $67 was charged on March 30. A payment of $200 was made on March 22. Find the average daily balance for the April 10 bill.

Solution

It is convenient to keep the results in a table. You may need a calendar at hand as well. The first purchase was made March 15. That is 5 days after the due date of March 10. A 5 is entered in Column 4. The next transaction is March 22, 7 days after March 15. Enter a 7 in Column 4. The charge on March 30 remains until April 10, the next due date. That is 11 days after March 30.

The calculation of the average daily balance begins by determining the amount owed each day of the month. For instance, consider March 15, the first purchase in the new billing cycle. On March 15, $620 has been owed for 5 days and $214 is owed for 1 day. The total for those 5 days is $5 \cdot \$620 + \$214 = \$3314$.

Repeat this calculation for each day on which a payment or purchase was made. For the last purchase in the billing cycle on March 30, there were 11 days remaining in the billing cycle and the amount owed for those 11 days was $701. The total for those days is $11 \cdot \$710 = \$7711$.

Date	Payments or Purchases ($)	Balances Each Day ($)	Number of Days Until Balance Changes	Unpaid Balance Times Number of Days ($)
March 10		620		
March 15	214	620 + 214 = 834	5	5 · 620 + 214 = 3314
March 22	−200	834 − 200 = 634	7	7 · 834 − 200 = 5638
March 30	67	634 + 67 = 701	8	8 · 634 + 67 = 5139
			11	11 · 701 = 7711
			Total	21,802

Now add the totals owed each day. $\$3314 + \$5638 + \$5139 + \$7711 = \$21,802$. The average daily balance is that sum divided by the number of days in the billing cycle.

$$\text{Average daily balance} = \frac{\text{Sum of total amounts owed each day of the month}}{\text{Number of days in billing cycle}}$$

$$= \frac{21,802}{31}$$

$$\approx 703.29$$

The average daily balance was $703.29.

Finance Charges

The finance charge on a credit card bill is calculated in various ways. Most credit card companies use the average daily balance, the number of days in the billing cycle, and the daily interest rate.

Example 2 **Calculate a Finance Charge Based on Average Daily Balance**

An unpaid credit card balance for $711 had a due date of January 8. Purchases of $175 were made on January 15, $108 was charged on January 17, and $72 was charged on February 2. A payment of $350 was made on January 25. Find the finance charge due on February 8 if the annual interest rate is 19.6%.

Solution

Prepare a table similar to Example 1 and calculate the average daily balance.

Date	Payments or Purchases ($)	Balances Each Day ($)	Number of Days Until Balance Changes	Unpaid Balance ($) Times Number of Days
1/8		711		
1/15	175	886	7	7 · 711 + 175 = 5152
1/17	108	994	2	2 · 886 + 108 = 1880
1/25	−350	644	8	8 · 994 − 350 = 7602
2/2	72	716	8	8 · 644 + 72 = 5224
			6	6 · 716 = 4296
			Total ($)	24,154

$$\text{Average daily balance} = \frac{\text{Sum of the total amounts owed for each day of the month}}{\text{Number of days in the billing cycle}}$$

$$= \frac{24,154}{31}$$

• There are 31 days from January 9 through February 8.

$$\approx 779.16$$

The monthly finance charge is annual interest rate divided by 365 (number of days in a year) times the number of days in the billing cycle times the average daily balance.

$$\text{Finance charge} = \text{Average daily balance} \cdot \frac{\text{Annual interest rate}}{365} \cdot \text{Number of days in billing cycle}$$

$$= 779.16 \cdot \frac{0.196}{365} \cdot 31$$

$$\approx 12.97$$

The finance charge was $12.97.

Calculating a finance charge manually as we have already done could be laborious if there were many transactions each month. A spreadsheet is helpful for these situations. For help calculating a credit card finance charge, you can access an Excel template from your digital resources.

Example 3 **Using a Spreadsheet to Calculate a Credit Card Finance Charge**

A credit card had an unpaid balance of $858.62 on July 1. The next due date was August 1. The following table shows purchases and payments made during that time. Calculate the finance charge using the average daily balance and an annual interest rate of 20.8%.

Date	Purchase or Payment Amount ($)	Date	Purchase or Payment Amount ($)
7/2	118.10	7/16	77.58
7/3	82.29	7/16	33.25
7/4	74.14	7/17	119.50
7/5	95.93	7/18	87.37
7/5	73.34	7/18	59.49
7/7	101.56	7/18	42.17
7/8	106.77	7/19	38.32
7/12	89.27	7/22	113.89
7/12	48.26	7/27	102.41
7/14	−750.00	7/27	51.98
7/16	100.73	7/28	36.34

Solution

Enter the dates and purchases or payments into the Excel. For help calculating an average daily balance, you can access an Excel template from your digital resources. Your result should look similar to the following.

Date payment is due	7/1/18	
Date of next payment	8/1/18	• Enter the payment due date
Days in billing cycle	31	• Enter the next payment due date
Annual Interest Rate	20.80%	• Enter the annual interest rate
Beginning Balance	858.62	• Enter the beginning balance
		• Enter the date of a transaction and its amount
		• Do not enter values in any shaded cell.

Date of Next Transaction	Purchase or Payment ($)	Balance ($)	Number of Days Until Balance Changes	Unpaid Balance Times Number of Days ($)
7/1/18		858.62		
7/2/18	118.10	976.72	1	976.72
7/3/18	82.29	1059.01	1	1059.01
7/4/18	74.14	1133.15	1	1133.15

(continued)

7/5/18	95.93	1229.08	1	1229.08
7/5/18	73.34	1302.42	1	73.34
7/7/18	101.56	1403.98	2	2706.40
7/8/18	106.77	1510.75	1	1510.75
7/12/18	89.27	1600.02	4	6132.27
7/12/18	48.26	1648.28	1	48.26
7/14/18	−750.00	898.28	2	2546.56
7/16/18	100.73	999.01	2	1897.29
7/16/18	77.58	1076.59	1	77.58
7/16/18	33.25	1109.84	1	33.25
7/17/18	119.50	1229.34	1	1229.34
7/18/18	87.37	1316.71	1	1316.71
7/18/18	59.49	1376.20	1	59.49
7/18/18	42.17	1418.37	1	42.17
7/19/18	38.32	1456.69	1	1456.69
7/22/18	113.89	1570.58	3	4483.96
7/27/18	102.41	1672.99	5	7955.31
7/27/18	51.98	1724.97	1	51.98
7/28/18	36.34	1761.31	1	1761.31
			4	7045.24
			Total	$44,825.86
			Average daily balance	$1,446.00
			Finance charge	$25.54

The finance charge was $25.54.

If a credit card bill is not paid in full when due, a minimum monthly payment is due. The minimum payment depends on the credit card issuer. A typical minimum is stated as the greater of 4% of the amount owed or $25.

Example 4 **Calculate the Minimum Due on a Credit Card Debt**

A lathe operator owes $638.45 on a credit card bill. If the minimum monthly payment is the greater of 4% of the amount owed or $25, what is the minimum payment the lathe operator can make?

Solution

Find 4% of the amount owed.

$$0.04 \cdot 638.45 = 25.538 \approx 25.54$$

The greater of $25.54 and $25 is $25.54.

The lathe operator's minimum payment is $25.54.

Pay Off Credit Card Debt

If someone wants to pay off a credit card debt by making equal monthly payments, then the time to repay the debt is found by solving the formula for the present value of an ordinary annuity for n. That formula is repeated here for convenience.

> ### Formula for the Present Value of an Ordinary Annuity
>
> $$PV = PMT\left[\frac{1 - (1 + i)^{-n}}{i}\right]$$
>
> where PMT is the cash flow amount, n is the number of payments, and i is the
>
> interest rate per period, $i = \dfrac{\text{Annual interest rate as a decimal}}{\text{Number of compounding periods per year}}$.

Example 5 **Calculate the Time to Repay Credit Card Debt**

A physical therapist accrued $5000 in credit card debt. If the therapist makes a monthly payment of $200 (and makes no additional charges on the account), how many months will it take to repay the debt if the annual interest rate on the credit card is 18.6%? Round up to the next month.

Solution

This problem requires financial software to solve. The calculation using Excel and the **NPER** function follows. The format is **NPER(rate, pmt, pv, [fv], [type])**. This is an ordinary annuity (payments at the end of the month), so enter a 0 in cell B6.

	B7	fx =NPER(B1/B2,B3,B4,B5,B6)	
	A	B	C
1	Annual interest rate	18.6%	
2	Compounding periods per year	12	
3	Payment	-$200.00	
4	Present value	$5,000.00	
5	Future value	$0.00	
6	Type	0	
7	Time to repay	31.87069011	

It will take 32 months to repay the credit card debt.

The solution to Example 5 is based on making the same monthly payment each month. This payment was based on 4% of the *original* debt amount, not 4% of the current debt amount.

This is important because when the therapist makes the first payment of $200, some of that payment goes to interest and some goes to reducing the debt.

$$I = Prt$$

$$\text{Amount to interest} = 5000\left(\frac{0.186}{12}\right)(1) = 77.50 \quad \bullet \; P = 5000, r = \frac{0.186}{12}, t = 1 \text{ (1 month)}$$

$$\text{Amount to principal} = 200.00 - 77.50 = 122.50.$$

The remaining credit card debt is now $5000.00 − 122.50 = $4877.50. If the therapist makes the minimum payment of 4% of the new balance, that payment is $195.10. This will make the repayment time much longer. Determining how much longer is one of the projects at the end of this section.

All of what we have discussed in this section gives you the power to understand credit card debt. Just one last comment: credit card debt is dangerous. Carefully monitor the use of credit cards.

4.4 Exercise Set

Think About It

1. If a credit card company charges an annual interest rate of 18%, what is the monthly interest rate?

2. Suppose you have a credit card balance of $200 on April 30, the last day of the billing cycle. If you have a $10 finance charge, what is your credit card balance on May 1, assuming you do not make any additional charges?

Preliminary Exercises

For Exercises P1 to P5, if necessary, use the referenced example following the exercise for assistance.

P1. A credit card had a balance of $620 due on July 1. Purchases of $315 were made on July 7, and $410 was charged on July 22. A payment of $400 was made on July 15. Find the average daily balance for the August 1 bill. **Example 1**

P2. An unpaid credit card bill for $652.95 had a due date of May 8. Purchases of $325 were made on May 15, $114.78 on May 17, and $69.67 was charged on June 2. A payment of $350 was made on May 25. Find the finance charge due on June 8 if the interest rate on the average daily balance is 1.8%. **Example 2**

P3. A credit card had a balance of $915.22 due on September 3. The next due date was October 3. The following table shows purchases and payments made during that time. Calculate the finance charge based on the average daily balance and an annual interest rate of 20.8%. **Example 3**

Date	Purchase or Payment	Date	Purchase or Payment
9/4	162.35	9/23	48.77
9/7	174.80	9/24	137.19
9/9	130.93	9/25	155.49
9/11	154.21	9/26	86.22
9/12	149.97	9/27	183.97
9/14	−425.00	9/28	70.77
9/17	156.97	9/29	70.52
9/19	170.52	9/30	40.50
9/20	28.15	10/1	53.15
9/21	16.54	10/2	163.66

P4. A credit card company calculates its minimum monthly payment as 3% of the credit card balance or $30, whichever is greater. If Emile Fritz has a credit balance of $365.48, what is his minimum monthly payment? **Example 4**

P5. Michela Navarro has $3856.71 in credit card debt. If Michela makes a monthly payment of $175 (and makes no additional charges on the account), how many months will it take to repay the debt if the annual interest rate on the credit card is 22.1%? Round up to the next month. **Example 5**

Section Exercises

1. **Minimum Payments** A machine shop has a credit card that offers rebates on purchases. At the end of May, the company had a credit card bill of $18,456.32. If the company must make a minimum payment that is the greater of $500 or 4.5% of the credit card bill, what is the minimum monthly payment for May?

2. **Minimum Payments** A printing company has a business credit card it uses for various purchases. At the end of June, the company had $11,209.16 in credit card debt. If the company must make a minimum payment that is the greater of $250 or 5% of the credit card debt, what is the minimum monthly payment for June?

3. **Average Daily Balance** A credit card had an unpaid balance of $653 on April 14. Purchases of $155.56 were made on April 19; $22.45 was charged on April 21, and $201.36 was charged on April 28. A payment of $500 was made on April 25. Find the average daily balance if the next due date is May 14.

4. **Average Daily Balance** A credit card had an unpaid balance of $1085.23 on October 1. Purchases of $42.57 were made on October 5, $130.25 on October 8, and $59.33 on October 28. A payment of $750 was made on October 25. Find the average daily balance if the next due date is November 1.

5. **Finance Charges** A credit card had an unpaid balance of $562 on September 14. Purchases of $286 were made on September 19, and $15 was charged on September 28. A payment of $350 was made on September 25. The annual interest on the average daily balance is 20.5%. Find the finance charge due on the October 14 bill.

6. **Finance Charges** A credit card had an unpaid balance of $874.25 on February 10. The following purchases were made: $187.69 on February 15, $404.56 on February 16, $12.90 on February 18, and $63.21 on February 25. A payment of $375 was made on March 2. The annual interest on the average daily balance is 18.8%. Find the finance charge on the March 10 bill. Assume it is not a leap year.

7. **Length of Debt Repayment** A landscape architect accrued $7841.36 in credit card debt. If the architect makes a monthly payment of $500 (and makes no additional charges on the account), how many months will it take to repay the debt if the annual interest rate on the credit card is 19.6%? Round up to the next month.

8. **Length of Debt Repayment** A career counselor decides to make monthly payments of $150 on credit card debt of $3489.64 and discontinue using that credit card. Assuming the annual interest rate is 22.2%, it will take the counselor approximately 31 months to repay the debt. How many fewer months would it take to repay the debt if the counselor makes monthly payments of $200? Round to the next month.

9. **Finance Charges** A credit card had an unpaid balance of $856.25 on July 15. The next due date was August 15. The following table shows purchases and payments made during that time. Calculate the finance charge based on the average daily balance and an annual interest rate of 20.5%. For help with this calculation, you can access an Excel template from your digital resources.

Date	Purchase or Payment ($)	Date	Purchase or Payment ($)
7/19	115.59	8/6	126.99
7/24	29.19	8/7	59.95
7/25	110.56	8/7	107.55
7/26	36.07	8/8	141.30
7/26	53.59	8/8	51.91
7/27	86.46	8/9	142.09
7/29	−450.00	8/10	48.14
7/30	29.12	8/11	85.65
8/4	73.01	8/13	144.49
8/5	21.62	8/14	105.58

10. **Finance Charges** A credit card had an unpaid balance of $721.04 on April 2. The next due date was May 2. The following table shows purchases and payments made during that time. Calculate the finance charge based on the average daily balance and an annual interest rate of 20.6%. For help with this calculation, you can access an Excel template from your digital resources.

Date	Purchase or Payment ($)	Date	Purchase or Payment ($)
April 6	199.58	April 17	−400.00
April 7	22.74	April 20	188.43
April 8	27.22	April 20	202.49
April 8	84.67	April 20	192.09
April 8	186.03	April 22	189.87
April 10	88.38	April 22	198.17
April 11	77.74	April 23	102.37
April 11	163.25	April 24	108.02
April 15	195.52	April 24	163.47
April 15	46.05	April 29	116.11

11. **Debt Repayment** Jason has $3540.85 in credit card debt. The annual interest rate on the unpaid balance is 19.6% compounded monthly. If Jason wants to pay off this credit card debt in 2 years and makes no additional purchases, what are the monthly payments? [Suggestion: use the formula for the present value of an ordinary annuity.]

12. **Debt Repayment** Aretha has $2634.28 in credit card debt. The annual interest rate on the unpaid balance is 21.5% compounded monthly. If Aretha wants to pay off this credit card debt in 3 years and makes no additional purchases, what are the monthly payments? [Suggestion: use the formula for the present value of an ordinary annuity.]

13. **Transferring Credit Card Debt** You will need a calculator such as a TI-84 or a spreadsheet program for this exercise. For help with this calculation, you can access an Excel template from your digital resources. Some websites offering financial information suggest that people with large credit card debts transfer their balances to a new credit card that offers an introductory rate of 0%. Suppose Jon Richards has $8000 in credit card debt on the A&L credit card for which the annual interest rate is 21.3%. Jon decides to transfer his debt to the School Credit Union credit card that offers 0% interest for 15 months.

   a. If Jon makes a payment of $400 per month on the A&L credit card, what will the balance be at the end of 15 months? Assume that no additional charges are made to the card. [Suggestion: You are trying to find the future value of a debt that has a present value of $8000. Use the formula for the future value of an ordinary annuity. Because the $8000 is owed, enter it as a negative number.]

   b. If Jon makes a payment of $400 per month on the School Credit Union credit card, what will the balance be at the end of 15 months? Assume that no additional charges are made to the card. [Suggestion: none of the payment goes to interest on the debt.]

   c. How much lower is the balance due after 15 months on the School Credit Union Card? Note: some credit card companies may charge a transfer fee that will affect the benefit of a 0% introductory rate.

14. **Transferring Credit Card Debt** You will need a calculator such as a TI-84 or a spreadsheet program for this exercise. For help with this calculation, you can access an Excel template from your digital resources. Suppose Olivia has $6480 in credit card debt on the VB credit card, which has an annual interest rate of 18.2%. Olivia decides to transfer her debt to the LA credit card that offers 0% interest for 18 months.

   a. If Olivia makes a payment of $360 per month on the VB credit card, what will the balance be at the end of 18 months? Assume that no additional charges are made to the card. [Suggestion: You are trying to find the future value of a debt that has a present value of $6480. Use the formula for the future value of an ordinary annuity. Because the $6480 is owed, enter it as a negative number.]

   b. If Olivia makes a payment of $360 per month on the LA credit card, what will the balance be at the end of 18 months? Assume that no additional charges are made to the card. [Suggestion: none of the payment goes to interest on the debt.]

   c. How much lower is the balance due after 18 months on the LA credit card? Note: some credit card companies may charge a transfer fee that will affect the benefit of a 0% introductory rate.

15. **Deferring Credit Card Payments** Richard has a credit card that allows deferred credit card payments for 1 year if Richard becomes unemployed. The interest on the card debt continues to accrue, but there are no late payment penalties. Suppose Richard has $1576.57 in credit card debt and the annual interest rate is 21.5% compounded monthly. How much will be owed after 1 year of this option is used? Assume no other purchases are made with the card.

16. **Deferring Credit Card Payments** Yolanda's credit card allows deferred credit card payments for 1 year if the cardholder becomes unemployed. The interest on the card debt continues to accrue, but there are no late payment penalties. Suppose Yolanda has $2503.14 in credit card debt and the annual interest rate is 18% compounded monthly. How much is owed after 1 year if the deferred option is used? Assume no other purchases are made with the card.

17. **Minimum Payment** The minimum amount due on credit card debt varies by issuer. However, most credit card companies use the larger of $25 or 2% of the current balance.

   a. Suppose a credit card company used a minimum payment that is the larger of $25 or 2% of the current balance and your current balance was $2500 on March 1. Suppose the annual percentage rate on the credit card bill was 25% compounded monthly, you did not make additional purchases with this card, and you make the minimum monthly payment for March. What would your credit card balance be on April 1?

b. Explain why the practice in part a is against the law.

18. Credit Card Teaser Rates Some credit card companies offer a "teaser" rate, an initial interest rate that is exceptionally low for a short period of time (usually less than 1 year) and then adjusts to a much higher rate. For borrowers with good credit scores (above 670), the interest rate ranges from 0% to 5.5%. Suppose Janice has two credit cards, one with a balance of $1563.60 and the second with a balance of $1125.73. If she transfers both of her card balances to a credit card with an annual interest rate of 4.5% and wants to repay her entire debt in 1 year, what are her monthly payments?

▶ **Investigation**

Understanding Debt Repayment In Example 5, we calculated the time to repay a credit card debt by making the same monthly payment ($200) each month. However, because of the way minimum monthly payments are calculated, the credit card bill would show a different amount each month for the minimum monthly payment. As shown just after Example 5, the minimum monthly payment the second month would be $195.10. Paying $195.10 instead of $200 will extend the repayment period. Create a spreadsheet that will determine the number of months required to repay the debt in Example 5, if the monthly payment is the greater of 4% of the current debt or $25. For help with this calculation, you can access an Excel template from your digital resources.

Section 4.5 Student Loans

Learning Objectives:

- Estimate the cost of college.
- Calculate Stafford Student Loans.
- Calculate PLUS Student Loans.

Estimating the Cost of College

Many factors affect the cost of attending college. There are direct costs such as tuition, books, and supplies. Then there are indirect costs such as food, housing, transportation, insurance, and cell phones. To pay those expenses, you must have saved the money, pay for it by working during the school year, obtain some kind of loan, or a combination of these.

Here is a screenshot of a spreadsheet that can be used to estimate cost. For help with this calculation, you can access an Excel template from your digital resources. Because this is a public spreadsheet, DO NOT enter your personal information into this sheet. This is a guide for the type of spreadsheet you can create to monitor expenses.

	A	B	C	D	E	F	G	H	I	J	K
1	Income	22000		School expenses	9105		Living Expenses	13280		Balance	−385
2	Source	Amount		Source	Amount		Source	Amount			
3	Your income	15000		Tuition	7500		Rent	5000			
4	Student loans	4000		Books	500		Utilities	500			
5	Scholarships	0		Insurance	250		Telephone	1400			
6	Income from savings	1000		Student activity fee	325		Internet	480			
7	Other income	2000		Lab fees	380		Credit card	0			
8				Supplies	150		Food	2500			
9							Insurance	0			
10							Entertainment	1200			
11							Transportation	1500			
12							Personal Expenses	200			
13							Miscellaneous	500			

The balance in the spreadsheet shows −$385. This means that this student has insufficient income for school and living expenses as entered into the spreadsheet.

You can download this spreadsheet and save it as an Excel file. Once you have done that, you can open it with Excel or with most other spreadsheet programs such as Numbers (an Apple program), Google Sheets, or Calc from OpenOffice.org. You can then use and edit the spreadsheet for your personal situation.

Stafford Student Loans

Under income in the spreadsheet above, there is an entry for student loans. The student loan program that benefits many students today was part of Title IV of the Higher Education Act of 1965. In 1988, Congress renamed the student loan program after Senator Robert T. Stafford, who authored a number of bills in support of students seeking postsecondary education. Stafford Loans, as they are now called, are available to students through the U.S. Department of Education. Because the federal government guarantees the repayment of these loans to lenders, the interest rate on Stafford Loans is usually lower than the rate on private student loans.

There are basically two types of Stafford Loans: subsidized and unsubsidized. For a subsidized loan, the interest on the loan is paid by the government while the student is in school. For unsubsidized loans, the student is responsible for all the interest on the loan while in school. In 2011, Congress eliminated subsidized loans for students in any graduate school program.

When a student secures a loan, there is a fee—called a *loan origination fee* or a *discount fee*—that is deducted from the loan amount. Although the student receives less money than borrowed, the student must repay the entire borrowed amount. This affects the APR of the loan.

Example 1   Calculate a Loan Fee for a Student Loan

A student obtains a loan for $3500. How much money does the student actually receive assuming a loan fee of 1.069%?

Solution

$$\text{Loan fee} = 0.01069 \cdot 3500 = 37.415$$

The loan fee is 37.42.

$$\text{Amount the student receives} = 3500 - 37.42 = 3462.58$$

The student received $3462.58.

There are many different repayment plans for student loans. We will look at payment plans that are fixed for the entire term of the loan. With this assumption, the loan payment is calculated using the formula for the present value of an ordinary annuity. We repeat the formula here.

Formula for the Present Value of an Ordinary Annuity

$$PV = PMT\left[\frac{1 - (1 + i)^{-n}}{i}\right]$$

where PMT is the cash flow amount, n is the number of payments, and i is the interest rate per period, $i = \dfrac{\text{Annual interest rate as a decimal}}{\text{Number of compounding periods per year}}$.

Example 2   Calculate the Monthly Payment of a Subsidized Student Loan

Kalie received a 9-year subsidized student loan of $32,000 at an annual interest rate of 4.5%. Assuming that Kalie graduates in 3 years, determine the monthly payment on the loan.

Solution

Because Kalie has a subsidized loan, interest does not accrue on the loan during the time payments are not made. Therefore, the payment can be calculated using the formula for the present value of an ordinary annuity. We will use a TI-84 calculator. N = 9 · 12 = 108, I% = 4.5, PV = −32,000, skip PMT for the moment, FV = 0, and P/Y and C/Y are 12. Now move the cursor to PMT and press ALPHA ENTER.

```
N=108
I%=4.5
PV=-32000
■PMT=360.8829634
FV=0
P/Y=12
C/Y=12
PMT: END BEGIN
```

The monthly payment is $360.88.

For Example 2, the loan was a subsidized loan. If Kalie were a graduate student, the loan would be a nonsubsidized loan, and the monthly payment at graduation would be different, as shown in Example 3.

Example 3   **Calculate the Monthly Payment of a Nonsubsidized Student Loan**

Taylor received a 9-year nonsubsidized student loan of $32,000 at an annual interest rate of 4.5%. Determine Taylor's monthly payment on the loan when payments begin 3 years later.

Solution

Because Taylor has a nonsubsidized loan, interest must be paid on the loan during the period of time that payments are not being made. The interest is

$$I = Prt$$
$$= 32{,}000(0.045)3 = 4320$$

Amount owed $= 32{,}000 + 4320 = 36{,}320$

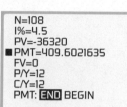

This changes the *PV* in the calculation of Example 2 to $36,320. See the calculation at the right using a TI-84 calculator.

The monthly payment is $409.60.

For the preceding two examples, the loan payment was calculated to repay the loan in the prescribed time, 9 years. It is possible to arrange *income-adjusted* repayment plans that are based on a student's ability to pay based on annual income after graduation. The entire amount borrowed must be repaid but at a lower monthly payment. This will extend the loan-repayment period.

Example 4   **Calculate the Number of Months to Repay an Income-Adjusted Student Loan**

Sydney received a 9-year subsidized student loan of $32,000 at an annual interest rate of 4.5%. If Sydney qualifies for an income-adjusted repayment plan of $250 per month, how long will it take to repay the loan?

Solution

In this case, the *PMT* is given as $250. Using that payment amount, determine the number, N, of months required to repay the loan. Enter the known values, then move the cursor to N. Press **ALPHA** and then **ENTER**.

```
■N=174.7071506
 I%=4.5
 PV=-32000
 PMT=250
 FV=0
 P/Y=12
 C/Y=12
 PMT: END BEGIN
```

It will take approximately 175 months to repay the loan.

Comparing the results from Example 2 and Example 4, the difference in repayment time, in months, is $175 - 108 = 67$. This means it takes about 5.6 years longer to repay the loan. With the increase in time comes an increase in the amount of interest paid.

For Example 2, the interest paid is

Interest paid = Number of payments · Payment amount − Loan amount

$$= 108 \cdot 360.88 - 32{,}000 = 6975.04$$

For Example 4, the interest paid is

Interest paid = Number of payments · Payment amount − Loan amount

$$= 175 \cdot 250 - 32{,}000 = 11{,}750$$

Using an income-adjusted payment, this student pays almost $5000 more in interest.

PLUS Loans

A PLUS Loan is a federal loan offered to parents of students enrolled at least half-time or to graduate or professional students to cover the cost for college. The original but now out-dated meaning of the acronym was Parent Loan for Undergraduate Students. Today, graduate students seeking advanced degrees or students attending professional schools (medical, dental, law, and others) can apply for the loan as well.

PLUS Loans differ from Stafford Loans in two substantial ways. The discount fee is higher than that of a Stafford Loan. Monthly loan payments usually begin within 6 months of the date the loan is disbursed. If the loan has a deferred payment plan, then interest accrues on the loan in the same way as for a nonsubsidized Stafford Loan.

> **Example 5** **Calculate a Payment for a PLUS Loan**

Linh received a 10-year PLUS Loan of $25,000 to attend the last year of medical school. If the current interest rate is 6.31% and Linh starts repayments 1 year after graduation, what are the monthly payments?

Solution

Because interest on a PLUS Loan accrues from the time it is disbursed, Linh owes interest on the $25,000 for 2 years (the last year of medical school plus the 1 year deferred payments after graduation).

```
N=120
I%=6.31
PV=28155
■PMT=-316.97922...
FV=0
P/Y=12
C/Y=12
PMT: END BEGIN
```

$$\text{Deferred interest} = Prt$$
$$= 25{,}000 \cdot 0.0631 \cdot 2 = 3155$$

The deferred interest is $3155. This amount is added to the $25,000 Linh borrowed to give $28,155. The monthly payment is based on this amount. Use the present value of an ordinary annuity formula. We will use the finance app from the TI-84 calculator.

Linh's monthly payment is $316.98.

As previously mentioned, a discount fee is applied to PLUS Loans. The amount as of this writing was 4.76%. For the $25,000 loan in Example 5, this means Linh receives 4.76% less than $25,000.

$$\text{Linh receives} = 25{,}000 - 25{,}000 \cdot 0.0476$$
$$= 25{,}000 - 1190$$
$$= 23{,}810$$

This affects the APR of the loan.

> **Example 6** **Calculate the APR on a Student Loan**

Calculate the APR on the loan in Example 5. Round to the nearest hundredth of a percent.

Solution

There are two steps to the calculation.

- Calculate the monthly payment based on the loan amount and the deferred interest. From Example 5, the monthly payment is $316.98.
- Replace the present value by the amount received, $23,810. Now calculate the interest rate (I%). You will need a calculator or a spreadsheet to do this. Here is the result using a TI-84 calculator.

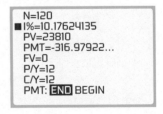

```
N=120
■I%=10.17624135
PV=23810
PMT=-316.97922...
FV=0
P/Y=12
C/Y=12
PMT: END BEGIN
```

The APR is 10.18%.

For more information on student loans, go to https://studentloans.gov.

4.5   Exercise Set

▶ Think About It

1. What is the difference between subsidized and nonsubsidized student loans?

2. Suppose a student secures a subsidized loan of $10,000 at an annual interest rate of 5%, 2 years before graduation. This same student also secures a nonsubsidized loan of $10,000 at an annual interest rate of 5%, 1 year before graduation. If the student begins repayment of both loans 3 years after graduating, for which loan will the student pay more interest?

3. What is the discount fee for a student loan?

▶ Preliminary Exercises

For Exercises P1 to P6, if necessary, use the referenced example following the exercise for assistance.

P1. Morgan obtains a loan for $5300. How much money does she actually receive assuming a loan fee of 0.95%? **Example 1**

P2. Peyton received a 10-year subsidized student loan of $27,000 at an annual interest rate of 4.1%. Assuming Peyton graduates in 4 years, determine Peyton's monthly payment on the loan. **Example 2**

P3. Mackenzie received a 4.75% nonsubsidized 15-year student loan for $22,000. Assuming she graduates in 4 years, determine Mackenzie's monthly payment on the loan. **Example 3**

P4. Ryan obtained a 9-year subsidized student loan of $28,000 at an annual interest rate of 4%. If Ryan qualifies for an income-adjusted repayment plan of $325 per month, how long will it take her to repay the loan? **Example 4**

P5. Theodora received an 8-year PLUS Loan of $15,000 to attend law school for her last 2 years. If the current interest rate is 5.68% and she starts her repayments 1 year after graduation, what are her monthly payments? **Example 5**

P6. Cheng obtains an 8-year nonsubsidized loan of $5000 at an annual interest rate of 3.9% and must pay a loan fee of 0.98%. What is the APR for Cheng's loan? **Example 6**

▶ Section Exercises

Solve.

1. **Student Loan Fees** A student obtains a loan for $7250. How much money does the student actually receive assuming a loan fee of 1.01%?

2. **Student Loan Fees** The loan fee for a student loan is 0.86%. If Shelley requests a loan for $6500, how much does Shelley actually receive?

3. **Student Loan Monthly Payments** Angelica received a 12-year nonsubsidized student loan of $18,000 at an annual interest rate of 5.3%. Determine her monthly payment on the loan after she graduates in 4 years.

4. **Student Loan Monthly Payments** Windsor received a 6-year nonsubsidized student loan of $15,000 at an annual interest rate of 3.75%. Determine his monthly payment on the loan after he graduates in 4 years.

5. **Student Loan Monthly Payments** Briana received a 10-year subsidized student loan of $21,000 at an annual interest rate of 4.125%. Determine her monthly payment on the loan after she graduates in 2 years.

6. **Student Loan Monthly Payments** Lois received a 9-year subsidized student loan of $32,000 at an annual interest rate of 4.875%. Determine her monthly payment on the loan after she graduates in 3 years.

7. **Length of Student Loan Repayment** Jefferson qualifies for an income-adjusted monthly payment of $425. If Jefferson has a subsidized student loan of $42,000 at an annual interest rate of 4%, how many months are required to repay the loan?

8. **Length of Student Loan Repayment** Victor received an 8-year subsidized student loan of $25,000 at an annual interest rate of 4.75%. If Victor qualifies for an income-adjusted repayment plan of $400 per month, how long will it take him to repay the loan?

9. **Student Loan Monthly Payments** Janice received a 10-year subsidized student loan of $5000 at the beginning of each of Janice's 4 years of college. The interest rate on each loan was 3.75%. What is the monthly payment for these loans after graduation?

10. **Student Loan Monthly Payments** Eunice received an 8-year subsidized student loan of $3500 at the beginning of each of Eunice's 4 years of college. The interest rate on each loan was 4.5%. What is the monthly payment for these loans after graduation?

11. **Student Loan Options** Katrina has the option of an 8-year nonsubsidized student loan of $28,000 at an annual interest rate of 3.5% or an 8-year subsidized loan of $28,000 at an annual interest rate of 4.5%. Determine for which loan Katrina will pay less interest over the term of the loan if repayment begins 2 years after obtaining the loan.

12. **Student Loan Options** Nicholas has the option of a 10-year nonsubsidized student loan of $25,000 at an annual interest rate of 3% or a 10-year subsidized loan of $25,000 at an annual interest rate of 4%. Determine for which loan Nicholas will pay less interest over the term of the loan if repayment begins 4 years after obtaining the loan.

13. **Student Loan Monthly Payments** A chiropractic student receives a 10-year PLUS Loan for $50,000 to complete the last 2 years of the program. If the interest rate of the loan is 5.78% and the student begins repaying the loan 2 years after graduation, what will the student's monthly payments be?

14. **Student Loan Monthly Payments** Jan borrows $12,000 through the PLUS program to complete the last year of architecture school. If the annual interest rate for an 8-year loan is 4.71% and Jan begins repayment of the loan 1 year after graduation, what is Jan's monthly payment?

15. **Student Loan Monthly Payments** Calculate the APR for the loan in Exercise 13.

16. **Student Loan Monthly Payments** What is the APR for Jan's loan in Exercise 14?

17. **Student Loan Consolidation** Keith has two student loans. One is for $25,000 at an APR of 6% for 12 years; the second is for $15,000 at an interest rate of 7% for 8 years. Keith is considering consolidating the loans and has found a bank that will loan him $40,000 for 10 years at an annual interest rate of 6.5%. If Keith is trying to pay off the loans and pay the least amount of interest, which consolidation option should be selected? Assume that interest is compounded monthly and payments are made on a monthly basis. Suggestion: Find the monthly payment for each loan. Based on this, determine the amount of interest paid for each loan. Then answer the question.

18. Student Loan Consolidation Davadene has two student loans. One is for $8000 at an APR of 5% for 10 years; the second is for $15,000 at an interest rate of 6% for 12 years. Davadene is considering consolidating the loans and has found a bank that will loan her $23,000 for 8 years at an annual interest rate of 5.5%. If Davadene is trying to pay off the loans and pay the least amount of interest, which consolidation option should be selected? Assume that interest is compounded monthly and payments are made on a monthly basis. Suggestion: Find the monthly payment for each loan. Based on this, determine the amount of interest paid for each loan. Then answer the question.

Investigation

Understanding Student Loan Repayment Nonsubsidized Stafford and PLUS Loans accrue interest from the time they are disbursed. When loan repayment begins, the accrued interest is added to the loan amount to determine the monthly payment. This results in *paying interest on interest*, a practice generally frowned on by financial advisers. One way to escape this is to pay the interest on the loan while in school. For the purposes of this problem, suppose Erin receives an 8-year, nonsubsidized student loan for $10,000 at an annual interest rate of 5.6% compounded monthly. Erin begins repayment of the loan 3 years after the loan is disbursed.

1. If Erin pays the interest on the loan each month for 3 years, how much interest will Erin pay over 3 years?

2. Because Erin has paid the interest on the loan, the monthly payment is based on the amount borrowed, $10,000. Calculate Erin's monthly payment.

3. How much interest will Erin pay on the loan? Include the interest in part 1 and the interest owed from 8 years of monthly payments.

4. If Erin does not make monthly interest payments, what is the monthly payment? [Suggestion: Add the interest owed during the time Erin is not making payments to $10,000. Then calculate the monthly payment on that amount.]

5. How much interest will be paid during the 8-year term of the loan if Erin does not make the initial monthly interest payments?

6. How much does Erin save over the term of the loan by making monthly interest payments?

Financial Considerations of Car Ownership

Learning Objectives:

- Calculate a monthly car payment.

- Calculate the cost to operate a car.

- Calculate a car lease payment.

Auto Loans

A loan for the purchase of a car is repaid in equal installments at the end of the installment period. Therefore, the present value of an ordinary annuity formula is used to calculate a car loan payment. The **present value** is the **loan amount** for the car.

> **Formula for the Present Value of an Ordinary Annuity**
>
> $$PV = PMT\left[\frac{1 - (1 + i)^{-n}}{i}\right]$$
>
> where PMT is the cash flow amount, n is the number of payments, and i is the interest rate per period, $i = \dfrac{\text{Annual interest rate as a decimal}}{\text{Number of compounding periods per year}}$.

Example 1 **Calculate a Monthly Car Payment**

An interior designer purchases a car for total cost, including tax and license, of $42,523.98. The designer obtains a 5-year loan at an annual interest rate of 2.79% compounded monthly. If the designer makes a 20% down payment, what is the monthly car payment?

Solution

First, calculate the amount financed, which is the purchase price minus the down payment.

Amount financed = Purchase price − 20% of purchase price

Amount financed = 42,523.98 − 42,523.98 · 0.20

= 34,019.184

The amount financed is $34,019.18.

```
N=60
I%=2.79
PV=34019.18
■PMT=-608.110607
FV=0
P/Y=12
C/Y=12
PMT: END BEGIN
```

Calculate the monthly payment solving the present value of an annuity formula for payment. We will use the finance app on a TI-84 calculator. See the screenshot on the previous page.

The monthly car payment is $608.11.

Example 2 **Calculate a Monthly Car Payment Including Tax and License**

A web page designer negotiates a purchase price of $23,250 for a car.

a. If the sales tax is 7.25% of the purchase price, what is the sales tax?

b. The registration and license fee is 1.2% of the purchase price. What is the registration and license fee?

c. The designer makes a down payment of $3000 and gets a loan at an interest rate of 3.5% compounded monthly for 5 years. Calculate the monthly car payment.

Solution

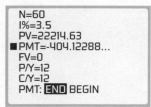

a. Sales tax = $23,250(0.0725) = 1685.625

The sales tax is $1685.63.

b. Registration and license fee = $23,250(0.012) = 279.00

The registration and license fee is $279.00.

c. The amount financed is the sum of the purchase price, the sales tax, and registration and license fee minus the down payment.

$$\text{Amount financed} = \$23,250.00 + \$1685.63 + \$279.00 - \$3000.00$$
$$= \$22,214.63$$

To find the monthly payment, use a calculator or a spreadsheet. The calculation using a TI-84 calculator is shown in the calculator screen image above.

The monthly payment is $404.12.

Example 3 **Calculate the Payoff for a Car Loan**

A firefighter for the National Park Service has a 5-year car loan for which the monthly payment is $750.46 and the annual interest rate of 3.75% is compounded monthly. After making 36 payments, the firefighter decides to trade in the car for a new car. Calculate the amount the firefighter still owes on the car.

Solution

The loan payoff is the present value of the remaining payments. Because 36 payments have been made, there are 60 − 36 = 24 payments remaining. Using the TI-84 finance app, N = 24, I% = 3.75, PMT = −750.46, P/Y and C/Y equal 12, and select END because payments are made at the end of the month (an ordinary annuity). Move the cursor to PV. Press ALPHA SOLVE.

```
N=24
I%=3.75
■PV=17326.14339
PMT=-750.46
FV=0
P/Y=12
C/Y=12
PMT: END BEGIN
```

The loan payoff is $17,326.14.

An important consideration when buying a car is deciding how much you can afford for a car payment.

| Example 4 | **Determine Car Affordability** |

From a budget that William has created, William can afford a monthly payment of $350 for a car. If William can obtain a 5-year loan at an annual interest rate of 4.3%, what is the maximum amount that can be financed?

Solution

To determine the amount William can finance, solve the present value of an ordinary annuity formula for *PV*. We will use a TI-84 calculator. N = 60 (5 years), I% = 4.3, skip PV, PMT = -350 (the amount William can afford), P/Y and C/Y equal 12 (monthly payments), and select END because payments are made at the end of the month (an ordinary annuity). Move the cursor to PV and press ALPHA SOLVE.

```
N=60
I%=4.3
■PV=18865.67271
PMT=-350
FV=0
P/Y=12
C/Y=12
PMT: END BEGIN
```

William can afford to finance $18,865.67.

The answer to Example 4 is the amount William can *finance*, not the actual price of the car. Taxes and license fees must be added to the purchase price. And William can make a down payment that would reduce the amount financed.

Cost to Operate a Car

There are many different fuel options for a car such as gas, electricity, natural gas, diesel, and hydrogen. Some hybrids use combinations of these fuels. Because of these options the miles per gallon (mpg) standard is no longer a good way to estimate fuel efficiency between cars using different fuels.

To compare the cost of operating a car with these different fuels, the energy required to drive 100 miles is frequently used. These fuel ratings are based on testing done by the Environmental Protection Agency (EPA).

| Example 5 | **Calculate Gallons per 100 Miles** |

The EPA rating of a car is 24 mpg. Find the gallons per 100 miles used by this car.

Solution

The mpg fraction is $\dfrac{24 \text{ miles}}{1 \text{ gallon}}$. The reciprocal of that fraction is $\dfrac{1 \text{ gallon}}{24 \text{ miles}}$. Now solve a proportion.

$$\frac{1 \text{ gallon}}{24 \text{ miles}} = \frac{x \text{ gallons}}{100 \text{ miles}}$$

$$100 \cdot \frac{1 \text{ gallon}}{24 \text{ miles}} = 100 \cdot \frac{x \text{ gallons}}{100 \text{ miles}}$$

$$4.17 \approx x$$

The car has a rating of 4.17 gallons per 100 miles.

If the cost of fuel is known in Example 5, then you can calculate how much it costs to drive 100 miles. For instance, suppose the gasoline cost is $2.48 per gallon. Then the cost to drive this car is

$$\frac{4.17 \text{ gallons}}{100 \text{ miles}} \cdot \frac{\$2.48}{1 \text{ gallon}} = \$10.3416$$

It costs approximately $10.34 per 100 miles driven for fuel.

One reason to calculate the cost per 100 miles is to compare the cost of operating cars that use different fuels. For instance, suppose the energy to drive an all-electric vehicle 100 miles is 25 kilowatt-hours (kWh). The energy to drive a gasoline powered car 100 miles is 3.1 gallons of gasoline. If electricity costs $0.21 per kWh and gasoline costs $2.33 per gallon, then

$$\text{Cost of electric car} = \text{Number of kWh} \cdot \text{cost per kWh} = 25 \cdot 0.21 = 5.25$$

$$\text{Cost of gas car} = \text{Number of gallons} \cdot \text{cost per gallon} = 3.1 \cdot 2.33 = 7.223$$

Because $5.25 < 7.223$, the electric car is more fuel efficient. Note, however, that other factors must be considered when determining total cost.

The single greatest cost involved in owning a new car is depreciation. Some companies that specialize in car valuations estimate that the *minute* a car is driven off the lot it depreciates (loses value) by 9% to 11%. By the end of the first year, it depreciates another 10%. These values are for cars that hold their value well. Some cars depreciate much more quickly. Information on depreciation rates can be found in various places such as Kelley Blue Book and Edmonds.com.

Example 6   **Calculate Car Depreciation**

Suppose a new car is purchased for $28,000 and depreciates by 21% over the first year of ownership. What is its value after 1 year?

Solution

$$\text{Depreciation} = \$28,000 \cdot 0.21 = 5880$$

The car depreciates by $5880.

$$\text{Value after 1 year} = 28,000 - 5880 = 22,120$$

The value of the car 1 year later is $22,120.

Besides the monthly car payment, depreciation, and fuel costs, other costs of ownership include insurance, maintenance, repairs, and license fees.

Example 7   **Calculate the Annual Cost of Owning a Car**

A tennis coach purchased a new car for $32,500 and financed $26,000 for 4 years at an annual interest rate of 3.75%. The monthly car payment is $584.15. Complete the following table to determine the cost of owning the car for the first year. Assume the car has an average fuel efficiency of 24 miles per gallon.

License and registration	1.2% of purchase price
Sales tax	6.75% of purchase price
Depreciation	22% of purchase price
Fuel cost driving 14,000 miles	Average cost of gasoline: $2.56 per gallon
Year 1 interest payment	Amount paid in interest on the car loan for the first year of ownership
Insurance	$832
Maintenance	$592

Solution

$$\text{License and registration} = 32,500 \cdot 0.012 = 390$$

$$\text{Sales tax} = 32,500 \cdot 0.0675 = 2193.75$$

$$\text{Depreciation} = 32,500 \cdot 0.22 = 7150.00$$

$$\text{Fuel cost} = \frac{\text{Miles in 1 year}}{\text{Fuel efficiency}} \cdot \text{Cost per gallon}$$

$$= \frac{14,000}{24} \cdot 2.56 \approx 1493.33$$

To find the year 1 interest payment, use the cumulative interest function (CUMINT) in a spreadsheet program. For help calculating Present Value of Annuity Formulas, you can access an Excel template from your digital resources.

The cumulative interest for year 1 is $870.19.

	A	B
	B9 ⋮ ⊘ ⊘ ∧ *fx* 870.19	
1	Calculate Cumulative Interest Paid for a Loan	
2	Annual interest rate	3.75%
3	Payments per year	12
4	Number of payments	48
5	Present Value	$26,000.00
6	Start period	1
7	End period	12
8	Ordinary or due annuity	0
9	Cumulative Interest	870.19

Enter the calculated values in the table and find the total.

License and registration	1.2% of purchase price	390.00
Sales tax	6.75% of purchase price	2193.75
Depreciation	22% of purchase price	7150.00
Fuel cost driving 14,000 miles	Average cost of gasoline: $2.56 per gallon	1493.33
Year 1 interest payment	Amount paid in interest on the car loan for first year of ownership	870.19
Insurance	$832	832.00
Maintenance	$592	592.00
	Total	13,521.27

The total cost of ownership for the first year is $13,521.27.

●

The year 1 ownership cost in Example 7 is extreme, partly because of the large depreciation cost and the sales tax. In year 2, the sales tax would not factor into the cost of ownership. The average depreciation in year 2 is about 18% of the current value of the car. Applying that to Example 7, the depreciation in year 2 is

$$\text{Depreciation in year 2} = (\text{Purchase price} - \text{Year 1 depreciation}) \cdot 0.18$$

$$= (32,500 - 7150) \cdot 0.18$$

$$= (25,350)0.18 = 4563$$

In year 2, the cost of ownership table would be similar to the following.

License and registration	Slight decrease from year 1.	340.00
Sales tax	None	0.00
Year 2 interest payment		635.95
Depreciation	18% of current value	4563.00
Fuel cost driving 14,000 miles	Average cost of gasoline: $2.56 per gallon	1493.33
Insurance	$832	832.00
Maintenance	$592	592.00
	Total	8456.28

The cost of ownership for year 2 is $8456.28.

Auto Leases

Leasing rather than buying a car is an option for some people. Leasing a car may result in lower monthly car payments. The trade-off of lower monthly payments is that, at the end of the lease, you do not own the car.

The value of the car at the end of the lease term is called its **residual value**. The residual value of a car is frequently based on a percentage of the manufacturer's suggested retail price (MSRP) and normally varies between 40% and 60% of the MSRP, depending on the type of lease.

For instance, suppose the MSRP of a car is $18,500, and the residual value is 45% of the MSRP. Then

$$\text{Residual value} = 18,500(0.45) = 8325$$

Take Note: The 2400 in the money factor is a sort of average monthly interest rate. For instance, suppose at the beginning of the year you owe $4000 on a car loan and at the end of the year you owe $2000. The average amount owed is

$$\frac{1}{2}(4000 + 2000) = 3000.$$

There are 12 months in a year, so each month you owe $\frac{1}{12}$ of the interest on $3000. To change an interest rate (a percent) to a fraction, multiply by $\frac{1}{100}$. Putting all this together, we have

$$\frac{1}{2} \cdot \frac{1}{12} \cdot \frac{1}{100} = \frac{1}{2400}.$$

The residual value is $8325. This is the amount the dealer thinks the car will be worth at the end of the lease period. The lessee (the person leasing the car) usually has the option of purchasing the car at that price at the end of the lease. The residual value of the car is a negotiable item. There is no rule that sets the residual value at a certain number.

In addition to the residual value of the car, the monthly lease payment for a car takes into consideration *net capitalized cost*, the *money factor, average monthly finance charge,* and *average monthly depreciation*. Definitions of these terms follow.

$$\textbf{Net capitalized cost} = \text{Negotiated price} - \text{Down payment} - \text{Trade-in value}$$

$$\textbf{Money factor} = \frac{\text{Annual interest rate}}{2400}$$

Note: Many lease agreements show just the money factor, not the interest rate. This formula can be used to find the interest rate. Lease agreements do not fall under the federal Truth in Lending Act. One other note: money factors are negotiable.

$$\textbf{Average monthly finance charge} = (\text{Net capitalized cost} + \text{Residual value}) \cdot \text{Money factor}$$

$$\textbf{Average monthly depreciation} = \frac{\text{Net capitalized cost} - \text{Residual value}}{\text{Number of months in the lease}}$$

Monthly Lease Payment Formula

The monthly lease payment, P, is given by

$$P = F + D$$

where F is the average monthly finance charge and D is the average monthly depreciation.

Example 8 **Calculate a Monthly Lease Payment**

The director of human resources for a company decides to lease a car for 30 months. Suppose the annual interest rate is 8.4%, the negotiated price is $29,500, there is no trade-in, and the down payment is $5000. Find the monthly lease payment. Assume that the residual value is 55% of the MSRP of $33,400.

Solution

Use the monthly lease payment formula. Calculate F and D. F is the average monthly finance charge. The money factor is given, so calculate the net capitalized cost and the residual value.

$$\text{Net capitalized cost} = \text{Negotiated price} - \text{Down payment} - \text{Trade-in value}$$

$$= 29,500 - 5000 - 0 = 24,500$$

Residual value = $33,400 \cdot 0.55 = 18,370$

F = (Net capitalized cost + Residual value) · Money factor

$= (24,500 + 18,370) \cdot \dfrac{8.4}{2400}$

$= (42,870) \cdot \dfrac{8.4}{2400}$

$= 150.045$

The average monthly finance charge, F, is \$150.05.

$D = \dfrac{\text{Net capitalized cost} - \text{Residual value}}{\text{Number of months in the lease}}$

$= \dfrac{24,500 - 18,370}{30}$

$= \dfrac{6130}{30}$

≈ 204.333

The average monthly depreciation, D, is \$204.33.

$P = F + D$

$P = 150.05 + 204.33 = 354.38$

The monthly lease payment is \$354.38.

As you see, there are quite a number of steps to calculate a monthly lease payment. For help with this calculation, you can access an Excel template from your digital resources.

One factor that was not considered in Example 8 is sales tax. When you purchase a car, the sales tax is due at the time of the purchase. When you lease a car, the sales tax is added to the lease payment each month.

For Example 8, if the sales tax for the car lease is 6.5%, then your monthly payment is \$354.38 plus 6.5% of that amount.

Monthly payment = $354.38 + 354.38 \cdot 0.065$

$= 354.38 + 23.0347$

$= 377.41$ • Rounded to the nearest cent

The actual monthly payment is \$377.41.

Buying or leasing a car comes down to personal circumstances. Here are some advantages and disadvantages of each.

Advantages of Leasing

- The down payment may not be as large. This can vary among leasing companies.
- When the lease agreement ends, usually in 3 or 4 years, you can turn in the car and get a new one.
- Lower maintenance cost because your lease time is during the manufacturer's warranty.

Disadvantages of Leasing

- You do not own the car, so you cannot make changes to it.
- There are mileage restrictions (typically 12,000 miles per year; 36,000 miles for a 3-year lease). If you exceed those miles, there is an additional cost per mile you must pay the leasing company.
- Leasing can be more expensive. If you purchase a car and keep it for 6 years, it may be less expensive than leasing two cars, each for 3 years.

- If you decide you do not like the car or would just like to get another one, the cost of terminating a lease can be quite high.

Advantages of Buying

- You own the car, so you can make changes to it.
- There are no mileage restrictions.
- You own the car, so you can sell it.
- If you intend to keep the car for a long time (more than 6 years), owning can be more economical.

Disadvantages of Buying

- The down payment can be quite large.
- The car depreciates as soon as you drive it off the lot.
- Because monthly lease payments are based on the residual value of the car, the monthly lease payment is usually less than the monthly car payment.
- If you keep the car past the warranty period, you are responsible for any repairs.

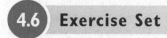

4.6 Exercise Set

Think About It

1. If a car depreciates 20% of its current value each year, does this mean it is worth $0 after 5 years? Why or why not?

2. One car has an EPA rating of 4 gallons of fuel per 100 miles driven, and a second car has a rating of 5 gallons of fuel per 100 miles driven. Which car has the best miles per gallon rating? Try to answer this without doing any calculations.

3. Suppose you are going to lease a car. Is it better for the residual value to be 45% of the cost of the car or 50% of the cost of the car?

Preliminary Exercises

For Exercises P1 to P8, if necessary, use the referenced example following the exercise for assistance.

P1. A metallurgist purchases a car for total cost including tax and license of $39,895.66. If the metallurgist obtains a 4-year loan at an annual interest rate of 4.9% compounded monthly, what is the monthly car payment? **Example 1**

P2. A dog trainer purchases a car for $26,500. **Example 2**

    a. The sales tax is 7.75% of the purchase price. Calculate the sales tax.

    b. The registration and license fee is 0.9% of the purchase price. What is the registration and license fee?

    c. If the trainer makes a down payment of $5000 and gets a loan for the purchase price, sales tax, and registration fee at an interest rate of 3.5% compounded monthly for 5 years, what is the monthly car payment?

P3. A firefighter for the National Park Service has a 5-year car loan for which the monthly payment is $750.46 with an annual interest rate of 3.75% compounded monthly. After making 36 payments, the firefighter decides to trade in the car for a new car. Calculate the amount the firefighter still owes on the car. **Example 3**

P4. Kato can afford a monthly payment of $475 for a car. If Kato can obtain a 4-year car loan at an annual interest rate of 3.9%, what is the maximum amount Kato can finance? **Example 4**

P5. The EPA rating of a car is 35 mpg. Find the gallons per 100 miles used by this car. **Example 5**

P6. Suppose a new car is purchased for $64,000 and depreciates by 24% over the first year of ownership. If the car is driven 14,200 miles in that year, what is the cost, in dollars, per mile for depreciation? Round to the nearest cent. **Example 6**

P7. A mechanical engineer purchased a minivan for $48,000 and financed $38,400 for 5 years at an annual interest rate of 4.2%. Complete the following table to determine the cost of owning the minivan for the first year of ownership. Assume the minivan has an average fuel efficiency of 28 miles per gallon. **Example 7**

License and registration	0.8% of purchase price	
Sales tax	4.5% of purchase price	
Depreciation	20% of purchase price	
Fuel cost driving 12,000 miles	Average cost of gasoline: $2.97 per gallon	
Year 1 interest payment	Amount paid in interest on the car loan for the first year of ownership	
Insurance	$789	
Maintenance	$425	
	Total	

P8. A real estate broker decides to lease a car for 36 months. Suppose the annual interest rate is 7.8%, the negotiated price is $48,000, there is no trade-in, and the down payment is $4000. Find the monthly lease payment. Assume that the residual value is 48% of the MSRP of $51,100. **Example 8**

Section Exercises

Solve.

1. Car Payments A computer chip designer purchased a car for $53,976.47, including sales tax and registration. If the designer obtains a 5-year loan for the total amount at an annual interest rate of 5.1% compounded monthly, what is the monthly car payment?

2. **Car Payments** A car dealer offered an end-of-year clearance that was 20% off the MSRP of $34,500. Janice Ng agrees to that price plus additional tax and license fees of $1785. If she obtains a 3-year loan at an annual interest rate of 5.75% compounded monthly, what is her monthly payment?

3. **Fuel Costs** The EPA rating of a car is 23 mpg. If this car is driven 1000 miles in 1 month and the price of gasoline remained constant at $3.06 per gallon, calculate the fuel cost for this car for 1 month. Round to the nearest cent.

4. **Mpg** Suppose a car has a rating of 3.8 gallons per 100 miles. What is the miles-per-gallon rating of this car? Round to the nearest tenth.

5. **Car Payments** A bank teller negotiated a purchase price of $22,350 for a car. In addition to the purchase price, the teller had to pay sales tax that was 6.25% of the purchase price and a $452 license fee. If the teller makes a down payment of 10% of the total cost of the car and secures a 4-year loan at an annual interest rate of 4.9% compounded monthly, what is the monthly car payment?

6. **Choosing a Term Length** A special education teacher is deciding between two options for a car purchase that have a total price of $34,769.31. Option 1 is a 4-year loan at an annual interest rate of 5% compounded monthly. Option 2 is a 5-year loan at an annual interest rate of 4% compounded monthly. Assuming the teacher keeps the car for 5 years, for which option will the total interest paid be less?

7. **Paying Off a Car Loan** An orthodontist has a 4-year car loan for which the monthly payment is $1472.43 with an annual interest rate of 4.95% compounded monthly. After making 30 payments, the orthodontist decides to trade in the car for a new vehicle. Calculate the amount the orthodontist still owes on the car.

8. **Present Value of a Car** A car dealer offers to buy your car for $9500 so that you can purchase a new one. If your monthly payment on a 4-year loan at an annual interest rate of 4% compounded monthly is $554.23 and you have 20 more payments remaining, is the present value of your car more or less than the amount the dealer is offering? By how much?

9. **Financing a Car** Abigail can afford a monthly payment of $475 for a car. If Abigail can obtain a 5-year car loan at an annual interest rate of 2.75%, what is the maximum Abigail can finance?

10. **Financing a Van** A florist needs to purchase another van to deliver flowers. If the florist can secure a 5-year loan at an annual interest rate of 3.6% and does not want a monthly payment more than $650, what is the maximum the florist can finance for another van? Round to the nearest hundred dollars.

11. **Calculating a Monthly Car Payment** A car dealer offers a 5-year loan that costs $18.416522 per month for each $1000 borrowed. What is the monthly payment for a car that is financed for $35,600?

12. **Calculating a Monthly Car Payment** A credit union offers a 5-year loan that costs $23.029293 per month for each $1000 borrowed. What is the monthly payment for a car that is financed for $52,550?

13. **Effects of Aggressive Driving on Mpg** The EPA estimates that aggressive driving (speeding and rapid acceleration or braking) during highway driving can lower gas mileage by as much as 33%. Suppose a person whose car has an EPA

highway rating of 28 miles per gallon (mpg) engages in aggressive driving that causes a 30% reduction in gas mileage.

   a. Calculate the miles per gallon during this behavior. Round to the nearest mpg.

   b. If this person drives in this manner for 50 miles and gasoline costs $2.89 per gallon, what is the increased cost for driving the 50 miles? Round to the nearest dollar.

14. **Effects of a Cargo Box on Mpg** Cargo boxes on a car roof can reduce fuel economy by 6% to 17% while driving at highway speeds.

   a. If a car normally gets 31 mpg but is carrying a cargo box on its roof that reduces fuel economy by 15%, what is the mpg for the car with the cargo box attached? Round to the nearest mpg.

   b. If this person drives 100 miles with the cargo box and gasoline costs $3.13 per gallon, what is the increased cost with the cargo box? Round to the nearest dollar.

15. **Car Depreciation** Suppose a new car is purchased for $48,357 and depreciates by 22% over the first year of ownership. If the car is driven 12,470 miles in that year, what is the cost, in dollars per 100 miles driven, for depreciation? Round to the nearest cent.

16. **Car Depreciation** You purchase a used car for $15,200, and it depreciates by 16% over the next year. If the car is driven 15,360 miles in that year, what is the cost, in dollars per 100 miles, for depreciation? Round to the nearest cent.

17. **Leasing a Car** A marketing manager leases a car for 24 months after agreeing to a negotiated price of $41,250 and makes a down payment that is 20% of the negotiated price. Find the monthly lease payment if the money factor is 0.0026. Assume that the residual value is 52% of the MSRP of $45,900 and there is no trade-in.

18. **Leasing a Car** The manager of a hotel negotiates a purchase price of $51,000 for a car whose MSRP is $53,300 and decides to lease the car for 30 months. Find the monthly lease payment if the annual interest rate is 4.35%, there is no trade-in, the manager makes a 10% down payment, and the residual value is 48% of the MSRP.

19. **Paying Off a Loan** Dagmar has a 4-year car loan at an annual interest rate of 5.5%. She has made 28 payments of $976.77. If Dagmar decides to pay off her loan, what is her payoff amount?

20. **Paying Off a Loan** Lourdes has a 5-year car loan with a monthly loan payment of $535.88 at an annual interest rate of 2.75%. Lourdes wants to purchase a new car and decides to pay off her loan after 37 payments. What is her payoff amount?

21. **Cost of Owning a Car** A cloud storage engineer purchased a new car for $29,900 and financed $25,000 for 4 years at an annual interest rate of 4.3%. The monthly payment is $567.84. Complete the following table to determine the cost of owning the car for the first year of ownership. Assume the car has an average fuel efficiency of 28 miles per gallon.

License and registration	0.75% of purchase price	
Sales tax	6.25% of purchase price	
Depreciation	21% of purchase price	
Fuel cost driving 14,000 miles	Average cost of gasoline: $2.74 per gallon	
Year 1 interest payment	Amount paid in interest on the car loan for first year of ownership	

(continued)

Insurance	$712	
Maintenance	$356	
	Total	

22. Cost of Owning a Car A receptionist purchased a new car for $18,500 and financed $15,000 for 4 years at an annual interest rate of 3.5%. Complete the following table to determine the cost of owning the car for the first year of ownership. Assume the car has an average fuel efficiency of 34 miles per gallon. The monthly payment is $335.34.

License and registration	0.9% of purchase price	
Sales tax	7.75% of purchase price	
Depreciation	19% of purchase price	
Fuel cost driving 14,000 miles	Average cost of gasoline: $3.05 per gallon	
Year 1 interest payment	Amount paid in interest on the car loan for first year of ownership	
Insurance	$694	
Maintenance	$411	
	Total	

23. Fuel Efficiency Two people are considering a new car that is more fuel efficient than their current car. One person is considering trading in a car that gets 20 mpg for one that gets 30 mpg, a 10 mpg increase. The second person is considering trading in a car that gets 35 mpg for one that gets 50 mpg, a 15 mpg increase. If gasoline costs $3.12 per gallon and each person plans on driving 10,000 miles in the next year, which person will see the greater decrease in fuel costs?

24. Fuel Efficiency Annabel and Evan are considering a new car that is more fuel efficient than their current car. Annabel is considering trading in a car that gets 24 mpg for one that gets 32 mpg. Evan is considering trading in a car that gets 32 mpg for one that gets 42 mpg. If gasoline costs $3.26 per gallon and each person plans on driving 12,000 miles in the next year, which person will see the greater percentage decrease in fuel costs?

25. Rolling Over a Car Loan A car dealership is offering to roll over the amount owed on your car into a new loan so you can purchase a new car. Suppose you owe $5606.77 on your current car and have agreed to a new car whose total cost, including tax and license, is $28,905.87. You are getting a 5-year loan at an annual interest rate of 4.25%. What are your new monthly payments?

26. Rolling Over a Car Loan Uvaldo is looking to purchase a new car and finds one whose total cost, including tax and license, is $45,099.82. Uvaldo owes $9854.87 on his current car and secures a 5-year loan that combines the cost of the new car and the amount he owes on his current car. The annual interest rate on the new loan is 3.875%. What are Uvaldo's monthly payments on the new car?

27. Leasing a Car Marisa leases a car that has a purchase price of $62,500 with an MSRP of $65,950 and decides to lease the car for 36 months. Find the monthly lease payment if the annual interest rate is 4.5%, the trade-in value of her car is $25,000, she makes a $5000 down payment, and the residual value is 45% of the MSRP. Include a sales tax of 7.25%.

28. Leasing a Car Julius negotiates a purchase price of $29,995 for a car whose MSRP is $34,800 and decides to lease the car for 24 months. Find the monthly lease payment if the annual interest rate is 4.15%, there is no trade-in, Julius makes a $1000 down payment, and the residual value is 41% of the MSRP. Include a sales tax of 6.25%.

29. Choosing a Car You are considering two cars that you plan on keeping for 5 years. One has an EPA combined city and highway rating of 32 mpg. The second has an EPA rating of 36 mpg. Suppose gasoline costs $3.41 per gallon and you drive 10,000 miles each year.
  a. How much will you save in 1 year by purchasing the 36 mpg car?
  b. If you deposit that amount at the end of each year for the 5 years of ownership into an account that earns 4.2% compounded annually, how much you will save over 5 years?

30. Choosing a Car A car has an EPA combined city and highway rating of 26 mpg. The second car has an EPA rating of 38 mpg. Suppose gasoline costs $3.08 per gallon and you drive 12,000 miles each year.
  a. If you keep the car for 4 years, how much will you save in 1 year by purchasing the 38 mpg car?
  b. If you deposit that amount at the end of each year for the 4 years of ownership into an account that earns 5.5% compounded annually, how much will you have saved over 4 years?

31. Calculating Annual Interest Rate A possible issue to offering a customer a loan based on dollars per $1000 borrowed as in Exercise 11 is that the annual interest rate is not obvious. Calculate the annual interest rate for the loan in Exercise 11. Round to the nearest tenth of a percent.

32. Calculating Annual Interest Rate A possible issue to offering a customer a loan based on dollars per $1000 borrowed as in Exercise 12 is that the annual interest rate is not obvious. Calculate the annual interest rate for the loan in Exercise 12. Round to the nearest tenth of a percent.

33. Hybrid Fuel Costs The energy needed to drive a certain plug-in hybrid 100 miles using only the electric motor is 39 kWh. The EPA estimates that using only the gasoline engine yields 33 miles per gallon. Suppose the owner of this car uses the battery alone for 40% of the miles driven and the gasoline engine for the remaining 60%. Find the cost of fuel to drive this car 12,000 miles in 1 year. Assume that electricity costs $0.18 per kWh and gas costs $3.10 per gallon.

34. Hybrid Fuel Costs The energy needed to drive a certain plug-in hybrid 100 miles using only the electric motor is 42 kWh. When using only the gasoline engine, the EPA estimate is 41 miles per gallon. Suppose the owner of this car drives 14,000 miles in 1 year and uses the battery alone for 50% of the miles driven and the gasoline engine for the remaining 50%. If the cost of electricity is $0.19 per kWh and gas costs $2.89 per gallon, which is greater—the cost for electricity or the cost for gas?

▶ Investigation

Buying a Car: New Versus Used Suppose you are going to purchase a car and have decided to keep it for 4 years. Estimate the difference in cost per 100 miles between buying a new car and purchasing essentially the same car but the previous year's model. In your calculations, consider how many miles per year you drive, depreciation, fuel costs, the interest rate on the car loan (which may be different depending on a new car versus a 1-year-old car), sales tax, license fee, and other factors that affect the cost of ownership. For help with this calculation, you can access an Excel template from your digital resources. You will need to modify the sheet for your *car* and make changes to calculate the values for a used car.

Financial Considerations of Home Ownership

Learning Objectives:

- Calculate fees to buy or sell a house.
- Calculate a fixed-rate mortgage payment.
- Calculate loan origination fees and points.
- Calculate an adjustable-rate mortgage payment.

Buying and Selling a Home

Probably the best place to start when deciding on the purchase of a home is how much you can afford. Lending institutions use guidelines to help determine the monthly payment a borrower can afford. One such guideline is the **debt-to-income (DTI) ratio**. DTI expresses, as a percent, how much of your gross monthly income is spent on servicing liabilities, such as auto loans, credit cards, mortgage payments or rent, insurance, property taxes, and HOA fees. This is calculated by adding all current monthly debts (not things like utilities, cell phone, gas for the car, and other recurring bills) for a borrower and then dividing by the borrower's gross (before taxes) monthly income. Then the result is written as a percentage.

A DTI ratio between 36 and 45 is acceptable to most lenders. The decision of acceptable ratios usually depends on the credit worthiness of the borrower (the borrower's credit score) or the amount of the down payment. The larger the down payment, the lower the acceptable DTI.

Example 1 **Calculate the DTI Ratio**

Each month Tessa Neiman has a minimum credit card payment of $150, a car payment of $325.36, a car insurance payment of $125, and Tessa is applying for a loan for which the monthly payment will be $1253.56. If Tessa has a gross monthly income of $5050, what is Tessa's DTI? Round to the nearest percent.

Solution

$$\text{DTI} = \frac{\text{Monthly payments}}{\text{Gross monthly income}}$$

$$\frac{150 + 325.36 + 125 + 1253.56}{5050} = \frac{1853.92}{5050}$$

$$\approx 0.3671$$

The DTI is 37%.

After you buy a home, you may later want to sell it. There are various costs in selling a house, including:
- real estate commissions (the amounts paid to real estate companies involved in the sale),
- title insurance (an insurance policy that guarantees the buyer that the seller actually owns the home), and
- an escrow fee (payment to a neutral third party that ensures that all payments are made and that all documents have been completed).

Example 2 **Calculate the Fees to Sell a Home**

Madison decides to sell a home for $275,000. The real estate selling agent charges a commission of 5% of the selling price; title insurance, which is 1.1% of the selling price; and an escrow fee of $750. What amount does Madison receive after fees?

Solution

$$\text{Real estate commission} = 275{,}000 \cdot 0.05 = 13{,}750$$

$$\text{Title insurance} = 275{,}000 \cdot 0.011 = 3025$$

$$\text{Madison receives} = 275{,}000 - 13{,}750 - 3025 - 750 = 257{,}475$$

Madison receives \$257,475.00.

The amount Madison received is approximately 6% less than the selling price. A rule of thumb used by real-estate agents is that the cost of selling a home is between 6% and 8% of the selling price.

Fixed-Rate Mortgages

A **mortgage** is a loan to buy a home. Many types of mortgages are available to home buyers today, so the terms of mortgages differ considerably. A **fixed-rate** mortgage, or **conventional** mortgage, is one in which the interest rate on the loan remains the same throughout the life of the mortgage. Because the interest rate is fixed, the amount of the monthly payment also remains unchanged throughout the term of the loan.

The term of a mortgage also can vary. Terms of 15, 20, 25, and 30 years are most common. The monthly payment on a mortgage is the **mortgage payment**. Because mortgage payments are a series of equal payments at equal time intervals and due at the end of a compounding period (usually 1 month), the payment for a mortgage is based on the present value of an ordinary annuity. The formula is restated here for convenience.

Formula for the Present Value of an Ordinary Annuity

$$PV = PMT \left[\frac{1 - (1 + i)^{-n}}{i} \right]$$

where *PMT* is the cash flow amount, *n* is the number of payments, and *i* is the interest rate per period, $i = \dfrac{\text{Annual interest rate as a decimal}}{\text{Number of compounding periods per year}}$.

Example 3   Calculate a Mortgage Payment

Janice Rigas secures a home loan for \$230,000 at an annual interest rate of 4.5%, compounded monthly, for 30 years.

a. Find the monthly mortgage payment.

b. If Janice keeps the home for 30 years, how much interest will be paid over that time?

Solution

a. To find the monthly payment, use a calculator or a spreadsheet. We will use the Excel *PMT* function. The present value is the loan amount; the future value is \$0 (the loan is paid off).

B7	▲▼	× ✓	f_x =PMT(B1/12,B2*B3,B4,B5,B6)	
	A		B	C
1	Annual interest rate		4.50%	
2	Paynments per year		12	
3	Number of years		30	
4	Present value (loan amount)		\$230,000.00	
5	Future value		\$0.00	
6	Type		0	
7	Monthly payment		(\$1165.38)	

The monthly mortgage payment is \$1165.38.

b. Interest paid = Monthly payment · Number of payments − Loan amount

$$= 1165.38 \cdot 360 - 230{,}000$$

$$= 419{,}536.80 - 230{,}000 = 189{,}536.80$$

The total interest repaid is $189,536.80.

●

Lenders know from past history that borrowers who make less than a 20% down payment for a home are more apt to default on their loans. To protect themselves, lenders require these borrowers to purchase a *private mortgage insurance* (PMI) policy. **Private mortgage insurance** is an insurance policy required of borrowers who make a down payment that is less than 20% of the purchase price of a home. The monthly PMI premium is added to the borrower's monthly payment. The PMI collected goes into a pool that lenders use to recover any loss that may result from a defaulted loan.

Example 4   **Calculate PMI for a Loan**

A lender requires PMI that is 0.65% of the loan amount of $325,000. How much will this add to the borrower's monthly payment?

Solution

PMI = Interest rate · Loan amount

$$= 0.0065 \cdot 325{,}000 = 2112.50$$

Because PMI is an annual fee, divide it by 12.

$$\text{Monthly PMI} = \frac{2112.50}{12} \approx 176.0417$$

PMI adds $176.04 to the monthly payment.

●

Loan Origination Fee and Points

Lenders charge a few fees to help borrowers secure home loans. One is a **loan origination fee**, which is used to cover costs to the lender. The fee is also used to verify the borrower's assets, income, employment history, and other factors that are part of determining the credit worthiness of the borrower. This fee is usually between 0.5% and 1% of the loan amount. For instance, if the loan origination fee is 0.75% on a $250,000 loan, then

Loan origination fee = $0.0075 \cdot 250{,}000 = 1875$

The loan origination fee is $1875.

Many factors influence the interest rate on a mortgage. **Discount points**, usually referred to as just **points**, are fees that may be required of a borrower to secure a loan. One point is 1% of the loan amount.

This fee increases the APR on the loan above the rate on which the monthly payment is calculated. For instance, a loan may be stated with a rate of 4% but an APR of 4.1%. The monthly payment is based on 4%; 4.1% is the interest rate on the loan when all fees to acquire the loan are considered. This is the interest rate required by the Truth in Lending Act.

A borrower can *buy down* the mortgage interest rate by agreeing to a loan with more points. Basically, the buyer is paying an upfront fee to have a lower interest rate. For instance, a borrower may opt for an interest rate of 5.125% on a loan and pay no points or a 4.875% loan and pay 1 point. The more paid in points, the lower the interest rate on the loan.

Example 5   Calculate Points on a Loan

Suppose a lender charges 1.75 points for a home buyer to obtain a loan of $230,000. Calculate the discount points.

Solution

$$\text{Discount points} = \text{Loan amount} \cdot \text{Points as a decimal}$$
$$= \$230{,}000 \cdot 0.0175 = \$4025$$

The fee for points is $4025.

Paying down the mortgage interest rate using points should be considered carefully. There are a few factors to consider, especially how long you plan to keep the loan.

Example 6   Determine the Impact of Points on a Loan

A bank offers Salvador Jenkins a home loan of $328,000 at an annual interest rate of 3.25% with 0.75 point or an interest rate of 3% and 1.65 points. If Salvador chooses the second loan, how long will it take in months to recoup the additional cost in points? Round to the nearest month.

Solution

First, calculate the difference in the cost of points.

$$\text{Difference in cost} = 328{,}000(0.0165) - 328{,}000(0.0075)$$
$$= 2952$$

Salvador must pay an additional $2952 in fees for the lower interest rate.

Now calculate the difference in monthly mortgage payment. We will use a TI-84 calculator to calculate the mortgage payments.

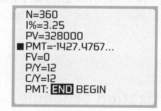 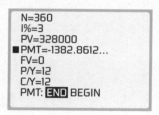

The difference in monthly payments is 1427.48 − 1382.86 = 44.62.

The number of months it takes to recoup the additional cost for points is

Difference in cost of points ÷ Difference in monthly payment
= Months to recoup cost of points

$$2952 \div 44.62 \approx 66$$

It will take approximately 66 months to recoup the additional cost in points.

Adjustable-Rate Mortgages

Some mortgages are **adjustable-rate mortgages** (ARMs). The interest rate charged on an ARM is adjusted periodically to more closely reflect current interest rates. Adjustable-rate mortgages are usually shown as 3/1, 5/1, 7/1, or 10/1. The first number indicates the number of years the rate remains fixed. The second number indicates how often the rate can change after the fixed period. For instance, a loan stated as 5/1 means the interest rate is fixed for 5 years and then can change every 1 year from then on.

The amount the interest rate can change each year once the fixed period has ended is also part of the loan agreement. A statement such as **2%/2%/5%** means that the interest rate can change by a maximum of 2% in the first year after the fixed period, a maximum of 2% in the second year, and a maximum of 5% over the entire life of the loan.

For instance, consider a 3/1 ARM with an initial rate of 2.65% and a 2%/2%/5% clause. In year 4 (1 year after the fixed period) the maximum interest on the loan can be 4.65%. In year 5, the maximum interest rate can be 6.65%. The maximum interest rate for the life of the loan is 7.65%. The Federal Reserve Board has a handbook on ARMs at http://files .consumerfinance.gov/f/201204_CFPB_ARMs-brochure.pdf.

Take Note: The initial monthly payment ($1086.58) is based on a 30-year mortgage even though the rate can change after 3 years. This is typical of ARMs.

| Example 7 | **Calculate a Change in Monthly Payment for an ARM** |

Michelle Li obtains a $275,000 loan for a condominium and secures a 3/1 ARM at an initial interest rate of 2.5%. The initial monthly payment is $1086.58. After 3 years, the interest rate on the loan changes to 3.25%. Calculate the new monthly payment in year 4 of the loan.

Solution

Because the monthly payment is based on the amount owed after 3 years, first calculate the amount she owes after making 36 payments. This is the present value of the remaining 324 payments. We will use the finance app in a TI-84 calculator.

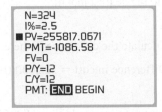
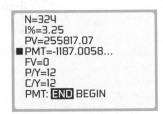

After making payments for 3 years, Michelle owes $255,817.07.

Now calculate the new monthly payment based on 324 more payments (she has already made 36 payments) and the new interest rate of 3.25%.

Michelle's new monthly payment is $1187.01.

Amortization Schedule

Amortization refers to the process of paying off a loan. An **amortization schedule** is a table showing the details of each periodic payment of a mortgage. These details show how much of each payment is interest, how much goes to repaying the loan, and the new loan balance.

For instance, consider the loan in Example 3. The monthly payment is $1165.38 on a 4.5% loan of $230,000 for 30 years. Suppose the loan begins on May 1. Then on June 1, the first payment is due. Janice has had the loan for 1 month, so she owes interest for that month. To find the interest owed, use $I = Prt$, where P is the amount of the loan, r is the monthly interest rate, and $t = 1$ (1 month).

$$\text{Month 1 interest} = 230,000 \cdot \frac{0.045}{12} = 862.50 \quad \bullet \text{ The interest rate for 1 month is } \frac{0.045}{12}.$$

The monthly payment is $1165.38. Of that, $862.50 was interest. The remaining amount goes to repaying the loan.

$$\text{Month 1 principal} = 1165.38 - 862.50 = 302.88$$

The amount owed on the loan after 1 month is the loan amount minus the principal payment.

$$\text{Loan balance after month 1} = 230,000.00 - 302.88 = 229,697.12$$

The loan balance after the first month of the loan is $229,697.12.

Now repeat the calculations for month 2.

$$\text{Month 2 interest} = \$229,697.12 \cdot \frac{0.045}{12} \approx 861.36$$

• The interest rate for 1 month is $\frac{0.045}{12}$.

The monthly payment is $1165.38. Of that, $861.36 was interest. The remaining amount goes to repaying the loan.

$$\text{Month 2 principal} = 1165.38 - 861.36 = 304.02$$

The amount owed on the loan after 1 month is the loan amount minus the principal payment.

$$\text{Loan balance after month 2} = 229,697.12 - 304.02 = 229,393.10$$

The loan balance after the second month of the loan is $229,393.10.

An amortization schedule shows the result of repeating these calculations for the number of months of the loan. The following amortization schedule was created using Excel for the loan in Example 3. We have only shown the first 12 months of the schedule and the last 12 months. For help creating the full Amortization schedule, you can access an Excel template from your digital resources.

Take Note: Look at the amortization schedule and observe that values for months 1 and 2 are the same as the calculations we did before the schedule. Using formulas within the cells of the spreadsheet, the calculations can be automated for the entire loan.

	A	B	C	D
	F367			
1	Annual interest rate	4.50%		
2	Monthly payment	$1,165.38		
3				
4		Interest paid	Principal paid	New Principal
5	Loan amount			$230,000.00
6	Month 1	$862.50	$302.88	$229,697.12
7	Month 2	$861.36	$304.02	$229,393.10
8	Month 3	$860.22	$305.16	$229,087.94
9	Month 4	$859.08	$306.30	$228,781.64
10	Month 5	$857.93	$307.45	$228,474.19
11	Month 6	$856.78	$308.60	$228,165.59
12	Month 7	$855.62	$309.76	$227,855.83
354	Month 349	$51.18	$1,114.20	$12,532.69
355	Month 350	$47.00	$1,118.38	$11,414.31
356	Month 351	$42.80	$1,122.58	$10,291.73
357	Month 352	$38.59	$1,126.79	$9,164.94
358	Month 353	$34.37	$1,131.01	$8,033.93
359	Month 354	$30.13	$1,135.25	$6,898.68
360	Month 355	$25.87	$1,139.51	$5,759.17
361	Month 356	$21.60	$1,143.78	$4,615.39
362	Month 357	$17.31	$1,148.07	$3,467.32
363	Month 358	$13.00	$1,152.38	$2,314.94
364	Month 359	$8.68	$1,156.70	$1,158.24
365	Month 360	$4.34	$1,161.04	–$2.80

The last value of −$2.80 indicates that the loan was overpaid by $2.80. This occurs because of rounding in the calculations. This amount would be deducted from the last payment.

Example 8   **Calculate a Mortgage Payoff**

The Lightfoot family has a 30-year mortgage at an annual interest rate of 6%. After making payments of $2315.69 for 6 years, the Lightfoots would like to pay off their loan so as to purchase a new home. Calculate the loan payoff.

Solution

Because they have made payments for 6 years (72 months), they have $360 - 72 = 288$ payments remaining. Use the present value of an annuity formula to calculate their payoff.

We will use the finance app available with the TI-84 calculator. Enter $N = 288$, $I\% = 6$, skip PV for now, $PMT = -2315.69$, and $FV = 0$.

Move the cursor to PV and press **ALPHA SOLVE**.

The loan payoff is $353,013.33.

```
N=288
I%=6
■PV=353013.3275
 PMT=-2315.69
 FV=0
 P/Y=12
 C/Y=12
 PMT: END BEGIN
```

The federal Truth in Lending Act requires that credit customers be made aware of the cost of credit. This applies to home mortgages as well. This means considering the other fees such as points and the loan origination fee that are part of acquiring a mortgage. For instance, a lender may offer a 4.375% mortgage rate but the APR is 4.478%. The difference is the APR adjusts for the points and the loan origination fee that are required to secure the loan.

Example 9   Calculate the APR of a Mortgage

A violinist for a symphony orchestra obtains a 30-year mortgage for $320,000 at an annual interest rate of 2.875% with 1.75 points. The loan origination fee is 0.6 point of the loan amount. Find the APR for the loan. Round to the nearest thousandth of a percent.

Solution

There is a series of steps to this calculation.

- Find the fee for points and the loan origination fee.

    Fee for points = $320,000 · 0.0175 = $5600

    Loan origination fee = $320,000 · 0.006 = $1920

- Add these amounts to the loan amount to give an adjusted loan balance.

    Adjusted loan balance = $320,000 + $5600 + $1920 = $327,520

```
N=360
I%=2.875
PV=327520
■PMT=-1358.8558...
 FV=0
 P/Y=12
 C/Y=12
 PMT: END BEGIN
```

- Find the monthly payment based on the adjusted balance. We will use a TI-84 here. The monthly payment is $1358.86.

- Replace the adjusted loan amount with the actual loan amount, $320,000, and then solve for $I\%$. The result follows.

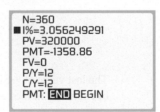

```
N=360
■I%=3.056249291
 PV=320000
 PMT=-1358.86
 FV=0
 P/Y=12
 C/Y=12
 PMT: END BEGIN
```

The APR for the loan is 3.056%.

The term **closing costs** refers to all of the fees involved in securing a home loan. These fees are collected upfront and are not part of the mortgage. Closing costs include real estate fees (the most costly of which generally range from 4% to 6% of the selling price), title insurance (an insurance policy that guarantees the homeowner actually owns the house), escrow fees (an independent agent who is paid to ensure all the details of a sale are done correctly), home repairs the buyer may require before purchasing the home, and a home warranty (an insurance policy that covers repair costs for major appliances such as a stove, refrigerator, heating and air-conditioning, and the water heater).

Buying versus Renting

Whether to rent or buy is a tough decision, and there are many questions to answer.

- How long do you plan to stay in the home (house, condo, townhouse, co-op)?
- How much do you have for a down payment?
- How much will it cost to buy a house?
- What is your income tax bracket?
- What is your consumer debt?
- How much can you afford for a monthly payment?
- What are the property taxes?
- What is the expected appreciation rate of your home's value?
- What is the expected growth rate of rents?

There are other considerations as well. To access your situation, look at some of the renting versus buying websites. Here are a few suggestions: zillow.com, realtor.com, trulia.com.

4.7 Exercise Set

▶ Think About It

1. Suppose a lender uses a debt-to-income ratio of 35 as reference for approving a home loan. If Howard's debt-to-income ratio is 40, would he be approved for a loan?

2. What is the difference between a fixed-rate mortgage and an adjustable-rate mortgage?

3. Janice makes a monthly mortgage payment of $1200 in March of which $500 goes to repaying principal and $700 goes to paying interest. When she makes her $1200 mortgage payment in April, will the amount going to principal be more or less than $500?

4. If a buyer is charged 1.5 points of a loan amount, what percentage of the loan amount is the buyer charged?

▶ Preliminary Exercises

For Exercises P1 to P9, if necessary, use the referenced example following the exercise for assistance.

P1. Calculate the DTI ratio for a borrower who has a gross monthly income of $4875 and has a minimum credit card bill of $125, a car lease payment of $256.78, and a student loan payment of $216, and who is applying for a loan with a monthly payment of $2031.79. **Example 1**

P2. The Tomlinsons decide to sell their home for $365,000. They are charged a real estate commission of 4.5% of the selling price, title insurance that is 1% of the selling price, and an escrow fee of $900. What amount do the Tomlinsons receive after fees? What percentage of the selling price was fees? Round to the nearest tenth of a percent. **Example 2**

P3. A hair stylist obtains a home loan for $411,000 at an annual interest rate of 5.25% compounded monthly for 30 years. Find the monthly mortgage payment. **Example 3**

P4. A lender requires PMI that is 0.8% of the loan amount of $410,000. How much will this add to the borrower's monthly payment? **Example 4**

P5. A savings and loan charges 2.125 points for a home buyer to obtain a loan of $245,000. Calculate the discount points. **Example 5**

P6. A credit union is offering a $485,000 30-year loan with an annual rate of 4.1% with 1.5 points or 3.8% with 1.85 points. If a borrower chooses the second option, how many months will it take to recoup the additional cost in points? Round to the nearest month. **Example 6**

P7. Zoey Bettincourt gets a $400,000 loan for a townhouse. She secures a 5/1 ARM at an initial interest rate of 2.75%. The initial monthly payment is $1632.96. After 5 years, the interest rate on the loan changes to 3.5%. Calculate the new monthly payment in year 6 of the loan. **Example 7**

P8. A Navy pilot has a 25-year mortgage at an annual interest rate of 4.875%. After making payments of $2453.65 for 4 years, the pilot would like to pay off the loan so as to purchase a new home. Calculate the loan payoff. **Example 8**

P9. A botanist secures a mortgage for $512,000 at an annual interest rate of 4.125% with 1.4 points. The loan origination fee is 0.75 point of the loan amount. Find the APR for the loan. Round to the nearest thousandth of a percent. **Example 9**

▶ Section Exercises

Solve.

1. **Calculating DTI** Calculate the DTI ratio for a borrower who has a gross monthly income of $5630 and has a minimum credit card bill of $250.00, a car lease payment of $364.97, and a student loan payment of $187, and who is applying for a loan for which the monthly payment will be $2784.64. Round to the nearest whole number.

2. **Calculating DTI** Josefina has a monthly income of $5500, and has a monthly car payment of $356.14, a monthly credit card bill of $175, and a $150 per month payment on a home entertainment system. Josefina is applying for a loan for which the monthly payment will be $2120.65. What is Josefina's DTI ratio? Round to the nearest whole number.

3. **Selling a Home** The Bourassas decide to sell a home for $490,000. They are charged a real estate commission of 6%

of the selling price, title insurance that is 1.2% of the selling price, and an escrow fee of $875.

a. What amount do the Bourassas receive after fees?

b. What percentage of the selling price were fees? Round to the nearest tenth of a percent.

4. **Selling an Apartment Building** The owner of an apartment building decides to sell the building for $2,500,000. The owner is charged a real estate commission of 4% of the selling price, title insurance that is 1.2% of the selling price, and an escrow fee of $1000.

a. What amount does the owner receive after fees?

b. What percentage of the selling price were fees? Round to the nearest tenth of a percent.

5. **Discount Points** A savings and loan charges 1.25 points for a home buyer to obtain a loan of $322,000. Calculate the discount points.

6. **Discount Points** A mortgage company charges 1.875 points for a home buyer to obtain a loan of $312,000. Calculate the discount points.

7. **Effect of PMI on a Loan Payment** A lender requires PMI that is 0.8% of the loan amount of $410,000. How much will this add to the borrower's cost to obtain the loan?

8. **Effect of PMI on a Loan Payment** A multimedia artist obtains a $390,000 home loan on which the lender requires PMI that is 0.75% of the loan amount. How much will this add to the artist's cost to obtain the loan?

9. **Monthly Mortgage Payment** A personal trainer obtains a home loan for $295,000 at an annual interest rate of 4.25% compounded monthly for 30 years. Find the monthly mortgage payment.

10. **Monthly Mortgage Payment** An orthodontist obtains a 30-year home loan for $575,000 at an annual interest rate of 4.125% compounded monthly. Find the monthly mortgage payment.

11. **Refinancing a Mortgage** Doreen Midcap has made payments for 5 years on a 30-year mortgage. Because interest rates have declined since the home was purchased, Doreen wants to refinance the mortgage interest rate from 6.5% to 4.5%. If the current mortgage payment is $1673.05 and Doreen will refinance the unpaid balance for 30 years, what is the new monthly payment?

12. **Refinancing a Mortgage** After owning his home for 4 years, Braxton wants to refinance the mortgage interest rate from 4.5% to 3.75%. If the current mortgage payment on the 20-year mortgage is $2236.72 and Braxton will refinance the unpaid balance for 30 years, what is Braxton's new monthly payment?

13. **Monthly Mortgage Payment** Remy obtains a mortgage in the amount of $625,000 for a co-op and secures a 7/1 ARM at an initial interest rate of 3%. The initial monthly payment is $2635.03. After 7 years, the interest rate on the loan changes to 3.5%. Calculate the new monthly payment in year 8 of the loan.

14. **Interest Rate Changes** Henry secures a $350,000 3/1 ARM loan at an initial interest rate of 3.125%. The monthly payment is $1499.31. After 3 years, the interest rate on Henry's loan changes to 3.75%. Calculate the new monthly payment.

15. **Discount Points** A credit union is offering 30-year home loan rates of 4.1% with 1.5 points or 3.8% with 1.85 points. If a borrower chooses the second option for a loan of $350,000, how many months will it take to recoup the additional cost in points? Round up to the nearest month.

16. **Discount Points Options** RGB Savings and Loan offers two 30-year home loan options on a $195,000 home loan. Option 1 is an interest rate of 3% with 1 point. Option 2 is an interest rate of 2.5% with 1.5 points. In how many months will option 1 be more expensive than option 2? Round to the nearest month.

17. **Mortgage APR** A computer scientist secures a 25-year mortgage for $675,000 at an annual interest rate of 3.625% with 1.25 points. The loan origination fee is 0.50 point of the loan amount. Calculate the APR for the loan. Round to the nearest thousandth of a percent.

18. **Mortgage APR** A high school music teacher obtains a 25-year mortgage for $410,000 at an annual interest rate of 4.875%. The teacher must pay 1.8 points and a loan origination fee of 1 point. Calculate the APR for the loan. Round to the nearest thousandth of a percent.

19. **Effect of Term Length on a Mortgage** Decreasing the number of years of a loan decreases the amount of interest repaid over the term of the loan. Suppose a dental hygienist has the option of a 30-year loan of $375,000 at an annual interest rate of 4.25% or 4.25% for 25 years.

a. Calculate the monthly payment for each loan.

b. Calculate the savings in interest by using the 25-year loan.

20. **Interest Rates** For Exercise 19, the interest rate for the two loans was the same. This is not typical. Usually, the shorter the term the smaller the interest rate. This has a substantial impact on the interest paid on a loan over the term of the loan. Suppose a 30-year loan of $250,000 has an interest rate of 4% while a 20-year loan has an interest rate of 3.75%. Find the difference between the amount of interest paid over the term of the 30-year loan and the 20-year loan.

21. **Effect of Term Length on a Mortgage** A restaurant manager has the option of a 30-year loan of $415,000 at an annual interest rate of 3.95% or the same interest rate but on a loan for 15 years.

a. Calculate the monthly payment for each loan.

b. Calculate the savings in interest by using the 15-year loan.

c. The term of the 15-year loan is one-half the term of the 30-year loan. Is the monthly payment for the 30-year loan one-half of that of the 15-year loan?

d. Assuming each loan is held to maturity, is the interest savings for the 15-year loan more or less than one-half of the interest paid on the 30-year loan?

22. **Total Monthly Home Payment** A veterinary ophthalmologist obtains a 20-year loan of $515,950 at an annual interest rate of 5.2%. Her annual property tax bill is $4250, and her annual fire insurance premium is $981. Find the total monthly payment for the mortgage, property tax, and fire insurance. Round to the nearest cent.

23. **Understanding Amortization** Using a spreadsheet program, create an amortization schedule for a 30-year mortgage of $500,000 at an annual interest rate of 4.5%. For help with this calculation, you can access an Excel template from your digital resources.

a. In which month does the amount of principal in a monthly payment first exceed the amount of interest?

b. How much interest is repaid for the term of the loan?

c. If the loan amount were $750,000 instead of $500,000, would the month in which the amount of principal in a monthly payment first exceeded the amount of interest change?

24. **Understanding Amortization** Using a spreadsheet program, create an amortization schedule for a 20-year mortgage of

$500,000 at an annual interest rate of 4.25%. For help with this calculation, you can access an Excel template from your digital resources.

a. In which month does the amount of principal in a monthly payment first exceed the amount of interest?

b. How much interest is repaid for the term of the loan?

c. If the loan amount was $750,000 instead of $500,000, would the month in which the amount of principal in a monthly payment first exceeded the amount of interest change?

▶ Investigations

Understanding Home Mortgages   Suppose you purchase a home for $300,000 and secure a 30-year loan for $240,000 at an interest rate of 4%. There are no points or PMI for your mortgage.

1. What is your monthly payment?

2. Assume your home appreciates at 3% per year. What is the value of your home after 5 years?

3. After the 5 years, you sell your home. When the sale is complete, you receive the difference between the selling price minus the loan payoff amount minus selling fees of 7% of the selling price. How much will you receive when escrow closes?

Understanding Mortgage Repayment Options   When repaying a mortgage, making an additional principal payment each month can significantly reduce the amount of interest paid over the term of the loan. In your digital resources, there is an amortization Excel template that will allow you to investigate various options for making additional principal payments.

Use the spreadsheet to experiment with various scenarios. Here is a suggestion to get you started. Suppose you purchase a home for $250,000 and obtain a 30-year loan at an annual interest rate of 3.75%. The answers to these questions can be found using the spreadsheet.

1. What is the monthly payment? How much interest will you pay on this loan over the 30-year period?

2. Suppose you add an additional $50 per month principal payment. How does this affect the term of the loan? How much interest will you pay over the term of the loan?

3. Suppose you add an additional $600 at the end of each year as a principal payment. How does this affect the term of the loan? How much interest will you pay over the term of the loan?

4. Explain why the answers to parts 2 and 3 are different even though in each case you have added $600 per year in extra principal payments.

Section 4.8

Stocks, Bonds, and Mutual Funds

Learning Objectives:

- Calculate ratios related to stocks.
- Calculate charges related to bonds.
- Calculate charges related to mutual funds.
- Calculate the effects of inflation on an investment.

Stocks

Stocks, bonds, and mutual funds are investment vehicles, but they differ in nature.

When owners of a company want to raise money, generally to expand their business, they may decide to sell part of the company to investors. An investor who purchases a part of the company is said to own stock in the company. Stock is measured in shares; a **share of stock** in a company is a certificate that indicates partial ownership in the company.

The owners of the certificates are called **stockholders** or **shareholders**. As owners, they share in the profits or losses of the corporation. A company may distribute profits to its shareholders in the form of dividends. A **dividend** is usually expressed as a per-share amount—for example, $0.07 per share.

Example 1   **Calculate Dividends Paid to a Shareholder**

Lockwood Design Science pays an annual dividend of $0.84 per share. Calculate the dividends paid to a shareholder who owns 200 shares of the company's stock.

Solution

$$\text{Dividends} = \text{Dividend per share} \cdot \text{Number of shares}$$
$$= 0.84 \cdot 200 = 168$$

The shareholder earned $168 in dividends.

The **dividend yield**, which is used to compare companies' dividends,

$$\text{Dividend yield} = \frac{\text{Dividend}}{\text{Stock price}} \times 100\%$$

Example 2   **Calculate a Dividend Yield**

A stock pays an annual dividend of $1.75 per share. The stock is trading at $70. Find the dividend yield.

Solution

$$\text{Dividend yield} = \frac{\text{Dividend}}{\text{Stock price}} \times 100\%$$

$$= \frac{\$1.75}{\$70} \times 100\% = 2.5\%$$

The dividend yield is 2.5%.

The **market value** of a share of stock is the price for which a stockholder is willing to sell a share of the stock and a buyer is willing to purchase it. Shares are always sold to the highest bidder. A **brokerage firm** is a dealer of stocks that acts as your agent when you want to buy or sell shares of stock. The brokers in the firm charge commissions for their service.

Most trading of stocks happens on a stock exchange. Stock exchanges are businesses whose purpose it is to bring together buyers and sellers of stock. The largest stock exchange in the United States is the New York Stock Exchange. There are, however, other exchanges such as the NASDAQ and the American Stock Exchange.

Information on the market value of a stock, its dividend yield, and other information can be found in newspapers and websites.

Here is a typical table with information about Target stock. This information can be obtained in real time from various Internet sites. Depending on the site, other metrics (information about a stock) may be available.

Open	Last	High	Low	52-Week Range	P/E	Dividend	Yield	EPS	Volume
163.07	161.71	168.00	155.20	155.20–268.98	11.46	3.61	2.23%	$3.22	50.8

This table gives Target's stock open (first bid of the day), its last (the price of the last trade), its high price of the day, its low price of the day, its trading range for the preceding 52 weeks, its price-to-earnings ratio (PE, discussed shortly), its dividend, its yield (the dividend yield discussed in Example 2), its earnings per share (EPS, also discussed shortly), and the volume (the number of shares that have been bought and sold during the day in millions).

The **earnings per share (EPS)** of a stock is the annual earnings of the company divided by the number of outstanding shares of stock that are currently available for purchase.

Example 3   **Calculate the EPS for a Company's Stock**

Suppose a company earned $1.5 billion during 1 year and there were 400 million outstanding shares of stock that year. Find the EPS for this company.

Solution

$$\text{EPS} = \frac{\text{Earnings}}{\text{Number of shares}} = \frac{1,500,000,000}{400,000,000} = 3.75$$

The EPS for the company was $3.75.

For Example 3, the $3.75 number means that the company earned $3.75 for each share of outstanding stock.

The **P/E** of a stock is the ratio of its price to its earnings per share.

Example 4   Calculate the P/E of a Company's Stock

The EPS of a company is $1.17, and its current stock price is $42.76. Find the company's P/E. Round to the nearest tenth.

Solution

$$P/E = \frac{\text{Stock price}}{\text{EPS}} = \frac{42.76}{1.17} \approx 36.547$$

The P/E is 36.5.

Bonds

When a corporation issues stock, it is *selling* part of the company to stockholders. A **bond** is a debt security that a corporation sells to **bondholders**, the investors who purchase the bonds. The corporation is *borrowing* money from investors. Corporations, the federal government, government agencies, states, and cities all issue bonds. These entities need money to operate—for example, to fund the federal deficit, repair roads, or build a new factory—so they borrow money from the public by issuing bonds.

Bonds are usually issued in units of $1000. The **par value** or **face value** of a bond is its value at maturity. The issuer promises to repay the bondholder on a particular day, called the **maturity date**, at a given rate of interest called the **coupon**. This rate is established at the time the bond is issued.

If a bondholder has bonds with a par value of $10,000 and a 3% coupon, then the annual interest paid to the bondholder is

$$I = Prt$$

$$= 10,000(0.03)(1) = 300$$

The annual interest is $300.

A key difference between owning a stock versus owning a bond is that stocks have no promises about dividends or returns, whereas the issuer of a bond guarantees that it will pay back the face value of the bond plus interest provided the issuer remains solvent.

Interest rates do not remain constant over time. The state of the economy and other factors influence interest rates. This in turn affects the current value of a bond.

$$\text{Current value of bond} = \frac{\text{Coupon rate of bond}}{\text{Current interest rate}} \cdot \text{Par value of the bond}$$

Example 5   Calculate the Current Value of a Bond

Suppose a bond with a par value of $1000 has a coupon of 4%. What is the current value of the bond if current interest rates are 5%?

Solution

$$\text{Current value of bond} = \frac{\text{Coupon rate of bond}}{\text{Current interest rate}} \cdot \text{Par value of the bond}$$

$$= \frac{0.04}{0.05} \cdot \$1000$$

$$= 800$$

The current value of the bond is $800.

Example 5 shows that although the par value of a bond is $1000, its current value to an investor is $800. The bond is said to be **trading at a discount**. If an investor owned the bond, the investor could sell it for $800, not the $1000 purchase price.

In Example 5, interest rates increased from 4% to 5%. If interest rates had decreased from, say, 4% to 3%, the current value of the bond would increase and be larger than the par value. The calculation is similar to Example 5.

$$\text{Current value of bond} = \frac{\text{Coupon rate of bond}}{\text{Current interest rate}} \cdot \text{Par value of the bond}$$

$$= \frac{0.04}{0.03} \cdot \$1000$$

$$\approx \$1333$$

In this case, the nominal value of the bond is $1333, and the bond is said to be **trading at a premium**.

Mutual Funds

A **mutual fund** is formed by a group of people who pool their money for a financial manager, frequently called the *fund manager*, to invest. Usually, the investments are highly specific. For instance, the mutual fund may invest only in healthcare stocks, technology stocks, apartment buildings, or commercial real estate. Sometimes the mandate of the fund is to buy large capitalization stocks (companies with a total value of $10 billion or more), mid-capitalization stocks (between $2 billion and $10 billion), or small capitalization stocks ($300 million to $2 billion). These are somewhat arbitrary divisions, and fund managers may use some discretion in selecting stocks.

Other types of mutual funds called **index funds** only invest in stocks that belong to a certain stock group. Probably the most notable is the Dow Jones Industrial Average, a group of 30 stocks. Other stock indexes are the S&P 500 Index (500 companies whose capitalization ranges from approximately $3 billion to $700 billion) and the Russel 2000 Index (2000 companies), among many others.

When a mutual fund is formed, say, for energy stocks, the fund manager purchases stocks using the pooled money from investors. Now some people may have invested $10,000 and others $5000. To make the arrangement equitable to all, some arbitrary number is chosen as the base value for owning part of the mutual fund, say $10. The person who invested $10,000 gets 1000 shares ($10,000 ÷ $10 = 1000); the person who invested $5000 gets 500 shares. The typical mutual fund has millions of shares.

The value of the original base ($10 in our example) will change over time as the value of the stocks purchased by the fund manager change in value. The **net asset value** for a share of stock in a mutual fund is the value of all the shares owned by the mutual fund, divided by the number of shares the fund issued.

Example 6   **Find the Net Asset Value of a Share in a Mutual Fund**

Suppose a mutual fund has a portfolio of stocks that have a market value of $2.5 billion and the company has 80,000,000 shares of stock. What is the net asset value of a share of the mutual fund?

Solution

$$\text{Net asset value of one share} = \frac{2,500,000,000}{80,000,000} = 31.25$$

The net asset value of one share is $31.25.

●

If the original investors in the mutual fund in Example 6 only paid $10 for a share in the fund, they would have had a good return on their investment.

For mutual funds, unlike stocks, an investor usually purchases a certain dollar amount, not a certain number of shares. The number of shares purchased is rounded to the nearest thousandth. For instance, if the net asset value of one share of a mutual fund is $21.47 and an investor wished to purchase for $3500 of this fund, then

$$\text{Number of shares purchased} = \frac{\$3500}{\$21.47} \approx 163.01816$$

The fund manager is compensated by fees charged to the people who buy shares in the mutual fund. There are basically two fee structures: load funds and no-load funds.

In a load fund, a person who purchases a share of the fund is charged a fee, generally between 4% and 8% of the purchase amount. In a no-load fund, fees are collected as a percentage of the total value of the fund. This fee can range from around 0.1% to 2%, depending on the fund.

Example 7 **Calculate a Fee for a Load Mutual Fund**

A physical therapist invests $5000 into a mutual fund for which the transaction fee is 4%. If the net asset value per share of the fund is $42.56, how many shares can the trainer purchase? Round to the nearest thousandth.

Solution

The fee is 4% of $5000.

Fee = 0.04 · $5000 = $200

The fee is subtracted from the amount invested so that $5000 − $200 = $4800 is available to purchase shares of the fund,

$$\text{Number of shares purchased} = \frac{\$4800}{\$42.56} \approx 112.78195$$

The physical therapist can purchase 112.782 shares of the fund.

Our discussion of fees is fairly general, and other fee structures are used by mutual funds. It is important to know those fees because they affect the value of your investment. For instance, the net asset value per share of the fund in Example 7 must increase to $43.33 before the value of the investment is $5000, the amount paid for the shares.

For a no-load fund, the fee is a percentage of total assets. This fee is deducted from the value of the fund before the net asset value per share is calculated. For instance, suppose a mutual fund has $90 billion in assets and the expense fee is 0.85%. Then the net asset value per share of stock is based on the value of the assets minus 0.85% of that value. If the fund has 1.5 billion shares outstanding, then the net asset value per share is

$$\text{Net asset value per share} = \frac{\$90 \text{ billion} - (0.0085 \cdot \$90 \text{ billion})}{1.5 \text{ billion}} = \$59.49$$

Effects of Inflation

The effects of inflation are important to investors as well. There are two interest rates that are important to investors: the *nominal* interest rate and the *inflation* rate. The **nominal rate** is the stated interest rate on an investment. It is the rate the investor expects to realize on the investment. Using the nominal rate and the inflation rate, an investor can determine what is called the **real rate of return**, the annual interest rate the investor is actually receiving.

For instance, suppose an investor pays $100 for an investment that earns an annual interest rate of 10%, the nominal rate. In 1 year, the investment will be worth

$$FV = PV(1 + r)$$
$$= 100(1 + 0.10) = 100(1.10)$$
$$= 110$$

The future value is $110, which is $10 more than the original investment. However, if the rate of inflation is, say, 4%, the present value of $110 is

$$PV = \frac{FV}{\left(1 + \dfrac{r}{n}\right)^{nt}}$$

$$= \frac{110}{\left(1 + \dfrac{0.04}{1}\right)^{1(1)}} \qquad • FV = 110, \ r = 0.04, \ n = 1, \ t = 1.$$

$$= \frac{110}{1.04}$$

$$\approx 105.7692308$$

Considering inflation, the investment increased by approximately $5.77, not the apparent $10 when inflation is not considered. Using the simple interest formula, the *real rate of return* is

$$I = Prt$$

$$5.77 = 100r(1) \qquad • I = 5.77, \ P = 100, \ t = 1.$$

$$0.0577 = r$$

The real rate of return is 5.77%.

If we repeat this calculation in a general way, we can derive a formula for real rate of return.

Real Rate of Return Formula

$$\text{Real rate of return} = \frac{1 + \text{nominal rate}}{1 + \text{inflation rate}} - 1$$

Real rate of return is usually written as a percentage.

Example 8 **Find the Real Rate of Return for an Investment**

A nurse expects to earn an annual interest rate of 6.5% on an investment of $10,000. If the current rate of inflation is 1.8%, what is the real rate of return on the investment? Round to the nearest hundredth of a percent.

Solution

Use the real rate of return formula.

$$\text{Real rate of return} = \frac{1 + \text{nominal rate}}{1 + \text{inflation rate}} - 1$$

$$= \frac{1 + 0.065}{1 + 0.018} - 1$$

$$\approx 1.046168959 - 1$$

$$\approx 0.04617$$

The real rate of return is 4.62%.

4.8 Exercise Set

Think About It

1. If a stock pays a dividend of $2.50 and the stock increases in value, does the dividend yield go up or down?

2. If the EPS of a stock increased and the price of the stock remained the same, would the P/E of the stock increase or decrease?

3. If the par value of a bond is $5000 and the nominal value of the bond is $4700, is the current interest rate less than or greater than the interest rate at the time the bond was issued?

4. If the par value of a bond is $1000 and the nominal value of the bond is $900, is the bond trading at a discount or a premium?

5. What is the difference between the nominal rate of return and the real rate of return on an investment?

Preliminary Exercises

For Exercises P1 to P8, if necessary, use the referenced example following the exercise for assistance.

P1. A stock pays an annual dividend of $2.03 per share. Calculate the dividends paid to a shareholder who has 500 shares of the company's stock. **Example 1**

P2. A stock pays an annual dividend of $2.035 per share. If the stock is trading at $68.35, what is the dividend yield? Round to the nearest tenth of a percent. **Example 2**

P3. Suppose a company earned $3.25 billion during 1 year and for that year there were 850 million outstanding shares of stock. Find the EPS for this company. Round to the nearest cent. **Example 3**

P4. The EPS of a company is $5.06, and its current stock price is $86.29. Find the company's P/E. Round to the nearest tenth. **Example 4**

P5. Suppose a bond with a par value of $1000 has coupon of 2.75%. What is the current value of the bond if the current interest rate is 3.25%? Round to the nearest dollar. **Example 5**

P6. Suppose a mutual fund has a portfolio of stocks that have a market value of $3.25 billion dollars, and the company has 120,000,000 shares of stock. What is the net asset value of a share of the mutual fund? Round to the nearest cent. **Example 6**

P7. Carley Montoya invests $7500 into a mutual fund for which the transaction fee is 5%. If the net asset value per share of the fund is $38.26, how many shares can Carley purchase? Round to the nearest thousandth. **Example 7**

P8. A music composer deposits $5000 into an investment that earns an annual interest rate of 5.6%. If the current rate of inflation is 1.3%, what is the real rate of return on the investment? Round to the nearest hundredth of a percent. **Example 8**

Section Exercises

Solve.

1. **Real Rate of Return** A dog trainer expects to earn an annual interest rate of 3.4% on an investment of $25,000. If the current rate of inflation is 1.2%, what is the real rate of return on the investment? Round to the nearest hundredth of a percent.

2. **Real Rate of Return** Heritage Associates LLC offers a 10-year investment into a shopping mall with projected annual returns of 6.5%. If Francois invests $30,000 into this investment and the current rate of inflation is 2.4%, what is the real rate of return? Round to the nearest hundredth of a percent.

3. **Dividend Yield** A stock pays an annual dividend of $1.27 per share. If the stock is trading at $41.05, what is the dividend yield?

4. **Dividend Yield** The stock of a fast-food restaurant chain pays a dividend of $0.63. If the share price for this company is $40.89, what is the dividend yield for the stock?

5. **Net Asset Value of Shares** Suppose a mutual fund has a portfolio of stocks that have a market value of $10.25 billion and the company has 900 million shares of stock. What is the net asset value of a share of the mutual fund? Round to the nearest cent. Disregard the fees that may be charged by the company.

6. **Net Asset Value of Shares** A recent value of the Wellington Fund's portfolio of assets was approximately $18.9 billion. If the Wellington Fund has 460 million outstanding shares, what is the net asset value of a share of the fund? Round to the nearest cent. Disregard the fees that may be charged by the company.

7. **Interest Payments** A bond with a $5000 par value has a 5.5% coupon and a 7-year maturity date. Calculate the total of the interest payments paid to the bondholder for the 7-year period.

8. **Interest Payments** A junk bond is one in which there is substantial doubt that the company will be able to repay investors holding that bond. Suppose a junk bond with a $10,000 par value has a 6% coupon and a 4-year maturity date. How much interest must this company pay over the 4-year period?

9. **Calculating EPS** A company earned $2.2 billion during 1 year, with 1.02 billion outstanding shares of stock. Find the EPS for this company. Round to the nearest cent.

10. **Calculating EPS** Apple had earnings of $45,687 million during a year when there were 5.5 billion shares of outstanding stock. Find the EPS for Apple during that year. Round to the nearest cent.

11. **A Company's P/E** A recent EPS for IBM was $13.89 when its stock price was $167.13. Find IBM's P/E. Round to the nearest tenth.

12. **A Company's P/E** The EPS of a company is $4.15, and its current stock price is $95.34. Find the company's P/E. Round to the nearest tenth.

13. **Nominal Value of a Bond** If a bond has a par value of $1000 and a coupon of 4%, what is the nominal value of the bond if the current interest rate is 3.5%? Round to the nearest dollar.

14. **Nominal Value of a Bond** A corporate bond has a par value of $1000 and a coupon of 5.75%. What is the nominal value of the bond if current interest rates are 5%?

15. **Dividend Payments** A stock pays an annual dividend of $1.966 per share (some companies pay dividends in tenths or hundredths of a cent—and sometimes even smaller portions of a cent). Calculate the dividends paid to a shareholder who has 500 shares of the company's stock.

16. **Dividend Payments** Cisco Systems paid an annual dividend of $0.84 per share in a recent year. Calculate the dividends paid to a shareholder who has 1000 shares of Cisco stock.

17. **Purchasing Shares** Seth Jefferson invests $10,000 into a mutual fund for which the transaction fee is 6.25%. If the net asset value per share of the fund is $42.08, how many shares can Seth purchase? Round to the nearest thousandth.

18. **Purchasing Shares** A diesel mechanic invests $4500 into a mutual fund for which the transaction fee is 6%. If the net asset value per share of the fund is $25.37, how many shares can the mechanic purchase? Round to the nearest thousandth.

19. **Net Asset Value** A no-load mutual fund specializing in healthcare stocks has a portfolio of stocks with a market value of $11.3 billion and 750 million shares of stock. What is the net asset value of a share of the mutual fund if the expense fee is 0.24%? Round to the nearest cent.

20. **Net Asset Value** The Horton no-load mutual fund, specializing in renewable energy stocks, has a portfolio of stocks with a market value of $9.6 billion and has 450 million shares of stock. What is the net asset value of a share of the mutual fund if the expense fee is 1.28%? Round to the nearest cent.

21. **A Company's P/E** A recent EPS for Johnson and Johnson was $6.04 when its price was $146.92. Find Johnson and Johnson's P/E. Round to the nearest tenth.

22. **Net Asset Value** A global real estate fund has a portfolio of buildings that have a market value of $239 million and the company has 12,500,000 shares of stock. What is the net asset value of a share of the mutual fund? Round to the nearest cent.

23. **Dividend Payments** During a recent year, Emerson Electric paid an annual dividend of $0.66 per share. Calculate the dividends paid to a shareholder who has 450 shares of Emerson Electric stock.

24. **Nominal Value of a Bond** A bond has a par value of $1000 and a coupon of 4.5%. What is the nominal value of the bond if current interest rates are 4.75%? Round to the nearest dollar.

25. **A Company's EPS** Home Depot had earnings of $7.1 billion during a year when it had 1.3 billion shares of outstanding stock. Find the EPS for Home Depot during that year.

26. **Purchasing Shares** A chef invests $7500 into a mutual fund for which the transaction fee is 3.5%. If the net asset value per share of the fund is $33.62, how many shares can the chef purchase? Round to the nearest thousandth.

27. **Calculating Dividend Yield** WindSolar stock is trading at $119.30 and pays a dividend of $3.08. What is the dividend yield for this stock?

28. **Interest Payments** A bond with a $5000 par value has a 4.25% coupon and a 10-year maturity date. Calculate the total of the interest payments paid to the bondholder over the 10 years.

▶ Investigation

Understanding Stocks and Mutual Funds Suppose you are given $50,000 to invest. Create a portfolio of stocks of your choosing using $25,000. Invest the remaining $25,000 in a diversified mutual fund. Follow the stocks in your portfolio and the value of the mutual fund for 2 weeks. After that time, calculate the percentage gain or loss you would have incurred from your portfolio and the mutual fund. Be sure to include whatever dividends you may have received.

Test Prep Managing Your Finances

Section 4.1 | Simple Interest

Simple Interest Formula   The simple interest formula is $I = Prt$, where I is the interest, P is the principal, r is the interest rate, and t is the time period.

See **Examples 1, 2,** and **3** on pages 144 and 145, and then try Exercises 1, 2, and 3 on page 213.

Maturity Value or Simple Interest Formulas for Simple Interest   The future or maturity value formulas for simple interest are

$$A = P + I$$
$$A = P + Prt$$
$$A = P(1 + rt)$$

where A is the amount after the interest (I) has been added to the principal (P), r is the interest rate, and t is the time.

    Each formula gives exactly the same result. The choice of which formula to use is just a matter of convenience.

See **Examples 4** and **5** on pages 146 and 147, and then try Exercises 4 and 5 on page 213.

Treasury bills, or **T-bills**, are U.S. government securities that have a *term* (the time a purchaser holds the T-bill) that is less than 1 year. T-bills are typically sold at a *discount* from the *par amount* (also called *face value* or *maturity value*).

See **Example 6** on page 147 and then try Exercise 6 on page 213.

Section 4.2 | Compound Interest

The **future value, FV**, of an amount deposited into compound interest account is given by

$$FV = PV\left(1 + \frac{r}{n}\right)^{nt}$$

where PV is the amount of money deposited (the present value), r is the annual interest rate as a decimal, n is the number of compounding periods per year, and t is the number of years.

See **Examples 1** and **2** on pages 151 and 153, and then try Exercises 8, 9, and 10 on page 213.

The **present value, PV**, of an amount deposited into a compound interest account is given by

$$PV = \frac{FV}{\left(1 + \frac{r}{n}\right)^{nt}}$$

where FV is the value of money expected to be received in the future (the future value), r is the annual interest rate as a decimal, n is the number of compounding periods per year, and t is the number of years.

See **Example 3** on pages 153 and 154 and then try Exercises 11 and 12 on page 213.

Inflation is an economic condition during which there are increases in the costs of goods and services.

See **Example 4** on page 155, and then try Exercise 13 on page 213.

The **consumer price index (CPI)** is a measure of the average change over time in the prices paid by consumers for a market basket of consumer goods and services.

    The CPI is used to convert **current dollars** (also called **nominal dollars**) into **constant dollars** (also called **real dollars**). Constant dollars are used to compare current salaries and prices to an equivalent salary or price at a different time.

See **Examples 5** and **6** on page 156 and then try Exercises 14, 15, and 16 on page 213.

The **annual percentage yield (APY)** is the simple interest rate that would yield the same amount of interest after 1 year as the nominal (current) rate. APY is also called **effective annual rate**.

$$APY = \left(1 + \frac{APR}{n}\right)^{n} - 1$$

where APR is the annual percentage rate as a decimal, n is the number of compounding periods per year, and APY is the annual percentage yield as a decimal.

See **Example 7** on page 157 and then try Exercise 17 on page 213.

Section 4.3 Future and Present Value of an Annuity

An **annuity** is a series of equal payments (sometimes called **cash flows**) for a given length of time. A **due annuity** is one for which the series of payments is made at the *beginning* of each payment period. An **ordinary annuity** is one for which the series of payments is made at the *end* of each payment period.

The formula for the future value of a due annuity is

$$FV = PMT\left[\frac{(1 + i)^n - 1}{i}\right](1 + i)$$

where *PMT* is the cash flow amount, *n* is the number of payments, and *i* is the interest rate per period:

$$i = \frac{\text{Annual interest rate as a decimal}}{\text{Number of compounding periods per year}}$$

The formula for the future value of an ordinary annuity is

$$FV = PMT\left[\frac{(1 + i)^n - 1}{i}\right]$$

where *PMT* is the cash flow amount, *n* is the number of payments, and *i* is the interest rate per period:

$$i = \frac{\text{Annual interest rate as a decimal}}{\text{Number of compounding periods per year}}$$

See **Examples 1, 2,** and **3** on pages 161 to 163, and then try Exercises 19, 20, and 21 on page 213.

A **sinking fund** is an annuity established to repay a debt such as a bond that a state or company may have sold or to save for future capital expenses such as a new computer system.

See **Example 4** on page 163 and then try Exercise 22 on page 213.

The **present value of an annuity** is the sum of the present values of all expected cash flows.

The formula for the present value of an ordinary annuity is

$$PV = PMT\left[\frac{1 - (1 + i)^{-n}}{i}\right]$$

where *PMT* is the cash flow amount, *n* is the number of payments, and *i* is the interest rate per period:

$$i = \frac{\text{Annual interest rate as a decimal}}{\text{Number of compounding periods per year}}$$

The formula for the present value of a due annuity is

$$PV = PMT\left[\frac{1 - (1 + i)^{-n}}{i}\right](1 + i)$$

where *PMT* is the cash flow amount, *n* is the number of payments, and *i* is the interest rate per period:

$$i = \frac{\text{Annual interest rate as a decimal}}{\text{Number of compounding periods per year}}$$

See **Examples 5, 6,** and **7** on pages 165 to 166, and then try Exercises 23, 24, and 25 on page 213.

Monthly payment for an installment loan The payment due on most consumer loans is based on the formula for the present value of an ordinary annuity.

See **Examples 8** and **9** on page 167 and then try Exercises 26 and 27 on page 213.

Determine a Loan Payoff To calculate a loan payoff for an APR installment loan, use the formula for the present value of an ordinary annuity. The value of *n* is the number of *remaining* payments to be made.

See **Example 10** on page 168 and then try Exercise 28 on page 213.

Section 4.4 Credit Cards

A **finance** charge is an amount paid in excess of the cash price; it is the cost to the customer for the use of credit.

The **average daily balance** of credit card charges and payments is calculated by dividing the sum of the total amounts owed each day of the billing cycle by the number of days in the billing cycle.

See **Example 1** on pages 170 and 171, and then try Exercise 30 on page 214.

Most finance charges on credit card bills are calculated as a percentage of the average daily balance.	See **Example 2** on pages 171 and 172, and then try Exercises 31 and 32 on page 214.
Because of the number of transactions for most credit card customers, it is necessary to use software to calculate a finance charge.	See **Example 3** on pages 172 and 173, and then try Exercises 31 and 32 on page 214.
A **minimum payment** due for a monthly credit card payment differs from company to company. However, minimum payments are a percentage of the due amount.	See **Example 4** on page 173, and then try Exercise 33 on page 214.
If a person wants to pay off their credit card debt, the formula for the present value of an ordinary annuity is used.	See **Example 5** on page 174, and then try Exercise 34 on page 214.

Section 4.5 Student Loans

Besides the interest rate on which a student loan repayment plan is based, student loans have a loan origination fee.	See **Example 1** on page 178, and then try Exercise 35 on page 214.
Stafford student loans are available through the U.S. Department of Education. A **subsidized** loan is one for which the interest on the loan is paid by the government. An **unsubsidized** loan is one for which the interest on the loan is paid by the borrower. To calculate the monthly payment to repay the loan, use the formula for the present value of an ordinary annuity.	See **Examples 2** and **3** on pages 178 and 179, and then try Exercises 36 and 37 on page 214.
An **income adjusted** student loan repayment plan that is based on the student's ability to pay, based on annual income earned after graduation.	See **Example 4** on page 179, and then try Exercise 38 on page 214.
PLUS Loans are offered to parents of students enrolled at least half-time or to graduate or professional students to cover the cost for college.	See **Example 5** on page 180, and then try Exercise 39 on page 214.
Because there are discount fees associated with a student loan, the APR of the loan is different from the interest rate on which monthly payments are calculated.	See **Example 6** on page 180, and then try Exercise 40 on page 214.

Section 4.6 Financial Considerations of Car Ownership

A monthly car payment is calculated using the formula for the present value of an ordinary annuity. This formula is also used to calculate the payoff amount for a car loan.	See **Examples 1, 2,** and **3** on pages 182 and 183, and then try Exercises 41, 42, and 43 on page 214.
An important consideration when buying a car is deciding how much a person can afford for a monthly payment.	See **Example 4** on page 184 and then try Exercise 44 on page 214.
Besides the monthly car payment, owning and operating a car has ongoing costs. Some of these expenses are fuel, insurance, license fees, and depreciation.	See **Examples 5, 6,** and **7** on pages 184 to 186, and then try Exercises 45, 46, and 47 on page 214.
An alternative to owning a car is leasing a car. Monthly lease payments may be less than monthly car payments when buying a car.	See **Example 8** on pages 187 and 188, and then try Exercise 48 on page 214.

Section 4.7 Financial Considerations of Home Ownership

To determine whether a borrower can afford to purchase a home, lenders use the **debt-to-income (DTI) ratio**.	See **Examples 1** and **2** on pages 192 and 193, and then try Exercise 49 on page 214.
A **mortgage** is a loan to buy a home. Many types of mortgages are available to home buyers today, so the terms of mortgages differ considerably. A **fixed-rate** mortgage, or **conventional** mortgage, is one in which the interest rate on the loan remains the same throughout the life of the mortgage. A monthly mortgage payment is calculated by using formula for the present value of an ordinary annuity.	See **Example 3** on pages 193 and 194, and then try Exercise 51 on page 214.
Private mortgage insurance (PMI) is an insurance policy required of borrowers who make a down payment that is less than 20% of the purchase price of a home. The premium for this insurance is added to the borrower's monthly payment.	See **Example 4** on page 194, and then try Exercise 52 on page 215.

A lender may charge a **loan origination fee** to cover costs to the lender. This fee is usually a percentage of the loan amount. **Discount points**, usually referred to as just **points**, is a fee that may be required of a borrower. One point is 1% of the loan amount.	See **Examples 5** and **6** on page 195, and then try Exercises 53 and 55 on page 215.
Some mortgages are **adjustable rate mortgages** (ARMs). The interest rate charged on an ARM is adjusted periodically to more closely reflect current interest rates.	See **Example 7** on page 196, and then try Exercise 56 on page 215.
An **amortization schedule** is a table showing the details of each periodic payment of a mortgage. These details show how much of each payment is interest, how much goes to repaying the loan, and the new loan balance. The schedule can be used to determine the amount owed on the loan at any time. A calculator can also be used to determine a loan payoff.	See **Example 8** on pages 197 and 198, and then try Exercise 57 on page 215.
Fees such as a loan origination fee and points can affect the APR of a home loan.	See **Example 9** on page 198, and then try Exercise 57 on page 215.

Section 4.8   Stocks, Bonds, and Mutual Funds

A **share of stock** in a company is a certificate that indicates partial ownership in the company.
 	
    The owners of the certificates are called **stockholders** or **shareholders**. As owners, the stockholders share in the profits or losses of the corporation. A company may distribute profits to its shareholders in the form of dividends.
 	
    The **dividend yield**, which is used to compare companies' dividends, is the amount of the dividend divided by the stock price and is expressed as a percentage.
	See **Examples 1** and **2** on pages 201 and 202, and then try Exercises 59 and 60 on page 215.

The **earnings per share (EPS)** of a stock is the annual earnings of the company divided by the number of outstanding shares of stock that are currently available for purchase.	See **Example 3** on page 202, and then try Exercise 61 on page 215.
The **P/E** of a stock is the ratio of its price to its earnings per share.	See **Example 4** on page 203, and then try Exercise 62 on page 215.
A **bond** is a debt security that a corporation sells to **bondholders**, the investors who purchase the bonds. Bonds are usually issued in units of $1000. The **par value** or **face value** of a bond is its value at maturity. The issuer promises to repay the bondholder on a certain day, called the **maturity date**, at a given rate of interest, called the **coupon**. This rate is established at the time the bond is issued.	See **Example 5** on page 203, and then try Exercise 63 on page 215.

A **mutual fund** is formed by a group of people who pool their money for a financial manager, frequently called the *fund manager*, to invest.
 	
    The **net asset value** for a share of stock in a mutual fund is the value of all the shares owned by the mutual fund, divided by the number of shares the fund issued.
 	
    There are **no-load** funds where fees are charged as a percentage of the net asset value and **load** funds where a fee is charged to the purchaser.
	See **Examples 6** and **7** on pages 204 and 205 and then try Exercises 64 and 65 on page 215.

Two interest rates are important to investors: the *nominal* interest rate and the *inflation* rate. The **nominal rate** is the stated interest rate on an investment of the interest rate the investor expects to realize on the investment. Using the nominal rate and the inflation rate, an investor can determine what is called the **real rate of return**, the annual interest rate the investor is actually receiving. The formula for real rate of return is $$\text{real rate of return} = \frac{1 + \text{nominal rate}}{1 + \text{inflation rate}} - 1$$	See **Example 8** on page 206, and then try Exercise 66 on page 215.

Review Exercises

Use 365 days in one year.

1. **Interest on a Loan** Calculate the simple interest due on a
   a. 3-year loan of $2750 if the annual interest rate is 6.75%.
   b. 3-month loan if the monthly interest rate is 0.9%.

2. **Interest on a Loan** Find the simple interest due on an 8-month loan of $8500 if the annual interest rate is 9%.

3. **Annual Interest Rate** The simple interest charged on a $500 loan of 14 days was $25. Find the annual interest rate. Use a 365-day year. Round to the nearest tenth of a percent.

4. **Maturity Value** Calculate the maturity value of a simple interest, 108-day loan of $7000 if the interest rate is 10.4%.

5. **Simple Interest Rate** The simple interest charged on a 3-month loan of $6800 is $127.50. Find the simple interest rate.

6. **Treasury Bills** Morris purchases a T-bill with a face value of $5000 for $4950. If the term of the T-bill is 3 months, what is the annual simple interest rate earned by the investor? Round to the nearest tenth of a percent.

7. **Compound Interest Rate** Calculate the compound amount when $3000 is deposited in an account earning 6.6% interest, compounded monthly, for 3 years.

8. **Compound Interest Rate** What is the compound amount when $6400 is deposited in an account earning an interest rate of 6%, compounded quarterly, for 10 years?

9. **Future Value** Find the future value of $6000 earning 9% interest, compounded daily, for 3 years.

10. **Interest Earned** Calculate the amount of interest earned in 4 years on $600 deposited in an account paying 7.2% interest, compounded daily.

11. **Savings Planning** How much money should be invested in an account that earns 8% interest, compounded semiannually, to have $18,500 in 7 years?

12. **Planning for College** A couple plans to save for their child's college education. What principal must be deposited by the parents when their child is born to have $80,000 when the child reaches the age of 18? Assume the money earns 8% interest, compounded quarterly.

13. **Purchasing Power** Cyrilla purchases an insurance policy today that will provide Cyrilla with $250,000 at retirement in 10 years. Assuming an annual inflation rate of 3.1%, what will be the purchasing power of $250,000 in 10 years?

14. **Inflation Rates** Two years ago, the consumer price index was 228.936. Today the CPI is 231.258. Find the inflation rate between the two years. Round to the nearest tenth.

15. **Inflation Rates** The CPI for housing in 2020 was 264.257, and in 2022 it was 274.561. If a house cost $275,000 in 2020, what would an equivalent house cost in 2022?

16. **Salaries in Constant Dollars** Leroy had a salary of $79,000 in 2021 when the CPI was 268.241 and a salary of $84,000 in 2023 when the CPI was 291.507. In constant 2023 dollars, has Leroy's salary increased between the two years?

17. **APY** An investment in a shopping center offers an annual interest rate of 8%, compounded monthly. Find the annual percentage yield. Round to the nearest tenth of a percent.

18. **Understanding Yield** Which has the higher annual yield—5.2% compounded quarterly or 5.4% compounded semiannually?

19. **Investment Value over Time** A sous chef starts a retirement savings plan by depositing $250 per month at the beginning of each month into an account that earns an annual interest rate of 4.5% compounded monthly. Find the value of this investment after 20 years.

20. **401(k) Value over Time** A cosmetologist begins a 401(k) program by investing $50 at the end of each week into an account that earns an annual interest rate of 5.25% compounded weekly. Find the value of the 401(k) after 15 years.

21. **Planning for Graduate School** Zelda wants to have $8000 in 3 years for tuition to begin a master's degree in business. How much should Zelda deposit at the beginning of each month into an account that earns 3.125% compounded monthly to reach the goal?

22. **Sinking Funds** The water district for a county issues $50 million in bonds that will mature in 25 years. How much money must be deposited at the end of each 6 months into a sinking fund that earns 6.5% compounded semiannually to retire the bonds?

23. **Present Value of a Structured Settlement** Rene has a structured settlement that pays $2500 at the end of each month for 10 years. If the current interest rate is 4.875% compounded monthly, what is the present value of the payments?

24. **Present Value** A clothing company pays an athlete $1250 at the beginning of each month to wear the company's sportswear during competitive events. If the contract is for 8 years at an annual interest rate of 4.75% compounded monthly, what is the present value of the payments?

25. **Choosing a Bonus Payout** As a signing bonus, a baseball player is given the option of $1,000,000 now or a series of payments of $25,000 per month for 20 years. If the player estimates that the average annual rate of return over the next 20 years will be 4.25%, which option has the greater present value?

26. **Financing Kitchen Appliances** Michel purchases all-new appliances for a kitchen at a cost of $8500. If Michel makes a down payment of 10% and finances the remaining balance for 3 years at an annual interest rate of 5.5% compounded monthly, what are the monthly payments?

27. **Financing New Windows** Olivia begins a remodel of a home by replacing all the windows with new energy efficient ones at a cost of $45,000. If Olivia makes a 25% down payment and finances the remainder at 3.8% compounded monthly for 8 years, what are the monthly payments? How much interest is paid over the term of the loan?

28. **Paying Off a Car Loan** Dasan has a $32,500, 5-year loan for a hybrid car at an annual interest rate of 4.8%. The monthly payment is $650.80. After 3 years, Dasan decides to purchase a new car. What is the payoff on the loan?

29. **Financing a Camera** Photo Experts offers a Nikon camera for $999, including taxes. If you finance the purchase of this

camera for 2 years at an annual interest rate of 8.5%, what is the monthly payment?

30. Average Daily Balance on a Credit Card A credit card account had an unpaid balance of $423.35 on March 11. Purchases of $145.50 were made on March 18, $212.94 on March 20, and $132.89 on April 1. A payment of $250 was made on March 29. Find the average daily balance if the billing date is April 11.

31. Determining a Finance Charge On September 10, a credit card account had an unpaid balance of $450. Purchases of $47.32 were made on September 12, $157.55 was charged on September 20, and $56.73 was charged on September 30. A payment of $175 was made on September 28. The annual interest on the average daily balance is 21.25%. Find the finance charge on the October 10 bill.

32. Determining a Finance Charge On February 3, a credit card account had an unpaid balance of $1023. Purchases of $68.54 were made on February 12, $52.33 was charged on February 20, and $124.75 was charged on February 22. A payment of $450 was made on February 28. The annual interest on the average daily balance is 19.6%. Find the finance charge on the March 3 bill. Assume it is not a leap year.

33. Minimum Payments A medical assistant owes $750.25 on a credit card bill. If the minimum monthly payment is the greater of 4% of the amount owed or $25, what is the minimum payment due?

34. Paying Off Debt An elite sports trainer has $4253.67 in credit card debt. If the trainer makes a monthly payment of $450 (and makes no additional charges on the account), how many months will it take to repay the debt if the annual interest rate on the credit card is 18.5%? Round up to the nearest month.

35. Student Loans Geraldine gets a Stafford student loan for $15,000 and must pay a loan fee of 1.1%. How much money does Geraldine receive?

36. Monthly Student Loan Payments A student receives a subsidized Stafford Loan of $17,000 at an annual interest rate of 4.1% for 6 years. What are the monthly payments on the loan when the student graduates 2 years later?

37. Monthly Student Loan Payments Pytor obtains a nonsubsidized student loan of $9000 for 6 years at an annual interest rate of 4.9%. What are the monthly payments when Pytor begins making payments 4 years later?

38. Student Loans Francis received a 10-year subsidized student loan of $27,000 at an annual interest rate of 5.2%. If Francis qualifies for an income-adjusted repayment plan of $325 per month, how long will it take to repay the loan?

39. Monthly Student Loan Payments Sabrina received a 5-year PLUS Loan of $36,000 to attend the last year of pharmacy school. If the current interest rate is 5.95% and Sabrina starts making repayments one year after graduation, what are the monthly payments?

40. APR on a Student Loan Hana received a nonsubsidized student loan of $7500 at an annual interest rate of 4.89% for 7 years. If Hana begins repaying the loan 5 years later, what is the APR on the loan? Round to the nearest tenth of a percent.

41. Monthly Car Payments Suppose that you decide to purchase a new car. You go to a credit union to get preapproval for your loan. The credit union offers you an annual interest rate of 3.25% for 3 years. The purchase price of the car you select

is $28,450, including taxes, and you make a 20% down payment. What is your monthly payment?

42. Purchasing a Car A cement contractor purchases a car for a negotiated price of $55,750 plus $558 in registration and license fees.

   a. The sales tax is 6.75% of the negotiated price. Calculate the sales tax.

   b. If the contractor makes a 10% down payment on the total cost of the car, including sales tax and license fees, what is the down payment?

   c. The contractor finances the remaining balance for 4 years at an annual interest rate of 4.85% compounded monthly. Find the monthly car payment.

43. Paying Off a Car Loan Rosina has a monthly car payment of $721.52 on a 4-year car loan at an annual interest rate of 3.15%. After making 32 payments, Rosina decides to buy a new car. What is Rosina's car loan payoff?

44. Buying a Car to Fit Your Budget Zia can afford a monthly car payment of $425. If Zia can obtain a 5-year loan at an annual interest rate of 4% compounded monthly, what is the maximum loan amount Zia can obtain?

45. Gas Mileage The EPA rating of a car is 33 mpg. Find the gallons per 100 miles used by this car. Round to the nearest tenth.

46. Value of a Car According to a car valuation website, a certain car will depreciate by 24% over the first year of ownership. If the purchase price of the car is $43,000, what is its value after 1 year?

47. Cost of Owning a Car A reporter purchased a car for $47,250. Determine the cost of owning the car for the first year assuming the car has an average fuel efficiency of 24 miles per gallon, the reporter drives 15,000 miles, the sales tax is 7.5%, the license and registration fee is 1% of the purchase price, depreciation is 18% of the purchase price, a 5-year loan of $40,000 is obtained at an annual interest rate of 4.625% with a monthly payment of $748.00, the annual insurance premium is $950, and gasoline costs $3.03 per gallon.

48. Leasing a Car An orthopedic nurse practitioner leases a car for 30 months. Suppose the annual interest rate is 6.8%, the negotiated price is $33,950, there is no trade-in, and the down payment is $7500. Find the monthly lease payment. Assume that the residual value is 55% of the MSRP of $35,400.

49. Calculating DTI Each month Naomi has a minimum credit card payment of $325, a car payment of $452.36, a car insurance payment of $107, and monthly gross income of $6125. If she is applying for a loan for which the monthly payment will be $2010.23, what is her debt-to-income ratio? Round to the nearest percent.

50. Purchasing a Condo Suppose you purchase a condominium and obtain a 30-year loan of $255,800 at an annual interest rate of 6.75%.

   a. What is the mortgage payment?

   b. What is the total of the payments over the life of the loan?

   c. Find the amount of interest paid on the mortgage loan over the 30 years.

51. Mortgages Shuman secures a 30-year mortgage for $329,900 and an annual interest rate of 4.875% compounded monthly.

   a. Find the monthly mortgage payment.

b. If Shuman keeps the loan for 30 years, how much interest is repaid over that time?

52. **Effect of PMI on a Mortgage Payment** A lender requires private mortgage insurance that is 0.8% of the loan amount of $415,000. How much will this add to the borrower's monthly payment?

53. **Discount Points** A private mortgage lender charges 2.25 points for a home buyer to obtain a loan of $362,000. Calculate the discount points.

54. **Recouping Points** A bank offers Samuel Nguyen a home loan of $258,790 at an annual interest rate of 4.25% with 1 point or an interest rate of 4% and 1.75 points. If Samuel chooses the second loan, how long will it take, in months, to recoup the additional cost in points? Round to the nearest month.

55. **Interest Rate Changes** Camelia Lopez obtains a $595,000 loan for a condominium and secures a 5/1 ARM at an initial interest rate of 3.125%. The initial monthly payment is $2548.83. After 5 years, the interest rate on the loan changes to 4.75%. Calculate the new monthly payment in year 6 of the loan.

56. **Paying Off a Condo** Garth Santacruz purchased a condominium and obtained a 25-year loan of $189,000 at an annual interest rate of 7.5%.

    a. What is the mortgage payment?

    b. After making payments for 10 years, Garth decides to sell the home. What is the loan payoff?

57. **APR for a Home Loan** A real estate attorney obtains a 30-year mortgage for $375,000 at an annual interest rate of 3.5% compounded monthly with 0.75 point. The loan origination fee is 0.5 point of the loan amount. Find the APR for the loan. Round to the nearest thousandth of a percent.

58. **Calculating the Total Monthly Home Payment** Geneva Goldberg obtains a 15-year loan of $278,950 at an annual interest rate of 7%. The annual property tax bill is $1134, and the annual fire insurance premium is $681. Find the total monthly payment for the mortgage, property tax, and fire insurance.

59. **Dividends** A corporation pays an annual dividend of $1.28 per share. Calculate the dividends paid to a shareholder who owns 700 shares of this stock.

60. **Dividend Yield** The Danko Box Company pays an annual dividend of $2.31 per share. If the stock is currently trading at $75.61 per share, what is the dividend yield? Round to the nearest tenth.

61. **A Company's EPS** Ming Engineering earned $2.2 billion during one year, and for that year there were 500 million outstanding shares of stock. Find the EPS for this company.

62. **A Company's P/E** A corporation's stock is trading at $61.34 and has earnings per share of $2.96. What is the company's P/E?

63. **Nominal Value of a Bond** The bonds issued by a school district have a par value of $1000 and a coupon of 4.5%. What is the nominal value of the bond if the current interest rates are 4%? Round to the nearest dollar.

64. **Net Asset Value** A mutual fund that specializes in oil and gas exploration has a portfolio of stocks that have a market value of $5.6 billion. If there are 62,000,000 shares of stock, what is the net asset value of a share of the mutual fund?

65. **Purchasing Shares** The data manager for a cloud-based software company invests $10,000 into a mutual fund for which the transaction fee is 3.25%. If the net asset value per share of the fund is $51.86, how many shares can the manager purchase? Round to the nearest thousandth.

66. **Real Rate of Return** The chief financial officer of a furniture company expects to earn an annual interest rate of 7% on an investment of $25,000. If the current rate of inflation is 2.1%, what is the real rate of return on the investment? Round to the nearest tenth of a percent.

Project 1

Dream Versus Reality: Financing a Car

You may have a dream car you would like to drive, but just how practical is it to own? For this project, you will compare monthly payments and interest on a dream car and those of a more practical choice. You will then compare your results to see which option best fits your budget.

1. Think of your dream car. Think big, think luxury, and don't skimp on the features. Once you decide on a make and model, research the cost of it (with all desirable features included, e.g., stereo, leather interior, heated seats). Splurge away! Choose a car that has a price between $85,000 and $125,000.

   Make and model of car: _____

   Price (without tax and license) _____

2. Now think of a more practical car choice. Do you have or plan to have a family? Choose one that will accommodate them. Do you have or plan to have a pet? Figure that into your decision. Choose a car that has a price between $10,000 and $50,000.

   Make and model of car: _____

   Price (without tax and license) _____

3. Using your current bank, information from a car dealership, or doing an Internet search for car loan interest rates, determine the annual interest rate for a car loan. Some financial advisers recommend the 20/4/10 rule when purchasing a car.

   • Make a down payment that is at least 20% of the cost of the car, excluding tax and license fees.

   • Do not finance the car for more than 4 years or 48 months.

   • Your monthly car payment should not exceed 10% of your gross monthly income.

   Based on these guidelines, determine each of the following. Use the formula for the present value of an ordinary annuity in Section 4.6 to calculate the monthly car payment.

   • Cost of dream car: _____
     ○ Down payment: _____
     ○ Amount to finance: _____
     ○ Loan interest rate: _____
     ○ Term of loan: _____
     ○ Monthly payment: _____

- Cost of practical car: _____
  - o Down payment: _____
  - o Amount to finance: _____
  - o Loan interest rate: _____
  - o Term of loan: _____
  - o Monthly payment: _____

4. Assuming you have a monthly income of $5000 (the approximate U.S. median income), does the monthly payment of your dream car satisfy the 10% rule recommended by some financial advisers? If not, determine how much your monthly income would have to be to meet the 10% requirement.

5. Assuming you have a monthly income of $5000 (the approximate U.S. median income), does the monthly payment of your practical car satisfy the 10% rule recommended by some financial advisers? If so, how much monthly income remains after making your car payment?

Project 2

Student Loan Debt

Financial advisers recommend that the monthly payment for student loan debt should not exceed 8% of anticipated gross monthly earnings after graduation. Based on this recommendation, complete this project.

1. What is your anticipated annual income at graduation?

   _____

   Enter a value between $25,000 and $80,000. As a guide, the median annual income of recent graduates is approximately $52,000.

2. What is the maximum recommended monthly payment for your student loan debt? _____

3. Complete the following to calculate the maximum amount of student debt you should incur.

   a. Expected salary from Exercise 1: _____

   b. Monthly payment from Exercise 2: _____

   c. Annual interest rate: _____

As a guide, current interest rates for student loans are between 5% and 7%.

Repayment period in years: _____

As a guide, the recommended repayment period should not exceed 10 years.

Maximum student loan debt: _____

4. Calculate the amount of interest repaid over the time of the loan.

5. Some lenders offer a 0.25% reduction in the annual interest rate if you use autopay, which automatically deducts your monthly payment from your bank account. Rework Exercises 3 and 4 based on the interest rate deduction.

6. What is the difference between the amount of interest paid over the term of the loan between Exercise 4 and Exercise 5?

Math in Practice

Saving money can provide impressive results. It just takes a little planning. Whether you decide to eat out less, use coupons, take advantage of sales, or employ other cost-cutting techniques, by setting money aside now, "future you" will appreciate this preparation.

Andrey_Popov/Shutterstock.com

Part 1

First, let's explore coffee consumption. You purchase an $8 coffee for 5 days a week for 50 weeks.

a. What is the total cost after 4 weeks? After 50 weeks?

b. If you purchase an $8 coffee 1 day a week and make coffee at home for 4 days a week at the cost of $1.20 per day, how much will you save after 4 weeks? After 50 weeks?

c. Suppose at the end of the year you have saved $1300 and then invest it at 3% interest compounded monthly. What is the value after 5 years? After 30 years? Round to the nearest cent.

Part 2

In preparation for buying a new car, you decide to make car payments to yourself. You decide to save $297 per month.

a. Without earning any interest, how much have you saved after 3 years? After 5 years?

b. If you invested $297 per month at the beginning of the month at 4% annual interest rate, compounded monthly, how much will you have after 3 years? After 5 years?

c. How much interest did you earn after 3 years? After 5 years?

Part 3

The financial institution holding the mortgage on your house allows you to make additional payments toward the principal. Suppose you have made on time monthly payments of $1457 for 15 years for a 30-year fixed-rate mortgage, with an annual interest rate of 5.875%. Your financial institution is offering a 15-year fixed-rate mortgage with refinance rate of 4.125% with refinance fees of $2500. Consider seeking the advice of a professional financial advisor before making major financial decisions.

a. If you make the same monthly payment of $1457, what is the total paid to the financial institution after 30 years?

b. If the original amount of the mortgage was $257,313, what is the total interest paid after 30 years?

c. What options do you have to reduce the total interest paid to your financial institution?

Modeling

Linear Functions

Section	How This Is Used . . .
5.1 Introduction to Functions	Functions are mathematical models of a realistic situation that imitate a real-world situation. Functions are used extensively in all branches of science from anthropology to zoology. Spreadsheet programs are essentially a large number of functions gathered together to allow analysts to evaluate how changes in one quantity affect another one.
5.2 Properties of Linear Functions	A linear function is one that can be graphed as a straight line. Although fairly simple functions, these functions apply to a variety of situations: from straight-line depreciation to creating simple approximations of cost of production of a product to the distance an object travels at a constant rate of speed.
5.3 Linear Models	One application of mathematics is to create a function that establishes how one quantity is related to another. The function that is created is called a *mathematical model* (or simply a *model*) of the situation. The model serves as a theoretical description of the relationship between quantities. For instance, if you work for an hourly wage, there is a linear relationship between the amount of money you earn and the number of hours you work.

Math in Practice:
Nursing, p. 258

Applications of Linear Functions

Linear functions are the simplest models of the relationship between two quantities. Nonetheless, these models have many applications. If you are participating in a road race at a constant rate of 7 miles per hour, the distance you run can be described by a linear model. Economists use linear models to predict the market price of a commodity given a certain demand by consumers. The pressure on a diver as the diver descends below the ocean's surface can be modeled by a linear function. These are just some of the applications of linear models. Exercise 21, Section 5.3, asks you to find a linear model that predicts the amount of water a professional tennis player loses during a match.

Introduction to Functions

Learning Objectives:

- Evaluate a function.
- Graph a function.
- Find the average rate of change of a function.

Be Prepared: Prep Test 5 reviews concepts used in this chapter. This can be found in your Companion Site at Cengage.com.

Introduction to Functions

One compelling application of mathematics is using and creating *functions*. Functions are one of the cornerstones of the applications of mathematics.

- A consumer may be interested in how the monthly cost of Internet service **depends** on the download speed of the service. This is stated as monthly cost, C, *is a function of* download speed, s, and can be written $C = f(s)$. The letter f is the name of the function.
- A physician wants to know how the dose of a medication **depends** on a patient's height, h, and weight, w. This is stated as dose *is a function* of height and weight, and can be written as $d = B(h, w)$. The letter B is the name of the function.
- An electrician wants to know how the monthly payment, P, of a new work van **depends** on the interest rate, r, the number of months, n, of the loan, and the loan amount, A. This is stated as monthly payment *is a function of* interest rate, months of the loan, and loan amount and can be written as $P = PMT(r, n, A)$. *PMT* is the name of the function. As in this case, the name of the function helps suggest what the function does.

For each of the cases above,

- Monthly Internet service cost, C, *depends* on download speed, s. C is called the **dependent variable**; s is the **independent variable**.
- Dose, d, *depends* on height, h, and weight, w. d is the **dependent variable**; s and h are the **independent variables**.
- Payment, P, *depends* on interest rate, r, the number of the months, n, of the loan, and the loan amount, A. P is the **dependent variable**; r, n, and A are the **independent variables**.

Each of $C = f(s)$, $d = B(h, w)$, and $P = PMT(r, n, A)$ are examples of **function notation**. The variables in parentheses are the independent variables; the variable to the left of the equal sign is the dependent variable.

To make all this useful, there must be some way to calculate the value of the dependent variable for the given values of the independent variables.

Take Note: Whether we write $C = 0.04s + 34$ or $f(s) = 0.04s + 34$, we are describing the same relationship. For the first equation, we give a name, C, to the dependent variable. In the second case, the emphasis is on function notation.

Suppose $f(s) = 0.04s + 34$ is a model of the monthly cost, in dollars, one Internet service provider (ISP) charges for connection speeds in megabits per second (mbps). From $C = f(s)$, we can write $C = 0.04s + 34$ replacing $f(s)$ with $0.04s + 34$. Using this model, a consumer can estimate the monthly cost of Internet service. The following table shows possible costs.

Speed (s)	$C = 0.04s + 34$
25	$C = 0.04(25) + 34 = 35$
50	$C = 0.04(50) + 34 = 36$
100	$C = 0.04(100) + 34 = 38$
500	$C = 0.04(500) + 34 = 54$
1000	$C = 0.04(1000) + 34 = 74$

From this table, a consumer can expect to pay \$54 per month for download speeds of 500 mbps.

Finding the value of the dependent variable for a given value of the independent variable is called **evaluating the function**. The *value of the function* is \$54 when the value of the independent variable is 500. If we were using function notation in the table, we would show evaluating the function as $f(500) = 0.04(500) + 34 = 54$.

Function

A **function** describes a relationship between independent variables and a dependent variable such that for given values of the independent variables, there is *one and only one* possible value of the dependent variable. The **domain** of a function is all the possible values for the independent variable. The **range** of a function is all the values of the dependent variable.

Example 1 Evaluate a Function

One model of the distance in feet a car will skid before coming to a stop after the brakes are suddenly applied is $skid(v) = 0.05v^2 - 0.4v + 9$, where v is the speed of the car in miles per hour. How far will a car skid when the brakes are suddenly applied at 45 mph?

Solution

The value of the independent variable, v, is given as 45. The name of the function is *skid*, and $skid(v)$ is the distance a car traveling v mph will skid after the brakes are applied. Replace v with 45. Then simplify.

$$skid(v) = 0.05v^2 - 0.4v + 9$$

$$skid(45) = 0.05(45)^2 - 0.4(45) + 9 \qquad \bullet \text{ Replace } v \text{ with } 45.$$

$$= 0.05(2025) - 18 + 9$$

$$= 101.25 - 9$$

$$= 92.25$$

A car traveling 45 mph will skid 92.25 feet after the brakes are applied.

The domain of a function is chosen to make sense for the application. For instance, if we allowed v (the speed) of the car to be zero (the car is not moving), then

$$skid(v) = 0.05v^2 - 0.4v + 9$$

$$skid(0) = 0.05(0)^2 - 0.4(0) + 9 \qquad \bullet \text{ Replace } v \text{ with } 0.$$

$$= 0 - 0 + 9$$

$$= 9$$

The model predicts the car would skid 9 feet when the speed is 0. That clearly does not make sense. The domain of a mathematical model is dictated by the values of the independent variable that were used to create the model. The model $skid(v) = 0.05v^2 - 0.4v + 9$ was based on speeds between 30 mph and 80 mph. Therefore, the domain of the function is $30 \le v \le 80$; that is, speeds between 30 mph and 80 mph.

In some cases, independent and dependent variables may be chosen differently. For instance, for the skid function $skid(v) = 0.05v^2 - 4v + 9$ from Example 1, we choose to find the length of the skid mark as a function of the velocity v of the car after the brakes were applied. Engineers who design tires may be interested in this function. However, an accident investigator might want to know a car's velocity as a function of skid distance. Now the skid distance is the independent variable and velocity is the dependent variable.

Example 2 Evaluate a Function

The velocity v, in miles per hours, of a car just before its brakes are applied can be estimated from the length L, in feet, of the skid mark using the equation $v = S(L) = 4 + 2\sqrt{5L - 41}$. Approximate the velocity of a car that leaves a 60-foot skid mark after the brakes are applied.

Solution

For this equation, the velocity of the car is a function of length L of the skid mark. Evaluate the function when $L = 60$.

$$S(L) = 4 + 2\sqrt{5L - 41}$$
$$S(60) = 4 + 2\sqrt{5(60) - 41}$$
$$= 4 + 2\sqrt{259}$$
$$\approx 36$$

The car was traveling approximately 36 miles per hour.

Because functions involve pairs of data, a rectangular coordinate system can be used to display those pairs. Here is a brief review of a rectangular coordinate system.

Rectangular Coordinate System Review

A **rectangular coordinate system** is formed by two number lines, one horizontal and one vertical, that intersect at the zero point of each line. The point of intersection is called the **origin**. The two number lines are called the **axes**. The horizontal axis normally represents the values of the independent variable, and the vertical axis normally represents the values of the dependent variable.

A pair of numbers called an **ordered pair** identifies each point in the plane. The first number of the ordered pair measures a horizontal change from zero. The second number measures a vertical change from zero. The ordered pairs $(3, -4)$ and $(-4, 3)$ are shown in Figure 5.1. Note that these ordered pairs identify different points. The *order* of the numbers in an ordered pair is important.

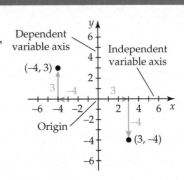

Figure 5.1

Table 5.1

Speed, v (mph)	Skid Distance, $skid(v)$, (feet)	$(v, skid(v))$
30	42	(30, 42)
40	73	(40, 73)
50	114	(50, 114)
60	165	(60, 165)
70	226	(70, 226)
80	297	(80, 297)

Graph of a Function

A **graph** of a function is a display of all the ordered pairs that belong to the function. To produce the graph, evaluate the function for selected values of the independent variable. Plot the resulting ordered pairs and then draw a smooth curve through the points. To create an accurate graph may require plotting many points.

Table 5.1 shows selected values of the independent variable (speed) and the resulting value of the dependent variable, skid distance. We've rounded the distance to the nearest whole number for ease in plotting the points.

Figure 5.2 shows the points and then a smooth curve drawn through the points. The blue curve is the graph of $skid(v) = 0.05v^2 - 0.4v + 9$.

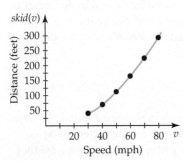

Figure 5.2

Example 3 **Graph a Function**

A homeowner has installed rain barrels to capture rainwater to be used for irrigation. On a day the homeowner wants to irrigate some fruit trees, a water pump is attached to a completely full 80-gallon rain barrel. If the pump drains the tank at a rate of 5 gallons per minute, then the amount of rainwater remaining in the barrel is a function of the time t the pump has been running and is given by $A(t) = 80 - 5t$. Complete the following table and draw a graph of $A(t) = 80 - 5t$.

Time, t (minutes)	Water Remaining $A(t)$ (gallons)	$(t, A(t))$
0	80	(0, 80)
4		
8	40	(8, 40)
12	20	(12, 20)
16		

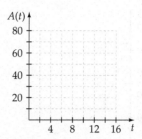

Solution

To complete the table, evaluate $A(t) = 80 - 5t$ when $t = 4$ and $t = 16$.

$$A(t) = 80 - 5t$$
$$A(4) = 80 - 5(4)$$
$$= 80 - 20$$
$$= 60$$

$$A(t) = 80 - 5t$$
$$A(16) = 80 - 5(16)$$
$$= 80 - 80$$
$$= 0$$

The ordered pair is (4, 60).
The following shows the completed table.

The ordered pair is (16, 0).

Plot the points and then draw a smooth curve through the points.

Time, t (minutes)	Water Remaining, $A(t)$ (gallons)	$(t, A(t))$
0	80	(0, 80)
4	60	(4, 60)
8	40	(8, 40)
12	20	(12, 20)
16	0	(16, 0)

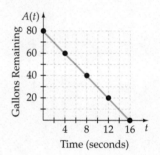

For Example 3, we emphasized function notation. It is also appropriate to choose a variable for $A(t)$—say g for gallons—and write $g = 80 - 5t$. This indicates that g is a *function* of t. The table and graph would now look like Table 5.2 and Figure 5.3, respectively. Note that the vertical axis is labeled g rather than $A(t)$.

Table 5.2

Time, t (minutes)	Water Remaining, g (gallons)	(t, g)
0	80	(0, 80)
4	60	(4, 60)
8	40	(8, 40)
12	20	(12, 20)
16	0	(16, 0)

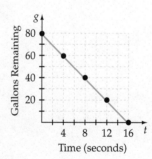

Figure 5.3

Using the letter g or function notation $A(t)$ is a matter of choice. The notation $A(t)$ provides a connection to the independent variable t.

Earlier, we were given a value of the independent variable and evaluated the function to find the corresponding value of the dependent variable. Sometimes we might want to reverse the procedure: Given a value of the dependent variable, find a corresponding value of the independent variable. Given the graph of a function, we can estimate the value.

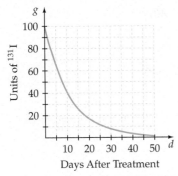

Figure 5.4

Example 4 **Estimate a Value of the Independent Variable**

Iodine-131, ^{131}I, is a radioactive isotope of iodine that is used in medical diagnostics and treatments. Figure 5.4 shows the amount of ^{131}I remaining in a person d days after receiving an ^{131}I treatment.

Using the graph in Figure 5.4, about how many days after treatment begins is the amount of ^{131}I 50 units?

Solution

The value of the dependent variable is given as 50. Draw a line from the 50-unit mark on the iodine axis to the graph. Then draw a line to the days after treatment axis. The line touches the axis somewhere between 5 and 10 days and closer to 10 days than 5 days. An estimate is that after 8 days, there are 50 units of ^{131}I in the bloodstream.

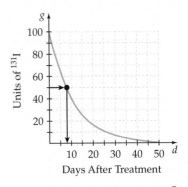

Average Rate of Change

The rate at which something is increasing or decreasing is important in many applications. For instance, if you have an investment that is increasing at a rate of 3% a year, that is good. If it is increasing at a rate of 10% per year, that is even better. Similarly, if the value of a car is decreasing at a rate of 10% per year, that is not good. If it is decreasing at a rate of 20% per year, that is worse.

Recall that a unit rate is the change in one quantity divided by the change in a second quantity. For instance, if a car travels 544 miles and uses 16 gallons of gasoline, then $\frac{544 \text{ miles}}{16 \text{ gallons}} = 34$ miles/gallon. The car *averaged* 34 miles/gallon (mpg). At times during the 544-mile drive, the car got more than 34 mpg, at other times less than 34 mpg.

Following similar reasoning, we can define the average rate of change for a function.

Average Rate of Change of a Function

$$\text{Average rate of change} = \frac{\text{Change in dependent variable}}{\text{Change in independent variable}}, \text{ where}$$

change in the dependent variable = value after change − value before change
change in the independent variable = value after change − value before change

---

Figure 5.5a again shows the amount of ^{131}I in the bloodstream d days after receiving treatment. Note the change of the independent variable from approximately 64 units in the bloodstream after 5 days to approximately 18 units in the bloodstream after 20 days.

Change in dependent variable = Value before change − Value after change

$$= 64 \text{ units} - 18 \text{ units} = 46 \text{ units}$$

Change in independent variable = Value before change − Value after change

$$= \text{day } 5 - \text{day } 20 = -15 \text{ days}$$

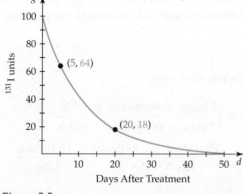

Figure 5.5a

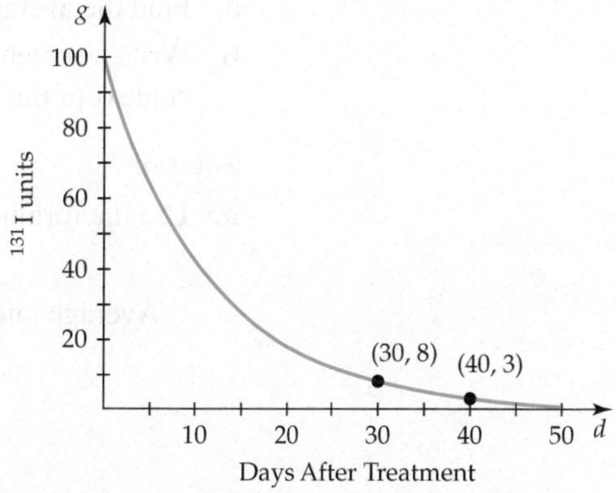

Figure 5.5b

$$\text{Average rate of change} = \frac{\text{Change in dependent variable}}{\text{Change in independent variable}}$$

$$= \frac{64 \text{ units} - 16 \text{ units}}{5 \text{ days} - 20 \text{ days}} = \frac{46 \text{ units}}{-15 \text{ days}}$$

$$\approx -3.06 \text{ units/day}$$

The average rate of change was approximately -3.06 units/day. The negative sign indicates that between day 5 and day 20, the amount of iodine in the bloodstream decreased by approximately 3.06 units each day.

If we had chosen two different points on the graph, say (30, 8) and (40, 3), (see Figure 5.5b) then

$$\text{Average rate of change} = \frac{\text{Change in dependent variable}}{\text{Change in independent variable}}$$

$$= \frac{8 \text{ units} - 3 \text{ units}}{\text{day } 30 - \text{day } 40} = \frac{5 \text{ units}}{-10 \text{ days}}$$

$$= -0.5 \text{ unit/day}$$

The number of units of iodine is decreasing at an average rate of -0.5 unit per day between day 30 and day 40 after treatments begin. This is a slower rate than the rate between day 10 and day 15.

If we calculated the average rate of change between other points, we could conclude that the rate at which iodine is eliminated from the bloodstream decreases over time. This kind of information is important to medical practitioners who use radioactive isotopes to diagnose and treat diseases.

Example 5   **Find Average Rate of Change**

Figure 5.6 shows the growth of yeast over a 10-day period.

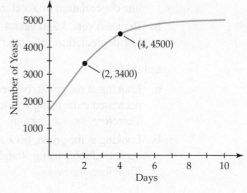

Figure 5.6

a. Find the average rate of change of the yeast population between day 2 and day 4.

b. Write a sentence that explains the meaning of the average rate of change in the context of this problem.

Solution

a. Use the formula for average rate of change.

$$\text{Average rate of change} = \frac{\text{Change in dependent variable}}{\text{Change in independent variable}}$$

$$= \frac{4500 \text{ yeast} - 3400 \text{ yeast}}{\text{day } 4 - \text{day } 2} = \frac{1100 \text{ yeast}}{2 \text{ days}}$$

$$= 550 \text{ yeast/day}$$

b. The yeast population is increasing at an average rate of 550 yeast per day between 2 and 4 days after measurements begin.

●

If we calculated the average rate of change between other points, we could conclude that the population is increasing over time. This could be important to a biologist who studies population growth.

> **Example 6** **Estimate Rate of Decrease**

Business owners may use accelerated or straight-line types of depreciation for equipment. The graphs in Figure 5.7 show the depreciated value in thousands of dollars of a company-wide computer system. Graph A is a straight-line depreciation; graph B is accelerated depreciation.

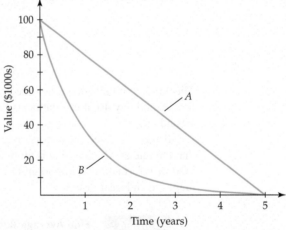

Figure 5.7

a. Between year 0 and year 1, was the average rate of depreciation greater for straight-line depreciation or accelerated depreciation?

b. Between year 3 and year 4, was the average rate of depreciation greater for straight-line depreciation or accelerated depreciation?

Solution

a. Looking at the graph, between years 0 and 1, the value of the computer system decreased more using accelerated depreciation than straight-line depreciation. Therefore, the average rate of depreciation was greater for accelerated depreciation.

b. Looking at the graph, between years 3 and 4, the value of the computer system decreased more using straight-line depreciation than accelerated depreciation. Therefore, the average rate of depreciation was greater for straight-line depreciation.

●

5.1 Exercise Set

Think About It

1. Suppose $z = F(x)$ and $F(-3) = 7$. What is the value of z?
2. If $y = f(x)$ and $(-2, 5)$ is an ordered pair of the function, what is the value of x and what is the value of y?
3. If f is a function, is it possible for $f(2) = 3$ and $f(2) = 4$?
4. If f is a function, is it possible for $f(3) = 4$ and $f(2) = 4$?
5. If $y = f(x)$, what is the independent variable, the dependent variable, and the name of the function?
6. The _____ of a function is the set of all first components of the ordered pairs of the function. The _____ of a function is the set of all second components of the ordered pairs of the function.

Preliminary Exercises

For Exercises P1 to P6, if necessary, use the referenced example following the exercise for assistance.

P1. The height, in feet, of a ball thrown vertically upward t seconds after it is released is given by $s(t) = -16t^2 + 64t + 4$. Find the height of the ball 3 seconds after it is released. **Example 1**

P2. The distance, in feet, from a new wall to an 8 foot grounded brace is given by $D(h) = \sqrt{h^2 + 64}$, where h is the wall brace height in feet. Find the grounded brace distance for a wall brace height of 7 feet. **Example 2**

P3. A medium-sized shopping mall is open from 8:00 A.M. to 6:00 P.M. The number of parked cars in a mall is a function of the number of hours after opening, 8:00 A.M., and is given by $C(t) = -50t^2 + 460t + 300$. **Example 3**

a. Complete the following table, which uses N to represent $C(t)$.

Time (hours), t	Number, N	(t, N)
0	300	(0, 300)
1	710	(1, 710)
2	1020	(2, 1020)
3		
4	1340	(4, 1340)
5		
6	1260	(6, 1260)
7	1070	(7, 1070)
8		
9	390	(9, 390)

b. Plot some points and then draw the graph. When you are plotting the points, you may have to do your best to approximate the location of the point. That is fine.

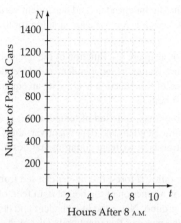

Hours After 8 A.M.

P4. Using the graph in Figure 5.4, about how many days after treatment begins is there 20 units of ^{131}I in the bloodstream? **Example 4**

P5. The following graph shows the decreasing value of a car over time. **Example 5**

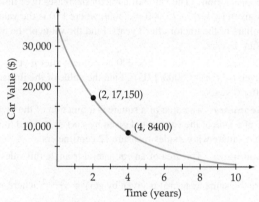

a. Write a sentence that explains the meaning of the ordered pair (4, 8400).
b. Find the average rate of depreciation for the given points.

P6. The following graph shows the speed of a falling object after being released at some point above the ground. **Example 6**

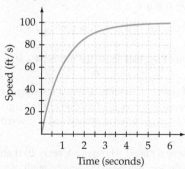

a. As time increases, does the speed increase or decrease?
b. Is the average rate of change of speed between 1 second and 2 seconds less than or greater than the average rate of change between 2 seconds and 3 seconds?

Section Exercises

1. The speed, s, of a ball released from a point above ground when it impacts the ground is a function of the height, h, from which it falls. Name the independent and dependent variables. The name of this function is Sp; write an equation for the function.

2. The perimeter, P, of a square is a function of the length of a side, s, of the square. Name the independent and dependent variables. The name of this function is G; write an equation that describes the function.

3. The speed, s, of sound is a function of temperature, T, of the air. Name the independent and dependent variables. The name of this function is v; write an equation that describes the function.

4. The distance, d, in miles, a forest fire ranger can see from an observation tower is a function of the height h, in feet, of the tower above level ground. Name the independent and dependent variables. If the name of this function is $Dist$, write an equation that describes the function.

For Exercises 5 to 18, evaluate the function to answer the question.

5. **Depreciation** The value of a tractor decreases over time and is given by $V(t) = 75,000 - 7500t$, where $V(t)$ is the value in dollars of the tractor after t years. Find the value of the tractor after 4 years.

6. **Finance** The value of a \$5000 investment after n years is given by $f(n) = 5000(1.05)^n$. Find the value of the investment after 6 years.

7. **Geometry** The area of a square is a function of the measure, s, of a side of the square and given by $f(s) = s^2$. Find the area of a square whose sides measure 12 centimeters.

8. **Geometry** The area of an equilateral triangle (all sides are the same measure) is given by $g(x) = \dfrac{\sqrt{3}}{4}x^2$, where x is the measure of the sides of the triangle. Find the area of an equilateral triangle with sides measuring 3 meters. Round the answer to the nearest hundredth.

9. **Sports** The height h, in feet, of a ball that is released 4 ft above the ground with an initial upward velocity of 80 ft/s is a function of the time t, in seconds, the ball is in the air and is given by

$$h(t) = -16t^2 + 80t + 4, 0 \le t \le 5.05$$

   a. Find the height of the ball above the ground 2 s after it is released.

   b. Find the height of the ball above the ground 4 s after it is released.

10. **Forestry** The distance d, in miles, a forest fire ranger can see from an observation tower is a function of the height h, in feet, of the tower above level ground and is given by $d(h) = 1.5\sqrt{h}$.

   a. Find the distance a ranger whose eye level is 20 ft above level ground can see. Round to the nearest tenth of a mile.

   b. Find the distance a ranger whose eye level is 35 ft above level ground can see. Round to the nearest tenth of a mile.

11. **Sound** The speed s, in feet per second, of sound in air depends on the temperature t of the air in degrees Celsius and is given by

$$s(t) = \frac{1087\sqrt{t + 273}}{16.52}$$

   a. What is the speed of sound in air when the temperature is 0°C (the temperature at which water freezes)? Round to the nearest foot per second.

   b. What is the speed of sound in air when the temperature is 25°C? Round to the nearest foot per second.

12. **Mixtures** The percent concentration P of salt in a particular salt water solution depends on the number of grams x of salt that are added to the solution and is $P(x) = \dfrac{100x + 100}{x + 10}$.

   a. What is the original percent concentration of salt?

   b. What is the percent concentration of salt after 5 more grams of salt is added?

13. **Pendulums** The time T, in seconds, it takes a pendulum to make one swing depends on the length of the pendulum and is given by $T(L) = 2\pi \sqrt{\dfrac{L}{32}}$, where L is the length of the pendulum in feet.

   a. Find the time it takes the pendulum to make one swing if the length of the pendulum is 3 ft. Round to the nearest hundredth of a second.

   b. Find the time it takes the pendulum to make one swing if the length of the pendulum is 9 in. Round to the nearest tenth of a second.

14. **Botany** Botanists have determined that some species of weed grow in a circular pattern. For one such species, the area A, in square meters, can be approximated by $A(t) = 0.005\pi t^2$, where t is the time in days after the growth of the weed first can be observed. Find the area this weed will cover 100 days after the growth is first observed. Round to the nearest square meter.

15. **Sports Training** The Karvonen formula is one formula used to determine the target heart rate for a person training for fitness. The formula is given by $T = p(220 - A) + H(1 - p)$, where T is the target heart rate, A is the age of the person, H is the resting heart rate, and p is the percent intensity (expressed as a decimal) of the workout.

   a. What are the independent variables?

   b. What is the dependent variable?

   c. A trainer has determined that a client who is 31 years old and has a resting heart rate of 70 beats per minute should train at an intensity of 65%. What is the target heart rate for this person? Round to the nearest whole number.

16. **Maximum of Two Numbers** Use the equation $Max = \dfrac{1}{2}(x + y + |x - y|)$ to answer the following questions. Recall that $|a|$ means the absolute value of a number.

   a. Evaluate the equation for $x = 6$ and $y = 10$.

   b. Evaluate the equation for $x = 12$ and $y = 5$.

c. Evaluate the equation for $x = -4$ and $y = -7$.

d. Evaluate the equation for $x = -8$ and $y = 7$.

e. Explain why the title of this exercise is maximum of two numbers.

17. Basal Metabolic Rate Basal metabolic rate (BMR) is the number of kilocalories (a kilocalorie is the number printed on nutrition labels as Calorie) required to keep your body functioning at rest. One formula for an adult man is $BMR = 13.4w + 4.8h - 5.7a + 88.4$, where w is the weight of the person in kilograms, h is the height of the person in centimeters, and a is the person's age in years.

a. What are the independent variables?

b. What is the dependent variable?

c. What is the BMR for a man who weighs 77 kilograms, is 170 centimeters tall, and is 28 years old? Round to the nearest whole number.

18. Basal Metabolic Rate As noted, BMR is the number of kilocalories required to keep your body functioning at rest. One formula for an adult woman is $BMR = 9.2w + 3.1h - 4.3a + 447.6$, where w is the weight of the person in kilograms, h is the height of the person in centimeters, and a is the person's age in years.

a. What are the independent variables?

b. What is the dependent variable?

c. What is the BMR for a woman who weighs 65 kilograms, is 165 centimeters tall, and is 35 years old? Round to the nearest whole number.

19. Download Times The time t, in seconds, it takes to download a file from a server is a function of the size s, in gigabytes, of the file and is given by $t = f(s) = 2.5s$.

a. Find the ordered pair to complete the following table.

s	t	(s, t)
8	20	(8, 20)
16	40	(16, 40)
	60	
32	80	(32, 80)
40	100	(40, 100)

b. Write a sentence that explains the meaning of the ordered pair (32, 80) in the context of this problem.

c. Use the following coordinate grid to graph the function for s between 8 and 40 gigabytes.

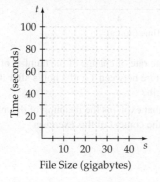

File Size (gigabytes)

20. Candles The height h, in inches, of a candle is a function of the length of time t, in minutes, it has been burning and is given by $h = f(t) = 10 - 0.1t$.

a. Find the ordered pair to complete the following table.

t	h	(t, h)
0	10	(0, 10)
20	8	(20, 8)
40	6	(40, 6)
60		
80	2	(80, 2)
100	0	(100, 0)

b. Write a sentence that explains the meaning of the ordered pair (100, 0) in the context of this problem.

c. Use the following coordinate grid to graph the function for t between 0 minutes and 100 minutes.

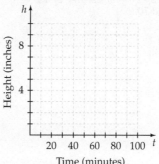

Time (minutes)

21. Video Games Through market research, a business analyst has determined that the number N (in thousands) of units of a new video game the company can sell is a function of the price p (in dollars) of the game and is given by $N = -0.07p^2 + 0.8p + 90$.

a. Find the ordered pair to complete the following table. Round to the nearest whole number.

p	N	(p, N)
10	91	(10, 91)
15	86	(15, 86)
20		
25	66	(25, 66)
30	51	(30, 51)
35	32	(35, 32)
40	10	(40, 10)

b. Write a sentence that explains the meaning of the ordered pair (25, 66) in the context of this problem.

c. Use the following coordinate grid to graph the function for p between $10 and $40.

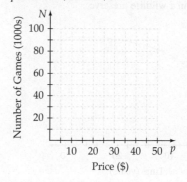

Price ($)

22. Forestry The distance d, in miles, a forest fire ranger can see from an observation tower is a function of the height h, in feet, of the tower above ground and is given by $d(h) = 1.5\sqrt{h}$.

a. Find the ordered pair to complete the following table. Round to the nearest tenth.

d	h	(d, h)
10	4.7	(10, 4.7)
20	6.7	(20, 6.7)
30	8.2	(30, 8.2)
40	9.5	(40, 9.5)
50		
60	11.6	(60, 11.6)
70	12.5	(70, 12.5)

b. Write a sentence that explains the meaning of the ordered pair (40, 9.5) in the context of this problem.

c. Use the following coordinate grid to graph the function for h between 10 and 70.

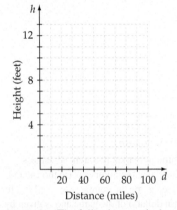

Distance (miles)

23. Basketball The following graph shows the arc of a basketball player's shot after it is released. Approximately how far from the shooter does the ball enter a basketball hoop situated 11 feet above the ground? Assume the ball enters the basket during the downward portion of the arc.

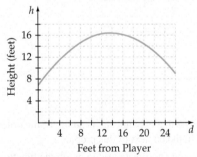

Feet from Player

24. Fox Population The following graph shows the growth of a fox population in a wildlife preserve.

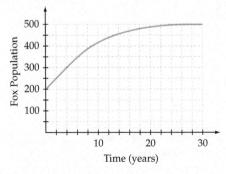

Time (years)

Estimate in how many years the fox population will be 400 foxes.

25. Golf The following graph shows the flight of a golf ball.

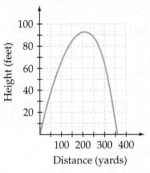

Distance (yards)

Use the graph to estimate the distance at which the ball reaches its maximum height.

26. Manufacturing Manufacturers of soda cans want to make a can that holds 350 milliliters of soda and uses the least amount of material. The amount of material of such a can is a function of the radius of the can. The graph is shown as follows.

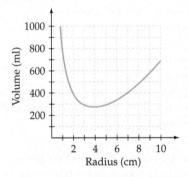

Radius (cm)

From the graph, estimate the radius of the can that will require the least amount of material.

27. Biology The following graph shows the decrease in the number of bacteria in a culture over time.

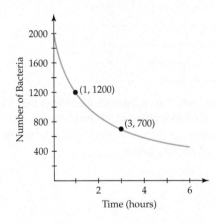

Time (hours)

a. What is the average rate of change of the number of bacteria in the culture between 1 and 3 hours after the measurements begin?

b. Write a sentence that explains the meaning of the average rate of change in the context of this exercise.

28. Life Science The following graph shows the decrease in the number of fish in a lake over time.

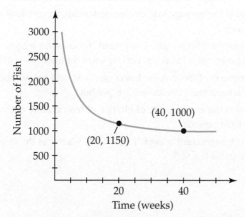

a. What is the average rate of change of the number of fish in the lake between 20 and 40 weeks after the measurements begin?

b. Write a sentence that explains the meaning of the average rate of change in the context of this exercise.

29. Disease Spread The following graph shows the increase in the number of flu cases in a small city over time.

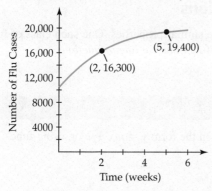

a. What is the average rate of change of the number of flu cases between 2 weeks and 5 weeks?

b. Write a sentence that explains the meaning of the average rate of change in the context of this exercise.

30. Physics The following graph shows the velocity of an object dropped from a place 200 feet above ground.

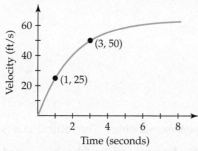

a. What is the average rate of change of velocity between 1 second and 3 seconds after the object is released?

b. Write a sentence that explains the meaning of the average rate of change in the context of this exercise.

31. Do the values of y increase or decrease as the values of x increase?

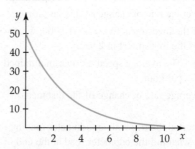

32. Do the values of y increase or decrease as the values of x increase?

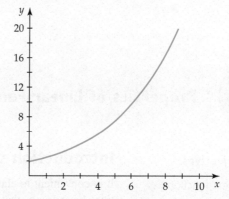

33. Which of the following graphs shows a more positive average rate of change between $x = 3$ and $x = 5$?

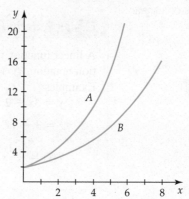

34. Which of the following graphs shows a more negative average rate of change between $x = 2$ and $x = 4$?

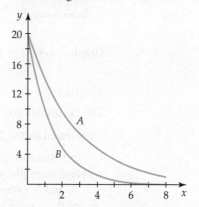

35. Finance Suppose you have an investment offer that guarantees an average investment gain of $1000 per year.

   a. What is the average rate of change of this investment?

   b. If the value of the investment today is $10,000, what will be the value of the investment in 2 years?

36. Driving Distance The average speed a driver maintained on a trip was 45 miles per hour.

   a. What is the average rate of change of the distance of this trip?

   b. If the driver started the trip with an odometer reading of 10,000 miles, what did the odometer read at the end of the trip 4 hours later?

37. Weight Loss A weight-loss program advertises a person can attain an average weight loss of 2 pounds per week.

   a. What is the average rate of change for this weight-loss program?

   b. If a person who weighs 180 pounds follows the weight-loss program, what is the person's weight after 5 weeks?

38. Temperature Between the hours of 1 A.M. and 6 A.M., the outside temperature decreased 2°F per hour.

   a. What is the average rate of change of temperature during this time period?

   b. If the temperature was 60°F at 1 A.M., what was the temperature at 6 A.M.?

Section 5.2

Properties of Linear Functions

Learning Objectives:

- Graph a linear function.
- Find the slope of a linear function.
- Find the intercepts of a linear function.

Introduction to Linear Functions

It is convenient to classify functions that have similar properties. One such class of functions have graphs that are straight lines. Such functions are *linear functions*.

Linear Function

A linear function is one that can be written in the form $y = mx + b$ or, using function notation, $f(x) = mx + b$.

Examples:

   $y = 3x - 2$ • $m = 3, b = -2$.

   $z = 3 - x$ • This is $-x + 3$ so $m = -1, b = 3$.

   $T(s) = -\dfrac{1}{2}s$ • $m = -\dfrac{1}{2}, b = 0$.

Example 1   **Graph a Linear Function**

Graph $s = -\dfrac{3}{2}t - 1$.

Solution

The horizontal axis is represented by the independent variable t. The vertical axis is represented by the dependent variable s. This is a linear function with $m = -\dfrac{3}{2}$ and $b = -1$. Because the graph is a straight line, we only need two points. However, we will find three just as a check. Choose three values for the independent variable, say, -2, 0, and 2. (We choose even numbers to make plotting the points easier.) We can record the results in a table.

t	$s = -\dfrac{3t}{2} - 1$		s	(s, t)
-2	$s = -\dfrac{3(-2)}{2} - 1$	$=$	2	$(-2, 2)$
0	$s = -\dfrac{3(0)}{2} - 1$	$=$	-1	$(0, -1)$
2	$s = -\dfrac{3(2)}{2} - 1$	$=$	-4	$(2, -4)$

Plot the ordered pairs and then draw a line through the points.

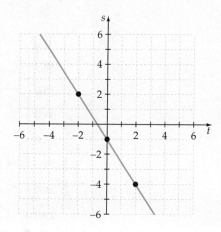

Slope

One characteristic of the graph of a line is its slant or *slope*. This characteristic has many applications.

The pressure on a scuba diver is a function of the diver's depth. Table 5.3 shows the approximate pressure in pounds per square foot on a scuba diver as the diver descends below the ocean's surface. The table also shows some of the ordered pairs of the function. The graph of the function is shown in Figure 5.8.

Table 5.3

Depth (ft), D	Pressure (lb/ft²), P	(D, P)
0	2088	(0, 2088)
5	2408	(5, 2408)
10	2728	(10, 2728)
15	3048	(15, 3048)
20	3368	(20, 3368)

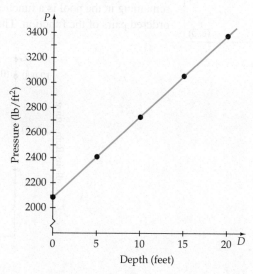

Figure 5.8

From Figure 5.9, we can calculate the rate at which the pressure on the diver is increasing as the depth of the diver increases, the rate of change of the function for the points (5, 2408) and (20, 3368).

$$\text{Average rate of change} = \frac{\text{Change in pressure}}{\text{Change in depth}}$$

$$= \frac{3368 \text{ lb/ft}^2 - 2408 \text{ lb/ft}^2}{20 \text{ ft} - 5 \text{ ft}}$$

$$= \frac{960 \text{ lb/ft}^2}{15 \text{ ft}}$$

$$= 64 \frac{\text{lb/ft}^2}{\text{ft}}$$

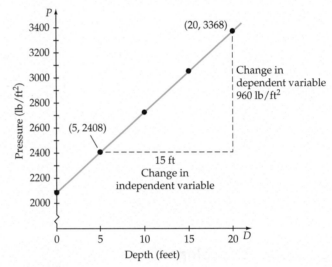

Figure 5.9

Table 5.4

Time (hours), t	Gallons Remaining, G	(t, G)
0	12,000	(0, 12,000)
2	9000	(2, 9000)
4	6000	(4, 6000)
6	3000	(6, 3000)
8	0	(8, 0)

For the diver, the average rate of change of pressure between the two depths is increasing $64 \dfrac{\text{lb/ft}^2}{\text{ft}}$; that is, with each 1-foot increase in depth, the pressure on the diver increases by 64 lb/ft². This is actually true for any two points on the graph.

Now consider a moderate size backyard pool that holds 12,000 gallons of water. A water pump is attached to a hose to drain the pool to make repairs. The amount of water remaining in the pool is a function of the time the pump runs. Table 5.4 shows some of the ordered pairs of the function. The graph of the function is shown in Figure 5.10.

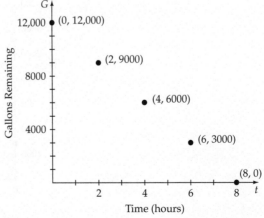

Figure 5.10

The points appear to lie on a line. If we connect the dots, we have the graph shown in Figure 5.11.

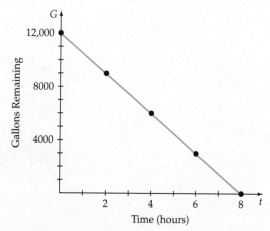

Figure 5.11

From Figure 5.12, we can calculate the average rate of change of the function for the points (2, 9000) and (6, 3000).

$$\text{Average rate of change} = \frac{\text{Change in gallons}}{\text{Change in time}}$$

$$= \frac{3000 \text{ gal} - 9000 \text{ gal}}{6 \text{ hr} - 2 \text{ hr}}$$

$$= \frac{-6000 \text{ gal}}{4 \text{ hr}}$$

$$= -1500 \text{ gal/hr}$$

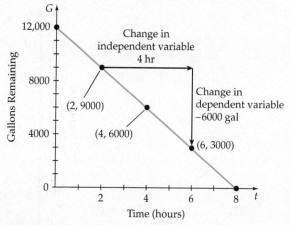

Figure 5.12

The amount of water remaining between the two times is decreasing at a rate of 1500 gallons per hour. In other words, with each 1-hour increase in time, the number of gallons in the pool decreases by 1500 gallons.

The examples of the pressure on the scuba diver and emptying the pool each resulted in a **constant** rate of change. Whatever two points are chosen on the graph, the ratio of the change between the two points was the same number. This is not always the case.

Table 5.5 gives the height in feet of a ball t, seconds, after it has been dropped from a building 500 feet tall. The points are plotted in Figure 5.13.

Table 5.5

Time, t (seconds)	Height, h (feet)	(t, h)
0	500	(0, 500)
1	484	(1, 484)
2	436	(2, 436)
3	356	(3, 356)
4	244	(4, 244)
5	100	(5, 100)
6	0	(5.59, 0)

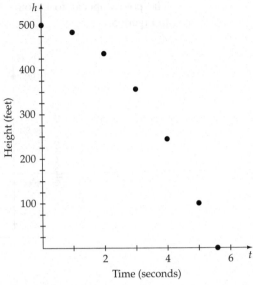

Figure 5.13

If we find the average rate of change of height to the change in time for the two points (1, 484) and (4, 244), we have

$$\frac{\text{Change in height}}{\text{Change in time}} = \frac{484 - 244}{1 - 4} = \frac{240 \text{ feet}}{-3 \text{ seconds}} = -80 \text{ feet/second}$$

Choose any two other points and find the ratio of the change in height to the change in time. We will use (2, 436) and (5, 100).

$$\frac{\text{Change in height}}{\text{Change in time}} = \frac{436 - 100}{2 - 5} = \frac{336 \text{ feet}}{-3 \text{ seconds}} = -112 \text{ feet/second}$$

The two rates are not equal. Figure 5.14 shows a graph of the plotted points. We can observe that the graph is not a straight line.

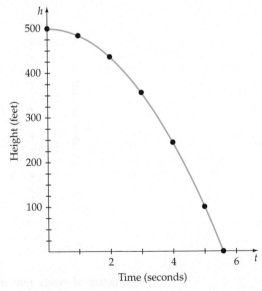

Figure 5.14

Constant Rate of Change

In the special case where the average rate of the change is the same constant for all points on the graph, the resulting graph is a straight line. The constant is called the *slope* of the line.

Take Note: For a straight line, the slope of a line and the average rate of change are equal.

Slope of a Line

Let (x_1, y_1) and (x_2, y_2) be the coordinates of two points on a nonvertical line. Then the **slope** of the line through the two points is the ratio of the change in the y-coordinates to the change in the x-coordinates.

$$m = \frac{\text{Change in } y}{\text{Change in } x} = \frac{y_2 - y_1}{x_2 - x_1}, x_1 \neq x_2$$

Example 2   **Find the Slope of a Line Between Two Points**

a. $(-4, -3)$ and $(-1, 1)$ **b.** $(-2, 3)$ and $(1, -3)$
c. $(-1, -3)$ and $(4, -3)$ **d.** $(4, 3)$ and $(4, -1)$

Solution

a. $(x_1, y_1) = (-4, -3)$, $(x_2, y_2) = (-1, 1)$

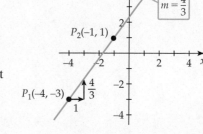

$$m = \frac{y_2 - y_1}{x_2 - x_1} = \frac{1 - (-3)}{-1 - (-4)} = \frac{4}{3}$$

The slope is $\frac{4}{3}$. A *positive* slope indicates that the line slopes *upward* to the right. For this particular line, the value of y *increases* by $\frac{4}{3}$ when x increases by 1.

b. $(x_1, y_1) = (-2, 3)$, $(x_2, y_2) = (1, -3)$

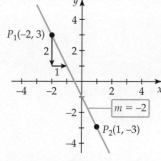

$$m = \frac{y_2 - y_1}{x_2 - x_1} = \frac{-3 - 3}{1 - (-2)} = \frac{-6}{3} = -2$$

The slope is -2. A *negative* slope indicates that the line slope *downward* to the right. For this particular line, the value of y *decreases* by 2 when x increases by 1.

c. $(x_1, y_1) = (-1, -3)$, $(x_2, y_2) = (4, -3)$

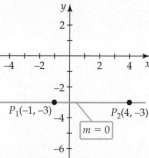

$$m = \frac{y_2 - y_1}{x_2 - x_1} = \frac{-3 - (-3)}{4 - (-1)} = \frac{0}{5} = 0$$

The slope is 0. A *zero* slope indicates that the line is *horizontal*. For this particular line, the value of y *stays the same* when x increases by any amount.

d. $(x_1, y_1) = (4, 3)$, $(x_2, y_2) = (4, -1)$

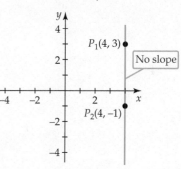

$$m = \frac{y_2 - y_1}{x_2 - x_1} = \frac{-1 - 3}{4 - 4} = \frac{-4}{0}$$ Division by 0 is undefined.

If the denominator of the slope formula is zero, the line has *no slope*. Sometimes we say that the slope of a vertical line is *undefined*.

Review of Slope

Let (x_1, y_1) and (x_2, y_2) be the coordinates of two points on a line.
- A line with positive slope slants upward to the right.
- A line with negative slope slants downward to the right.
- A horizontal line has 0 slope.
- A vertical line has no slope.

The concept of positive and negative slope has many applications. Recall the discussion of the pressure on a scuba diver. Figure 5.15 shows the relationship between pressure and depth.

The slope of the line is *positive*. Because the line has positive slope, pressure is *increasing* on the diver. For each 1-foot increase in depth, the pressure on the diver increases by $64 \dfrac{\text{lb}}{\text{ft}^2}$.

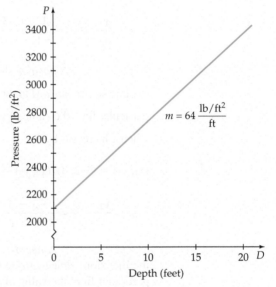

Figure 5.15

As discussed earlier in this chapter, the graph in Figure 5.16 shows the number of gallons of water in a pool over time. The slope of the line is negative. Because the line has a *negative* slope, the amount of water in the pool is *decreasing*. For each 1-hour increase in time, the amount of water in the pool decreases by 1500 gallons.

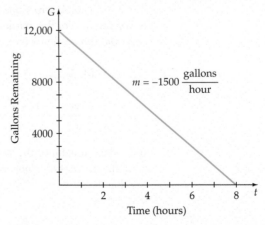

Figure 5.16

Table 5.6

Time, t (hours)	Distance, d (miles)	(t, d)
0.0	0	(0.0, 0)
0.5	3	(0.5, 3)
1.0	6	(1.0, 6)
1.5	9	(1.5, 9)
2.0	12	(2.0, 12)
2.5	15	(2.5, 15)
3.0	18	(3.0, 18)
3.5	21	(3.5, 21)

Table 5.6 and Figure 5.17 show the distance traveled by a jogger who is jogging at a rate of 6 mph. The ordered pairs from the table are shown on the graph.

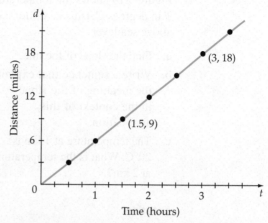

Figure 5.17

The slope of the line can be calculated using two points on the graph.

$$m = \frac{\text{Change in } d}{\text{Change in } t} = \frac{18\,\text{mi} - 9\,\text{mi}}{3\,\text{hr} - 1.5\,\text{hr}}$$

$$= \frac{9\,\text{mi}}{1.5\,\text{hr}} = 6\,\text{mph}$$

Note that the speed of the jogger and the slope of the line are the same. Speed is a real-world application of slope.

In economics, the cost for a manufacturer to produce one additional item of a product is called its *marginal cost*. As shown in the next example, marginal cost is another real-world application of slope.

Example 3   **Application of Slope to Business**

Figure 5.18 shows the total cost C in dollars for a company to make n wooden patio chairs.

a. Find the slope of the line.

b. Write a sentence that explains the meaning of the slope in the context of this application.

c. If the total cost to produce 12 chairs is $3200, what is the total cost to produce 13 chairs?

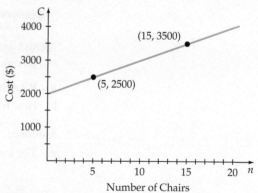

Figure 5.18

Solution

a. $m = \dfrac{\text{Change in } C}{\text{Change in } n} = \dfrac{\$3500 - \$2500}{15\,\text{chairs} - 5\,\text{chairs}}$

$= \dfrac{\$1000}{10\,\text{chairs}} = \$100/\text{chair}$

b. The slope means that total cost of producing one additional chair is $100.

c. From part b, the total cost of producing one additional chair is $100. If the total cost to produce 12 chairs is $3200, then the total cost to produce 13 chairs is $3300.

Take Note: Economists would say that the marginal cost to produce the next chair in the production line is $100/chair.

Example 4   **Application of Slope to Atmospheric Science**

Figure 5.19 shows the temperature
T in degrees Celsius at x kilometers
above sea level.

a. Find the slope of the line.
b. Write a sentence that explains
    the meaning of the slope
    in the context of this
    application.
c. The temperature at 1 km is
    20°C. What is the temperature
    at 2 km?

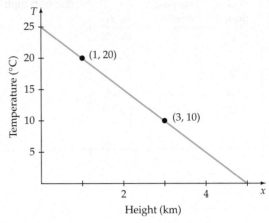

Solution   **Figure 5.19**

a. $m = \dfrac{\text{Change in } T}{\text{Change in } x} = \dfrac{10°C - 20°C}{3 \text{ km} - 1 \text{ km}}$

    $= \dfrac{-10°C}{2 \text{ km}} = -5°C/\text{km}$

b. The slope means that the temperature is decreasing (because the slope is negative)
    5°C for each 1-km increase in height above sea level.

c. From part b, the temperature is decreasing at a rate of 5°C per kilometer. If
    the temperature at 1 km is 20°C, then the temperature at 2 km is 5°C less or
    15°C.

Example 5   **Application of Slope to Child Development**

A child grows at an approximately constant rate from 80 centimeters at age 12 months
to 100 centimeters at age 36 months.

a. What is the average growth rate per month? Round to the nearest hundredth.
b. How many centimeters will this child grow in 4 months? Round to the nearest tenth.

Solution

a. The average growth rate is the slope of the line between (12, 80) and (36, 100).

    $m = \dfrac{\text{Change in height}}{\text{Change in months}}$

    $= \dfrac{100 \text{ cm} - 80 \text{ cm}}{36 \text{ months} - 12 \text{ months}} = \dfrac{20 \text{ cm}}{24 \text{ months}}$

    $\approx 0.83 \text{ cm/month}$

    The child is growing at an approximate rate of 0.8 cm/month.

b. To find the number of centimeters the child grows in 4 months, multiply the rate per
    month (slope) by 4 months.

    $0.83 \text{ cm/month} \cdot 4 \text{ months} = 3.32 \text{ cm}$

    The child grows 3.3 cm in 4 months.

Intercepts

An infinite number of parallel lines have the same slope. Figure 5.20 shows just two examples.

$$m_{\text{top line}} = \frac{6 - 2}{3 - (-3)} = \frac{4}{6} = \frac{2}{3}$$

$$m_{\text{bottom line}} = \frac{2 - (-4)}{6 - (-3)} = \frac{6}{9} = \frac{2}{3}$$

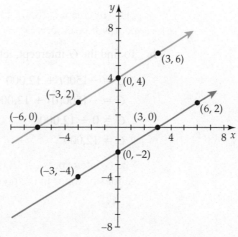

Figure 5.20

What distinguishes the two lines is the point at which they cross the axes. The top line crosses the y-axis at **(0, 4)** and the x-axis at **(−6, 0)**. The bottom line crosses the y-axis at **(0, −2)** and the x-axis at **(3, 0)**.

Intercepts of a Graph

An *x*-**intercept** of a graph in an *xy*-coordinate plane is a point at which the graph crosses the *x*-axis. **For an x-intercept, the y-coordinate of the point is 0.** A *y*-**intercept** of a graph in an *xy*-coordinate plane is a point at which the graph crosses the *y*-axis. **For a y-intercept, the x-coordinate of the point is 0.**

Intercepts have real-world applications. In Example 6, we revisit the function of the amount of water remaining in a pool.

Example 6 **Find and Interpret Intercepts**

A model of the amount of water G in gallons remaining in a 12,000 gallon pool after t hours is given by $G = -1500t + 12{,}000$. The graph of the equation is shown in Figure 5.21.

a. Find the t-intercept. Write a sentence that explains the intercept in the context of this application.

b. Find the G-intercept. Write a sentence that explains the intercept in the context of this application.

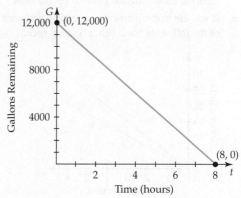

Figure 5.21

Solution

Although we could read the intercepts from the graph, we will give an algebraic solution to use in those cases where it is not possible to read the intercepts from the graph.

a. To find the t-intercept, let $G = 0$ and solve for t.

$$G = -1500t + 12{,}000$$
$$0 = -1500t + 12{,}000 \qquad \bullet \text{Let } G = 0.$$
$$-12{,}000 = -1500t \qquad \bullet \text{Solve for } t.$$
$$8 = t$$

The t-intercept is $(8, 0)$. In 8 hours, there will be 0 gallons of water in the pool; the pool will be empty.

b. To find the G-intercept, let $t = 0$ and solve for G.

$$G = -1500t + 12{,}000$$
$$G = -1500(0) + 12{,}000 \qquad \bullet \text{Let } t = 0.$$
$$G = 0 + 12{,}000 \qquad \bullet \text{Solve for } G.$$
$$G = 12{,}000$$

The G-intercept is $(0, 12{,}000)$. When $t = 0$ (the pump is turned on), there are 12,000 gallons of water in the pool.

5.2 Exercise Set

Think About It

1. Identify which of the following are linear functions. If it is a linear function, identify m and b.

   a. $y = \dfrac{1}{x}$ **b.** $y = x$

   c. $F(v) = 4 - 0.256v$ **d.** $B(n) = 3n - n^2$

2. What is the difference between a line with 0 slope and one with no slope?

3. The coordinates of two points are $(0, -3)$ and $(2, 0)$. Which ordered pair is the x-intercept, and which ordered pair is the y-intercept?

4. Suppose two different lines are parallel. Are the slopes of the lines the same? Are the y-intercepts the same?

5. If you are participating in a marathon and want to win, which of the following has a better running speed?

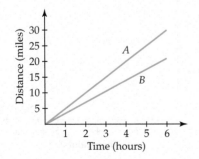

6. The following graph shows the depreciation of two cars that originally cost \$30,000. Which of the two cars is depreciating more quickly?

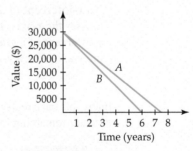

7. Which of the following are not examples of slope?

   a. 4 centimeters/hour

   b. 7 minutes

   c. \$3 per hour

   d. 5 kilometers

   e. 7 cars per day

Preliminary Exercises

For Exercises P1 to P6, if necessary, use the referenced example following the exercise for assistance.

P1. Graph $f(v) = \dfrac{2}{3}v + 1$. **Example 1**

P2. Find the slope of the line between the points with coordinates $(3, 1)$ and $(-1, 3)$. **Example 2**

P3. The following graph shows the amount *A* of gasoline remaining in a gas tank after a car is driven *d* miles. **Example 3**

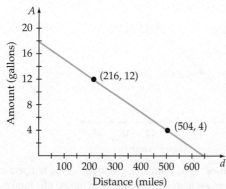

a. Find the slope of the line.

b. Write a sentence that explains the meaning of the slope in the context of this application.

P4. The following graph shows the distance *s* in meters a swimmer has traveled in *t* seconds. **Example 4**

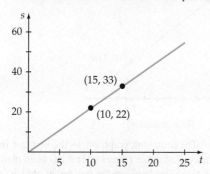

a. Find the slope of the line.

b. Write a sentence that explains the meaning of the slope in the context of this application.

c. At 20 seconds, the swimmer has traveled 44 meters. How far will the swimmer travel in 21 seconds?

P5. Human hair grows at an approximate rate of 0.6 inch per month. **Example 5**

a. What is the average growth rate per day?

b. How many inches will human hair grow in 15 days?

P6. A model of the total cost *C*, in dollars, to produce *n* cell phones is given by $C = 210n + 1000$. **Example 6**

a. Find the *C*-intercept. Write a sentence to explain the meaning of the intercept in the context of this exercise.

b. Find the *n*-intercept. Write a sentence to explain the meaning of the intercept in the context of this exercise.

▶ **Section Exercises**

For Exercises 1 to 4, graph the linear function.

1. $y = 2x - 3$

2. $y = -\dfrac{x}{2} + 2$

3. $B(n) = 3 - \dfrac{n}{4}$

4. $f(t) = -3t + 3$

For Exercises 5 to 8, find the slope of the line between the two points.

5. $(-2, 4); (4, -1)$

6. $(3, 0); (2, -5)$

7. $(5, 7); (-1, 7)$

8. $(-3, 4); (-3, -4)$

9. **Water** The number of gallons *g* of water in a cistern depends on the time *t*, in minutes, the water has been flowing into the cistern. Three ordered pairs of this function are (1, 2), (2, 5), and (4, 11). Is it possible that these three points lie on the same line? Explain.

10. **Landing a Plane** The altitude, in feet, of a plane descending for a landing is a function of the time, in minutes, the plane has been descending. Three ordered pairs of this function are (20, 10,000), (22, 7000), and (23, 5500). Is it possible that these three points lie on the same line? Explain.

11. **Throwing a Ball** The height of a ball thrown directly upward with a certain speed is a function of the time after the ball has been released. Three ordered pairs of this function are (1, 85), (2, 133), and (3, 149). Is it possible that these three points lie on the same line? Explain.

12. **Value of an Investment** The value of an investment is a function of the number of months the investment has been held. Three ordered pairs of this function are (5, 1051), (10, 1104), and (12, 1127). Is it possible that these three points lie on the same line? Explain.

13. **Travel** The following graph shows the relationship between the distance traveled by a motorist and the time of travel. Find the slope of the line between the two points shown on the graph. Write a sentence that states the meaning of the slope in the context of this application.

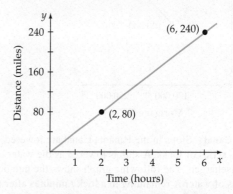

14. **Depreciation** The following graph shows the depreciated value of a building allowed for income tax purposes. Find the slope of the line between the two points shown on the graph. Write a sentence that states the meaning of the slope in the context of this application.

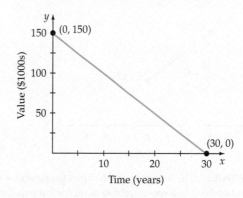

15. **Aviation** The following graph shows the height of a plane above an airport during its 30-minute descent from cruising

altitude to landing. Find the slope of the line. Write a sentence that explains the meaning of the slope.

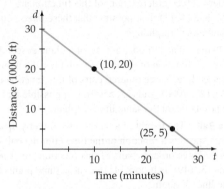

Time (minutes)

16. **Mortgages** The following graph shows the relationship between the monthly payment on a mortgage and the amount of the mortgage. Find the slope of the line between the two points shown on the graph. Write a sentence that states the meaning of the slope in the context of this application.

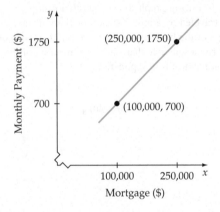

17. **Panama Canal** Ships in the Panama Canal are lowered through a series of locks. A ship is lowered as the water in a lock is discharged. The following graph shows the number of gallons of water N remaining in a lock t minutes after the valves are opened to discharge the water. Find the slope of the line. Write a sentence that explains the meaning of the slope.

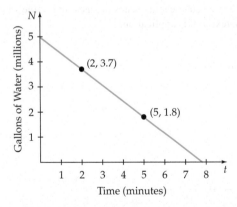

18. **Website Traffic** The following graph shows the number n of people subscribing to a sports website of increasing popularity m months after the site was created. Find the slope of the line between the two points shown on the graph. Write a sentence that states the meaning of the slope.

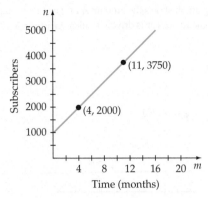

19. **Temperature** The following graph shows the temperature T in an oven t minutes after it has been turned off. Find the slope of the line between the two points shown on the graph. Write a sentence that states the meaning of the slope.

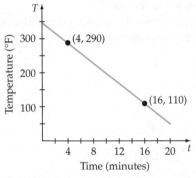

20. **Tree Growth** The following graph shows the increase in the circumference C of a tree t years after it has been planted. Find the slope of the line between the two points shown on the graph. Write a sentence that states the meaning of the slope.

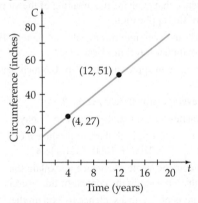

21. **Hiking** On a hike, Parker noticed that the temperature at the base of a mountain was 62°F. At a point on the mountain that was 2000 feet above the base, the temperature was 56°F. What is the average rate of change of temperature? How much does the temperature change when the altitude changes by 100 feet?

22. **Road Construction** A highway is slightly raised in the center so that rain will run off to collection areas on the side of the road. The edge of the highway is 3 feet above the collection area. At the center of the road, 40 feet from the edge, the road is 3.2 feet above the collection area. What is the slope of the road?

23. **Wheelchair Ramp** The recommended slope for a wheelchair ramp is a height-to-distance ratio of no more than

1 to 12 feet. Does a ramp that has a slope of 0.09 satisfy the recommendation?

24. Sports   The rate at which a professional tennis player used carbohydrates during a strenuous workout was found to be 1.5 grams per minute. If a line were graphed showing time on the horizontal axis and carbohydrates used on the vertical axis, what would be the slope of the line? How many carbohydrates would the athlete use in 20 minutes?

25. Evaporation   The amount of water that evaporates from a pool in a one-hour period depends of the surface area of the pool. For a pool with a surface area of 100 square feet, 25 gallons of water will evaporate in 1 hour. For a pool with 400 square feet of surface area, 60 gallons of water will evaporate.

  a. What is the evaporation rate per square foot of surface area? Round to the nearest thousandth.

  b. How many gallons of water will evaporate from a pool of 200 square feet?

26. Chemistry   The temperature in degrees Celsius of an endothermic (heat-absorbing) chemical reaction decreases at an approximately constant rate. Two measurements from this reaction found a 20°C temperature 5 minutes after the reaction began and an 18°C temperature 4 minutes later.

  a. What is the rate of change of temperature?

  b. Write a sentence that explains the result in part a in the context of this problem.

  c. What was the temperature 11 minutes after the reaction begins?

27. Business   Marketing surveys conducted by the sales manager for a company determined that 2000 power drills could be sold for $50 each but 3000 drills could be sold for $40 each.

  a. If the two data points were plotted on a coordinate system with the number of drills sold on the horizontal axis and the price of each drill on the vertical axis, what is the slope of the line between the two points?

  b. Write a sentence that explains the result in part a in the context of this problem.

  c. Assuming the trend in part a continues, how many drills can be sold for $38?

28. Business   A production engineer has determined that the total cost to produce 500 tractor tires is $110,000 and producing 600 tractor tires costs $120,000.

  a. If the two data points were plotted on a coordinate system and the number of tires sold was plotted on the horizontal axis and the price of each tire on the vertical axis, what is the slope of the line between the two points?

  b. Write a sentence that explains the result in part a in the context of this problem.

29. Basketball Camp   The director of a basketball camp estimates that the number N of students who will enroll in a basketball camp can be modeled by $N = -0.3t + 270$, where t is the tuition for the camp in dollars.

  a. What is the N-intercept?

  b. Write a sentence that explains the answer to part a in the context of the problem.

  c. What is the t-intercept?

  d. Write a sentence that explains the answer to part c in the context of the problem.

30. Skiing   The elevation E, in feet, of a skier is a function of the time t, in seconds, the skier has been gliding down a ski run and can be modeled by $E = -20t + 10{,}000$.

  a. What is the E-intercept?

  b. Write a sentence that explains the answer to part a in the context of the problem.

  c. What is the t-intercept?

  d. Write a sentence that explains the answer to part c in the context of the problem.

31. Business   The profit P for selling n smartwatches can be modeled by $P = 25n - 1000$.

  a. What is the P-intercept?

  b. Write a sentence that explains the answer to part a in the context of the problem.

  c. What is the n-intercept?

  d. Write a sentence that explains the answer to part c in the context of the problem.

32. Gardening   The growth G in inches of a tomato plant can be modeled by $G = 5d + 36$, where d is the number of days after the first measurement is taken.

  a. What is the G-intercept?

  b. Write a sentence that explains the answer to part a in the context of the problem.

  c. What is the d-intercept?

  d. Assuming a constant growth rate from planting to maturity, what is one possible interpretation of the answer to part c?

Investigation

Hiking Distance   The following graph represents the distance a hiker is from home t hours after leaving home.

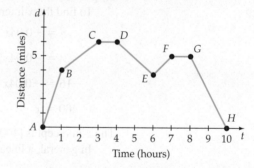

For each of the following problems, the vertical axis represents the distance from home. The horizontal axis is time.

1. Write a sentence that explains what is happening between A and B.

2. Between which two points—A and B or C and D—is the rate at which the hiker is traveling greater?

3. Write a sentence that explains what is happening between C and D.

4. Between D and E, is the hiker walking away from home or walking toward home?

5. Write a sentence that explains what is happening between G and H.

6. Between which two points—D and E or G and H—is the rate at which the hiker is walking greater?

7. How long was the hiker gone from home?

Section 5.3 Linear Models

Learning Objective:

• Find linear models.

Find Linear Models

Suppose that a car burns 0.04 gallon of gas per mile driven and that the fuel tank, which holds 16 gallons of gas, is full. Using this information, we can determine a linear model for the amount of fuel remaining in the gas tank after driving x miles.

$$\begin{array}{ccccc} \text{Gallons of gas} \\ \text{remaining} \end{array} = \begin{array}{c} \text{Rate of} \\ \text{consumption} \end{array} \cdot \begin{array}{c} \text{Miles} \\ \text{driven} \end{array} + \begin{array}{c} \text{Initial number} \\ \text{of gallons} \end{array}$$

$$g = m \cdot x + b$$

$$g = -0.04 \cdot x + 16$$

A linear model for gas consumption is $g = -0.04x + 16$. A graph of the model is shown in Figure 5.22. The slope of the line is the rate gas is consumed, -0.04 gallon per mile. The rate is negative because the amount of gas in the tank is decreasing. We can determine the intercepts as well.

To find the g-intercept, let $x = 0$ and solve for g.

$$g = -0.04x + 16$$

$$= -0.04(0) + 16 \qquad \bullet \text{ Replace } x \text{ with } 0.$$

$$= 16$$

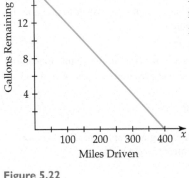

Figure 5.22

The g-intercept is $(0, 16)$. Before driving any miles, $x = 0$, there were 16 gallons of gas in the gas tank.

To find the x-intercept, let $g = 0$ and solve for x.

$$g = -0.04x + 16$$

$$0 = -0.04x + 16 \qquad \bullet \text{ Replace } g \text{ with } 0.$$

$$-16 = -0.04x \qquad \bullet \text{ Solve for } x.$$

$$400 = x$$

The x-intercept is $(400, 0)$. The car can be driven 400 miles before the tank is empty, $g = 0$. In general, a linear model for any situation will have the form

Linear model

$$\begin{array}{c} \text{Dependent} \\ \text{variable} \end{array} = \begin{array}{c} \text{Rate of} \\ \text{change} \end{array} \cdot \begin{array}{c} \text{Independent} \\ \text{variable} \end{array} + \begin{array}{c} \text{Initial} \\ \text{state} \end{array}$$

$$y = m \cdot x + b$$

Take Note: We choose to use function notation for Example 1. We could have selected a name for the dependent variable, say G, and written the model as $G = 0.2t + 2$. Either method is fine.

Example 1 **Find a Linear Model Given the Rate of Change and the Initial State**

Suppose a 20-gallon gas tank contains 2 gallons of gas when a motorist decides to fill the tank. If the gas pump fills the tank at a rate of 0.2 gallon per second:

a. Find a linear function that models the amount of fuel in the tank t seconds after fueling begins.

b. Find the number of gallons in gas in the tank 40 seconds after fueling begins.

Take Note: The following calculation is the same as the one in part b of Example 1 but with the units included.

$$f(40 \text{ seconds}) = 0.2\frac{\text{gallon}}{\text{second}}(40 \text{ seconds})$$
$$+ 2 \text{ gallons}$$
$$= 8 \text{ gallons} + 2 \text{ gallons}$$
$$= 10 \text{ gallons}$$

Note that result has units of gallons, as it should. Being aware of units can help when solving application problems.

Solution

a. The slope is the rate at which fuel is being added to the tank. Because the amount of fuel in the tank is increasing, the slope is positive and we have $m = 0.2$ gallon/second. The initial state is when fueling begins at $t = 0$. When fueling begins, there are 2 gallons of gas in the tank, $b = 2$ gallons. Use these values to write the linear function. We will use function notation for this example.

$$f(t) = mt + b$$
$$f(t) = 0.2t + 2 \qquad \bullet \, m = 0.2, b = 2.$$

The linear function is $f(t) = 0.2t + 2$, where $f(t)$ is the number of gallons of fuel in the tank t seconds after fueling begins.

b. To find the number of gallons of gas in the tank after 40 seconds, evaluate the function for $t = 40$.

$$f(t) = mt + b$$
$$f(40) = 0.2(40) + 2 \qquad \bullet \, t = 40.$$
$$= 8 + 2 = 10$$

There are 10 gallons of gas in the tank after 40 seconds.

For Example 1, the initial state (when the independent variable = 0) was given. That value is the value of b in the model $y = mx + b$. When some condition other than the initial state is given, we use the fact that the condition must be an ordered pair of the model.

Example 2 **Find a Linear Model Given the Rate of Change and a Data Point**

A zoologist found that during the first 5 years after birth, an elephant gains weight at a rate of 1.9 pounds per day. The weight of the baby elephant 10 days after birth was 250 pounds. Find a linear model that gives the weight of the elephant d days after birth.

Solution

Let w represent the weight of the elephant d days after birth. Then

Dependent variable	=	Rate of change	·	Independent variable	+	Initial state
weight		1.9 lb/day		days after birth		unknown

$$w = m \cdot d + b$$

We are given the weight of the elephant is 250 pounds after 10 days. Replace w, m, and d with their values and solve for b.

$$w = m \cdot d + b$$
$$250 = 1.9 \cdot 10 + b \qquad \bullet \, w = 250, m = 1.9, d = 10$$
$$250 = 19 + b$$
$$231 = b$$

The weight of the elephant at birth was 231 pounds. Using that information and the rate at which the elephant is growing, 1.9 pounds per day, write a linear model for the weight of the elephant d days after birth.

A linear model for the weight of an elephant d days after birth is $w = 1.9d + 231$.

Before leaving Example 2, we could check that our model is correct. This means checking that the given data point satisfies the model.

$$w = 1.9d + 231$$
$$250 \overset{?}{=} 1.9(10) + 231 \qquad \bullet \text{ Given } w \text{ is } 250 \text{ when } d \text{ is } 10.$$
$$250 \overset{?}{=} 19 + 231$$
$$250 = 250 \qquad \bullet \text{ Check}$$

If the rate is unknown, we can use two given data points and the slope formula to find the rate.

Example 3 Find a Linear Model Given Two Data Points

The boiling point of water changes with altitude. The boiling point is 93°C at an altitude of 2 kilometers and 86°C at an altitude of 4 kilometers.

a. Find a linear model that gives the boiling point as a function of altitude.

b. Use your model to find the boiling point of water at an altitude of 1.5 kilometers.

Solution

a. To find a linear model, we need to know the slope, m, and the initial state, b. Let a represent the altitude (independent variable) and T represent the temperature (dependent variable). We are given two data points (2 km, 93°C) and (4 km, 86°C). Use the slope formula to find the slope the line between the two points.

$$m = \frac{\text{Change in dependent variable}}{\text{Change in independent variable}}$$

$$= \frac{86°C - 93°C}{4\text{ km} - 2\text{ km}} = \frac{-7°C}{2\text{ km}}$$

$$= -3.5°C/\text{km}$$

The slope of the line is -3.5°C/km.

Now use the method of Example 2 to find b. We will use the data point (2 km, 93°C). It does not matter which data point is selected.

$$T = m \cdot a + b$$
$$93 = -3.5 \cdot 2 + b$$
$$93 = -7 + b$$
$$100 = b$$

Using $m = -3.5$ and $b = 100$, we can write a linear model.

A linear model that gives the boiling point of water as a function of altitude is $T = -3.5a + 100$.

b. To find the boiling point of water at an altitude of 1.5 km, replace a with 1.5 and solve for T.

$$T = m \cdot a + b$$
$$T = -3.5 \cdot 1.5 + 100$$
$$T = -5.25 + 100$$
$$T = 94.75$$

The boiling point of water at an altitude of 1.5 km is 94.75°C.

Sometimes there is a need to approximate a data set by a linear function. There are a few ways to do this, and we look at one method here.

Example 4 Find a Linear Model from a Data Set

Photographers use sodium thiosulfate to develop some types of film. The amount of this chemical that will dissolve in water depends on the temperature of the water. Table 5.7 gives the number of grams of sodium thiosulfate that will dissolve in 100 milliliters of water at various temperatures.

Table 5.7

Temperature, x, in Degrees Celsius	20	35	50	60	75	90	100
Sodium Thiosulfate Dissolved, y, in Grams	50	80	120	145	175	205	230

a. Use the points (20, 50) and (100, 230) to find a linear model that will predict the number of grams of sodium thiosulfate that will dissolve in water at a temperature of x degrees Celsius.

b. How many grams of sodium thiosulfate does the model predict will dissolve in 100 milliliters of water when the temperature is 70°C?

Solution

a. To find a linear model, we need to know the slope, m, and the initial state, b. The independent variable is x (temperature), and the dependent variable is y grams of sodium thiosulfate. We have to find the model using data points (20°C, 50 g) and (100°C, 230 g). Use the slope formula to find the slope of the line between the two points.

$$m = \frac{\text{Change in dependent variable}}{\text{Change in independent variable}}$$

$$= \frac{230 \text{ g} - 50 \text{ g}}{100°\text{C} - 20°\text{C}} = \frac{180 \text{ g}}{80°\text{C}}$$

$$= 2.25 \text{ g/°C}$$

The slope of the line is 2.25 g/°C.

   Now use the method of Example 2 to find b. We will use the data point (100°C, 230 g). It does not matter which two given data points are selected.

$$y = m \cdot x + b$$
$$230 = 2.25 \cdot 100 + b$$
$$230 = 225 + b$$
$$5 = b$$

Using $m = 2.25$ and $b = 5$, we can write a linear model.

A linear model that gives the grams of sodium thiosulfate that can be dissolved in water of a given temperature is $y = 2.25x + 5$.

b. To find the number of grams of sodium thiosulfate that can be dissolved in water at 70°C, replace x with 70 and solve for y.

$$y = 2.25 \cdot x + 5$$
$$y = 2.25 \cdot 70 + 5$$
$$y = 157.5 + 5$$
$$y = 162.5$$

The model predicts that 162.5 g of sodium thiosulfate will dissolve in 100 ml of water at a temperature of 70°C.

The graph of the linear model and the data points in the table are shown in Figure 5.23. From the graph, we can see that the data points do not lie on a line but the linear model is a good approximation of the data.

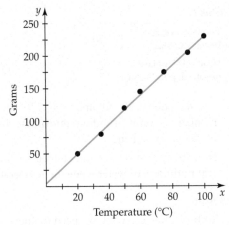

Figure 5.23

5.3 Exercise Set

Think About It

1. For each of the following would the graph of the function increase or decrease as the values of the independent variable increase?

   a. $f(x) = -x + 3$ **b.** $y = x$

   c. $F(v) = 4 - 0.256v$ **d.** $B(n) = \dfrac{2n}{3} - 4$

2. A model of the temperature T, in °F, of a cup of tea is given by $T = -2x + 130$, where x is the number of minutes after the tea was poured into the cup. What are the units of the slope?

3. Two buckets have holes in their bottoms and are leaking water. A model of the amount of water in bucket A is $y = -2x + 15$ while a model for bucket B is $y = -3x + 12$. In each case, y is the number of gallons of water remaining in the bucket and x is the number of hours after 1 P.M. Which bucket is leaking water at a greater rate?

4. Using the model $y = -3x + 12$ from Exercise 3, when $x = 4$, $y = 0$. Write a sentence that explains the meaning of this in the context of the problem.

Preliminary Exercises

For Exercises P1 to P4, if necessary, use the referenced example following the exercise for assistance.

P1. During a snowstorm at a ski resort, snow was falling at a rate of 2 inches per hour. When the snow began, 20 inches of snow were already on the ground. Find a linear model of the snow on the ground after t hours. **Example 1**

P2. Whales, dolphins, and porpoises communicate using high-pitched sounds that travel through the water. The speed at which sound travels depends on many factors, one of which is the depth of the water. At approximately 1000 meters below sea level, the speed of sound is 1480 meters per second. The speed of sound increases at a constant rate of 0.017 meter per second for each additional meter below 1000 meters. Find a linear function for the speed of sound in terms of the number of meters below sea level. Use the function to approximate the speed of sound 2500 meters below sea level. Round to the nearest whole number. **Example 2**

P3. The resale value of a sports car declines as the number of miles driven increases. Data from a website that values used cars show that a car with 25,000 miles had a value of $34,400. A similarly equipped car with 50,000 miles was valued at $33,000. Find a linear function for the value of the car in terms of the number of miles driven. Use the function to approximate the value of a car that has been driven 45,000 miles. **Example 3**

P4. The following table shows the circumference in centimeters of a baby's head d days after birth.

Days After Birth	0	5	10	15	20	25	30
Circumference (cm)	13.50	13.96	14.44	14.45	14.79	15.28	15.50

Using the data points (0, 13.50) and (30, 15.50), find a linear model that approximates the data. Use the model to predict the circumference of a baby's head 18 days after birth. Round to the nearest tenth. **Example 4**

Section Exercises

1. **Sales** A sales executive for a financial firm earns a base salary of $4000 per month plus a commission of 2% on all sales s in dollars. Find a linear model for the total monthly salary T earned by the executive. How much will the executive earn in a month for which sales were $80,000?

2. **Home Heating** A propane tank for a home can hold 420 gallons of propane. When a refueling truck arrives, there are 100 gallons of propane in the tank. If the refueling truck delivers 32 gallons of propane per minute into the tank, write a linear function for the amount of propane in the tank t minutes after refueling begins. How much propane is in the tank after 3 minutes?

3. **Aviation** Once an airplane has reached its cruising altitude, its computer shows the destination to be 2700 miles. If the plane is traveling at a rate of 500 miles per hour, write a linear function of the distance d the plane is from its destination in t hours after reaching cruising altitude. How far is the plane from its destination 2 hours after the flight begins?

4. **Earth Science** Sediment buildup behind a dam can adversely affect a reservoir's storage capacity. Measurements of a reservoir in 1988 showed it had a capacity of 400,000 acre-feet of water. Measurements in 2018 showed that the reservoir's capacity has been decreasing at a rate of 1500 acre-feet per year. Find a linear model of the number of acre-feet of water in the reservoir n years after 1988. Use the model to predict the capacity of the reservoir in 2020.

5. **Aviation** A passenger jet passed through an altitude of 12,000 feet 8 minutes after takeoff and was climbing at a rate of 1250 feet per minute. Find a linear model for the altitude of the jet t minutes after takeoff. Use your model to predict the altitude of the jet 20 minutes into the flight.

6. **Forestry** From previous studies, a biologist has determined that the circumference of a tree in its early growth stages increases at a rate of approximately 3 inches per year. A 4-year-old tree has a circumference of 27 inches. Find a linear model of the circumference of this tree after n years. Use the model to predict the circumference of the tree in 12 years.

7. **Walking** Some pedometers have a default setting that assumes that the number of steps a person must take to walk 1 mile decreases by 34 steps per 1 inch increase in height. Suppose a person 68 inches tall makes 2250 steps per mile. Find a linear model of the number of steps per mile as a function of height. Use the model to predict the number of steps per mile a person who is 60 inches tall will take.

8. **Weather Balloons** A weather balloon is descending at a constant rate of 50 feet per minute. If the initial height of the balloon was 12,000 feet, find a linear model of the balloon's altitude t minutes after it begins its descent. Use your model to find the height, h, of the balloon 2 hours after it begins its descent.

9. **Tree Growth** A red maple sapling was 3 feet tall when planted in 2010. Six years later, the tree was 24 feet tall. The growth rate of the tree is constant over time. Find a linear model for the growth of the red maple n years after 2010. What is the expected height of the red maple in 2020? Let $n = 0$ represent 2010.

10. **Business** A financial analyst has determined that the total cost to produce 3-foot-diameter inflatable exercise balls can be modeled by a linear function. The total for producing 1000 balls is $7500, and producing 2000 balls is $14,500. Find a linear model for the total cost to produce n balls. Use the model to find the total cost to produce 3500 balls.

11. **Automotive Engineering** An automotive engineer studied the effect of car weight in tons on fuel efficiency, which is measured in miles per gallon (mpg). The data from two cars in the study showed that a 2-ton car had a fuel efficiency of 33 mpg and a 2.5-ton car had a fuel efficiency of 29 mpg. Find a linear model that gives the fuel efficiency of a car as a function of its weight. Use the model to find the fuel efficiency of a car that weighs 3 tons.

12. **Social Media** A technology specialist studied the relationship between time spent on social media and age. One observation found that a 50-year-old person spends 1 hour per week on social media. A second observation showed that a 20-year-old person spends 13 hours per week on social media. Assuming time spent per week on social media is a linear function of age, find a linear model of this event. Use the model to find the age of a person who spends 10 hours per week on social media.

13. **Telecommunications** Taylor wants to buy a brand new cell phone, which has a retail cost of $749. A cell phone dealer tells Taylor that instead of paying that entire amount, there is a 2-year plan requiring a down payment of $99 and $25 per month.

    a. Write a linear function that models the total amount of money Taylor has paid for the phone as a function of months of ownership. Use the function to determine the total amount paid toward ownership of the phone after the 2 years.

    b. After 2 years, has the total amount of money Taylor spent on the phone been more or less than the original retail price?

14. **Hotel Industry** The operator of a hotel estimates that 500 rooms per night will be rented if the room rate per night is $100. For each $10 increase in the price of a room, six fewer rooms per night will be rented. Determine a linear function that will predict the number of rooms that will be rented per night for a given price per room. Use this model to predict the number of rooms that will be rented if the room rate is $150/night.

15. **Construction** A general building contractor estimates that the cost to build a new home is $30,000 plus $85 for each square foot of floor space in the house. Determine a linear function that gives the cost of building a house that contains x square feet of floor space. Use this model to determine the cost to build a house that contains 1800 ft^2 of floor space.

16. **Compensation** An account executive receives a base salary plus a commission. On $20,000 in monthly sales, an account executive would receive compensation of $2700. On $50,000 in monthly sales, an account executive would receive compensation of $4500. Determine a linear function that yields the compensation of an account executive for x dollars in monthly sales. Use this model to determine the compensation of an account executive who has $85,000 in monthly sales.

17. **Car Sales** A manufacturer of economy cars has determined that 50,000 cars per month can be sold at a price of $18,000 per car. At a price of $17,500, the number of cars sold per month would increase to 55,000. Determine a linear function that predicts the number of cars that will be sold at a price of x dollars. Use this model to predict the number of cars that will be sold at a price of $17,000.

18. **Calculator Sales** A manufacturer of graphing calculators has determined that 10,000 calculators per week will be sold at a price of $95 per calculator. At a price of $90, it is estimated that 12,000 calculators will be sold. Determine a linear function that predicts the number of calculators that will be sold per week at a price of x dollars. Use this model to predict the number of calculators that will be sold at a price of $75.

19. **Test Scores** The data in the following table show five students' reading test grades and final exam grades in a history class.

Reading Test Score, x	8.5	9.4	10.0	11.4	12.0
History Final Exam Grade, y	64	68	76	87	92

a. Use the points (8.5, 64) and (10.0, 76) to find a linear model that will predict a student's final exam grade given his or her reading test score.

b. Use your linear model to estimate a student's final exam grade in the history class given that the student's reading test grade was 10.5. Round to the nearest whole number.

20. Stress A research hospital did a study on the relationship between stress and diastolic blood pressure. The results from eight patients are given in the following table. Blood pressure values are measured in millimeters of mercury.

Stress Test Score, x	55	62	58	78	92	88	75	80
Blood Pressure, y	70	85	72	85	96	90	82	85

a. Use the points (55, 70) and (92, 96) to find a linear model that will predict the blood pressure of a patient for a given stress test score. Round the slope to the nearest ten-thousandth.

b. Use your linear model to estimate the diastolic blood pressure of a person whose stress test score was 85. Round to the nearest whole number.

21. Sports The data in the following table show the amount of water in milliliters a professional tennis player loses for various times, in minutes, of play during a tennis match.

Time of Workout (in minutes), x	10	20	30	40	50	60
Water Lost (in milliliters), y	600	900	1200	1500	2000	2300

a. Use the points (10, 600) and (60, 2300) to find a linear model that will predict the water lost by a player who has been playing for x minutes.

b. Use your linear model to estimate the amount of water lost after playing a tennis match for 25 minutes. Round to the nearest milliliter.

22. Fuel Efficiency An automotive engineer studied the relationship between the speed of a car and the number of miles traveled per gallon of fuel consumed at that speed. The results of the study are shown in the following table.

Speed (in miles per hour), x	40	25	30	50	60	80	55	35	45
Consumption (in miles traveled per gallon), y	26	27	28	24	22	21	23	27	25

a. Use the points (25, 27) and (60, 22) to find a linear model that will predict the fuel consumption of a car traveling at x miles per hour.

b. Use your linear model to estimate the expected number of miles traveled per gallon of fuel consumed for a car traveling at 65 mph. Round to the nearest mile per gallon.

23. Meteorology A meteorologist studied the maximum temperatures at various latitudes for January of a certain year. The results of the study are shown in the following table.

Latitude (in °N), x	22	30	36	42	56	51	48
Maximum Temperature (in °F), y	80	65	47	54	21	44	52

a. Use the points (22, 80) and (56, 21) to find a linear model that will predict the maximum temperature at a latitude of x degrees north of the equator.

b. Use your linear model to estimate the expected maximum temperature in January at a latitude of 45°N. Round to the nearest degree.

24. Zoology A zoologist studied the running speeds of animals in terms of the animals' body lengths. The results of the study are shown in the following table.

Body Length (in centimeters), x	1	9	15	16	24	25	60
Running Speed (in meters per second), y	1	2.5	7.5	5	7.4	7.6	20

a. Use the points (9, 2.5) and (60, 20) to find a linear model that will predict the running speed of an animal with a body length of x centimeters.

b. Use your linear model to estimate the expected running speed of a deer mouse, whose body length is 10 centimeters. Round to the nearest tenth of a meter per second.

Investigations

A Linear Business Model Two people decide to open a business reconditioning toner cartridges for copy machines. They rent a building for $7000/year and estimate that building maintenance, taxes, and insurance will cost $6500/year. Each person wants to make $12/hr in the first year and will work 10 hr per day for 260 days of the year. Assume that it costs $28 to restore a cartridge and that the restored cartridge can be sold for $45.

1. Write a linear function for the total cost C to operate the business and restore n cartridges during the first year, not including the hourly wage the owners wish to earn.
2. Write a linear function for the total revenue R the business will earn during the first year by selling n cartridges.
3. How many cartridges must the business restore and sell annually to break even, not including the hourly wage the owners wish to earn?
4. How many cartridges must the business restore and sell annually for the owners to pay all expenses and earn the hourly wage they desire?
5. Suppose the entrepreneurs are successful in their business and are restoring and selling 25 cartridges each day of the 260 days the business is open. What will be their hourly wage for the year if the profit is shared equally?
6. As the company becomes successful and is selling and restoring 25 cartridges each day of the 260 days it is open, the entrepreneurs decide to hire a part-time employee. The employee works 4 hr per day for 260 days and is paid $8/hr. How many additional cartridges must be restored and sold each year just to cover the cost of the new employee? You can ignore employee costs such as social security, worker's compensation, and other benefits.
7. Suppose the company decides that it could increase its business by advertising. Answer Exercises 1, 2, 3, and 5 if the owners decide to spend $400/month on advertising.

Understanding Sales Revenue For Section Exercise 1, it is difficult to know, without more information, how many sales the executive could have in a month. Suppose a reasonable domain for monthly sales is $0 to $250,000. What is the range for this function?

Test Prep 5 — Linear Functions

Section 5.1 — Introduction to Functions

Function A **function** describes a relationship between independent variables and a dependent variable such that for given values of the independent variables, there is *one and only one* value of the dependent variable. The **domain** of a function is all the possible values for the independent variable. The **range** of a function is all the values of the dependent variable.	See **Example 1** on page 221, and then try Exercise 1 on page 254.
Evaluate a Function To evaluate a function means to replace the independent variable with a given value and then simplify the resulting expression. The result is the value of the corresponding dependent variable.	See **Example 2** on page 221, and then try Exercises 2, 3, and 4 on page 254.
Graph of a Function A **graph** of a function is a display of all of the ordered pairs that belong to the function. To produce the graph, evaluate the function for selected values of the independent variable. Plot the resulting ordered pairs and then draw a smooth curve through the points.	See **Examples 3** and **4** on pages 222 and 224, and then try Exercises 5, 6, and 7 on page 254.
Average Rate of Change of a Function $$\text{Average rate of change} = \frac{\text{Change in dependent variable}}{\text{Change in independent variable}},$$ where Change in the dependent variable = Value after change – Value before change Change in the independent variable = Value after change – Value before change	See **Examples 5** and **6** on pages 225 and 226, and then try Exercises 8 through 11 on pages 254 and 255.

Section 5.2 — Properties of Linear Functions

Linear Function A *linear function* is one that can be written in the form $y = mx + b$ or, using function notation, $f(x) = mx + b$.	See **Example 1** on page 232, and then try Exercises 12 and 13 on page 255.
Slope of a Line Let (x_1, y_1) and (x_2, y_2) be the coordinates of two points on a nonvertical line. Then the **slope** of the line through the two points is the ratio of the change in the y-coordinates to the change in the x-coordinates. $$m = \frac{\text{Change in } y}{\text{Change in } x} = \frac{y_2 - y_1}{x_2 - x_1}, x_1 \neq x_2$$	See **Examples 2, 3, 4,** and **5** on pages 237, 239, and 240, and then try Exercises 14, 16, 18, and 19 on page 255.
Intercepts An *x*-intercept of a graph in an *xy*-coordinate plane is a point at which the graph crosses the *x*-axis. **For an *x*-intercept, the *y*-coordinate of the point is 0.** A *y*-intercept of a graph in an *xy*-coordinate plane is a point at which the graph crosses the *y*-axis. **For a *y*-intercept, the *x*-coordinate of the point is 0.**	See **Example 6** on page 241, and then try Exercise 21 on page 256.

Section 5.3 — Linear Models

A **linear model** is a linear function that models a real-world application.	See **Examples 1, 2, 3,** and **4** on pages 246, 247, and 248, and then try Exercises 22, 23, 24, and 25 on page 256.

Review Exercises

1. **Uploading Files** The time t it takes to upload a file to a web server is a function of the size s of the file. Name the independent and dependent variables. The name of the function is f. Write an equation for the function.

2. **Cameras** The number of cameras that can be sold by a company is a function of the price of the camera and can be approximated by $N(p) = 10,000 - 5p$, where $N(p)$ is the number sold for price p in dollars. How many cameras does the company expect to sell for a camera priced at $1000?

3. **Footballs** The height of a football after the ball has been kicked can be approximated by $h(t) = -16t^2 + 70t + 5$, where $h(t)$ is the height, in feet, of the football t seconds after being kicked. Find the height of the football 3 seconds after it is kicked.

4. **Soup** The temperature of a cup of soup can be approximated by $T(m) = 70 + 30 \cdot 0.7^m$, where $T(m)$ is the temperature of the soup in degrees Fahrenheit m minutes after it has been poured into a bowl. Find the temperature of the soup 2 minutes after being poured into a bowl. Round to the nearest degree.

5. **Springs** The length of a hanging spring is a function of the weight attached to the end of the spring and can be approximated by $L = f(x) = 3 + 0.25x$, where L is the length, in inches, of the spring when a weight of x ounces is attached to the spring.

   a. Find the ordered pair to complete the following table.

x	L	(x, L)
0	3	(0, 3)
4	4	(4, 4)
8		
12	6	(12, 6)

   b. Write a sentence that explains the meaning of the ordered pair (12, 6) in the context of this problem.

   c. Use the following coordinate grid to graph the function for x between 0 ounces and 12 ounces.

   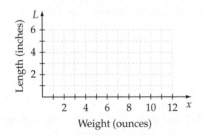

   Weight (ounces)

6. **Business** The profit a company earns to produce a certain machine part is a function of the number of parts produced and can be approximated by $P = g(x) = -0.01x^2 + 4x - 100$, where P is the profit in thousands of dollars and x is the number of parts produced in thousands.

   a. Find the ordered pair to complete the following table.

x	P	(x, P)
50	75	(50, 75)
100	200	(100, 200)
150	275	(150, 275)
200		
250	275	(250, 275)
300	200	(300, 200)
350	75	(350, 75)

   b. Write a sentence that explains the meaning of the ordered pair (250, 275) in the context of this problem.

   c. Use the following coordinate grid to graph the function for x between 50,000 parts and 350,000 parts.

   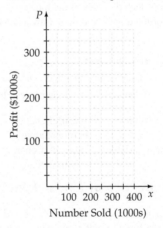

   Number Sold (1000s)

7. **Car Payments** The amount of a monthly payment for a car loan for a fixed time period and interest rate is a function of the amount borrowed. Use the following graph to estimate the amount borrowed if the car payment is $500.

   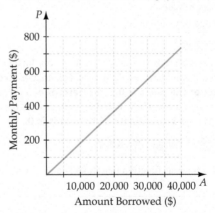

   Amount Borrowed ($)

8. **Throwing a Ball** The following graph shows the decrease in the velocity v, in feet per second, of a ball thrown upward as its distance d, in feet, above ground increases.

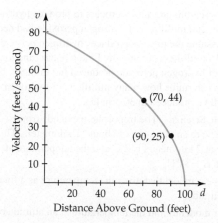

a. Find the average rate of change of velocity between the two given points.

b. Write a sentence that explains the meaning of the average rate of change in the context of this problem.

9. **Throwing a Ball**   The following graph shows the decrease in the velocity v, in feet per second, of a ball thrown upward as its distance d, in feet above ground, increases.

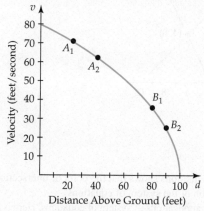

Between which two points is the velocity decreasing the fastest?

10. **Spreading a Rumor**   The following graph shows the spread of a rumor through a small town; d is the number of days after the rumor starts and N is the number of people who have heard the rumor.

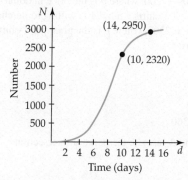

a. Find the average rate of change of the number of people who have heard the rumor between the two given points. Round to the nearest whole number.

b. Write a sentence that explains the meaning of the average rate of change in the context of this problem.

11. **Spreading a Rumor**   The following graph shows the spread of a rumor through a small town; d is the number of days

after the rumor starts and N is the number of people who have heard the rumor. See Exercise 10.

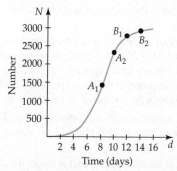

Is the average rate of change between A_1 and A_2 less than or greater than the average rate of change between B_1 and B_2?

12. Graph $f(x) = -\dfrac{3}{4}x + 1$.

13. Graph $z = \dfrac{2}{3}t - 3$.

For Exercises 14 to 17, find the slope of the line between the points with the given coordinates.

14. $(3, 2), (2, -3)$
15. $(-1, 4), (-3, -1)$
16. $(2, -5), (-4, -5)$
17. $(5, 2), (5, 7)$

18. **Fuel Consumption**   The following graph shows how the amount of gas in the tank of a car decreases as the car is driven. Find the slope of the line. Write a sentence that explains the meaning of the slope.

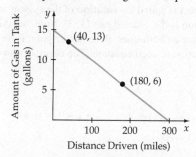

19. **Roses**   The following graph shows the profit P for selling x rose bushes.

a. Find the slope of the line.

b. Write a sentence that explains the meaning of the slope in the context of this application.

c. If the total profit for selling 300 rose bushes is $300, what is the total profit to sell 350 rose bushes?

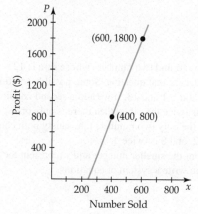

20. **Car Slowing Down**   A car is slowing for a stop sign from 40 feet per second to 0 feet per second at a constant rate over 10 seconds. What is the average rate of change per second?

21. **Value of a Delivery Van**   The accountant for a small business uses the model $V(t) = 25,000 - 5000t$ to approximate the value, $V(t)$, of a delivery van t years after its purchase. Find and discuss the meaning of the intercepts on both vertical and horizontal axes in the context of this application.

22. **Installing Carpet**   A dentist's office is being carpeted. The cost to install the new carpet is $100 plus $12 per square foot of carpeting.

    a. Determine a linear function for the cost to carpet the office.

    b. Use this function to determine the cost to carpet 288 square feet of floor space.

23. **Chemistry**   Ten minutes into a process to produce hydrogen peroxide, 200 milliliters of hydrogen peroxide had been produced. Assume the process produces hydrogen peroxide at a rate of 15 milliliters per minute. Find a linear function of the amount of hydrogen peroxide produced in t minutes. Use the model to determine how many milliliters of hydrogen peroxide will be produced in 30 minutes.

24. **Atmospheric Science**   The troposphere extends from Earth's surface to an elevation of about 11 kilometers. The temperature at 2 kilometers is 2°C and the temperature at 8 kilometers is −34°C.

    a. Find a linear model that gives the temperature as a function of altitude.

    b. Use your model to find the temperature at an altitude of 7 kilometers.

Project 1

Systems of Equations

A system of linear equations is a set of two or more equations taken together. For instance,

$$\begin{cases} y = 2x + 1 \\ y = -x + 4 \end{cases}$$

A graph of the two equations is shown below. The point where the two graphs intersect is called the **solution** of the system of equations. The solution is the ordered pair (1, 3). Note that this is the only ordered pair on both lines. Because the point is on both lines, it is a solution to each equation. The check is shown as follows.

$$y = 2x + 1 \qquad y = -x + 4$$

$$3 \overset{?}{=} 2(1) + 1 \qquad 3 \overset{?}{=} -(1) + 4 \qquad \bullet\, x = 1, y = 3$$

$$3 = 3 \qquad\qquad 3 = 3$$

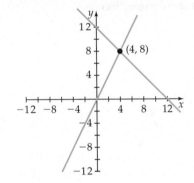

Suppose we asked you to find two numbers whose sum is 12. There are many answers to that question. Some possible solutions are 3 and 9, 4 and 8, and −1 and 13. If we add a second condition such as one number is twice the other, then there is only one solution: 4 and 8 are the only two numbers that satisfy both conditions. Their sum is 12, and 8 is twice 4.

If we let x represent the smaller number and y represent the larger number, we can write a system of equations.

$$\begin{cases} x + y = 12 \\ y = 2x \end{cases}$$

Graphing the two equations shows that the point of intersection, (4, 8), is the solution of the system of equations.

There are many applications of system of equations. Here is one from economics. The **equilibrium price** of a commodity is the price a manufacturer is willing to accept and a buyer is willing to buy.

Example

Suppose that the number x of T-shirts a manufacturer is willing to sell is given by $x = 100p - 200$, where p is the price in dollars of a T-shirt. The number x of T-shirts a retailer is willing to purchase is given by $x = -150p + 2300$, where p is the price per T-shirt. Find the equilibrium price.

Solution

The system of equation is

$$\begin{cases} x = 100p - 200 \\ x = -150p + 2300 \end{cases}$$

Graph the equations and determine the point of intersection.

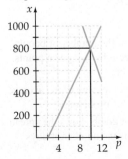

From the graph, (10, 800) is the solution of the system of equations. The equilibrium price is $10. This means that the manufacturer is willing to sell some number of T-shirts for that price, and the retailer is willing to buy the same number of T-shirts at that price.

There are other methods to solve systems of equations but this project will just focus on the ones that can be done by graphing.

Project Exercises

1. Suppose that the number x of digital cameras a manufacturer is willing to sell is given by $x = 25p - 600$, where p is the price in dollars of a camera. The number x of cameras a retailer is willing to purchase is given by $x = -5p + 1800$, where p is the price per camera. Find the equilibrium price.

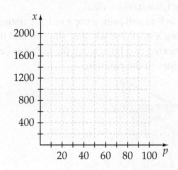

2. Suppose that the number x of smartphones a manufacturer is willing to sell is given by $x = 25p - 450$, where p is the price in dollars of a smartphone. The number x of smartphones a retailer is willing to purchase is given by $x = -5p + 1800$, where p is the price per smartphone. Find the equilibrium price.

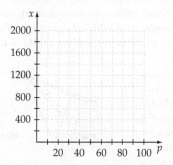

Project 2

Dilations of a Geometric Figure

A **dilation** of a geometric figure changes the size of the figure by either enlarging it or reducing it. This is accomplished by multiplying the coordinates of the figure by a positive number called the **dilation constant**. When the dilation constant is greater than 1, the geometric figure is enlarged. When the dilation constant is between 0 and 1, the geometric figure is reduced. Examples of enlarging and reducing a geometric figure are shown as follows.

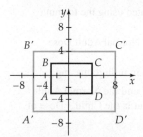

ABCD was enlarged by multiplying its x- and y-coordinates by 2. The result is $A'B'C'D'$.

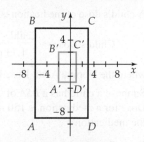

ABCD was reduced by multiplying its x- and y-coordinates by $\frac{1}{3}$. The result is $A'B'C'D'$.

Take Note: Photocopy machines have reduction and enlargement features that function essentially as constants of dilation. The numbers are usually expressed as percentages. A copier selection of 50% reduces the size of the object being copied by 50%. A copier selection of 125% increases the size of the object being copied by 25%.

When each coordinate of a figure is multiplied by the same number to produce a dilation, the **center of dilation** will be the origin of the coordinate system. For triangle *ABC* at the right, a constant of dilation of 3 was used to produce triangle $A'B'C'$. Note that lines through the corresponding vertices of the two triangles intersect at the origin, the center of dilation.

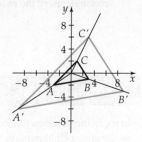

For rectangle *ABCD* at the right, a constant of dilation of $\frac{1}{3}$ was used to produce rectangle $A'B'C'D'$. Lines through the corresponding vertices of the two rectangles intersect at the origin, the center of dilation.

For a figure in space, or one not oriented in the *xy*-plane, the center of dilation can be any point. See Exercises 3 and 4 that follow.

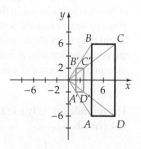

Project Exercises

1. A dilation is performed on the figure with vertices $A(-2, 0)$, $B(2, 0)$, $C(4, -2)$, $D(2, -4)$, and $E(-2, -4)$.

   a. Draw the original figure and a new figure using 2 as the dilation constant.

   b. Draw the original figure and a new figure using $\frac{1}{2}$ as the dilation constant.

2. Because each coordinate of a geometric figure is multiplied by a number, the lengths of the sides of the figure will change. It is possible to show that the lengths change by a factor equal to the constant of dilation. In this exercise, you will examine the effect of a dilation on the angles of a geometric figure. Draw some figures and then draw a dilation of each figure using the origin as the center of dilation. Using a protractor, determine whether the measures of the angles of the dilated figure are different from the measures of the corresponding angles of the original figure.

3. Graphic artists use centers of dilation to create three-dimensional (3-D) effects. Consider the block letter A shown at the left. Draw another block letter A by changing the center of dilation to see how it affects the 3-D look of the letter. You determine how thick to draw the letter. To enhance the 3-D look, make the face of the letter darker than the rest of it. Programs such as PowerPoint use these methods to create various shading options for design elements in a presentation.

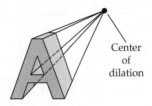

Center of dilation

4. Draw an enlargement and a reduction of the quadrilateral at the right for the given center of dilation P.

5. On a blank piece of paper, draw a rectangle 4 in. by 6 in. with the center of the rectangle in the center of the paper. Make various photocopies of the rectangle using the reduction and enlargement settings on a copy machine. Where is the center of dilation for the copy machine?

6. **Perspective in Art** Many paintings we see today have a 3-D quality, even though they are painted on a flat surface. This was not always the case. Not until the Renaissance did artists start to paint "in perspective." Using lines is one way to create this perspective. Here is a simple example.

Draw a dot, called the *vanishing point*, and a rectangle on a piece of paper. Draw windows as shown. To keep the perspective accurate, the lines through opposite corners of the windows should be parallel. A table in proper perspective is created in the same way. This method of creating perspective was employed by Leonardo da Vinci.

Search online for to find and print a copy of his painting *The Last Supper*. Using a ruler, see whether you can find the vanishing point by drawing two lines along the top edges of the windows on the sides of the painting.

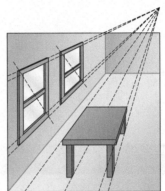

Math in Practice

In medicine, it seems like common sense that the amount of medication a child requires is much smaller than the amount of medication an adult requires. In fact, the value for a child's dose of medication is based in their body surface area, *BSA*, which is given in square meters. Some nurses are given the responsibility of calculating a child's dose of medication. The formula for finding the body surface area is based on height and weight, which are measured by a nurse at a medical appointment.

Part 1

The formula for body surface area, *BSA*, for a child weighing 101 pounds is

$$BSA = \sqrt{\frac{h}{31}},$$

where *h* is the child's height in inches.
Suppose a child weighing 101 pounds is 60 inches tall. What is the child's *BSA*?

Part 2

A child's dose of medication is calculated using the formula

$$\text{Child's Dose} = \frac{\text{Child's } BSA}{1.73 \text{ m}^2} \times \text{Normal Adult Dose}$$

where the Normal Adult Dose is given in milligrams, mg.

Suppose a child has a *BSA* of 0.76 m² and the Normal Adult Dose for a medication is 150 mg. What is the Child's Dose of the medication?

Part 3

The medical office has many pieces of equipment that requires replacement every few years. The office manager keeps track of these pieces of equipment and their depreciation. There are several different methods of depreciation. The simplest is called the straight-line depreciation method. The formula for straight-line depreciation is

$$\text{Depreciation per year} = \frac{\text{Asset Cost } - \text{ Salvage Value}}{\text{Useful life}}$$

The first year's end of year book value is calculated by

Asset Cost − Deprecation per year

SofikoS/Shutterstock.com

The remaining years' end of year book value is calculated by

Previous year's end of book value − Deprecation per year

Suppose a piece of equipment in the medical office cost $14,500, has a useful life of 4 years, and a salvage value of $1300.

a. Use the straight-line depreciation method to find the depreciation per year.

b. Using $14,500 for year 0, find the asset value for years 1, 2, 3, and 4.

c. Make a depreciation graph with years on the horizontal axis, and asset value on the vertical axis.

d. What is the slope of the graph?

e. What is the intercept on the vertical axis? What does this value represent?

f. Will the graph ever reach the intercept on the horizontal axis? Why or why not?

Nonlinear Functions

Section	How This Is Used . . .
6.1 Exponential Functions	Exponential functions have applications in finance to calculate compound interest, in biology to determine the growth of a bacteria colony, in archeology to determine the age of a rock or bone, and in other areas of the sciences.
6.2 Logarithmic Functions	Logarithms can be used to solve some exponential equations discussed in Section 6.1 and to model natural phenomena such as the intensity of earthquakes and the intensity of a sound.
6.3 Quadratic Functions	The path of a pitched ball, the path of a punted football, or the arc of a cable supporting a bridge are only some of the applications of a quadratic function.

Math in Practice:
Music, p. 297

Applications of Nonlinear Functions

Radioactive isotopes have many applications in the sciences. One such application is determining the age of an artifact. By measuring the amount of a certain isotope of carbon in an artifact, it is possible to determine its approximate age. Using its age and other evidence, archaeologists can determine the state of the environment thousands and sometimes millions of years ago.

In Exercise 24, Section 6.2, logarithms are used to determine the age of an artifact.

Section 6.1 Exponential Functions

Learning Objectives:

- Find an exponential growth model.
- Find an exponential decay model.
- Find the half-life of an event.

Be Prepared: Prep Test 6 reviews concepts used in this chapter. This can be found in your Companion Site at Cengage.com.

Recall that a linear function has a constant rate of change. For a one-unit increase in x, the y-value increases (positive slope) or decreases (negative slope) by the *same amount*. For instance, if premium chocolate costs \$3 per ounce, then the cost increases by \$3 for every 1 ounce of chocolate purchased. The resulting graph is a straight line, as shown in Figure 6.1. The amount increases by \$3 for each additional pound of chocolate purchased.

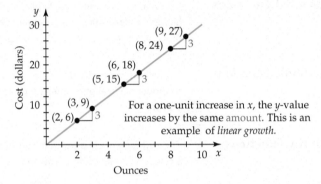

Figure 6.1

Take Note: One gigawatt is enough power for about 100 homes for one year.

In this section, we look at another type of function where, for a one-unit increase in x, the y-value increases or decreases by the *same percent*.

By some estimates, the growth in solar power will increase by approximately 15% each year for the next few years. The graph in Figure 6.2 shows this increase. This is an example of *exponential growth*.

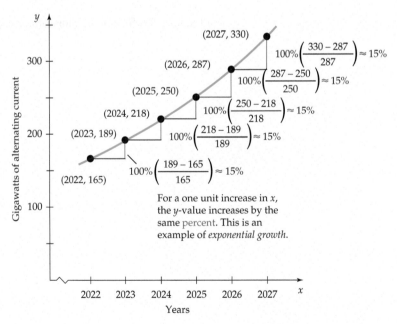

Figure 6.2

Xenon-133 is a radioactive substance that is used to diagnose pulmonary function in patients. The amount of Xenon-133 remaining in a patient decreases by about 13% per hour. The graph in Figure 6.3 shows this decrease. This is an example of *exponential decay*.

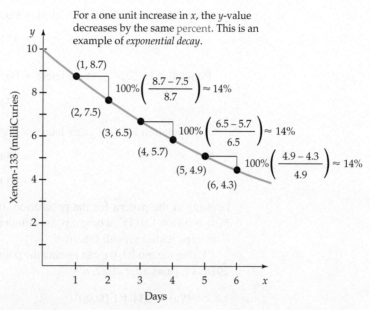

For a one unit increase in x, the y-value decreases by the same percent. This is an example of *exponential decay*.

Figure 6.3

Exponential Function

An **exponential function** is one that can be written in the form

$$y = b^x, b > 0, b \neq 1$$

If $0 < b < 1$, the function is called an **exponential decay** function. If $b > 1$, then the function is an **exponential growth** function.

Example 1   **Identify the Type of Function**

For each of the following, identify the function as exponential growth, exponential decay, or neither.

a. $f(x) = 3^x$ **b.** $g(x) = x^2$ **c.** $F(x) = \left(\dfrac{2}{3}\right)^x$ **d.** $v(x) = 1.02^x$

Solution

a. Exponential growth, $b = 3$, which is greater than 1.

b. Neither. Exponential growth or decay has the variable as an exponent.

c. Exponential decay, $b = \frac{2}{3}$, which is between 0 and 1.

d. Exponential growth, $b = 1.02$, which is greater than 1.

Exponential Growth Models

The U.S. population is estimated to be increasing at a rate of 0.58% per year. Because the rate is increasing by a percent each year, we can model the growth with an exponential growth function.

To find this model, we begin with a base year—say, 2022—when the U.S. population was approximately 334.4 million.

Population in 2023 (1 year later) = Population in 2022 $\cdot$ 1.0058

$$= 334.4(1.0058)^1$$

Population in 2024 (2 years later) = Population in 2023 $\cdot$ 1.0058

$$= 334.4(1.0058)(1.0058) = 334.4(1.0058)^2$$

Population 2025 (3 years later) = Population in 2024 $\cdot$ 1.0058

$$= 334.4(1.0058)^2(1.0058)$$

$$= 334.4(1.0058)^3$$

Looking at the pattern for the population, the world's population appears to be given by $P(n) = 334.4(1.011)^n$, where n is the number of years after 2022. Because $1.0058 > 1$, this is an exponential growth function.

Using the model, we can predict the population in some future year, say, 2028. Because 2028 is 6 years after 2022, $n = 6$.

$$P(n) = 334.4(1.0058)^n$$

$$P(11) = 334.4(1.0058)^6 \approx 346.2$$

The U.S. population in 2028 will be approximately 346.2 million people.

Before we leave the population model, we want to note that it is impossible to sustain exponential population growth. Eventually, there would not be enough natural resources such as land, water, and food to sustain a large population. Ecologists refer to Earth's *carrying capacity* as the largest population that can be sustained.

We can follow the same strategy that we used for the population model for any situation for which a quantity is increasing at $r\%$. Instead of having $1 + 0.0058$ as we did above, we have $1 + r$, where r is written as a decimal.

Model for Exponential Growth

Suppose a quantity is increasing at a rate of $r\%$ per unit time. Let A represent the amount of a quantity at some future time T and A_0 the amount when $T = 0$. Then

$$A = A_0(1 + r)^T$$

is a model for exponential growth.

Note: The phrase "rate of $r\%$ per unit time" means that if something is increasing at $r\%$ per minute, then T must be in minutes; if the growth rate is $r\%$ per year, then T must be in years.

Example 2   Determine an Exponential Growth Model

The month over month increase of hits to a website went from 20,000 hits in March 2022 to 20,500 hits in April 2022. Assuming that this percent increase continues, write an exponential function for the number of hits on the website as a function of t months after March 2022.

Solution

To use the equation $A = A_0(1 + r)^T$, we need to find A_0 and r. $A_0 = 20{,}000$, the number of hits in March 2022. The value of r is the percent increase from March to April.

$$\text{Percent increase} = \frac{\text{Increase from March to April}}{\text{Number of hits in March}}$$

$$r = \frac{20{,}500 - 20{,}000}{20{,}000} = \frac{500}{20{,}000} = 0.025$$

Find a model. The units of the growth (months) and the time (months after March 2022) are the same. Therefore, a model of the growth of hits to the website is

$$A = A_0(1 + r)^t$$

$$A = 20,000(1.025)^t$$

An exponential model of traffic to the website is $A = 20,000(1.025)^t$, where t is in months.

●

Example 3 **Find an Exponential Growth Model for the Coronavirus**

Early in 2020, when COVID-19 was first noticed, the models of how fast the virus was spreading depended on the country. For the United States, on October 1, 2020, there were approximately 39,000 cases and the virus was spreading at a rate of about 7% per week. Find an exponential growth model for the spread of the virus t days after October 1, 2020. Use the model to predict the number of cases of COVID-19 on December 31, 2020.

Solution

On October 1, 2020, the initial number of cases was 39,000. Therefore, $A_0 = 39,000$. The rate is given as 7% per week but we are asked to find a model in terms of days.

$$T = \frac{t \text{ days}}{1 \text{ week}} = \frac{t \text{ days}}{7 \text{ days}} = \frac{t}{7} \qquad \bullet \text{ Change weeks to days.}$$

The rate is given as 7%. Therefore, $r = 7\% = 0.07$.

$$A = A_0(1 + r)^T$$

$$A = 39,000(1 + 0.07)^{t/7} \qquad \bullet A_0 = 39,000; \ r = 0.07; \ T = t/7.$$

$$A = 39,000(1.07)^{t/7}$$

An exponential model for the number of cases is $A = 39,000(1.07)^{t/7}$.

To find the number of cases on December 31, find the number of days between October 1, 2020 and December 31, 2020. There are 91 days between the two dates. Use this number as the value of t.

$$A = 39,000(1.07)^{t/7}$$

$$= 39,000(1.07)^{91/7}$$

$$\approx 93,984$$

There were approximately 94,984 cases on December 31, 2020.

The exponential growth of a virus cannot continue forever. For instance, if we tried to use this model through December 31, 2022, the model would predict over 100 million coronavirus infections in the United States. That did not happen because of mitigation efforts by states to slow the spread of the virus.

●

Compound interest is another example of exponential growth. **Compound interest** is interest paid on the original principal and on any interest previously earned on the principal. It is sometimes referred to as "interest on interest."

Example 4 **Find an Exponential Growth Model for Compound Interest**

An investor deposits $10,000 into an account that earns an annual interest rate of 5% compounded daily. Find an exponential growth model for the value of the account after t years. What is the value of the account after 3 years? Round to the nearest cent.

Solution

The interest rate is given as an annual rate, but the time is in days (compounded daily). As mentioned in the note following the model for exponential growth, r and T must have the same units. Change 5% per year to an interest rate per day.

$$r = \frac{0.05}{365} \approx 0.000136986$$

Using 365 days in one year, there are $T = 365t$ days in t years. Now we have r as an interest rate per *day* and time in *days*.

Write the model.

$$A = A_0(1 + r)^T$$
$$A = 10{,}000(1 + 0.000136986)^{365t} \qquad • A_0 = 10{,}000, \; r \approx 0.000136986, \; T = 365t.$$
$$A = 10{,}000(1.000136986)^{365t}$$

To find the value of the account after 3 years, evaluate the function when $t = 3$ years.

$$A = 10{,}000(1.000136986)^{365t}$$
$$A = 10{,}000(1.000136986)^{365 \cdot 3}$$
$$= 10{,}000(1.000136986)^{1095}$$
$$\approx 11{,}618.219$$

The value of the account after 3 years is $11,618.22.

Some bacteria, such as *Salmonella*, can be present in some vegetables and undercooked meat. The growth of the number of bacteria is sometimes measured in **doubling time**, which is defined as the time it takes the number of bacteria to double in size. For *Salmonella*, the doubling time is 30 minutes. This means that every 30 minutes there are twice the number of bacteria in the food.

Doubling in size every 30 minutes means that the number of bacteria in the food is increasing by 100% every 30 minutes, so $r = 100\%$ or 1 (as a decimal number) in the exponential growth equation. One doubling time is 30 minutes so 1 minute is $\frac{1}{30}$ of a doubling time. Therefore, there are $\frac{t}{30}$ doubling times in t minutes. In the exponential growth equation, $T = \frac{t}{30}$. A model of the number of bacteria in a sample that originally contained 100 bacteria is given by

$$A = A_0(1 + r)^T$$
$$A = 100(1 + 1)^{t/30} \qquad • A_0 = 100, \; r = 1, \; T = \frac{t}{30}.$$
$$A = 100(2)^{t/30}$$

Using this model, we can predict the number of *Salmonella* in the food after, say, 45 minutes.

$$A = A_0(2)^{t/d}$$
$$A = 100(2)^{45/30} \qquad • A_0 = 100, \; t = 45, \; d = 30.$$
$$A \approx 283$$

There would be approximately 283 *Salmonella* in the food 45 minutes later.

Doubling Time Equation

$$A = A_0(2)^{t/d}$$

where A is the amount of a substance after time t, A_0 is the initial amount of substance, and d the doubling time of the process.

Example 5 **Use a Doubling Time Equation**

A sample of water from a stagnant pond showed 50,000 *Escherichia coli* (*E. coli*) bacteria. Further testing of the sample showed that the *E. coli* bacteria was doubling in size every 3 days. Find a model for the growth of the bacteria. Use the model to find the amount of *E. coli* in the sample after 7 days.

Solution

Use the Doubling Time Equation.

$$A = A_0(2)^{t/d}$$

$$A = 50,000(2)^{t/3} \qquad \bullet A_0 = 50,000; d = 3.$$

A model of the growth of the bacteria is $A = 50,000(2)^{t/3}$

Use the model to find the population of *E. coli* after 7 days.

$$A = 50,000(2)^{t/3}$$

$$A = 50,000(2)^{7/3} \qquad \bullet t = 7.$$

$$A \approx 251,984$$

To the nearest thousand, there are approximately 252,000 *E. coli* in the sample.

Exponential Decay Models

Exponential decay occurs when a quantity decreases by a certain percent per unit of time.

Model for Exponential Decay

Let A represent the amount of a quantity that is decreasing at a rate that is $r\%$ of the amount present. Then

$$A = A_0(1 - r)^T$$

where A_0 is the initial amount when $T = 0$ is a model for exponential decay.

Example 6 **Write an Exponential Decay Function for Deforestation**

One estimate is that deforestation in Africa is occurring at a rate of 0.7% per year. If there were 700,000,000 hectares (1 hectare $\approx$ 2.5 acres) of forest in Africa in 1990, write an exponential function for the number of hectares of forest in Africa t years after 1990. Use your model to predict the number of hectares of forest in Africa in 2030.

Solution

$A_0 = 700,000,000$. Deforestation is occurring at 0.7% per year. Therefore, $r = 0.007$. The rate is measured in years, and the model is to be written in years. This is exponential decay because the amount of forest is less each year. A model for deforestation is

$$A = A_0(1 - r)^t$$

$$A = 700,000,000(0.993)^t \qquad \bullet 1 - 0.007 = 0.993, A_0 = 700,000,000.$$

A deforestation model for Africa is $A = 700,000,000(0.993)^t$.

To predict the number of hectares of forest in 2030, evaluate the model when $t = 40$ (2030 is 40 years after 1990).

$$A = 700,000,000(0.993)^t$$

$$A = 700,000,000(0.993)^{40}$$

$$\approx 529,000,000$$

In 2030, the model predicts there will be approximately 529,000,000 hectares of forest.

The next example shows that time is not always a variable in exponential decay.

> **Example 7** **Write an Exponential Decay Function for Light Absorption**
>
> The percentage of light that penetrates the ocean decreases by 11% for each 5 meters in depth. Find an exponential function for the percent of light arriving x meters below the surface of the ocean. Photosynthesis will not occur if the amount of available light is less than 1% of the amount striking the ocean surface. Use your model to determine whether photosynthesis will occur at 200 meters below the surface.
>
> **Solution**
>
> The percent of light that hits the surface is 100. Therefore, $A_0 = 100$. The percentage of light at a depth d meters is decreasing at a rate of 11% each 5 meters; $r = 0.11$. The model is to be written in terms of meters; r is given in terms of 5-meter intervals. Because 1 meter is $\frac{1}{5}$ of a depth interval, there are $\frac{d}{5}$ intervals in d meters: $T = \frac{d}{5}$.
>
> $$A = A_0 (1 - r)^T$$
> $$A = 100(1 - 0.11)^{d/5} \qquad \bullet\, A_0 = 100, r = 0.11, T = \frac{d}{5}.$$
> $$A = 100(0.89)^{d/5}$$
>
> An exponential decay model for the absorption of light is $A = 100(0.89)^{d/5}$. Using the model, the percent of light reaching 200 meters is given by
>
> $$A = 100(0.89)^{d/5}$$
> $$= 100(0.89)^{200/5}$$
> $$\approx 0.9\%$$
>
> Because $0.9\% < 1\%$, photosynthesis will not occur at 200 meters.

Many radioactive materials decrease in mass exponentially over time. This decrease, called *radioactive decay*, is measured in terms of **half-life**, which is defined as the time required for the disintegration of 50% or $\frac{1}{2}$ of the atoms in a sample of a radioactive substance.

For instance, naturally occurring carbon has various forms. Two of these are denoted by ^{12}C and ^{14}C. The 12 and 14 refer to the sum of the number of protons and neutrons in the nucleus. For ^{12}C, there are six protons and six neutrons; for ^{14}C there are six protons and eight neutrons. The two extra neutrons cause ^{14}C to be unstable, and a sample of ^{14}C decays to become nitrogen in a predictable manner. Measuring the time it takes ^{14}C to decay to nitrogen has shown that the half-life of ^{14}C is approximately 5730 years. This means that every 5730 years, the amount of ^{14}C in a sample decreases by one-half of the amount currently in the sample.

Because the amount is decreasing by 50% each half-life, $r = 0.50$. The model is to be written in terms of years; r is given in terms of 5370-year intervals (the half-life). Because 1 year is $\frac{1}{5370}$ of a half-life, there are $\frac{t}{5370}$ half-lives in t years: $T = \frac{t}{5370}$. A model of the amount of ^{14}C remaining in a 20-mg sample is given by

$$A = A_0(1 - r)^T$$
$$A = 20(1 - 0.5)^{t/5730} \qquad \bullet\, A_0 = 20, r = 0.5, T = \frac{t}{5730}.$$
$$A = 20(0.5)^{t/5730}$$

Using the model, we can predict the amount of ^{14}C in a fossil that is 8000 years old.

$$A = 20(0.50)^{t/5730}$$
$$A = 20(0.50)^{8000/5730}$$
$$\approx 7.60$$

There would be approximately 7.60 mg of ^{14}C in a fossil that is 8000 years old.

The equation for a half-life process is given in the following.

> **Half-Life Equation**
>
> The half-life equation is
>
> $$A = A_0\left(\frac{1}{2}\right)^{t/h}$$
>
> where A_0 is the initial amount of a substance, h is the half-life of the substance, and A is the amount of the substance remaining after time t.

Example 8 **Use the Half-Life Equation for an Insecticide**

Methoprene is an insecticide used to kill fleas, among other insects. When sprayed on soil, it has a half-life of approximately 10 days. If 300 milligrams of methoprene is spayed on a lawn, how many milligrams of methoprene remain after 15 days? Round to the nearest milligram.

Solution

The initial amount sprayed on the soil is $A_0 = 300$. The half-life is 10 days, $h = 10$. The exponential decay equation is

$$A = A_0(0.5)^{t/h}$$
$$A = 300(0.5)^{t/10} \qquad \bullet\ A_0 = 300,\ h = 10.$$

To find how many milligrams remain after 15 days, evaluate the equation when $t = 15$.

$$A = 300(0.5)^{t/10}$$
$$A = 300(0.5)^{15/10} \qquad \bullet\ t = 15.$$
$$A \approx 106.07$$

There are approximately 106 milligrams of methoprene remaining after 15 days.

6.1 Exercise Set

Think About It

1. Suppose the exponential function for a process is given by $A = 15(1.03)^t$, where t is in minutes. Does this represent exponential growth or exponential decay? What is the percent growth or decay?

2. Suppose the exponential function for a process is given by $A = 15(0.75)^t$, where t is in years. Does this represent exponential growth or exponential decay? What is the percent growth or decay?

3. Suppose two compound interest accounts are opened for $100. For one account the interest rate is 2%, and for the other the interest rate is 4%.

   True or False: After 5 years, the amount of interest in the 4% account is twice the amount in the 2% account.

4. A process is decaying according to the model $A = 25\left(\frac{1}{2}\right)^{t/50}$, where t is measured in seconds. What is the half-life of this process?

5. A process is growing according to the model $A = 3.2(2)^{t/5.5}$, where t is measured in hours. What is the doubling time of this process?

> **Preliminary Exercises**

For Exercises P1 through P8, if necessary, use the referenced example at the end of the exercise for assistance.

P1. For each of the following, determine whether the equation represents exponential growth, exponential decay, or neither. **Example 1**

a. $y = 2 \cdot x^{1/2}$ **b.** $y = 3 \cdot \left(\dfrac{2}{3}\right)^x$ **c.** $s = \dfrac{2}{3} \cdot 2^t$

d. $s = t^2$ **e.** $y = 0.5 \cdot 1.8^{z/5}$ **f.** $v = 4 \cdot 0.8^{5s}$

P2. A colony of bacteria is increasing at the rate of 15% each hour. There are 2500 bacteria in the colony at the time observations begin. Find an exponential growth model for the colony t hours after the first observation. Use the model to determine the number of bacteria in the colony 10 hours after the initial observation. **Example 2**

P3. The value of a certain rare silver dollar was $50,000 in 1980. A price history of the silver dollar shows that it has been doubling in value every 20 years. Find an exponential growth model for the value of the coin t years after 1980. Use your model to estimate the value of the coin in 2025. Round to the nearest thousand. **Example 3**

P4. A restaurant owner deposits $5000 into an account that earns an annual interest rate of 6% compounded monthly. Find an exponential growth model for the value of the account after t years. What is the value of the account after 5 years? Round to the nearest cent. **Example 4**

P5. Listeriosis is a foodborne bacterial infection caused by *Listeria* and usually found in food. A test of contaminated milk found 1500 *Listeria* in 1 quart of milk. Additional testing found that the *Listeria* population was doubling every 50 minutes. Find a model for the number of *Listeria* in the sample t hours after the initial test. Use the model to predict the number of *Listeria* in the sample after 200 minutes. **Example 5**

P6. A desalinization process is eliminating 20% of the salt from seawater every 10 minutes. There are initially 350 g of salt in 10 liters of water. Find an exponential function for the grams of salt in the 10 liters of seawater t minutes after the process has begun. Use the model to predict the amount of salt in the seawater after 45 minutes. Round to the nearest whole number. **Example 6**

P7. A cup of coffee contains approximately 90 milligrams of caffeine. The time it takes a healthy person to metabolize one-half of the consumed caffeine is approximately 5 hours. Find an exponential function that gives the number of milligrams of caffeine in the body t hours after drinking a cup of coffee. **Example 7**

P8. Ibuprofen is a nonsteroidal anti-inflammatory drug with a half-life of approximately 90 minutes. If 200 milligrams of ibuprofen is taken at 2:00 P.M., how many milligrams of ibuprofen remain in the bloodstream at 6:00 P.M.? Round to the nearest milligram. **Example 8**

> **Section Exercises**

1. **Finance**   A financial adviser recommends that a client deposit $2500 into a fund that earns 7.5% annual interest compounded monthly. What will be the value of the investment after 3 years? Round to the nearest cent.

2. **Finance**   An investment broker deposits $1000 into an account that earns 8% annual interest compounded quarterly. What is the value of the investment after 2 years? Round to the nearest dollar.

3. **Population Growth**   Suppose that the population of a certain country grows at an annual rate of 1.5%. If the current population is 3 million, what will the population be in 10 years? Round to the nearest hundred-thousand.

4. **Tuition**   The tuition in the school year 2016–2017 at a certain university was $12,000. For the school year 2021–2022, the tuition was $13,320. Find an exponential growth function for tuition at this university t years after the 2016–2017 school year. Assuming increases at the same annual rate, use the function to predict the tuition in the 2024–2025 school year.

5. **Oil Spill**   When first observed, an oil spill covers 2 square miles. Measurements show that the area is tripling every 8 hours. Find an exponential model for the area of the oil spill as a function of time from the beginning of the spill.

6. **Forestry**   Fire rangers observe a fire that has consumed 10 acres and is doubling in size every 3 hours. Find an exponential model for the area of the fire as a function of time from when firefighters first arrive.

7. **Light Intensity**   The intensity of light below the surface of a clear lake depends on many factors. One model shows that 19% of light is absorbed for each 1-meter increase in depth. Find an exponential model for the percentage of light that reaches a depth of m meters. Use your model to predict what percent of light will reach a depth of 7 meters.

8. **Safety**   To test the effectiveness of a carbon monoxide (CO) alarm, a company begins a controlled experiment that delivers carbon monoxide to a closed room at the rate of 1.5% of the current amount in the room every 2 hours. Find an exponential model for the amount of CO in the room as a function of time from the beginning of the experiment when there are 40 parts per million CO in the room. Use your model to predict the amount of CO in the room 15 hours after the experiment begins.

9. **Computer Science**   The cost of 1 gigabyte (GB) of storage in 1980 was approximately $300,000. From that time, the cost of 1 GB of storage has been decreasing at a rate of approximately 39% every 2 years. Find an exponential model for the cost of 1 GB of storage t years after 1980. Using the model, find the cost of 1 GB of storage in 2025.

10. **Computer Science**   A law for the density of hard drive storage (how much information can be stored on the drive) was suggested by Mark Kryder. He proposed that storage density was doubling every 13 months. The density of a hard drive in January 2005 was 150 kilobits per square millimeter. Find an exponential model for the density of a hard drive n months after January 2005. Assuming Kryder's law continues to be true, what was the storage density of a hard drive in January 2022?

11. **Computer Science**   An Internet analytics company measured the number of people watching a video posted on a social media platform. The company found 127 people had watched the video and that the number of people watching it was increasing by 20% every 2 hours.

a. Write an exponential function for the number of people watching the video n hours after the initial observation.

b. A video is said to go "viral" if the number of people who have watched the video exceeds 5 million within 5 days (120 hours). Would this video be considered to have gone viral?

12. Medicine Researchers studied the replication behavior of a certain virus. From an initial 1000 copies of the virus, the researchers found that the virus population was increasing at a rate of 16% every 3 hours. Write an exponential model of the number of viruses t hours after the study began with the 1000 copies of the virus. Use your model to predict the number of viruses in 20 hours. Round to the nearest whole number.

13. Internet of Things Internet of Things (IoT) consists of appliances like home voice controllers, dash cams, smart light switches, doorbells, thermostats, and many, many other devices. Each device requires a unique IP (Internet Protocol) address. In 2022, there were approximately 12.3 billion IoT devices connected to the Internet. It is estimated that the doubling time for connected IoT devices is about 4 years. Find a model for the number of IoT devices t years after 2022. Use the model to predict the number of IoT devices in 2030.

14. Smart Watches The value of the smart watch market in the United States in 2019 was approximately $26.3 billion and is expected double in size every 3 years. Find an exponential growth model for the value of smart watches t years after 2019. Use the model to predict the value of the smart watch market in 2025.

15. PET Scans Positron emission tomography (PET) scans are used to diagnose some cancers. In some cases, Fluorine-18 (^{18}F), which has a half-life of about 110 minutes, is used. If a patient receives an injection of 5 micrograms of ^{18}F, find an exponential decay function that gives the number of micrograms of ^{18}F t minutes after receiving the injection. Use the model to determine the amount of ^{18}F 45 minutes after the injection.

16. Light Absorption The intensity of sunlight decreases by 50% for each 45 meters below the surface of a certain lake. Find an exponential decay model for the intensity of light m meters below the surface. Use the model to find the intensity of light 100 meters below the surface.

17. Dentistry A lead vest used for patients having dental x-rays taken is being tested to determine what percent P of the x-rays penetrate the vest for various thicknesses of lead inside the vest. The current model shows that 25% fewer x-rays penetrate the vest for each 4-millimeter (mm) increase in thickness of the lead in the vest. Find an exponential model for the percent of x-rays that penetrate the vest as a function of the thickness of the vest in mm. Use your model to predict the percent of x-rays that will penetrate a vest 100 mm thick. Round to the nearest tenth of a percent.

18. Optometry Photochromatic lenses for glasses will darken as the intensity of ultraviolet (UV) rays striking the lenses increases. The UV index scale measures the intensity of the UV rays. An index of 0 is extremely low UV intensity; a value of 10 shows extremely high UV intensity. An increase of 2 on the UV scale results in a 75% decrease in transparency. Find an exponential model for the percentage transparency of a lens as a function of the UV index, with a UV index of 0 corresponding to 100% transparency. Use your model to predict the percent transparency of a lens when the UV index is 3.5. Round to the nearest percent.

19. Education A grandparent puts $5000 into a college education fund for a grandchild. If the fund earns 3.75% annual interest compounded daily, what is the value of the account after 18 years? Assume all years have 365 days.

20. Finance A financial adviser recommends that a client deposit $7500 into a real estate investment trust that earns 4.5% annual interest compounded monthly. What will be the value of the investment after 5 years?

21. Finance On January 1, 2020, $1000 is placed in an account that earns 3% annual interest compounded quarterly. On January 1, 2021, another $1000 is placed in the same account. What is the value of the account on December 31, 2022?

22. Rumors A person begins a fake news story by telling it to four people. Call this generation 1. Each of those four people tells the story to four different people. Call this generation 2. Now each of the people in generation 2 tells the story to four different people. This is generation 3. For each of the following parts, assume no one tells the story to a person who has already heard the story and that once a person tells the story to four other people, that person does not tell the story again.

a. How many people will hear the story in generation 1?

b. How many people will hear the story in generation 2?

c. How many people will hear the story in generation 3?

d. What is the percent increase in the number of people who will hear the story from one generation to another?

e. Write an exponential function for the number of people who will hear the story in the nth generation.

23. Rumors Two people start a rumor. Each person tells the rumor to three other people. Call this the *first telling* of the rumor. Those six people each tell the rumor to three other people. This is the second telling of the rumor. The telling of the rumor is then repeated in a similar manner. Assuming no one tells the rumor to a person who has already heard the rumor and that once a person tells the rumor to three other people, that person does not tell the rumor again, find an exponential function for the number of people who will hear the rumor in the nth telling.

24. Finance You are offered a one-month job that will pay you $0.01 on day 1, $0.02 on day 2, $0.04 on day 3, and so on, doubling the amount of the previous day for 30 consecutive days. The following table shows the results for 10 days.

n	1	2	3	4	5	6	7	8	9	10	...
Earnings of Days n	0.01	0.02	0.04	0.08	0.16	0.32	0.64	1.28	2.56	5.12	...
Total Earnings	0.01	0.03	0.07	0.15	0.31	0.63	1.27	2.55	5.11	10.23	...

a. Write an exponential function for the amount you will earn on day n.

b. Write an exponential function for the total amount you will earn through day n.

c. Will you make more or less than $10 million for the 30-day job?

25. **Optometry** A lens used to observe a solar eclipse will filter 68% of the sunlight entering the lens for each 10 millimeters in thickness. Find an exponential function for the percentage of sunlight passing through the lens as a function of the thickness of the lens. If a safe percent of sunlight passing through the lens is 10%, would a 25-mm-thick lens be considered safe?

26. **Physics** A spring is stretched by a 2-pound weight and brought to rest. It is then given an additional pull of 12 inches and released. Because of frictional forces, the spring rebounds 80% of its distance past the equilibrium (the place where the weight is at rest) at each bounce. This process continues until the weight comes to rest. Find an exponential function for the distance the weight is from the equilibrium after n passes through the equilibrium.

27. **Physics** The half-life of carbon-14, ^{14}C, is approximately 5730 years. A bone fragment is estimated to have originally contained 5 milligrams of ^{14}C. How many milligrams of ^{14}C should be in the bone fragment 10,000 years later? Round to the nearest tenth.

28. **Archeology** The age of some ancient rocks is determined using uranium-235 (^{235}U), which has a half-life of approximately 700 million years. Suppose a rock is determined to have originally 2 milligrams of ^{235}U. Find an exponential decay model for the amount of ^{235}U in t years. Use the model to determine the amount of ^{235}U in the rock in 1 billion years. Round to the nearest hundredth.

29. **Newton's Law of Cooling** When an oven is turned off, its interior temperature is given by $T = 70 + 280(0.8)^t$, where T is the temperature (in °F) in the oven and t is the time in minutes after the oven is turned off. For each of the following, assume room temperature remains constant.
    a. What is the oven temperature after 10 minutes?
    b. Is it possible for the oven temperature to become less than 70°F? Explain. (*Suggestion:* Examine values of T when t is 20, 30, and 40 minutes.)

30. **Bacterial Growth** The growth of a bacterium in a petri dish is given by $A = 5000 - 4000(0.8)^t$, where A is the number of bacteria in the petri dish after t hours.
    a. How many bacteria are in the petri dish after 3 hours?
    b. Is it possible that the bacteria population will ever exceed 5000? Explain. (*Suggestion:* Examine values of A when t is 20, 30, and 40 hours.)

31. **Atmospheric Science** The atmospheric pressure on Earth's surface is approximately 15 pounds per square inch (psi) and decreases at a rate of about 20% per mile above Earth's surface. Find an exponential decay model of the atmospheric pressure n feet above Earth. Use the model to predict the atmospheric pressure at 49,500 feet above Earth, the approximate altitude of some spy satellites.

32. **Decline of Desktop Computers** In 2018 approximately 94 million desktop computers were sold. Estimates are that sales of desktop computers will decline around 4% per year for the next 10 years. Find an exponential decay model for the sales of desktop computers. Use the model to predict the number of desktop computers that will be sold in 2028.

33. **Employment Opportunities** According to the Bureau of Labor Statistics (BLS), in January 2020, there were approximately 33,000 people employed in the United States as Computer and Information Research Scientists. The BLS estimates that job growth opportunities will grow at a rate of 2% per year through 2030. Find an exponential growth function to model this growth. Use the model to predict how many people will be employed as Computer and Information Research Scientists in 2030.

34. **Tablet Computers** The global market for tablet computers in 2020 was $1.2 billion and is estimated to grow by 11% every two years. Find an exponential growth model of the tablet computer market t years after 2020. Use the model to predict the value of the tablet computer market in 2024.

▶ Investigation

Chess and Exponential Growth According to legend, when Sissa Ben Dahir of India invented the game of chess, King Shirham was so impressed with the game that he offered Sissa Ben Dahir the reward of his choosing. Sissa Ben Dahir pointed to the chessboard and requested one grain of wheat on the first square, two grains of wheat on the second square, four grains of wheat on the third square, eight grains on the fourth square, and so on until all 64 squares on the chessboard were covered. The king considered this a modest reward and said he would grant the inventor's wish. (*Suggestion:* You may want to review Exercise 24 before doing this project.)

1. The following table shows the number of grains of wheat on the board for five squares. Complete the table for squares 6, 7, and 8.

Square n	1	2	3	4	5	6	7	8
Grains on Square n	1	2	4	8	16			
Total Number of Grains on the Board	1	3	7	15	31			

2. Write an exponential function for the grains of wheat on square n.

3. Write an exponential function for the total number of grains on the board after square n is filled.

4. How many grains of wheat would be on the board after square 64 is filled?

5. The annual world production of wheat is approximately 24 billion bushels. Each bushel contains approximately 900,000 wheat seeds. How many years of wheat production would be required to fulfill Sissa Ben Dahir's request? Round to the nearest 1000 years.

Logarithmic Functions

Learning Objectives:

- Evaluate a logarithmic function.
- Apply logarithmic functions.
- Solve logarithmic equations.
- Find doubling time or the half-life of an event.

Introduction to Logarithmic Functions

Consider the equation $10x = 30$. To find the solution of this equation, we need the answer to "10 *times* what number equals 30?" We can find that number by dividing each side of the equation by 10.

$$10x = 30$$
$$\frac{10x}{10} = \frac{30}{10}$$
$$x = 3$$

• Divide each side of the equation by 10.

Suppose, however, that we have the *exponential equation* $10^x = 30$, where the variable is an exponent. We need to answer the question "10 to what *power* equals 30?" The answer to that question is a *logarithm* of 30.

Take Note: The important point to the definition of common logarithm is that **a common logarithm is *always* an exponent on 10**. For instance, $10^{1.2354}$ means that 1.2354 is the logarithm of some number. Using the 10^x key on a calculator, $10^{1.2354} \approx 17.19491$. Therefore, $\log 17.19491 \approx 1.2354$.

Definition of Common Logarithm

The **common logarithm*** of a positive number a is the number b such that $10^b = a$. In this case, we write $\log a = b$.

The **exponential form** of $\log a = b$ is $10^b = a$. The **logarithmic form** of $10^b = a$ is $\log a = b$.

Examples

1. $\log 100 = 2$ because $10^2 = 100$.
2. Because $10^4 = 10{,}000$, $\log 10{,}000 = 4$.
3. $\log 1 = 0$ because $10^0 = 1$.
4. Because $10^{-3} = 0.001$, $\log 0.001 = -3$.
5. $\log 5 \approx 0.69897$ because $10^{0.69897} \approx 5$.
6. Because $\log 30 \approx 1.47712$, $10^{1.47712} \approx 30$.

*There are other types of logarithms. For our purposes, we will use common logarithms. Rather than always writing common logarithm, we will just write *logarithm*.

```
log(5)
        .6989700043
log(30)
        1.477121255
■
```

Figure 6.5

The values for examples 5 and 6 in the definition above were found by using the **log** key from a calculator. The results are shown in Figure 6.5. To find the logarithm of most numbers, a calculator is required.

Example 1 **Evaluate a Logarithm**

a. Use the definition of logarithm to find each of the following.

   i. $\log 10$

   ii. $\log 0.1$

   iii. $\log 100{,}000$

b. Use a calculator to find each of the following. Round to the nearest one-hundred thousandth.

   i. $\log 9$

   ii. $\log 658$

   iii. $\log 0.125$

Solution

a. From the definition, $\log a = b$ if and only if $10^b = a$.

  i. $\log 10 = 1$ because $10^1 = 10$

  ii. $\log 0.1 = -1$ because $10^{-1} = \dfrac{1}{10} = 0.1$

  iii. $\log 100{,}000 = 5$ because $10^5 = 100{,}000$

b. Use a calculator. Round to the nearest one-hundred thousandth.

  i. $\log 9 \approx 0.95424$

  ii. $\log 658 \approx 2.81823$

  iii. $\log 0.125 \approx -0.90309$

Logarithmic Scales

Logarithmic functions often are used to *scale* extremely large (or extremely small) numbers into numbers that are easier to comprehend. For instance, the hydrogen ion concentration, [H^+], in a solution affects its alkalinity (think baking soda, lye, drain cleaners) and acidity (think vinegar, orange juice, stomach acid). The typical concentrations can range from 0.00000000000001 to 0.1. These numbers (especially the small ones) are difficult to read. To make reading and other operations easier, a logarithm *scale* is used.

$$\log 0.00000000000001 = -14 \quad \text{and} \quad \log 0.1 = -1$$

Instead of having to deal with extremely small numbers, we have numbers between -14 and -1. This scale is modified by putting a negative in front of each number so the scale ranges from 1 to 14. The resulting scale is called the *pH scale*.

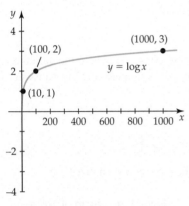

Graph of the equation $y = \log x$. Logarithmic growth works in such a way that as the x value increases by 10-fold (multiplied by 10) the y-value increases by 1. For the graph to exceed 4 on the y-axis, we would have to extend the x-axis to 10,000.

> ### pH Scale
>
> The pH of a solution is given by $\text{pH} = -\log [H^+]$ or $10^{-\text{pH}} = [H^+]$, where [H^+] is the concentration of hydrogen ions in moles per liter.
>   Pure water has a pH of 7. An acid has a pH that is less than 7. A base (alkaline solution) has a pH that is greater than 7.

Example 2   Use the pH Formulas

a. When rain reacts with air pollutants such as sulfur dioxide, the result is acid rain. If a sample of acid rain has a pH of 4.5, what is the hydrogen ion concentration?

b. The hydrogen ion concentration for a certain cola soda is 0.0032 mole per liter. What is the pH of the soda? Round to the nearest tenth. Is the soda an acid or a base?

c. The hydrogen ion concentration for a certain bottled water is 0.0000000056 mole per liter. What is the pH of the water? Is the water an acid or a base?

Solution

a. Use the equation $10^{-\text{pH}} = [H^+]$.

$$10^{-\text{pH}} = [H^+]$$
$$10^{-4.5} = [H^+]$$
$$0.000032 \approx [H^+]$$

The hydrogen ion concentration is approximately 0.000032 mole per liter.

b. Use the equation $\text{pH} = -\log[H^+]$.

$$\text{pH} = -\log[H^+]$$
$$\text{pH} = -\log 0.0032$$
$$\approx 2.5$$

The pH is 2.5. Because $2.5 < 7$, the soda is an acid.

c. Use the equation pH $= -\log[\text{H}^+]$.

$$\text{pH} = -\log[\text{H}^+]$$
$$\text{pH} = -\log 0.0000000056$$
$$\approx 8.3$$

The pH is 8.3. Because $8.3 > 7$, the bottled water is a base.

It is sometimes useful to measure a quantity *relative* to some base value or standard. As an example, an economic study may use the federal minimum wage as a base value. Then each city's minimum wage is reported as a multiple of the federal minimum wage. For instance, Chicago's minimum wage is approximately 1.8 times the federal minimum wage. In this case, we may not know the federal minimum wage but do know that Chicago's minimum wage is almost twice the federal minimum wage.

Some logarithmic scales use *relative* scales. A base value is chosen, and all other measurements are a multiple of the base value. The *Richter scale* is one such scale. It uses a logarithmic function to convert the **relative intensity** I of an earthquake's shock waves into a number M called the **magnitude** of the earthquake. The base intensity, I_0, of the Richter scale is the smallest wave that can be detected by a seismograph stationed 100 kilometers from the epicenter of the earthquake.

Richter Scale

The magnitude M of an earthquake that has an intensity that is I times the base intensity earthquake I_0 is given by $M = \log I$.

> **Example 3**   **Use the Richter Scale to Find the Magnitude of an Earthquake**

Find the magnitude of an earthquake that was 900,000 times the intensity of the base intensity earthquake. Round to the nearest tenth.

Solution

$$M = \log I$$
$$M = \log(900{,}000)$$
$$\approx 5.954$$

The Richter magnitude was approximately 6.0.

Because the Richter scale is a logarithmic scale, an increase of 1 on the scale corresponds to a 10-fold increase in intensity. For instance, suppose one earthquake measures 4 on the Richter scale and another earthquake measures 5. Then $4 = \log I_1$ and $5 = \log I_2$. Writing the exponential form of each logarithm,

$$I_1 = 10^4 \quad \text{and} \quad I_2 = 10^5$$

From these equations, $I_2 = 10I_1$; that is, an increase of 1 on Richter scale results in a 10 times increase in intensity.

> **Example 4**   **Compare the Intensities of Two Earthquakes**

An earthquake in Chile in 1960 measured 9.5 on the Richter scale. An earthquake in San Francisco in 1989 measured 7.1 on the Richter scale. Compare the intensities of the two earthquakes. Round to the nearest whole number.

Solution

For the Chile earthquake, $9.5 = \log I_1$. For the San Francisco earthquake, $7.1 = \log I_2$. Write each of these in exponential form:

$$I_1 = 10^{9.5} \qquad I_2 = 10^{7.1}$$

To compare the intensities, find the ratio of the two earthquakes.

$$\frac{I_1}{I_2} = \frac{10^{9.5}}{10^{7.1}}$$

$$= 10^{9.5-7.1} \qquad \bullet \text{ To divide exponential expressions}$$
$$= 10^{2.4} \approx 251 \qquad \text{with the same base, subtract the exponents.}$$

The earthquake in Chile was approximately 251 times as intense as the earthquake in San Francisco.

Sound intensity level can be expressed with a logarithmic scale. In this case, the base intensity, I_0, is the threshold of hearing. Because the ability to hear noises varies from person to person, an international standard has been set for I_0. It is 10^{-12} watt/meter^2. To give a perspective, think about the intensity of a 10-watt light bulb one meter (about three feet) from you. 10^{-12} watt is *extremely* small in comparison.

Decibel Scale

Sound intensity level is measured by decibels (dB) and is given by

$$dB(I) = 10 \log\left(\frac{I}{I_0}\right),$$

where I_0 is the threshold of hearing and I is the intensity of the sound.

Example 5 **Find the Decibel Level for Normal Conversation**

Normal conversation has an intensity that is 1.5×10^6 times the threshold level, I_0. Find the decibel value for normal conversation. Round to the nearest decibel.

Solution

The intensity I is 1.5×10^6 times the threshold level, I_0. Therefore,

$$dB(I) = 10 \log\left(\frac{I}{I_0}\right)$$

$$dB(1.5 \times 10^6 I_0) = 10 \log\left(\frac{1.5 \times 10^6 I_0}{I_0}\right) = 10 \log(1.5 \times 10^6)$$

$$\approx 61.8$$

The decibel level of normal conversation is approximately 62 dB.

Example 6 **Decibel Level of Trumpets**

The relative intensity level of a single note from a trumpet when it reaches an orchestra conductor is 76 dB. What relative intensity level would the conductor perceive if 3 trumpeters played at the same volume? Assume that the three trumpeters are at the same distance from the conductor.

Solution

Let I be the intensity of the sound from one trumpet. The relative intensity level for the conductor is 76 dB. Therefore, $76 = 10 \log\left(\frac{I}{I_0}\right)$. Solve this equation for I.

$$76 = 10 \log\left(\frac{I}{I_0}\right)$$

$$7.6 = \log\left(\frac{I}{I_0}\right) \qquad \text{• Divide each side of the equation by 10.}$$

$$10^{7.6} = \frac{I}{I_0} \qquad \text{• Use } \log a = b \text{ is equivalent to } 10^b = a.$$

$$10^{7.6} I_0 = I \qquad \text{• Solve for } I.$$

The intensity of the sound of one trumpet is $10^{7.6} I_0$. Therefore, the intensity of three trumpet notes reaching the conductor is $3 \cdot 10^{7.6} I_0$. The decibel level is

$$dB(I) = 10 \log\left(\frac{I}{I_0}\right)$$

$$dB(3 \cdot 10^{7.6} I_0) = 10 \log\left(\frac{3 \cdot 10^{7.6} I_0}{I_0}\right)$$

$$= 10 \log(3 \cdot 10^{7.6})$$

$$\approx 80.8$$

The relative intensity level is 80.8 decibels.

Solving Exponential Equations

Logarithms can be used to solve exponential equations that occur in certain applications. For instance, suppose an investment of \$1000 has been growing at a rate of 10% each year. An exponential equation for the growth of the investment is $A = 1000(1.1)^t$. Using this equation, we can input the time we have owned the investment, and the output is its value at that time. For instance, if we wanted to know the value of the investment after 5 years, replace t by 5. Then evaluate the expression.

$$A = 1000(1.1)^t$$

$$A = 1000(1.1)^5$$

$$= 1610.51$$

The investment will be worth \$1610.51 in 5 years.

Now suppose we want to know how long it will take for the investment to reach a value of \$1750. In this case, we use the same equation.

$$A = 1000(1.1)^t$$

$$1750 = 1000(1.1)^t \qquad \text{• } A = 1750.$$

Now we need to solve for t with the help of the following.

Solution of the Exponential Equation $A = A_0(p)^t$

The solution of the equation $A = A_0(p)^t$ for t is

$$t = \frac{\log\left(\dfrac{A}{A_0}\right)}{\log(p)}$$

where $p = 1 + r$ for exponential growth, $p = 1 - r$ for exponential decay, and r is the rate of exponential growth or decay.

Use the Solution of an Exponential Equation to solve $1750 = 1000(1.1)^t$ for t.

$$1750 = 1000(1.1)^t$$

$$t = \frac{\log\left(\dfrac{A}{A_0}\right)}{\log(p)}$$

$$= \frac{\log\left(\dfrac{1750}{1000}\right)}{\log(1.1)} \qquad \bullet A = 1750, A_0 = 1000, p = 1.1.$$

$$\approx 5.87$$

It will take about 5.87 years (about 5 years 10 months) to attain a value of $1750.

Example 7 **Solve an Exponential Equation**

A process involves running pure water into a tank, thoroughly mixing a solution, and then draining some of the solution from the tank. This process removes 8% of salt from the tank for each hour. If there were initially 10 pounds of salt in the solution, in how many hours will there be 3 pounds of salt in the water? Round to the nearest tenth.

Solution

Because salt is removed at a rate of 8% per hour, the process is exponential decay and $r = 0.08$. Therefore, $p = 1 - 0.08 = 0.92$. The solution is

$$t = \frac{\log\left(\dfrac{A}{A_0}\right)}{\log(p)}$$

$$t = \frac{\log\left(\dfrac{3}{10}\right)}{\log(0.92)} \qquad \bullet A = 3, A_0 = 10, p = 0.92.$$

$$\approx 14.439$$

It will take 14.4 hours to reduce the solution to 3 pounds of salt.

The half-life for an exponential decay process can be determined by solving the half-life equation for h.

Equation for the Half-Life of a Process

$$h = \frac{t \cdot \log\left(\dfrac{1}{2}\right)}{\log\left(\dfrac{A}{A_0}\right)}$$

where A is the amount of a substance after time t, A_0 is the initial amount of substance, and h is the half-life of the process.

Example 8 **Find the Half-Life of an Exponential Decay Process**

Iodine-131 is a radioisotope used in the treatment of some thyroid cancers. A measurement found 10 mg of iodine-131 in a sample. A second measurement taken 8 hours later showed 9.7152 mg of iodine-131 in the sample. Find the half-life, in hours, of iodine-131. Round to the nearest whole number.

Solution

The equation for an exponential decay process is $A = A_0\left(\frac{1}{2}\right)^{t/h}$, where A is the amount of iodine-131 remaining in a sample after t hours, A_0 is the initial amount of iodine-131, and h is the half-life.

$$h = \frac{t \cdot \log\left(\frac{1}{2}\right)}{\log\left(\frac{A}{A_0}\right)}$$

$$h = \frac{8 \cdot \log\left(\frac{1}{2}\right)}{\log\left(\frac{9.7152}{10}\right)} \qquad \bullet\ A = 9.7152 \text{ when } t = 8 \text{ and } A_0 = 10.$$

$$h \approx 191.918 \qquad \bullet \text{ Evaluate the expression.}$$

The half-life of iodine-131 is approximately 192 hours.

Population ecologists, financial analysts, and others want to determine the doubling time of a process.

Equation for the Doubling Time of a Process

$$d = \frac{t \cdot \log(2)}{\log\left(\frac{A}{A_0}\right)}$$

where A is the amount of a substance after time t, A_0 is the initial amount of substance, and d is the doubling time of the process.

Take Note: The doubling time of an investment is such a popular statistic that an approximate formula is used to compute the time.

Rule of 72 The time it takes an investment that earns a compounded annual interest rate of $r\%$ is

$$\text{Doubling time} \approx \frac{72}{r}.$$

Using this formula for Example 9, we have, to the nearest tenth, $\frac{72}{7} \approx 10.3$. This formula is reasonably accurate for interest rates between 6% and 10%.

Example 9   Find the Doubling Time of an Investment

A financial advisor can offer a client an investment that returns a compounded annual return of 7%. Find the time, in years, it will take the investment to double in value.

Solution

Let A_0 be the initial amount of the investment. Because the investment earns 7% per year, the value of the investment after 1 year is $A = 1.07A_0$. Use the Equation for the Doubling Time of a Process.

$$d = \frac{t \cdot \log(2)}{\log\left(\frac{A}{A_0}\right)}$$

$$= \frac{1 \cdot \log(2)}{\log\left(\frac{1.07A_0}{A_0}\right)} \qquad \bullet\ t = 1, A = 1.07A_0.$$

$$d \approx 10.2$$

It will take a little more than 10 years for the investment to double in value.

| Example 10 | Find the Doubling Time for a Population |

A biologist measured the growth of a certain haploid yeast by placing 1000 cells into a synthetic medium. Thirty minutes later, the biologist found 1260 cells in the medium. Find the doubling time of this yeast. Round to the nearest minute.

Solution

The equation for an exponential doubling process is $A = A_0(2)^{t/h}$, where A is the amount of yeast in the sample after t minutes, A_0 is the initial amount of yeast, and h is the doubling time.

$$h = \frac{t \cdot \log(2)}{\log\left(\dfrac{A}{A_0}\right)}$$

$$h = \frac{30 \cdot \log(2)}{\log\left(\dfrac{1260}{1000}\right)} \qquad \bullet\ A = 1260 \text{ when } t = 30 \text{ and } A_0 = 1000.$$

$$h \approx 89.98 \qquad\qquad \bullet \text{ Evaluate the expression.}$$

The doubling time is approximately 90 minutes.

6.2   Exercise Set

Think About It

1. What is the logarithmic form of $10^x = 7$?

2. What is the exponential form of $\log 9 = y$?

3. If earthquake A has a magnitude of 2 on the Richter scale and earthquake B has a magnitude of 4 on the Richter scale, is the intensity of B twice that of A?

4. If the pH of a liquid is 4, is it acidic or alkaline?

5. What is the doubling time of an investment?

Preliminary Exercises

For Exercises P1 through P10, if necessary, use the referenced example following the exercise for assistance.

P1. Use the definition of logarithm to find each of the following. If necessary, round to the nearest ten-thousandth. **Example 1**

   a. $\log x = -4$ **b.** $\log x = -0.25$

   c. $\log 100{,}000 = x$ **d.** $\log\left(\dfrac{5}{2}\right) = x$

P2. The hydrogen ion concentration for a brand of canned tomato sauce is 0.000051 mole per liter. What is the pH of the tomato sauce? Round to the nearest tenth. Is the sauce an acid or a base? **Example 2**

P3. Find the magnitude of an earthquake that was 750,000 times the intensity of the base intensity earthquake. Round to the nearest tenth. **Example 3**

P4. A 1964 earthquake in Alaska measured 9.2 on the Richter scale. An earthquake in Assam, Tibet, in 1950 measured 8.6 on the Richter scale. Compare the intensities of the two earthquakes. Round to the nearest whole numbers. **Example 4**

P5. A harp seal's call has a sound intensity that is 9.2×10^{12} times greater than the threshold level. Find the decibel value for a harp seal. Round to the nearest decibel. **Example 5**

P6. A home theater system has a speaker playing background music at 75 dB. If a second 75 dB speaker is added, what is the decibel level of the two speakers? **Example 6**

P7. The breaking strength s in tons of a steel cable of diameter d in inches is given by $s = 5(6.2)^d$. If an engineer wants a cable to have a breaking strength of 20 tons, what diameter cable is needed? Round to the nearest hundredth. **Example 7**

P8. Technetium-99 is a radioisotope used in brain scans. A measurement found 5 mg of technetium-99 in a sample. A second measurement taken 2 hours later showed 3.9685 mg of technetium-99 in the sample. Find the half-life in hours of technetium-99. Round to the nearest tenth. **Example 8**

P9. An investment is advertised giving a 5% compounded annual return. Find the time, in years, for the investment to double in value. **Example 9**

P10. Using historical data, a forester found that the population of a certain tree grew from 100 trees to 122 trees in 10 years. Find the doubling time for the growth of this tree population. Round to the nearest year. **Example 10**

Section Exercises

Solve.

1. Solve for x.

   a. $\log x = 5$ **b.** $\log x = -1$

   c. $\log 10{,}000 = x$ **d.** $\log\left(\dfrac{1}{100}\right) = x$

2. Solve for x. If necessary, round to the nearest ten-thousandth.

    a. $\log x = \dfrac{3}{4}$ **b.** $\log \dfrac{3}{4} = x$

    c. $\log x = \sqrt{2}$ **d.** $\log \sqrt{2} = x$

3. **A Purring Cat** The purr of a cat has a sound intensity that is 320 times greater than the threshold level. Find the decibel value for this cat's purr. Round to the nearest decibel.

4. **A Barking Dog** A dog's bark can have a sound intensity that is 3.2×10^8 times greater than the threshold level. Find the decibel value for this dog's bark. Round to the nearest decibel.

5. **Car Stereos** The average car stereo system has a maximum volume of approximately 100 decibels. How many times greater than the threshold level is the maximum volume of an average car stereo?

6. **Sound of a Pin Dropping** When a pin is dropped from 1 inch above a surface, the sound is approximately 12 decibels. How many times greater than the threshold sound level is the sound of a pin drop? Round to the nearest whole number.

7. **Earthquake Magnitude** Find the Richter magnitude of an earthquake that occurred off the coast of Chile in 2010 that was 630,000,000 times the intensity of the base intensity earthquake. Round to the nearest tenth.

8. **Earthquake Magnitude** The Great Alaska Earthquake in 1964 was 1,600,000,000 times the intensity of the base intensity earthquake. Find the Richter magnitude of the earthquake. Round to the nearest tenth.

9. **Blood Samples** Sodium-24 (^{24}Na) is a radioisotope used to study circulatory dysfunction. A measurement found 4 micrograms of ^{24}Na in a blood sample. A second measurement taken 5 hours later showed 3.18 micrograms of ^{24}Na in a blood sample. Find the half-life in hours of ^{24}Na. Round to the nearest tenth.

10. **Deep Space Equipment** Polonium-210 (^{210}Po) is used as a heat source for some deep space equipment. An experiment on a sample of ^{210}Po originally contained 10 g of ^{210}Po. Five days later, the sample contained 9.75 g of ^{210}Po. Find the half-life in days of ^{210}Po. Round to the nearest day.

11. **Bacterial Growth** During an experiment, there were 500,000 *Escherichia coli* (*E. coli*) bacteria in broth. A second measurement 10 minutes later showed 710,000 bacteria in the broth. Find the doubling time in minutes of the *E. coli* bacteria. Round to the nearest minute.

12. **Sour Milk** *Streptococcus lactis* is a milk contaminant that leads to souring. A test of a milk sample found 200,000 *S. lactis* in the sample. Five minutes later, a second test showed 230,000 *S. lactis* in the sample. Find the doubling time in minutes of the bacteria. Round to the nearest minute.

13. **Population Growth** From 2015 to 2022, the population of the United States grew at a rate of approximately 0.4% per year. Assuming the U.S. population continues to grow at that rate, what is the doubling time of the U.S. population?

14. **CD Investment Growth** How long will it take to triple your money if it is invested in a CD that pays 5% annual interest compounded daily? Round to the nearest tenth of a year.

15. **Stereo Speakers** Three car stereo speakers each produce 70 decibels. What is the combined decibel level of the three speakers? Round to the nearest whole number.

16. **Sound Intensity** A speaker is playing music at 80 dB. If you added nine more 80 dB speakers, would the total decibel level be less than or greater than 100 dB? (*Suggestion:* See Exercise P6.)

17. **Oil Leak** Crude oil is leaking from a tank at the rate of 10% of the tank volume every 2 hours. If the tanker originally contained 400,000 gallons of oil, how many gallons of oil remain in the tank after 3 hours? Round to the nearest gallon.

18. **Cooking** In the initial stages of cooking, the temperature inside a beef roast rises 30% of its current temperature every 20 minutes. Assuming the initial temperature of the roast is 45°F, what is the internal temperature of the roast after 2 hours? Round to the nearest degree.

19. **Earthquake Intensity** An earthquake had a Richter scale magnitude of 7.2. Its aftershock had a Richter scale magnitude of 3.7. How many times more was the intensity of the original earthquake than the aftershock? Round to the nearest whole number.

20. **Earthquake Intensity** On March 28, 1964, an earthquake of magnitude 9.2 on the Richter scale struck Prince William Sound, Alaska. On May 2, 2008, an earthquake of magnitude 6.6 on the Richter scale struck the Aleutian Islands in Alaska. Compare the intensity of the larger earthquake to the intensity of the smaller earthquake by finding the ratio of the larger intensity to the smaller intensity. Round to the nearest whole number.

21. **LED Lights** A manufacturer of light-emitting diode (LED) lights has tested its lights and found that the percent chance P that a light will fail after t hours of use can be approximated by

$$P = 100\left(\frac{1}{2}\right)^{t/50,000}$$

    a. What is the percent chance that one of the company's LED lights will fail after 20,000 hours? Round to the nearest tenth.

    b. What is the half-life of the process?

22. **Small Town Population Growth** A recent survey conducted in 2022 by the Census Bureau determined that the average growth rate for a small town with a population of 35,240 in the northeast United States was approximately 0.75% per year.

    a. Assuming that growth rate continues, write an equation that will give the population in t years after 2022.

    b. What is the doubling time of the population? Round to the nearest tenth of a year.

23. **Sterilizing Medical Supplies** Cobalt-60 (^{60}Co) is used to sterilize some medical supplies such as syringes and surgical gloves. In the laboratory, an original 8-milligram sample of ^{60}Co decayed to 7 milligrams of ^{60}Co in 1 year. Find the half-life in years of ^{60}Co. Round to the nearest tenth.

24. **Age of an Artifact** The half-life of ^{14}C is approximately 5730 years. what percent of the original ^{14}C remains after 10,000 years? Round to the nearest tenth.

25. **Liver Diseases** Gold-198 (^{198}Au) is used in the diagnosis of some liver diseases. A 5-milligram sample of ^{198}Au will decay to 3 milligrams in 2 days. Find the half-life in days of ^{198}Au. Round to the nearest tenth.

26. **Bacterial Growth** *Staphylococcus epidermidis* is a bacterium that can grow on surgical implants and is extremely difficult to treat. An observation of a laboratory colony of

S. epidermidis determined that it grew from 0.25 millimeter to 1.3 millimeters in 10 hours. Find the doubling time of *S. epidermidis*. Round to the nearest hour.

27. **Smart Watches**   The value of the smart watch market in the United States in 2019 was approximately $26.3 billion and is expected double in size every 3 years. Find an exponential growth model for the value of smart watches *t* years after 2019. Use the model to predict the value of the smart watch market in 2025.

28. **Inflation Rate Growth**   In 2022, the inflation rate for consumption goods was approximately 8.5%, the highest in more than 30 years. If that rate continues, how long will it take for the cost of consumption goods to double? Round to the nearest year.

29. **Processing Sugar**   Inversion is one step in processing raw sugar. In this process, 20% of the available raw sugar is changed to a refined sugar every 5 hours. If a process begins with 1000 pounds of raw sugar, write an exponential equation for the amount of raw sugar *t* hours after the process begins. How many hours must elapse before there is 50 pounds of raw sugar? Round to the nearest tenth.

30. Bacteria can grow on warm food left in a pot or pan. To minimize this danger, a chef may take gravy prepared for later use and place it in an ice bath to quickly cool the gravy. The goal is to cool the gravy to less than 40°F, a temperature at which the bacteria normally cannot grow. Suppose the container of gravy has a temperature of 165°F when removed from the stove. The gravy pot is placed in an ice bath and the temperature 2 hours later is 100°F. Find an exponential decay equation for the cooling of the gravy. What is the half-life of the process?

31. The population of an ant colony is doubling every 12 weeks. If the colony currently consists of 1000 ants, find an exponential growth equation for the colony *t* days from now. Use the equation to find the number of ants in the colony 200 days from now.

Investigation

Earthquakes   Seismologists generally determine the Richter scale magnitude of an earthquake by examining a seismogram.

The magnitude of an earthquake cannot be determined just by examining the amplitude of a seismogram because this amplitude decreases as the distance between the epicenter of the earthquake and the observation station increases. To account for the distance between the epicenter and the observation station, a seismologist examines a seismogram for both small waves, called *p-waves*, and larger waves, called *s-waves*. Charles Richter developed the Amplitude-Time-Difference Formula to determine the magnitude *M* of an earthquake from the data in a seismogram: $M = \log A + 3 \log 8t - 2.92$, where *A* is the amplitude (in millimeters) of the s-waves on a seismogram and *t* is the time (in seconds) between the s-waves and p-waves.

1. Find the Richter scale magnitude of the earthquake that produces a seismogram with an s-wave with an amplitude of 18 mm occurring 31 seconds after the p-wave. Round to the nearest tenth.

2. Find the Richter scale magnitude of the earthquake that produced the following seismograph. Round to the nearest tenth.

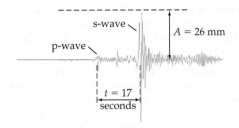

Section **6.3**

Quadratic Functions

Learning Objectives:

- Identify a quadratic function.
- Find the vertex of a quadratic function.
- Find the intercepts of the graph of a function.
- Find the maximum or minimum of a quadratic function.
- Solve application problems involving maximum or minimum of a quadratic function.

Introduction to Quadratic Functions

For linear functions, a one-unit change in *x* produced a *constant* change in *y*. For exponential functions, a one-unit change in *x* produced a *constant percent* change in *y*. Other types of functions have neither of these qualities.

In an experiment in the early days of the U.S. space program, astronauts rode in a balloon to extreme altitudes (more than 100,000 feet) and then parachuted from the gondola. The astronaut would begin free fall. Because of the height, there was minimal air resistance during the first part of the fall.

Figure 6.6 shows the distance the astronaut falls for each second of free fall. Looking at the graph, for each one-unit increase in time, the *change* in distance is not constant. The function is not linear (which is evident from its graph). For each one-unit increase in time, the *percent change* in distance is not the same, so the graph does not represent an exponential function. This graph is an example of the graph of a *quadratic function*.

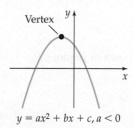

Figure 6.6

Definition of a Quadratic Function

A **quadratic function** is one that can be expressed by the equation

$$y = ax^2 + bx + c$$

where a, b, and c are real numbers and $a \neq 0$.

Examples

$y = 2x^2 - 3x + 1$	• $a = 2, b = -3, c = 1$
$s = -t^2 - 3$	• $a = -1, b = 0, c = -3$
$p = q^2 + 5q$	• $a = 1, b = 5, c = 0$

> **Example 1** **Determine Whether an Equation Represents a Quadratic Function**

Determine which of the following is a quadratic function. For a quadratic function, identify a, b, and c.

a. $y = x^2$ **b.** $y = 2^x + 3x - 4$ **c.** $s = 0.06t^2 - \dfrac{t}{2} + 1$

Solution

a. Quadratic function; $a = 1, b = 0, c = 0$

b. Not a quadratic function

c. Quadratic function; $a = 0.06, b = -\dfrac{1}{2}, c = 1$

●

For the astronaut in free fall, using some principles from physics, the equation for the distance traveled for given values of time is $s = 16t^2$, where t is the time in seconds the astronaut falls and s is the distance the astronaut falls during that time.

Vertex of a Parabola

The graph of a quadratic function is a **parabola**. The following are two examples of curves that are part of a parabola. A basketball shot travels along a parabola. The arc of a suspension bridge is part of a parabola.

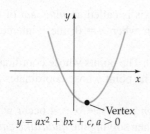

Vertex

$y = ax^2 + bx + c, a < 0$

Figure 6.7

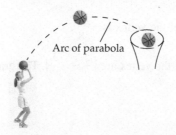

Arc of parabola

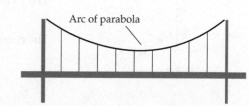

Arc of parabola

The graph of a parabola either **opens down** (like the arc of a basketball) when $a < 0$ (see Figure 6.7) or **opens up** (like the arc of the suspension bridge) when $a > 0$ (see Figure 6.8). The **vertex** of a parabola is the highest point on a parabola that opens down or the lowest point on a parabola that opens up.

In terms of the basketball shot, the vertex of the shot is the highest point the basketball reaches; for the suspension bridge, the vertex is the lowest point on the support cable. These particular points on a parabola have real-world applications. Because of these applications, it is convenient to have a formula to find the vertex of a parabola. These formulas can be derived by using a technique called *completing the square* and can be found in most college algebra texts.

$y = ax^2 + bx + c, a > 0$

Vertex

Figure 6.8

> ## Vertex Formula
>
> The x-coordinate of the vertex of $y = ax^2 + bx + c$ is given by $x = -\dfrac{b}{2a}$. To find
> the y-coordinate of the vertex, evaluate $y = ax^2 + bx + c$ at the x-coordinate of the
> vertex.

Example 2   Find the Vertex of a Parabola

Determine whether the graph of $y = 2x^2 - 5x - 6$ opens up or down. Then find the
coordinates of the vertex.

Solution

For $y = 2x^2 - 5x - 6$, $a = 2$, $b = -5$, and $c = -6$. Because $a > 0$, the parabola opens up.

Find the x-coordinate of the vertex.

$$x = -\frac{b}{2a}$$

$$= -\frac{-5}{2(2)} = \frac{5}{4}$$

The x-coordinate of the vertex is $\dfrac{5}{4}$.

To find the y-coordinate of the vertex, evaluate $y = 2x^2 - 5x - 6$ when $x = \dfrac{5}{4}$.

$$y = 2x^2 - 5x - 6$$

$$= 2\left(\frac{5}{4}\right)^2 - 5\left(\frac{5}{4}\right) - 6 \qquad \text{• Replace } x \text{ by } \frac{5}{4}.$$

$$= 2\left(\frac{25}{16}\right) - 5\left(\frac{5}{4}\right) - 6$$

$$= \frac{25}{8} - \frac{25}{4} - 6 = \frac{25}{8} - \frac{50}{8} - \frac{48}{8} \qquad \text{• The common denominator is 8.}$$

$$= -\frac{73}{8}$$

The y-coordinate of the vertex is $-\dfrac{73}{8}$.

The coordinates of the vertex are $\left(\dfrac{5}{4}, -\dfrac{73}{8}\right)$. The graph is shown in Figure 6.9.

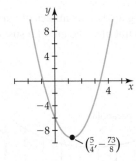

Figure 6.9

Intercepts of a Quadratic Function

Recall that a point at which a graph crosses the x- or y-axis is called an *intercept* of the
graph. The x-intercepts of the graph of an equation occur when $y = 0$; the y-intercepts
occur when $x = 0$.

The graph of $y = x^2 + 3x - 4$ is shown in Figure 6.10. The points whose coordinates
are $(-4, 0)$ and $(1, 0)$ are the **x-intercepts** of the graph. The point whose coordinates are
$(0, -4)$ is the **y-intercept** of the graph.

The x-intercepts of the graph of the *quadratic function* $y = x^2 + 3x - 4$ occur when
$y = 0$. The x-intercepts are found by replacing y by 0 and then solving the *quadratic equation* $0 = x^2 + 3x - 4$.

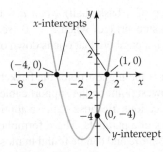

Figure 6.10

Review of Solving a Quadratic Equation

The solutions of the quadratic equation $ax^2 + bx + c = 0$ can be found by using the quadratic formula

$$x = \frac{-b \pm \sqrt{b^2 - 4ac}}{2a}$$

Example

Solve $2x^2 - x - 4 = 0$.

Use the quadratic formula with $a = 2$, $b = -1$, and $c = -4$.

$$x = \frac{-b \pm \sqrt{b^2 - 4ac}}{2a}$$

$$= \frac{-(-1) \pm \sqrt{(-1)^2 - 4(2)(-4)}}{2(2)}$$

$$= \frac{1 \pm \sqrt{1 + 32}}{4} = \frac{1 \pm \sqrt{33}}{4}$$

$$\approx \frac{1 \pm 5.7446}{4}$$

$$x \approx \frac{1 + 5.7446}{4} \approx 1.686 \quad \text{or} \quad x \approx \frac{1 - 5.7446}{4} \approx -1.186$$

The solutions are approximately 1.686 and -1.186.

Example 3   **Find the Intercepts for a Graph**

Find the x- and y-intercepts of the graph of $y = x^2 + 3x - 4$.

Solution

Before we begin the solution, note that the equation is the same as the one graphed in Figure 6.10, which shows the intercepts. This is an algebraic verification of the graph.

• Find the y-intercept.

To find the y-intercept, replace x with 0 and simplify.

$$y = x^2 + 3x - 4$$
$$= 0^2 + 3(0) - 4$$
$$= -4$$

The y-intercept is $(0, -4)$.

• Find the x-intercepts.

To find the x-intercept, replace y with 0 and then solve for x.

$$0 = x^2 + 3x - 4$$

This is a quadratic equation. Solve for x by using the quadratic formula.

$$x = \frac{-b \pm \sqrt{b^2 - 4ac}}{2a}$$

$$= \frac{-(3) \pm \sqrt{(3)^2 - 4(1)(-4)}}{2(1)} \qquad \bullet\, a = 1, b = 3, c = -4.$$

$$= \frac{-3 \pm \sqrt{9 + 16}}{2} = \frac{-3 \pm \sqrt{25}}{2}$$

$$= \frac{-3 \pm 5}{2}$$

$$x = \frac{-3 + 5}{2} = 1 \quad \text{or} \quad x = \frac{-3 - 5}{2} = -4$$

The x-intercepts are $(1, 0)$ and $(-4, 0)$.

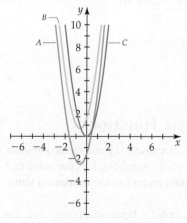

Graphs of quadratic functions can have two x-intercepts like curve A, no x-intercepts like B, or one x-intercept like C. In all cases, however, parabolas all have only one y-intercept.

There are applications of intercepts to real-world situations.

> **Example 4**   **Calculate the Airtime for a Snowboarder's Jump**

The height y in feet of a snowboarder t seconds after beginning a certain jump can be approximated by $y = -16t^2 + 22.9t + 9$. The snowboarder lands at a point that is 3 feet below the base of the jump (see Figure 6.11). Determine the air time (the time the snowboarder is in the air) for this jump. Round to the nearest tenth of a second.

9 ft

3 ft

Figure 6.11

Solution

From Figure 6.11, y represents the height of the snowboarder t seconds after the beginning of the jump, the snowboarder lands when $y = -3$, 3 feet *below* the base of the jump.

$$y = -16t^2 + 22.9t + 9$$

$$-3 = -16t^2 + 22.9t + 9 \qquad \bullet \text{ Replace } y \text{ by } -3.$$

$$0 = -16t^2 + 22.9t + 12 \qquad \bullet \text{ Add 3 to each side. To use the quadratic formula, the equation must be set equal to zero.}$$

$$t = \frac{-22.9 \pm \sqrt{22.9^2 - 4(-16)(12)}}{2(-16)}$$

$$= \frac{-22.9 \pm \sqrt{1292.41}}{-32}$$

$$\approx \frac{-22.9 \pm 35.9501}{-32}$$

$$t \approx \frac{-22.9 - 35.9501}{-32} \approx 1.8 \quad \text{or} \quad t \approx \frac{-22.9 + 35.9501}{-32} \approx -0.4$$

Because a negative time is not possible, the airtime for this jump is approximately 1.8 seconds.

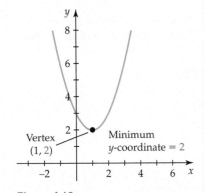

Figure 6.12

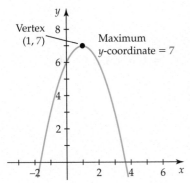

Figure 6.13

Maximum or Minimum of a Quadratic Function

The graph of $f(x) = x^2 - 2x + 3$ is shown in Figure 6.12. Because a is positive, the parabola opens up. The vertex of the parabola is the lowest point on the parabola. It is the point that has the minimum y-coordinate. The y-coordinate of this point represents the **minimum value of the function.**

The graph of $f(x) = -x^2 + 2x + 6$ is shown in Figure 6.13. Because a is negative, the parabola opens down. The vertex of the parabola is the highest point on the parabola. It is the point that has the maximum y-coordinate. The y-coordinate of this point represents the **maximum value of the function.**

To find the minimum or maximum value of a quadratic function, first find the x-coordinate of the vertex. Then evaluate the function at that value.

Example 5   **Find the Minimum or Maximum of a Quadratic Function**

Determine whether $y = 2x^2 - 3x + 1$ has a maximum or minimum. Then determine the maximum or minimum.

Solution

• Determine whether the function has a maximum or minimum.

If the graph of the parabola opens up, the function has a minimum; if the graph opens down, the function has a maximum. Because $a = 2 > 0$, the graph opens up. The function has a minimum.

• Find the minimum.

The minimum occurs at the vertex. Use the procedure to find the vertex of a parabola.

$$x = -\frac{b}{2a}$$

$$= -\frac{-3}{2(2)} = \frac{3}{4} \qquad \bullet\ a = 2, b = -3.$$

The x-coordinate of the vertex is $\frac{3}{4}$.

Evaluate the function when $x = \frac{3}{4}$.

$$y = 2x^2 - 3x + 1$$

$$y = 2\left(\frac{3}{4}\right)^2 - 3\left(\frac{3}{4}\right) + 1$$

$$= 2\left(\frac{9}{16}\right) - 3\left(\frac{3}{4}\right) + 1 = \frac{9}{8} - \frac{9}{4} + 1$$

$$= \frac{9}{8} - \frac{18}{8} + \frac{8}{8} = -\frac{1}{8}$$

The minimum value of the function is $-\frac{1}{8}$.

Example 6   **Solve a Projectile Motion Problem**

A ball is thrown vertically upward with an initial velocity of 48 feet per second. If the ball started its flight at a height of 8 feet, then its height s in feet t seconds after it is released can be determined by $s = -16t^2 + 48t + 8$.

a. Determine the time it takes the ball to attain its maximum height.

b. What is the maximum height attained by the ball?

Solution

a. The graph of $s = -16t^2 + 48t + 8$ is a parabola that opens downward. Therefore, s will attain its maximum value at the vertex of its graph. Using the vertex formula with $a = -16$ and $b = 48$, we have

$$t = -\frac{b}{2a}$$

$$= -\frac{48}{2(-16)} = \frac{3}{2}$$

The ball attains its maximum height in $\frac{3}{2}$ seconds.

b. When $t = \dfrac{3}{2}$ seconds,

$$s = -16t^2 + 48t + 8$$

$$= -16\left(\dfrac{3}{2}\right)^2 + 48\left(\dfrac{3}{2}\right) + 8$$

$$= -16\left(\dfrac{9}{4}\right) + 48\left(\dfrac{3}{2}\right) + 8 = -36 + 72 + 8$$

$$= 44$$

The maximum height attained by the ball is 44 feet.

Example 7 **Solve an Application Problem**

A fence is being constructed to enclose a rectangular picnic area next to a stream. If 200 feet of fence are available and there is no fencing along the stream, what dimensions of the rectangle will produce the maximum area? What is the maximum area?

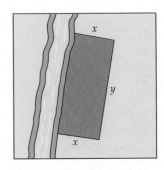

Solution

There are 200 feet of fencing available. Therefore, $200 = 2x + y$. Let $A = xy$ be the area of the enclosure. The goal is to maximize A. Rewrite A in terms of a single variable by solving $200 = 2x + y$ for y.

$$200 = 2x + y$$

$$200 - 2x = y$$

Substitute $200 - 2x$ for y in $A = xy$. Then simplify.

$$A = xy$$

$$A = x(200 - 2x)$$

$$A = -2x^2 + 200x$$

The equation $A = -2x^2 + 200x$ is a quadratic function whose graph opens down (coefficient of x^2 is -2). Find the x-coordinate of the vertex.

$$x = -\dfrac{b}{2a} = -\dfrac{200}{2(-2)} = 50$$

The x-coordinate of the vertex is 50.

To find the A-coordinate of the vertex, replace x in $200 - 2x$ by 50.

$$200 - 2x = A$$

$$200 - 2(50) = 100$$

The A-coordinate of the vertex is 100.

The dimensions of the rectangle that will give the maximum area are 50 feet by 100 feet. To find the maximum area, replace x and y in $A = xy$. Then evaluate.

$$A = xy$$

$$A = 50(100) = 5000$$

The maximum area of the enclosure is 5000 ft^2.

Quadratic functions can be applied to economics. A **demand function** gives the relationship between the price at which an item can be sold and the number of items that can be sold at that price. For instance, suppose $p = 500 - 5x$ gives the price p, in dollars, at which x espresso machines can be sold. If $x = 10$, then $p = 500 - 5(10) = 450$. This means that 10 espresso machines can be sold for $450. If $x = 50$, then $p = 500 - 5(50) = 250$. This means that 50 espresso machines can be sold for $250. This is typical of a demand function. More items can be sold as the price goes down.

Revenue is the amount of money taken in for selling x items:

Revenue = Price per item × Number of items sold

Because the demand function gives the price per item,

Revenue = Demand · x

The process of producing and selling items can be expensive and is given by a **cost function**, which describes the total cost to produce x items. The profit for selling items is revenue minus cost:

Profit = Revenue − Cost

In Example 8, we will continue with the espresso machine.

Example 8   **Solve an Application to Economics**

The demand function for an espresso machine is given by $p = 500 - 5x$, where p is the price in dollars that x espresso machines can be sold. The cost function for the espresso machines is $C = 100x + 1000$.

a. Find the revenue function.
b. Find the profit function.
c. How many machines should be sold to maximize profit?
d. What is the maximum profit?

Solution

a. $R = px$
    $R = (500 - 5x)x$
    $R = -5x^2 + 500x$

b. $P = R - C$
    $P = -5x^2 + 500x - (100x + 1000)$
    $P = -5x^2 + 400x - 1000$

c. The profit function is a quadratic function. To find how many machines to sell to maximize profit, find the x-coordinate of the vertex.

$$x = -\frac{b}{2a}$$

$$= -\frac{400}{2(-5)}$$

$$= 40$$

Maximum profit occurs by selling 40 espresso machines.

d. To find the maximum profit, evaluate the profit function for $x = 40$.
    $P = -5x^2 + 400x - 1000$
    $= -5(40)^2 + 400(40) - 1000$
    $= 7000$
    The maximum profit is $7000.

6.3 Exercise Set

Think About It

1. Does the graph of $y = -0.5x^2 + 3x + 2$ open up or down?

2. Do all quadratic functions have either a maximum or minimum? Explain.

3. Identify each of the following as either an exponential function or a quadratic function.
   a. $y = x^2$ **b.** $z = 2^x$

4. Fill in the blank: The maximum or minimum of a quadratic function occurs at the _____ of the graph of the function.

Preliminary Exercises

For Exercises P1 to P8, if necessary, use the referenced example following the exercise for assistance.

P1. Determine which of the following is a quadratic function. For a quadratic function, identify a, b, and c. **Example 1**
   a. $y = 2x^3 + 3x^2 + 4$
   b. $s = v^2 - 4v + 1$
   c. $s = 1 + 32t - 16t^2$

P2. Determine whether the graph of $y = -x^2 + 3x + 4$ opens up or down. Then find the coordinates of the vertex. **Example 2**

P3. Find the x- and y-intercepts of the graph of $y = x^2 - 3x - 2$. Round the intercepts to the nearest hundredth. **Example 3**

P4. The height y, in feet, of a baseball t seconds after being thrown upward can be approximated by $y = -16t^2 + 96t + 5$. How many seconds after the ball is released is it 20 feet above the ground? Round to the nearest tenth of a second. **Example 4**

P5. Determine whether $y = -\dfrac{x^2}{2} + 4x - 6$ has a maximum or minimum. Then determine the maximum or minimum. **Example 5**

P6. The cables on a suspension bridge can be modeled by $h = 0.0088x^2 - 0.88x + 40$, where h is the height in feet of the cable above the roadway x feet from the start of the bridge. What is the minimum height of the cables above the roadway? **Example 6**

P7. A courtyard at the corner of two buildings is to be enclosed using 100 ft of redwood fencing. Find the dimensions of the courtyard that will maximize the area. **Example 7**

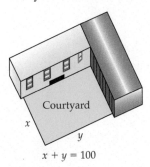

$x + y = 100$

P8. The demand function for a child's tricycle is given by $p = 137 - 4x$, where p is the price in dollars at which x tricycles can be sold. The cost function to produce tricycles is $C = 25x + 200$. **Example 8**
   a. Find the revenue function.
   b. Find the profit function.
   c. What is the maximum profit?

Section Exercises

In Exercises 1 to 10, determine whether the graph opens up or down and find the vertex of the graph.

1. $f(x) = x^2 - 10x$
2. $f(x) = x^2 - 6x$
3. $f(x) = x^2 - 10$
4. $f(x) = x^2 - 4$
5. $f(x) = -x^2 + 6x + 1$
6. $f(x) = -x^2 + 4x + 1$
7. $f(x) = 2x^2 - 3x + 7$
8. $f(x) = 3x^2 - 10x + 2$
9. $f(x) = -4x^2 + x + 1$
10. $f(x) = -5x^2 - 6x + 3$

In Exercises 11 to 20, find the maximum or minimum of the function. State whether this value is a maximum or minimum.

11. $y = x^2 + 8x$
12. $y = -x^2 - 6x$
13. $y = 3x^2$
14. $y = -x^2$
15. $y = -x^2 + 8x - 1$
16. $y = x^2 + 10x - 3$
17. $s = 2t^2 - 8t - 1$
18. $v = -2z^2 - 4z + 3$
19. $m = -2n^2 + 10n + 4$
20. $y = 2p^2 + 6p - 5$

21. **Astronaut Training** To prepare astronauts for the experience of zero gravity (technically, microgravity) in space, NASA uses a specially designed jet. A pilot accelerates the plane upward to an altitude of approximately 9000 meters and then reduces power. During the time of reduced power, the plane is in freefall and the astronauts experience microgravity. The altitude $A(t)$, in meters, of the plane t seconds after power was reduced can be approximated by $A(t) = -4.9t^2 + 90t + 9000$. The graph is shown as follows.

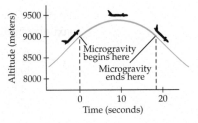

If the pilot increases power when the plane descends to 9000 meters, ending microgravity, find the time the astronauts experience microgravity during one of the maneuvers. Round to the nearest tenth of a second.

22. **Soccer Ball Kick** The height $h(t)$, in meters, above the ground of a certain soccer ball kick t seconds after the ball is kicked can be approximated by $h(t) = -4.9t^2 + 12.8t$. Determine the time for which the ball is in the air. Round to the nearest tenth of a second.

23. **Building a Garden Bed** A homeowner wants to build a raised rectangular garden bed. The homeowner has enough lumber to create a perimeter of 100 feet. What dimensions will create the garden bed of maximum area?

24. **Building a Fenced Enclosure** A rectangular enclosure for animals is fenced to produce three separate feeding areas as shown in the following diagram. If 800 feet of fencing is available, what dimensions of the enclosure will maximize the feeding area? What is the maximum area?

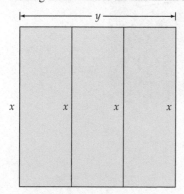

   a. Fill in the following equation.

      __ x + __ y = 800

   b. Solve the equation in part a for y.

      y = __ + __ x

   c. Using $A = x \cdot y$, write an equation for the area A in terms of x.

   d. What values of x and y will maximize the area of the enclosures?

   e. What is the maximum area?

25. **Softball Fields** A large lot in a park is going to be split into two softball fields, and each field will be enclosed with a fence. The parks and recreation department has 2100 feet of fencing to enclose the fields. What dimensions will enclose the greatest area?

   a. Fill in the following equation.

      __ x + __ y = 2100

   b. Solve the equation in part a for y.

      y = __ + __ x

   c. Using $A = x \cdot y$, write an equation for the area A in terms of x.

   d. What values of x and y will maximize the area of the enclosures?

   e. What is the maximum area?

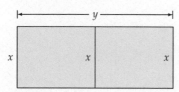

26. **Larvae Survival** Soon after insect larvae are hatched, they must begin to search for food. The survival rate of the larvae depends on many factors, but the temperature of the environment is one of the most important. For a certain species of insect, a model of the number of larvae, $N(T)$, that survive this searching period is given by $N(T) = -0.6T^2 + 32.1T - 350$ where T is the temperature in degrees Celsius.

   a. At what temperature will the maximum number of larvae survive? Round to the nearest degree.

   b. What is the maximum number of surviving larvae? Round to the nearest integer.

   c. Find the x-intercepts to the nearest integer for the graph of this function.

   d. Write a sentence that describes the meaning of the x-intercepts in the context of this problem.

27. **Temperature Fluctuations** The temperature $T(t)$, in degrees Fahrenheit, during the day can be modeled by the equation $T(t) = -0.7t^2 + 9.4t + 59.3$, where t is the number of hours after 6 A.M.

   a. At what time is the temperature a maximum? Round to the nearest hour.

   b. What is the maximum temperature? Round to the nearest degree.

28. **Geology** In June 2001, Mt. Etna in Sicily, Italy, erupted, sending volcanic bombs (masses of molten lava ejected from the volcano) into the air. A model of the height h, in meters, of a volcanic bomb above the crater of the volcano t seconds after the eruption is given by $h(t) = -9.8t^2 + 100t$. Find the maximum height of a volcanic bomb above the crater for this eruption. Round to the nearest meter.

29. **Projectile** If the initial velocity of a projectile is 128 feet per second, then its height in feet is a function of time t in seconds given by the equation $h(t) = -16t^2 + 128t$.

   a. Find the time t when the projectile achieves its maximum height.

   b. Find the maximum height of the projectile.

   c. Find the time t when the projectile hits the ground.

30. **Projectile** The height in feet of a projectile with an initial velocity 64 feet per second and an initial height of 80 feet is a function of time t in seconds given by

$$h(t) = -16t^2 + 64t + 80$$

   a. Find the maximum height of the projectile.

   b. Find the time t when the projectile achieves its maximum height.

   c. Find the time t when the projectile has a height of 0 feet.

31. **Economics** The demand function for small high-definition televisions is given by $p = 6000 - 45x$, where p is the price, in dollars, that x televisions can be sold. The cost function to produce televisions is $C = 1500x + 3000$.

   a. Find the revenue function.

   b. Find the profit function.

   c. Find how many televisions should be sold to maximize profit.

   d. What is the maximum profit?

32. **Economics** The demand function for golf carts is given by $p = 9000 - 50x$, where p is the price in dollars of x golf carts that can be sold. The cost function to produce golf carts is $C = 200x + 2000$.

   a. Find the revenue function.

   b. Find the profit function.

   c. Find how many golf carts should be sold to maximize profit.

   d. What is the maximum profit?

Investigation

Understanding Nonlinear Functions   A Norman window has the shape of a rectangle surmounted by a semicircle, as shown in the following figure. The exterior perimeter of the window in the figure is 48 feet. Find the height h and radius r that will maximize the area of the window. [*Suggestion:* The area of the window is the area of the rectangular portion of the window, plus the area of the semicircle. Area of the rectangle is $A_1 = 2r \cdot h$. The area of the semicircle is $A_2 = \frac{1}{2}\pi r^2$. The total area is $A = 2r \cdot h + \frac{1}{2}\pi r^2$. The perimeter of the window is $48 = 2h + 2r + \frac{1}{2}(2\pi r)$. Solve this equation for h and then substitute into the equation for A. Now use the strategies of this section to find the value of r that maximizes the area.]

Test Prep 6 — Nonlinear Functions

Section 6.1 — Exponential Functions

Exponential Function An **exponential function** is one that can be written in the form $y = b^x$, $b > 0$, $b \neq 1$. If $0 < b < 1$, the function is called an **exponential decay** function. If $b > 1$, the function is an **exponential growth** function.	See **Example 1** on page 263, and then try Exercises 1 to 4 on page 295.
Model for Exponential Growth Suppose a quantity is increasing at a rate of $r\%$ per unit time. Let A represent the amount of a quantity at some future time T and A_0 the amount when $T = 0$. Then $$A = A_0(1 + r)^T$$ is a model for exponential growth.	See **Examples 2** and **3** on pages 264 and 265, and then try Exercises 7 and 8 on page 295.
Compound Interest **Compound interest** is interest paid on the original principal and on any interest previously earned on the principal. It is sometimes referred to as "interest on interest."	See **Example 4** on pages 265 and 266, and then try Exercises 10 and 11 on page 295.
Model for Exponential Decay Suppose a quantity is decreasing at a rate of $r\%$ per unit time. Let A represent the amount of a quantity at some future time T and A_0 the amount when $T = 0$. Then $$A = A_0(1 - r)^T$$ is a model for exponential growth.	See **Example 6** on page 267, and then try Exercises 15 and 16 on page 295.
Half-life The **half-life** of a an event is the time it takes a quantity to decrease to one-half its current size.	See **Examples 7** and **8** on pages 268 and 269, and then try Exercises 18 and 19 on page 295.
Doubling time The **doubling time** of an event is the time it takes a quantity to decrease to one-half its current size.	See **Example 5** on page 267, and then try Exercises 13 and 14 on page 295.

Section 6.2 — Logarithmic Functions

Common Logarithm The **common logarithm** of a positive number a is the number b such that $10^b = a$. In this case, we write $\log a = b$. The **exponential form** of $\log a = b$ is $10^b = a$. The **logarithmic form** of $10^b = a$ is $\log a = b$.	See **Example 1** on page 273, and then try Exercises 20 and 21 on page 296.
pH Scale The pH of a solution is given by $\text{pH} = -\log[\text{H}^+]$ or $10^{-\text{pH}} = [\text{H}^+]$, where $[\text{H}^+]$ is the concentration of hydrogen ions in moles per liter.	See **Example 2** on page 274, and then try Exercises 22 and 23 on page 296.
Richter Scale The magnitude M of an earthquake that has an intensity that is I times the base intensity earthquake I_0 is given by $M = \log I$.	See **Examples 3** and **4** on pages 275 and 276, and then try Exercises 24 and 25 on page 296.
Decibel Scale Sound intensity level I is measured in decibels (dB) and is given by $\text{dB}(I) = 10 \log\left(\dfrac{I}{I_0}\right)$, where I_0 is the standard intensity threshold for a sound to be heard.	See **Examples 5** and **6** on pages 276 and 277, and then try Exercises 26 and 27 on page 296.
Solution of the Exponential Equation $A = A_0 p^t$ The solution of $A = A_0 p^t$ for t is $t = \dfrac{\log\left(\dfrac{A}{A_0}\right)}{\log p}$, where $p = 1 + r$ for exponential growth and $p = 1 - r$ for exponential decay and r is the rate of exponential growth or decay.	See **Example 7** on page 278, and then try Exercises 28 and 29 on page 296.

Half-life Equation and Its Solution Let A be the amount of a substance after time t, A_0 the initial amount of a substance, and h the half-life of the process. Then **Half-life equation:** $A = A_0\left(\dfrac{1}{2}\right)^{t/h}$ **Solution of the half-life equation for half-life h:** $$h = \frac{t \cdot \log\left(\dfrac{1}{2}\right)}{\log\left(\dfrac{A}{A_0}\right)}$$	See **Example 8** on page 279, and then try Exercise 30 on page 296.
Doubling Time Equation and Its Solution Let A be the amount of a substance after time t, A_0 the initial amount of a substance, and h the doubling time of the process. Then **Doubling time equation:** $A = A_0(2)^{t/h}$ **Solution of the doubling time equation for doubling time d:** $$d = \frac{t \cdot \log(2)}{\log\left(\dfrac{A}{A_0}\right)}$$	See **Examples 9** and **10** on pages 279 and 280, and then try Exercise 31 on page 296.

Section 6.3 Quadratic Functions

Quadratic Function A **quadratic function** is one that can be expressed by the equation $y = ax^2 + bx + c$, where a, b, and c are real numbers and $a \neq 0$.	See **Example 1** on page 283, and then try Exercise 32 on page 296.
Vertex Formula The x-coordinate of the vertex of $y = ax^2 + bx + c$ is given by $x = -\dfrac{b}{2a}$. To find the y-coordinate of the vertex, evaluate $y = ax^2 + bx + c$ at the x-coordinate of the vertex.	See **Example 2** on page 284, and then try Exercise 33 on page 296.
Intercepts of the Graph of a Quadratic Function The x-intercepts of the graph of an equation occur when $y = 0$; the y-intercepts occur when $x = 0$. The quadratic formula can be used to find the x-intercepts of a quadratic equation. $$x = \frac{-b \pm \sqrt{b^2 - 4ac}}{2a}$$	See **Examples 3** and **4** on pages 285 and 286, and then try Exercise 35 on page 296.
Minimum or Maximum of a Quadratic Function The vertex of a parabola that opens up is the lowest point on the parabola. It is the point that has the minimum y-coordinate. That y-coordinate is the minimum value of the function. The vertex of a parabola that opens down is the highest point on the parabola. It is the point that has the maximum y-coordinate. That y-coordinate is the maximum value of the function.	See **Examples 5, 6, 7,** and **8** on pages 287 through 289, and then try Exercises 36, 37, 38, 39, and 40 on page 296.

Review Exercises

For Exercises 1 to 4, identify whether the equation represents exponential growth, exponential decay, or neither.

1. $y = 0.6(1.025)^x$

2. $s = 7.5(0.9)^{t/5}$

3. $y = \dfrac{1}{2}x^2$

4. $y = \dfrac{4}{3} \cdot 2^{5x}$

5. Frequency The frequency f in vibrations per second of the sound of a struck piano key can be expressed as the exponential equation $f = 27.5 \cdot 4^{(n-1)/24}$, where n is the number of the key on the standard 88-key piano, with the leftmost key being key 1.

   a. What is the frequency of middle C, the 40th key on the keyboard?

   b. If the ratio of the frequencies of two keys is 2:1, the span of keys is called an *octave*. Show that the ratio of middle C (the 40th key) and the next C above middle C (key 52) is 2:1.

6. Powering Satellites A radioactive isotope is used as a power supply for certain satellites. The equation for the number of watts produced by the power supply is given by $E = 25(0.997)^{t/24}$, where t is the number of hours the power supply has been operating. How many watts will the power supply produce 20 days (480 hours) after it is activated?

7. Bacterial Growth Initially, a bacterial colony contained 50,000 cells. One hour later, the colony had 60,000. Assuming that this percentage increase continues, write an exponential function for the number of bacteria in the colony t hours after the initial observation. Use the model to determine the number of bacteria in the colony 3.5 hours after the initial observation.

8. Predicting Height For approximately the first 24 months of a child's life, they grow approximately 5% taller each month. Write an exponential function for the growth of a child t months after birth if the birth height of the child is 50 centimeters. Use the model to determine the height of the child after 12 months.

9. Predicting Weight In the first 3 years of a child's life, the child's weight will double approximately every 15 months. Assume a child has a birth weight of 7.2 pounds. Find an exponential function for the weight of the child t months after birth. Use the model to predict the child's weight 3 years after birth.

10. Investment Growth A dog trainer deposits $5000 into an account that earns 5% annual interest compounded monthly. Find an exponential growth model for the value of the account after t years. What is the value of the account after 7 years?

11. Investment Growth A bank offers an account that earns 3.75% annual interest compounded daily. Assume an investor opens one of these accounts with a deposit of $10,000. Write an exponential function for the value of the account t years after it is opened. What is the value of the account in 8 years?

12. Social Media Advertising An advertising company retweeted an advertising piece as a way to reach potential customers. When the company first measured the effectiveness of the campaign, there were 2000 tweets for the product, and the number of retweets was increasing at the rate of 12% every 4 hours.

   a. Write an exponential expression for the number of people who have retweeted as a function of the number of hours after the initial measurement.

   b. Assume this rate continues and use your model to determine how many people have received the tweet after 40 hours. Round to the nearest whole number.

13. Bacterial Growth *Streptococcus lactis* is a bacterium used in the production of some cheeses and buttermilk and has a doubling time of 30 minutes. If a sample of *Streptococcus lactis* original contains 25,000 bacteria, find a model for the growth of the bacteria. Use the model to find the amount of *Streptococcus lactis* in the sample in 75 minutes. Round to the nearest thousand.

14. Baking Bread *Saccharomyces cerevisiae* is a yeast used when baking bread and has a doubling time of 90 minutes. If 6 grams of *Streptococcus lacitis* is added to a batch of bread dough, find a model for the growth of the yeast. Use the model to find the grams of *Saccharomyces cerevisiae* in the dough in 2 hours. Round to the nearest whole number.

15. Water Mixtures A solution of salt and water originally contains 25 pounds of salt. Pure water is being added to the tank and mixed with the saltwater solution, and then some of the saltwater solution is drained from the tank. The result is that the number of pounds of salt in the tank is decreasing at a rate of 30% of the current amount of salt in the tank each 30 minutes. Find an exponential decay model for the amount of salt in the tank t minutes after the process begins. Use your model to find the number of pounds of salt in the solution 2 hours after the process begins.

16. Disinfecting a Room One process used to disinfect airborne pathogens in an enclosed room is called *continuous ultraviolet germicidal irradiation* with ultraviolet light used to kill pathogens. A certain room originally contains 500,000 pathogens when an ultraviolet light is turned on. Measurements of the number of pathogens t seconds later shows that the number is decreasing at a rate of 5% every 5 seconds. Write an exponential function for the number of pathogens t seconds after the light is turned on. Use the model to predict the number of pathogens after 60 seconds. Round to the nearest thousand.

17. Bouncing a Ball A high-elasticity rubber ball is dropped from a height of 10 feet. The ball rebounds to a height of 90% of its current height on each bounce. Find an exponential function for the height of the ball after the nth bounce.

18. Nuclear Fission Strontium-90, ^{90}Sr, is a byproduct of nuclear fission and has a half-life of approximately 28.8 years. If a sample originally contains 75 g of ^{90}Sr, find an exponential function for the number of grams of ^{90}Sr t years after the measurement of the sample. How many grams will remain after 50 years? Round to the nearest tenth.

19. DDT Dichlorodiphenyltrichloroethane (DDT) is an insecticide that has been banned in the United States because of its environmental impacts, but it is still used in some parts of the world. The half-life of DDT is approximately 15 years. Assume 50 g of DDT is sprayed on a field. Find an exponential function for the number of grams of DDT t years after it has been sprayed. How many grams of DDT remain after 6 months? Round to the nearest tenth.

20. Solve for x.

a. $\log x = 4$

b. $\log x = -3$

c. $\log(10,000,000) = x$

d. $\log\left(\dfrac{1}{10,000}\right) = x$

21. Solve for x. If necessary, round to the nearest ten-thousandth.

a. $\log x = \dfrac{4}{5}$

b. $\log x = -1.2$

c. $\log 23 = x$

d. $\log x = 0.15$

22. **Maple Syrup** The hydrogen ion concentration for a maple syrup is 0.0000002 mole per liter. What is the pH of the maple syrup? Round to the nearest tenth. Is the syrup an acid or a base?

23. **Carrot Juice** The pH of a carrot juice is 5.2. What is the hydrogen ion concentration of the carrot juice in moles per liter?

24. **Earthquake Magnitude** Find the magnitude of an earthquake that is 800,000 times the intensity of the base intensity earthquake. Round to the nearest tenth.

25. **Earthquake Magnitude** One earthquake measured 7.7 on the Richter scale and a second earthquake measured 4.3. How many times more intense was the magnitude 7.7 than the magnitude 4.3 earthquake? Round to the nearest whole number.

26. **Blue Whales** A blue whale can make a sound with an intensity that is 6.3×10^{18} times the threshold intensity. Find the decibel level of the sound. Round to the nearest whole number.

27. **Meows Versus Barks** A typical cat meow has a sound intensity level of 45 decibels, whereas a typical dog bark is about 90 decibels. Is the intensity of a dog's bark twice that of a cat's meow? If not, how many more times is the intensity of the dog's bark than the cat's meow? Round to the nearest whole number.

28. **Making a Vacuum Chamber** To prepare a vacuum chamber for an experiment, a pump is attached to the chamber to extract the air in the chamber. The pump extracts 30% of the available air every 50 seconds. When the pump is started, there are 5000 liters of air. Write an exponential decay function for the amount of air in the chamber t seconds after the pump is turned on. How many seconds are required to reduce the amount of air in the chamber to 100 liters? Round to the nearest second.

29. **Value of a Sculpture** The original selling price of a sculpture was $5000. Over the years, as the sculptor became more famous, the value of the sculpture increased about 7% each year. Assuming this trend continues, in how many years will the sculpture be worth $7500? Round to the nearest whole number.

30. **Insulin Therapy** To test the effectiveness of a new insulin therapy, a patient is given 25 milligrams of the new drug. Four hours later, 20 milligrams of the drug were in the patient's bloodstream. What is the half-life, in hours, of the drug? Round to the nearest tenth.

31. **Investment Growth** Suppose the current rate on a savings account is 2.1% compounded semiannually. If $5000 is placed in this account today, how many years will it take for the value of the account to double in value? Round to the nearest year.

32. Determine whether the equation represents a quadratic function. If so, state the values of a, b, and c.

a. $y = \dfrac{2}{x^2} - \dfrac{3}{x} + 4$

b. $s = \dfrac{1}{2}t^2 + 3$

c. $y = 2x - 3$

d. $y = 1 - x - x^2$

33. Determine whether the graph of the equation opens up or down. Then find the coordinates of the vertex.

a. $y = x^2 - 8x - 6$

b. $y = -x^2 + 3x$

c. $y = -2z^2 - 4z - 3$

d. $y = 3 - x + x^2$

34. Find the x- and y-intercepts of the graph of the equation.

a. $y = x^2 - 2x - 8$

b. $y = -x^2 + 3x + 18$

35. **Football** Assuming no air resistance, the height y of a football punt t seconds after being kicked can be approximated by $y = -16t^2 + 64.952t + 5$. If the football is allowed to hit the ground ($y = 0$), what is the "hang time" for the punt? Round to the nearest tenth of a second.

36. **Football** Assuming no air resistance, the height y of a football punt x feet from the spot it is kicked can be approximated by $y = -0.0113778x^2 + 1.73205x + 5$. If the football is allowed to hit the ground ($y = 0$), how many feet will the ball have traveled? Round to the nearest whole number.

37. Determine whether the graph has a maximum or a minimum. Then determine the maximum or minimum value.

a. $y = x^2 + 4x - 1$

b. $y = -2x^2 + 6x + 3$

38. **Height of a Throw** A ball is thrown vertically upward with an initial velocity of 64 feet per second. If the ball started its flight at a height of 6 feet, then its height s in feet at t seconds after it is released can be determined by $s = -16t^2 + 64t + 6$.

a. Determine the time it takes the ball to attain its maximum height.

b. What is the maximum height attained by the ball?

39. **Enclosing a Garden** Parker has 72 feet of fencing to surround a garden that is bounded on one side by the wall of a garden shed. What are the dimensions of the largest rectangular area that Parker can enclose?

40. **Economics** The demand function for a washing machine is given by $p = 2700 - 30x$, where p is the price in dollars that x washing machines can be sold. The cost function to produce washing machines is $C = 600x + 2200$. Find the number of machines that should be sold to maximize profit. What is the maximum profit?

Project I

Wildlife Populations

We mentioned earlier that an exponential growth model for population growth is unsustainable because natural resources will eventually run out. A population model that considers that population growth will be curbed by the lack of natural resources is called a **logistic growth model**. One form of this model is given by

$$P(t) = \frac{c}{1 + a \cdot 2^{-kt}}$$

where c is the **carrying capacity** (the maximum population that can be supported by available resources as time t increases, and k is a positive constant called the *growth rate constant*). The **initial population** is $P_0 = P(0)$. The constant a is related to the initial population P_0 and the carrying capacity c by the formula

$$a = \frac{c - P_0}{P_0}$$

Consider a wildlife preserve where the manager is interested in the growth of the elephant population. At the beginning of 2016, the elephant population in the preserve was estimated at 200. The preserve manager estimates that the carrying capacity of the preserve is 500 elephants and the growth constant is 0.3.

1. Use the information to find a.

2. Write the logistic model for this elephant population.

3. Use the logistic model to predict the number of elephants in the preserve in 2020.

4. Find the expected average rate of change of growth of the elephant population for each of the following time periods.

   a. Between 2016 and 2021.

   b. Between 2021 and 2026.

   c. Between 2026 and 2031.

   d. Between 2031 and 2036.

5. Between 2016 and 2036, does the average rate of change in growth of the elephant population increase or decrease?

6. Assuming this model is good for many years, what will the elephant population be in 2066?

7. Explain why this model shows that the carrying capacity is 500 elephants.

Project 2

Biological Diversity

Not all logarithms are base 10. There are many applications of logarithms base 2. In this case,

$$\log_2 x = y \quad \text{is equivalent to} \quad 2^y = x$$

For instance, $\log_2 8 = 3$ because $2^3 = 8$

Unfortunately, calculators do not generally have a logarithm base 2 key. Fortunately, there is a way to use the **LOG** key to find base two logarithms using the following formula:

$$\log_2 x = \frac{\log x}{\log 2}$$

For instance, $\log_2 17 = \dfrac{\log 17}{\log 2} \approx \dfrac{1.2304}{0.3010} \approx 4.0875$

To discuss the variety of species that live in a certain environment, a biologist needs a precise definition of *diversity*. Let $p_1, p_2, \ldots, p_n$ be the proportions of n species that live in an environment. The biological diversity, D, of this system is

$$D = -(p_1 \log_2 p_1 + p_2 \log_2 p_2 + \cdots + p_n \log_2 p_n)$$

The larger the value of D, the greater the diversity of the system. Suppose an ecosystem has exactly five different varieties of grass: rye (R), Bermuda (B), blue (L), fescue (F), and St. Augustine (A).

1. Calculate the diversity of this ecosystem if the proportions are as shown in Table 1.

2. Because Bermuda and St. Augustine are virulent grasses, after a time the proportions are as shown in Table 2. Does this system have more or less diversity than the one given in Table 1?

3. After an even longer period, the Bermuda and St. Augustine completely overrun the environment, and the proportions are as in Table 3. Calculate the diversity of this system. (*Note:* for purposes of the diversity definition, $0 \log_2 0 = 0$.) Does it have more or less diversity than the system given in Table 2?

4. Finally, the St. Augustine overruns the Bermuda, and the proportions are as in Table 4. Calculate the diversity of this system. Write a sentence that describes your answer.

Table 1

R	B	L	F	A
$\frac{1}{5}$	$\frac{1}{5}$	$\frac{1}{5}$	$\frac{1}{5}$	$\frac{1}{5}$

Table 2

R	B	L	F	A
$\frac{1}{8}$	$\frac{3}{8}$	$\frac{1}{16}$	$\frac{1}{8}$	$\frac{5}{16}$

Table 3

R	B	L	F	A
0	$\frac{1}{4}$	0	0	$\frac{3}{4}$

Table 4

R	B	L	F	A
0	0	0	0	1

Math in Practice

Zheltyshev/Shutterstock.com

In music, an instrument is kept in tune by adjusting the instrument's ability to play at certain frequencies. Tuning an instrument usually happens before any kind of concert or performance. Of the many different kinds of instruments, the piano has keys and notes that are easy to identify. There are 88 keys on a piano. When a piano key is played, a sound is produced with a very specific frequency. Some keys are white and some are black, and they follow a very special pattern. The first note on a piano is at the left end of the piano, a white key, also called A0, which has a frequency of 27.5 Hertz. The 88th note is at the right end of the piano, a white key, also called C8, which has a frequency of 4186 Hertz. Starting at the left end of the piano, the white keys have note names A0, B0, C1, D1, E1, F1, G1, A1, B1, and then C2. This pattern repeats until note name C8. There are five black keys from C1 and C2: one between C1 and D1, one between D1 and E1, one between F1 and G1, one between G1 and A1, and one between A1 and B1. A semitone is an interval between two adjacent notes. From C1 to C2 there are 12 semitones, which is also called an octave.

Part 1

The relationship between a note's frequency, in Hertz, and a specific number of semitones above or below that note is calculated using a logarithmic formula. Inversely, we can calculate the number of semitones from a note's frequency using an exponential formula. Before a concert, a guitar's strings must be correctly tuned, which means playing at a very specific frequency. Let's find some of those frequencies.

a. Suppose we know that the frequency for A3 is 220.0 Hertz. Using the formula $f(n) = (220.0)2^{n/12}$, where n is the number of semitones above A3, find the frequency of D4, which is five semitones above A3. Round to the hundredth.

b. Suppose we know that the frequency for A4 is 440.0 Hertz. Find the frequency of G4, which is two semitones below A4. Using the formula $f(n) = (440.0)2^{n/12}$, where n is the number of semitones above A4, if n is positive. If n is negative, then n represents the number of semitones below A4. Round to the hundredth.

c. Suppose we know that the frequency for A4 is 440.0 Hertz. Using the formula

$$n = \frac{12 \log\left(\dfrac{f}{440}\right)}{\log 2}$$

where n is the number of semitones above A4, find the number of semitones above A4, which has frequency 659.26 Hertz.

d. Suppose we know that the frequency for A4 is 440.0 Hertz. Using the formula

$$n = \frac{12 \log\left(\dfrac{f}{440}\right)}{\log 2}$$

where n is the number of semitones above A4, find the number of semitones above A4, which has frequency 246.94 Hertz. If n is negative, then n represents the number of semitones below A4.

Part 2

With electronic music, a MIDI number is given for each pitch. For example, for the piano key A0, the MIDI number is 21. For the piano key A4, the MIDI number is 69.

To find M, the MIDI number for a given f, the frequency:

$$M = \frac{12 \log\left(\dfrac{f}{440}\right)}{\log 2} + 69$$

To find f, the frequency in Hertz, for a given M, the MIDI number:
$f_M = (440)2^{(M-69)/12}$

Suppose you have an electronic keyboard and you want to tune a bass guitar before a concert.

a. Find the frequencies for MIDI numbers 38 and 43. Round to the ten-thousandth.

b. Find the MIDI numbers for frequencies 82.407 Hertz and 110.000 Hertz.

Analyzing Data

Introduction to Probability

Section	How This Is Used . . .
7.1 The Counting Principle	If you have ever purchased a lottery ticket, you may have wondered how many different lottery tickets are possible. For the Powerball lottery in 2018, there are 292,201,338. That number is found by using the counting principle.
7.2 Permutations and Combinations	Consider the letters S T A H Y. There are 120 *permutations* (arrangements) of those letters but only one leads to the word HASTY. Now suppose you are dealt two cards, the ace of diamonds and the king of hearts. The arrangements ace then king or king then ace are essentially the same. The order in which the cards are received does not matter. This is called a *combination*. There are many other applications of these two concepts.
7.3 Probability and Odds	When a meteorologist reports that there is a 25% chance of rain, the report is a probability. In the past, when weather patterns were as they are today, it rained 25% of the time. If we say there is a 50-50 chance a team will win a football game, we are stating the odds of winning. In this case, the odds are a toss-up; either team is equally like to win.
7.4 Addition and Complement Rules	Some probability events are difficult to compute and require some helpful shortcuts that are provided by probability addition and complement rules.
7.5 Conditional Probability	Some probability problems depend on a previous event. An example is determining the probability of getting the flu given that you got a flu shot last week. This is a conditional probability.
7.6 Expectation	The cost of car insurance (and all other types of insurance) is based on the insurance company's *expectation* they will have to pay a claim and the amount of that claim.

Math in Practice:
Fundraiser, p. 353

Applications of Combinatorics and Probability

The Human Genome Project, a project to completely map the genetic make-up of *Homo sapiens*, was completed in 2003. In Example 5 on p. 322, we show a calculation of the probability a child may have cystic fibrosis based on the genotype of the parents.

The Counting Principle

Learning Objectives:

- Count the number of outcomes of an experiment.
- Use the counting principle.
- Count the outcomes of an event with or without replacement.

Be Prepared: Prep Test 7 reviews concepts used in this chapter. This can be found in your Companion Site at Cengage.com.

Counting by Making a List

Combinatorics is the study of counting the different outcomes of some task. For example, if a coin is flipped, the side facing upward will be a head or a tail. The outcomes can be listed as {H, T}. There are two possible outcomes.

If a regular six-sided die is rolled, the possible outcomes are ⚀, ⚁, ⚂, ⚃, ⚄, ⚅. The outcomes can also be listed as {1, 2, 3, 4, 5, 6}. There are six possible outcomes.

Example 1   **Counting by Forming a List**

List and then count the number of different outcomes that are possible when one letter from the word *Tennessee* is chosen.

Solution

The possible outcomes are {T, e, n, s}. There are four possible outcomes.

In combinatorics, an **experiment** is an activity with an observable outcome. The set of all possible outcomes of an experiment is called the **sample space** of the experiment. Flipping a coin, rolling a die, and choosing a letter from the word *Tennessee* are experiments. The sample spaces are {H, T}, {1, 2, 3, 4, 5, 6}, and {T, e, n, s}, respectively.

An **event** is one or more of the possible outcomes of an experiment. Flipping a coin and having a head show on the upward face, rolling a 5 when a die is tossed, and choosing a T from one of the letters in the word *Tennessee* are all examples of events. An event is a *subset* of the sample space.

Example 2   **Listing the Elements of an Event**

One number is chosen from the sample space

$S = \{1, 2, 3, 4, 5, 6, 7, 8, 9, 10, 11, 12, 13, 14, 15, 16, 17, 18, 19, 20\}$

List the elements in the following events.
a. The number is even.
b. The number is divisible by 5.
c. The number is a prime number.

Solution
a. {2, 4, 6, 8, 10, 12, 14, 16, 18, 20}
b. {5, 10, 15, 20}
c. {2, 3, 5, 7, 11, 13, 17, 19}

Counting by Making a Table

Each of the experiments given previously illustrates a *single-stage experiment*. A **single-stage experiment** is an experiment for which there is a single outcome. Experiments that have two, three, or more stages are called **multi-stage experiments**. To count the number of outcomes of such an experiment, a systematic procedure is helpful. Using a table to record results is one such procedure.

Consider the two-stage experiment of rolling two dice, one white and one blue. How many different outcomes are possible? To determine the number of outcomes, make a table with the different outcomes of rolling the white die across the top and the different outcomes of rolling the blue die down the side.

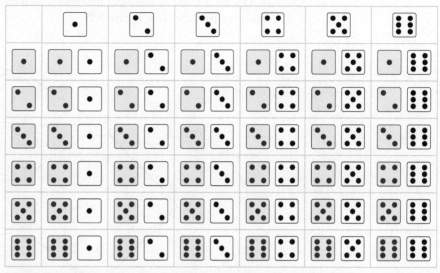

Figure 7.1 Outcomes of rolling two dice

By counting the number of entries in Figure 7.1, we see that there are 36 different outcomes of the experiment of rolling two dice. The sample space is

$$\{\boxed{\cdot}\,\boxed{\cdot}, \boxed{\cdot\cdot}\,\boxed{\cdot}, \boxed{\cdot\cdot}\,\boxed{\cdot}, \dots, \boxed{\cdot\cdot}\,\boxed{::}, \boxed{::}\,\boxed{::}\}$$

From Figure 7.1, several different events can be discussed.

- The sum of the pips (dots) on the upward faces is 7. There are six outcomes of this event. They are $\{\boxed{\cdot}\,\boxed{::}, \boxed{\cdot\cdot}\,\boxed{::}, \boxed{\cdot\cdot}\,\boxed{\cdot\cdot}, \boxed{::}\,\boxed{\cdot\cdot}, \boxed{::}\,\boxed{\cdot\cdot}, \boxed{::}\,\boxed{\cdot}\}$.
- The sum of the pips on the upward faces is 11. There are two outcomes of this event. They are $\{\boxed{::}\,\boxed{::}, \boxed{::}\,\boxed{::}\}$.
- The numbers of pips on the upward faces are equal. There are six outcomes of this event. They are $\{\boxed{\cdot}\,\boxed{\cdot}, \boxed{\cdot\cdot}\,\boxed{\cdot\cdot}, \boxed{\cdot\cdot}\,\boxed{\cdot\cdot}, \boxed{::}\,\boxed{::}, \boxed{::}\,\boxed{::}, \boxed{::}\,\boxed{::}\}$.

Example 3   Counting Using a Table

Two-digit numbers are formed from the digits 1, 3, and 8. Find the sample space and determine the number of elements in the sample space.

Solution

Use a table to list all the different two-digit numbers that can be formed by using the digits 1, 3, and 8.

	1	3	8
1	11	13	18
3	31	33	38
8	81	83	88

The sample space is {11, 13, 18, 31, 33, 38, 81, 83, 88}. There are nine two-digit numbers that can be formed from the digits 1, 3, and 8.

Counting by Using a Tree Diagram

A **tree diagram** is another way to organize the outcomes of a multi-stage experiment. To illustrate the method, consider a computer store offering special prices on its most popular laptop models. A customer can choose from two sizes of RAM, three screen sizes, and two preloaded application packages. How many different laptops can customers choose?

We can organize the information by letting M_1 and M_2 represent the two sizes of RAM; S_1, S_2, and S_3 represent the three screen sizes; and A_1 and A_2 represent the two application packages (see Figure 7.2).

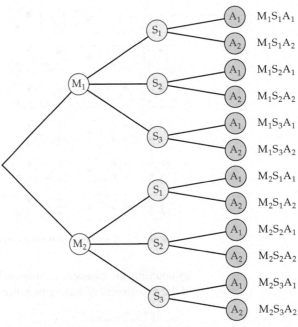

Figure 7.2 There are 12 possible laptops.

| Example 4 | **Counting Using a Tree Diagram** |

A true/false test consists of 10 questions. Draw a tree diagram to show the number of ways to answer the first three questions.

Solution

See the tree diagram in Figure 7.3. There are eight possible ways to answer the first three questions.

Question 1 Question 2 Question 3

T
- T
  - T → TTT
  - F → TTF
- F
  - T → TFT
  - F → TFF

F
- T
  - T → FTT
  - F → FTF
- F
  - T → FFT
  - F → FFF

Figure 7.3

The Counting Principle

For each of the previous problems, the possible outcomes were listed and then counted to determine the number of different outcomes. However, it is not always possible or practical to list and count outcomes. For example, the number of different 5-card poker hands that can be drawn from a standard deck of 52 playing cards is 2,598,960. Trying to create a list of these hands would be quite time consuming.

Consider again the problem of selecting a laptop. By using a tree diagram, we listed the 12 possible laptops. Another way to arrive at this result is to find the product of the numbers of choices available for RAM sizes, screen sizes, and application packages.

$$\begin{bmatrix} \text{number of} \\ \text{RAM sizes} \end{bmatrix} \times \begin{bmatrix} \text{number of} \\ \text{screen sizes} \end{bmatrix} \times \begin{bmatrix} \text{number of} \\ \text{application packages} \end{bmatrix} = \begin{bmatrix} \text{number of} \\ \text{laptops} \end{bmatrix}$$
$$2 \quad\times\quad 3 \quad\times\quad 2 \quad=\quad 12$$

For the example of tossing two dice, there were 36 possible outcomes. We can arrive at this result without listing the outcomes by finding the product of the number of possible outcomes of rolling the white die and the number of possible outcomes of rolling the blue die.

$$\begin{bmatrix} \text{outcomes} \\ \text{of white die} \end{bmatrix} \times \begin{bmatrix} \text{outcomes} \\ \text{of blue die} \end{bmatrix} = \begin{bmatrix} \text{number of} \\ \text{outcomes} \end{bmatrix}$$

$$\quad\quad 6 \quad\quad \times \quad\quad 6 \quad\quad = \quad\quad 36$$

This method of determining the number of outcomes of a multi-stage experiment without listing them is called the **counting principle**.

Counting Principle

Let E be a multi-stage experiment. If $n_1, n_2, n_3, \ldots, n_k$ are the number of possible outcomes of each of the k stages of E, then there are $n_1 \cdot n_2 \cdot n_3 \cdot \ldots \cdot n_k$ possible outcomes for E.

Example 5   **Counting by Using the Counting Principle**

In horse racing, betting on a *trifecta* refers to choosing the exact order of the first three horses across the finish line. If there are eight horses in a race, how many trifectas are possible, assuming there are no ties?

Solution

Any one of the eight horses can be first, so $n_1 = 8$. Because a horse cannot finish both first and second, there are seven horses that can finish second; thus $n_2 = 7$. Similarly, there are six horses that can finish third; $n_3 = 6$. By the counting principle, there are $8 \cdot 7 \cdot 6 = 336$ possible trifectas.

Counting With and Without Replacement

Consider an experiment in which three balls colored red, blue, and green are placed in a box. A person reaches into the box and repeatedly pulls out a colored ball, keeping note of the color picked. The sequence of colors will depend on whether the balls are returned to the box after each pick. We say that the experiment can be performed *with replacement* or *without replacement*.

Consider the following two situations.

1. How many four-digit numbers can be formed from the digits 1 through 9 if no digit can be repeated?
2. How many four-digit numbers can be formed from the digits 1 through 9 if a digit can be used repeatedly?

In the first case, there are nine choices for the first digit ($n_1 = 9$). Because a digit cannot be repeated, the first digit chosen cannot be used again. Thus there are only eight choices for the second digit ($n_2 = 8$). Because neither of the first two digits can be used as the third digit, there are only seven choices for the third digit ($n_3 = 7$). Similarly, there are six choices for the fourth digit ($n_4 = 6$). By the counting principle, there are $9 \cdot 8 \cdot 7 \cdot 6 = 3024$ four-digit numbers in which no digit is repeated.

In the second case, there are nine choices for the first digit ($n_1 = 9$). Because a digit can be used repeatedly, the first digit chosen can be used again. Thus there are nine choices for the second digit ($n_2 = 9$) and, similarly, nine choices for the third and fourth digits ($n_3 = 9, n_4 = 9$). By the counting principle, there are $9 \cdot 9 \cdot 9 \cdot 9 = 6561$ four-digit numbers when digits can be used repeatedly.

The set of four-digit numbers created without replacement includes numbers such as 3867, 7941, and 9128. In these numbers, no digit is repeated. However, numbers such as 6465, 9911, and 2222, each of which contains at least one repeated digit, can be created only with replacement.

Example 6 — Counting With and Without Replacement

From the letters a, b, c, d, and e, how many four-letter groups can be formed if

a. a letter can be used more than once?

b. each letter can be used exactly once?

Solution

a. Because each letter can be repeated, there are $5 \cdot 5 \cdot 5 \cdot 5 = 625$ possible four-letter groups.

b. Because each letter can be used only once, there are $5 \cdot 4 \cdot 3 \cdot 2 = 120$ four-letter groups in which no letter is repeated.

7.1 Exercise Set

▶ Think About It

1. In combinatorics, what is an experiment?
2. In combinatorics, what is a sample space?
3. What is the relationship between an event and a sample space?
4. Does a multi-stage experiment performed with replacement generally have more or fewer possible outcomes than the same experiment performed without replacement?

▶ Preliminary Exercises

For Exercises P1 to P6, if necessary, use the referenced example following the exercise for assistance.

P1. List and then count the number of different outcomes that are possible when one letter is chosen from the word *Mississippi*. **Example 1**

P2. One digit is chosen from the digits 0 through 9. The sample space S is $\{0, 1, 2, 3, 4, 5, 6, 7, 8, 9\}$. List the elements in the following events. **Example 2**

    a. The number is odd.

    b. The number is divisible by 3.

    c. The number is greater than 7.

P3. A die is tossed and then a coin is flipped. Find the sample space and determine the number of elements in the sample space. **Example 3**

P4. A two-letter code is formed from the letters a, b, and c. Draw a tree diagram to determine the elements of the sample space. Assume the same letter can occur in the two-letter code. **Example 4**

P5. Nine runners are entered in a 100-meter dash for which a gold, silver, and bronze medal will be awarded for first, second, and third place finishes, respectively. In how many possible ways can the medals be awarded? (Assume that there are no ties.) **Example 5**

P6. In how many ways can three awards be given to five students if

    a. each student may receive more than one award?

    b. each student may receive no more than one award? **Example 6**

▶ Section Exercises

In Exercises 1 to 10, list the elements of the sample space defined by each experiment.

1. Select an even single-digit whole number.
2. Select an odd single-digit whole number.
3. Select one day from the days of the week.
4. Select one month from the months of the year.
5. Toss a coin twice.
6. Toss a coin three times.
7. Roll a single die and then toss a coin.
8. Toss a coin and then choose a digit from the digits 1 through 4.
9. Choose a complete dinner from a dinner menu that allows a customer to choose from two salads, three entrees, and two desserts.
10. Choose a car during a new car promotion that allows a buyer to choose from three body styles, two radios, and two interior color schemes.

For Exercises 11 and 12, list the sample space of paths that start at *A* and pass through each vertex of the figure exactly once.

11. Square *ABCD*

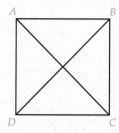

12. Pentagon *ABCDE*

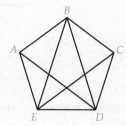

In Exercises 13 to 18, use the counting principle to determine the number of elements in the sample space.

13. Two digits are selected without replacement from the digits 1, 2, 3, and 4.

14. Two digits are selected with replacement from the digits 1, 2, 3, and 4.

15. The possible ways to complete a multiple-choice test consisting of 20 questions, with each question having four possible answers (a, b, c, or d).

16. The possible ways to complete a true–false examination consisting of 25 questions.

17. The possible four-digit telephone number extensions that can be formed if 0, 8, and 9 are excluded as the first digit.

18. The possible six-character passwords that can be formed using a letter from a to h and five numbers from 1 to 9. Assume a letter or number cannot be used more than once in any six-character password.

In Exercises 19 to 24, use the following experiment. Two-digit numbers are formed, with replacement, from the digits 0–9.

19. How many two-digit numbers are possible?

20. How many two-digit even numbers are possible?

21. How many numbers are divisible by 5?

22. How many numbers are divisible by 3?

23. How many numbers are greater than 37?

24. How many numbers are less than 59?

In Exercises 25 to 28, use the following experiment. Four cards labeled A, B, C, and D are randomly placed in four boxes labeled A, B, C, and D. Each box receives exactly one card.

25. In how many ways can the cards be placed in the boxes?

26. Count the number of elements in the event that no box contains a card with the same letter as the box.

27. Count the number of elements in the event that *at least* one card is placed in the box with the corresponding letter.

28. If you add the answer for Exercise 26 and the answer for Exercise 27, is the sum the answer for Exercise 25? Why or why not?

In Exercises 29 to 32, use the following experiment. A state lottery game consists of choosing one card from each of the four suits in a standard deck of playing cards. (There are 13 cards in each suit.)

29. Count the number of elements in the sample space.

30. Count the number of elements in the event that an ace, a king, a queen, and a jack are chosen.

31. Count the number of ways in which four aces can be chosen.

32. Count the number of ways in which four cards, each of a different face value, can be chosen.

33. Write a lesson that you could use to explain the meanings of the words *experiment, sample space,* and *event* as they apply to combinatorics.

34. Explain how a tree diagram is used to count the number of ways an experiment can be performed.

▶ **Investigation**

Decision Trees **Decision trees** are tree diagrams that are used to solve problems that involve many choices. To illustrate, suppose we are given eight coins, one of which is counterfeit and slightly heavier than the other seven. Using a balance scale, we must find the counterfeit coin.

Designate the coins as c_1, c_2, c_3, c_4, c_5, c_6, c_7, and c_8. One way to determine the counterfeit coin is to weigh coins in pairs. This method is illustrated by the decision tree in Figure 7.4. In this case, it would take from one to four weighings to determine the counterfeit coin.

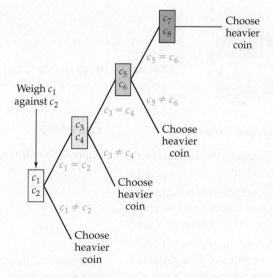

Figure 7.4 Using this method, it may take up to four weighings to determine which is the counterfeit coin.

A second method is to divide the coins into two groups of four coins each, place each group on a pan of the balance scale, and take the coins from the side that goes down. Now divide these four coins into two groups of two coins each and weigh them on the balance scale. Again keep the coins from the heavier side. Weighing one of these coins against the other will reveal the counterfeit coin. The decision tree for this procedure is shown in Figure 7.5. Using this method, the counterfeit coin will be found in three weighings.

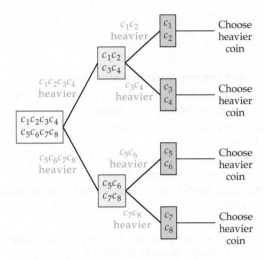

Figure 7.5 Using this method, it will always take three weighings to determine which is the counterfeit coin.

A third method is even more efficient. Divide the coins into three groups. Two of the groups contain three coins, and the third group contains two coins. Place each of the three-coin groups on the balance scale. If they balance, the counterfeit coin is in the

third group. Placing a coin from the third group on each of the balance pans will determine the counterfeit coin. If the three coin groups do not balance, then take two of the coins from the pan that goes down and weigh them against each other. If these balance, the third coin is the counterfeit. If not, the counterfeit is the coin on the pan that goes down. This method requires only two weighings and is shown in the decision tree in Figure 7.6.

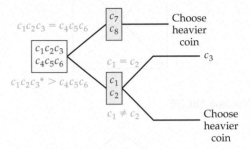

*We are assuming that $c_1 c_2 c_3$ is heavier than $c_4 c_5 c_6$. If $c_4 c_5 c_6$ is heavier, use those coins in the final weighing.

Figure 7.6 Using this method, it will take only two weighings to determine which is the counterfeit coin.

For each of the following problems, draw a decision tree and determine the minimum number of weighings necessary to identify the counterfeit coin.

1. In a stack of 12 identical-looking coins, one is counterfeit and is lighter than the remaining 11 coins.

2. In a stack of 13 identical-looking coins, one is counterfeit and is heavier than the remaining 12 coins.

Section 7.2

Permutations and Combinations

Learning Objectives:

- Compute the factorial of a number.

- Count the number of permutations of an experiment consisting of different objects.

- Count the number of permutations of an experiment when some objects occur more than once.

- Count the number of combinations of an experiment.

Factorial

Suppose four different squares are arranged in a row. One possibility is shown in Figure 7.7.

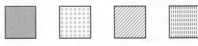

Figure 7.7

How many different ways are there to order the squares? There are four choices for the first square, three choices for the second square, two choices for the third square, and one choice for the fourth square. By the counting principle, there are $4 \cdot 3 \cdot 2 \cdot 1 = 24$ different arrangements of the four squares. Note from this example that the number of arrangements equals the product of the natural numbers n through 1, where n is the number of objects. This product is called a *factorial*.

n Factorial

***n* factorial** is the product of the natural numbers *n* through 1 and is symbolized by ***n*!**.

$$n! = n \cdot (n-1) \cdot (n-2) \cdot \ldots \cdot 3 \cdot 2 \cdot 1$$

Here are some examples:

$$5! = 5 \cdot 4 \cdot 3 \cdot 2 \cdot 1 = 120$$

$$8! = 8 \cdot 7 \cdot 6 \cdot 5 \cdot 4 \cdot 3 \cdot 2 \cdot 1 = 40,320$$

$$1! = 1$$

On some occasions it will be necessary to use 0! (zero factorial). Because it is impossible to define zero factorial in terms of a product of natural numbers, a standard definition is used.

Zero Factorial

$$0! = 1$$

A factorial can be written in terms of smaller factorials. This is useful when calculating large factorials. For example,

$$10! = 10 \cdot 9!$$

$$10! = 10 \cdot 9 \cdot 8!$$

$$10! = 10 \cdot 9 \cdot 8 \cdot 7!$$

Example I Simplify Factorials

Evaluate: **a.** $5! - 3!$ **b.** $\dfrac{9!}{6!}$

Solution

a. $5! - 3! = (5 \cdot 4 \cdot 3 \cdot 2 \cdot 1) - (3 \cdot 2 \cdot 1) = 120 - 6 = 114$

b. $\dfrac{9!}{6!} = \dfrac{9 \cdot 8 \cdot 7 \cdot \cancel{6!}}{\cancel{6!}} = 9 \cdot 8 \cdot 7 = 504$

Permutations

Determining the number of possible ordered arrangements of a group of distinct objects, as we did with the squares earlier, is one application of the counting principle. Each arrangement of this type is called a *permutation*.

Permutation

A **permutation** is an arrangement of objects in a definite order.

For example, abc and cba are two different permutations of the letters a, b, and c. As a second example, 122 and 212 are two different permutations of one 1 and two 2s.

The counting principle is used to count the number of different permutations of any set of objects. We will begin our discussion using sets of *distinct* objects; that is, sets in which no two objects are the same. For instance, the objects a, b, c, d are distinct, whereas the objects □, ☆, ○, ☆ are not.

Most music players allow the user to create a playlist, which is a list of songs that can be played on the device. Many of these players have a *shuffle* feature that plays the songs in a playlist in a different order each time.

Suppose a playlist consists of two songs, a rock song and a reggae song. If the shuffle feature is used, the songs could be played in two orders:

rock then reggae   or   reggae then rock

Thus there are two choices for the first song (rock or reggae) and there is one choice for the second song (whichever song was not played first). By the counting principle, there are $2 \cdot 1 = 2! = 2$ permutations or orders in which to play the songs.

With three songs in a playlist, one rock, one reggae, and one country, there are three choices for the first song, two choices for the second song, and one choice for the third song. By the counting principle, there are $3 \cdot 2 \cdot 1 = 3! = 6$ permutations in which the three songs could be played as shown in Table 7.1.

Table 7.1

Permutation 1	Permutation 2	Permutation 3	Permutation 4	Permutation 5	Permutation 6
Rock	Rock	Reggae	Reggae	Country	Country
Reggae	Country	Rock	Country	Rock	Reggae
Country	Reggae	Country	Rock	Reggae	Rock

With four songs in a playlist, there are $4 \cdot 3 \cdot 2 \cdot 1 = 4! = 24$ orders in which the songs could be played. In general, if there are n songs in a playlist, then there are $n!$ permutations, or orders, in which the songs could be played.

Suppose now that you have a playlist that consists of eight songs but you have time to listen to only three of the songs. In shuffle mode, any one of the eight songs could play first, then any one of the seven remaining songs could play second, and any one of the remaining six songs could play third. By the counting principle, there are $8 \cdot 7 \cdot 6 = 336$ permutations in which the songs could be played.

The following formula can be used to determine the number of permutations of n distinct objects (the songs in the example above), of which k are selected.

Permutation Formula for Distinct Objects

The number of permutations of n distinct objects selected k at a time is

$$P(n, k) = \frac{n!}{(n - k)!}$$

Applying this formula to the situation above in which there were eight songs ($n = 8$), of which only three songs ($k = 3$) could be played, we have

$$P(8, 3) = \frac{8!}{(8 - 3)!} = \frac{8!}{5!} = \frac{8 \cdot 7 \cdot 6 \cdot 5!}{5!} = 8 \cdot 7 \cdot 6 = 336$$

There are 336 permutations of playing the songs. This is the same answer we obtained using the counting principle.

Example 2 **Counting Permutations**

A university tennis team consists of six players who are ranked from 1 through 6. If a tennis coach has 10 players from which to choose, how many different ranked tennis teams can the coach select?

Solution

Because the players on the tennis team are ranked from 1 through 6, a team with player A in position 1 is different from a team with player A in position 2. Therefore, the number of different teams is the number of permutations of 10 players selected 6 at a time.

$$P(10, 6) = \frac{10!}{(10 - 6)!} = \frac{10!}{4!} = \frac{10 \cdot 9 \cdot 8 \cdot 7 \cdot 6 \cdot 5 \cdot 4!}{4!}$$
$$= 10 \cdot 9 \cdot 8 \cdot 7 \cdot 6 \cdot 5 = 151{,}200$$

There are 151,200 possible tennis teams.

Example 3 **Counting Permutations**

In 2015, 18 horses were entered in the Kentucky Derby. How many different finishes of first through fourth place were possible?

Solution

Because the order in which the horses finish the race is important, the number of possible finishes of first through fourth place is $P(18, 4)$.

$$P(18, 4) = \frac{18!}{(18 - 4)!} = \frac{18!}{14!} = \frac{18 \cdot 17 \cdot 16 \cdot 15 \cdot 14!}{14!}$$
$$= 18 \cdot 17 \cdot 16 \cdot 15 = 73{,}440$$

There were 73,440 possible finishes of first through fourth places.

Applying Several Counting Techniques

The permutation formula is derived from the counting principle. This formula is a convenient way of expressing the number of ways in which the items in an ordered list can be arranged. Both the permutation formula and the counting principle are needed to solve some counting problems.

Example 4 **Counting Using Several Methods**

Five white cubes numbered 1 to 5 and four red cubes numbered 1 to 4 are placed on a shelf. How many different arrangements are possible if

a. there are no restrictions on the arrangements?

b. the white cubes are together and the red cubes are together?

Solution

Because these arrangements have a definite order, they are permutations.

a. If there are no restrictions on the arrangements, then the number of arrangements is $P(9, 9)$.

$$P(9, 9) = \frac{9!}{(9 - 9)!} = \frac{9!}{0!} = 9! = 362{,}880$$

There are 362,880 arrangements.

b. This is a multi-stage experiment, so both the permutation formula and the counting principle will be used. There are 5! ways to arrange the white cubes and 4! ways to arrange the red cubes. We must also consider that either the white cubes or the red cubes could be arranged at the beginning of the row. There are two ways to do this. By the counting principle, there are $2 \cdot 5! \cdot 4!$ ways to arrange the white cubes together and the red cubes together.

$$2 \cdot 5! \cdot 4! = 5760$$

There are 5760 arrangements in which white cubes are together and red cubes are together.

Permutations of Indistinguishable Objects

Up to this point we have been counting the number of permutations of *distinct* objects. We now look at the situation of arranging objects when some of them are identical. In the case of identical or indistinguishable objects, a modification of the permutation formula is necessary. The general idea is to count the number of permutations as if all of the objects were distinct and then remove the permutations that look alike.

Consider the permutations of the letters *bbbcc*. We first assume all the letters are different by labeling them as $b_1b_2b_3c_1c_2$. Using the permutation formula, there are $5! = 120$ permutations. Now we need to remove repeated permutations. Note that

$$b_1b_2b_3c_1c_2 \quad b_1b_3b_2c_1c_2 \quad b_2b_1b_3c_1c_2 \quad b_2b_3b_1c_1c_2 \quad b_3b_2b_1c_1c_2 \quad b_3b_1b_2c_1c_2$$

are all distinct permutations that end with c_1c_2. However, if we replace each b_1, b_2, and b_3 with b, all six of these permutations will look the same. Thus there are $3! = 6$ times too many arrangements for each arrangement of c_1 and c_2. A similar argument applies to the c's. There are $2! = 2$ times too many permutations of the c's for each arrangement of b's.

Combining the results above, the number of permutations of *bbbcc* is

$$\frac{5!}{3! \cdot 2!} = \frac{5 \cdot 4 \cdot 3 \cdot 2 \cdot 1}{(3 \cdot 2 \cdot 1) \cdot (2 \cdot 1)} = 10$$

There are 10 distinct permutations of *bbbcc*.

Permutations of Objects, Some of Which Are Identical

The number of distinguishable permutations of n objects of r different types, where k_1 identical objects are of one type, k_2 are of another, and so on, is given by

$$\frac{n!}{k_1! \cdot k_2! \cdot \ldots \cdot k_r!}$$

where $k_1 + k_2 + \cdots + k_r = n$.

Example 5 **Permutations of Identical Objects**

A password requires 7 characters. If a person who lives at 155 Nunn Road wants a password to be an arrangement of the characters 155NUNN, how many different passwords are possible?

Solution

We are looking for the number of permutations of the characters 155NUNN. With $n = 7$ (number of characters), $k_1 = 1$ (number of 1's), $k_2 = 2$ (number of 5's), $k_3 = 3$ (number of N's), and $k_4 = 1$ (number of U's) we have

$$\frac{7!}{1! \cdot 2! \cdot 3! \cdot 1!} = \frac{7 \cdot 6 \cdot 5 \cdot 4 \cdot 3!}{2 \cdot 3!} = 420$$

There are 420 possible passwords.

Combinations

For some arrangements of objects, the order in which the objects are arranged is important. These are permutations. If a telephone extension is 2537, then the digits must be dialed in exactly that order. On the other hand, if you were to receive a $1 bill, a $5 bill, and a $10 bill, you would have $16 regardless of the order in which you received the bills. A **combination** is a collection of objects in which the order of the objects is not important. The three-letter sequences acb and bca are *different* permutations but the *same* combination.

The formula for finding the number of combinations is derived in much the same manner as the formula for finding the number of permutations of identical objects. Consider the problem of finding the number of possible combinations when choosing three letters from the letters a, b, c, d, and e, without replacement. For each choice of three letters, there are 3! permutations. For example, choosing the letters a, d, and e gives the following six permutations.

ade   aed   dea   dae   ead   eda

Because there are six permutations and each permutation represents the *same* combination, the number of permutations is six times the number of combinations. This is true each time three letters are selected. Therefore, to find the number of combinations of five objects chosen three at a time, divide the number of permutations by $3! = 6$. The number of combinations of five objects chosen three at a time is

$$\frac{P(5, 3)}{3!} = \frac{5!}{3! \cdot (5 - 3)!} = \frac{5!}{3! \cdot 2!} = \frac{5 \cdot 4 \cdot 3!}{3! \cdot 2!} = \frac{5 \cdot 4}{2 \cdot 1} = 10$$

Combination Formula

The number of combinations of n objects chosen k at a time is

$$C(n, k) = \frac{P(n, k)}{k!} = \frac{n!}{k! \cdot (n - k)!}$$

In Section 7.1, we stated that there were 2,598,960 possible 5-card poker hands. This number was calculated using the combination formula. Because the 5-card hand ace of hearts, king of diamonds, queen of clubs, jack of spades, 10 of hearts is exactly the same as the 5-card hand king of diamonds, jack of spades, queen of clubs, 10 of hearts, ace of hearts, the order of the cards is not important, and therefore the number of hands is a combination. The number of different 5-card poker hands is the combination of 52 cards chosen 5 at a time, which is given by $C(52, 5)$.

$$C(52, 5) = \frac{52!}{5! \cdot (52 - 5)!} = \frac{52!}{5! \cdot 47!} = \frac{52 \cdot 51 \cdot 50 \cdot 49 \cdot 48 \cdot 47!}{5! \cdot 47!}$$

$$= \frac{52 \cdot 51 \cdot 50 \cdot 49 \cdot 48}{5 \cdot 4 \cdot 3 \cdot 2 \cdot 1} = 2,598,960$$

Example 6   **Counting Using the Combination Formula**

An emergency room at a hospital has 11 nurses on staff. Each night a team of 5 nurses is on duty. In how many different ways can the team of 5 nurses be chosen?

Solution

This is a combination problem, because the order in which the nurses are chosen is not important. The 5 nurses N_1, N_2, N_3, N_4, N_5 are the same as the 5 nurses N_3, N_5, N_1, N_2, N_4.

$$C(11, 5) = \frac{11!}{5! \cdot (11 - 5)!} = \frac{11!}{5! \cdot 6!} = \frac{11 \cdot 10 \cdot 9 \cdot 8 \cdot 7 \cdot 6!}{5! \cdot 6!}$$

$$= \frac{11 \cdot 10 \cdot 9 \cdot 8 \cdot 7}{5 \cdot 4 \cdot 3 \cdot 2 \cdot 1} = 462$$

There are 462 possible teams of 5 nurses.

Example 7 **Counting Using the Combination Formula and the Counting Principle**

A committee of 5 is chosen from 5 mathematicians and 6 economists. How many different committees are possible if the committee must include 2 mathematicians and 3 economists?

Solution

Because a committee of professors A, B, C, D, and E is exactly the same as a committee of professors B, D, E, A, and C, choosing a committee is an example of choosing a combination. There are 5 mathematicians from whom 2 are chosen, which is equivalent to $C(5, 2)$ combinations. There are 6 economists from whom 3 are chosen, which is equivalent to $C(6, 3)$ combinations. Therefore, by the counting principle, there are $C(5, 2) \cdot C(6, 3)$ ways to choose 2 mathematicians and 3 economists.

$$C(5, 2) \cdot C(6, 3) = \frac{5!}{2! \cdot 3!} \cdot \frac{6!}{3! \cdot 3!} = 10 \cdot 20 = 200$$

There are 200 possible committees consisting of 2 mathematicians and 3 economists.

A standard deck of playing cards contains four suits: spades, hearts, diamonds, and clubs. Each suit has 13 cards: 2 through 10, jack, queen, king, and ace.

Example 8 **Counting Problems with Cards**

From a standard deck of playing cards, 5 cards are chosen. How many 5-card combinations contain

a. 2 kings and 3 queens?

b. 5 hearts?

c. 5 cards of the same suit?

Solution

a. There are $C(4, 2)$ ways of choosing 2 kings from 4 kings and $C(4, 3)$ ways of choosing 3 queens from 4 queens. By the counting principle, there are $C(4, 2) \cdot C(4, 3)$ ways of choosing 2 kings and 3 queens.

$$C(4, 2) \cdot C(4, 3) = \frac{4!}{2! \cdot 2!} \cdot \frac{4!}{3! \cdot 1!} = 6 \cdot 4 = 24$$

There are 24 ways of choosing 2 kings and 3 queens.

b. There are $C(13, 5)$ ways of choosing 5 hearts from 13 hearts.

$$C(13, 5) = \frac{13!}{5! \cdot 8!} = 1287$$

There are 1287 ways to choose 5 hearts from 13 hearts.

c. From part b, there are also 1287 ways of choosing 5 spades from 13 spades, 5 clubs from 13 clubs, or 5 diamonds from 13 diamonds. Because there are 4 suits from which to choose and $C(13, 5)$ ways of choosing 5 cards from a suit, by the counting principle there are $4 \cdot C(13, 5)$ ways to choose 5 cards of the same suit.

$$4 \cdot C(13, 5) = 4 \cdot 1287 = 5148$$

There are 5148 ways of choosing 5 cards of the same suit from a standard deck of playing cards.

7.2 Exercise Set

Think About It

1. What is the value of 0!?
2. What is the difference between a permutation and a combination?
3. For each of the following, determine whether the result is calculated as a permutation or combination. Do not perform the calculation. In each case, assume that the experiment is done without replacement.

    a. The number of ways 10 people can stand in a line to purchase tickets to a concert.

    b. Two people are chosen from 10 people. One of the chosen individuals will stand on your left, the other on your right.

    c. You choose three individuals from a group of 10 people to form a three-person team.

    d. To complete your degree program, you must take two of five courses.

Preliminary Exercises

For Exercises P1 to P8, if necessary, use the referenced example following the exercise for assistance.

P1. Evaluate: **a.** $7! + 4!$ **b.** $\dfrac{8!}{4!}$ **Example 1**

P2. A college golf team consists of five players who are ranked from 1 through 5. If a golf coach has eight players from which to choose, how many different ranked golf teams can the coach select? **Example 2**

P3. There were 43 cars entered in the 2015 Daytona 500 NASCAR race. In how many different ways could the first, second, and third place prizes have been awarded? **Example 3**

P4. There are seven tutors, three juniors and four seniors, who must be assigned to the 7 hours that a math center is open each day. If each tutor works 1 hour per day, how many different tutoring schedules are possible if

    a. there are no restrictions?

    b. the juniors tutor during the first 3 hr and the seniors tutor during the last 4 hr? **Example 4**

P5. Eight coins—three pennies, two nickels, and three dimes—are placed in a single stack. How many different stacks are possible if

    a. there are no restrictions on the placement of the coins?

    b. the dimes must stay together? **Example 5**

P6. A restaurant employs 16 waitstaff. In how many ways can a group of 9 waitstaff be chosen for the lunch shift? **Example 6**

P7. An IRS auditor randomly chooses 5 tax returns to audit from a stack of 10 tax returns, 4 of which are from corporations and 6 of which are from individuals. In how many different ways can the auditor choose the tax returns if the auditor wants to include 3 corporate and 2 individual returns? **Example 7**

P8. Five cards are chosen from a standard deck of playing cards. How many five-card combinations contain four cards of the same suit? **Example 8**

Section Exercises

In Exercises 1 to 30, evaluate the expression.

1. $8!$
2. $5!$
3. $9! - 5!$
4. $(9 - 5)!$
5. $(8 - 3)!$
6. $8! - 3!$
7. $P(8, 5)$
8. $P(7, 2)$
9. $P(9, 7)$
10. $P(10, 5)$
11. $P(8, 0)$
12. $P(7, 0)$
13. $P(8, 8)$
14. $P(10, 10)$
15. $P(8, 2) \cdot P(5, 3)$
16. $\dfrac{P(10, 4)}{P(8, 4)}$
17. $\dfrac{P(6, 0)}{P(6, 6)}$
18. $\dfrac{P(6, 3) \cdot P(5, 2)}{P(4, 3)}$
19. $C(9, 2)$
20. $C(8, 6)$
21. $C(12, 0)$
22. $C(11, 11)$
23. $C(6, 2) \cdot C(7, 3)$
24. $C(8, 5) \cdot C(9, 4)$
25. $\dfrac{C(10, 4) \cdot C(5, 2)}{C(15, 6)}$
26. $\dfrac{C(4, 3) \cdot C(5, 2)}{C(9, 5)}$
27. $\dfrac{C(9, 7) \cdot C(5, 3)}{C(14, 10)}$
28. $3! \cdot C(8, 5)$
29. $4! \cdot C(10, 3)$
30. $5! \cdot C(18, 0)$

In Exercises 31 to 34, how many combinations are possible? Assume that the items are distinct.

31. 7 items chosen 5 at a time
32. 8 items chosen 3 at a time
33. 12 items chosen 7 at a time
34. 11 items chosen 11 at a time
35. Is it possible to calculate $C(7, 9)$? Think of your answer in terms of 7 items chosen 9 at a time.
36. Is it possible to calculate $C(n, k)$ where $k > n$? See Exercise 35 for some help.
37. **Music Downloads** A student downloads five music files to a portable music player. In how many different orders can the songs be played?
38. **Elections** The board of directors of a corporation must select a president, a secretary, and a treasurer. In how many possible ways can this be accomplished if there are 20 members on the board of directors?
39. **Elections** A committee of 16 students must select a president, a vice-president, a secretary, and a treasurer. In how many possible ways can this be accomplished?
40. **The Olympics** A gold, a silver, and a bronze medal are awarded in an Olympic event. In how many possible ways can the medals be awarded for a 200-meter sprint in which there are nine runners?
41. **Music Festival** Six country music bands and three rock bands are signed up to perform at an all-day festival. How many different orders can the bands play in if

    a. there are no restrictions on the order?

    b. all the bands of each type must perform in a row?

42. Radio Show Five rock songs and six rap songs are on a disc jockey's playlist for a radio show. How many different orders can the songs be played in if
  a. there are no restrictions on the order?
  b. all the rap songs are played consecutively?

43. Passwords A password requires eight characters. If a soccer player whose jersey number is 77 wants his password to be an arrangement of the numbers and letters in SOCCER77, how many passwords are possible?

44. Gardening A gardener is planting a row of tulip bulbs. She has a mix of bulbs for 10 yellow tulips and 6 red tulips. How many color arrangements are possible for the row of tulips?

45. Firefighters At a certain fire station, one team of firefighters consists of 8 firefighters. If there are 24 firefighters qualified for a team, how many different teams are possible?

46. Platoons A typical platoon consists of 20 soldiers. If there are 30 soldiers available to create a platoon, how many different platoons are possible?

47. Exam Questions A professor gives a class seven essay questions to prepare for an exam. Only three of the questions will actually appear on the exam. How many different exams are possible?

48. Test Banks A math quiz is generated by randomly choosing 5 questions from a test bank consisting of 50 questions. How many different quizzes are possible?

49. Committee Selection A committee of six people is chosen from eight managers and eight supervisors. How many different committees are possible that consist of three managers and three supervisors?

50. Quality Control In a shipment of 20 smartphones, 2 are defective. How many ways can a quality control inspector randomly test 5 smartphones, of which 2 are defective?

51. Football In the National Football Conference (NFC), the NFC East division has four teams. During the regular season, each NFC East team plays each of the other NFC East teams twice. How many games are played between teams of the NFC East during a regular season?

52. Basketball In the Eastern Conference of the National Basketball Association (NBA), the Atlantic, Southeast, and Central Divisions have five teams each. During the regular season, each team plays every other team in its own division four times a year, plays six of the teams from the other two divisions four times a year, plays the remaining four teams from the other two divisions three times a year, and plays each team in the Western conference (which has 15 teams) twice. How many games are played by a team in the Eastern Conference during the regular season?

53. Geometry A hexagon is a six-sided plane figure. A diagonal is a line segment connecting any two nonadjacent vertices. How many diagonals are possible?

54. Geometry Seven distinct points are drawn on a circle. How many different triangles can be drawn in which each vertex of the triangle is at one of the seven points?

55. Softball Eighteen people decide to play softball. In how many ways can the 18 people be divided into 2 teams of 9 people?

56. Bowling Fifteen people decide to form a bowling league. In how many ways can the 15 people be divided into 3 teams of 5 people each?

57. Signal Flags The Coast Guard uses signal flags as a method of communicating between ships. If four different flags are available and the order in which the flags are raised is important, how many different signals are possible? Assume that four flags are raised.

58. Pizza Toppings A restaurant offers a special pizza with any 5 toppings. If the restaurant has 12 toppings from which to choose, how many different special pizzas are possible?

59. Letter Arrangements How many different letter arrangements are possible using all the letters of the word *committee*?

60. Color Arrangements A set of 12 plates came with 4 red, 4 green, and 4 yellow plates, but 2 red plates have been broken. The remaining plates are stacked on a shelf. How many different color arrangements are possible for the stack of plates?

61. Coin Tosses Ten identical coins are tossed. How many possible arrangements of the coins include five heads and five tails?

62. Coin Tosses Twelve identical coins are tossed. How many possible arrangements of the coins include eight heads and four tails?

63. Concerts Three groups will perform at a choral concert. One group will sing three pieces, one group will sing four pieces, and one group will sing two pieces. In how many possible orders can the nine pieces be performed, assuming each group performs all its songs consecutively?

64. Morse Code In 1835, Samuel Morse, a professor of art at New York University, devised a code that could be transmitted over a wire by using an electric current. This was the invention of the telegraph. The code used is called Morse code. It consists of a dot or a dash or a combination of up to five dots and/or dashes. For instance, the letter *c* is represented by — • — •, and • — represents the letter *a*. The numeral 0 is represented by five dashes, — — — — —. How many different symbols can be represented using Morse code?

Exercises 65 to 70 refer to a standard deck of playing cards. Assume that 5 cards are randomly chosen from the deck.

65. How many hands contain four aces?

66. How many hands contain two aces and two kings?

67. How many hands contain exactly three jacks?

68. How many hands contain exactly three jacks and two queens?

69. How many hands contain exactly two 7s?

70. How many hands contain exactly two 7s and two 8s?

▶ **Investigation**

Choosing Numbers in Keno A popular gambling game called keno, first introduced in China over 2000 years ago, is played in many casinos. In keno, there are 80 balls numbered from 1 to 80. The casino randomly chooses 20 balls from the 80 balls. These are "lucky balls" because if a gambler chooses some of the numbers on these balls, there is a possibility of winning money. The amount that is won depends on the number of lucky numbers the gambler has selected. The number of ways in which a casino can choose 20 balls from 80 is

$$C(80, 20) = \frac{80!}{20! \cdot 60!} \approx 3{,}535{,}000{,}000{,}000{,}000{,}000$$

Once the casino chooses the 20 lucky balls, the remaining 60 balls are unlucky for the gambler. A gambler who chooses 5 numbers will have from 0 to 5 lucky numbers.

Let's consider the case in which 2 of the 5 numbers chosen by the gambler are lucky numbers. Because 5 numbers were chosen, there must be 3 unlucky numbers among the 5 numbers. The number of ways of choosing 2 lucky numbers from 20 lucky numbers is $C(20, 2)$. The number of ways of choosing 3 unlucky numbers from 60 unlucky numbers is $C(60, 3)$. By the counting principle, there are $C(20, 2) \cdot C(60, 3) = 190 \cdot 34,220 = 6,501,800$ ways to choose 2 lucky and 3 unlucky numbers.

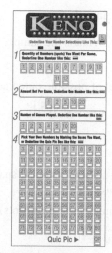

A keno card is used to mark the numbers chosen.

For each of the following exercises, assume that a gambler playing keno has randomly chosen four numbers.

1. In how many ways can the gambler choose no lucky numbers?

2. In how many ways can the gambler choose exactly one lucky number?

3. In how many ways can the gambler choose exactly two lucky numbers?

4. In how many ways can the gambler choose exactly three lucky numbers?

5. In how many ways can the gambler choose exactly four lucky numbers?

Section 7.3 Probability and Odds

Learning Objectives:

- Find the sample space for an experiment.
- Find an event in a sample space.
- Find the probability of an event by counting.
- Calculate an empirical probability.
- Apply probability of genetics.
- Calculate the odds of an event.

Introduction to Probability

The official state lottery of Indiana is the Hoosier Lottery where a player chooses 6 numbers from the numbers 1 to 46. The chances of matching the 6 numbers drawn by the lottery is approximately 1 in 9.4 million. In contrast, the likelihood of being struck by lightning is about 1 chance in 500,000. Comparing the likelihood of winning the Hoosier Lottery to the likelihood of being struck by lightning, you are about 20 times more likely to be struck by lightning than to pick the winning lottery numbers.

The likelihood of the occurrence of a particular event is described by a number between 0 and 1. (You can think of this as a percentage between 0% and 100%, inclusive.) This number is called the **probability** of the event. An event that is not very likely has a probability close to 0; an event that is very likely has a probability close to 1 (100%). For instance, the probability of being struck by lightning is close to 0. However, if you randomly choose a car in Indiana and observe the state that issued the license plate the chance the state was Indiana is quite high, so the probability is close to 1.

Because any event has from 0% to 100% chance of occurring, probabilities are always between 0% and 1%, inclusive or between 0 and 1. If an event *must* occur, its probability is 1. If an event *cannot* occur, its probability is 0.

Probabilities can be calculated by considering the outcomes of experiments. Here are some examples of experiments:

- Flip a coin and observe the outcome as a head or a tail.
- Select a company and observe its annual profit.
- Record the time a person spends at the checkout line in a supermarket.

The sample space of an experiment is the set of all possible outcomes of the experiment. For example, consider tossing a coin three times and observing the outcome as a head or a tail. Using H for head and T for tail, the sample space is

$$S = \{\text{HHH, HHT, HTH, HTT, THH, THT, TTH, TTT}\}$$

Note that the sample space consists of *every* possible outcome of tossing three coins.

Example 1 | **Find a Sample Space**

A single die is rolled once. What is the sample space for this experiment?

Solution

The sample space is the set of possible outcomes of the experiment.

$S = \{\boxed{\cdot}, \boxed{\because}, \boxed{\therefore}, \boxed{::}, \boxed{\vdots}, \boxed{:::}\}$

●

Formally, an event is a subset of a sample space. Using the sample space of Example 1, here are some possible events:

- There are an even number of pips (dots) facing up. The event is

$E_1 = \{\boxed{\because}, \boxed{::}, \boxed{:::}\}$.

- The number of pips facing up is greater than 4. The event is $E_2 = \{\boxed{\vdots}, \boxed{:::}\}$.
- The number of pips facing up is less than 20. The event is

$E_3 = \{\boxed{\cdot}, \boxed{\because}, \boxed{\therefore}, \boxed{::}, \boxed{\vdots}, \boxed{:::}\}$.

Because the number of pips facing up is always less than 20, this event will always occur. The event and the sample space are the same.

- The number of pips facing up is greater than 15. The event is $E_4 = \varnothing$, the empty set. This is an impossible event; the number of pips facing up cannot be greater than 15.

Outcomes of some experiments are **equally likely**, which means that the chance of any one outcome is just as likely as the chance of any other. For instance, if four balls of the same size but different colors—red, blue, green, and white—are placed in a box and a blindfolded person chooses one ball, the chance of choosing a green ball is the same as the chance of choosing any other color ball.

In the case of equally likely outcomes, the probability of an event is based on the number of elements in the event and the number of elements in the sample space. We will use $n(E)$ to denote the number of elements in the event E and $n(S)$ to denote the number of elements in the sample space S.

Probability of an Event

For an experiment with sample space S of *equally likely outcomes,* the probability $P(E)$ of an event E is given by

$$P(E) = \frac{n(E)}{n(S)} = \frac{\text{number of elements in } E}{\text{total number of elements in sample space } S}$$

Because each outcome of rolling a fair die is equally likely, the probability of the events E_1 through E_4 following Example 1 can be determined from the formula for the probability of an event.

$P(E_1) = \dfrac{3}{6}$ ←Number of elements in E_1
←Number of elements in the sample space

$= \dfrac{1}{2}$

The probability of rolling an even number of pips on a single roll of one die is $\frac{1}{2}$ (or 50%).

$P(E_2) = \dfrac{2}{6}$ ←Number of elements in E_2
←Number of elements in the sample space

$= \dfrac{1}{3}$

The probability of rolling a number greater than 4 on a single roll of one die is $\frac{1}{3}$.

$$P(E_3) = \frac{6}{6} \quad \begin{array}{l} \leftarrow \text{Number of elements in } E_3 \\ \leftarrow \text{Number of elements in the sample space} \end{array}$$

$$= 1$$

The probability of rolling a number less than 20 on a single roll of one die is 1 (or 100%). Recall that the probability of any event that is certain to occur is 1.

$$P(E_4) = \frac{0}{6} \quad \begin{array}{l} \leftarrow \text{Number of elements in } E_4 \\ \leftarrow \text{Number of elements in the sample space} \end{array}$$

$$= 0$$

The probability of rolling a number greater than 15 on a single roll of one die is 0. (It is not possible to roll any number greater than 6.)

Example 2   Probability of Equally Likely Outcomes

A fair coin—one for which it is equally likely that heads or tails will result from a single toss—is tossed three times. What is the probability that two heads and one tail are tossed?

Solution

Determine the number of elements in the sample space. The sample space must include every possible toss of a head or a tail (in order) in three tosses of the coin.

$$S = \{\text{HHH, HHT, HTH, HTT, THH, THT, TTH, TTT}\}$$

The elements in the event are $E = \{\text{HHT, HTH, THH}\}$.

$$P(E) = \frac{n(E)}{n(S)} = \frac{3}{8}$$

The probability is $\frac{3}{8}$.

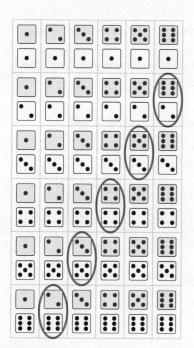

Figure 7.8 Outcomes of the roll of two dice

In Example 2, we calculated the probability as $\frac{3}{8}$. However, we could have expressed the probability as a decimal, 0.375, or as a percent, 37.5%. A probability can always be expressed as a fraction, a decimal, or a percent.

Example 3   Calculating Probabilities With Dice

Two fair dice are tossed. What is the probability that the sum of the pips on the upward faces of the two dice equals 8?

Solution

The dice must be considered as distinct, so there are 36 possible outcomes. (⚀⚁ and ⚁⚀ are considered different outcomes.) Therefore, $n(S) = 36$. The sample space is shown at the left. Let E represent the event that the sum of the pips on the upward faces is 8. These outcomes are circled in Figure 7.8. By counting the number of circled pairs, $n(E) = 5$.

$$P(E) = \frac{n(E)}{n(S)} = \frac{5}{36}$$

The probability that the sum of the pips is 8 is $\frac{5}{36}$.

Table 7.2

Point up	15
Side	85
Total	100

Empirical Probability

Probabilities such as those calculated in the preceding examples are sometimes referred to as **theoretical probabilities**. In Example 2, we assumed that, in theory, we had a perfectly balanced coin, and we calculated the probability based on the fact that each outcome was equally likely. Similarly, we assumed the dice in Example 3 were equally likely to land with any of the six faces upward.

When a probability is based on data gathered from an experiment, it is called an **experimental** or **empirical probability**. For instance, if we tossed a thumbtack 100 times and recorded the number of times it landed "point up," the results might be as shown in Table 7.2. From this experiment, the empirical probability of "point up" is

$$P(\text{point up}) = \frac{15}{100} = 0.15$$

Empirical Probability of an Event

If an experiment is performed repeatedly and the occurrence of the event E is observed, the probability $P(E)$ of the event E is given by

$$P(E) = \frac{\text{number of times event } E \text{ occurred}}{\text{number of times the experiment was performed}}$$

Example 4 **Calculate an Empirical Probability**

A survey showed the following information on the ages and party affiliations of registered voters in a certain city.

a. If one voter is chosen at random from this survey, what is the probability that the voter is a Republican?

b. If one voter is chosen at random from this survey, what is the probability the age of the voter is between the age of 18 to 28?

c. If one voter is chosen at random from this survey, what is the probability the age of the voter is greater than 38?

d. If one voter is chosen at random from this survey, what is the probability the voter is an Independent and voter's age is between 29 and 38?

Age	Republican	Democrat	Independent	Other	Total
18–28	205	432	98	112	847
29–38	311	301	109	83	804
39–49	250	251	150	122	773
≥ 50	272	283	142	107	804
Total	1038	1267	499	424	3228

Solution

a. Let R represent the event that a Republican is selected. Then

$$P(R) = \frac{1038}{3228} \leftarrow \text{Number of Republicans in the survey}$$
$$\phantom{P(R) = \frac{1038}{3228}} \leftarrow \text{Total number of people surveyed}$$
$$\approx 0.322$$

The probability that the selected person is a Republican is approximately 0.32.

b. Let A represent the event the voter's age is between 18 and 28. Then

$$P(A) = \frac{847}{3228}$$
$$\approx 0.26$$

The probability the selected voter's age is between 18 and 28 is approximately 0.26.

c. Let C represent the event the age of the voter is greater than 38. These are the voters in the group from 39 to 49 (773) and the group whose age is greater than or equal to 50 (804).

$$P(C) = \frac{773 + 804}{3228} = \frac{1577}{3228}$$
$$\approx 0.49$$

The probability the selected voter's age is greater than 38 is approximately 0.49.

d. Let D represent the event the voter is an Independent and the voter's age is between 29 and 38. From the table, there are 109 voters in that group.

$$P(D) = \frac{109}{3228}$$
$$\approx 0.03$$

The probability the selected voter is an Independent and the voter's age is between 29 and 38 is approximately 0.03.

Application to Genetics

Completed in April 2003, the Human Genome Project was a 13-year-long project designed to completely map the genetic make-up of *Homo sapiens*. Researchers hope to use this information to treat and prevent certain hereditary diseases.

The concept behind this project began with Gregor Mendel and his work on flower color and how it is transmitted from generation to generation. From his studies, Mendel concluded that flower color seems to be predictable in future generations by making certain assumptions about a plant's color "determiner." He concluded that red was a *dominant* determiner of color and that white was a *recessive* determiner. Today, geneticists talk about the *gene* for flower color and the *allele* of the gene. A gene consists of two dominant alleles (two red), a dominant and a recessive allele (red and white), or two recessive alleles (two white). Because red is the dominant allele, a flower will be white only if no dominant allele is present.

Later work by Reginald Punnett (1875–1967) showed how to determine the probability of certain flower colors by using a **Punnett square** (see Table 7.3). Using a capital letter for a dominant allele (R for red) and the corresponding lowercase letter for a recessive allele (r for white), Punnett arranged the alleles of the parents in a square. A parent could be RR, Rr, or rr.

Suppose that the genotype (genetic composition) of one parent is rr and the genotype of the other is Rr. The first parent is represented in the left column of the square, and the second parent is represented in the top row. The genotypes of the offspring are shown in the body of the table and are the result of combining one allele from each parent.

Because each of the genotypes of the offspring is equally likely, the probability that a flower will be red is $\frac{1}{2}$ (two Rr genotypes of the four possible genotypes), and the probability that a flower will be white is $\frac{1}{2}$ (two rr genotypes of the four possible genotypes).

Table 7.3

Parents	R	r
r	Rr	rr
r	Rr	rr

Example 5 **Probability Using a Punnett Square**

A child will have cystic fibrosis if the child inherits the recessive gene from both parents. Using F for the normal allele and f for the mutant recessive allele, suppose a parent who is Ff (said to be a *carrier*) and a parent who is FF (does not have the mutant allele) decide to have a child.

a. What is the probability that the child will have cystic fibrosis? (To have the disease, the child must be ff.)

b. What is the probability that the child will be a carrier?

Solution

Make a Punnett square.

Parents	F	F
F	FF	FF
f	Ff	Ff

a. To have the disease, the child must be ff. From the table, there is no combination of the alleles that will produce ff. Therefore, the child cannot have the disease, and the probability is 0.

b. To be a carrier, exactly one allele must be f. From the table, there are two cases out of four of a genotype with one f. The probability that the child will be a carrier is $\frac{2}{4} = \frac{1}{2}$.

Calculating Odds

Statistics kept by some fantasy sports enthusiasts show that the *odds in favor* of a professional basketball player making a free throw are 9 to 2. These odds indicate that out of every 11 (9 + 2) free throws, a player is expected to make 9 baskets.

A **favorable outcome** of an experiment is one that satisfies some event. For instance, in the case of free-throw attempts by professional basketball players, you can assume that out of 11 free-throw attempts, there will be 9 favorable outcomes (9 baskets). The opposite event, a missed basket, is an **unfavorable outcome**. Odds are frequently expressed in terms of favorable and unfavorable outcomes.

Odds of an Event

Let E be an event in a sample space of equally likely outcomes. Then

$$\textbf{Odds in favor of } E = \frac{\text{number of favorable outcomes}}{\text{number of unfavorable outcomes}}$$

$$\textbf{Odds against } E = \frac{\text{number of unfavorable outcomes}}{\text{number of favorable outcomes}}$$

When the odds of an event are written in fractional form, the fraction bar is read as the word *to*. Thus odds of $\frac{3}{2}$ are read as "3 to 2." We can also write odds of $\frac{3}{2}$ as $3:2$. This form is also read as "3 to 2."

Example 6 **Calculate Odds**

If a pair of fair dice is rolled once, what are the odds in favor of rolling a sum of 7?

Solution

Let E be the event of rolling a sum of 7. From Figure 7.8 on page 319, the 6 favorable outcomes are $E = \{$ [⚀][⚅], [⚁][⚄], [⚂][⚃], [⚃][⚂], [⚄][⚁], [⚅][⚀] $\}$. The unfavorable outcomes are the remaining 30 possibilities. (Because there are 36 possible outcomes when tossing 2 dice and 6 of them are favorable, there are $36 - 6 = 30$ unfavorable outcomes.)

$$\text{Odds in favor of } E = \frac{\text{number of favorable outcomes}}{\text{number of unfavorable outcomes}} = \frac{6}{30} = \frac{1}{5}.$$

The odds in favor of rolling a sum of 7 are 1 to 5.

Odds express the likelihood of an event and are therefore related to probability. When the odds of an event are known, the probability of the event can be determined. Conversely, when the probability of an event is known, the odds of the event can be determined.

The Relationship Between Odds and Probability

1. Suppose E is an event in a sample space and that the *odds in favor* of E are $\frac{a}{b}$. Then $P(E) = \frac{a}{a + b}$.
2. Suppose E is an event in a sample space. Then the *odds in favor* of E are $\frac{P(E)}{1 - P(E)}$.

Example 7   Determine Probability from Odds

In 2022, the racehorse Rick Strike won the Kentucky Derby. The odds against Rick Strike winning the race were 80 to 1. What was the probability of Rick Strike winning the race?

Solution

Because the odds *against* Rick Strike winning the race were 80 to 1, the odds *in favor* of Rick Strike winning the race were 1 to 80, or, as a fraction, $\frac{1}{80}$. Now use the formula for calculating the probability of an event when the odds in favor are known.

$$P(E) = \frac{a}{a + b}$$

$$P(E) = \frac{1}{1 + 80} = \frac{1}{81} \qquad \bullet \, a = 1, b = 80$$

The probability of Rick Strike winning the race was $\frac{1}{80}$, or 1.25%.

7.3   Exercise Set

Think About It

1. Is it possible that the probability of an event could be 1.23?

2. A box contains a red pencil, a green pencil, an orange pencil, and a blue pencil. Without looking in the box, a pencil is removed and its color recorded. Is the color of the pencil an equally likely event?

3. A pair of dice are rolled once. Are the sums of the pips of the upward faces equally likely events?

4. If the odds of some event are 1 to 1, what is the probability of the event?

5. A box contains five red balls and five green balls. Suppose a red ball is randomly selected from the box and set aside. Then a second ball is drawn. Is the probability of a red ball

less than, equal to, or greater than the probability that the first ball was red?

▶ Preliminary Exercises

For Exercises P1 to P7, if necessary, use the referenced example following the exercise for assistance.

P1. A coin is tossed twice. What is the sample space for this experiment? **Example 1**

P2. If a fair die is rolled once, what is the probability that an odd number will show on the upward face? **Example 2**

P3. Two fair dice are tossed. What is the probability that the sum of the pips on the upward faces of the two dice equals 7? **Example 3**

P4. Using the data from Example 4, what is the probability that a randomly selected person is between the ages of 39 and 49? **Example 4**

P5. For a certain type of hamster, the color cinnamon, C, is dominant and the color white, c, is recessive. If both parents are Cc, what is the probability that an offspring will be white? **Example 5**

P6. If three red, four white, and five blue balls are placed in a box and one ball is randomly selected from the box, what are the odds against the ball being blue? **Example 6**

P7. A report issued by the Southern California Earthquake Data Center estimates that the probability of an earthquake of magnitude 6.7 or greater within 30 years in the San Francisco Bay Area is about 60%. What are the odds in favor of such an earthquake occurring in that region in the next 30 years? **Example 7**

▶ Section Exercises

In Exercises 1 to 6, list the elements of the sample space for each experiment.

1. A coin is flipped three times.
2. An even number between 1 and 11 is selected at random.
3. One day in the first 2 weeks of November is selected.
4. A current U.S. coin is selected from a piggy bank.
5. A state is selected from the U.S. states whose names begin with the letter A.
6. A month is selected from the months that have exactly 30 days.

In Exercises 7 to 15, assume a coin is tossed three times.

7. List the elements of the sample space.
8. List the elements of the event of 2 heads and 1 tail.
9. List the elements of the event of at least 2 tails.
10. List the elements of the event no tails.
11. List the elements of the event at least 1 tail.
12. Compute the probability of 2 heads and 1 tail.
13. Compute the probability of at least 2 tails.
14. Compute the probability of no tails.
15. Compute the probability of at least 1 tail.

In Exercises 16 to 18, a coin is tossed four times. Assuming the coin is equally likely to land on heads or tails, compute the probability of each event occurring.

16. 2 heads and 2 tails
17. 1 head and 3 tails
18. All 4 coin tosses are identical.

In Exercises 19 to 22, a dodecahedral die (one with 12 sides numbered from 1 to 12) is tossed once. Find each of the following probabilities.

19. The number on the upward face is 12.
20. The number on the upward face is not 10.
21. The number on the upward face is divisible by 4.
22. The number on the upward face is less than 5 or greater than 9.

In Exercises 23 to 32, two regular six-sided dice are tossed. Compute the probability that the sum of the pips on the upward faces of the two dice is each of the following. (See Figure 7.8 on page 319 for the sample space of this experiment.)

23. 6	**24.** 11
25. 2	**26.** 12
27. 1	**28.** 14
29. At least 10	**30.** At most 5
31. An even number	**32.** An odd number

33. If two dice are rolled, compute the probability of rolling doubles (both dice show the same number of pips).
34. If two dice are rolled, compute the probability of *not* rolling doubles.

In Exercises 35 to 38, a card is selected at random from a standard deck of playing cards.

35. Compute the probability that the card is a 9.
36. Compute the probability that the card is a face card (jack, queen, or king).
37. Compute the probability that the card is between 5 and 9, inclusive.
38. Compute the probability that the card is between 3 and 6, inclusive.

In Exercises 39 to 44, use the data given in Example 4, page 320, to compute the probability that a randomly chosen voter from the survey will satisfy the following. Round to the nearest hundredth.

39. The voter is a Democrat.
40. The voter is not a Republican.
41. The voter is 50 years old or older.
42. The voter is under 39 years old.
43. The voter is between 39 and 49 and is registered as an Independent.
44. The voter is under 29 and is registered as a Democrat.

In Exercises 45 to 48, a survey asked 850 respondents about their highest levels of completed education. The results are given in the following table.

Education Completed	Number of Respondents
No high school diploma	52
High school diploma	234
Associate's degree or 2 years of college	274
Bachelor's degree	187
Master's degree	67
Ph.D. or professional degree	36

If a respondent from the survey is selected at random, compute the probability of each of the following.

45. The respondent did not complete high school.

46. The respondent has an associate's degree or 2 years of college (but not more).

47. The respondent has a Ph.D. or professional degree.

48. The respondent has a degree beyond a bachelor's degree.

A random survey asked respondents about their current annual salaries. The results are given in the following table. Use the table for Exercises 49 to 52.

Salary Range	Number of Respondents
Below $28,000	24
$28,000–$37,999	41
$38,000–$45,999	52
$46,000–$55,999	58
$56,000–$69,999	43
$70,000–$89,999	39
$90,000–$109,999	22
$110,000 or more	14

If a respondent from the survey is selected at random, compute the probability of the following.

49. The respondent earns from $46,000 to $55,999 annually.

50. The respondent earns from $70,000 to $89,999 per year.

51. The respondent earns at least $90,000 per year.

52. The respondent earns less than $46,000 annually.

53. Dreams Three groups of student volunteers participated in a study of dream recall. During a one-week period, each morning every student was asked how many dreams were remembered. The results are shown in the following table.

	None	1 or 2	3 or 4	5 or More	Total
Group A	12	45	56	6	119
Group B	17	53	48	9	127
Group C	9	42	35	2	88
Total	38	140	139	17	334

a. If a student is randomly chosen from this study, what is the probability the student recalls 5 or more dreams each week.

b. If a student is randomly chosen from this study, what is the probability the student is Group C.

c. If a student is randomly chosen from this study, what is the probability the student recalls 4 or fewer dreams per week?

d. If a student is randomly chosen from this study, what is the probability the student recalls 3 or 4 dreams and comes from Group B.

54. Blood Type There are four major blood types, A, B, O, and AB. Each of these blood types may contain what is called the rhesus antigen. The presence of the rhesus antigen is designated as Rh+; the omission of the rhesus is designated by Rh–. A sample of population and the presence of a blood type is given in the table below.

Type	A	B	O	AB	Total
Rh+	170	45	190	15	420
Rh–	30	10	35	5	80
Total	200	55	225	20	500

a. If a person is selected from this group, what is the probability that person has blood type O?

b. If a person is selected from this group, what is the probability that person is Rh+?

c. If a person is selected from this group, what is the probability that person has blood type AB and is Rh–?

55. Genotypes The following Punnett square for flower color shows two parents of genotype Rr, where R corresponds to the dominant red flower allele and r represents the recessive white flower allele. (See Example 5.)

Parents	R	r
R	RR	Rr
r	Rr	rr

What is the probability that the offspring of these parents will have white flowers?

56. Genotypes One parent plant with red flowers has genotype RR and the other with white flowers has genotype rr, where R is the dominant allele for a red flower and r is the recessive allele for a white flower. Compute the probability of one of the offspring having white flowers. *Hint:* Draw a Punnett square.

57. Genotypes The eye color of mice is determined by a dominant allele E, corresponding to black eyes, and a recessive allele e, corresponding to red eyes. If two mice, one of genotype EE and the other of genotype ee, have offspring, compute the probability of one of the offspring having red eyes. *Hint:* Draw a Punnett square.

58. Genotypes The height of a certain plant is determined by a dominant allele T corresponding to tall plants and a recessive allele t corresponding to short (or dwarf) plants. If both parent plants have genotype Tt, compute the probability that the offspring plants will be tall. *Hint:* Draw a Punnett square.

In Exercises 59 to 64, the odds in favor of an event are given. Compute the probability of the event.

59. 1 to 2

60. 1 to 4

61. 3 to 7

62. 3 to 5

63. 8 to 5

64. 11 to 9

In Exercises 65 to 70, the probability of an event is given. Find the odds in favor of the event.

65. 0.2   66. 0.6

67. 0.375   68. 0.28

69. 0.55   70. 0.81

71. **Game Shows** The game board for the television show "Jeopardy" is divided into six categories, with each category containing five answers. In the Double Jeopardy round, there are two hidden Daily Double squares. What are the odds in favor of choosing a Daily Double square on the first turn?

72. **Game Shows** The spinner for the television game show "Wheel of Fortune" is shown below. On a single spin of the wheel, what are the odds against stopping on $300?

73. If a single fair die is rolled, what are the odds in favor of rolling an even number?

74. If a card is randomly pulled from a standard deck of playing cards, what are the odds in favor of pulling a heart?

75. A coin is tossed four times. What are the odds against the coin showing heads all four times?

For Exercises 76 to 78, use the following table that shows the approximate probability or odds of an earthquake of various magnitudes in the next 30 years for two regions of California.

Earthquake of Magnitude . . .	Probability or Odds in San Francisco Area	Probability or Odds in Los Angeles Area
6.7	72%	Odds in favor: 3 to 5
7.0	Odds against: 27 to 50	51%
7.5	20%	Odds in favor: 3 to 10

Source: https://www.usgs.gov/faqs/what-probability-earthquake-will-occur-los-angeles-area-san-francisco-bay-area?qt-news_science_products=7#qt-news_science_products

76. Is the probability of a magnitude 6.7 earthquake greater in the San Francisco area or the Los Angeles area?

77. What is the probability of a magnitude 7.0 earthquake in the San Francisco area?

78. What is the probability of a magnitude 7.5 earthquake in the Los Angeles area?

79. **Football** An online betting site offered 8 to 3 odds *against* a particular football team winning its next game. What is the probability, in the online betting site's view, of the team winning?

80. **Candy Colors** A snack-size bag of M&M's candies contains 12 red candies, 12 blue, 7 green, 13 brown, 3 orange, and 10 yellow. If a candy is randomly picked from the bag, compute

   a. the odds of getting a green M&M.

   b. the probability of getting a green M&M.

▶ **Investigation**

Roulette Roulette is played by spinning a wheel with 38 numbered slots. The numbers 1 through 36 appear on the wheel, half of them colored black and half colored red. Two slots, numbered 0 and 00, are colored green. A ball is placed on the spinning wheel and allowed to come to rest in one of the slots. Bets are placed on where the ball will land.

1. You can place a bet that the ball will stop in a black slot. If you win, the casino will pay you $1 for each dollar you bet. What is the probability of winning this bet?

2. You can bet that the ball will land on an odd number. If you win, the casino will pay you $1 for each dollar you bet. What is the probability of winning this bet?

3. You can bet that the ball will land on any number from 1 to 12. If you win, the casino will pay you $2 for each dollar you bet. What is the probability of winning this bet?

4. You can bet that the ball will land on any particular number. If you win, the casino will pay you $35 for each dollar you bet. What is the probability of winning this bet?

5. You can bet that the ball will land on one of 0 or 00. If you win, the casino will pay you $17 for each dollar you bet. What is the probability of winning this bet?

6. You can bet that the ball will land on certain groups of 6 numbers (such as 1 to 6). If you win, the casino will pay you $5 for each dollar you bet. What is the probability of winning this bet?

Addition and Complement Rules

Learning Objectives:

- Apply the Addition Rule for Probabilities.

- Find the complement of an event.

- Calculate probabilities using counting techniques.

The Addition Rule for Probabilities

Suppose you draw a single card from a standard deck of playing cards. The sample space S consists of the 52 cards of the deck. Therefore, $n(S) = 52$. Now consider the events

$$E_1 = \text{a 4 is drawn} = \{\spadesuit 4, \heartsuit 4, \diamondsuit 4, \clubsuit 4\}$$

$$E_2 = \text{a spade is drawn}$$

$$= \{\spadesuit A, \spadesuit 2, \spadesuit 3, \spadesuit 4, \spadesuit 5, \spadesuit 6, \spadesuit 7, \spadesuit 8, \spadesuit 9, \spadesuit 10, \spadesuit J, \spadesuit Q, \spadesuit K\}$$

It is possible, on one draw, to satisfy the conditions of both events: the ♠4 could be drawn. This card is an element of both E_1 and E_2.

Now consider the events

$$E_3 = \text{a 5 is drawn} = \{♠5, ♥5, ♦5, ♣5\}$$

$$E_4 = \text{a king is drawn} = \{♠K, ♥K, ♦K, ♣K\}$$

In this case, it is not possible to draw one card that satisfies the conditions of both events. There are no elements common to both sets. Two events that cannot both occur at the same time are called **mutually exclusive events**. The events E_3 and E_4 are mutually exclusive events, whereas E_1 and E_2 are not.

Mutually Exclusive Events

Two events A and B are mutually exclusive if they cannot occur at the same time. That is, A and B are mutually exclusive when $A \cap B = \varnothing$.

The probability of either of two mutually exclusive events occurring can be determined by adding the probabilities of the individual events.

Probability of Mutually Exclusive Events

If A and B are two mutually exclusive events, then the probability of A or B occurring is

$$P(A \text{ or } B) = P(A) + P(B)$$

Example 1   **Probability of Mutually Exclusive Events**

Suppose a single card is drawn from a standard deck of playing cards. Find the probability of drawing a 5 or a king.

Solution

Let $A = \{♠5, ♥5, ♦5, ♣5\}$ and $B = \{♠K, ♥K, ♦K, ♣K\}$. There are 52 cards in a standard deck of playing cards; thus $n(S) = 52$. Because the events are mutually exclusive, we can use the formula for the probability of mutually exclusive events.

$$P(A \text{ or } B) = P(A) + P(B) \qquad \bullet \text{ Formula for the probability of mutually exclusive events}$$

$$= \frac{1}{13} + \frac{1}{13} = \frac{2}{13} \qquad \bullet \, P(A) = \frac{4}{52} = \frac{1}{13}, P(B) = \frac{4}{52} = \frac{1}{13}$$

The probability of drawing a 5 or a king is $\frac{2}{13}$.

Consider the experiment of rolling two dice. Let A be the event of rolling a sum of 8 and let B be the event of rolling a double (the same number on both dice).

$$A = \{ \boxed{∴}\boxed{⁚⁚},\ \boxed{⁚⁚}\boxed{∴},\ \boxed{∵}\boxed{∷},\ \boxed{∷}\boxed{∵},\ \boxed{⁚⁚}\boxed{⁚⁚} \}$$

$$B = \{ \boxed{·}\boxed{·},\ \boxed{∴}\boxed{∴},\ \boxed{∵}\boxed{∵},\ \boxed{⁚⁚}\boxed{⁚⁚},\ \boxed{∷}\boxed{∷},\ \boxed{⁚⁚⁚}\boxed{⁚⁚⁚} \}$$

These events are *not* mutually exclusive because it is possible to satisfy the conditions of each event on one toss of the dice—a $\boxed{∵}\boxed{∷}$ could be rolled. Therefore, $P(A \text{ or } B)$, the probability of a sum of 8 or a double, cannot be calculated using the formula for the probability of mutually exclusive events. However, a modification of that formula can be used.

Take Note: The $P(A \text{ and } B)$ term in the Addition Rule for Probabilities is subtracted to compensate for the overcounting of the first two terms of the formula. If two events are mutually exclusive, then $A \cap B = \varnothing$. Therefore, $n(A \cap B) = 0$ and $P(A \text{ and } B) = \dfrac{n(A \cap B)}{n(S)} = 0$. For mutually exclusive events, the Addition Rule for Probabilities is the same as the formula for the probability of mutually exclusive events.

Take Note: Recall that the probability of an event A is $P(A) = \dfrac{n(A)}{n(S)}$. Therefore,

$$P(A \text{ and } B) = \dfrac{n(A \cap B)}{n(S)}.$$

> **Addition Rule for Probabilities**
>
> If A and B are two events in a sample space S, then
>
> $$P(A \text{ or } B) = P(A) + P(B) - P(A \text{ and } B)$$

Using this formula with

$$A = \left\{ \boxed{\cdot\ \ \text{::}}, \boxed{\text{::}\ \cdot}, \boxed{\cdot\ \text{::}}, \boxed{\text{::}\ \cdot}, \boxed{\text{::}\ \cdot}, \boxed{\text{::}\ \text{::}} \right\}$$

$$B = \left\{ \boxed{\cdot\ \cdot}, \boxed{\cdot\ \cdot}, \boxed{\cdot\ \cdot}, \boxed{\text{::}\ \text{::}}, \boxed{\text{::}\ \text{::}}, \boxed{\text{::}\ \text{::}} \right\}$$

$$A \cap B = \left\{ \boxed{\text{::}\ \text{::}} \right\}$$

the probability of A or B can be calculated.

$$P(A \text{ or } B) = P(A) + P(B) - P(A \text{ and } B)$$

$$= \frac{5}{36} + \frac{6}{36} - \frac{1}{36} \qquad \bullet\ P(A) = \frac{5}{36},\ P(B) = \frac{6}{36},\ P(A \cap B) = \frac{1}{36}$$

$$= \frac{10}{36} = \frac{5}{18}$$

On a single roll of two dice, the probability of rolling a sum of 8 or a double is $\frac{5}{18}$.

Example 2 **Use the Addition Rule for Probabilities**

Table 7.4 shows data from an experiment conducted to test the effectiveness of a flu vaccine. If one person is selected from this population, what is the probability that the person was vaccinated or contracted the flu?

Solution

Let $V = \{$people who were vaccinated$\}$ and $F = \{$people who contracted the flu$\}$. These events are not mutually exclusive because there are 21 people who were vaccinated and who contracted the flu. The sample space S consists of the 490 people who participated in the experiment. From the table, $n(V) = 219$, $n(F) = 97$, and $n(V \text{ and } F) = 21$.

$$P(V \text{ or } F) = P(V) + P(F) - P(V \text{ and } F)$$

$$= \frac{219}{490} + \frac{97}{490} - \frac{21}{490}$$

$$= \frac{295}{490} \approx 0.602$$

The probability of selecting a person who was vaccinated or who contracted the flu is approximately 60.2%.

Table 7.4

	F	No F	Total
V	21	198	219
No V	76	195	271
Total	97	393	490

V: Vaccinated
F: Contracted the flu

The Complement of an Event

Consider the experiment of tossing a single die once. The sample space is

$$S = \left\{ \boxed{\cdot}, \boxed{\cdot\cdot}, \boxed{\cdot\cdot\cdot}, \boxed{\text{::}}, \boxed{\text{::}\cdot}, \boxed{\text{:::}} \right\}$$

Now consider the event $E = \left\{ \boxed{\cdot\cdot\cdot} \right\}$, that is, the event of tossing a $\boxed{\cdot\cdot\cdot}$. The probability of E is

$$P(E) = \frac{1}{6} \quad \begin{array}{l} \leftarrow \text{Number of elements in } E \\ \leftarrow \text{Number of elements in the sample space} \end{array}$$

The **complement** of an event E, symbolized by E', is the "opposite" event of E. The complement includes all those outcomes of S that are not in E and excludes the outcomes in E. For the event E above, E' is the event of not tossing a ⚃. Thus

$$E' = \{⚀, ⚁, ⚂, ⚄, ⚅\}$$

Note that because E and E' are opposite events, they are mutually exclusive, and their union is the entire sample space S. Thus $P(E) + P(E') = P(S)$. But $P(S) = 1$, so $P(E') = 1 - P(E)$.

> ### Probability of the Complement of an Event
>
> If E is an event and E' is the complement of the event, then
>
> $$P(E') = 1 - P(E)$$

Continuing our example, the probability of not tossing a ⚃ is given by

$$P\!\left(\text{not a } ⚃\right) = 1 - P\!\left(⚃\right)$$

$$= 1 - \frac{1}{6} = \frac{5}{6}$$

You can also verify the probability of E' directly:

$$P\!\left(\text{not a } ⚃\right) = \frac{5}{6} \quad \begin{array}{l}\leftarrow \text{Number of elements in } E' \\ \leftarrow \text{Number of elements in the sample space}\end{array}$$

Example 3 Find a Probability by the Complement Rule

The probability of tossing a sum of 11 on the toss of 2 dice is $\frac{1}{18}$. What is the probability of not tossing a sum of 11 on the toss of 2 dice?

Solution

Use the formula for the probability of the complement of an event.
Let $E = \{$toss a sum of 11$\}$. Then $E' = \{$toss a sum that is not 11$\}$.

$$P(E') = 1 - P(E)$$

$$= 1 - \frac{1}{18} = \frac{17}{18} \qquad \bullet \, P(E) = \frac{1}{18}$$

The probability of not tossing a sum of 11 is $\frac{17}{18}$.

Take Note: The phrase "at least one" means 1 or more. Tossing a coin 3 times and asking the probability of getting at least 1 head means calculating the probability of getting 1, 2, or 3 heads.

Suppose we toss a coin three times and want to calculate the probability of having heads occur *at least once*. We could list all the possibilities of tossing three coins, as shown, and then find the ones that contain at least one head.

$$\underbrace{\{\text{HHH, HHT, HTH, HTT, THH, THT, TTH,}}_{\text{at least one head}} \text{TTT}\}$$

The probability of at least 1 head is $\frac{7}{8}$.

Another way to calculate this result is to use the formula for the probability of the complement of an event. Let $E = \{$at least 1 head$\}$. From the list above, note that E contains every outcome except TTT (no heads). Thus $E' = \{$TTT$\}$ and we have

$$P(E) = 1 - P(E')$$

$$= 1 - \frac{1}{8} = \frac{7}{8} \qquad \bullet \, P(E') = \frac{n(E')}{n(S)} = \frac{1}{8}$$

This is the same result that we calculated above. As we will see, sometimes working with a complement is much less work than proceeding directly.

Combinatoric Formulas and Probability

In many cases, the principles of counting that were discussed in Sections 7.1 and 7.2 are part of the process of calculating a probability.

Example 4 **Find a Probability Using the Complement Rule**

A die is tossed 4 times. What is the probability that a ⚅ will show on the upward face at least once?

Solution

Let E = {at least one 6}. Then E' = {no 6s}. To calculate the number of elements in the sample space (all possible outcomes of tossing a die 4 times) and the number of items in E', we will use the counting principle.

Because on each toss of the die there are 6 possible outcomes,

$$n(S) = 6 \cdot 6 \cdot 6 \cdot 6 = 1296$$

On each toss of the die there are 5 numbers that are not 6s. Therefore,

$$n(E') = 5 \cdot 5 \cdot 5 \cdot 5 = 625$$

$$P(E) = 1 - P(E')$$

$$= 1 - \frac{625}{1296} = \frac{671}{1296}$$

$$\approx 0.518$$

When a die is tossed 4 times, the probability that a ⚅ will show on the upward face at least once is approximately 0.518.

In the next example, we use the combination formula $C(n, r) = \dfrac{n!}{r!(n - r)!}$ to find the number of ways in which r objects can be chosen from n objects.

Example 5 **Find a Probability Using the Combination Formula**

Suppose a manufacturing process for tableware produces 40 dinner plates, of which 3 are defective. If 5 plates are randomly selected from the 40, what is the probability that at least one is defective?

Solution

Let E = {at least one plate is defective}. It is easier to work with the complement event, E' = {no plates are defective}. To calculate the number of elements in the sample space (all possible outcomes of choosing 5 plates from 40), use the combination formula with $n = 40$ (the total number of plates) and $r = 5$ (the number of plates that are chosen). Then

$$n(S) = C(40, 5) = \frac{40!}{5!\,(40 - 5)!} \qquad \bullet\, n = 40, r = 5$$

$$= \frac{40!}{5!\,35!} = 658{,}008$$

To find the number of outcomes that contain no defective plates, all of the plates chosen must come from the 37 nondefective plates. Therefore, we must calculate the number of ways in which 5 objects can be chosen from 37. Thus $n = 37$ (the number of nondefective plates) and $r = 5$ (the number of plates chosen).

$$n(E') = C(37, 5) = \frac{37!}{5!\,(37 - 5)!} \qquad \bullet\, n = 37, r = 5$$

$$= \frac{37!}{5!\,32!} = 435{,}897$$

$$P(E) = 1 - P(E')$$

$$= 1 - \frac{435{,}897}{658{,}008} = \frac{222{,}111}{658{,}008}$$

$$\approx 0.338$$

The probability is approximately 0.338, or 33.8%.

7.4 Exercise Set

Think About It

1. A die is rolled once. Let E be the event that an even number is rolled and let O be the event that an odd number is rolled. Are the events E and O mutually exclusive?

2. A single number is chosen from the natural numbers from 1 to 10, inclusive. Let P be the event a prime number is selected and let O be the event an odd number is selected. Are the events P and O mutually exclusive?

3. If the probability of an event is 0.3, what is the probability of the complement of the event?

4. An experiment consists of tossing a coin four times. Let L be the event of observing at least three heads. List the elements of L.

5. An experiment consists of tossing a coin four times. Let M be the event of observing at most three heads. List the elements of M.

Preliminary Exercises

For Exercises P1 to P5, if necessary, use the referenced example following the exercise for assistance.

P1. Two fair dice are tossed once. What is the probability of rolling a 7 or an 11? For the sample space for this experiment, see Figure 7.1 on page 303. **Example 1**

P2. The data in Table 7.5 show the starting salaries of college graduates with selected degrees. If one person is chosen from this population, what is the probability that the person has a degree in business or has a starting salary between $20,000 and $24,999? **Example 2**

Table 7.5

Salary (in $)	Engineering	Business	Chemistry	Psychology
Less than 20,000	0	4	1	12
20,000–24,999	4	16	3	16
25,000–29,999	7	21	5	15
30,000–34,999	12	35	5	7
35,000 or more	12	22	4	5

(Degree header spans Engineering, Business, Chemistry, Psychology columns)

P3. The probability that a person has type A blood is 34%. What is the probability that a person does not have type A blood? **Example 3**

P4. A pair of dice is rolled 3 times. What is the probability that a sum of 7 will occur at least once? **Example 4**

P5. The winner of a contest will be blindfolded and then allowed to reach into a hat containing 31 $1 bills and 4 $100 bills. The winner can remove 4 bills from the hat and keep the money. Find the probability that the winner will pull out at least 1 $100 bill. **Example 5**

Section Exercises

In Exercises 1 to 4, first verify that the compound event consists of two mutually exclusive events, and then compute the probability of the compound event occurring.

1. A single card is drawn from a standard deck of playing cards. Find the probability of drawing a 4 or an ace.

2. A single card is drawn from a standard deck of playing cards. Find the probability of drawing a heart or a club.

3. Two dice are rolled. Find the probability of rolling a 2 or a 10.

4. Two dice are rolled. Find the probability of rolling a 7 or an 8.

5. If $P(A) = 0.2$, $P(B) = 0.5$, and $P(A \text{ and } B) = 0.1$, find $P(A \text{ or } B)$.

6. If $P(A) = 0.6$, $P(B) = 0.4$, and $P(A \text{ and } B) = 0.2$, find $P(A \text{ or } B)$.

7. If $P(A) = 0.3$, $P(B) = 0.8$, and $P(A \text{ or } B) = 0.9$, find $P(A \text{ and } B)$.

8. If $P(A) = 0.7$, $P(A \text{ and } B) = 0.4$, and $P(A \text{ or } B) = 0.8$, find $P(B)$.

In Exercises 9 to 12, suppose you ask a friend to randomly choose an integer between 1 and 10, inclusive.

9. What is the probability that the number will be more than 6 or odd?

10. What is the probability that the number will be less than 5 or even?

11. What is the probability that the number will be even or prime?

12. What is the probability that the number will be prime or greater than 7?

In Exercises 13 to 18, two dice are rolled. Determine the probability of each of the following. ("Doubles" means that both dice show the same number.)

13. Rolling a 6 or doubles

14. Rolling a 7 or doubles

15. Rolling an even number or doubles

16. Rolling a number greater than 7 or doubles

17. Rolling an odd number or a number less than 6

18. Rolling an even number or a number greater than 9

In Exercises 19 to 24, a single card is drawn from a standard deck. Find the probability of each of the following events.

19. Drawing an 8 or a spade

20. Drawing an ace or a red card

21. Drawing a jack or a face card

22. Drawing a red card or a face card

23. Drawing a diamond or a black card

24. Drawing a spade or a red card

In Exercises 25 to 28, use the data in the following table, which shows the employment status of individuals in a particular town by age group.

Age	Full-Time	Part-Time	Unemployed
0–17	24	164	371
18–25	185	203	148
26–34	348	67	27
35–49	581	179	104
≥50	443	162	173

25. If a person is randomly chosen from the town's population, what is the probability that the person is aged 26 to 34 or is employed part-time?

26. If a person is randomly chosen from the town's population, what is the probability that the person is at least 50 years old or unemployed?

27. If a person is randomly chosen from the town's population, what is the probability that the person is under 18 or employed part-time?

28. If a person is randomly chosen from the town's population, what is the probability that the person is 18 or older or employed full-time?

29. **Contests** If the probability of winning a particular contest is 0.04, what is the probability of not winning the contest?

30. **Weather** Suppose the probability that it will rain tomorrow is 0.38. What is the probability that it will not rain tomorrow?

The National Collegiate Athletic Association (NCAA) keeps statistics on the number of seniors playing on NCAA teams who become professionals after graduation. Exercises 31 and 32 use some of these statistics.

31. Suppose there is a 1 in 75 chance that a senior on an NCAA basketball team will play professional basketball. What is the probability that a senior NCAA basketball player will not play professional basketball?

32. Suppose the odds in favor of a senior on an NCAA golf team playing professional golf are 1 to 25. What is the probability that a senior NCAA golfer will not play professional golf?

In Exercises 33 to 44, use the formula for the probability of the complement of an event.

33. Two dice are tossed. What is the probability of not tossing a 7?

34. Two dice are tossed. What is the probability of not getting doubles?

35. Two dice are tossed. What is the probability of getting a sum of at least 4?

36. Two dice are tossed. What is the probability of getting a sum of at most 11?

37. A single card is drawn from a deck. What is the probability of not drawing an ace?

38. A single card is drawn from a deck. What is the probability of not drawing a face card?

39. A coin is flipped four times. What is the probability of getting at least one tail?

40. A coin is flipped four times. What is the probability of getting at least two heads?

41. A single die is rolled three times. What is the probability that a 1 will show on the upward face at least once?

42. A single die is rolled four times. Find the probability that a 5 will be rolled at least once.

43. A pair of dice is rolled three times. What is the probability that a sum of 8 on the two dice will occur at least once?

44. A pair of dice is rolled four times. Compute the probability that a sum of 11 on the two dice will occur at least once.

45. **Magic Trick** A magician shuffles a standard deck of playing cards and allows an audience member to pull out a card, look at it, and replace it in the deck. Two additional people do the same. Find the probability that of the three cards drawn, at least one is a face card.

46. **Face Cards** If a person draws three cards from a standard deck (without replacing them), what is the probability that at least one of the cards is a face card?

47. **E-Readers** An electronics store receives a shipment of 30 e-readers. Unbeknownst to the store, 4 of the e-readers are defective. If the store sells 12 of these e-readers the first day, what is the probability that at least 1 of the 12 buyers will get a defective e-reader?

48. **Blu-ray Players** An electronics store currently has 28 new Blu-ray players in stock, of which 5 are defective. If customers buy 3 Blu-ray players, what is the probability that at least 1 of them will be defective?

49. **Prize Drawing** Three employees of a restaurant each contributed one business card for a random drawing to win

a prize. Forty-two business cards were received in all, and three cards will be drawn for prizes. Determine the probability that at least one of the restaurant employees will win a prize.

50. **Coins** A bag contains 44 U.S. quarters and 6 Canadian quarters. (The coins are identical in size.) If 5 quarters are randomly picked from the bag, what is the probability of getting at least 1 Canadian quarter?

Investigations

Blackjack In the game blackjack, a player is dealt two cards from a standard deck of playing cards. The player has a blackjack if one card is an ace and the other card is a 10, a jack, a queen, or a king. In some casinos, blackjack is played with more than one standard deck of playing cards. Does using two decks of cards change the probability of getting a blackjack? Give a reason for your answer.

Door Codes A planned community has 300 homes, each with an automatic garage door opener operated by a code of 8 num-

bers. The homeowner sets each of the 8 numbers to be 0 or 1. For example, a door opener code might be 01101001. Assuming all the homes in the community are sold, what is the probability that at least 2 homeowners will set their door openers to use the same code and will therefore be able to open each other's garage doors?

Poker In 5-card stud poker, a hand containing 5 cards of the same suit from a standard deck is called a flush. In this exercise, you will compute the probability of getting a flush.

1. How many 5-card poker hands are possible?
2. How many 5-card poker hands are possible that contain all spades?
3. What is the probability of getting a 5-card poker hand containing all spades?
4. What is the probability of getting a 5-card poker hand containing all hearts? All diamonds? All clubs?
5. Are the events of getting 5 spades, 5 hearts, 5 diamonds, or 5 clubs mutually exclusive?
6. What is the probability of getting a flush in 5-card stud poker?

Section 7.5 Conditional Probability

Learning Objectives:

- Apply conditional probability.
- Find the probability that two or more events occur.
- Determine whether events are dependent or independent.

Table 7.6

	F	No F	Total
V	21	198	219
No V	76	195	271
Total	97	393	490

V: Vaccinated
F: Contracted the flu

Table 7.7

	F	No F	Total
V	21	198	219

Conditional Probability

In the preceding section, we discussed the effectiveness of a flu vaccination in preventing the onset of the flu. Table 7.6 shows the data from that discussion. From the table, we can calculate the probability that one person randomly selected from this population will contract the flu.

$$P(F) = \frac{n(F)}{n(S)} = \frac{97}{490} \approx 0.198$$ • $n(F) = 97, n(S) = 490$
(S denotes the sample space)

Now consider a slightly different situation. We could ask, "What is the probability that a person randomly chosen from this population will contract the flu *given* that the person received the flu vaccination?"

In this case, we know that the person received a flu vaccination, and we want to determine the probability that the person will contract the flu. Therefore, the only part of the table that is of concern to us is the top row, as shown in Table 7.7. In this case, we have

$$P(F \text{ given } V) = \frac{21}{219} \approx 0.096$$

Thus the probability that an individual will contract the flu given that the individual has been vaccinated is about 0.096.

The probability of an event B occurring given that some other event A has already occurred is called a **conditional probability** and is denoted $P(B|A)$.

Conditional Probability Formula

If A and B are two events in a sample space S, then the conditional probability of B given that A has occurred is

$$P(B|A) = \frac{P(A \text{ and } B)}{P(A)}$$

The symbol $P(B|A)$ is read "the probability of B given A."

To see how this formula applies to the flu data on the preceding page, let

$S = \{\text{all people participating in the test}\}$

$F = \{\text{people who contracted the flu}\}$

$V = \{\text{people who were vaccinated}\}$

Then $F \cap V = \{\text{people who contracted the flu } and \text{ were vaccinated}\}$.

$$P(F|V) = \frac{P(F \text{ and } V)}{P(V)} = \frac{\dfrac{21}{490}}{\dfrac{219}{490}}$$

- $P(F \text{ and } V) = \dfrac{n(F \cap V)}{n(S)} = \dfrac{21}{490}$

- $P(V) = \dfrac{n(V)}{n(S)} = \dfrac{219}{490}$

$$= \frac{21}{219} \approx 0.096$$

The probability that one person selected from this population contracted the flu given that the person received the vaccination, $P(F|V)$, is about 0.096. Our answer agrees with the calculation we performed directly from the table, but the Conditional Probability Formula enables us to find conditional probabilities even when we cannot compute them directly.

Example 1 **Determine a Conditional Probability**

The data in Table 7.8 show the results of a survey used to determine the number of adults who received financial help from their parents for certain purchases.

Table 7.8

Age	College Tuition	Buy a Car	Buy a House	Total
18–28	405	253	261	919
29–39	389	219	392	1000
40–49	291	146	245	682
50–59	150	71	112	333
≥60	62	15	98	175
Total	1297	704	1108	3109

If one person is selected from this survey, what is the probability that the person received financial help to purchase a home, given that the person is between the ages of 29 and 39?

Solution

Let $B = \{$adults who received financial help for a home purchase$\}$ and
$A = \{$adults between 29 and 39$\}$. From the table, $n(A \cap B) = 392$, $n(A) = 1000$, and
$n(S) = 3109$. Using the Conditional Probability Formula, we have

$$P(B \mid A) = \frac{P(A \text{ and } B)}{P(A)} = \frac{\dfrac{392}{3109}}{\dfrac{1000}{3109}} \qquad \bullet P(A \text{ and } B) = \frac{n(A \cap B)}{n(S)} = \frac{392}{3109}$$

$$\bullet P(A) = \frac{n(A)}{n(S)} = \frac{1000}{3109}$$

$$= \frac{392}{1000} = 0.392$$

The probability that a person received financial help to purchase a home, given that the
person is between the ages of 29 and 39, is 0.392.

Product Rule for Probabilities

Suppose that two cards are drawn, without replacement, from a standard deck of playing
cards. Let A be the event that an ace is drawn on the first draw and B the event that an ace
is drawn on the second draw. Then the probability that an ace is drawn on the first *and* second draws is $P(A \text{ and } B)$. To find this probability, we can solve the Conditional Probability
Formula for $P(A \text{ and } B)$.

$$\frac{P(A \text{ and } B)}{P(A)} = P(B \mid A)$$

$$P(A) \cdot \frac{P(A \text{ and } B)}{P(A)} = P(A) \cdot P(B \mid A) \qquad \bullet \text{Multiply each side by } P(A).$$

$$P(A \text{ and } B) = P(A) \cdot P(B \mid A)$$

This is called the Product Rule for Probabilities.

Product Rule for Probabilities

If A and B are two events in a sample space S, then

$$P(A \text{ and } B) = P(A) \cdot P(B \mid A)$$

For the problem just given, $P(A \text{ and } B)$ is the product of $P(A)$, the probability that the
first card drawn is an ace, and $P(B \mid A)$, the probability of an ace on the second draw *given*
that the first card drawn was an ace.

The tree diagram in Figure 7.9 shows the possible outcomes of drawing 2 cards from
a deck without replacement. On the first draw, there are 4 aces in the deck of 52 cards.
Therefore, $P(A) = \dfrac{4}{52} = \dfrac{1}{13}$. On the second draw, there are only 51 cards remaining and
only 3 aces (an ace was drawn on the first draw). Therefore, $P(B \mid A) = \dfrac{3}{51} = \dfrac{1}{17}$. Putting
these calculations together, we have

$$P(A \cap B) = P(A) \cdot P(B \mid A)$$

$$= \frac{1}{13} \cdot \frac{1}{17} = \frac{1}{221}$$

The probability of drawing an ace on the first and second draws is $\dfrac{1}{221}$.

The Product Rule for Probabilities can be extended to more than two events. The probability that a certain sequence of events will occur in succession is the product of the probabilities of each of the events *given* that the preceding events have occurred.

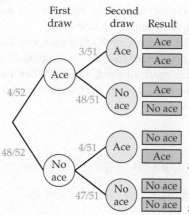

First draw, Second draw, Result

Figure 7.9

Probability of Successive Events

The probability of two or more events occurring in succession is the product of the conditional probabilities of each of the events.

Example 2   **Find the Probability of Successive Events**

A box contains four red, three white, and five green balls. Suppose that three balls are randomly selected from the box in succession, without replacement.

a. What is the probability that first a red, then a white, and then a green ball are selected?

b. What is the probability that two white balls followed by one green ball are selected?

Take Note: In part a, there are originally 12 balls in the box. After a red ball is selected, there are only 11 balls remaining, of which 3 are white. Thus $P(B|A) = \frac{3}{11}$. After a red ball and a white ball are selected, there are 10 balls left, of which 5 are green. Thus $P(C|A \text{ and } B) = \frac{5}{10}$.

In part b, we have a similar situation. However, after a white ball is selected, there are 11 balls remaining, of which only 2 are white. Therefore, $P(B|A) = \frac{2}{11}$.

Solution

a. Let $A = \{$a red ball is selected first$\}$, $B = \{$a white ball is selected second$\}$, and $C = \{$a green ball is selected third$\}$. Then

$$P(A \text{ followed by } B \text{ followed by } C) = P(A) \cdot P(B|A) \cdot P(C|A \text{ and } B)$$

$$= \frac{4}{12} \cdot \frac{3}{11} \cdot \frac{5}{10}$$

$$= \frac{1}{22}$$

The probability of choosing a red, then a white, then a green ball is $\frac{1}{22}$.

b. Let $A = \{$a white ball is selected first$\}$, $B = \{$a white ball is selected second$\}$, and $C = \{$a green ball is selected third$\}$. Then

$$P(A \text{ followed by } B \text{ followed by } C) = P(A) \cdot P(B|A) \cdot P(C|A \text{ and } B)$$

$$= \frac{3}{12} \cdot \frac{2}{11} \cdot \frac{5}{10}$$

$$= \frac{1}{44}$$

The probability of choosing two white balls followed by one green ball is $\frac{1}{44}$.

Independent Events

Earlier in this section we considered the probability of drawing two aces in a row from a standard deck of playing cards. Because the cards were drawn without replacement, the probability of an ace on the second draw *depended* on the result of the first draw.

Now consider the case of tossing a coin twice. The outcome of the first coin toss has no effect on the outcome of the second toss. So the probability of the coin flipping to a head or a tail on the second toss is not affected by the result of the first toss. When the outcome of a first event does not affect the outcome of a second event, the events are called *independent*.

Independent Events

If A and B are two events in a sample space and $P(B|A) = P(B)$, then A and B are called **independent events**.

For a mathematical verification, consider tossing a coin twice. We can compute the probability that the second toss comes up heads, given that the first coin toss came up heads. If A is the event of a head on the first toss, then $A = \{HH, HT\}$. Let B be the event of a head on the second toss. Then $B = \{HH, TH\}$. The sample space is $S = \{HH, HT, TH, TT\}$. The conditional probability $P(B|A)$ (the probability of a head on the second toss given a head on the first toss) is

$$P(B|A) = \frac{P(A \text{ and } B)}{P(A)} = \frac{\frac{1}{4}}{\frac{1}{2}} = \frac{1}{2}$$

$$\bullet\ P(A \text{ and } B) = \frac{n(A \cap B)}{n(S)} = \frac{1}{4}$$

$$\bullet\ P(A) = \frac{n(A)}{n(S)} = \frac{2}{4} = \frac{1}{2}$$

Thus $P(B|A) = \frac{1}{2}$. Note, however, that $P(B) = \frac{n(B)}{n(S)} = \frac{2}{4} = \frac{1}{2}$. Therefore, in the case of tossing a coin twice, the probability of the second event does not depend on the outcome of the first event, and we have $P(B|A) = P(B)$.

In general, this result enables us to simplify the product rule when two events are independent; the probability of two independent events occurring in succession is simply the product of the probabilities of the individual events.

Take Note: The Product Rule for Independent Events can be extended to more than two events. If E_1, E_2, E_3, and E_4 are independent events, then the probability that all four events will occur is $P(E_1) \cdot P(E_2) \cdot P(E_3) \cdot P(E_4)$.

> **Product Rule for Independent Events**
>
> If A and B are two independent events from the sample space S, then
> $P(A \text{ and } B) = P(A) \cdot P(B)$.

Example 3   **Find the Probability of Independent Events**

A pair of dice is tossed twice. What is the probability that the first roll is a sum of 7 and the second roll is a sum of 11?

Solution

Take Note: See page 319 for all the possible outcomes of the roll of two dice.

The rolls of a pair of dice are independent; the probability of a sum of 11 on the second roll does not depend on the outcome of the first roll. Let $A = \{\text{sum of 7 on the first roll}\}$ and $B = \{\text{sum of 11 on the second roll}\}$. Then

$$P(A \text{ and } B) = P(A) \cdot P(B) = \frac{6}{36} \cdot \frac{2}{36} = \frac{1}{108}$$

Applications of Conditional Probability

Conditional probability is used in many real-world situations, such as to determine the efficacy of a drug test, to verify the accuracy of genetic testing, and to analyze evidence in legal proceedings.

Example 4   **Drug Testing and Conditional Probability**

Suppose a company claims it has a test that is 95% effective in determining whether an athlete is using a steroid. That is, if an athlete is using a steroid, the test will be positive 95% of the time. In the case of a negative result, the company says its test is 97% accurate. That is, even if an athlete is not using steroids, it is possible that the test will be positive in 3% of the cases. Such an occurrence is called a **false positive**. Suppose this test is given to a group of athletes in which 10% of the athletes are using steroids. What is the probability that a randomly chosen athlete actually uses steroids, given that the athlete's test is positive?

Solution

Let S be the event that an athlete uses steroids and let T be the event that the test is positive. Then the probability we wish to determine is $P(S\,|\,T)$. Using the Conditional Probability Formula, we have

$$P(S\,|\,T) = \frac{P(S \text{ and } T)}{P(T)}$$

A tree diagram, as shown in Figure 7.10, can be used to calculate this probability. A positive test result can occur in two ways: either an athlete using steroids correctly tests positive, or an athlete not using steroids incorrectly tests positive. The probability of a positive test result, $P(T)$, corresponds to an athlete following path ST or path $S'T$ in the tree diagram. (S' symbolizes no steroid use and T' symbolizes a negative result.) $P(S \text{ and } T)$, the probability of using steroids and getting a positive test result, is path ST. Thus,

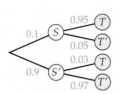

Figure 7.10

$$P(S\,|\,T) = \frac{P(S \text{ and } T)}{P(T)}$$

$$= \frac{(0.1)(0.95)}{(0.1)(0.95) + (0.9)(0.03)} \approx 0.779$$

Given that an athlete tests positive, the probability that the athlete actually uses steroids is approximately 77.9%.

This result is probably lower than you might expect considering that the manufacturer claims its test is 95% accurate! In fact, the prevalence of false-positive test results can be much more dramatic in cases of conditions that are present in a small percentage of the population. (See Exercise 59, page 340.)

7.5   Exercise Set

Think About It

1. If A is the event a number is odd and B is the event that a number is prime, translate the symbol $P(A\,|\,B)$.

2. If $P(A\,|\,B) = P(A)$, are A and B independent events?

3. If two events are mutually exclusive, are they dependent or independent events?

4. If A and B are independent events, what is the probability of the events A and B occurring?

Preliminary Exercises

For Exercises P1 to P4, if necessary, use the referenced example following the exercise for assistance.

P1. Two dice are tossed, one after the other. What is the probability that the result is a sum of 6, given that the first toss is not a 3? **Example 1**

P2. A standard deck of playing cards is shuffled and three cards are dealt. Find the probability that the cards dealt are a spade followed by a heart followed by another spade. **Example 2**

P3. A coin is tossed three times. What is the probability that heads appears on all three tosses? **Example 3**

P4. A pharmaceutical company has a test that is 95% effective in determining whether a person has a certain genetic defect.

However, the test may give a false positive result in 4% of cases. Suppose this particular genetic defect occurs in 2% of the population. Given that a person tests positive, what is the probability that the person actually has the defect? **Example 4**

Section Exercises

In Exercises 1 to 4, compute the conditional probabilities $P(A\,|\,B)$ and $P(B\,|\,A)$.

1. $P(A) = 0.7$, $P(B) = 0.4$, $P(A \text{ and } B) = 0.25$

2. $P(A) = 0.45$, $P(B) = 0.8$, $P(A \text{ and } B) = 0.3$

3. $P(A) = 0.61$, $P(B) = 0.18$, $P(A \text{ and } B) = 0.07$

4. $P(A) = 0.2$, $P(B) = 0.5$, $P(A \text{ and } B) = 0.2$

In Exercises 5 to 8, use the data in the following table, which shows the employment status of individuals in a particular town by age group.

	Full-Time	Part-Time	Unemployed
0–17	24	164	371
18–25	185	203	148
26–34	348	67	27
35–49	581	179	104
≥50	443	162	173

5. If a person in this town is selected at random, find the probability that the individual is employed part-time, given that they are between the ages of 35 and 49.

6. If a person in the town is randomly selected, what is the probability that the individual is unemployed, given that they are 50 years old or older?

7. A person from the town is randomly selected; what is the probability that the individual is employed full-time, given that they are between 18 and 49 years of age?

8. A person from the town is randomly selected; what is the probability that the individual is employed part-time, given that they are at least 35 years old?

In Exercises 9 to 12, use the data in the following table, which shows the results of a survey of 2000 gamers concerning their favorite home video game systems, organized by age group. If a survey participant is selected at random, determine the probability of each of the following. Round to the nearest hundredth.

	Nintendo Switch	Microsoft Xbox Series S/X	Steam Deck	Sony PS5
0–12	63	84	55	51
13–18	105	139	92	113
19–24	248	217	83	169
≥25	191	166	88	136

9. The participant prefers the Nintendo Switch system.

10. The participant prefers the Xbox S/X, given that the person is between the ages of 13 and 18.

11. The participant prefers Steam Deck, given that the person is between the ages of 13 and 24.

12. The participant is under 12 years of age, given that the person prefers the PS5 system.

13. A pair of dice is tossed. Find the probability that the sum on the two dice is 8, given that the sum is even.

14. A pair of dice is tossed. Find the probability that the sum on the two dice is 12, given that doubles are rolled.

15. A pair of dice is tossed. What is the probability that doubles are rolled, given that the sum on the two dice is less than 7?

16. A pair of dice is tossed. What is the probability that the sum on the two dice is 8, given that the sum is more than 6?

17. What is the probability of drawing two cards in succession (without replacement) from a standard deck and having them both be face cards? Round to the nearest thousandth.

18. Two cards are drawn from a standard deck without replacement. Find the probability that both cards are hearts. Round to the nearest thousandth.

19. Two cards are drawn from a standard deck without replacement. What is the probability that the first card is a spade and the second card is red? Round to the nearest thousandth.

20. Two cards are drawn from a standard deck without replacement. What is the probability that the first card is a king and the second card is not? Round to the nearest thousandth.

In Exercises 21 to 24, a snack-size bag of M&M's candies is opened. Inside, there are 12 red candies, 12 blue, 7 green, 13 brown, 3 orange, and 10 yellow. Three candies are pulled from the bag in succession, without replacement.

21. Determine the probability that the first candy drawn is blue, the second is red, and the third is green.

22. Determine the probability that the first candy drawn is brown, the second is orange, and the third is yellow.

23. What is the probability that the first two candies drawn are green and the third is red?

24. What is the probability that the first candy drawn is orange, the second is blue, and the third is orange?

In Exercises 25 to 28, three cards are dealt from a shuffled standard deck of playing cards.

25. Find the probability that the first card dealt is red, the second is black, and the third is red.

26. Find the probability that the first two cards dealt are clubs and the third is a spade.

27. What is the probability that the three cards dealt are, in order, an ace, a face card, and an 8? (A face card is a jack, queen, or king.)

28. What is the probability that the three cards dealt are, in order, a red card, a club, and another red card?

In Exercises 29 to 32, the probability that a student enrolled at a local high school will be absent on a particular day is 0.04, assuming that the student was in attendance the previous school day. However, if a student is absent, the probability that they will be absent again the following day is 0.11. For each exercise, assume that the student was in attendance the previous day.

29. What is the probability that a student will be absent 3 days in a row?

30. What is the probability that a student will be absent 2 days in a row but then show up on the third day?

31. Find the probability that a student will be absent, attend the next day, but then be absent again the third day.

32. Find the probability that a student will be absent 4 days in a row.

In Exercises 33 to 36, determine whether the events are independent.

33. A single die is rolled and then rolled a second time.

34. Numbered balls are pulled from a bin one by one to determine the winning lottery numbers.

35. Numbers are written on slips of paper in a hat; one person pulls out a slip of paper without replacing it, then a second person pulls out a slip of paper.

36. In order to determine who goes first in a game, one person picks a number between 1 and 10, then a second person picks a number from the remaining 9 numbers.

In Exercises 37 to 42, a pair of dice is tossed twice.

37. Find the probability that both rolls give a sum of 8.

38. Find the probability that the first roll is a sum of 6 and the second roll is a sum of 12.

39. Find the probability that the first roll is a total of at least 10 and the second roll is a total of at least 11.

40. Find the probability that both rolls result in doubles.

41. Find the probability that both rolls give even sums.

42. Find the probability that both rolls give at most a sum of 4.

43. A fair coin is tossed four times in succession. Find the probability of getting two heads followed by two tails.

44. **Monopoly** In the game of Monopoly, a player is sent to jail if they roll doubles with a pair of dice three times in a row. What is the probability of rolling doubles three times in succession?

45. Find the probability of tossing a pair of dice three times in succession and getting a sum of at least 10 on all three tosses.

46. Find the probability of tossing a pair of dice three times in succession and getting a sum of at most 3 on all three tosses.

In Exercises 47 to 52, a card is drawn from a standard deck and replaced. After the deck is shuffled, another card is pulled.

47. What is the probability that both cards pulled are aces?

48. What is the probability that both cards pulled are face cards?

49. What is the probability that the first card drawn is a spade and the second card is a diamond?

50. What is the probability that the first card drawn is an ace and the second card is not an ace?

51. Find the probability that the first card drawn is a heart and the second card is a spade.

52. Find the probability that the first card drawn is a face card and the second card is black.

53. A standard deck of playing cards is shuffled, and three people each choose a card. Find the probability that the first two cards chosen are diamonds and the third card is black if

    a. the cards are chosen *with* replacement.

    b. the cards are chosen *without* replacement.

54. A standard deck of playing cards is shuffled and three people each choose a card. Find the probability that all three cards are face cards if

    a. the cards are chosen *with* replacement.

    b. the cards are chosen *without* replacement.

55. A bag contains five red marbles, four green marbles, and eight blue marbles. Find the probability of pulling two red marbles followed by a green marble if the marbles are pulled from the bag

    a. with replacement.

    b. without replacement.

56. A box contains three medium t-shirts, five large t-shirts, and four extra-large t-shirts. If someone randomly chooses three t-shirts from the box, find the probability that the first t-shirt is large, the second is medium, and the third is large if the shirts are chosen

    a. with replacement.   b. without replacement.

57. **Drug Testing** A company that performs drug testing guarantees that its test determines a positive result with 97% accuracy. However, the test also gives 6% false positives. If 5% of those being tested actually have the drug present in their bloodstream, find the probability that a person testing positive has actually been using drugs. Round to the nearest thousandth.

58. **Genetic Testing** A test for a genetic disorder can detect the disorder with 94% accuracy. However, the test will incorrectly report positive results for 3% of those without the disorder. If 12% of the population has the disorder, find the probability that a person testing positive actually has the genetic disorder. Round to the nearest thousandth.

59. **Disease Testing** A pharmaceutical company has developed a test for a rare disease that is present in 0.5% of the population. The test is 98% accurate in determining a positive result, and the chance of a false positive is 4%. What is the probability that someone who tests positive actually has the disease? Round to the nearest thousandth.

▶ Investigation

Random Walk Suppose you are standing at a street corner and flip a coin to decide whether you will go north or south from your current position. When you reach the next intersection, you repeat the procedure. This problem is a simplified version of what is called a *random walk* problem. Problems of this type are important in economics, physics, chemistry, biology, and other disciplines.

1. After performing this experiment three times, what is the probability that you will be three blocks north of your original position?

2. After performing this experiment four times, what is the probability that you will be two blocks north of your original position?

3. After performing this experiment four times, what is the probability that you will be back at your original position?

To extend the problem on moving north or south, suppose you decide to go north, south, east, or west based on tossing a coin twice. If the result is HH, you move north; HT, you move east; TH, you move south, and TT, you move west. The probability of moving in any one direction is $\frac{1}{4}$ or 0.25.

4. If you perform this experiment four times, how many different walks can you take?

5. Of all the possible walks you can take, how many end with you back at the starting point?

6. What is the probability of returning to the starting point? Is this more than or less than the probability of ending at some other point other than the starting point?

Section
7.6

Expectation

Learning Objectives:

- Calculate the expectation of an event.
- Apply expectation to business applications.

Expectation

Suppose a barrel contains a large number of balls, half of which have the number 1000 painted on them and the other half of which have the number 500 painted on them. As the grand prize winner of a contest, you get to reach into the barrel (blindfolded, of course) and select 10 balls. Your prize is the sum of the numbers on the balls in cash.

If you are very lucky, all of the balls will have 1000 painted on them, and you will win $10,000. If you are very unlucky, all of the balls will have 500 painted on them, and you will win $5000. Most likely, however, approximately one-half of the balls will have 1000 painted on them and one-half will have 500 painted on them. The amount of your winnings in this case will be 5(1000) + 5(500), or $7500. Because 10 balls are drawn, your amount of winnings per ball is $\dfrac{\$7500}{10} = \$750$.

The number $750 is called the *expected value* or *expectation* of the game. You cannot win $750 on one draw, but if given the opportunity to draw many times, you will win, on average, $750 per ball.

We can also calculate expectation by using probabilities. For the game above, one-half of the balls have the number 1000 painted on them and one-half have the number 500 painted on them. Therefore, $P(1000) = \frac{1}{2}$ and $P(500) = \frac{1}{2}$. The expectation is calculated as follows.

Expectation

$$= (\text{probability of winning } \$1000) \cdot \$1000 + (\text{probability of winning } \$500) \cdot \$500$$

$$= P(1000) \cdot \$1000 + P(500) \cdot \$500$$

$$= \frac{1}{2} \cdot \$1000 + \frac{1}{2} \cdot \$500 = \$500 + \$250 = \$750$$

The general result for experiments with numerical outcomes follows.

Expectation

Let $S_1, S_2, S_3, \ldots, S_n$ be the possible numerical outcomes of an experiment, and let $P(S_1), P(S_2), P(S_3), \ldots, P(S_n)$ be the probabilities of those outcomes. Then the **expectation** of the experiment is

$$P(S_1) \cdot S_1 + P(S_2) \cdot S_2 + P(S_3) \cdot S_3 + \cdots + P(S_n) \cdot S_n$$

That is, to find the expectation of an experiment, multiply the probability of each outcome of the experiment by the outcome and then add the results.

Example 1   **Expectation in Gambling**

One of the wagers in roulette is to place a bet on 1 of the numbers from 0 to 36 or on 00. If that number comes up, the player wins 35 times the amount bet (and keeps the original bet). Suppose a player bets $1 on a number. What is the player's expectation?

Take Note: In Example 1, suppose the player bets $5 instead of $1. The payoff is then $175 ($35 \cdot 5$) if the player wins and $-$5$ if the player loses. The expectation is $\frac{1}{38}(175) + \frac{37}{38}(-5) = -\frac{5}{19}$. Note that the bet is 5 times greater and the expectation, $-\frac{5}{19}$, is 5 times the expectation when $1 is bet. Thus a player who makes $5 bets can expect to lose 5 times as much money as a player who makes $1 bets.

Solution

Let S_1 be the event that the player's number comes up and the player wins $35. Because there are 38 numbers from which to choose, $P(S_1) = \frac{1}{38}$. Let S_2 be the event that the player's number does not come up and the player therefore loses $1. Then $P(S_2) = 1 - \frac{1}{38} = \frac{37}{38}$.

$$\text{Expectation} = P(S_1) \cdot S_1 + P(S_2) \cdot S_2$$

$$= \frac{1}{38}(35) + \frac{37}{38}(-1) = -\frac{1}{19}$$

• The amount the player can win is entered as a positive number. The amount that can be lost is entered as a negative number.

$$\approx -0.053$$

The player's expectation is approximately $-$0.053$. This means that, on average, the player will lose about $0.05 every time this bet is made.

In Example 1, the fact that the player is losing approximately 5 cents on each dollar bet means that the casino's expectation is positive 5 cents; it is earning (on average) 5 cents for every dollar spent making that particular bet at the roulette wheel. An individual player may get lucky, but over time the casino can plan on a predictable profit.

Example 2   **Use Excel to Find Expected Value**

Historical records for the number goals scored by a hockey team and the probability of scoring those goals are shown in the following table.

Goals	Probability	Goals	Probability
0	0.05	5	0.06
1	0.19	6	0.04
2	0.23	7	0.02
3	0.22	8	0.01
4	0.17	9	0.01

Use Excel to find the expected value of the number of goals scored by the team.

Solution

Step 1: Enter the values in two columns

	A	B
1	Goals	Probability
2	0	0.05
3	1	0.19
4	2	0.23
5	3	0.22
6	4	0.17
7	5	0.06
8	6	0.04
9	7	0.02
10	8	0.01
11	9	0.01

Step 2: Find the product of the number of goals and the probability of that number of goals for the second row.

	A	B	C
1	Goals	Probability	Goals x Probability
2	0	0.05	=A2*B2

Step 3: Drag the formula in C2 through cell C11.

	A	B	C
1	Goals	Probability	Goals x Probability
2	0	0.05	0
3	1	0.19	0.19
4	2	0.23	0.46
5	3	0.22	0.66
6	4	0.17	0.68
7	5	0.06	0.3
8	6	0.04	0.24
9	7	0.02	0.14
10	8	0.01	0.08
11	9	0.01	0.09
12			

Step 4: Sum the values from C2 to C11. Place the result in C12.

	A	B	C
1	Goals	Probability	Goals x Probability
2	0	0.05	0
3	1	0.19	0.19
4	2	0.23	0.46
5	3	0.22	0.66
6	4	0.17	0.68
7	5	0.06	0.3
8	6	0.04	0.24
9	7	0.02	0.14
10	8	0.01	0.08
11	9	0.01	0.09
12			2.84

=SUM(C2:C11)

=SUM(C2:C11)

The team scores 2.84 goals per game.

Business Applications of Expectation

When an insurance company sells a life insurance policy, the premium (the cost to purchase the policy) is based to a large extent on the probability that the insured person will outlive the term of the policy. Such probabilities are found in *mortality tables,* which give the probability that a person of a certain age will live 1 more year. The insurance company wants to know its expectation on a policy—that is, how much it will have to pay out, on average, for each policy it writes.

Example 3   Expectation in Insurance

According to mortality tables published in the National Vital Statistics Report, the probability that a 21-year-old will die within 1 year is approximately 0.000962. Suppose that the premium for a 1-year, $25,000 life insurance policy for a 21-year-old is $32. What is the insurance company's expectation for this policy?

Solution

Let S_1 be the event that the person dies within 1 year. Then $P(S_1) = 0.000962$, and the company must pay out $25,000. Because the company charged $32 for the policy, the company's actual loss is $24,968. Let S_2 be the event that the policy holder does not die during the year of the policy. Then $P(S_2) = 0.999038$, and the company keeps the premium of $32. The expectation is

$$\text{Expectation} = P(S_1) \cdot S_1 + P(S_2) \cdot S_2$$

$$= 0.000962(-24{,}968) + 0.999038(32)$$

$$= 7.95$$

• The amount the company pays out is entered as a negative number. The amount the company receives is entered as a positive number.

The company's expectation is $7.95, so the company earns, on average, $7.95 for each policy sold.

Expectation is also used when a company bids on a project. The company must try to predict the costs and amount of work involved if it is to make a bid that will ensure a profit. At the same time, if the bid is too high, the client may reject the offer. Because it is impossible to predict in advance the exact requirements of a job, probabilities can be used to analyze the likelihood of making a profit.

Example 4   Expected Company Profits

Suppose a software company bids on a project to update the database program for an accounting firm. The software company assesses its potential profit as shown in Table 7.9.

Table 7.9

Profit/Loss	Probability
$75,000	0.10
$50,000	0.25
$20,000	0.50
−$10,000	0.10
−$25,000	0.05

What is the profit expectation for the company?

Solution

The company's expected profit is

Expectation

$$= 0.10(75{,}000) + 0.25(50{,}000) + 0.50(20{,}000) + 0.10(-10{,}000) + 0.05(-25{,}000)$$

$$= 7500 + 12{,}500 + 10{,}000 - 1000 - 1250 = 27{,}750$$

The company's expected profit is $27,750.

7.6 Exercise Set

Think About It

1. Can the expectation of an event be less than zero?
2. If the expectation of an event is −$2, write a sentence that explains the meaning of this expectation.
3. Suppose a wager of $5 has an expectation of $4.80. If $10 is wagered on the same bet, what is the expectation?

Preliminary Exercises

For Exercises P1 to P3, if necessary, use the referenced example following the exercise for assistance.

P1. In roulette it is possible to place a wager that 1 of the numbers between 1 and 12 (inclusive) will come up. If it does, the player wins twice the amount bet. Suppose a player bets $5 that a number between 1 and 12 will come up. What is the player's expectation? **Example 1**

P2. Entrepreneurship Suppose you are considering quitting your day job and opening a new online business. Your current annual salary is $75,000. The following table shows the probability of earning a certain annual salary from your new business. **Example 2**

Annual Salary ($)	Probability	Annual Salary ($)	Probability
45,000	0.09	90,000	0.08
60,000	0.37	100,000	0.03
70,000	0.26	125,000	0.02
85,000	0.14	150,000	0.01

Use Excel to determine whether your expected annual salary from your new business is more than or less than your current annual salary.

P3. The probability that an 18-year-old will die within 1 year is approximately 0.000753. Suppose that the premium for a 1-year, $10,000 life insurance policy for an 18-year-old is $45. What is the insurance company's expectation for this policy? **Example 3**

P4. A road construction company bids on a project to build a new freeway. The company estimates its potential profit as shown in Table 7.10. What is the profit expectation for the company? **Example 4**

Table 7.10

Profit/Loss	Probability
$500,000	0.05
$250,000	0.30
$150,000	0.35
−$100,000	0.20
−$350,000	0.10

Section Exercises

1. The outcomes of an experiment and the probability of each outcome are given in the following table. Compute the expectation for this experiment.

Outcome	Probability
30	0.15
40	0.2
50	0.4
60	0.05
70	0.2

2. The outcomes of an experiment and the probability of each outcome are given in the following table. Compute the expectation for this experiment.

Outcome	Probability
5	0.4
6	0.3
7	0.1
8	0.08
9	0.07
10	0.05

3. **Roulette** One of the wagers in the game of roulette is to place a bet that the ball will land on a black number. (Eighteen of the numbers are black, 18 are red, and 2 are green.) If the ball lands on a black number, the player wins the amount bet. If a player bets $1, find the player's expectation.

4. **Roulette** One of the wagers in roulette is to bet that the ball will stop on a number that is a multiple of 3. (Both 0 and 00 are not included.) If the ball stops on such a number, the player wins double the amount bet. If a player bets $1, compute the player's expectation.

Many casinos have a game called the Big Six Money Wheel, which has 54 slots in which are displayed a Joker, the casino logo, and various dollar amounts, as shown in the following table. Players may bet on the Joker, the casino logo, or one or more dollar denominations. The wheel is spun and if the wheel stops on the same place as the player's bet, the player wins that amount for each dollar bet. Exercises 5 to 8 use this game.

Denomination	Number of Slots
$40 (Joker)	1
$40 (Casino logo)	1
$20	2
$10	4
$5	7
$2	15
$1	24

5. If a player bets $1 on the Joker denomination, find the player's expectation.

6. If a player bets $1 on the $20 denomination, find the player's expectation.

7. If a player bets $1 on the $5 denomination, find the player's expectation.

8. If a player bets $1 on the $2 denomination, find the player's expectation.

Exercises 9 to 14 use data taken from mortality tables published in the National Vital Statistics Report.

9. The probability that a 22-year-old female in the United States will die within 1 year is approximately 0.000487. If an insurance company sells a 1-year, $25,000 life insurance policy to such a person for $75, what is the company's expectation?

10. The probability that a 28-year-old male in the United States will die within 1 year is approximately 0.001362. If an insurance company sells a 1-year, $20,000 life insurance policy to such a person for $155, what is the company's expectation?

11. The probability that an 80-year-old male in the United States will die within 1 year is approximately 0.070471. If an insurance company sells a 1-year, $10,000 life insurance policy to such a person for $495, what is the company's expectation?

12. The probability that an 80-year-old female in the United States will die within 1 year is approximately 0.050409. If an insurance company sells a 1-year, $15,000 life insurance policy to such a person for $860, what is the company's expectation?

13. The probability that a 30-year-old male in the United States will die within 1 year is about 0.001406. An insurance company is preparing to sell a 30-year-old male a 1-year, $30,000 life insurance policy. How much should it charge for its premium in order to have a positive expectation for the policy?

14. The probability that a 25-year-old female in the United States will die within 1 year is about 0.000509. An insurance company is preparing to sell a 25-year-old female a 1-year, $75,000 life insurance policy. How much should it charge for its premium in order to have a positive expectation for the policy?

15. **Construction** A construction company has been hired to build a custom home. The builder estimates the probabilities of potential profit (or loss) as shown in the following table. What is the profit expectation for the company?

Profit/Loss	Probability
$100,000	0.10
$60,000	0.40
$30,000	0.25
$0	0.15
−$20,000	0.08
−$40,000	0.02

16. **Painting** A professional painter has been hired to paint a commercial building for $18,000. From this fee, the painter must buy supplies and pay employees. The painter estimates

the potential profit as shown in the following table. What is the profit expectation for the painter?

Profit/Loss	Probability
$10,000	0.15
$8,000	0.35
$5,000	0.2
$3,000	0.2
$1,000	0.1

17. **Design Consultant** A consultant has been hired to redesign a company's production facility. The consultant estimates the probabilities of her potential profit as shown in the following table. What is her profit expectation?

Profit/Loss	Probability
$40,000	0.05
$30,000	0.2
$20,000	0.5
$10,000	0.2
$5,000	0.05

18. **Office Rentals** A real estate company has purchased an office building with the intention of renting office space to small businesses. The company estimates the probabilities of potential profit (or loss) as shown in the following table. What is the profit expectation for the company?

Profit/Loss	Probability
$700,000	0.15
$400,000	0.25
$200,000	0.25
$50,000	0.20
−$100,000	0.10
−$250,000	0.05

19. **Overbooking** Airlines typically overbook a flight because historical data shows that there are no-shows for flights. Suppose an airline has historical data, shown in the following table, that shows the probability the airline will overbook a flight by a certain number of seats and that the airline will have to pay $250 to compensate each passenger in the overbooked situation.

Seats Overbooked	Cost to Airline ($)	Probability of Overbooking
1	250	0.0013
2	500	0.0052
3	750	0.0095
4	1000	0.0195
5	1250	0.0295
6	1500	0.0481
7	1750	0.1050
8	2000	0.1419
9	2250	0.2480
10	2500	0.3920

Use Excel to find the expected value the airline will have to pay for overbooking this flight. You can assume that the flight never has more than 10 overbooked seats.

20. **ATM** Suppose an ATM dispenses only $20 bills. A bank teller must fill the machine for Sunday when the bank is not open. The following table is based on historical records from the bank showing the probability of a certain number of $20 bills withdrawn for each transaction from the ATM on a Sunday. You may assume there is a limit of 10 $20 bills that can be withdrawn in a single transaction.

No. of $20 Bills	Probability
1	0.039
2	0.091
3	0.149
4	0.172
5	0.187
6	0.145
7	0.101
8	0.053
9	0.025
10	0.038

Use Excel to find the expected number of $20 bills withdrawn per transaction.

Investigations

Sicherman Dice   Two dice, one labeled 1, 2, 2, 3, 3, 4 and the other labeled 1, 3, 4, 5, 6, 8, are rolled once. Use the formula for expectation to determine the expected sum of the numbers on the upward faces of the two dice. Dice such as these are called *Sicherman dice*.

Efron's Dice   Suppose you are offered one of two pairs of dice, a red pair or a green pair, that are labeled as follows:

Red die 1: 0, 0, 4, 4, 4, 4

Red die 2: 2, 3, 3, 9, 10, 11

Green die 1: 3, 3, 3, 3, 3, 3

Green die 2: 0, 1, 7, 8, 8, 8

After you choose, your friend will receive the other pair. Which pair should you choose if you are going to play a game in which each of you rolls your dice and the player with the higher sum wins? Dice such as these are part of a set of four pairs of dice called *Efron's dice*.

Test Prep Introduction to Probability

Section 7.1   The Counting Principle

Sample Spaces An experiment is an activity with an observable outcome. The sample space of an experiment is the set of all possible outcomes. A table or a tree diagram can be used to list all the outcomes in the sample space of a multi-stage experiment.	See **Examples 3** and **4** on pages 303 and 304, and then try Exercises 3 and 4 on page 350.
The Counting Principle Let E be a multi-stage experiment. If $n_1, n_2, n_3, \ldots, n_k$ are the numbers of possible outcomes of each of the k stages of E, then there are $n_1 \cdot n_2 \cdot n_3 \cdot \ldots \cdot n_k$ possible outcomes for E.	See **Example 5** on page 305, and then try Exercises 5 and 7 on page 350.
Counting With and Without Replacement Some multi-stage experiments involve repeatedly choosing an element from a given set. If an element is returned to the set and can be chosen again, the experiment is performed *with replacement*. If an element is not returned to the set and cannot be chosen again, the experiment is performed *without replacement*.	See **Example 6** on page 306, and then try Exercises 1 and 2 on page 350.

Section 7.2   Permutations and Combinations

n factorial n factorial is the product of the natural numbers n through 1. $$n! = (n-1) \cdot (n-2) \cdot \ldots \cdot 3 \cdot 2 \cdot 1$$ $$0! = 1$$	See **Example 1** on page 309, and then try Exercises 9 to 11 on page 350.
Permutation Formula for Distinct Objects A permutation is an arrangement of objects in a definite order. The number of permutations of n distinct objects selected k at a time is $P(n, k) = \dfrac{n!}{(n-k)!}$	See **Examples 2** and **3** on page 311, and then try Exercises 15 and 20 on page 350.
Permutations of Objects, Some of Which Are Identical The number of permutations of n objects of r different types, where k_1 identical objects are of one type, k_2 are of another, and so on, is $\dfrac{n!}{k_1! \cdot k_2! \cdot \ldots \cdot k_r!}$, where $k_1 + k_2 + \cdots + k_r = n$.	See **Example 5** on page 312, and then try Exercise 18 on page 350.
Combination Formula A combination is a collection of objects for which the order is not important. The number of combinations of n objects chosen k at a time is $$C(n, k) = \dfrac{P(n, k)}{k!} = \dfrac{n!}{k! \cdot (n-k)!}$$	See **Example 6** on page 313, and then try Exercises 21 and 22 on page 350.
Applying Several Counting Techniques Some counting problems require using the counting principle along with a permutation or combination formula.	See **Examples 4, 7,** and **8** on pages 311 and 314, and then try Exercises 23 to 25 on page 350.

Section 7.3   Probability and Odds

Theoretical Probability of an Event For an experiment with sample space S of equally likely outcomes, the theoretical probability $P(E)$ of an event E is given by $$P(E) = \frac{n(E)}{n(S)} = \frac{\text{number of elements in } E}{\text{number of elements in } S}$$	See **Examples 2** and **3** on page 319, and then try Exercises 27 and 31 on pages 350 and 351.
Empirical Probability of an Event If an experiment is performed repeatedly and the occurrence of the event E is observed, the empirical probability $P(E)$ of the event E is given by $$P(E) = \frac{\text{number of times event } E \text{ occurred}}{\text{number of times the experiment was performed}}$$	See **Example 4** on page 320, and then try Exercise 29 on page 351.

Punnett Square Using a capital letter to represent a dominant allele and a lowercase letter to represent a recessive allele, a Punnett square shows all possible genotypes of the offspring of two parents with given genotypes.	See **Example 5** on page 322, and then try Exercise 44 on page 351.
Odds of an Event Let E be an event in a sample space of equally likely outcomes. Then Odds in favor of $E = \dfrac{\text{number of favorable outcomes}}{\text{number of unfavorable outcomes}}$ Odds against $E = \dfrac{\text{number of unfavorable outcomes}}{\text{number of favorable outcomes}}$	See **Example 6** on pages 322 to 323, and then try Exercises 41 and 42 on page 351.
Relationship Between Odds and Probability If the odds in favor of event E are $\dfrac{a}{b}$, then $P(E) = \dfrac{a}{a+b}$. This relationship is also expressed by the equation Odds in favor of $E = \dfrac{P(E)}{1 - P(E)}$	See **Example 7** on page 323, and then try Exercise 43 on page 351.

Section 7.4   Addition and Complement Rules

Probability of Mutually Exclusive Events Two events A and B are mutually exclusive if they cannot occur at the same time. In this case, the probability of A or B occurring is $P(A \text{ or } B) = P(A) + P(B)$	See **Example 1** on page 327, and then try Exercises 27 and 31 on pages 350 and 351.
Addition Rule for Probabilities If A and B are two events in a sample space S, then $P(A \text{ or } B) = P(A) + P(B) - P(A \text{ and } B)$	See **Example 2** on page 328, and then try Exercise 38 on page 350.
Probability of the Complement of an Event The complement of an event E in a sample space S includes all those outcomes of S that are not in E and excludes the outcomes in E. The symbol for the complement of E is E'. To find the probability of E', use the relationship $P(E') = 1 - P(E)$	See **Example 3** on page 329, and then try Exercise 39 on page 350.
Using Combinatorics Formulas to Find Probabilities In many cases, the process of finding a probability involves using the counting principle and/or a permutation or combination formula.	See **Examples 4** and **5** on pages 330 to 331, and then try Exercises 60 and 63 on pages 351 and 352.

Section 7.5   Conditional Probability

Conditional Probability Formula If A and B are two events in a sample space S, then the conditional probability of B given that A has occurred is $P(B \mid A) = \dfrac{P(A \text{ and } B)}{P(A)}$	See **Example 1** on pages 334 to 335, and then try Exercises 35 and 36 on page 351.
Product Rule for Probabilities If A and B are two events in a sample space S, then $P(A \text{ and } B) = P(A) \cdot P(B \mid A)$ **Probability of Successive Events** The probability of two or more events occurring in succession is the product of the conditional probabilities of each of the events.	See **Example 2** on page 336, and then try Exercises 51 and 57 on page 351.
Product Rule for Independent Events If A and B are two events in a sample space S and $P(B \mid A) = P(B)$, then A and B are independent events. In this case $P(A \text{ and } B) = P(A) \cdot P(B)$	See **Example 3** on page 337, and then try Exercise 39 on page 351.
Using Several Formulas to Find Probabilities The process of finding a conditional probability may involve using other probability formulas.	See **Example 4** on pages 337 to 338, and then try Exercise 61 on page 351.

Section 7.6 Expectation

Expectation If $S_1, S_2, S_3, \ldots, S_n$ are the possible numerical outcomes of an experiment and $P(S_1), P(S_2), P(S_3), \ldots, P(S_n)$ are the probabilities of those outcomes, then the expectation of the experiment is $$P(S_1) \cdot S_1 + P(S_2) \cdot S_2 + P(S_3) \cdot S_3 + \cdots + P(S_n) \cdot S_n$$	See **Examples 1** and **2** on pages 341 to 343, and then try Exercises 64 and 65 on page 352.
Business Applications of Expectation Insurance companies can use expectation to help determine life insurance premiums. Other businesses can use expectation to determine potential profits.	See **Examples 3** and **4** on page 344, and then try Exercises 68 and 70 on page 352.

Review Exercises

In Exercises 1 and 2, list the elements of the sample space for the given experiment.

1. Two-digit numbers are formed, with replacement, from the digits 1, 2, and 3.

2. Two-digit numbers are formed, without replacement, from the digits 2, 6, and 8.

3. Use a tree diagram to list all possible outcomes that result from tossing four coins.

4. Use a table to list all possible two-character codes that can be formed from one of the digits 7, 8, or 9 followed by one of the letters A or B.

5. An athletic shoe store sells running shoes in three styles that come in four colors. Each color comes in six sizes. How many distinct shoes are available?

6. The combination for a lock to a bicycle chain contains four numbers chosen from the numbers 0 through 9. How many different lock combinations are possible? Assume that a number can be used more than once.

7. **Serial Numbers** In the 1970s and 1980s, the Conn music company assigned the first four characters in serial numbers of instruments in the following way. The first character is one of the letters G or H and indicates the decade in which the instrument was made: "G" for 1970s and "H" for 1980s. The second character indicates the month of the year in which the instrument was made: "A" for January, "B" for February, and so on. The third character is a number from 0 to 9 indicating the year of the decade in which the instrument was made, and the fourth character is a number from 0 to 9 indicating the type of instrument: 1 = cornet, 2 = trumpet, 3 = alto horn, 4 = French horn, 5 = mellophone, 6 = valve trombone, 7 = slide trombone, 8 = euphonium, 9 = tuba, 0 = sousaphone. How many four-character sequences are possible?

8. **Codes** A *biquinary code* is a code that consists of two different binary digits (a binary digit is a 0 or a 1) followed by five binary digits for which there are no restrictions. How many biquinary codes are possible?

In Exercises 9 to 14, evaluate each expression.

9. $7!$

10. $8! - 4!$

11. $\dfrac{9!}{2! \, 3! \, 4!}$

12. $P(10, 6)$

13. $P(8, 3)$

14. $\dfrac{C(6, 2) \cdot C(8, 3)}{C(14, 5)}$

15. In how many different ways can seven people arrange themselves in a line to receive service from a bank teller?

16. A matching test has seven definitions that are to be paired with seven words. Assuming each word corresponds to exactly one definition, how many different matches are possible by random matching?

17. A matching test has seven definitions to be matched with five words. Assuming each word corresponds to exactly one definition, how many different matches are possible by random matching?

18. How many distinct arrangements are possible using the letters of the word *letter*?

19. Twelve identical coins are tossed. How many distinct arrangements are possible consisting of four heads and eight tails?

20. **Work Shifts** Three positions are open at a manufacturing plant: the day shift, the swing shift, and the night shift. In how many different ways can five people be assigned to the three shifts?

21. A professor assigns 25 homework problems, of which 10 will be graded. How many different sets of 10 problems can the professor choose to grade?

22. **Stock Portfolios** A stockbroker recommends 11 stocks to a client. If the client will invest in 3 of the stocks, how many different 3-stock portfolios can be selected?

23. **Quality Control** A quality control inspector receives a shipment of 15 computer monitors, of which 3 are defective. If the inspector randomly chooses 5 monitors, how many different sets can be formed that consist of 3 nondefective monitors and 2 defective monitors?

24. In how many ways can nine people be seated in nine chairs if two of the people refuse to sit next to each other?

25. How many 5-card poker hands consist of 4 of a kind (4 aces, 4 kings, 4 queens, and so on)?

26. For two-factor authentication, a six-digit code is sent to a user. What is the probability that the code has 2 even numbers and 4 odd numbers? Remember that 0 is an even number.

27. If a coin is tossed three times, what is the probability of getting one head and two tails?

28. A large university currently employs 5739 teaching assistants and 7290 professors. If an employee is selected at random, what is the probability that the employee is a professor?

In Exercises 29 and 30, use the table below, which shows the number of students at a university who are currently in each class level.

Class Level	Number of Students
First year	642
Sophomore	549
Junior	483
Senior	445
Graduate student	376

29. If a student is selected at random, what is the probability that the student is an upper-division undergraduate student (junior or senior)?

30. If a student is selected at random, what is the probability that the student is not a graduate student?

In Exercises 31 to 36, a pair of dice is tossed.

31. Find the probability that the sum of the pips on the two upward faces is 9.

32. Find the probability that the sum on the two dice is not 11.

33. Find the probability that the sum on the two dice is at least 10.

34. Find the probability that the sum on the two dice is an even number or a number less than 5.

35. What is the probability that the sum on the two dice is 9, given that the sum is odd?

36. What is the probability that the sum on the two dice is 8, given that doubles were rolled?

In Exercises 37 to 40, a single card is selected from a standard deck of playing cards.

37. What is the probability that the card is a heart or a black card?

38. What is the probability that the card is a heart or a jack?

39. What is the probability that the card is not a 3?

40. What is the probability that the card is red, given that it is not a club?

41. If a pair of dice is rolled, what are the odds in favor of getting a sum of 6?

42. If one card is drawn from a standard deck of playing cards, what are the odds that the card is a heart?

43. If the odds against an event occurring are 4 to 5, compute the probability of the event occurring.

44. **Genotypes** The hair length of a particular rodent is determined by a dominant allele H, corresponding to long hair, and a recessive allele h, corresponding to short hair. Draw a Punnett square for parents of genotypes Hh and hh, and compute the probability that the offspring of the parents will have short hair.

45. Two cards are drawn, without replacement, from a standard deck of playing cards. The probability that exactly one card is an ace is 0.145. The probability that exactly one card is a face card (jack, queen, or king) is 0.362, and the probability that a selection of two cards will contain an ace or a face card is 0.471. Find the probability that the two cards are an ace and a face card.

46. **Surveys** A recent survey asked 1000 people whether they liked cheese-flavored corn chips (642 people),

jalapeño-flavored chips (487 people), or both (302 people). If 1 person is chosen from this survey, what is the probability that the person does not like either of the 2 flavors?

In Exercises 47 to 51, a box contains 24 different colored chips that are identical in size. Five are black, 4 are red, 8 are white, and 7 are yellow.

47. If a chip is selected at random, what is the probability that the chip will be yellow or white?

48. If a chip is selected at random, what are the odds in favor of getting a red chip?

49. If a chip is selected at random, find the probability that the chip is yellow given that it is not white.

50. If five chips are randomly chosen, without replacement, what is the probability that none of them is red?

51. If three chips are chosen without replacement, find the probability that the first one is yellow, the second is white, and the third is yellow.

In Exercises 52 to 56, use the following table, which shows the number of voters in a city who voted for or against a proposition (or abstained from voting) according to their party affiliations. Round answers to the nearest hundredth.

	For	Against	Abstained
Democrat	8452	2527	894
Republican	2593	5370	1041
Independent	1225	712	686

52. If a voter is chosen at random, compute the probability that the person voted against the proposition.

53. If a voter is chosen at random, compute the probability that the person is a Democrat or an Independent.

54. If a voter is randomly chosen, what is the probability that the person abstained from voting on the proposition and is not a Republican?

55. A voter is randomly selected. What is the probability that the individual voted for the proposition, given that the voter is a registered Independent?

56. A voter is randomly selected. What is the probability that the individual is registered as a Democrat, given that the person voted against the proposition?

57. A single die is rolled three times in succession. What is the probability that each roll gives a 6?

58. A single die is rolled five times in succession. Find the probability that a 6 will be rolled at least once.

59. A single die is rolled five times in succession. Find the probability that exactly two of the rolls give a 6.

60. A person draws a card from a standard deck and replaces it; she then does this three more times. What is the probability that she drew a spade at least once?

61. **Disease Testing** A veterinarian uses a test to determine whether a dog has a disease that affects 7% of the dog population. The test correctly gives a positive result for 98% of dogs that have the disease, but gives false positives for 4% of dogs that do not have the disease. If a dog tests positive, what is the probability that the dog has the disease?

62. **Weather** Suppose that in your area in the wintertime, if it rains one day, there is a 65% chance that it will rain the next day. If it does not rain on a given day, there is only a 15% chance that it will rain the following day. What is the probability that, if it didn't rain today, it will rain the next 2 days but not the following 2?

63. **Batteries** About 1.2% of AA batteries produced by a particular manufacturer are defective. If a consumer buys a box of 12 of these batteries, what is the probability that at least one battery is defective?

64. Suppose it costs $4 to play a game in which a single die is rolled and you win the amount of dollars that the die shows. What is the expectation for the game?

65. I will flip two coins. If both coins come up tails, I will pay you $5. If one shows heads and one shows tails, you will pay me $2. If both coins come up heads, we will call it a draw. What is your expectation for this game?

66. **Raffle Tickets** For a fundraiser, an elementary school is selling 800 raffle tickets for $1 each. From these, 5 tickets will be drawn. One of the winners gets $200 and the others each get $75. If you buy 1 raffle ticket, what is your expectation?

67. If a pair of dice is rolled 65 times, on how many rolls can we expect to get a total of 4?

68. **Life Insurance** The probability that a 29-year-old female in the United States will die within 1 year is approximately 0.000595. If an insurance company sells a 1-year, $40,000 life insurance policy to a 29-year-old for $320, what is the company's expectation?

69. **Life Insurance** The probability that a 19-year-old male in the United States will die within 1 year is approximately 0.001188. If an insurance company sells a 1-year, $25,000 life insurance policy to a 19-year-old for $795, what is the company's expectation?

70. **Construction** A construction company has bid on a building renovation project. The company estimates the probabilities of potential profit (or loss) as shown in the table below. What is the profit expectation for the company?

Profit/Loss	Probability
$25,000	0.20
$15,000	0.25
$10,000	0.20
$5,000	0.15
$0	0.10
−$5,000	0.10

71. **Hurricane insurance** An insurance company is planning to issue a policy to an apartment building owner for potential damage due to a hurricane. The following table shows the amount of potential loss, in millions of dollars, and the probability of that loss. Find the expectation for the insurance company.

Loss ($, in millions)	2.0	1.75	1.50	1.0	0.75	0.50	0.25	0.15
Probability of Loss	0.010	0.023	0.031	0.059	0.075	0.121	0.240	0.441

Project 1

Keno

The casino game keno is played by choosing numbers from 1 to 80 with hopes that the casino will draw balls with the same numbers. A player can choose exactly 1 number or as many as 20. The casino will then pick 20 numbered balls from the 80 possible; if enough of the player's numbers match the lucky numbers the casino chooses, the player wins money. The amount won varies according to how many numbers were chosen and how many are "lucky."

Project Exercises

1. A gambler playing keno randomly chooses five numbers. What is the probability that the gambler will match at least one lucky number?

2. If five numbers are chosen, compute the probability of matching fewer than five lucky numbers.

3. If the keno player chooses 15 numbers, bets $1, and matches 13 of the lucky numbers, the gambler will be paid $12,000. What is the probability of this occurring?

4. If the keno player chooses 15 numbers and matches 5 or 6 of the lucky numbers, the gambler gets the bet back but is not paid any extra. What is the probability of this occurring?

5. Some casinos will let you choose up to 20 numbers. In this case, if you don't match any of the lucky numbers, the casino pays you! Although this may seem unusual, it is actually more difficult not to match any of the lucky numbers than it is to match a few of them. Compute the probability of not matching any of the lucky numbers at all, and compare it to the probability of matching five lucky numbers.

6. If 20 numbers are chosen, find the probability of matching *at least one* lucky number.

Project 2

Sharing Birthdays

Have you ever been introduced to someone at a party or other social gathering and discovered that you share the same birthday? It seems like an amazing coincidence, but exactly how rare is this coincidence?

As an example, suppose four people have gathered for a dinner party. We can determine the probability that at least two of the guests have the same birthday. (For simplicity, we will ignore the

February 29th birthday from leap years.) Let E be the event that at least two people share a birthday. In this case, it is easier to look at the complement E', the event that no one shares the same birthday.

If we start with one of the guests, then the second guest cannot share the same birthday, so that person has 364 possible dates for his or her birthday from a total of 365. Thus the conditional probability that the second guest has a different birthday, given that we know the first person's birthday, is $\frac{364}{365}$. Similarly, the third person has 363 possible birthday dates that do not coincide with those of the first two guests. So the conditional probability that the third person does not share a birthday with either of the first two guests, given that we know the birthdays of the first two people, is $\frac{363}{365}$. The probability of the fourth guest having a distinct birthday is, similarly, $\frac{362}{365}$. We can use the Product Rule for Probabilities to find the probability that all of these conditions are met; that is, none of the four guests share a birthday.

$$P(E') = \frac{364}{365} \cdot \frac{363}{365} \cdot \frac{362}{365} \approx 0.984$$

Then $P(E) = 1 - P(E') \approx 0.016$, so there is about a 1.6% chance that in a group of 4 people, 2 or more will have the same birthday.

It would require 366 people gathered together to *guarantee* that 2 people in the group will have the same birthday. But how many people would be necessary to guarantee that the chance of at least 2 of them sharing a birthday is at least 50/50? Make a guess before you proceed through the exercises. The results may surprise you!

Project Exercises

1. If eight people are present at a meeting, find the probability that at least two share a common birthday.
2. Compute the probability that at least 2 people among a group of 15 have the same birthday.
3. If 23 people are in attendance at a party, what is the probability that at least 2 share a birthday?
4. In a group of 40 people, what would you estimate to be the probability that at least 2 people share a birthday? If you have the patience, compute the probability to check your guess.

Math in Practice

At a fundraiser, a raffle is being held for forty bottles of wine. Each bottle of wine is inside a brown paper bag, so no one can see what kind of wine it is.

Part 1

There are 500 tickets available for the raffle. You buy 5 tickets.
a. What is the probability that you win the first bottle?
b. If forty winning tickets are drawn, what is the probability that you win the first two bottles?

Part 2

Your friend is a member of the fundraising committee, who tells you that there are 26 bottles of red wine and 14 bottles of white wine. After winning a bottle, you get to select one from the brown bags.
a. What is the probability of selecting a bottle of red wine?
b. What is the probability of selecting a bottle of white wine?

Part 3

The following table gives probabilities of winning a certain bottle of wine. Suppose you have one winning ticket, so you get to select one brown bagged bottle.

Value ($)	Probability
4	0.275
12	0.250
15	0.175
20	0.100
25	0.125
30	0.050
60	0.025

a. What is the probability of winning a $12 bottle or a $15 bottle?
b. What is the probability of winning a bottle less than $23?
c. What is the probability of not winning a $30 bottle?
d. What is the expected value?

Shippee/Shutterstock.com

Introduction to Statistics

Section	How This Is Used . . .
8.1 Measures of Central Tendency	You might hear expressions such as "the average daily temperature is 68°F" or "the average rent in a neighborhood is $2100 per month." In this section we refine the concept of "average" because there are quite a few of these that are used in various situations.
8.2 Measures of Dispersion	If you had four test scores of 90, 70, 100, and 60, you would have an average score of 80. Another student might have scores of 79, 80, 80, and 81 and also have an average score of 80. However, the dispersion of the scores is quite different. Knowing the dispersion of a data set has implications in many areas, one of which is making precision tools.
8.3 Measures of Relative Position	Many times the position of a data value relative to some other data value is important. For instance, giving the price of a stock as $30 is more useful if at one time it was $15 versus at one time being $50.
8.4 Normal Distributions	The normal distribution approximates many natural phenomena, such as the weights of candy bars produced by a machine, crop yields, the fluctuations of stock prices, many biological and chemical processes, and a host of other important applications.
8.5 Linear Regression and Correlation	When researchers gather data, they will plot the data on a coordinate system. Looking at the graphed data, researchers may ask, "Is there a model (an equation) that relates one variable to another?" One way to answer the question is using linear regression.

Math in Practice: Real Estate, p. 405

Applications of Statistics

Statistics is used to predict weather, if a law will pass, the on-time performance of an airline, the efficacy of a medical procedure, and countless other applications. In Exercise 8, on 391, we look at a biologist who is studying the wingspan of monarch butterflies.

Measures of Central Tendency

Learning Objectives:

- Find the mean of a set of data.
- Find the median of a set of data.
- Find the mode of a set of data.
- Find the weighted mean of a set of data.

Be Prepared: Prep Test 8 reviews concepts used in this chapter. This can be found in your Companion Site at Cengage.com

The Arithmetic Mean

Statistics involves the collection, organization, summarization, presentation, and interpretation of data. The branch of statistics that involves the collection, organization, summarization, and presentation of data is called **descriptive statistics**. The branch that interprets and draws conclusions from data is called **inferential statistics**.

One of the most basic statistical concepts involves finding a *measure of central tendency* of a set of numerical data. This is a single number that, in some sense, is the center of the data set.

For instance, suppose a senior is graduating with a degree in landscape architecture. Data from the college's career services department shows that landscape architecture graduates from last year's senior class received job offers with the following annual salaries.

$63,750 $59,500 $58,000 $61,250 $64,000

One "central" number around which the salaries cluster is the *arithmetic mean*, more commonly called the average.

$$\text{Average} = \frac{\$63,750 + \$59,500 + \$58,000 + \$61,250 + \$64,000}{5}$$

$$= \frac{\$306,500}{5} = \$61,300$$

This average suggests that the graduate can reasonably expect to find a position as a landscape architect that pays around $61,300 annually. Because, as will be shown shortly, there are various "central" numbers, we will use *arithmetic mean* (sometimes, just mean) for this average.

The **arithmetic mean** for the salaries is $61,300.

Arithmetic Mean

The **arithmetic mean** (or just mean) of n numbers is the sum of the numbers divided by n.

$$\text{Arithmetic mean} = \frac{\text{Sum of data values}}{n}$$

The Excel function AVERAGE can be used to find the arithmetic mean. For the salary data, enter

=AVERAGE(63750,59500,58000,61250,64000)

in a cell. Note that although the data values contain commas, do not use commas when entering the numbers. Excel uses a comma to separate data elements.

A1	▲▼	✗ ✓	*fx*	=AVERAGE(63750,59500,58000,61250,64000)		
	A	B	C	D	E	F	G
1	61300						

Statisticians often collect data from small portions of a large group in order to determine information about the group. In such situations the entire group under consideration is known as the **population**, and any subset of the population is called a **sample**. It is traditional to denote the mean of a *sample* by $\bar{x}$ (which is read as "x bar") and to denote the mean of a *population* by the Greek letter μ (lowercase mu).

Example 1 Find a Mean

Six friends in a biology class of 20 students receive test grades of

> 92, 84, 65, 76, 88, and 90

Find the mean of these test scores.

Solution

The six friends are a sample of the population of 20 students. Use $\bar{x}$ to represent the mean.

$$\bar{x} = \frac{92 + 84 + 65 + 76 + 88 + 90}{6} = \frac{495}{6} = 82.5$$

The mean of these test scores is 82.5.

The Median

Another type of average is the *median*. Essentially, the median is the *middle number* or the *mean of the two middle numbers* in a list of numbers that have been arranged in numerical order from smallest to largest or largest to smallest. Any list of numbers that is arranged in numerical order from smallest to largest or largest to smallest is a **ranked list**.

Median

The median of a ranked list of n numbers is:

- the middle number if n is odd.
- the mean of the two middle numbers if n is even.

Example 2 Find a Median

a. The hourly wage for 13 business major interns is shown below. Find the median hourly wage for these interns.

> 17.50, 14.50, 21.75, 20.50, 15.50, 16.75, 22.50, 19.25, 17.25, 22.25, 15.00, 17.75, 22.50

b. The number of hours 12 commercial pilots flew in a one-month period are shown below. Find the median number of hours flown by these pilots.

> 75, 78, 79, 73, 74, 69, 67, 80, 64, 72, 61, 74

Solution

a. Arrange the data from smallest to largest or largest to smallest.

> 14.50, 15.00, 15.50, 16.75, 17.25, 17.50, 17.75, 19.25, 20.50, 21.75, 22.25, 22.50, 22.50

Because there are an odd number of data, the median is the middle number. The median hourly wage was $17.75.

b. Arrange the data from smallest to largest or largest to smallest.

> 61, 64, 67, 69, 72, **73, 74,** 74, 75, 78, 79, 80

$$\frac{73 + 74}{2} = 73.5$$

Because there are an even number of data, the median is the average of the two middle numbers, 73 and 74. The median number of hours flown was 73.5 hours.

Excel can be used to find the median of data. Here is part b of Example 2. It is not necessary to enter the data in ranked form.

A2		×	✓	f_x	=MEDIAN(A1:L1)

	A	B	C	D
1	75	78	79	
2	73.5			

The Mode

A third type of average is the *mode*.

Mode

The **mode** of a list of numbers is the number that occurs most frequently.

Some lists of numbers do not have a mode. For instance, in the list 1, 6, 8, 10, 32, 15, 49, each number occurs exactly once. Because no number occurs more often than the other numbers, there is no mode.

A list of numerical data can have more than one mode. For instance, in the list 4, 2, 6, 2, 7, 9, 2, 4, 9, 8, 9, 7, the number 2 occurs three times and the number 9 occurs three times. Each of the other numbers occurs less than three times. Thus 2 and 9 are both modes for the data.

Example 3 **Find a Mode**

Find the mode of the data in the following lists.
a. 18, 15, 21, 16, 15, 14, 15, 21   **b.** 2, 5, 8, 9, 11, 4, 7, 23

Solution

a. In the list 18, 15, 21, 16, 15, 14, 15, 21, the number 15 occurs more often than the other numbers. Thus 15 is the mode.
b. Each number in the list 2, 5, 8, 9, 11, 4, 7, 23 occurs only once. Because no number occurs more often than the others, there is no mode.

●

Example 4 **Solve an Application**

A veterinarian researcher measured the weights, in pounds, of 20 adult golden retrievers. The weights were 66, 54, 66, 66, 49, 70, 57, 62, 71, 66, 74, 66, 72, 76, 59, 68, 69, 59, 80, and 67. Find the arithmetic mean, median, and mode of this data set.

Solution

Using Excel, enter the data. Then use the AVERAGE, MEDIAN, and MODE.SNGL functions to calculate the arithmetic mean, median, and mode. We have entered the data in a row just to conserve space. You can enter the data in a column as well.

B2		×	✓	f_x	=AVERAGE(A1:T1)

	A	B	C	D	E	F	G	H	I	J	K	L	M	N	O	P	Q	R	S	T
1	66	54	66	66	49	70	57	62	71	66	74	66	72	76	59	68	69	59	80	67
2	Mean	65.85																		

The arithmetic mean is 65.85 pounds.

The median is the middle number when the data is arranged from smallest to largest or largest to smallest. It is not necessary to enter the data in a ranked format. Excel can still calculate the value.

The median is 66 pounds.

The mode is the most frequently occurring value.

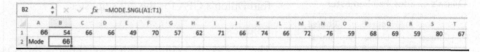

The mode is 66 pounds.

The mean, the median, and the mode are all averages; however, they are generally not equal. When a data set has one or more extreme values that are very different from most data values, the mean will not necessarily be a good indicator of an average value.

For instance, suppose the annual salaries of the 8 employees of a company are

$60,000 $60,000 $60,000 $60,000 $60,000 $100,000 $250,000
$500,000

The mean is $143,750; the median is $60,000. If the names of all the employees were put in a hat and you randomly selected one name, it is more likely that person's salary would be closer to $60,000 than $143,750. The median is a better estimate of the average salary.

The Weighted Mean

A value called the *weighted mean* is often used when some data values are more important than others. For instance, many professors determine a student's course grade from the student's tests and the final examination. Consider the situation in which a professor counts the final examination score as two test scores. To find the weighted mean of the student's scores, the professor first assigns a weight to each score. In this case the professor could assign each of the test scores a weight of 1 and the final exam score a weight of 2. A student with test scores of 65, 70, and 75 and a final examination score of 90 has a weighted mean of

$$\frac{(65 \times 1) + (70 \times 1) + (75 \times 1) + (90 \times 2)}{5} = \frac{390}{5} = 78$$

Note that the numerator of this weighted mean is the sum of the products of each test score and its corresponding weight. The number 5 in the denominator is the sum of all the weights $(1 + 1 + 1 + 2 = 5)$. The procedure for finding the weighted mean can be generalized as follows.

The Weighted Mean

The **weighted mean** of the n numbers $x_1, x_2, x_3, \ldots, x_n$ with the respective assigned weights $w_1, w_2, w_3, \ldots, w_n$ is

$$\text{Weighted mean} = \frac{\text{sum of each data value times its assigned weight}}{\text{sum of the weights}}$$

Many colleges use the 4-point grading system:

$$A = 4, B = 3, C = 2, D = 1, F = 0$$

A student's grade point average (GPA) is calculated as a weighted mean, where the student's grade in each course is given a weight equal to the number of units (or credits) that course is worth. Use this 4-point grading system for Example 5.

Example 5 Find the GPA

Table 8.1
Dillon's Grades, Fall Semester

Course	Course Grade	Course Units
English	B	4
History	A	3
Chemistry	D	3
Algebra	C	4

Table 8.1 shows Dillon's fall semester course grades. Use the weighted mean formula to find Dillon's GPA for the fall semester.

Solution

The B is worth 3 points, with a weight of 4; the A is worth 4 points, with a weight of 3; the D is worth 1 point, with a weight of 3; and the C is worth 2 points, with a weight of 4. The sum of all the weights is $4 + 3 + 3 + 4$, or 14.

$$\text{Weighted mean} = \frac{(3 \times 4) + (4 \times 3) + (1 \times 3) + (2 \times 4)}{14}$$

$$= \frac{35}{14} = 2.5$$

Dillon's GPA for the fall semester is 2.5.

Excel can be used to find a weighted mean using the SUMPRODUCT function and SUM function.

$$\text{Weighted mean} = \frac{\text{SUMPRODUCT(data, weights)}}{\text{SUM(weights)}}$$

where data is a row or column of the data values and weights is the corresponding row or column of the weights.

Here is the calculation in Example 5. The course grade has been replaced their associated values, $A = 4, B = 3, C = 2, D = 1$.

C6	▲▼	✕ ✓	fx	=SUMPRODUCT(B2:B5,C2:C5)/SUM(C2:C5)

	A	B	C	D	E	F
1	Course	Course Grade	Course Units			
2	English	3	4			
3	History	4	3			
4	Chemistry	1	3			
5	Algebra	2	4			
6		GPA	2.5			

Using percents as weights is a common application of weighted averages.

Example 6 Calculate a Course Grade

A sociology professor uses the following scale to compute a student's grade for a course.

Average of midterms: 25%   Term paper: 20%   Research paper: 20%

Final exam: 25%   Average of quizzes: 10%

Ariel's scores for the course were:

Average of midterms: 86	Term paper: 91	Research paper: 94
Final exam: 88	Average of quizzes: 85	

Find Ariel's course grade.

Solution

Record the information in a table. Then find the products of a score and its weight, as a decimal. Add the results. In this case, the sum of the weights is 1.

	Score	Weight	Score × Weight
Average of midterms	86	0.25	21.5
Term paper	91	0.20	18.2
Research paper	94	0.20	18.8
Final Exam	88	0.25	22
Average of quizzes	85	0.10	8.5
	Total	1.00	89

Ariel's course grade is 89.

The Excel function SUMPRODUCT can be used to calculate a weighted mean. For Example 6, enter the scores in one column and the weights in a second column.

B7		f_x =SUMPRODUCT(A2:A6,B2:B6)					
	A	**B**	**C**	**D**	**E**	**F**	**G**
1	Score	Weight					
2	86	0.25					
3	91	0.20					
4	94	0.20					
5	88	0.25					
6	85	0.10					
7	Course grade	89					

Data that have not been organized or manipulated in any manner are called **raw data**. A large collection of raw data may not provide much readily observable information. A **frequency distribution**, which is a table that lists observed events and the frequency of occurrence of each observed event, is often used to organize raw data. For instance, consider Table 8.2, which lists the number of laptop computers owned by families in each of 40 homes in a subdivision.

Table 8.2
Number of Laptop Computers per Household

2	0	3	1	2	1	0	4
2	1	1	7	2	0	1	1
0	2	2	1	3	2	2	1
1	4	2	5	2	3	1	2
2	1	2	1	5	0	2	5

The frequency distribution in Table 8.3 was constructed using the data from Table 8.2. The first column of the frequency distribution consists of the numbers 0, 1, 2, 3, 4, 5, 6, and 7. The corresponding frequency of occurrence, f, of each of the numbers in the first column is listed in the second column.

Table 8.3
A Frequency Distribution for Table 8.2

Observed Event Number of Laptop Computers, x	Frequency Number of Households, f, with x Laptop Computers
0	5
1	12
2	14
3	3
4	2
5	3
6	0
7	1

This row indicates that there are 14 households with 2 laptop computers.

The formula for a weighted mean can be used to find the mean of the data in a frequency distribution. The only change is that the weights $w_1, w_2, w_3, \ldots, w_n$ are replaced with the frequencies $f_1, f_2, f_3, \ldots, f_n$. This procedure is illustrated in the next example.

Example 7   Find the Mean of Data Displayed in a Frequency Distribution

Find the mean of the data in Table 8.3.

Solution

The solution, using Excel, is shown at the right. The data values are the frequencies (number of computers). The weights are the number of households.

B3		f_x	=SUMPRODUCT(B1:I1,B2:I2)/SUM(B2:I2)						
	A	B	C	D	E	F	G	H	I
1	Number of laptops	0	1	2	3	4	5	6	7
2	Number of households	5	12	14	3	2	3	0	1
3	Weighted Average	1.975							

The mean number of laptop computers per household for the homes in the subdivision is 1.975.

8.1   Exercise Set

Think About It

1. What is the difference between descriptive and inferential statistics?

2. Suppose 2500 is the largest number in a data set consisting of 100 numbers. If that number is replaced by 2700, does the median of the data set change? Does the mean of the data set change?

3. If a data set has a mode, must that number belong to the data set?

4. Suppose a data set consists of the values 1, 2, 3, 4, and 20. Without doing a calculation, which has the greater value, the median or the mean?

5. What is a weighted mean?

Preliminary Exercises

For Exercises P1 to P7, if necessary, use the referenced example following the exercise for assistance.

P1. A doctor ordered four separate blood tests to measure a patient's total blood cholesterol levels. The test results were

245,   235,   220,   and   210

Find the mean of the blood cholesterol levels. **Example 1**

P2. Find the median of the data in the following lists. **Example 2**

a. 14, 27, 3, 82, 64, 34, 8, 51

b. 21.3, 37.4, 11.6, 82.5, 17.2

P3. Find the mode of the data in the following lists. **Example 3**

a. 3, 3, 3, 3, 3, 4, 4, 5, 5, 5, 8

b. 12, 34, 12, 71, 48, 93, 71

P4. The points scored in hockey for the home team in the last 20 games were 1, 5, 3, 2, 2, 3, 4, 1, 0, 2, 1, 1, 4, 2, 1, 5, 0, 3, 1, and 4. Find the arithmetic mean, median, and mode of this data set. **Example 4**

P5. The following table shows a student's spring semester course grades. Use the weighted mean formula to find the student's GPA for the spring semester. Use the scale A = 4, B = 3, C = 2, D = 1, F = 0. Round to the nearest hundredth. **Example 5**

Course	Course Grade	Course Units
Biology	A	4
Statistics	B	3
Business	C	3
Psychology	F	2
CAD	B	2

P6. A biology professor uses the following scale to compute a student's grade for a course.

Average of exams: 20%
Lab experiments: 25%
Average of quizzes: 15%
Research paper: 15%
Final exam: 25%

Shasta's scores for the course were:
Average of exams: 85
Lab experiments: 98
Average of quizzes: 90
Research paper: 85
Final exam: 93

Find Shasta's course grade. **Example 6**

P7. A housing subdivision consists of 45 homes. The following frequency distribution shows the number of homes of a certain size, in square feet, in a new subdivision. Find the weighted mean number of square feet for the 45 homes. **Example 7**

Number of Square Feet	Number of Homes
1589	5
1705	25
1847	10
2164	5

Section Exercises

In Exercises 1 to 4, find the mean, median, and mode(s), if any, for the given data. Round noninteger means to the nearest tenth.

1. 2, 7, 5, 7, 14

2. 8, 3, 3, 17, 9, 22, 19

3. 11, 8, 2, 5, 17, 39, 52, 42

4. 5, 5, 5, 5, 5, 5, 5, 5, 5, 5, 5, 5, 5

5. Ages   In a survey the ages of 11 students living in student housing were 21, 22, 25, 26, 21, 25, 18, 21, 23, 29, and 21. Find the mean, median, and mode for this data.

6. Ant Colonies   A research entomologist estimates from 9 fire ant colonies determined there were 320,000, 350,000, 320,000, 260,000, 380,000, 340,000, 310,000, 330,000, and 310,000 ants in each colony. Find the mean, median, and mode for this data.

7. CPA Salaries   The starting salary for 10 certified public accounts (CPA) were $84,500, $83,500, $82,500, $78,000, $89,500, $80,000, $81,000, $72,500, $85,500, and $84,000. Find the mean, median, and mode for this data.

8. Robo Calls   Twelve people volunteered to count the number of robocalls received in a one-month period. The results were 90, 87, 87, 88, 82, 67, 71, 94, 79, 92, 84, and 85. Find the mean, median, and mode for this data.

9. Winter Temperatures   The temperatures at noon for a city in November for a 14-day period were 56, 61, 60, 60, 56, 58, 59, 59, 60, 59, 59, 57, 54, and 58. Find the mean, median, and mode for this data.

10. Corn Yield   The corn yield, in bushels, per acre for a 20-acre farm were 158, 171, 177, 172, 177, 170, 158, 182, 186, 164, 188, 174, 190, 174, 185, 171, 169, 180, 165, and 176. Find the mean, median, and mode for this data.

11. Tire Wear   Twenty-five tires were tested as to useful miles before the tire was unsafe to drive on. The test results were 60,000, 62,000, 58,000, 57,000, 61,000, 56,000, 62,000, 62,000, 65,000, 58,000, 62,000, 67,000, 72,000, 54,000, 58,000, 56,000, 54,000, 54,000, 65,000, 60,000, 61,000, 59,000, 56,000, 57,000, and 53,000. Find the mean, median, and mode for this data.

12. Airplane Fuel Usage   The number of gallons of fuel used by 25 passenger jets from Los Angeles to Miami were recorded as 14,750, 14,750, 13,750, 14,000, 12,750, 14,000, 14,250, 14,250, 13,250, 13,750, 16,000, 13,000, 13,750, 13,500, 15,000, 13,250, 15,000, 12,750, 12,500, 12,750, 16,750,

11,750, 14,500, 14,500, and 14,250. Find the mean, median, and mode for this data.

13. **Blood Pressure** The diastolic blood pressure for 20 patients were 67, 86, 77, 48, 90, 98, 88, 96, 73, 68, 102, 93, 92, 96, 76, 80, 92, 90, 90, and 86. Find the mean, median, and mode for this data.

14. **Shoe Sizes** A podiatrist measured the shoe size of 15 patients so they could be fitted for orthotics. The shoe sizes were 9.5, 9, 8, 11.5, 11.5, 11.5, 9, 9, 10, 10.5, 10, 9.5, 11, 8, and 11. Find the mean, median, and mode for this data.

In Exercises 15 to 18, use the grading system to find each student's GPA. Round to the nearest hundredth.

Grade Point Average In some 4.0 grading systems, a student's grade point average (GPA) is calculated by assigning letter grades the following numerical values.

A = 4.00	B− = 2.67	D+ = 1.33
A− = 3.67	C+ = 2.33	D = 1.00
B+ = 3.33	C = 2.00	D− = 0.67
B = 3.00	C− = 1.67	F = 0.00

15. Jerry's Grades, Fall Semester

Course	Course Grade	Course Units
English	A	3
Anthropology	A	3
Chemistry	B	4
French	C+	3
Theatre	B−	2

16. Rhonda's Grades, Spring Semester

Course	Course Grade	Course Units
English	C	3
History	D+	3
Computer science	B+	2
Calculus	B−	3
Photography	A−	1

17. Tessa's cumulative GPA for three semesters was 3.24 for 46 course units. Tessa's fourth semester GPA was 3.86 for 12 course units. What is Tessa's cumulative GPA for all four semesters?

18. Richard's cumulative GPA for three semesters was 2.0 for 42 credits. The fourth semester GPA for Richard was 4.0 for 14 course units. What is Richard's cumulative GPA for all four semesters?

19. **Calculate a Course Grade** A professor grades students on five tests, a project, and a final examination. Each test counts as 10% of the course grade. The project counts as 20% of the course grade. The final examination counts as 30% of the course grade. If a student has test scores of 70, 65, 82, 94, and 85; a project score of 92; and a final

examination score of 80, find the student's average for the course.

20. **Calculate a Course Grade** A professor grades students on four tests, a term paper, and a final examination. Each test counts as 15% of the course grade. The term paper counts as 20% of the course grade. The final examination counts as 20% of the course grade. Suppose a student has test scores of 80, 78, 92, and 84; an 84 on the term paper; and a final examination score of 88. Use the weighted mean formula to find the student's average for the course.

In Exercises 21 to 24, find the mean, for the given frequency distribution.

21. Points Scored by Loba

Points Scored in a Basketball Game	Frequency
2	6
4	5
5	6
9	3
10	1
14	2
19	1

22. Mystic Pizza Company

Hourly Pay Rates for Employees	Frequency
$8.00	14
$11.50	9
$14.00	8
$16.00	5
$19.00	2
$22.50	1
$35.00	1

23. Quiz Scores

Scores on a 10-Point Biology Quiz	Frequency
2	1
4	2
6	7
7	12
8	10
9	4
10	3

24. Ages of Science Fair Contestants

Age	Frequency
7	3
8	4
9	6
10	15
11	11
12	7
13	1

In Exercises 25 to 28, use the following information about another measure of central tendency for a set of data, called the midrange.

Meteorology The **midrange** is defined as the value that is halfway between the minimum data value and the maximum data value. That is,

$$\text{Midrange} = \frac{\text{minimum value} + \text{maximum value}}{2}$$

The midrange is often stated as the *average* of a set of data in situations in which there are a large amount of data and the data are constantly changing. Many weather reports state the average daily temperature of a city as the midrange of the temperatures achieved during that day. For instance, if the minimum daily temperature of a city was 60° and the maximum daily temperature was 90°, then the midrange of the temperatures is $\frac{60° + 90°}{2} = 75°$.

25. Find the midrange of the following daily temperatures, which were recorded at 3-hour intervals.

52°, 65°, 71°, 74°, 76°, 75°, 68°, 57°, 54°

26. Find the midrange of the following daily temperatures, which were recorded at 3-hour intervals.

−6°, 4°, 14°, 21°, 25°, 26°, 18°, 12°, 2°

27. During a 24-hour period on January 23–24, 1916, the temperature in Browning, Montana, decreased from a high of 44°F to a low of −56°F. Find the midrange of the temperatures during this 24-hour period.

28. During a 2-minute period on January 22, 1943, the temperature in Spearfish, South Dakota, increased from a low of −4°F to a high of 45°F. Find the midrange of the temperatures during this 2-minute period.

▶ **Investigation**

Geometric Mean The *geometric mean* is average that is used when trying to find an average percent change. For instance, suppose an investor starts with $1000 and earns a 5% return after one year, a 3% return after the second year, and a 1% loss after the third year.

Beginning value: $1000

After year 1:

$1000 + 5% of $1000 = $1000 + $50 = $1050

This is the same as $1000(1.05) = $1050

After year 2:

$1050 + 3% of $1050 = $1050 + $31.50 = $1081.50

This is the same as $1050(1.03) = $1081.50

After year 3:

$1081.50 − 1% of $1081.50 = $1081.50 − $10.82 = $1070.68

This is the same as $1081.50(0.99) = $1070.68

The question is "What single percent applied to $1000 would result in $1070.68 after 3 years?" The answer uses the geometric mean.

Geometric Mean

The **geometric mean** of the n numbers $x_1, x_2, x_3, \ldots, x_n$ is $\sqrt[n]{x_1 \cdot x_2 \cdot x_3 \cdot \ldots \cdot x_n}$.

In words, the geometric mean of n numbers is the nth root of the product of the n numbers.

For the $1000 investment,

$$\text{geometric mean} = \sqrt[3]{1.05 \cdot 1.03 \cdot 0.99} \approx 1.02303$$

The single percent $r = 1.02303 − 1 = 0.02303$ or 2.303%. Using the compound interest formula $P = A(1 + r)^n$, we have

$$P = A(1 + r)^n$$
$$= 1000(1 + 0.02303)^3$$
$$= 1000(1.02303)^3 \approx 1070.69$$

This shows, within 1 cent, that a 2.303% increase per year yields the same as a 5% increase, 3% increase, and 1% decrease over 3 years.

1. The annual percent gains for an investment are shown in the table below. Find the geometric mean of the returns. What was the percent increase in the investment over the 6-year period?

Year	1	2	3	4	5	6
Percent change	5	8	−10	12	−7	4

2. The percent increases and decreases in a city's population are shown in the following table.

Year	1	2	3	4	5
Percent change	8	−1	3	−5	−3

Find the geometric mean to find the average percent change in the city's population over the 5-year period.

Section

8.2

Section 8.2 Measures of Dispersion

Learning Objectives:

- Find the range of a set of data.
- Find the standard deviation of a set of data.
- Find the variance of a set of data.

Table 8.4
Soda Dispensed (ounces)

Machine 1	Machine 2
9.52	8.01
6.41	7.99
10.07	7.98
5.85	8.03
8.15	8.02
$\bar{x} = 8.0$	$\bar{x} = 8.0$

The Range

In the preceding section we introduced three types of average values for a data set—the mean, the median, and the mode. Some characteristics of a set of data may not be evident from an examination of averages. For instance, consider a soft-drink dispensing machine that should dispense 8 oz of your selection into a cup. Table 8.4 shows data for two of these machines.

The mean data value for each machine is 8 oz. However, look at the variation in data values for Machine 1. The quantity of soda dispensed is very inconsistent—in some cases the soda overflows the cup, and in other cases too little soda is dispensed. The machine obviously needs adjustment. Machine 2, on the other hand, is working just fine. The quantity dispensed is very consistent, with little variation.

This example shows that average values do not reflect the *spread* or *dispersion* of data. To measure the spread or dispersion of data, we must introduce statistical values known as the *range* and the *standard deviation*.

> **Range**
>
> The **range** of a set of data values is the difference between the greatest data value and the least data value.

Example 1 Find a Range

Find the range of the numbers of ounces dispensed by Machine 1 and Machine 2 in Table 8.4.

Solution

For Machine 1, the greatest number of ounces dispensed is 10.07 and the least is 5.85. The range of the numbers of ounces dispensed is $10.07 - 5.85 = 4.22$ oz.

For Machine 2, the greatest number of ounces dispensed is 8.03 and the least is 7.98. The range of the numbers of ounces dispensed is $8.03 - 7.98 = 0.05$ oz.

The Standard Deviation

The range of a set of data is easy to compute, but it can be deceiving. The range is a measure that depends only on the two most extreme values, and as such it is very sensitive. A measure of dispersion that is less sensitive to extreme values is the *standard deviation*. The standard deviation of a set of numerical data makes use of the amount by which each individual data value deviates from the mean.

Consider the two data sets

Set 1:   10   20   30   40   50

Set 2:   2   17   30   45   56

The mean for Set 1 is $\dfrac{10 + 20 + 30 + 40 + 50}{5} = \dfrac{150}{5} = 30$.

The mean for Set 2 is $\dfrac{2 + 17 + 30 + 45 + 56}{5} = \dfrac{150}{5} = 30$.

Both sets have the same mean. However, the data values in Set 1 range from 20 units below the mean to 20 units above the mean while the data values in Set 2 range from 28 units below the mean to 26 units above the mean. The standard deviation is used to measure this variability or *dispersion*.

Take Note: The sample standard deviation is often used to estimate the population standard deviation, and it can be shown mathematically that the use of $n - 1$ yields a better estimate.

Standard Deviations for Populations and Samples

If $x_1, x_2, x_3, \ldots, x_n$ is a *population* of n numbers with a mean of μ, then the **standard deviation** of the population is

$$\sigma = \sqrt{\frac{(x_1 - \mu)^2 + (x_2 - \mu)^2 + (x_3 - \mu)^2 + \cdots + (x_n - \mu)^2}{n}}.$$

If $x_1, x_2, x_3, \ldots, x_n$ is a *sample* of n numbers with a mean of $\bar{x}$, then the **standard deviation** of the sample is

$$s = \sqrt{\frac{(x_1 - \bar{x})^2 + (x_2 - \bar{x})^2 + (x_3 - \bar{x})^2 + \cdots + (x_n - \bar{x})^2}{n - 1}}.$$

Most statistical applications involve a sample rather than a population, which is the complete set of data values. Sample standard deviations are designated by the lowercase letter *s*. In those cases in which we *do* work with a population, we designate the standard deviation of the population by σ, which is the lowercase Greek letter *sigma*. We can use the following procedure to calculate the standard deviation of n numbers.

Procedure for Computing a Standard Deviation

1. Determine the mean of the n numbers.
2. For each number, calculate the deviation (difference) between the number and the mean of the numbers.
3. Calculate the square of each deviation and find the sum of these squared deviations.
4. If the data is a *population*, then divide the sum by n. If the data is a *sample*, then divide the sum by $n - 1$.
5. Find the square root of the quotient in Step 4.

For instance, to find the sample standard deviation of

2, 4, 7, 12, 15

Step 1: Find the mean of the sample.

$$\bar{x} = \frac{2 + 4 + 7 + 12 + 15}{5} = \frac{40}{5} = 8$$

Step 2: For each number, calculate the deviation between the number and the mean.

x	x − x̄
2	$2 - 8 = -6$
4	$4 - 8 = -4$
7	$7 - 8 = -1$
12	$12 - 8 = 4$
15	$15 - 8 = 7$

Step 3: Calculate the square of each deviation in Step 2, and find the sum of these squared deviations.

x	$x - \bar{x}$	$(x - \bar{x})^2$
2	$2 - 8 = -6$	$(-6)^2 = 36$
4	$4 - 8 = -4$	$(-4)^2 = 16$
7	$7 - 8 = -1$	$(-1)^2 = 1$
12	$12 - 8 = 4$	$4^2 = 16$
15	$15 - 8 = 7$	$7^2 = 49$
		118 $\leftarrow$ Sum of the squared deviations

Step 4: Because we have a sample of $n = 5$ values, divide the sum 118 by $n - 1$, which is 4.

$$\frac{118}{4} = 29.5$$

Step 5: The standard deviation of the sample is $s = \sqrt{29.5}$. To the nearest hundredth, the standard deviation is $s = 5.43$.

In most cases, a calculator or spreadsheet program is used to calculate standard deviation.

Example 2   Use Standard Deviations

A consumer group has tested a sample of ten size-D batteries from each of three companies. The results of the tests are shown in the following table. According to these tests, which company produces batteries for which the values representing hours of constant use have the smallest standard deviation?

Company	Hours of Constant Use Per Battery
EverSoBright	108, 101, 94, 98, 102, 99, 98, 104, 95, 101
Dependable	101, 97, 97, 103, 98, 98, 99, 106, 100, 101
Beacon	104, 95, 97, 101, 101, 105, 103, 95, 96, 103

Solution

The mean for each sample of batteries is 100 hr. Because we have a sample of battery lives from a population, calculate the sample standard deviation. Use the STDEV.S Excel function. STDEV.P is used to calculate the population standard deviation.

B2 $\times$ $\checkmark$ fx =STDEV.S(A1:J1)

	A	B	C	D	E	F	G	H	I	J
1	108	101	94	98	102	99	98	104	95	101
2	s1 =	4.1633								

The standard deviation is approximately 4.1633 hr.

B2 $\times$ $\checkmark$ fx =STDEV.S(A1:J1)

	A	B	C	D	E	F	G	H	I	J
1	101	97	97	103	98	98	99	106	100	101
2	s2 =	2.8674								

The standard deviation is approximately 2.8674 hr.

B2		$\times$ $\checkmark$ f_x	=STDEV.S(A1:J1)						
	A	B	C	D	E	F	G	H	I	J
1	104	95	97	101	101	105	103	95	96	103
2	s3 =	3.8873								

The standard deviation is approximately 3.8873 hr.

The batteries from Dependable have the smallest standard deviation. According to these results, the Dependable company produces the most consistent batteries with regard to life expectancy under constant use.

Many calculators have built-in statistics features for calculating the mean and standard deviation of a set of numbers. The next example illustrates these features on a TI-84 graphing calculator.

Example 3   **Use TI-84 to Calculate Statistics**

The length, in minutes, of 12 online philosophy lectures are shown in the table below. Use a graphing calculator to find the mean and population standard deviation for the lecture times.

Online lecture, in minutes.

52.0	51.08	49.29	48.88	48.83	48.65	48.83	48.25	49.11	49.41	49.62	49.55

Solution

On a TI-84 calculator, press STAT ENTER and then enter the above times into list L1. Press STAT ▷ ENTER ENTER ENTER ENTER. The calculator displays the mean and standard deviations shown below. Because we are working with a population, we are interested in the population standard deviation. From the calculator screen, $\bar{x} \approx 49.458$ min and $\sigma x \approx 1.021$ min.

Take Note: Because the calculations of the population mean and the sample mean are the same, a graphing calculator uses the same symbol $\bar{x}$ for both. The symbols for the population standard deviation, σx, and the sample standard deviation, Sx, are different.

TI-84 Display of List 1   TI-84 Display of $\bar{x}$, s, and σ

The Variance

A statistic known as *variance* is also used as a measure of dispersion. The **variance** for a given set of data is the square of the standard deviation of the data. The following chart shows the mathematical notations that are used to denote standard deviations and variances.

Notations for Standard Deviation and Variance

σ is the standard deviation of a population.
σ^2 is the variance of a population.
s is the standard deviation of a sample.
s^2 is the variance of a sample.

Example 4	Find the Variance

Find the variance for the sample of Dependable batteries from Example 2.

Solution

In Example 2, we found $s = 2.8674$. The variance is the square of the standard deviation. Thus the variance is $s^2 = (2.8674)^2 \approx 8.22198$.

Although the variance of a set of data is an important measure of dispersion, it has a disadvantage that is not shared by the standard deviation: the variance does not have the same unit of measure as the original data. For instance, if a set of data consists of times measured in hours, then the variance of the data will be measured in *square* hours. The standard deviation of this data set is the square root of the variance, and as such it is measured in hours, which is a more intuitive unit of measure.

8.2 Exercise Set

Think About It

1. What is the range of a data set?

2. What does the variance of a data set measure?

3. What is the relationship between the variance of a data set and the standard deviation of the data set?

4. If the standard deviation of a data set is zero, what can be said about the data set?

5. The data sets 2, 4, 6, 8, 10 and 4, 6, 6, 6, 8 have the same mean. Without doing a calculation, which of the two data sets has the larger standard deviation?

6. If you were in charge of a production line that filled large boxes of cereal, would you want the standard deviation of the weights the boxes were filled to be small or large?

Preliminary Exercises

For Exercises P1 to P4, if necessary, use the referenced example following the exercise for assistance.

P1. Find the range of the numbers of ounces dispensed by Machine 2 in Table 8.4. **Example 1**

P2. A consumer testing agency has tested the strengths of three brands of $\frac{1}{8}$-inch rope. The results of the tests are shown in the following table. According to the sample test results, which company produces $\frac{1}{8}$-inch rope for which the breaking point has the smallest standard deviation? **Example 2**

Company	Breaking Point of $\frac{1}{8}$-inch Rope in Pounds
Trustworthy	122, 141, 151, 114, 108, 149, 125
Brand X	128, 127, 148, 164, 97, 109, 137
NeverSnap	112, 121, 138, 131, 134, 139, 135

P3. The height, in inches, at birth of 30 giraffes is given in the table below. Use Excel to find the mean and the population standard deviation. **Example 3**

71.7	72.8	70.9	75.3	71.5	76.7	67.4	74.1	71.2	75.3
65.8	73.5	71.0	72.5	72.8	69.9	70.8	75.9	69.3	69.1
73.0	72.1	68.5	72.1	70.7	72.5	73.6	69.8	73.6	73.5

P4. Find the population variance for the measured speeds, in gigahertz, of five computers given as 3.1, 2.8, 3.2, 2.9, and 3.0. **Example 4**

Section Exercises

1. **Meteorology** During a 12-hour period on December 24, 1924, the temperature in Fairfield, Montana, dropped from a high of 63°F to a low of −21°F. What was the range of the temperatures during this period?

2. **Meteorology** During a 2-hour period on January 12, 1911, the temperature in Rapid City, South Dakota, dropped from a high of 49°F to a low of −13°F. What was the range of the temperatures during this period?

3. **Fuel Efficiency** The fuel efficiency, in miles per gallon, of 10 small utility trucks was measured. The results are recorded in the following table.

Fuel Efficiency (mpg)

22	25	23	27	15	24	24	32	23	22	25	22

Find the mean and sample standard deviation of these data. Round to the nearest hundredth.

4. **Waiting Times** A customer at a specialty coffee shop observed the amount of time, in minutes, that each of 20 customers spent waiting to receive an order. The results are recorded in the following table.

Time (min) to Receive Order

3.2	4.0	3.8	2.4	4.7	5.1	4.6	3.5	3.5	6.2
3.5	4.9	4.5	5.0	2.8	3.5	2.2	3.9	5.3	2.9

Find the mean and sample standard deviation of these data. Round to the nearest hundredth.

5. **Fast-food Calories** A survey of 10 fast-food restaurants noted the number of calories in a mid-sized hamburger. The results are given in the following table.

Calories in a Mid-sized Hamburger

514	507	502	498	496	506	458	478	463	514

Find the mean and sample standard deviation of these data. Round to the nearest hundredth.

6. **Energy Drinks** A survey of 16 energy drinks noted the caffeine concentration of each drink in milligrams per ounce. The results are given in the following table.

Concentration of Caffeine (mg/oz)

9.1	7.5	7.8	8.9	9.0	8.2	9.1	8.7
9.0	7.7	8.8	8.9	9.0	9.1	8.2	8.9

Find the mean and sample standard deviation of these data. Round to the nearest hundredth.

7. **Weekly Commute Times** A survey of 15 large cities noted the average weekly commute times, in hours, of the residents of each city. The results are recorded in the following table.

Weekly Commute Time (hr)

4.5	4.0	5.8	5.4	4.7
4.0	3.6	3.9	4.7	3.7
4.6	3.4	3.5	3.9	4.4

Find the mean and sample standard deviation of these data. Round to the nearest hundredth.

8. **Biology** Some studies show that the mean normal human body temperature is actually somewhat lower than the commonly given value of 98.6°F. This is reflected in the following data set of body temperatures.

Body Temperatures (°F) of 30 Healthy Adults

97.1	97.8	98.0	98.7	99.5	96.3
98.4	98.5	98.0	100.8	98.6	98.2
99.0	99.3	98.8	97.6	97.4	99.0
97.4	96.4	98.0	98.1	97.8	98.5
98.7	98.8	98.2	97.6	98.2	98.8

a. Find the mean and sample standard deviation of the body temperatures. Round each result to the nearest hundredth.

b. Are there any temperatures in the data set that do not lie within 2 standard deviations of the mean? If so, list them.

9. **Recording Industry** The following table shows a random sample of the lengths of songs in a playlist.

Lengths of Songs (minutes:seconds)

3:42	3:40	3:50	3:17	3:15	3:37
2:27	3:01	3:47	3:49	4:02	3:30

a. Find the mean and sample standard deviation of the song lengths, in seconds. Round each result to the nearest second.

b. Are there any song lengths in the data set that do not lie within 1 standard deviation of the mean? If so, list them.

10. **Data Analytics** A large data analytics company measured the time it takes new employees to learn its software system. The time, in minutes, for the training times of 25 new employees were 91, 94, 92, 85, 76, 96, 77, 78, 100, 73, 86, 95, 104, 87, 79, 97, 97, 95, 85, 91, 102, 93, 89, 99, and 97. Find the sample mean and standard deviation for this data.

11. **Heart Rate** An investigational drug to reduce heart rate was given to 20 patients. The change in heartrate, beats per minute (bps) before and after administration of the drug to the patients was recorded as 3, 0, −7, −3, −1, −1, −5, 1, −1, 5, 2, 6, −8, −1, 5, 2, 5, 2, −4, and 8. Find the sample mean and standard deviation for the data.

12. **Temperatures** The low temperatures, in Fahrenheit, at a ski lift over a 20-day period were recorded as 3, −1, −7, −1, −3, 3, −2, −11, 5, 5, 4, 0, 0, −3, 10, −2, −3, 1, 6, and 1. Find the sample standard deviation of the temperatures.

13. **Ophthalmology** Two eyedrop mediations were tested on 11 glaucoma patients. The standard deviation in reduction in eye pressure, in millimeters of mercury (mm Hg), for eyedrop medication 1 was 5.42 mm Hg. For eyedrop medication 2, the standard deviation for reduction in pressure was 3.78 mm Hg. Which of the two medications showed less variability in pressure reduction?

14. **Health Sciences** Neutrophils are a type of white blood cell that are the first line of defense against an infection. The neutrophil count, in neutrophils per milliliter of blood, for 15 patients were measured as 1180, 830, 1190, 1280, 1290, 1110, 1010, 1000, 930, 1200, 1090, 1210, 1190, 1160, and 900. Find the sample mean and standard deviation of neutrophil counts for these patients.

15. **Stock Volatility** One measure used by stock market analysts is *stock return volatility*. This is the standard deviation of the daily percent price changes of the stock price. The price of a stock with a smaller stock return volatility does not change as much over the measured time frame. The daily percent changes for two stocks are:

Stock 1: −1.9934, 1.3115, 0.6515, −0.3268, −1.3245, 1.6287, −1.6556, −0.6667, 1.6393, 0.6515

Stock 2: 2.2727, −1.3158, 0.9772, −0.9868, −0.3300, −0.3311, −0.3322, 2.9032, −0.6494, −0.6536

Using stock return volatility for the two stocks, which of the two stocks has less volatility?

16. **Stock Volatility** One measure used by stock market analysts is *stock return volatility*. This is the standard deviation of the daily percent price changes of the stock price. The price of a stock with a smaller stock return volatility does not change as much over the measured time frame. The daily percent changes for two stocks are:

Stock 1: −0.3976, 0.1984, −0.3984, 0.9862, −0.1976, −0.9980, 0.9881, 0.1972, 0.0000, −0.7952

Stock 2: −0.1969, 0.0000, −0.9940, 1.1788, −0.5929, −0.1980, −0.5976, 0.1988, 1.1788, 0.1961

Using stock return volatility for the two stocks, which of the two stocks has less volatility?

▶ Investigations

Coefficient of Variation Consider the following two data sets.

Set 1: 8, 9, 10, 11, 12

Set 2: 0, 5, 10, 15, 20

The mean of both sets is 10. The standard deviation of Set 1 is 1.58114. The standard deviation of Set 2 is 7.90569. Looking at the data, there is quite a difference in the variation of the data and reflected in the differences in the standard deviations.

Another measure of variation in data is the *coefficient of variation*.

Coefficient of Variation

The **coefficient of variation** (abbreviated *CV*) of a data set is the quotient of the standard deviation of the data set and the mean of the data set and written as a percent.

$$CV = \frac{\text{standard deviation}}{\text{mean}} \cdot 100\%$$

If we write a fraction as a percent, for instance, $\frac{3}{5} = 60\%$, it means that 3 is 60% of 5. This is the interpretation of the *CV*. The coefficient of variation is the percent that the standard deviation is of the mean.

For Set 1, $CV_1 = \frac{1.58114}{10} \cdot 100\% \approx 15.8\%$. The standard deviation is 15.8% of the mean.

For Set 2, $CV_2 = \frac{7.90569}{10} \cdot 100\% \approx 79.1\%$. The standard deviation is 79.1% of the mean.

Generally, a smaller *CV* is better. One use for *CV* is when analyzing data that uses two different units.

1. Suppose a group of 100 sixth-grade students has a mean height of 50 inches and a standard deviation of 2 inches. That same group of students has a mean weight of 80 pounds with a standard deviation of 7 pounds. Find *CV* for the two measures. Is weight more variable than height?

2. An arborist recorded the height and width of Eastern Pine trees. The mean height of the trees was 54.25 feet with a standard deviation of 4.94 feet. The trunk diameter measured three feet above the ground of each tree had a mean of 36.8 inches with a standard deviation of 4.3 inches. Find the *CV* for the two measurements. Is tree height more variable than trunk diameter?

Section 8.3

Measures of Relative Position

Learning Objectives:

- Find a z-score.
- Find percentiles.
- Find quartiles.
- Create a box-and-whisker plot.

z-Scores

Consider an Internet site that offers movie downloads. Based on data kept by the site, an estimate of the mean time to download a certain movie is 12 min, with a standard deviation of 4 min. When you download this movie, the download takes 20 min, which you think is an unusually long time for the download. On the other hand, when your friend downloads the movie, the download takes only 6 min, and your friend is pleasantly surprised at how quickly the movie downloaded. In each case, a data value far from the mean is unexpected.

The following graph shows the download times for this movie using two different measures: the number of *minutes* a download time is from the mean and the number of *standard deviations* the download time is from the mean.

Movie Download Times, in Minutes
$\bar{x} = 12, s = 4$

1.5 standard deviations below the mean   2 standard deviations above the mean

6 minutes below the mean   8 minutes above the mean

The number of standard deviations between a data value and the mean is known as the data value's *z-score* or *standard score*.

z-Score

The **z-score** for a given data value x is the number of standard deviations that x is above or below the mean of the data. The following formulas show how to calculate the z-score for a data value x in a population and in a sample.

$$\text{Population: } z_x = \frac{x - \mu}{\sigma} \qquad \text{Sample: } z_x = \frac{x - \bar{x}}{s}$$

In the next example, we use a student's z-scores for two tests to determine how well the student did on each test in comparison to the other students.

Example 1   Compare z-Scores

Blair has taken two chemistry tests. The score on the first test was 72, for which the mean of all scores was 65 and the standard deviation was 8. The score on the second test was 60, for which the mean of all scores was 45 and the standard deviation was 12. In comparison to the other students, did Blair do better on the first test or the second test?

Solution

Find the z-score for each test.

$$z_{72} = \frac{72 - 65}{8} = 0.875 \qquad z_{60} = \frac{60 - 45}{12} = 1.25$$

Blair scored 0.875 standard deviation above the mean on the first test and 1.25 standard deviations above the mean on the second test. These z-scores indicate that, in comparison to their classmates, Blair scored better on the second test than on the first test.

Example 2   Use z-Scores

A consumer group tested a sample of 100 light bulbs. It found that the mean life expectancy of the bulbs was 842 hr, with a standard deviation of 90. One particular light bulb from the DuraBright Company had a z-score of 1.2. What was the life span of this light bulb?

Solution

Substitute the given values into the z-score equation and solve for x.

$$z_x = \frac{x - \bar{x}}{s}$$

$$1.2 = \frac{x - 842}{90} \qquad \bullet \, z_x = 1.2, \bar{x} = 842, s = 90$$

$$108 = x - 842 \qquad \bullet \text{ Solve for } x.$$

$$950 = x$$

The light bulb had a life span of 950 hr.

Percentiles

Most standardized examinations provide scores in terms of *percentiles*, which are defined as follows:

pth Percentile

A value x is called the **pth percentile** of a data set provided $p\%$ of the data values are less than x.

Example 3 **Using Percentiles**

In a recent year, the median annual salary for a physical therapist was $74,480. If the 90th percentile for the annual salary of a physical therapist was $105,900, find the percent of physical therapists whose annual salary was

a. more than $74,480.
b. less than $105,900.
c. between $74,480 and $105,900.

Solution

a. By definition, the median is the 50th percentile. Therefore, 50% of the physical therapists earned more than $74,480 per year.

b. Because $105,900 is the 90th percentile, 90% of all physical therapists made less than $105,900.

c. From parts a and b, 90% − 50% = 40% of the physical therapists earned between $74,480 and $105,900.

The following formula can be used to find the percentile that corresponds to a particular data value in a set of data.

Percentile for a Given Data Value

Given a set of data and a data value x,

$$\text{Percentile of score } x = \frac{\text{number of data values less than } x}{\text{total number of data values}} \cdot 100$$

Example 4 **Find a Percentile**

On a reading examination given to 900 students, one student's score of 602 was higher than the scores of 576 of the students who took the examination. What is the percentile for the student's score?

Solution

$$\text{Percentile} = \frac{\text{number of data values less than } 602}{\text{total number of data values}} \cdot 100$$

$$= \frac{576}{900} \cdot 100$$

$$= 64$$

The student's score is at the 64th percentile.

When there is a large data set, a spreadsheet such as Excel, Numbers, or Google Sheets is used to calculate percentiles.

Example 5 **Find Percentiles**

A survey asked 50 college students how far from campus each of them live. The results, in miles, were 6.5, 6.1, 5.5, 5.8, 5.9, 7.2, 6.1, 6.4, 5.7, 6.3, 6.1, 6.5, 6.3, 4.8, 6.0, 5.9, 5.5, 5.8, 6.5, 5.6, 6.0, 6.7, 5.1, 6.4, 6.7, 5.1, 4.9, 6.1, 5.4, 5.4, 6.2, 5.1, 4.7, 5.8, 5.6, 6.4, 5.7, 5.4, 6.3, 5.7, 6.0, 7.4, 6.2, 5.1, 6.4, 5.9, 5.4, 4.9, 5.8, and 6.8.

a. Find the 95th percentile.
b. Find the 50th percentile.
c. Find the 22nd percentile.

Solution

Enter the data in a row or column in an Excel sheet. The Excel function to find a percentile is PERCENTILE.INC(data_array, percentile). The data do not have to be sorted. The percentile can be entered as a percent or decimal. For instance, for the 86th percentile, enter 86% or 0.86.

For each part of this example, only the first 6 data values are shown.

a.

A2		fx	=PERCENTILE.INC(A1:AX1,0.95)			
	A	B	C	D	E	F	G
1	6.5	6.1	5.5	5.8	5.9	7.2	6.1
2	6.755						

The 95th percentile is 6.755 miles. 95% of the students live less than 6.755 miles from campus.

b.

A2		fx	=PERCENTILE.INC(A1:AX1,0.5)			
	A	B	C	D	E	F	G
1	6.5	6.1	5.5	5.8	5.9	7.2	6.1
2	5.9						

The 50th percentile is 5.9 miles. 50% of the students live less than 5.9 miles from campus.

c.

A2		fx	=PERCENTILE.INC(A1:AX1,0.22)			
	A	B	C	D	E	F	G
1	6.5	6.1	5.5	5.8	5.9	7.2	6.1
2	5.4						

The 22nd percentile is 5.4 miles. 22% of the students live less than 5.4 miles from campus.

> **Take Note** The result in part a shows the 95th percentile is 6.755 miles. If you actually count the number of students living less than 6.755 miles from campus, you will find that it is 94%. The problem is that there is no **exact** 95th percentile so Excel estimates the value as best as possible. The given percentile value, 6.755, is the number where 95% of the data is *less than or equal* to 6.755.

In Example 5, part b, the 50th percentile is the median of the data. Applying median function to the data in Example 5 gives =MEDIAN(A1:AX1) = 5.9, as expected.

Quartiles

The three numbers Q_1, Q_2, and Q_3 that partition a ranked data set into four (approximately) equal groups are called the **quartiles** of the data. For instance, for the following data set, the values $Q_1 = 11$, $Q_2 = 29$, and $Q_3 = 104$ are the quartiles of the data.

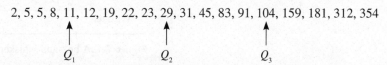

2, 5, 5, 8, 11, 12, 19, 22, 23, 29, 31, 45, 83, 91, 104, 159, 181, 312, 354

$$Q_1 \qquad Q_2 \qquad Q_3$$

The quartile Q_1 is called the *first quartile*. The quartile Q_2 is called the *second quartile*. It is the median of the data. The quartile Q_3 is called the *third quartile*. The following method of finding quartiles makes use of medians.

The Median Procedure for Finding Quartiles

1. Rank the data.
2. Find the median of the data. This is the second quartile, Q_2.
3. The first quartile, Q_1, is the median of the data values less than Q_2. The third quartile, Q_3, is the median of the data values greater than Q_2.

Example 6   **Use Medians to Find the Quartiles of a Data Set**

The following table lists the calories per 100 milliliters of 25 popular sodas. Find the quartiles for the data.

Calories, per 100 Milliliters, of Selected Sodas

43	37	42	40	53	62	36	32	50	49
26	53	73	48	45	39	45	48	40	56
41	36	58	42	39					

Solution

Step 1: Rank the data as shown in the following table.

1) 26	**2)** 32	**3)** 36	**4)** 36	**5)** 37	**6)** 39	**7)** 39	**8)** 40	**9)** 40
10) 41	**11)** 42	**12)** 42	**13)** 43	**14)** 45	**15)** 45	**16)** 48	**17)** 48	**18)** 49
19) 50	**20)** 53	**21)** 53	**22)** 56	**23)** 58	**24)** 62	**25)** 73		

Step 2: The median of these 25 data values has a rank of 13. Thus the median is 43. The second quartile Q_2 is the median of the data, so $Q_2 = 43$.

Step 3: There are 12 data values less than the median and 12 data values greater than the median. The first quartile is the median of the data values less than the median. Thus Q_1 is the mean of the data values with ranks of 6 and 7.

$$Q_1 = \frac{39 + 39}{2} = 39$$

The third quartile is the median of the data values greater than the median. Thus Q_3 is the mean of the data values with ranks of 19 and 20.

$$Q_3 = \frac{50 + 53}{2} = 51.5$$

Box-and-Whisker Plots

A **box-and-whisker plot** (sometimes called a **box plot**) is often used to provide a visual summary of a set of data. A box-and-whisker plot shows the median, the first and third quartiles, and the minimum and maximum values of a data set. See Figure 8.1.

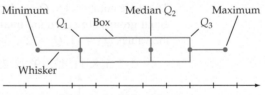

Figure 8.1 A box-and-whisker plot

Construction of a Box-and-Whisker Plot

1. Draw a horizontal scale that extends from the minimum data value to the maximum data value.
2. Above the scale, draw a rectangle (box) with its left side at Q_1 and its right side at Q_3.
3. Draw a vertical line segment across the rectangle at the median, Q_2.
4. Draw a horizontal line segment, called a whisker, that extends from Q_1 to the minimum and another whisker that extends from Q_3 to the maximum.

Example 7   **Construct a Box-and-Whisker Plot**

Construct a box-and-whisker plot for the data set in Example 6.

Solution

For the data set in Example 6, we determined that $Q_1 = 39$, $Q_2 = 43$, and $Q_3 = 51.5$. The minimum data value for the data set is 26, and the maximum data value is 73. Thus the box-and-whisker plot is as shown in the following figure.

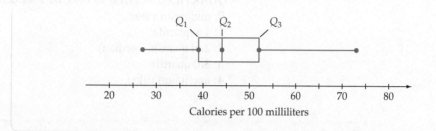

Calories per 100 milliliters

Box plots are popular because they are easy to construct and they illustrate several important features of a data set in a simple diagram. Note from the box plot in Example 7 that we can easily estimate

- the quartiles of the data.
- the range of the data.
- the position of the middle half of the data as shown by the length of the box.

Some graphing calculators can be used to produce box-and-whisker plots. For instance, on a TI-84, you enter the data into a list, as shown on the first screen in Figure 8.2. The **WINDOW** menu is used to enter appropriate boundaries that contain all the data. Use the key sequence **2nd [STAT PLOT] ENTER** and choose from the **Type** menu the box-and-whisker plot icon (see the third screen in Figure 8.2). The **GRAPH** key is then used to display the box-and-whisker plot. After the calculator displays the box-and-whisker plot, the **TRACE** key and the ▷ key enable you to view Q_1, Q_2, Q_3, and the minimum and maximum of your data set.

Take Note: The following data were used to produce the box plot shown in Figure 8.2.

21.2, 20.5, 17.0, 16.8, 16.8, 16.5, 16.2, 14.0, 13.7, 13.3, 13.1, 13.0, 12.4, 12.1, 12.0

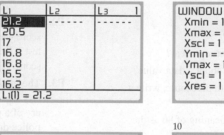

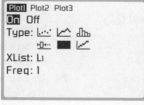

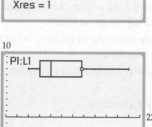

Figure 8.2 TI-84 screen displays

To create a box-and-whisker plot with Excel,

- Enter the data in a row or column. The image below shows the same data that was used for the TI-84 calculation.
- Select the data.

- On ribbon tab, click on the Insert tab.
- On Windows, click Insert > Insert Statistic Chart > Box and Whisker. On macOS, click the Statistical Chart icon, then select Box and Whisker.

The result is the image shown below. We have added the minimum value, 1st quartile, 2nd quartile (median), 3rd quartile, and maximum value using the QUARTILE.INC function. The "x" in the middle of the box is the mean of the data. Note that the box-and-whisker plot is shown vertically instead of horizontally as from the TI-84 calculator.

Here is the format for the QUARTILE.INC function.

$$= \text{QUARTILE.INC(array of data, number between 0 and 4)}$$

    0: minimum value

    1: 1st quartile

    2: 2nd quartile (median)

    3: 3rd quartile

    4: maximum value

8.3 Exercise Set

Think About It

1. What does the z-score of a data point represent?

2. What does a z-score of -1 indicate about a data value?

3. If a data value is 34 and is in the 58th percentile, write a sentence explaining what that means.

4. Write a sentence that explains the meaning of 86 as the value of the first quartile.

5. What percent of the data in a data set lies between Q_1 and Q_3?

Preliminary Exercises

For Exercises P1 to P6, if necessary, use the referenced example following the exercise for assistance.

P1. Cameron has taken two quizzes in a history class. The score of the first quiz was 15, for which the mean of all scores was 12 and the standard deviation was 2.4. The score on the second quiz, for which the mean of all scores was 11 and the standard deviation was 2.0, was 14. In comparison to Cameron's classmates, on which quiz did Cameron do better? **Example 1**

P2. Roland received a score of 70 on a test for which the mean score was 65.5. Roland has learned that the z-score for his test is 0.6. What is the standard deviation for this set of test scores? **Example 2**

P3. The median annual salary for a police dispatcher in a large city was $44,528. If the 25th percentile for the annual salary of a police dispatcher was $32,761, find the percent of police dispatchers whose annual salaries were

  a. less than $44,528.

  b. more than $32,761.

  c. between $32,761 and $44,528.

  Example 3

P4. On an examination given to 8600 students, Hal's score of 405 was higher than the scores of 3952 of the students who took the examination. What is the percentile for Hal's score? **Example 4**

P5. A survey asked 50 college students how much sleep they got last night. The results, in hours, were 6.8, 7.9, 8.2, 9.1, 5.5, 7.2, 8.0, 7.8, 6.2, 7.0, 7.5, 7.3, 6.5, 6.0, 7.4, 6.9, 7.3, 7.5, 8.1, 7.2, 6.6, 7.1, 7.6, 8.0, 7.7, 7.8, 7.5, 7.5, 9.0, 8.5, 8.2, 7.8, 7.5, 7.2, 7.0, 6.8, 6.5, 7.5, 7.8, 6.9, 7.2, 7.0, 8.5, 7.0, 8.2, 8.1, 7.8, and 7.5. Find the 90th, 50th, and 25th percentiles. **Example 5**

P6. The following table lists the weights, in ounces, of 15 avocados in a random sample. Find the quartiles for the data. Example 6

Weights, in Ounces, of Avocados

12.4	10.8	14.2	7.5	10.2	11.4	12.6	12.8	13.1	15.6
9.8	11.4	12.2	16.4	14.5					

P7. Construct a box-and-whisker plot for the following data. Example 7

Number of Rooms Occupied in a Resort During an 18-Day Period

86	77	58	45	94	96	83	76	75
65	68	72	78	85	87	92	55	61

▶ Section Exercises

In Exercises 1 to 4, round each z-score to the nearest hundredth.

1. A data set has a mean of $\bar{x} = 75$ and a standard deviation of 11.5. Find the z-score for each of the following.
  a. $x = 85$ **b.** $x = 95$
  c. $x = 50$ **d.** $x = 75$

2. A data set has a mean of $\bar{x} = 212$ and a standard deviation of 40. Find the z-score for each of the following.
  a. $x = 200$ **b.** $x = 224$
  c. $x = 300$ **d.** $x = 100$

3. A data set has a mean of $\bar{x} = 6.8$ and a standard deviation of 1.9. Find the z-score for each of the following.
  a. $x = 6.2$ **b.** $x = 7.2$
  c. $x = 9.0$ **d.** $x = 5.0$

4. A data set has a mean of $\bar{x} = 4010$ and a standard deviation of 115. Find the z-score for each of the following.
  a. $x = 3840$ **b.** $x = 4200$
  c. $x = 4300$ **d.** $x = 4030$

5. Blood Pressure A blood pressure test was given to 450 women ages 20 to 36. It showed that their mean systolic blood pressure was 119.4 mm Hg, with a standard deviation of 13.2 mm Hg.
  a. Determine the z-score, to the nearest hundredth, for a woman who had a systolic blood pressure reading of 110.5 mm Hg.
  b. The z-score for one woman was 2.15. What was her systolic blood pressure reading?

6. Fruit Juice A random sample of 1000 oranges showed that the mean amount of juice per orange was 7.4 fluid ounces, with a standard deviation of 1.1 fluid ounces.
  a. Determine the z-score, to the nearest hundredth, of an orange that produced 6.6 fluid ounces of juice.
  b. The z-score for one orange was 3.15. How much juice was produced by this orange? Round to the nearest tenth of a fluid ounce.

7. Cholesterol A test involving 380 patients ages 20 to 24 found that their blood cholesterol levels had a mean of 182 mg/dl and a standard deviation of 44.2 mg/dl.
  a. Determine the z-score, to the nearest hundredth, for one of the patients who had a blood cholesterol level of 214 mg/dl.
  b. The z-score for one patient was −1.58. What was this patient's blood cholesterol level? Round to the nearest hundredth.

8. Tire Wear A random sample of 80 tires showed that the mean mileage per tire was 41,700 mi, with a standard deviation of 4300 mi.
  a. Determine the z-score, to the nearest hundredth, for a tire that provided 46,300 mi of wear.
  b. The z-score for one tire was −2.44. What mileage did this tire provide? Round your result to the nearest hundred miles.

9. Test Scores Which of the following three test scores is the highest relative score?
  a. A score of 65 on a test with a mean of 72 and a standard deviation of 8.2
  b. A score of 102 on a test with a mean of 130 and a standard deviation of 18.5
  c. A score of 605 on a test with a mean of 720 and a standard deviation of 116.4

10. Physical Fitness Which of the following fitness scores is the highest relative score?
  a. A score of 42 on a test with a mean of 31 and a standard deviation of 6.5
  b. A score of 1140 on a test with a mean of 1080 and a standard deviation of 68.2
  c. A score of 4710 on a test with a mean of 3960 and a standard deviation of 560.4

11. Reading Test On a reading test, Shaylen's score of 455 was higher than the scores of 4256 of the 7210 students who took the test. Find the percentile, rounded to the nearest percent, for Shaylen's score.

12. Placement Exams On a placement examination, one student scored lower than 1210 of the 12,860 students who took the exam. Find the percentile, rounded to the nearest percent, for that score.

13. Test Scores Agan scored at the 65th percentile on a test given to 9840 students. How many students scored lower than Agan?

14. Test Scores Shion scored at the 84th percentile on a test given to 12,600 students. How many students scored higher than Shion?

15. Median Income In a recent year, the median family income in the United States was $66,650. If the 90th percentile for median four-person family income was $178,500, find the percentage of families whose income was
  a. more than $66,650.   **b.** more than $178,500.
  c. between $66,650 and $178,500.

16. Monthly Rents A recent survey determined that the median monthly housing rent was $708. If the first quartile for monthly housing rent was $570, find the percent of monthly housing rents that were
  a. more than $570.   **b.** less than $708.
  c. between $570 and $708.

17. Commute to School A survey was given to 18 students. One question asked about the one-way distance the student had to travel to attend college. The results, in miles, are shown in the following table. Use the median procedure for finding quartiles to find the first, second, and third quartiles for the data.

Miles Traveled to Attend College

12	18	4	5	26	41	1	8	10
10	3	28	32	10	85	7	5	15

18. **Prescriptions** The following table shows the number of prescriptions a doctor wrote each day for a 36-day period. Use the median procedure for finding quartiles to find the first, second, and third quartiles for the data.

Number of Prescriptions Written per Day

8	12	14	10	9	16
7	14	10	7	11	16
11	12	8	14	13	10
9	14	15	12	10	8
10	14	8	7	12	15
14	10	9	15	10	12

19. **Home Sales** The accompanying table shows the median selling prices of existing single-family homes in the United States in the four regions of the country for an 11-year period. Prices have been rounded to the nearest hundred. Use Excel to draw a box-and-whisker plot of the data for each of the four regions. Enter the data in columns. Then select all the data. Now create the box-and-whisker plot. Write a few sentences that explain any differences you found.

Median Prices of Homes Sold in the United States over an 11-year Period

Year	Northwest	Midwest	South	West
1	227,400	169,700	148,000	196,400
2	246,400	172,600	155,400	213,600
3	264,300	178,000	163,400	238,500
4	264,500	184,300	168,100	260,900
5	315,800	205,000	181,100	283,100
6	343,800	216,900	197,300	332,600
7	346,000	213,500	208,200	337,700
8	320,200	208,600	217,700	330,900
9	343,600	198,900	203,700	294,800
10	302,500	189,200	194,800	263,700
11	335,500	197,600	196,000	259,700

20. **Medicine** The blood lead concentrations, in micrograms per deciliter ($\mu g/dl$), of 20 children from two different neighborhoods were measured. The results are recorded in the table.

Neighborhood 1	3.97	3.91	3.98	3.70	4.13	3.97	4.01	3.88	4.11	3.70
	3.96	3.77	4.30	4.08	4.12	4.93	3.93	3.94	3.85	3.83
Neighborhood 2	4.31	4.22	3.78	4.10	4.34	4.20	4.35	4.20	4.01	4.04
	4.28	4.12	4.59	4.12	4.01	3.85	3.96	4.28	4.39	4.13

Using the same scale, draw a box-and-whisker plot for each of the two data sets. Considering that high blood lead concentrations are harmful to humans, in which of the two neighborhoods would you prefer to live?

21. **Farming** The following table shows the numbers of bushels of barley cultivated per acre for 12 one-acre plots of land for two different strains of barley, PHT-34 and CBX-21.

PHT-34	CBX-21
43	56
49	47
47	44
38	45
47	46
45	50
50	48
46	60
46	53
46	50
45	49
43	52

Using the same scale, draw a box-and-whisker plot for each of the two data sets. Write a valid conclusion based on the data.

▶ **Investigations**

Stem-and-Leaf Diagrams The relative position of each data value in a small set of data can be graphically displayed by using a *stem-and-leaf diagram*. For instance, consider the following history test scores:

65, 72, 96, 86, 43, 61, 75, 86, 49, 68, 98, 74, 84, 78, 85, 75, 86, 73

In the following stem-and-leaf diagram, we have organized the history test scores by placing all of the scores that are in the 40s in the top row, the scores that are in the 50s in the second row, the scores that are in the 60s in the third row, and so on. The tens digits of the scores have been placed to the left of the vertical line. In this diagram, they are referred to as *stems*. The ones digits of the test scores have been placed in the proper row to the right of the vertical line. In this diagram, they are the *leaves*. It is now easy to make observations about the distribution of the scores. Only two of the scores are in the 90s. Six of the scores are in the 70s, and none of the scores are in the 50s. The lowest score is 43, and the highest is 98.

A Stem-and-Leaf Diagram of a Set of History Test Scores

Stems	Leaves
4	3 9
5	
6	1 5 8
7	2 3 4 5 5 8
8	4 5 6 6 6
9	6 8

Legend: 8|6 represents 86

Steps in Construction of a Stem-and-Leaf Diagram

1. Determine the stems and list them in a column from smallest to largest or largest to smallest.
2. List the remaining digit of each stem as a leaf to the right of the stem.
3. Include a *legend* that explains the meaning of the stems and the leaves. Include a title for the diagram.

The choice of how many leading digits to use as the stem will depend on the particular data set. For instance, consider the following data set, in which a travel agent has recorded the amount spent by customers for a cruise.

Amount Spent for a Cruise

$3600	$4700	$7200	$2100	$5700	$4400	$9400
$6200	$5900	$2100	$4100	$5200	$7300	$6200
$3800	$4900	$5400	$5400	$3100	$3100	$4500
$4500	$2900	$3700	$3700	$4800	$4800	$2400

One method of choosing the stems is to let each thousands digit be a stem and each hundreds digit be a leaf. If the stems and leaves are assigned in this manner, then the notation 2|1, with a stem of 2 and a leaf of 1, represents a cost of $2100, and 5|4 represents a cost of $5400. A stem-and-leaf diagram can now be constructed by writing all of the stems in a column from smallest to largest to the left of a vertical line and writing the corresponding leaves to the right of the line. See the following diagram.

Amount Spent for a Cruise   Legend: 7|3 represents $7300

Stems	Leaves
2	1 1 4 9
3	1 1 6 7 7 8
4	1 4 5 5 7 8 8 9
5	2 4 4 7 9
6	2 2
7	2 3
8	
9	4

Sometimes two sets of data can be compared by using a *back-to-back stem-and-leaf diagram*, in which common stems are listed in the middle column of the diagram. Leaves from one data set are displayed to the right of the stems, and leaves from the other data set are displayed to the left. For instance, the following back-to-back stem-and-leaf diagram shows the test scores for two classes that took the same test. It is easy to see that the 8 A.M. class did better on the test because it had more scores in the 80s and 90s and fewer scores in the 40s, 50s, and 60s. The number of scores in the 70s was the same for both classes.

Biology Test Scores

8 A.M. Class		10 A.M. Class
2	4	5 8
7	5	6 7 9 9
5 8	6	2 3 4 8
1 2 3 3 3 7 8	7	1 3 3 5 5 6 8
4 4 5 5 6 8 8 9	8	2 3 6 6 6
2 4 5 5 8	9	4 5

Legend: 3|7 represents 73   Legend: 8|2 represents 82

Project Exercises

1. The following table lists the ages of customers who purchased a cruise. Construct a stem-and-leaf diagram for the data.

Ages of Customers Who Purchased a Cruise

32	45	66	21	62	68	72
61	55	23	38	44	77	64
46	50	33	35	42	45	51
51	28	40	41	52	52	33

2. Two groups of people were part of a test to determine how long, in seconds, it took to solve a logic problem when exposed to different ambient noise levels. Group 1 was given the problem in a room where a constant decibel (dB) level of 65 dB was maintained. For Group 2, the decibel level was maintained at 30 dB. The results are shown in the following stem-and-leaf diagram.

Group 1	Stem	Group 2
	3	3
4	4	2 2 6 8
8 3 1	5	0 4 4 4 5 8
6 2 2 2 1	6	2 3 5
5 3 2 1	7	4
6 1	8	

Legend:
8|5 represents 58 seconds

Legend:
5|8 represents 58 seconds

a. How many people in Group 1 required more than 60 s to solve the problem?
b. How many people in Group 2 required more than 60 s to solve the problem?
c. By just looking at the stem-and-leaf diagram, which group appears to have the larger mean time to solve the problem?

3. The exercise heart rate, in beats per minute (bpm), of 20 people was tested before and after a 10-week training program. The results are recorded in the following table.

Before	128	128	131	151	141	139	128	139	161	156
	136	134	134	136	116	174	158	148	156	144
After	125	107	121	140	150	149	126	119	134	138
	164	140	134	129	123	133	139	117	128	139

Draw a back-to-back stem-and-leaf diagram for these data.

Normal Distributions

Learning Objectives:

- Read data from a frequency distribution and histogram.
- Use the empirical rule for a normal distribution.
- Find probabilities using a normal distribution.

Frequency Distributions and Histograms

Large sets of data are often displayed using a *grouped frequency distribution* or a *histogram*. For instance, consider the following situation. An Internet service provider (ISP) has installed new computers. To estimate the new download times its subscribers will experience, the ISP surveyed 1000 of its subscribers to determine the time required for each subscriber to download a particular file from an Internet site. The results of that survey are summarized in Table 8.5.

Table 8.5
A Grouped Frequency Distribution with 12 Classes

Download Time (in seconds)	Number of Subscribers
0–5	6
5–10	17
10–15	43
15–20	92
20–25	151
25–30	192
30–35	190
35–40	149
40–45	90
45–50	45
50–55	15
55–60	10

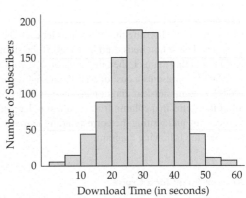

Figure 8.3 A histogram for the frequency distribution in Table 8.6

Table 8.5 is called a **grouped frequency distribution**. It shows how often (frequently) certain events occurred. Each interval, 0–5, 5–10, and so on, is called a **class**. This distribution has 12 classes. For the 10–15 class, 10 is the **lower class boundary** and 15 is the **upper class boundary**. Any data value that lies on a common boundary is assigned to the higher class. The *graph* of a frequency distribution is called a **histogram**. A histogram provides a pictorial view of how the data are distributed. In Figure 8.3, the height of each bar of the histogram indicates how many subscribers experienced the download times shown by the class on the base of the bar.

Examine the distribution in Table 8.6. It shows the *percent* of subscribers that are in each class, as opposed to the frequency distribution in Table 8.5, which shows the *number* of customers in each class. The type of frequency distribution that lists the *percent* of data in each class is called a **relative frequency distribution**. The **relative frequency histogram** in Figure 8.4 was drawn by using the data in the relative frequency distribution. It shows the *percent* of subscribers along its vertical axis.

One advantage of using a relative frequency distribution instead of a grouped frequency distribution is that there is a direct correspondence between the percent values of the relative frequency distribution and probabilities. For instance, in the relative frequency distribution in Table 8.6, the percent of the data that lies between 35 s and 40 s is 14.9%. Thus, if a subscriber is chosen at random, the probability that the subscriber will require at least 35 s but less than 40 s to download the file is 0.149.

Table 8.6
A Relative Frequency Distribution

Download Time (in seconds)	Percent of Subscribers
0–5	0.6
5–10	1.7
10–15	4.3
15–20	9.2
20–25	15.1
25–30	19.2
30–35	19.0
35–40	14.9
40–45	9.0
45–50	4.5
50–55	1.5
55–60	1.0

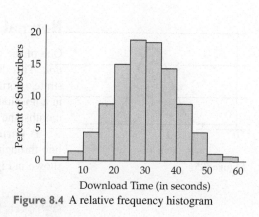

Figure 8.4 A relative frequency histogram

Example 1   **Use a Relative Frequency Distribution**

Use the relative frequency distribution in Table 8.6 to determine the

a. *percent* of subscribers who required at least 25 s to download the file.

b. *probability* that a subscriber chosen at random will require at least 5 s but less than 20 s to download the file.

Solution

a. The percent of data in all the classes with a lower boundary of 25 s or more is the sum of the percents printed in red in Table 8.7. Thus the percent of subscribers who required at least 25 s to download the file is 69.1%.

Table 8.7

Download Time (in seconds)	Percent of Subscribers	
0–5	0.6	
5–10	1.7	
10–15	4.3	Sum is
15–20	9.2	15.2%
20–25	15.1	
25–30	19.2	
30–35	19.0	
35–40	14.9	
40–45	9.0	Sum is
45–50	4.5	69.1%
50–55	1.5	
55–60	1.0	

b. The percent of data in all the classes with a lower boundary of 5 s and an upper boundary of 20 s is the sum of the percents printed in blue in Table 8.7. Thus the percent of subscribers who required at least 5 s but less than 20 s to download the file is 15.2%. The probability that a subscriber chosen at random will require at least 5 s but less than 20 s to download the file is 0.152.

Normal Distributions and the Empirical Rule

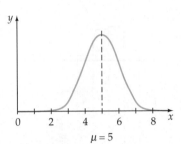

Figure 8.5

One of the most important statistical distributions of data is known as a *normal distribution*. This distribution occurs in a variety of applications. Types of data that may demonstrate a normal distribution include the lengths of leaves on a tree, the weights of newborns in a hospital, the lengths of time of a student's trip from home to school over a period of months, the SAT scores of a large group of students, and the life spans of light bulbs.

A **normal distribution** forms a bell-shaped curve that is symmetric about a vertical line through the mean of the data. A graph of a normal distribution with a mean of 5 is shown in Figure 8.5.

Properties of a Normal Distribution

Every normal distribution has the following properties.

- The graph is symmetric about a vertical line through the mean of the distribution.
- The mean, median, and mode are equal.
- The y-value of each point on the curve is the *percent* (expressed as a decimal) of the data at the corresponding x-value.
- Areas under the curve that are symmetric about the mean are equal.
- The total area under the curve is 1.

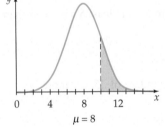

Figure 8.6

In the normal distribution shown in Figure 8.6, the area of the shaded region is 0.159 unit. This region represents the fact that 15.9% of the data values are greater than or equal to 10. Because the area under the curve is 1, the unshaded region under the curve has area $1 - 0.159$, or 0.841, representing the fact that 84.1% of the data are less than 10.

The following rule, called the Empirical Rule, describes the percents of data that lie within 1, 2, and 3 standard deviations of the mean in a normal distribution.

Empirical Rule for a Normal Distribution

In a normal distribution, approximately

- 68% of the data lie within 1 standard deviation of the mean.
- 95% of the data lie within 2 standard deviations of the mean.
- 99.7% of the data lie within 3 standard deviations of the mean.

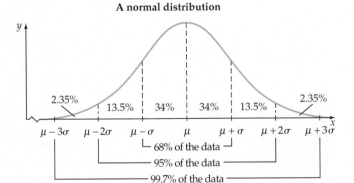

Example 2 | **Use the Empirical Rule to Solve an Application**

A survey of 1000 U.S. gas stations found that the price charged for a gallon of regular gas could be closely approximated by a normal distribution with a mean of $3.10 and a standard deviation of $0.18. How many of the stations charge

a. between $2.74 and $3.46 for a gallon of regular gas?

b. less than $3.28 for a gallon of regular gas?

c. more than $3.46 for a gallon of regular gas?

Solution

a. The $2.74 per gallon price is 2 standard deviations below the mean. The $3.46 price is 2 standard deviations above the mean. In a normal distribution, 95% of all data lie within 2 standard deviations of the mean. See Figure 8.7. Therefore, approximately

$$(95\%)(1000) = (0.95)(1000) = 950$$

of the stations charge between $2.74 and $3.46 for a gallon of regular gas.

b. The $3.28 price is 1 standard deviation above the mean. See Figure 8.8. In a normal distribution, 34% of all data lie between the mean and 1 standard deviation above the mean. Thus, approximately

$$(34\%)(1000) = (0.34)(1000) = 340$$

of the stations charge between $3.10 and $3.28 for a gallon of regular gasoline. Half of the 1000 stations, or 500 stations, charge less than the mean. Therefore, about $340 + 500 = 840$ of the stations charge less than $3.28 for a gallon of regular gas.

c. The $3.46 price is 2 standard deviations above the mean. In a normal distribution, 95% of all data are within 2 standard deviations of the mean. This means that the other 5% of the data will lie either more than 2 standard deviations above the mean or more than 2 standard deviations below the mean. We are interested only in the data that are more than 2 standard deviations above the mean, which is $\frac{1}{2}$ of 5%, or 2.5%, of the data. See Figure 8.9. Thus about $(2.5\%)(1000) = (0.025)(1000) = 25$ of the stations charge more than $3.46 for a gallon of regular gas.

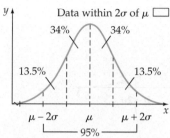

Figure 8.7

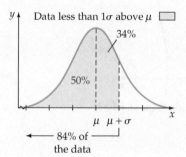

Figure 8.8

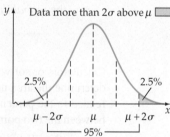

Figure 8.9

The Standard Normal Distribution

It is often helpful to convert data values x to z-scores, a measure of the number of standard deviations from the mean. A positive z-score is to the right of the mean; a negative z-score is the left of the mean.

$$z_x = \frac{x - \mu}{\sigma} \qquad \text{or} \qquad z_x = \frac{x - \bar{x}}{s}$$

If the original distribution of x values is a normal distribution, then the corresponding distribution of z-scores will also be a normal distribution. This normal distribution of *z*-scores is called the *standard normal distribution*. See Figure 8.10. It has a mean of 0 and a standard deviation of 1.

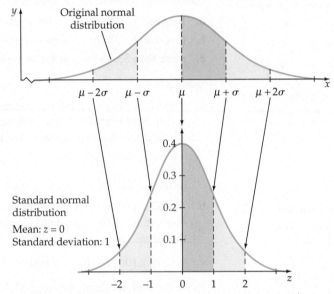

Figure 8.10 Conversion of a normal distribution to the standard normal distribution

---

The Standard Normal Distribution

The **standard normal distribution** is the normal distribution that has a mean of 0 and a standard deviation of 1.

In the standard normal distribution, the area of the distribution from $z = a$ to $z = b$ represents

- the *percentage* of *z*-values that lie in the interval from *a* to *b*.
- the *probability* that *z* lies in the interval from *a* to *b*.

Calculators or software programs like Excel, Numbers, or Google Sheets are used to determine the area under the normal distribution between two points. These areas are related to the percent of data between the two points or the probability that data lies between the two points. Here are some examples using TI-84 calculator and Excel. For each of these examples, we will use a normal distribution with a mean of $\mu = 500$ and a standard deviation of $\sigma = 100$.

TI-84

Find the probability that a value is less than 430.
Press **2ND DISTR**. Select normalcdf(

```
DISTR DRAW
1:normalpdf(
2:normalcdf(
3:invNorm(
4:invT(
5:tpdf(
6:tcdf(
7↓χ²pdf(
```

Enter the values for lower bound, upper bound, μ, and σ. Use a lower bound that is at least 4 standard deviations below the mean or 200 $(500 - 4 \cdot 100 = 100)$. We have used 100 for this example.

```
normalcdf
lower:100
upper:430
μ:500
σ:100
Paste
```

Scroll to Paste and press
ENTER twice. The probability is
approximately 0.2420.

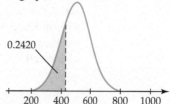

A graph is shown below.

0.2420

Find the probability that a value is
greater than 575.
Press **2ND DISTR**. Select
normalcdf(

```
DISTR DRAW
1:normalpdf(
2:normalcdf(
3:invNorm(
4:invT(
5:tpdf(
6:tcdf(
7↓χ²pdf(
```

Enter the values for
lower bound, upper
bound, μ, and σ. Use
an upper bound that is
at least 4 standard devi-
ations larger than mean.
$(500 + 4 \cdot 100 = 900)$

```
normalcdf
lower:575
upper:900
μ:500
σ:100
Paste
```

Scroll to Paste and press
ENTER twice. The probability is
approximately 0.2266.

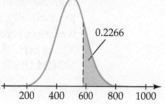

A graph is shown below.

0.2266

Find the probability that a value
lies between 400 and 650.
Press **2ND DISTR**. Select
normalcdf(

```
DISTR DRAW
1:normalpdf(
2:normalcdf(
3:invNorm(
4:invT(
5:tpdf(
6:tcdf(
7↓χ²pdf(
```

Enter the values for
lower bound, upper
bound, μ, and σ.

```
normalcdf
lower:400
upper:650
μ:500
σ:100
Paste
```

Scroll to Paste and press
ENTER twice. The probability is
approximately 0.7745.

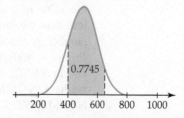

A graph is shown below.

0.7745

Excel

Excel uses the NORM.DIST function to calculate probabilities for a normal distribution. The format for this function is NORM.DIST(x,mean,std_deviation,TRUE).
This function returns the probability that a randomly chosen value is **_less than_** x. The
parameter TRUE tells Excel to find the probability and is required. Note the *less than*.
This is very important.

Here is a repeat of the calculations above using Excel. Recall that the mean of $\mu = 500$ and a standard deviation of $\sigma = 100$. To find the probability that a value x is less than 430, in Excel we use the function NORM.DIST(x,_mean,std_deviation,TRUE). Here we have entered mean = 500 in cell B1, standard deviation 100 in cell B2, and end value 430 in cell B3. Then, our Excel function is NORM.DIST(B3,B1,B2,TRUE).

B4		fx	=NORM.DIST(B3,B1,B2,TRUE)		
	A	B	C	D	E
1	Mean	500			
2	Std Deviation	100			
3	End value	430			
4	Probability	0.24196365			

Prob($x < 430$) ≈ 0.0.2420

Find the probability that a value is greater than 575. Because Excel gives the probability for a value *less than* a given value, by the complement rule, the probability that a value x is *greater than* a given value is 1 minus the probability the value *less than* the given value. Symbolically, *Prob*($x > 575$) = 1 − *Prob*($x < 575$). Excel gives *Prob*($x < 575$). In Excel we have entered mean = 500 in cell B1, standard deviation 100 in cell B2, and end value 575 in cell B3. Then our Excel function NORM.DIST(x,mean,std_deviation,TRUE) is NORM.DIST(B3,B1,B2,TRUE).

B4		fx	=1-NORM.DIST(B3,B1,B2,TRUE)		
	A	B	C	D	E
1	Mean	500			
2	Std Deviation	100			
3	End value	575			
4	Probability	0.22662735			

Prob($x > 575$) = 1 − *Prob*($x < 575$) ≈ 0.2266

Find the probability that a randomly chosen value is between 400 and 650. Because Excel gives the probability for a value *less than* a given value, the probability that the value of a single data element is between two numbers is the probability it is less than the larger value minus the probability the number is less than the smaller number. Symbolically, this is written *Prob*($400 < x < 650$) = *Prob*($x < 650$) − *Prob*($x < 400$). In Excel we have entered mean = 500 in cell B1, standard deviation 100 in cell B2, smaller value 400 in cell B3, and larger value 600 in cell B4. Because we have the difference of the larger value and the smaller value, we use the function NORM.DIST(B4,B1,B2,TRUE) − NORM.DIST(B3,B1,B2,TRUE).

B5		fx	=NORM.DIST(B4,B1,B2,TRUE)-NORM.DIST(B3,B1,B2,TRUE)					
	A	B	C	D	E	F	G	H
1	Mean	500						
2	Std Deviation	100						
3	Smaller value	400						
4	Larger value	650						
5	Probability	0.77453754						

Prob($400 < x < 650$) = *Prob*($x < 650$) − *Prob*($x < 400$)
≈ 0.7745

> **Example 3** Solve an Application Using a TI-84 Calculator

A soda machine dispenses soda into 12-ounce cups. Tests show that the actual amount of soda dispensed is normally distributed, with a mean of 12.01 oz and a standard deviation of 0.09 oz.

a. What percent of cups will receive less than 12 oz of soda?

b. What percent of cups will receive between 11.95 oz and 12.1 oz of soda?

c. If a cup is filled at random, what is the probability that the machine will dispense more 12.05 oz of soda?

Take Note The lower value should be at least 4 standard deviations *below* the mean.

$12.01 - 4 \cdot 0.09 = 12.07 - 0.36$
$= 11.65$

You could use any number less than 11.65.

Solution

This solution uses the TI-84. $\mu = 12.01$ oz and $\sigma = 0.09$ oz.

a. Use the TI-84 normalcdf function with lower = 11.65, upper = 12, $\mu = 12.01$, and $\sigma = 0.09$ oz. Approximately 45.6% of the cups will have less than 12 oz of soda.

```
  normalcdf
lower:11.65
upper:12
μ:12.01
σ:0.09
Paste
```
```
normalcdf(11.65,12,12.01, ▸
              0.4557323813
```

b. Use the TI-84 normalcdf function with lower = 11.95, upper = 12.1, $\mu = 12.01$, and $\sigma = 0.09$ oz. Approximately 58.9% of the cups will 11.95 oz and 12.1 oz of soda.

```
  normalcdf
lower:11.95
upper:12.1
μ:12.01
σ:0.09
Paste
```
```
normalcdf(11.95,12.1,12.0 ▸
              0.5888522734
```

Take Note The upper value should be at least 4 standard deviations *above* the mean,

$12.01 + 4 \cdot 0.09 = 12.01 + 0.36$
$= 12.37$

You could use any number greater than 12.37.

c. Use the TI-84 normalcdf function with lower = 12.05, upper = 12.37, $\mu = 12.01$, and $\sigma = 0.09$ oz. The probability that the machine will dispense more than 12.05 ounces of soda is approximately 0.328.

```
  normalcdf
lower:12.05
upper:12.37
μ:12.01
σ:0.09
Paste
```
```
normalcdf(12.05,12.37,12. ▸
              0.3283289806
```

> **Example 4** Solve an Application Using Excel

The weight of a large egg is normally distributed with a mean of 60.3 grams and a standard deviation of 1.167 grams.

a. What percent of eggs randomly selected from a basket of large eggs weigh less than 56.9 grams?

b. What is the probability that a randomly selected egg from a basket of large eggs weighs between 59.7 grams and 61.4 grams?

c. What is the probability that a randomly selected egg from a basket of large eggs weighs more than 63 grams?

Solution

a. Use the Excel NORM.DIST function. Because this part of the problem asks for a percent, find the probability. Then multiply by 100%.

Take Note The NORM.DIST function gives the probability of being less than a value.
$Prob(x < 56.9) = $ NORM.DIST(56.9,60.3,1.167,TRUE)

B4		f_x	=NORM.DIST(B3,B1,B2,TRUE)		
	A	B	C	D	E
1	Mean	60.3			
2	Std Deviation	1.167			
3	Smaller value	56.9			
4	Probability	0.00178728			
5	Percent	0.18%			

Approximately 0.18% of the eggs would weigh less than 56.9 grams.

b. Use the Excel NORM.DIST function. Find the probability that the egg weighs less than 61.4 grams minus the probability the egg weighs less than 59.7 grams. Symbolically, find

$$Prob(59.7 < x < 61.4) = Prob(x < 61.4) - Prob(x < 59.7)$$
$$= \text{NORM.DIST}(61.4, 60.3, 1.167, \text{TRUE})$$
$$- \text{NORM.DIST}(59.7, 60.3, 1.167, \text{TRUE})$$

B5	▲▼	× ✓	f_x	=NORM.DIST(B4,B1,B2,TRUE)-NORM.DIST(B3,B1,B2,TRUE)				
	A	B	C	D	E	F	G	H
1	Mean	60.3						
2	Std Deviation	1.167						
3	Smaller value	59.7						
4	Larger value	61.4						
5	Probability	0.523477						

The probability is approximately 0.523.

c. Because Excel gives the probability for a value *less than* a given value (63 in this case), the probability that a value is greater than 63 is 1 minus the probability the value is *less than* 63. Symbolically,

$$Prob(x > 63) = 1 - Prob(x < 63)$$
$$= 1 - \text{NORM.DIST}(63, 60.3, 1.167, \text{TRUE})$$

B4	▲▼	× ✓	f_x	=1-NORM.DIST(B3,B1,B2,TRUE)	
	A	B	C	D	E
1	Mean	60.3			
2	Std Deviation	1.167			
3	End value	63			
4	Probability	0.010344			

The probability is approximately 0.010.

8.4 Exercise Set

▶ Think About It

1. What is the standard normal distribution?

2. For data that are normally distributed, what percent of the data lie below the mean? What percent of the data lie above the median?

3. According to the Empirical Rule for normally distributed data, approximately how much of the data lie between plus or minus 1 standard deviation from the mean?

4. If a data value is randomly selected from normally distributed data, what is the probability that it will be more than 3 standard deviations from the mean?

▶ Preliminary Exercises

For Exercises P1 to P4, if necessary, use the referenced example following the exercise for assistance.

P1. Use the relative frequency distribution in Table 8.6 to determine the

  a. *percent* of subscribers who required less than 25 s to download the file.

  b. *probability* that a subscriber chosen at random will require at least 10 s but less than 30 s to download the file. **Example 1**

P2. A vegetable distributor knows that during the month of August, the weights of its tomatoes are normally distributed with a mean of 0.61 lb and a standard deviation of 0.15 lb. Use the Empirical Rule for parts a, b, and c. **Example 2**

  a. What percent of the tomatoes weigh less than 0.76 lb?

  b. In a shipment of 6000 tomatoes, how many tomatoes can be expected to weigh more than 0.31 lb?

  c. In a shipment of 4500 tomatoes, how many tomatoes can be expected to weigh from 0.31 lb to 0.91 lb?

P3. The mean score on the verbal reasoning portion of a recent Graduate Record Exam (GRE) is 150 with a standard deviation of 12.5.

  a. What percent of students taking the GRE had a score greater than 165 on the verbal reasoning portion? Round to the nearest tenth of a percent.

  b. What is the probability that a student taking the GRE scored between 140 and 155 on the verbal reasoning portion? Round to the nearest ten-thousandths.

c. What percent of students taking the GRE had a score less than 160 on the verbal reasoning portion? Round to the nearest tenth of a percent. **Example 3**

P4. A study shows that the lengths of the careers of professional football players are nearly normally distributed, with a mean of 6.1 years and a standard deviation of 1.8 years. **Example 4**

 a. What percent of professional football players have a career of more than 9 years?

 b. If a professional football player is chosen at random, what is the probability that the player will have a career of between 3 and 4 years?

▶ **Section Exercises**

1. **Weights of New-Born Elephants** The following relative frequency distribution shows the weights, in pounds, of new-born elephants.

Weight (in pounds)	Percent of Elephants
<175	5
176–210	20
211–245	30
246–280	27
281–315	15
>316	3

 a. What percent of the new-born elephants were at least 246 pounds?

 b. What is the probability one new-born elephants chosen at random had a weight that is at most 210 pounds?

2. **Biology** A biologist measured the lengths of hundreds of cuckoo bird eggs. Use the relative frequency distribution below to answer the questions that follow.

Lengths of Cuckoo Bird Eggs

Length (in millimeters)	Percent of Eggs
18.75–19.75	0.8
19.75–20.75	4.0
20.75–21.75	17.3
21.75–22.75	37.9
22.75–23.75	28.5
23.75–24.75	10.7
24.75–25.75	0.8

 a. What percent of the group of eggs was less than 21.75 mm long?

 b. What is the probability that one of the eggs selected at random was at least 20.75 mm long but less than 24.75 mm long?

In Exercises 3 to 8, use the Empirical Rule to answer each question.

3. In a normal distribution, what percent of the data lie

 a. within 2 standard deviations of the mean?

 b. more than 1 standard deviation above the mean?

 c. between 1 standard deviation below the mean and 2 standard deviations above the mean?

4. In a normal distribution, what percent of the data lie

 a. within 3 standard deviations of the mean?

 b. more than 2 standard deviations below the mean?

 c. between 2 standard deviations below the mean and 3 standard deviations above the mean?

5. **Shipping** During 1 week, an overnight delivery company found that the weights of its parcels were normally distributed, with a mean of 24 oz and a standard deviation of 6 oz.

 a. What percent of the parcels weighed between 12 oz and 30 oz?

 b. What percent of the parcels weighed more than 42 oz?

6. **Sports** A sports team finds that the attendance at its home games is normally distributed, with a mean of 16,000 and a standard deviation of 4000.

 a. What percent of the home games have an attendance between 12,000 and 20,000 people?

 b. What percent of the home games have an attendance of fewer than 8000 people?

7. **Traffic** A highway study of 8000 vehicles that passed by a checkpoint found that their speeds were normally distributed, with a mean of 61 mph and a standard deviation of 7 mph.

 a. How many of the vehicles had a speed of more than 68 mph?

 b. How many of the vehicles had a speed of less than 40 mph?

In Exercises 8 to 22, answer each question. Round z-scores to the nearest hundredth.

8. **Biology** A biologist found the wingspans of a group of monarch butterflies to be normally distributed, with a mean of 52.2 mm and a standard deviation of 2.3 mm. What percent of the butterflies had a wingspan

 a. less than 48.5 mm?

 b. between 50 and 55 mm?

9. **Cholesterol Levels** The cholesterol levels of a group of students at a university are normally distributed, with a mean of 185 and a standard deviation of 39. What percent of the students have a cholesterol level

 a. greater than 219?

 b. between 190 and 225?

10. **Heart Rates** The resting heart rates of a group of healthy adult men were found to have a mean of 73.4 beats per minute, with a standard deviation of 5.9 beats per minute. What percent of these men had a resting heart rate of

 a. greater than 80 beats per minute?

 b. between 70 and 85 beats per minute?

11. **Light Bulbs** A manufacturer of light bulbs finds that one light bulb model has a mean life span of 1025 hr with a standard deviation of 87 hr. What percent of these light bulbs will last

 a. at least 950 hr?

 b. between 800 and 900 hr?

12. **Telephone Calls** A telephone company has found that the lengths of its long distance telephone calls are normally distributed, with a mean of 225 s and a standard deviation of 55 s. What percent of its long distance calls are

 a. longer than 340 s?

 b. between 200 and 300 s?

13. **Cereal Weight** The weights of all the boxes of corn flakes filled by a machine are normally distributed, with a mean weight of 14.5 oz and a standard deviation of 0.4 oz. What percent of the boxes will

 a. weigh less than 14 oz?

 b. weigh between 13.5 oz and 15.5 oz?

14. **Tire Mileage** The mileage for WearEver tires is normally distributed, with a mean of 48,000 mi and a standard deviation

of 7400 mi. What is the probability that the WearEver tires you purchase will provide a mileage of

a. more than 60,000 mi?

b. between 40,000 and 50,000 mi?

15. Rope Strength The breaking point of a particular type of rope is normally distributed, with a mean of 350 lb and a standard deviation of 24 lb. What is the probability that a piece of this rope chosen at random will have a breaking point of

a. less than 320 lb?

b. between 340 and 370 lb?

16. IQ Tests A psychologist finds that the intelligence quotients of a group of patients are normally distributed, with a mean of 102 and a standard deviation of 16. Find the percent of the patients with IQs

a. above 114.

b. between 90 and 118.

17. Grocery Store Lines The amount of time customers spend waiting in line at a grocery store is normally distributed, with a mean of 2.5 min and a standard deviation of 0.75 min. Find the probability that the time a customer spends waiting is

a. less than 3 min.

b. less than 1 min.

18. The mean snowfall in Vail, CO in January is 48 inches with a standard deviation of 8 inches.

a. What is the probability that Vail will get more than 55 inches of snow in a given January?

b. What is the probability that Vail will get more than 30 inches of snow in a given January?

c. What is the probability that Vail will get between 45 and 57 inches of snow in a given January?

19. The mean weight of an adult Great Dane dog is 170 pounds with a standard deviation of 15 pounds,

a. What is the probability that a particular adult Great Dane will weigh more than 205 pounds?

b. What is the probability a particular adult Great Dane will weigh less than 180 pounds?

c. What is the probability that a particular adult Great Dane will weigh between 150 and 185 pounds?

20. A manufacturer of precision machine parts makes a metal plate that has a hole with a diameter of 4 centimeters. Because all machine processes have some variability, plates that have a diameter that is less than 3.995 centimeters or more than 4.005 centimeters are rejected. If the standard deviation of the process is 0.0025 centimeters, what is the probability that a plate will be rejected?

21. A manufacturer of LED light bulbs knows that bulbs will burn continuously with a mean of 60,000 hours and a standard deviation of 10,000 hours. If the manufacturer advertises that it will send a consumer a replacement bulb for one that lasts less than 45,000 hours, what is the probability that the manufacturer will have to replace a bulb?

22. The daily calls into a software help center are normally distributed with a mean of 900 calls with a standard deviation of 50 calls. What is the probability that on a particular day there would be between 825 calls and 1020 calls?

▶ **Investigations**

In Exercises 1 to 7, determine whether the given statement is true or false.

1. The standard normal distribution has a mean of 0.

2. Every normal distribution is a bell-shaped distribution.

3. In a normal distribution, the mean, the median, and the mode of the distribution are all located at the center of the distribution.

4. The mean of a normal distribution is always larger than the standard deviation of the distribution.

5. The standard deviation of the standard normal distribution is 1.

6. If a data value x from a normal distribution is positive, then its z-score also must be positive.

7. All normal distributions have a mean of 0.

8. a. Make a sketch of two normal distributions that have the same standard deviation but different means.

   b. Make a sketch of two normal distributions that have the same mean but different standard deviations.

9. Determine the approximate z-scores for the first quartile and the third quartile of the standard normal distribution.

Section
8.5 **Linear Regression and Correlation**

Learning Objectives:

• Find a linear regression equation.

• Find a linear correlation coefficient.

Linear Regression

In the last section, we used two points of a data set to create a linear model of the data. However, researchers prefer a more precise method called *linear regression*.

When performing research studies, scientists often wish to know whether two variables are related. If the variables are determined to be related, then a scientist may wish to find an equation that can be used to model the relationship. For instance, a geologist might want to know whether there is a relationship between the duration of an eruption of a geyser and the time between eruptions. A first step in this determination is to collect data. Data involving two variables are called *bivariate data*. Table 8.8 gives bivariate data showing the time between two eruptions and the duration of the second eruption for 10 eruptions of the geyser Old Faithful.

Table 8.8

Time Between Eruptions (in seconds), x	272	227	237	238	203	270	218	226	250	245
Duration of Eruption (in seconds), y	89	79	83	82	81	85	78	81	85	79

Once the data are collected, a **scatter diagram** or **scatter plot** can be drawn, as shown in Figure 8.11.

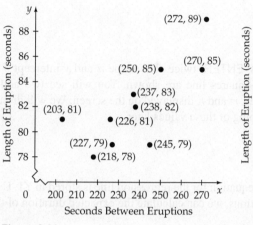

Figure 8.11

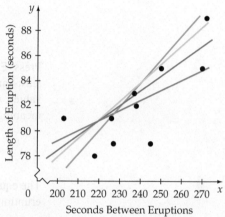

Figure 8.12

One way for the geologist to create a model of the relationship between the time between two eruptions and the duration of the second eruption is to find the equation of a line whose graph *approximates* the data points plotted in the scatter plot. There are many such lines that can be drawn, as shown in Figure 8.12.

Of all the possible lines that can be drawn, the one that is usually of most interest is called the *line of best fit* or the *least-squares regression line*.

The Least-Squares Regression Line

The **least-squares regression line** is a straight line that shows how a dependent variable changes for changes in an independent variable.

For the data in Table 8.8, the dependent variable is the duration of an eruption. The independent variable is the time between eruptions.

The formulas to calculate the equation of the regression line are programmed into spreadsheet programs and calculators. The following example shows the use of a TI-84 to find the equation of the regression line for the Old Faithful data.

Enter the data from Table 8.8 into L1 and L2, as shown at the right.

L1	L2	L3 1
272	89	------
227	79	
237	83	
238	82	
203	81	
270	85	
218	78	

L1(1) = 272

Press the **STAT** key. Tab to **CALC** , and scroll to 4:. Then press **ENTER** .

```
 EDIT  CALC  TESTS
 1: 1-Var Stats
 2: 2-Var Stats
 3: Med-Med
 4: LinReg(ax+b)
 5: QuadReg
 6: CubicReg
 7↓ QuartReg
```

Scroll to the Store RegEQ line. Press the **VARS** key and scroll to **Y-VARS**. Press **ENTER** twice.

```
    LinReg(ax+b)
 Xlist: L1
 Ylist: L2
 FreqList:
 Store RegEQ: Y1█
 Calculate
```

Press **ENTER** twice. The slope a and y-intercept b of the least-squares line are shown. You will see two additional values, r^2 and r, displayed on the screen. We will discuss the meaning of these values later.

```
        LinReg
 y=ax+b
 a=.1189559666
 b=53.81710637
 r²=.5819197284
 r=.7628366328
```

The equation for the regression line is stored in Y1. Using 200 seconds as the time between eruptions, we can calculate the expected duration of the eruptions as follows.

Press the **VARS** key and scroll to **Y-VARS** . Press **ENTER** twice. Now enter "(200)" and press **ENTER** . The predicted duration of the eruption is approximately 78 seconds.

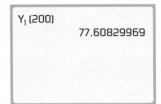

```
 Y₁(200)
              77.60829969
```

Here is an additional example of calculating regression lines. Professor R. McNeill Alexander wanted to determine whether the *stride length* of a dinosaur, as shown by its fossilized footprint, could be used to estimate the speed of the dinosaur. Stride length for an animal is defined as the distance x from a particular point on a footprint to that same point on the next footprint of the same foot. (See the figure at the left.) Because dinosaurs are extinct, Alexander and fellow scientist A. S. Jayes carried out experiments with many types of animals, including dogs, camels, ostriches, and elephants. Some of the results from these experiments are recorded in Table 8.9.

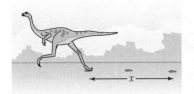

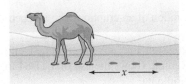

Table 8.9
Speeds of Camels for Selected Stride Lengths

Stride Length (m)	2.5	3.0	3.2	3.4	3.5	3.8	4.0	4.2
Speed (m/s)	2.3	3.9	4.1	5.0	5.5	6.2	7.1	7.6

Example 1 **Find the Equation of a Least-Squares Line**

Find the equation of the least-squares line that predicts speed of a camel for a certain stride length using the data in Table 8.9.

For this solution, Excel is used. Data is entered into columns with the dependent variable (speed) data in one column and the independent variable (stride length) data in a second column. Then use the SLOPE and INTERCEPT Excel function. For this example, the following formulas are placed in cells E2 and E3.

$$=\text{SLOPE(B2:B9,A2:A9)} = 3.164351852$$
$$=\text{INTERCEPT(B2:B9,A2:A9)} = -5.704513889$$

	A	B	C	D	E
1	Stride Length (meters)	Speed (meters/second)			
2	2.5	2.3		Slope	3.164351852
3	3.0	3.9		Intercept	-5.704513889
4	3.2	4.1			
5	3.4	5.0			
6	3.5	5.5			
7	3.8	6.2			
8	4.0	7.1			
9	4.2	7.6			

The regression equation is $\hat{y} = 3.164351852x - 5.704513889$

To get a graph of the regression line, ensure that one of the data cells is selected. Click on Insert and select Chart. Scroll to Scatter chart. This will give a scatterplot of the data points. Click on the Add Chart Element and select Trendline. This will draw the regression line. This is shown below. We have also selected other Add Chart elements options to add axes titles.

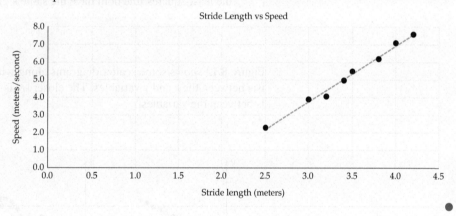

Example 2   Use a Least-Squares Line to Make a Prediction

Use the equation of the least-squares line from Example 1 to predict the average speed of camel for each of the following stride lengths. Round your results to the nearest tenth of a meter per second.

a. 2.8 meters   **b.** 3.9 meters

Solution

Use the FORECAST.LINEAR function in Excel. The format for the FORECAST.LINEAR function is =FORECAST.LINEAR(x-value,dependent variable data, independent variable data)

a. =FORECAST.LINEAR(2.8,B2:B9,A2:A9) = 3.155671296

The predicted average speed of camel with a stride length of 2.8 m is 3.2 m/s.

b. =FORECAST.LINEAR(3.9,B2:B9,A2:A9) = 6.636458333

The predicted average speed of a camel with a stride length of 3.9 m is 6.6 m/s.

For Example 2, if you are using a TI-84 calculator, store the stride lengths as L1 and the speed in L2. Select

STAT CALC 4: Scroll to Store RegEQ: Press **VARS** →
ENTER ENTER ENTER ENTER. Doing this will store the regression equation in **Y1**.

To find the value of the dependent variable when $x = 2.8$, press

   VARS → 1 1 (2.8) ENTER.

The value of the independent variable is given as 3.155671296.

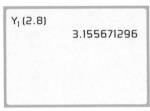

Linear Correlation Coefficient

To determine the strength of a linear relationship between two variables, statisticians use a statistic called the *linear correlation coefficient*, which is denoted by r.

Properties of the Linear Correlation Coefficient

- The linear correlation coefficient r is a real number between -1 and 1, inclusive.
  - If all of the ordered pairs lie on a line with positive slope, $r = 1$.
  - If all of the ordered pairs lie on a line with negative slope, $r = -1$.
  - If $r = 0$, it indicates that the two variables cannot be modeled by linear function.
- For any set of ordered pairs, the linear correlation coefficient r and the slope of the least-squares line both have the same sign.

Figure 8.13 shows some scatter diagrams along with the type of linear correlation that exists between the x and y variables. The closer $|r|$ is to 1, the stronger the linear relationship is between the variables.

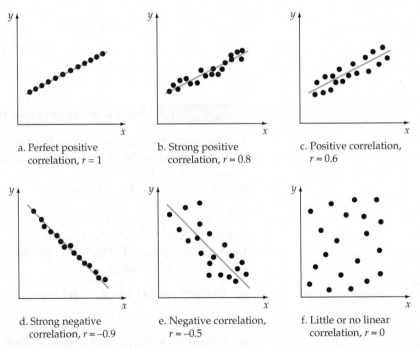

a. Perfect positive correlation, $r = 1$

b. Strong positive correlation, $r \approx 0.8$

c. Positive correlation, $r \approx 0.6$

d. Strong negative correlation, $r \approx -0.9$

e. Negative correlation, $r \approx -0.5$

f. Little or no linear correlation, $r \approx 0$

Figure 8.13 Linear correlation

Example 3   Find a Linear Correlation Coefficient

Find the linear correlation coefficient for stride length versus speed of a camel. Use the data in Table 8.9. Round your result to the nearest thousandth.

Solution

Using the spreadsheet from Example 1, the correlation coefficient r is shown in the spreadsheet below.

D2	f_x =CORREL(B2:B9,A2:A9)

	A	B	C	D
1	Stride Length (meters)	Speed (meters/second)		
2	2.5	2.3	r	0.9958553
3	3.0	3.9		
4	3.2	4.1		
5	3.4	5.0		
6	3.5	5.5		
7	3.8	6.2		
8	4.0	7.1		
9	4.2	7.6		

The linear correlation coefficient, rounded to the nearest thousandth, is 0.996.

The linear correlation coefficient indicates the strength of a linear relationship between two variables, but it does not indicate the presence of a *cause-and-effect relationship*. For instance, the data in Table 8.10 show the hours per week that a student spent playing pool and the student's weekly algebra test scores for those same weeks.

Table 8.10
Algebra Test Scores vs. Hours Spent Playing Pool

Hours per Week Spent Playing Pool	4	5	7	8	10
Weekly Algebra Test Score	52	60	72	79	83

The linear correlation coefficient for the ordered pairs in the table is $r \approx 0.98$. Thus, there is a strong positive linear relationship between the student's algebra test scores and the time the student spent playing pool. This does not mean that the higher algebra test scores were caused by the increased time spent playing pool. The fact that the student's test scores increased with the increase in the time spent playing pool could have been the result of many other factors, or it could just be a coincidence.

8.5   Exercise Set

Think About It

1. What is the purpose of finding a least-squares regression line?

2. If two variables have a correlation coefficient of –1, does this mean there is no relationship between the two variables? Explain.

3. Choose the correct word. Suppose x is the independent variable, y is the dependent variable, and $r = -0.8$. Then as x increases, y *increases* or *decreases*.

4. If the slope of the regression line is positive, is r positive or negative?

Preliminary Exercises

For Exercises P1 to P3, if necessary, use the referenced example following the exercise for assistance.

P1. Find the equation of the least-squares line for the stride length and speed of camels given in Table 8.9. **Example 1**

P2. Use the equation of the least-squares line from Exercise P1 to predict the average speed of a camel with a stride length of 3.6 meters. Round your results to the nearest tenth of a meter per second. **Example 2**

P3. Find the linear correlation coefficient for stride length versus speed of a camel as given in Table 8.9. Round your result to the nearest hundredth. **Example 3**

Section Exercises

1. Which of the following scatter diagrams suggests:
   a. a positive linear correlation between the x and y variables?
   b. a negative linear correlation between the x and y variables?

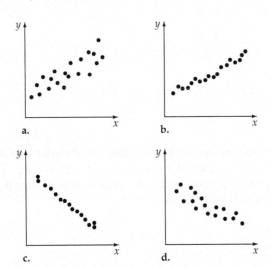

   a.   b.

   c.   d.

2. Which of the following scatter diagrams suggests:
   a. a nearly perfect positive linear correlation between the x and y variables?
   b. little or no linear correlation between the x and y variables?

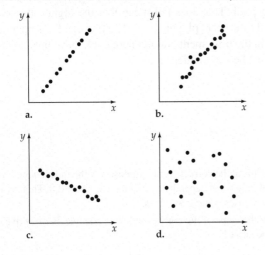

   a.   b.

   c.   d.

3. **Value of a Corvette**   The following table gives retail values of a 2020 Corvette for various odometer readings.

Odometer Reading	Retail Value ($)
13,000	52,275
18,000	51,525
20,000	51,200
25,000	50,275
29,000	49,625
32,000	49,075

   a. Find the equation of the least-squares line for the data.
   b. Use the equation from part a to predict the retail price of a 2020 Corvette with an odometer reading of 30,000. Round to the nearest $100.
   c. Find the linear correlation coefficient for these data.
   d. What is the significance of the fact that the linear correlation coefficient is negative for these data?

4. **Paleontology**   The following table shows the length, in centimeters, of the humerus and the total wingspan, in centimeters, of several pterosaurs, which are extinct flying reptiles.

Pterosaur Data

Humerus	Wingspan	Humerus	Wingspan
24	600	20	500
32	750	27	570
22	430	15	300
17	370	15	310
13	270	9	240
4.4	68	4.4	55
3.2	53	2.9	50
1.5	24		

Source: Southwest Educational Development Laboratory

   a. Find the equation of the least-squares line for the data. Round constants to the nearest hundredth.
   b. Use the equation from part a to determine, to the nearest centimeter, the projected wingspan of a pterosaur if its humerus is 54 cm.

5. **Oceanography**   An oceanographer measured the length, in meters, of a deepwater wave and its speed, in meters per second. The results are shown in the following table.

Wave Length (m)	Speed (m/s)
100	14.4
125	13.4
130	16.1
175	16.8
210	21.9
350	24.6
400	24.8

a. Find the equation of a linear regression line for the data where wave length is the independent variable and speed is the dependent variable.

b. Using the equation from part a, estimate the speed of a wave that is 200 m long.

6. **Astronomy** Astronomer Edwin Hubble postulated a relationship between the distance from Earth and the velocity at which a galaxy appears to be traveling away from Earth. The following table shows observations of seven galaxies. Distance is measured in megaparsecs (1 Mpc is approximately 3260 light-years), and velocity is measured in kilometers per second.

Distance (Mpc)	Velocity (km/s)
51.8	4545
12.2	1204
27.1	1766
46.2	3792
58.2	5133
46.2	3792
29.1	1734

a. Find the equation of linear regression line for the data where distance is the independent variable and velocity is the dependent variable.

b. Using the equation from part a, estimate the velocity at which a galaxy 100 Mpc from Earth is traveling.

7. **Internet Usage** A psychologist collected data on a person's age and the number of hours per week that person spent on the Internet. The data are shown in the following table.

Age (years)	73	67	63	56	49	46	39	32	25	21	15
Time (hours)	0	2	4	6	8	10	12	14	16	18	20

a. Find the equation of a linear regression line for the data where age is the independent variable and time is the dependent variable.

b. Using the equation from part a, estimate the number of hours a person 30 years old spends on the Internet.

c. Find the linear correlation coefficient.

d. Based on the value of the linear correlation coefficient, would you conclude, at the $|r| > 0.9$ level, that the data can be reasonably modeled by a linear function?

8. **Automotive Technology** The weight of a car affects its fuel efficiency. The following table shows the weight in pounds of a car and its fuel efficiency rating in miles per gallon.

Weight (lbs)	2000	2500	2890	3100	3400	3750	4100	4370	4400
Miles per Gallon	40	33	32	30	33	28	23	27	27

a. Find the equation of a linear regression line for the data where weight is the independent variable and miles per gallon is the dependent variable.

b. Using the equation from part a, estimate the miles per gallon a car weighing 3500 lbs is expected to get.

c. Find the linear correlation coefficient.

d. Based on the value of the linear correlation coefficient, would you conclude, at the $|r| > 0.9$ level, that the data can be reasonably modeled by a linear function?

9. **Tennis Serves** The speed of tennis ball as it leaves the racket depends on the speed of the racket when it hits the ball. The table below is data collected from 10 serves.

Racket Speed in Miles per Hour	68	41	58	40	61	55	43	65	40	50
Ball Speed in Miles per Hour	91	60	79	59	82	76	62	87	59	70

a. Find the equation of the least-squares line for the data.

b. Use the equation from part a to estimate the ball speed for a racket speed of 65 miles per hour.

10. **Freezing Point Depression** The point at which a sugar-water solution will freeze depends on the amount of sugar in the water. The table below shows the freezing point temperature, in Celsius, for various amounts, in grams, of sugar dissolved in 100 grams of water.

Sugar Grams	40	22	43	22	48	34	49	38	35	38
Freezing Temp Celsius	-7.8	-4.1	-8.3	-4.0	-9.3	-6.4	-9.5	-7.2	-6.6	-7.3

a. Find the equation of the least-squares line for the data.

b. Use the equation from part a to estimate the freezing temperature when 45 grams of sugar is dissolved in 100 grams of water.

11. **Exercise** An exercise physiologist wants to determine how speed on an exercise bike affects heart rate. The data for one person are shown in the following table.

Speed (mph)	0.0	0.5	1.0	1.5	2.0	2.5	3.0	3.5	4.0
Pulse (bpm)	61	65	69	74	78	82	87	91	96

a. Find the least-squares regression for these data where speed is the independent variable and pulse rate in beats per minute (bpm) is the dependent variable.

b. Assuming the regression line is accurate for higher speeds, what is the expected pulse rate for someone traveling 5 mph? Round to the nearest whole number.

c. At the level of $|r| > 0.9$, is there a strong correlation between speed and bpm?

12. **Fitness** An aerobic exercise instructor remembers the data given in the following table, which shows the recommended maximum exercise heart rates for individuals of the given ages.

Age (x years)	20	40	60
Maximum Heart Rate (y beats per minute)	170	153	136

a. Find the linear correlation coefficient for the data.

b. What is the significance of the value found in part a?

c. Find the equation of the least-squares line.

d. Use the equation from part c to predict the maximum exercise heart rate for a person who is 72.

▶ Investigation

Tuition   The following table shows the average annual tuition and fees at private and public 4-year colleges and universities for the school years 2015–2016 through 2021–2022.

Four-year Colleges and Universities Tuition and Fees

Year	Private	Public
2015–2016	24,476	7629
2016–2017	27,983	8276
2017–2018	28,989	8646
2018–2019	30,131	8895
2019–2020	31,283	9145
2020–2021	32,334	9417
2021–2022	33,479	9648

1. Using 1 for 2015–2016, 2 for 2016–2017, and so on, find the linear correlation coefficient and the equation of the least-squares line for the tuition and fees at private 4-year colleges and universities based on the year.

2. Using 1 for 2015–2016, 2 for 2016–2017, and so on, find the linear correlation coefficient and the equation of the least-squares line for the tuition and fees at public 4-year colleges and universities based on the year.

3. Based on the linear correlation coefficients you found in parts a and b, are the equations you wrote in Exercises 1 and 2 good models of the growth in tuition and fees at 4-year colleges and universities?

4. The equation of a least-squares line is written in the form $\hat{y} = ax + b$. Explain the meaning of the value of a for each equation you wrote in Exercises 1 and 2.

Test Prep

8 Introduction to Statistics

Section 8.1 — Measures of Central Tendency

Mean, Median, and Mode The *mean* of n numbers is the sum of the numbers divided by n. The *median* of a ranked list of n numbers is the middle number if n is odd, or the mean of the two middle numbers if n is even. The *mode* of a list of numbers is the number that occurs most frequently.

See **Examples 1, 2, 3,** and **4** on pages 357 to 359, and then try Exercise 1 on page 402.

The Weighted Mean The **weighted mean** of the n numbers $x_1, x_2, x_3, \ldots, x_n$ with the respective assigned weights $w_1, w_2, w_3, \ldots, w_n$ is

$$\text{Weighted mean} = \frac{\text{sum of each data value times its assigned weight}}{\text{sum of the weights}}$$

See **Examples 5** and **6** on pages 360 and 361, and then try Exercise 7 on pages 402 and 403.

Section 8.2 — Measures of Dispersion

Range The range of a set of data values is the difference between the greatest data value and the least data value.

See **Example 1** on page 366, and then try Exercise 5 on page 402.

Notations for Standard Deviation and Variance If $x_1, x_2, x_3, \ldots, x_n$ is a *population* of n numbers with mean μ, then the standard deviation of the population is

$$\sigma = \sqrt{\frac{(x_1 - \mu)^2 + (x_2 - \mu)^2 + (x_3 - \mu)^2 + \cdots + (x_n - \mu)^2}{n}}.$$

If $x_1, x_2, x_3, \ldots, x_n$ is a *sample* of n numbers with mean $\bar{x}$, then the standard deviation of the sample is

$$s = \sqrt{\frac{(x_1 - \bar{x})^2 + (x_2 - \bar{x})^2 + (x_3 - \bar{x})^2 + \cdots + (x_n - \bar{x})^2}{n - 1}}.$$

See **Examples 2** and **4** on pages 368 and 370, and then try Exercise 9 on page 403.

Section 8.3 — Measures of Relative Position

z-score The z-score for a given data value x is the number of standard deviations that x is above or below the mean.

z-score for a population data value: $z_x = \dfrac{x - \mu}{\sigma}$

z-score for a sample data value: $z_x = \dfrac{x - \bar{x}}{s}$

See **Example 1** on page 373, and then try Exercises 8a and 12 on page 403.

Percentiles A value x is called the *p*th percentile of a data set provided $p\%$ of the data values are less than x. Given a set of data and a data value x,

$$\text{Percentile score of } x = \frac{\text{number of data values less than } x}{\text{total number of data values}} \cdot 100$$

See **Example 4** on page 374, and then try Exercise 8b on page 403.

Quartiles and Box-and-Whisker Plots The quartiles of a data set are the three numbers Q_1, Q_2, and Q_3 that partition the ranked data into four (approximately) equal groups. Q_2 is the median of the data, Q_1 is the median of the data values less than Q_2, and Q_3 is the median of the data values greater than Q_2. A box-and-whisker plot is a display used to show the quartiles and the maximum and minimum values of a data set.

See **Examples 5** and **6** on pages 374 and 376, and then try Exercise 13 on page 403.

Section 8.4 **Normal Distributions**	
Frequency Distributions A frequency distribution displays a data set by dividing the data into intervals, or classes, and listing the number of data values that fall into each interval. A relative frequency distribution lists the percent of data in each interval.	See **Example 1** on pages 383 and 384, and then try Exercise 15 on page 403.
Normal Distributions and the Empirical Rule A normal distribution of data is a bell-shaped curve that is symmetric about a vertical line through the mean. The *y*-value of each point on the curve is the *percent* (expressed as a decimal) of the data at the corresponding *x*-value. The total area under the curve is 1. The Empirical Rule for a normal distribution states that approximately 68% of the data lie within 1 standard deviation of the mean, 95% of the data lie within 2 standard deviations of the mean, and 99.7% of the data lie within 3 standard deviations of the mean.	See **Example 2** on page 385, and then try Exercise 18 on page 404.
Using the Standard Normal Distribution The standard normal distribution is the normal distribution that has a mean of 0 and a standard deviation of 1. Any normal distribution can be converted into the standard normal distribution by converting data values to their *z*-scores. Then the percent of data values that lie in a given interval can be found as the area under the standard normal curve between the *z*-scores of the endpoints of the given interval.	See **Example 3** on page 389, and then try Exercise 19 on page 404.

Section 8.5 **Linear Regression and Correlation**	
Least-Squares Line Bivariate data are data given as ordered pairs. The least-squares regression line, or least-squares line, for a set of bivariate data is the line that minimizes the sum of the squares of the vertical deviations from each data point to the line. **Linear Correlation Coefficient** The linear correlation coefficient r measures the strength of a linear relationship between two variables. The closer $|r|$ is to 1, the stronger the linear relationship is between the variables.	See **Examples 1, 2,** and **3** on pages 394, 395, and 397, and then try Exercise 19 on page 404.

Review Exercises

1. Find the mean, median, mode, range, population variance, and population standard deviation for the following data. Round noninteger values to the nearest tenth.

   12, 17, 14, 12, 8, 19, 21

2. A set of data has a mean of 16, a median of 15, and a mode of 14. Which of these numbers must be a value in the data set?

3. Write a set of data with five data values for which the mean, median, and mode are all 55.

4. State whether the mean, median, or mode is being used.

   a. In 2002, there were as many people aged 25 and younger in the world as there were people aged 25 and older.

   b. The majority of full-time students carry a load of 15 credit hours per semester.

   c. The average annual return on an investment is 6.5%.

5. **Bridges**   The lengths of cantilever bridges in the United States are shown. Find the mean, median, mode, and range of the data.

Bridge Length (in feet)

Baton Rouge (Louisiana), 1235
Commodore Barry (Pennsylvania), 1644
Crescent City Connection (Louisiana), 1576
Longview (Washington), 1200
Patapsco River (Maryland), 1200
Queensboro (New York), 1182
Tappan Zee (New York), 1212
Transbay Bridge (California), 1400

6. **Average Speed**   Cleone traveled 45 mi to her sister's house in 1 hr. The return trip took 1.5 hr. What was Cleone's average rate for the entire trip?

7. **Grade Point Average**   In a 4.0 grading system, each letter grade has the following numerical value.

A = 4.00	B− = 2.67	D+ = 1.33
A− = 3.67	C+ = 2.33	D = 1.00
B+ = 3.33	C = 2.00	D− = 0.67
B = 3.00	C− = 1.67	F = 0.00

Use the weighted mean formula to find the grade point average for a student with the following grades. Round to the nearest hundredth.

Course	Credits	Grade
Mathematics	3	A
English	3	C+
Computers	2	B−
Biology	4	B
Art	1	A

8. **Test Scores** A teacher finds that the test scores of a group of 40 students have a mean of 72 and a standard deviation of 8.

   a. If a student has a test score of 82, what is their z-score?

   b. If a student's score is higher than that of 35 of the 40 students who took the test, find the percentile, rounded to the nearest percent, for the student's score.

9. **Airline Industry** An airline recorded the times it took for a ground crew to unload the baggage from an airplane. The recorded times, in minutes, were 12, 18, 20, 14, and 16. Find the *sample* standard deviation and the variance of these times. Round your results to the nearest hundredth of a minute.

10. **Ticket Prices** The following table gives the average annual admission prices to U.S. movie theaters for 10 recent years.

Average Annual Admission Price, 2006–2015

$6.55	$6.88	$7.18	$7.50	$7.89
$7.93	$7.96	$8.15	$8.17	$8.12

Source: Theatrical Market Statistics, 2015, Motion Picture Association of America

Find the mean, median, and standard deviation for this *sample* of admission prices. Round to the nearest cent.

11. A *population* data set has a mean of 81 and a standard deviation of 5.2. Find the z-scores for each of the following. Round to the nearest hundredth.

   a. $x = 72$ **b.** $x = 84$

12. **Cholesterol Levels** The cholesterol levels for 10 adults are shown in the following table. Draw a box-and-whisker plot of the data.

Cholesterol Levels

310	185	254	221	170
214	172	208	164	182

13. **Test Scores** The following histogram shows the distribution of the test scores for a history test.

   a. How many students scored at least 84 on the test?

   b. How many students took the test?

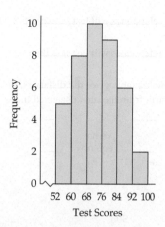

14. **Teacher Salaries** Use the following relative frequency distribution to determine the

   a. *percent* of the states that paid an average teacher salary of at least $48,000.

   b. *probability, as a decimal,* that a state selected at random paid an average teacher salary of at least $56,000 but less than $72,000.

Average Salaries of Public School Teachers

Average Salary, s	Number of States	Relative Frequency
$40,000 \le s < $44,000	2	4%
$44,000 \le s < $48,000	9	18%
$48,000 \le s < $52,000	14	28%
$52,000 \le s < $56,000	4	8%
$56,000 \le s < $60,000	9	18%
$60,000 \le s < $64,000	2	4%
$64,000 \le s < $68,000	4	8%
$68,000 \le s < $72,000	2	4%
$72,000 \le s < $76,000	3	6%
$76,000 \le s < $80,000	1	2%

15. **Waiting Time** The amount of time customers spend waiting in line at the ticket counter of an amusement park is normally distributed, with a mean of 6.5 min and a standard deviation of 1 min. Find the probability that the time a customer will spend waiting is:

   a. less than 8 min.   **b.** less than 6 min.

16. **Pet Food** The weights of all the sacks of dog food filled by a machine are normally distributed, with an average weight of 50 lb and a standard deviation of 0.5 lb. What percent of the sacks will

   a. weigh less than 49.5 lb?

   b. weigh between 49 and 51 lb?

17. **Telecommunication** A telephone manufacturer finds that the life spans of its telephones are normally distributed, with a mean of 6.5 years and a standard deviation of 0.5 year.

   a. What percent of its telephones will last at least 7.25 years?

b. What percent of its telephones will last between 5.8 years and 6.8 years?

c. What percent of its telephones will last less than 6.9 years?

18. Astronomy The following table gives the distances, in millions of miles, of Earth from the sun at selected times during the year.

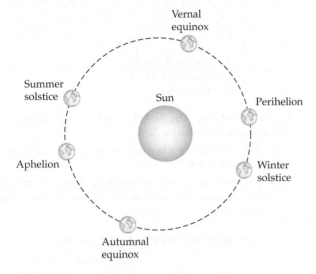

Position	Distance (millions of miles)
Perihelion	91.4
Vernal equinox	92.6
Summer solstice	94.5
Aphelion	94.6
Autumnal equinox	94.3
Winter solstice	91.5

On the basis of these data, what is the mean distance of Earth from the sun?

19. Springs A student has recorded data in the following table that show the distance a spring stretches in inches for a given weight in pounds.

Weight, x	80	100	110	150	170
Distance, y	6.2	7.4	8.3	11.1	12.7

a. Find the linear correlation coefficient.

b. Find the equation of the least-squares line.

c. Use the equation of the least-squares line from part b to predict the distance a weight of 135 lb will stretch the spring. Round to the nearest tenth.

Project 1

Linear Interpolation and Animation

Linear interpolation is a method used to find a particular number between two given numbers. For instance, if a table lists the two entries 0.3156 and 0.8248, then the value exactly halfway between the numbers is the mean of the numbers, which is 0.5702. To find the number that is 0.2 of the way from 0.3156 to 0.8248, compute 0.2 times the difference between the numbers and, because the first number is smaller than the second number, add this result to the smaller number.

$0.8248 - 0.3156 = 0.5092$ $\leftarrow$ Difference between the table entries

$0.2 \cdot (0.5902) = 0.10184$ $\leftarrow$ 0.2 of the above difference

$0.3156 + 0.10184 = 0.41744$ $\leftarrow$ Interpolated result, which is 0.2 of the way between the two table entries

The above linear interpolation process can be used to find an intermediate number that is any specified fraction of the difference between two given numbers.

Project Exercises

1. Use linear interpolation to find the number that is 0.7 of the way from 1.856 to 1.972.

2. Use linear interpolation to find the number that is 0.3 of the way from 0.8765 to 0.8652. Note that because 0.8765 is larger than 0.8652, three-tenths of the difference between 0.8765 and 0.8652 must be subtracted from 0.8765 to find the desired number.

3. A calculator shows that $\sqrt{2} \approx 1.414$ and $\sqrt{3} \approx 1.732$. Use linear interpolation to estimate $\sqrt{2.4}$. *Hint:* Find the number that is 0.4 of the difference between 1.414 and 1.732 and add this number to the smaller number, 1.414. Round your estimate to the nearest thousandth.

4. We know that $2^1 = 2$ and $2^2 = 4$. Use linear interpolation to estimate $2^{1.2}$.

5. At the present time, a football player weighs 325 lb. There are 90 days until the player needs to report to spring training at a weight of 290 lb. The player wants to lose weight at a constant rate. That is, the player wants to lose the same amount of weight each day of the 90 days. What weight, to the nearest tenth of a pound, should the player attain in 25 days?

Math in Practice

In real estate, the value of a property is generally given as a dollar value. However, that dollar amount doesn't provide a very specific description about the property or the house. Real estate agents use a different measure when determining the value of a property: price per square foot.

Suppose you are looking for a house in a town. Your real estate agent supplies you with a list of 20 houses by providing each property's price per square foot.

185   178   154   222   104   144   207   188   104   163

178   142   104   219   104   214   194   104   106   228

Part 1

a. Calculate the mean, standard deviation, and mode of the data.

b. Calculate the mode, range, first quartile, median, and third quartile.

Part 2

Create a box-and-whisker plot of the data.

Takasu/Shutterstock.com

Svetlana Khutornaia/Shutterstock.com

Part 3

You ask your real estate agent to give you more information about each of the houses. You are interested in not only the price per square foot, but also, the asking price, the square footage of the house, and the number of bedrooms and bathrooms. You asked for the data to be sorted from least to greatest by the size of the house, given in square feet.

a. From the table, are the largest houses the most expensive? Give examples or counterexamples.

b. From the table, are the smallest houses the least expensive? Give examples or counterexamples.

c. What kind of factors could cause similarly sized houses to vary in price?

$/sqft	Price	sqft	Bed Bath
194	$245,000	1269	4 bd 2 ba
185	$245,000	1324	3 bd 2 ba
222	$293,800	1324	3 bd 3 ba
188	$245,000	1324	3 bd 2 ba
214	$308,000	1440	4 bd 4 ba
228	$335,000	1470	3 bd 3 ba
154	$235,000	1530	3 bd 2 ba
144	$230,000	1595	2 bd 3 ba
178	$295,000	1660	3 bd 3 ba
163	$270,000	1660	3 bd 3 ba
178	$295,000	1660	3 bd 3 ba
207	$399,900	1930	4 bd 3 ba
219	$289,500	2324	4 bd 4 ba
142	$375,000	2648	4 bd 3 ba
104	$275,000	2656	3 bd 4 ba
104	$276,200	2656	3 bd 4 ba
104	$275,000	2656	3 bd 4 ba
104	$305,000	2944	3 bd 4 ba
104	$305,000	2944	3 bd 4 ba
106	$312,200	2944	3 bd 4 ba

Selected Topics

Voting and Apportionment

Section	How This Is Used . . .
9.1 Introduction to Apportionment	The U.S. Constitution, in Article I, Section 2, states in part that The House of Representatives shall be composed of members chosen every second year by the people of the several states, and the electors in each state shall have the qualifications requisite for electors of the most numerous branch of the state legislature. . . . Representatives and direct taxes shall be *apportioned* [emphasis added] among the several states which may be included within this union, according to their respective numbers. . . Apportioning the House is only one of many applications of apportionment. Today, it is used in many other instances. For instance, if a school is going to hire a teacher, apportionment may be used to help decide where the teacher should be assigned.
9.2 Introduction to Voting	Our system of government is based on participation from its citizens. Our right to vote is critical to the success of democracy. However, there are many different ways to set up a voting system that differ from the one person, one vote rule. Two of many instances are voting for the Oscars and voting in the United Nations Security Council.
9.3 Weighted Voting Systems	In a weighted voting system, some votes have more influence than others. In an extreme case, in a dictatorship, only one person's vote really matters.

Math in Practice: Coffee Shops, p. 454

Apportionment

As mentioned, the U.S. Constitution requires that representatives to the House be apportioned every 10 years. There are several formulas that can be used to decide how many representatives each state should have. As a matter of historical fact, the first presidential veto ever issued was by George Washington in 1792 to reject an apportionment plan chosen by the House of Representatives. The plan Washington preferred gave Virginia, his home state, one more representative than the proposed plan.

In Exercise 11, page 425, an example of using different apportionment plans is discussed.

Introduction to Apportionment

Learning Objectives:

- Use the Hamilton Plan for apportionment.
- Use the Jefferson Plan for apportionment.
- Use the Webster Plan for apportionment.
- Calculate average constituency.
- Calculate the absolute unfairness of an apportionment.
- Use the apportionment principle.
- Use the Huntington-Hill apportionment method.

Be Prepared: Prep Test 9 reviews concepts used in this chapter. This can be found in your Companion Site at Cengage.com.

Table 9.1 Andromeda

State	Population
Apus	11,123
Libra	879
Draco	3518
Cephus	1563
Orion	2917
Total	20,000

The mathematical investigation into **apportionment**, which is a method of dividing a whole into various parts, has its roots in the U.S. Constitution. Since 1790, when the House of Representatives first attempted to apportion itself, various methods have been used to decide how many voters would be represented by each member of the House. The two competing plans in 1790 were put forward by Alexander Hamilton and Thomas Jefferson.

To illustrate how the Hamilton and Jefferson plans were used to calculate the number of representatives each state should have, we will consider the fictitious country of Andromeda, with a population of 20,000 and five states. The population of each state is given in Table 9.1.

Andromeda's constitution calls for 25 representatives to be chosen from these states. (The United States has 435 representatives for the 50 states.) The number of representatives is to be apportioned according to the states' respective populations.

The Hamilton Plan

Under the Hamilton plan, the total population of the country (20,000) is divided by the number of representatives (25). This gives the number of citizens represented by each representative. This number is called the *standard divisor*.

Standard Divisor

$$\text{Standard divisor} = \frac{\text{total population}}{\text{number of people to apportion}}$$

For Andromeda, we have

$$\text{Standard divisor} = \frac{\text{total population}}{\text{number of people to apportion}} = \frac{20,000}{25} = 800$$

Now divide the population of each state by the standard divisor and round the quotient *down* to a whole number. For example, both 15.1 and 15.9 would be rounded to 15. Each whole number quotient is called a *standard quota*.

Standard Quota

The **standard quota** is the whole number part of the quotient of a population divided by the standard divisor.

Table 9.2

State	Population	Quotient	Standard Quota
Apus	11,123	$\dfrac{11,123}{800} \approx 13.904$	13
Libra	879	$\dfrac{879}{800} \approx 1.099$	1
Draco	3518	$\dfrac{3518}{800} \approx 4.398$	4
Cephus	1563	$\dfrac{1563}{800} \approx 1.954$	1
Orion	2917	$\dfrac{2917}{800} \approx 3.646$	3
		Total	22

Take Note: Andromeda is 3 representatives short so use the decimal remainders to determine which state receives an additional representative. The remainders are

0.904
0.099
0.398
0.954
0.646

Cephus has the largest decimal remainder (0.954) so it receives an additional representative bringing its total to 2. Apus has the next largest decimal remainder (0.904) so it receives an additional representative bringing its total to 14. Orion has the next largest remainder (0.646) so it receives an additional representative bringing its total to 4.

From the calculations in Table 9.2, the total number of representatives is 22, not 25 as required by Andromeda's constitution. When this happens, the Hamilton plan calls for revisiting the calculation of the quotients and assigning an additional representative to the state with the largest decimal remainder. This process is continued until the number of representatives equals the number required by the constitution. The final result for Andromeda is shown in Table 9.3.

Table 9.3

State	Population	Quotient	Standard Quota	Number of Representatives
Apus	11,123	$\dfrac{11,123}{800} \approx 13.904$	13	14
Libra	879	$\dfrac{879}{800} \approx 1.099$	1	1
Draco	3518	$\dfrac{3518}{800} \approx 4.398$	4	4
Cephus	1563	$\dfrac{1563}{800} \approx 1.954$	1	2
Orion	2917	$\dfrac{2917}{800} \approx 3.646$	3	4
		Total	22	25

The Jefferson Plan

As we saw with the Hamilton plan, dividing by the standard divisor and then rounding down does not always yield the correct number of representatives. In the previous example, we were three representatives short. The Jefferson plan attempts to overcome this difficulty by using a *modified standard divisor*. This number is chosen, by trial and error, so that the sum of the standard quotas is equal to the total number of representatives. In a specific apportionment calculation, there may be more than one number that can serve as the modified standard divisor. For instance, in the following apportionment calculation shown in Table 9.4, we have used 740 as our modified standard divisor. However, 741 can also be used as the modified standard divisor. The modified standard divisor will always be smaller than the standard divisor.

Table 9.4

State	Population	Quotient	Number of Representatives
Apus	11,123	$\frac{11{,}123}{740} \approx 15.031$	15
Libra	879	$\frac{879}{740} \approx 1.188$	1
Draco	3518	$\frac{3518}{740} \approx 4.754$	4
Cephus	1563	$\frac{1563}{740} \approx 2.112$	2
Orion	2917	$\frac{2917}{740} \approx 3.942$	3
		Total	25

Webster Apportionment

The Webster method of apportionment is similar to the Jefferson method except that quotas are rounded up when the decimal remainder is 0.5 or greater and quotas are rounded down when the decimal remainder is less than 0.5. This method of rounding is referred to as rounding to the nearest integer. For instance, using the Jefferson method, a quotient of 15.91 would be rounded to 15; using the Webster method, it would be rounded to 16. A quotient of 15.49 would be rounded to 15 in both methods.

To use the Webster method, you must still experiment to find a modified standard divisor for which the sum of the quotas rounded to the nearest integer equals the number of items to be apportioned. Although the Webster method is similar to the Jefferson method, the Webster method is generally more difficult to apply because the modified divisor may be less than, equal to, or more than the standard divisor.

We begin by using the standard divisor $= \dfrac{20{,}000}{25} = 800$

State	Population	Quotient	Standard Quota
Apus	11,123	$\frac{11{,}123}{800} \approx 13.91$	14
Libra	879	$\frac{879}{800} \approx 1.10$	1
Draco	3518	$\frac{3518}{800} \approx 4.40$	4
Cephus	1563	$\frac{1563}{800} \approx 1.95$	2
Orion	2917	$\frac{2917}{800} \approx 3.65$	4
		Total	25

The total number of representatives is 25 as required by the Andromeda constitution.

Table 9.5 shows how the results of the Hamilton, Jefferson, and Webster apportionment methods differ. Note that each method assigns a different number of representatives to certain states.

Table 9.5

State	Population	Hamilton Plan	Jefferson Plan	Webster Plan
Apus	11,123	14	15	14
Libra	879	1	1	1
Draco	3518	4	4	4
Cephus	1563	2	2	2
Orion	2917	4	3	4
Total		25	25	25

Although we have applied apportionment to allocating representatives to a congress, there are many other applications of apportionment. For instance, nurses can be assigned to hospitals according to the number of patients requiring care; police officers can be assigned to precincts based on the number of reported crimes; and math classes can be scheduled based on student demand for those classes.

Table 9.6 Ruben County

City	Population
Cardiff	7020
Solana	2430
Vista	1540
Pauma	3720
Pacific	5290

Example 1   Apportioning Board Members Using the Hamilton, Jefferson, and Webster Methods

Suppose the 18 members on the board of the Ruben County environmental agency are selected according to the populations of the five cities in the county, as shown in Table 9.6.

a. Use the Hamilton method to determine the number of board members each city should have.

b. Use the Jefferson method to determine the number of board members each city should have.

c. Use the Webster method to determine the number of board members each city should have.

Solution

a. First find the total population of the cities.

$$7020 + 2430 + 1540 + 3720 + 5290 = 20,000$$

Now calculate the standard divisor.

$$\text{Standard divisor} = \frac{\text{population of the cities}}{\text{number of board members}} = \frac{20,000}{18} \approx 1111.11$$

Use the standard divisor to find the standard quota for each city. See Table 9.7.

Table 9.7

City	Population	Quotient	Standard Quota	Number of Board Members
Cardiff	7020	$\frac{7020}{1111.11} \approx 6.318$	6	6
Solana	2430	$\frac{2430}{1111.11} \approx 2.187$	2	2
Vista	1540	$\frac{1540}{1111.11} \approx 1.386$	1	2
Pauma	3720	$\frac{3720}{1111.11} \approx 3.348$	3	3
Pacific	5290	$\frac{5290}{1111.11} \approx 4.761$	4	5
Total			16	18

Calculator Note: Using lists and the iPart function (which returns only the whole number part of a number) of a TI-84 calculator can be helpful when trying to find a modified standard divisor.

Press **STAT ENTER** to display the list editor. If a list is present in L1, use the up arrow key to highlight L1, then press **CLEAR ENTER**. Enter the populations of each city in L1.

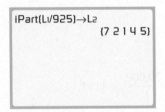

L1	L2	L3	1
7020	------	------	
2430			
1540			
3720			
5290			

L1(6) =

Press **2nd** [QUIT]. To divide each number in L1 by a modified divisor (we are using 925 in part b of this example), enter the following.

MATH ▶ 3 2nd [L1]

÷ 925)

STO 2nd [L2] ENTER

iPart(L1/925)→L2
{7 2 1 4 5}

The standard quotas are shown on the screen (and stored in the list L2). The sum of these numbers is 19, one more than the desired number of representatives.

Calculator Note: To repeat the computation of the number of board members using 950 as the modified standard divisor, press **2nd** [ENTRY] and use the arrow keys to change 925 to 950.

Then press **ENTER**.

From Table 9.7, the sum of the standard quotas is 16, so we must add 2 more members. The two cities with the largest decimal remainders are Pacific and Vista. Each of these two cities gets one additional board member. Thus the composition of the environmental board using the Hamilton method is Cardiff: 6, Solana: 2, Vista: 2, Pauma: 3, and Pacific: 5.

b. To use the Jefferson method, we must find a *modified* standard divisor that is less than the standard divisor we calculated in part a. We must do this by trial and error. For instance, if we choose 925 as the modified standard divisor, we have the results in Table 9.8.

Table 9.8

City	Population	Quotient	Number of Board Members
Cardiff	7020	$\frac{7020}{925} \approx 7.589$	7
Solana	2430	$\frac{2430}{925} \approx 2.627$	2
Vista	1540	$\frac{1540}{925} \approx 1.665$	1
Pauma	3720	$\frac{3720}{925} \approx 4.022$	4
Pacific	5290	$\frac{5290}{925} \approx 5.719$	5
		Total	19

This result yields too many board members. Thus we must increase the modified standard divisor. By experimenting with different divisors, we find that 950 gives the correct number of board members, as shown in Table 9.9.

Table 9.9

City	Population	Quotient	Number of Board Members
Cardiff	7020	$\frac{7020}{950} \approx 7.389$	7
Solana	2430	$\frac{2430}{950} \approx 2.558$	2
Vista	1540	$\frac{1540}{950} \approx 1.621$	1
Pauma	3720	$\frac{3720}{950} \approx 3.916$	3
Pacific	5290	$\frac{5290}{950} \approx 5.568$	5
		Total	18

Thus the composition of the environmental board using the Jefferson method is Cardiff: 7, Solana: 2, Vista: 1, Pauma: 3, and Pacific: 5.

c. For part c, we will use Excel. Spreadsheets that can assist with finding the correct apportionment can be found at REF.COM. This is the Webster Plan spreadsheet.

For the first attempt, the standard divisor is used. Note that the number of representatives is 17, not 18. The message in the pink cell indicates we need to decrease the standard divisor.

1	**Webster Apportionment Plan**				
2	To use this spreadsheet, enter your data for each state and its population. A 'state' is any quantity that is being apportioned. **You must begin entering states in A6 and population values in B6.** Enter the number of representatives to be apportioned in C3. When you have finished entering data, check that the number of representatives being apportioned in D51 eqauls the number in C3. If not, decrease the initial standard divisor and enter that number in E4 as the Modified divisor.				
3	Number of Representatives		18		
4	Standard Divisor		1111.111111	Modified divisor	
5	State	Population	Quota	Number of Representatives	Decrease the modified divisor
6	Cardiff	7,020	6.318	6	
7	Solana	2,430	2.187	2	
8	Vista	1,540	1.386	1	
9	Pauma	3,720	3.348	3	
10	Pacific	5,290	4.761	5	
51	Total Population	20,000	No. of Representatives	17	

Try a modified divisor and enter it into cell E4.

1	**Webster Apportionment Plan**				
2	To use this spreadsheet, enter your data for each state and its population. A 'state' is any quantity that is being apportioned. **You must begin entering states in A6 and population values in B6.** Enter the number of representatives to be apportioned in C3. When you have finished entering data, check that the number of representatives being apportioned in D51 eqauls the number in C3. If not, decrease the initial standard divisor and enter that number in E4 as the Modified divisor.				
3	Number of Representatives		18		
4	Standard Divisor		1111.111111	Modified divisor	1075
5	State	Population	Quota	Number of Representatives	You have the correct divisor.
6	Cardiff	7,020	6.530232558	7	
7	Solana	2,430	2.260465116	2	
8	Vista	1,540	1.43255814	1	
9	Pauma	3,720	3.460465116	3	
10	Pacific	5,290	4.920930233	5	
51	Total Population	20,000	No. of Representatives	18	

The green cell indicates that 1075 works as a modified divisor. Thus the composition of the environmental board using the Webster method is Cardiff: 7, Solana: 2, Vista: 1, Pauma: 3, and Pacific: 5.

Suppose that the environmental agency in Example 1 decides to add one more member to the board even though the population of each city remains the same. The total number of members is now 19, and we must determine how the members of the board will be apportioned.

The standard divisor is now $\frac{20,000}{19} \approx 1052.63$. Using Hamilton's method, the calculations necessary to apportion the board members are shown in Table 9.10.

Table 9.10

City	Population	Quotient	Standard Quota	Number of Board Members
Cardiff	7020	$\dfrac{7020}{1052.63} \approx 6.669$	6	7
Solana	2430	$\dfrac{2430}{1052.63} \approx 2.309$	2	2
Vista	1540	$\dfrac{1540}{1052.63} \approx 1.463$	1	1
Pauma	3720	$\dfrac{3720}{1052.63} \approx 3.534$	3	4
Pacific	5290	$\dfrac{5290}{1052.63} \approx 5.026$	5	5
		Total	17	19

Table 9.11 summarizes the number of board members each city would have if the board consisted of 18 members (Example 1) or 19 members.

Table 9.11

City	Hamilton Apportionment with 18 Board Members	Hamilton Apportionment with 19 Board Members
Cardiff	6	7
Solana	2	2
Vista	2	1
Pauma	3	4
Pacific	5	5
Total	18	19

Note that although one more board member was added, Vista lost a board member, even though the populations of the cities did not change. This is called the **Alabama paradox** where an increase in the total number of representatives being apportioned results in one state (or whatever is being apportioned) to lose a seat. In the interest of fairness, an apportionment method should not exhibit the Alabama paradox.

There are other paradoxes that involve apportionment methods. Two of them are the *population paradox* and the *new states paradox*. It is possible for the population of one state to be increasing at a rate that is faster than that of another state and for the state still to lose a representative. This is an example of the **population paradox**. For instance, consider the following situation of apportioning 11 representatives among the three states in the following table.

State	Mimas	Rhea	Hyperion
Initial population	600	1364	2073
Number of representatives	1	4	6
Revised population	676	1637	2379
Approximate percent increase	12.7	20.0	14.8
Number of representatives	2	4	5

Note that although Hyperion's population is increasing at a greater rate than that of Mimas, Hyperion lost a representative.

In 1907, when Oklahoma was added to the Union, the size of the House was increased by five representatives to accommodate Oklahoma's population. However, when the complete apportionment of Congress was recalculated, New York lost a seat and Maine gained a seat. This is an example of the **new states paradox**. Here is an example. Consider the two states Io and Ganymede, which have populations of 10,445 and 89,555, respectively.

There are 100 delegates to apportion between the two states. Using the Hamilton method, the apportionment is shown in the following table.

State	Population	Number of Representatives
Io	10,445	10
Ganymede	89,555	90

A new state, Callisto, is added to the confederation that has a population of 5240. Because the standard divisor is 1000, Callisto is entitled to $\frac{5240}{1000} \approx 5$ delegates. The new apportionment is shown in the following table.

State	Population	Number of Representatives
Io	10,445	11
Ganymede	89,555	89
Callisto	5240	5

Note that Ganymede lost a delegate and Io gained a delegate, an example of the new states paradox.

Fairness in Apportionment

As we have seen, the choice of apportionment method affects the number of representatives a state will have. Given that fact, mathematicians and others have tried to work out an apportionment method that is fair. The difficulty lies in trying to define what is fair, so we will try to state conditions by which an apportionment plan is judged fair.

One criterion of fairness for an apportionment plan is that it should satisfy the *quota rule*.

Quota Rule

The number of representatives apportioned to a state is the standard quota or one more than the standard quota.

We can show that the Jefferson plan, for instance, does not always satisfy the quota rule by calculating the standard quota of Apus (see page 411).

$$\text{Standard quota} = \frac{\text{population of Apus}}{\text{standard divisor}} = \frac{11,123}{800} \approx 13$$

The standard quota of Apus is 13. However, the Jefferson plan assigns 15 representatives to that state (see page 412), two more than its standard quota. Therefore, the Jefferson method violates the quota rule.

Another measure of fairness is *average constituency*. This is the population of a state divided by the number of representatives from the state and then rounded to the nearest whole number.

Average Constituency

$$\text{Average constituency} = \frac{\text{population of a state}}{\text{number of representatives from the state}}$$

Consider the two states Hampton and Shasta in Table 9.12.

Table 9.12

State	Population	Representatives	Average Constituency
Hampton	16,000	10	$\frac{16,000}{10} = 1600$
Shasta	8340	5	$\frac{8340}{5} = 1668$

Take Note: The idea of average constituency is an essential aspect of our democracy. To understand this, suppose state A has an average constituency of 1000 and state B has an average constituency of 10,000. When a bill is voted on in the House of Representatives, each vote has equal weight. However, a vote from a representative from state A would represent 1000 people, but a vote from a representative from state B would represent 10,000 people. Consequently, in this situation, we do not have "equal representation."

Because the average constituencies are approximately equal, it seems natural to say that both states are equally represented. See the Take Note at the left.

Now suppose that one representative will be added to one of the states. Which state is more deserving of the new representative? In other words, to be fair, which state should receive the new representative?

The changes in the average constituency are shown in Table 9.13.

Table 9.13

State	Average Constituency (old)	Average Constituency (new)
Hampton	$\frac{16,000}{10} = 1600$	$\frac{16,000}{11} \approx 1455$
Shasta	$\frac{8340}{5} = 1668$	$\frac{8340}{6} = 1390$

From the table, there are two possibilities for adding one representative. If Hampton receives the representative, its average constituency will be 1455 and Shasta's will remain at 1668. The difference in the average constituencies is $1668 - 1455 = 213$. This difference is called the *absolute unfairness of the apportionment*.

Absolute Unfairness of an Apportionment

The **absolute unfairness of an apportionment** is the absolute value of the difference between the average constituency of state A and the average constituency of state B.

$$|\text{Average constituency of A} - \text{average constituency of B}|$$

If Shasta receives the representative, its average constituency will be 1390 and Hampton's will remain at 1600. The absolute unfairness of apportionment is $1600 - 1390 = 210$. This is summarized in Table 9.14.

Table 9.14

	Hampton's Average Constituency	Shasta's Average Constituency	Absolute Unfairness of Apportionment
Hampton Receives the New Representative	1455	1668	213
Shasta Receives the New Representative	1600	1390	210

Because the smaller absolute unfairness of apportionment occurs if Shasta receives the new representative, it might seem that Shasta should receive the representative. However, this is not necessarily true.

To understand this concept, let's consider a somewhat different situation. Suppose an investor makes two investments, one of $10,000 and another of $20,000. One year later, the first investment is worth $11,000 and the second investment is worth $21,500. This is shown in Table 9.15.

Table 9.15

	Original Investment	One Year Later	Increase
Investment A	$10,000	$11,000	$1000
Investment B	$20,000	$21,500	$1500

Although there is a larger increase in investment B, the increase *per dollar* of the original investment is $\frac{1500}{20,000} = 0.075$. On the other hand, the increase per dollar of investment A is $\frac{1000}{10,000} = 0.10$. Another way of saying this is that each $1 of investment A produced a return of 10 cents ($0.10), whereas each $1 of investment B produced a return of 7.5 cents ($0.075). Therefore, even though the increase in investment A was less than the increase in investment B, investment A was more productive.

A similar process is used when deciding which state should receive another representative. Rather than look at the difference in the absolute unfairness of apportionment, we determine the *relative unfairness* of adding the representative.

Take Note: The average constituency in the denominator is computed *after* the new representative is added.

Relative Unfairness of an Apportionment

The **relative unfairness of an apportionment** is the quotient of the absolute unfairness of the apportionment and the average constituency of the state receiving the new representative.

$$\frac{\text{Absolute unfairness of the apportionment}}{\text{Average constituency of the state receiving the new representative}}$$

Example 2 **Determine the Relative Unfairness of an Apportionment**

Determine the relative unfairness of an apportionment that gives a new representative to Hampton rather than Shasta.

Solution

Using data for Hampton and Shasta shown in Table 9.14, we have

Relative unfairness of the apportionment

$$= \frac{\text{absolute unfairness of the apportionment}}{\text{average constituency of Hampton with a new representative}}$$

$$= \frac{213}{1455} \approx 0.146$$

The relative unfairness of the apportionment is approximately 0.146.

The following principle uses the relative unfairness of an apportionment to determine how to add a representative to an existing apportionment.

Apportionment Principle

When adding a new representative to a state, the representative is assigned to the state in such a way as to give the smallest relative unfairness of apportionment.

Although we have focused on assigning representatives to states, the apportionment principle can be used in many other situations.

Example 3 **Use the Apportionment Principle**

Table 9.16 shows the number of paramedics and the annual number of paramedic calls for two cities. If a new paramedic is hired, use the apportionment principle to determine to which city the paramedic should be assigned.

Table 9.16

	Paramedics	Annual Paramedic Calls
Tahoe	125	17,526
Erie	143	22,461

Solution

Calculate the relative unfairness of the apportionment that assigns the paramedic to Tahoe and the relative unfairness of the apportionment that assigns the paramedic to Erie. In this case, average constituency is the number of annual paramedic calls divided by the number of paramedics. This is shown in Table 9.17.

Table 9.17

	Tahoe's Annual Paramedic Calls per Paramedic	Erie's Annual Paramedic Calls per Paramedic	Absolute Unfairness of Apportionment
Tahoe Receives a New Paramedic	$\frac{17{,}526}{125 + 1} \approx 139$	$\frac{22{,}461}{143} \approx 157$	$157 - 139 = 18$
Erie Receives a New Paramedic	$\frac{17{,}526}{125} \approx 140$	$\frac{22{,}461}{143 + 1} \approx 156$	$156 - 140 = 16$

If Tahoe receives the new paramedic, the relative unfairness of the apportionment is

Relative unfairness of the apportionment

$$= \frac{\text{absolute unfairness of the apportionment}}{\text{Tahoe's average constituency with a new paramedic}}$$

$$= \frac{18}{139} \approx 0.129$$

If Erie receives the new paramedic, the relative unfairness of the apportionment is

Relative unfairness of the apportionment

$$= \frac{\text{absolute unfairness of the apportionment}}{\text{Erie's average constituency with a new paramedic}}$$

$$= \frac{16}{156} \approx 0.103$$

Because the smaller relative unfairness results from adding the paramedic to Erie, that city should receive the paramedic.

●

Huntington-Hill Apportionment Method

As we mentioned earlier, the members of the House of Representatives are apportioned among the states every 10 years. The present method used by the House is based on the apportionment principle and is called the *method of equal proportions* or the *Huntington-Hill method*. This method has been used since 1940.

The Huntington-Hill method is implemented by calculating what is called a *Huntington-Hill number*.

Huntington-Hill Number

The value of $\dfrac{(P_A)^2}{a(a+1)}$, where P_A is the population of state A and a is the current number of representatives from state A, is called the **Huntington-Hill number** for state A.

When the Huntington-Hill method is used to apportion representatives between two states, the state with the greater Huntington-Hill number receives the next representative. This method can be extended to more than two states.

Huntington-Hill Apportionment Principle

When there is a choice of adding one representative to one of several states, the representative should be added to the state with the greatest Huntington-Hill number.

| Example 4 | Use the Huntington-Hill Apportionment Principle |

Table 9.18 shows the numbers of lifeguards that are assigned to three different beaches and the numbers of rescues made by lifeguards at those beaches. Use the Huntington-Hill apportionment principle to determine to which beach a new lifeguard should be assigned.

Table 9.18

Beach	Number of Lifeguards	Number of Rescues
Mellon	37	1227
Donovan	51	1473
Ferris	24	889

Solution

Calculate the Huntington-Hill number for each of the beaches. In this case, the population is the number of rescues and the number of representatives is the number of lifeguards.

Mellon:

$$\frac{1227^2}{37(37 + 1)} \approx 1071$$

Donovan:

$$\frac{1473^2}{51(51 + 1)} \approx 818$$

Ferris:

$$\frac{889^2}{24(24 + 1)} \approx 1317$$

Ferris has the greatest Huntington-Hill number. Thus, according to the Huntington-Hill apportionment principle, the new lifeguard should be assigned to Ferris.

Now that we have looked at various apportionment methods, it seems reasonable to ask which is the best method. Unfortunately, all apportionment methods have some flaws. This was proved by Michael Balinski and H. Peyton Young in 1982.

Balinski-Young Impossibility Theorem

Any apportionment method either will violate the quota rule or will produce paradoxes such as the Alabama paradox.

Table 9.19 lists flaws that may occur in the application of different apportionment methods.

Table 9.19 Summary of Apportionment Methods and Possible Flaws

Apportionment Methods	Flaws			
	Violation of the Quota Rule	Alabama Paradox	Population Paradox	New States Paradox
Hamilton method	Cannot occur	May occur	May occur	May occur
Jefferson method	May violate	Cannot occur	Cannot occur	Cannot occur
Huntington-Hill method	May violate	Cannot occur	Cannot occur	Cannot occur
Webster method	May violate	Cannot occur	Cannot occur	Cannot occur

Although there is no perfect apportionment method, Balinski and Young went on to present a strong case that the Webster method (following Exercise 26 on page 426) is the system that most closely satisfies the goal of one person, one vote. However, political expediency sometimes overrules mathematical proof. Some historians have suggested that although the Huntington-Hill apportionment method was better than some of the previous methods, President Franklin Roosevelt chose this method in 1941 because it allotted one more seat to Arkansas and one less to Michigan. This essentially meant that the House of Representatives would have one more seat for the Democrats, Roosevelt's party.

9.1 Exercise Set

Think About It

1. When considering apportionment, suppose the standard divisor for a state is 800. Write a sentence that explains the meaning of that number.

2. The standard divisor is used in the _____ apportionment method; the modified standard divisor is used in the _____ apportionment method.

3. When considering apportionment, suppose the average constituency of a state is 25,000. Write a sentence that explains the meaning of that number.

4. What is the quota rule?

5. What is the Alabama paradox?

Preliminary Exercises

For Exercises P1 to P4, if necessary, use the referenced example following the exercise for assistance.

P1. Suppose the 20 members of a committee from five European countries are selected according to the populations of the five countries, as shown in the following table.

European Countries' Populations, 2020

Country	Population
France	65,721,000
Germany	82,850,000
Italy	59,132,000
Spain	46,330,000
Belgium	47,404,000

a. Use the Hamilton method to determine the number of representatives each country should have.

b. Use the Jefferson method to determine the number of representatives each country should have.

c. Use the Webster method to determine the number of representatives each country should have. **Example 1**

P2. Using the data in Table 9.14, determine the relative unfairness of an apportionment that gives a new representative to Shasta rather than Hampton. Based on the relative unfairness results from this exercise and result in Example 2, which state should receive the new representative? **Example 2**

P3. The following table shows the number of first and second grade teachers in a school district and the number of students in each of those grades. If a new teacher is hired, use the apportionment principle to determine to which grade the teacher should be assigned. **Example 3**

	Number of Teachers	Number of Students
First Grade	512	12,317
Second Grade	551	15,439

P4. A university has a president's council that is composed of students from each of the undergraduate classes. If a new student representative is added to the council, use the Huntington-Hill apportionment principle to determine which class the new student council member should represent. **Example 4**

Class	Number of Representatives	Number of Students
First year	12	2015
Second year	10	1755
Third year	9	1430
Fourth year	8	1309

Section Exercises

Spreadsheets are available on the Companion Website at Cengage.com to assist with the calculations of the Jefferson, Hamilton, and Huntington-Hill apportionment methods.

1. Explain how to calculate the standard divisor of an apportionment for a total population p with n items to apportion.

2. **Teacher Aides** A total of 25 teacher aides are to be apportioned among seven classes at a new elementary school. The enrollments in the seven classes are shown in the following table.

Class	Number of Students
Kindergarten	38
First grade	39
Second grade	35
Third grade	27
Fourth grade	21
Fifth grade	31
Sixth grade	33
Total	224

a. Determine the standard divisor. What is the meaning of the standard divisor in the context of this exercise?

b. Use the Hamilton method to determine the number of teacher aides to be apportioned to each class.

c. Use the Jefferson method to determine the number of teacher aides to be apportioned to each class. Is this apportionment in violation of the quota rule?

d. How do the apportionment results produced using the Jefferson method compare with the results produced using the Hamilton method?

3. In the Hamilton apportionment method, explain how to calculate the standard quota for a particular state (group).

4. What is the quota rule?

5. **Governing Boards** The following table shows how the average constituency changes for two regional governing boards, Joshua and Salinas, when a new representative is added to each board.

	Joshua's Average Constituency	Salinas's Average Constituency
Joshua Receives New Board Member	1215	1547
Salinas Receives New Board Member	1498	1195

a. Determine the relative unfairness of an apportionment that gives a new board member to Joshua rather than to Salinas. Round to the nearest thousandth.

b. Determine the relative unfairness of an apportionment that gives a new board member to Salinas rather than to Joshua. Round to the nearest thousandth.

c. Using the apportionment principle, determine which regional governing board should receive the new board member.

6. **Forest Rangers** The following table shows how the average constituency changes when two different national parks, Evergreen State Park and Rust Canyon Preserve, add a new forest ranger.

	Evergreen State Park's Average Constituency	Rust Canyon Preserve's Average Constituency
Evergreen Receives New Forest Ranger	466	638
Rust Canyon Receives New Forest Ranger	650	489

a. Determine the relative unfairness of an apportionment that gives a new forest ranger to Evergreen rather than to Rust Canyon. Round to the nearest thousandth.

b. Determine the relative unfairness of an apportionment that gives a new forest ranger to Rust Canyon rather than to Evergreen. Round to the nearest thousandth.

c. Using the apportionment principle, determine which national park should receive the new forest ranger.

7. **Sales Associates** The following table shows the number of sales associates and the average number of customers per day at a company's two department stores. The company is planning to add a new sales associate to one of the stores. Use the apportionment principle to determine which store should receive the new employee.

Shopping Mall Location	Number of Sales Associates	Average Number of Customers per Day
Summer Hill Galleria	587	5289
Seaside Mall Galleria	614	6215

8. **Hospital Interns** The following table shows the number of interns and the average number of patients admitted each day at two different hospitals. The hospital administrator is planning to add a new intern to one of the hospitals. Use the apportionment principle to determine which hospital should receive the new intern.

Hospital Location	Number of Interns	Average Number of Patients Admitted per Day
South Coast Hospital	128	518
Rainer Hospital	145	860

9. **House of Representatives** The U.S. House of Representatives currently has 435 members to represent the 331,108,434 citizens of the United States as determined by the 2020 census.

a. Calculate the standard divisor for the apportionment of these representatives and explain the meaning of this standard divisor in the context of this exercise.

b. According to the 2020 census, the population of Delaware was 990,837. Delaware currently has only one representative in the House of Representatives. Is Delaware currently overrepresented or underrepresented in the House of Representatives? Explain.

c. According to the 2020 census, the population of Vermont was 643,503. Vermont currently has only one representative in the House of Representatives. Is Vermont currently overrepresented or underrepresented in the House of Representatives? Explain.

10. **College Enrollment** The following table shows the enrollment for each of the four divisions of a college. The four divisions are liberal arts, business, humanities, and science. There are 180 new computers that are to be apportioned among the divisions based on the enrollments.

Division	Enrollment
Liberal arts	3455
Business	5780
Humanities	1896
Science	4678
Total	15,809

a. What is the standard divisor for an apportionment of the computers? What is the meaning of the standard divisor in the context of this exercise?

b. Use the Hamilton method to determine the number of computers to be apportioned to each division.

c. If the computers are to be apportioned using the Jefferson method, explain why neither 86 nor 87 can be used as a modified standard divisor. Explain why 86.5 can be used as a modified standard divisor.

d. Explain why the modified standard divisor used in the Jefferson method cannot be larger than the standard divisor.

e. Use the Jefferson method to determine the number of computers to be apportioned to each division. Is this apportionment in violation of the quota rule?

f. How do the apportionment results produced using the Jefferson method compare with the results produced using the Hamilton method?

11. Medical Care A hospital district consists of six hospitals. The district administrators have decided that 48 new nurses should be apportioned based on the number of beds in each of the hospitals. The following table shows the number of beds in each hospital.

Hospital	Number of Beds
Sharp	242
Palomar	356
Tri-City	308
Del Raye	190
Rancho Verde	275
Bel Air	410
Total	1781

a. Determine the standard divisor. What is the meaning of the standard divisor in the context of this exercise?

b. Use the Hamilton method to determine the number of nurses to be apportioned to each hospital.

c. Use the Jefferson method to determine the number of nurses to be apportioned to each hospital.

d. How do the apportionment results produced using the Jefferson method compare with the results produced using the Hamilton method?

12. What is the Alabama paradox?

13. What is the population paradox?

14. What is the new states paradox?

15. What is the Balinski-Young Impossibility Theorem?

16. Apportionment of Projectors Consider the apportionment of 27 digital projectors for a school district with four campus locations labeled A, B, C, and D. The following table shows the apportionment of the projectors using the Hamilton method.

Campus	A	B	C	D
Enrollment	840	1936	310	2744
Apportionment of 27 Projectors	4	9	1	13

a. If the number of projectors to be apportioned increases from 27 to 28, what will be the apportionment if the Hamilton method is used? Will the Alabama paradox occur? Explain.

b. If the number of projectors to be apportioned using the Hamilton method increases from 28 to 29, will the Alabama paradox occur? Explain.

17. Hotel Management A company operates four resorts. The CEO of the company decides to use the Hamilton method to apportion 115 new LCD television sets to the resorts based on the number of guest rooms at each resort.

Resort	A	B	C	D
Number of Guest Rooms	23	256	182	301
Apportionment of 115 Televisions	4	39	27	45

a. If the number of television sets to be apportioned by the Hamilton method increases from 115 to 116, will the Alabama paradox occur?

b. If the number of television sets to be apportioned by the Hamilton method increases from 116 to 117, will the Alabama paradox occur?

c. If the number of television sets to be apportioned by the Hamilton method increases from 117 to 118, will the Alabama paradox occur?

18. College Security A college apportions 40 security personnel among three education centers according to their enrollments. The following table shows the present enrollments at the three centers.

Center	A	B	C
Enrollment	356	1054	2590

a. Use the Hamilton method to apportion the security personnel.

b. After one semester, the centers have the following enrollments.

Center	A	B	C
Enrollment	370	1079	2600

Center A has an increased enrollment of 14 students, which is an increase of $\frac{14}{356} \approx 0.039 = 3.9\%$. Center B has an increased enrollment of 25 students, which is an increase of $\frac{25}{1054} \approx 0.024 = 2.4\%$.

If the security personnel are reapportioned using the Hamilton method, will the population paradox occur? Explain.

19. Management Scientific Research Corporation has offices in Boston and Chicago. The number of employees at each office is shown in the following table. There are 22 vice presidents to be apportioned between the offices.

Office	Boston	Chicago
Employees	151	1210

a. Use the Hamilton method to find each office's apportionment of vice presidents.

b. The corporation opens an additional office in San Francisco with 135 employees and decides to have a total of 24 vice presidents. If the vice presidents are reapportioned using the Hamilton method, will the new states paradox occur? Explain.

20. Education The science division of a college consists of three departments: mathematics, physics, and chemistry. The number of students enrolled in each department is shown in the following table. There are 19 clerical assistants to be apportioned among the departments.

Department	Math	Physics	Chemistry
Student Enrollment	4325	520	1165

a. Use the Hamilton method to find each department's apportionment of clerical assistants.

b. The division opens a new computer science department with a student enrollment of 495. The division decides to have a total of 20 clerical assistants. If the clerical assistants are reapportioned using the Hamilton method, will the new states paradox occur? Explain.

21. Elementary School Teachers The following table shows the number of fifth and sixth grade teachers in a school district and the number of students in the two grades. The number of teachers for each of the grade levels was determined by using the Huntington-Hill apportionment method.

	Number of Teachers	Number of Students
Fifth Grade	19	604
Sixth Grade	21	698

The district has decided to hire a new teacher for either the fifth or the sixth grade.

a. Use the apportionment principle to determine to which grade the new teacher should be assigned.

b. Use the Huntington-Hill apportionment principle to determine to which grade the new teacher should be assigned. How does this result compare with the result in part a?

22. Social Workers The following table shows the number of social workers and the number of cases (the case load) handled by the social workers for two offices. The number of social workers for each office was determined by using the Huntington-Hill apportionment method.

	Number of Social Workers	Case Load
Hill Street Office	20	584
Valley Office	24	712

A new social worker is to be hired for one of the offices.

a. Use the apportionment principle to determine to which office the social worker should be assigned.

b. Use the Huntington-Hill apportionment principle to determine to which office the new social worker should be assigned. How does this result compare with the result in part a?

c. The results of part b indicate that the new social worker should be assigned to the Valley office. At this moment the Hill Street office has 20 social workers and the Valley office has 25 social workers. Use the Huntington-Hill apportionment principle to determine to which office the *next* new social worker should be assigned. Assume the case loads remain the same.

23. Computer Usage The following table shows the number of computers that are assigned to four different schools and the number of students in those schools. Use the Huntington-Hill apportionment principle to determine to which school a new computer should be assigned.

School	Number of Computers	Number of Students
Rose	26	625
Lincoln	22	532
Midway	26	620
Valley	31	754

24. The population of California increased by over 2.4 million from 2010 to 2020, yet the state lost a seat in the House of Representatives. How must California's population increase have compared to the changes of other states' populations?

25. The population of Utah increased 18.4% from 2010 to 2020 and the state did not gain a representative. Texas's population increased at a lower rate, 15.9%, yet the state gained two representatives. Explain the seeming discrepancy.

26. House of Representatives Currently, the U.S. House of Representatives has 435 members who have been apportioned by the Huntington-Hill apportionment method. If the number of representatives were to be increased to 436, then, according to the 2020 census figures, North Carolina would be given the new representative. How must North Carolina's 2020 census Huntington-Hill number compare with the 2020 census Huntington-Hill numbers for the other 49 states? Explain.

27. Computer Usage Use the Webster method to apportion the computers in Exercise 10, page 424. How do the apportionment results produced using the Webster method compare with the results produced using the

a. Hamilton method?

b. Jefferson method?

28. **Demographics** The following table shows the populations of five European countries. A committee of 20 people from these countries is to be formed using the Webster method of apportionment.

Country	Population
France	66,550,000
Germany	80,850,000
Italy	61,860,000
Spain	48,150,000
Belgium	11,320,000

Total 268,730,000

Source: U.S. Census Bureau, International Data Base

a. Explain why 13,500,000 *cannot* be used as a modified standard divisor.

b. Explain why 13,750,000 *can* be used as a modified standard divisor.

c. Use the Webster apportionment method to determine the apportionment of the 20 committee members.

29. Which of the following apportionment methods can violate the quota rule?

   - Hamilton method
   - Jefferson method
   - Webster method
   - Huntington-Hill method

30. According to Michael Balinski and H. Peyton Young, which of the apportionment methods most closely satisfies the goal of one person, one vote?

31. What method is presently used to apportion the members of the U.S. House of Representatives?

▶ **Investigations**

In the Huntington-Hill method of apportionment, each state is first given one representative and then additional representatives are assigned, one at a time, to the state currently having the highest Huntington-Hill number. This way of implementing the Huntington-Hill apportionment method is time consuming. Another process for implementing the Huntington-Hill apportionment method consists of using a modified divisor and a special rounding procedure that involves the *geometric mean* of two consecutive integers.

The geometric mean of two consecutive natural numbers n and $n + 1$ is given by $\sqrt{n(n + 1)}$.

To use this Huntington-Hill method,

- Find the total population for all the states.
- Divide the total population by the number of seats to apportion. This is the standard divisor.

- Divide the population of each state by the standard divisor.
- The lower quotient is the whole number part of the quotient from the previous step.
- The upper quotient is one more than the lower quotient.
- Find the geometric mean of the two quotients.
- If the geometric mean is greater than the quotient from step 2, the state gets the *lower quotient*. If the geometric mean is less than the quotient from step 2, the state gets the *upper quotient*.
- Continue until all seats have been apportioned.
- Add the apportioned seats. If this number equals the number of seats to be apportioned, you're done. If not, by trial and error, use modified standard divisors until the two numbers are equal.

For assistance with this Investigation, you can access an Excel template from your digital resources for help.

Apply this process to apportion 22 new security vehicles to each of the following schools, based on their student populations.

School	Number of Students	Number of Security Vehicles
Del Mar	5230	?
Wheatly	12,375	?
West	8568	?
Mountain View	14,245	?

Deriving the Huntington-Hill Number The Huntington-Hill number is derived by using the apportionment principle. Let

P_A = population of state A
a = number of representatives from state A
P_B = population of state B
b = number of representatives from state B

Complete the following to derive the Huntington-Hill number.

1. Write the fraction that gives the average constituency of state A when it receives a new representative.

2. Write the fraction that gives the average constituency of state B without a new representative.

3. Express the relative unfairness of apportionment by giving state A the new representative in terms of the fractions from Exercises 1 and 2.

4. Express the relative unfairness of apportionment by giving state B the new representative.

5. According to the apportionment principle, state A should receive the next representative instead of state B if the relative unfairness of giving the new representative to state A is *less than* the relative unfairness of giving the new representative to state B. Express this inequality in terms of the expressions in Exercises 3 and 4.

6. Simplify the inequality and you will have the Huntington-Hill number.

Introduction to Voting

Learning Objectives:

- Use the plurality method of voting to determine the winner of an election.
- Use the Borda Count method of voting to determine the winner of an election.
- Use the plurality with elimination method of voting to determine the winner of an election.
- Use the pairwise comparison method of voting to determine the winner of an election.

Take Note: When an issue requires a **majority vote**, it means that more than 50% of the people voting must vote for the issue. This is not the same as a **plurality**, in which the person or issue with the most votes wins.

Take Note: A preference schedule lists the number of people who gave a particular ranking. For example, the column shaded green means that 3 voters ranked solid chocolate first, caramel centers second, almond centers third, toffee centers fourth, and vanilla centers fifth.

Plurality Method of Voting

One of the most revered privileges that those of us who live in a democracy enjoy is the right to vote for our representatives. Sometimes, however, we are puzzled by the fact that the best candidate did not get elected. Unfortunately, because of the way our *plurality* voting system works, it is possible to elect someone or pass a proposition that has less than *majority support*. As we proceed through this section, we will look at the problems with plurality voting and alternatives to this system. We start with a definition.

The Plurality Method of Voting

Each voter votes for one candidate, and the candidate with the most votes wins. The winning candidate does not have to have a majority of the votes.

Example 1 **Determine the Winner Using Plurality Voting**

Fifty people were asked to rank their preferences of five varieties of chocolate candy, using 1 for their favorite and 5 for their least favorite. This type of ranking of choices is called a **preference schedule**. The results are shown in Table 9.20.

Table 9.20

	Rankings					
Caramel Center	5	4	4	4	2	4
Vanilla Center	1	5	5	5	5	5
Almond Center	2	3	2	1	3	3
Toffee Center	4	1	1	3	4	2
Solid Chocolate	3	2	3	2	1	1
Number of voters:	17	11	9	8	3	2

According to this table, which variety of candy would win the taste test using the plurality voting system?

Solution

To answer the question, Table 9.21 shows the number of first-place votes for each candy.

Table 9.21

	First-Place Votes
Caramel Center	0
Vanilla Center	17
Almond Center	8
Toffee Center	$11 + 9 = 20$
Solid Chocolate	$3 + 2 = 5$

Because toffee centers received 20 first-place votes, this type of candy would win the plurality taste test.

Example 1 can be used to show the difference between plurality and majority. There were 20 first-place votes for toffee-centered chocolate, so it wins the taste test. However, toffee-centered chocolate was the first choice of only 40% $\left(\frac{20}{50} = 40\%\right)$ of the people voting. Thus less than half of the people voted for toffee-centered chocolate as number one, so it did not receive a majority vote.

Borda Count Method of Voting

The problem with plurality voting is that alternative choices are not considered. For instance, the result of the Minnesota governor's contest might have been quite different if voters had been asked, "Choose the candidate you prefer, but if that candidate does not receive a *majority* of the votes, which candidate would be your second choice?"

To see why this might be a reasonable alternative to plurality voting, consider the following situation. Thirty-six senators are considering an educational funding measure. Because the senate leadership wants an educational funding measure to pass, the leadership first determines that the senators prefer measure A for $50 million over measure B for $30 million. However, because of an unexpected dip in state revenues, measure A is removed from consideration and a new measure, C, for $15 million, is proposed. The senate leadership determines that senators favor measure B over measure C. In summary, we have

A majority of senators favor measure A over measure B.

A majority of senators favor measure B over measure C.

From these results, it seems reasonable to think that a majority of senators would prefer measure A over measure C. However, when the senators are asked about their preferences between the two measures, measure C is preferred over measure A. To understand how this could happen, consider the preference schedule for the senators shown in Table 9.22.

Table 9.22

	Rankings		
Measure A: $50 Million	1	3	3
Measure B: $30 Million	3	1	2
Measure C: $15 Million	2	2	1
Number of senators:	15	12	9

Notice that 15 senators prefer measure A over measure C, but $12 + 9 = 21$ senators, a majority of the 36 senators, prefer measure C over measure A. According to the preference schedule, if all three measures were on the ballot, A would come in first, B would come in second, and C would come in third. However, if just A and C were on the ballot, C would win over A. This paradoxical result was first discussed by Jean C. Borda in 1770.

In an attempt to remove such paradoxical results from voting, Borda proposed that voters rank their choices by giving each choice a certain number of points.

Take Note: Paradoxes occur in voting only when there are three or more candidates or issues on a ballot. If there are only two candidates in a race, then the candidate receiving the majority of the votes cast is the winner. In a two-candidate race, the majority and the plurality are the same.

The Borda Count Method of Voting

If there are n candidates or issues in an election, each voter ranks the candidates or issues by giving n points to the voter's first choice, $n - 1$ points to the voter's second choice, and so on, with the voter's least favorite choice receiving 1 point. The candidate or issue that receives the most total points is the winner.

Applying the Borda count method to the education measures, a measure receiving a first-place vote receives 3 points. (There are three different measures.) Each measure receiving a second-place vote receives 2 points, and each measure receiving a third-place vote receives 1 point. The calculations are shown.

Points per vote

Measure A:	15 first-place votes:	$15 \cdot 3 = 45$
	0 second-place votes:	$0 \cdot 2 = 0$
	21 third-place votes:	$21 \cdot 1 = 21$
	Total:	66
Measure B:	12 first-place votes:	$12 \cdot 3 = 36$
	9 second-place votes:	$9 \cdot 2 = 18$
	15 third-place votes:	$15 \cdot 1 = 15$
	Total:	69
Measure C:	9 first-place votes:	$9 \cdot 3 = 27$
	27 second-place votes:	$27 \cdot 2 = 54$
	0 third-place votes:	$0 \cdot 1 = 0$
	Total:	81

Using the Borda count method, measure C is the clear winner (even though it is not the plurality winner).

Example 2   Use the Borda Count Method

The members of a club are going to elect a president from four nominees using the Borda count method. If the 100 members of the club mark their ballots as shown in Table 9.23, who will be elected president?

Table 9.23

	Rankings					
Avalon	2	2	2	2	3	2
Branson	1	4	4	3	2	1
Columbus	3	3	1	4	1	4
Dunkirk	4	1	3	1	4	3
Number of voters:	30	24	18	12	10	6

Solution

Using the Borda count method, each first-place vote receives 4 points, each second-place vote receives 3 points, each third-place vote receives 2 points, and each last-place vote receives 1 point. The summary for each candidate is shown.

Avalon:	0 first-place votes	$0 \cdot 4 = 0$
	90 second-place votes	$90 \cdot 3 = 270$
	10 third-place votes	$10 \cdot 2 = 20$
	0 fourth-place votes	$0 \cdot 1 = 0$
	Total	290

Branson:	36 first-place votes	$36 \cdot 4 = 144$
	10 second-place votes	$10 \cdot 3 = 30$
	12 third-place votes	$12 \cdot 2 = 24$
	42 fourth-place votes	$42 \cdot 1 = 42$
		Total 240
Columbus:	28 first-place votes	$28 \cdot 4 = 112$
	0 second-place votes	$0 \cdot 3 = 0$
	54 third-place votes	$54 \cdot 2 = 108$
	18 fourth-place votes	$18 \cdot 1 = 18$
		Total 238
Dunkirk:	36 first-place votes	$36 \cdot 4 = 144$
	0 second-place votes	$0 \cdot 3 = 0$
	24 third-place votes	$24 \cdot 2 = 48$
	40 fourth-place votes	$40 \cdot 1 = 40$
		Total 232

Avalon has the largest total score. By the Borda count method, Avalon is elected president.

Take Note: Note in Example 2 that Avalon was the winner even though that candidate did not receive any first-place votes. The Borda count method was devised to allow voters to say, "If my first choice does not win, then consider my second choice."

Plurality with Elimination

A variation of the plurality method of voting is called *plurality with elimination*. Like the Borda count method, the method of plurality with elimination considers a voter's alternate choices.

Suppose that 30 members of a regional planning board must decide where to build a new airport. The airport consultants to the regional board have recommended four different sites. The preference schedule for the board members is shown in Table 9.24.

Take Note: When the second round of voting occurs, the two ballots that listed Bremerton as the first choice must be adjusted. The second choice on those ballots becomes the first, the third choice becomes the second, and the fourth choice becomes the third. The order of preference does not change. Similar adjustments must be made to the 12 ballots that listed Bremerton as the second choice. Because that choice is no longer available, Apple Valley becomes the second choice and Del Mar becomes the third choice. Adjustments must be made to the 11 ballots that listed Bremerton as the third choice. The fourth choice of those ballots, Del Mar, becomes the third choice. A similar adjustment must be made for the five ballots that listed Bremerton as the third choice.

Table 9.24

	Rankings			
Apple Valley	3	1	2	3
Bremerton	2	3	3	1
Coachella	1	2	4	2
Del Mar	4	4	1	4
Number of ballots:	12	11	5	2

Using the plurality with elimination method, the board members first eliminate the site with the fewest number of first-place votes. If two or more of these alternatives have the same number of first-place votes, all are eliminated unless that would eliminate all alternatives. In that case, a different method of voting must be used. From Table 9.24, Bremerton is eliminated because it received only two first-place votes. Now a vote is retaken using the following important assumption: *Voters do not change their preferences from round to round.* This means that after Bremerton is deleted, the 12 people in the first column would adjust their preferences so that Apple Valley becomes their second choice, Coachella remains their first choice, and Del Mar becomes their third choice. For the 11 voters in the second column, Apple Valley remains their first choice, Coachella remains their second choice, and Del Mar becomes their third choice. Similar adjustments are made by the remaining voters. The new preference schedule is shown in Table 9.25.

Table 9.25

	Rankings			
Apple Valley	2	1	2	2
Coachella	1	2	3	1
Del Mar	3	3	1	3
Number of ballots:	12	11	5	2

The board members now repeat the process and eliminate the site with the fewest first-place votes. In this case it is Del Mar. The new adjusted preference schedule is shown in Table 9.26.

Table 9.26

	Rankings			
Apple Valley	2	1	1	2
Coachella	1	2	2	1
Number of ballots:	12	11	5	2

From this table, Apple Valley has 16 first-place votes and Coachella has 14 first-place votes. Therefore, Apple Valley is the selected site for the new airport.

Example 3 **Use the Plurality with Elimination Voting Method**

A university wants to add a new sport to its existing program. To help ensure that the new sport will have student support, the students of the university are asked to rank the four sports under consideration. The results are shown in Table 9.27.

Table 9.27

	Rankings					
Lacrosse	3	2	3	1	1	2
Squash	2	1	4	2	3	1
Rowing	4	3	2	4	4	4
Golf	1	4	1	3	2	3
Number of ballots:	326	297	287	250	214	197

Use the plurality with elimination method to determine which of these sports should be added to the university's program.

Solution

Because rowing received no first-place votes, it is eliminated from consideration. The new preference schedule is shown in Table 9.28.

Table 9.28

	Rankings					
Lacrosse	3	2	2	1	1	2
Squash	2	1	3	2	3	1
Golf	1	3	1	3	2	3
Number of ballots:	326	297	287	250	214	197

Take Note: Remember to shift the preferences at each stage of the elimination method. The 297 students who chose rowing as their third choice and golf as their fourth choice now have golf as their third choice. Check each preference schedule and update it as necessary.

From this table, lacrosse has 464 first-place votes, squash has 494 first-place votes, and golf has 613 first-place votes. Because lacrosse has the fewest first-place votes, it is eliminated. The new preference schedule is shown in Table 9.29.

Table 9.29

	Rankings					
Squash	2	1	2	1	2	1
Golf	1	2	1	2	1	2
Number of ballots:	326	297	287	250	214	197

From this table, squash received 744 first-place votes and golf received 827 first-place votes. Therefore, golf is added to the sports program.

Pairwise Comparison Voting Method

The *pairwise comparison* method of voting is sometimes referred to as the "head-to-head" method. In this method, each candidate is compared one-on-one with each of the other candidates. A candidate receives 1 point for a win, 0.5 point for a tie, and 0 points for a loss. The candidate with the greatest number of points wins the election.

A voting method that elects the candidate who wins all head-to-head matchups is said to satisfy the Condorcet criterion.

Condorcet Criterion

A candidate who wins all possible head-to-head matchups should win an election when all candidates appear on the ballot.

This is one of the *fairness criteria* that a voting method should exhibit. We will discuss other fairness criteria later in this section.

Example 4   Use the Pairwise Comparison Voting Method

There are four proposals for the name of a new football stadium at a college: Panther Stadium, after the team mascot; Sanchez Stadium, after a large university contributor; Mosher Stadium, after a famous alumnus known for humanitarian work; and Fritz Stadium, after the college's most winning football coach. The preference schedule cast by alumni and students is shown in Table 9.30.

Table 9.30

	Rankings				
Panther Stadium	2	3	1	2	4
Sanchez Stadium	1	4	2	4	3
Mosher Stadium	3	1	4	3	2
Fritz Stadium	4	2	3	1	1
Number of ballots:	752	678	599	512	487

Use the pairwise comparison voting method to determine the name of the stadium.

Solution

We will create a table to keep track of each of the head-to-head comparisons. Before we begin, note that a matchup between, say, Panther and Sanchez is the same as the matchup between Sanchez and Panther. Therefore, we will shade the duplicate cells and the cells between the same candidates. This is shown in the following table.

Versus	Panther	Sanchez	Mosher	Fritz
Panther				
Sanchez				
Mosher				
Fritz				

To complete the table, we will place the name of the winner in the cell of each head-to-head match. For instance, for the Panther-Sanchez matchup,

✓ Panther was favored over Sanchez on $678 + 599 + 512 = 1789$ ballots.

Sanchez was favored over Panther on $752 + 487 = 1239$ ballots.

The winner of this matchup is Panther, so that name is placed in the Panther versus Sanchez cell. Do this for each of the matchups.

✓ Panther was favored over Mosher on $752 + 599 + 512 = 1863$ ballots.

Mosher was favored over Panther on $678 + 487 = 1165$ ballots.

Panther was favored over Fritz on $752 + 599 = 1351$ ballots.

✓ Fritz was favored over Panther on $678 + 512 + 487 = 1677$ ballots.

Sanchez was favored over Mosher on $752 + 599 = 1351$ ballots.

✓ Mosher was favored over Sanchez on $678 + 512 + 487 = 1677$ ballots.

Sanchez was favored over Fritz on $752 + 599 = 1351$ ballots.

✓ Fritz was favored over Sanchez on $678 + 512 + 487 = 1677$ ballots.

Mosher was favored over Fritz on $752 + 678 = 1430$ ballots.

✓ Fritz was favored over Mosher on $599 + 512 + 487 = 1598$ ballots.

Versus	Panther	Sanchez	Mosher	Fritz
Panther		Panther	Panther	Fritz
Sanchez			Mosher	Fritz
Mosher				Fritz
Fritz				

Fritz has three wins, Panther has two wins, and Mosher has one win. Using pairwise comparison, Fritz Stadium is the winning name.

Take Note: Although we have shown the totals for both Panther over Sanchez and Sanchez over Panther, it is only necessary to do one of the matchups. You can then just subtract that number from the total number of ballots cast, which in this case is 3028.

Panther over Sanchez: 1789

Sanchez over Panther:
$3028 - 1789 = 1239$

We could have used this calculation method for all of the other matchups. For instance, in the final matchup of Mosher versus Fritz, we have

Mosher over Fritz: 1430

Fritz over Mosher:
$3028 - 1430 = 1598$

Fairness of Voting Methods and Arrow's Theorem

Now that we have examined various voting options, we will stop to ask which of these options is the *fairest*. To answer that question, we must first determine what we mean by fair.

Fairness Criteria

1. *Majority criterion:* The candidate who receives a majority (more than 50%) of the first-place votes is the winner.
2. *Monotonicity criterion:* If candidate A wins an election, then candidate A will also win the election if the only change in the voters' preferences is that supporters of a different candidate change their votes to support candidate A.
3. *Condorcet criterion:* A candidate who wins all possible head-to-head matchups should win an election when all candidates appear on the ballot.
4. *Independence of irrelevant alternatives:* If a candidate wins an election, the winner should remain the winner in any recount in which losing candidates withdraw from the race.

There are other criteria, such as the *dictator criterion,* that will be discussed in the next section. The economist and Nobel laureate Kenneth Arrow was able to prove that no matter what kind of voting system we devise, it is impossible for it to satisfy the fairness criteria.

Arrow's Impossibility Theorem

There is no voting method involving three or more choices that satisfies all four fairness criteria.

By Arrow's Impossibility Theorem, none of the voting methods we have discussed is fair. Not only that, but we *cannot* construct a fair voting system for three or more candidates. We will now give some examples of each of the methods we have discussed and show which of the fairness criteria are not satisfied.

Example 5 **Show That the Borda Count Method Violates the Majority Criterion**

Suppose the preference schedule for three candidates, Alpha, Beta, and Gamma, is given by Table 9.31.

Table 9.31

	Rankings		
Alpha	1	3	3
Beta	2	1	2
Gamma	3	2	1
Number of ballots:	55	50	3

Show that using the Borda count method violates the majority criterion.

Solution

The calculations for Borda's method are shown.

Alpha

55 first-place votes	$55 \cdot 3 = 165$
0 second-place votes	$0 \cdot 2 = 0$
53 third-place votes	$53 \cdot 1 = 53$
	Total 218

Beta

50 first-place votes	$50 \cdot 3 = 150$
58 second-place votes	$58 \cdot 2 = 116$
0 third-place votes	$0 \cdot 1 = 0$
	Total 266

Gamma

3 first-place votes	$3 \cdot 3 =$	9
50 second-place votes	$50 \cdot 2 =$	100
55 third-place votes	$55 \cdot 1 =$	55
	Total	164

From these calculations, Beta should win the election. However, Alpha has the majority (more than 50%) of the first-place votes. This result violates the majority criterion.

Example 6   **Show That Plurality with Elimination Violates the Monotonicity Criterion**

Suppose the preference schedule for three candidates, Alpha, Beta, and Gamma, is given by Table 9.32.

Table 9.32

	Rankings			
Alpha	2	3	1	1
Beta	3	1	2	3
Gamma	1	2	3	2
Number of ballots:	25	20	16	10

a. Show that, using plurality with elimination voting, Gamma wins the election.

b. Suppose that the 10 people who voted for Alpha first and Gamma second changed their votes such that they all voted for Alpha second and Gamma first. Show that, using plurality with elimination voting, Beta will now be elected.

c. Explain why this result violates the monotonicity criterion.

Solution

a. Beta received the fewest first-place votes, so Beta is eliminated. The new preference schedule is shown in Table 9.33.

Table 9.33

	Rankings			
Alpha	2	2	1	1
Gamma	1	1	2	2
Number of ballots:	25	20	16	10

From this schedule, Gamma has 45 first-place votes and Alpha has 26 first-place votes, so Gamma is the winner.

b. If the 10 people who voted for Alpha first and Gamma second changed their votes such that they all voted for Alpha second and Gamma first, the preference schedule is given in Table 9.34.

Table 9.34

	Rankings			
Alpha	2	3	1	2
Beta	3	1	2	3
Gamma	1	2	3	1
Number of ballots:	25	20	16	10

From this schedule, Alpha has the fewest first-place votes and is eliminated. The new preference schedule is shown in Table 9.35.

Table 9.35

	Rankings			
Beta	2	1	1	2
Gamma	1	2	2	1
Number of ballots:	25	20	16	10

From this schedule, Gamma has 35 first-place votes and Beta has 36 first-place votes, so Beta is the winner.

c. This result violates the monotonicity criterion because Gamma, who won the first election, loses the second election even though Gamma received a larger number of first-place votes in the second election.

9.2 Exercise Set

Think About It

1. What is the difference between an issue winning by a majority of the votes cast and winning by a plurality of votes cast?

2. The following preference schedule shows the ranking of preferences of three colors. What does the number 5 represent in the row of numbers at the bottom of the table?

	Rankings		
Red	3	3	1
Green	1	2	2
Blue	2	1	3
Totals:	7	5	9

3. What is Arrow's Impossibility Theorem?
4. What is the Condorcet criterion?

Preliminary Exercises

For Exercises P1 to P6, if necessary, use the referenced example following the exercise for assistance.

P1. According to the table in Example 1 on page 428, which variety of candy would win second place using the plurality voting system? **Example 1**

P2. The preference schedule given in Example 1 for the 50 people who were asked to rank their preferences of five varieties of chocolate candy is shown again in the following table.

	Rankings					
Caramel Center	5	4	4	4	2	4
Vanilla Center	1	5	5	5	5	5
Almond Center	2	3	2	1	3	3
Toffee Center	4	1	1	3	4	2
Solid Chocolate	3	2	3	2	1	1
Number of voters:	17	11	9	8	3	2

Determine the taste test favorite using the Borda count method. **Example 2**

P3. A service club is going to sponsor a dinner to raise money for a charity. The club has decided to serve Italian, Mexican, Thai, Chinese, or Indian food. The members of the club were surveyed to determine their preferences. The results are shown in the following table.

	Rankings				
Italian	2	5	1	4	3
Mexican	1	4	5	2	1
Thai	3	1	4	5	2
Chinese	4	2	3	1	4
Indian	5	3	2	3	5
Number of ballots:	33	30	25	20	18

Use the plurality with elimination method to determine the food preference of the club members. **Example 3**

P4. One hundred restaurant critics were asked to rank their favorite restaurants from a list of four. The preference schedule for the critics is shown in the following table.

	Rankings				
Sanborn's Fine Dining	3	1	4	3	1
The Apple Inn	4	3	3	2	4
May's Steak House	2	2	1	1	3
Tory's Seafood	1	4	2	4	2
Number of ballots:	31	25	18	15	11

Use the pairwise voting method to determine the critics' favorite restaurant. **Example 4**

P5. Using Table 9.31 in Example 5 on page 435, show that the Borda count method violates the Condorcet criterion.
Example 5

P6. The following table shows the preferences for three new car colors.

	Rankings		
Radiant Silver	1	3	3
Electric Red	2	2	1
Lightning Blue	3	1	2
Number of votes:	30	27	2

Show that the Borda count method violates the independence of irrelevant alternatives criterion. (*Hint:* Reevaluate the vote if lightning blue is eliminated.)
Example 6

Section Exercises

1. **Breakfast Cereal**   Sixteen people were asked to rank three breakfast cereals in order of preference. Their responses are given in the following table.

Corn Flakes	3	1	1	2	3	3	2	2	1	3	1	3	1	2	1	2
Raisin Bran	1	3	2	3	1	2	1	1	2	1	2	2	3	1	3	3
Mini Wheats	2	2	3	1	2	1	3	3	3	2	3	1	2	3	2	1

If the plurality method of voting is used, which cereal is the group's first preference?

2. **Cartoon Characters**   A kindergarten class was surveyed to determine the children's favorite cartoon characters among Dora the Explorer, SpongeBob SquarePants, and Buzz Lightyear. The students ranked the characters in order of preference; the results are shown in the following preference schedule.

	Rankings					
Dora the Explorer	1	1	2	2	3	3
SpongeBob SquarePants	2	3	1	3	1	2
Buzz Lightyear	3	2	3	1	2	1
Number of students:	6	4	6	5	6	8

   a. How many students are in the class?

   b. How many votes are required for a majority?

   c. Using plurality voting, which character is the children's favorite?

3. **Catering**   A 15-person committee is having lunch catered for a meeting. Three caterers, each specializing in a different cuisine, are available. In order to choose a caterer for the group, each member is asked to rank the cuisine options in order of preference. The results are given in the following preference schedule.

	Rankings				
Italian	1	1	2	3	3
Mexican	2	3	1	1	2
Japanese	3	2	3	2	1
Number of votes:	2	4	1	5	3

Using plurality voting, which caterer should be chosen?

4. **Movies**   Fifty consumers were surveyed about their movie-watching habits. They were asked to rank the likelihood that they would participate in each listed activity. The results are summarized in the following table.

	Rankings				
Stream Online	2	3	1	2	1
Go to a Theater	3	1	3	1	2
Rent from a Kiosk	1	2	2	3	3
Number of votes:	8	13	15	7	7

Using the Borda count method of voting, which activity is the most popular choice among this group of consumers?

5. **Breakfast Cereal**   Use the Borda count method of voting to determine the preferred breakfast cereal in Exercise 1.

6. **Cartoons**   Use the Borda count method of voting to determine the children's favorite cartoon character in Exercise 2.

7. **Catering**   Use the Borda count method of voting to determine which caterer the committee should hire in Exercise 3.

8. **Class Election**   A senior high school class held an election for class president. Instead of just voting for one candidate, the students were asked to rank all four candidates in order of preference. The results are shown in the following table.

	Rankings					
Rith Lee	2	3	1	3	4	2
Sabana Brewer	4	1	3	4	1	3
Elaine Garcia	1	2	2	2	3	4
Michael Turley	3	4	4	1	2	1
Number of votes:	36	53	41	27	31	45

Using the Borda count method, which student should be class president?

9. **Cell Phone Usage**   A journalist reviewing various cellular phone services surveyed 200 customers and asked each one to rank four service providers in order of preference. The group's results are shown in the following table.

	Rankings				
AT&T	3	4	2	3	4
Sprint	1	1	4	4	3
Verizon	2	2	1	2	1
T-Mobile	4	3	3	1	2
Number of votes:	18	38	42	63	39

Using the Borda count method, which provider is the favorite of these customers?

10. **Baseball Uniforms** A Little League baseball team must choose the colors for its uniforms. The coach offered four different choices, and the players ranked them in order of preference, as shown in the following table.

	Rankings			
Red and White	2	3	3	2
Green and Yellow	4	1	4	1
Red and Blue	3	4	2	4
Blue and White	1	2	1	3
Number of votes:	4	2	5	4

Using the plurality with elimination method, what colors should be used for the uniforms?

11. **Radio Stations** A number of college students were asked to rank four radio stations in order of preference. The responses are given in the following table.

	Rankings				
WNNX	3	1	1	2	4
WKLS	1	3	4	1	2
WWVV	4	2	2	3	1
WSTR	2	4	3	4	3
Number of votes:	57	72	38	61	15

Use plurality with elimination to determine the students' favorite radio station among the four.

12. **Class Election** Use plurality with elimination to choose the class president in Exercise 8.

13. **Cell Phone Usage** Use plurality with elimination to determine the preferred cellular phone service in Exercise 9.

14. **Campus Club** A campus club has money left over in its budget and must spend it before the school year ends. The members arrive at five different possibilities, and each member ranks them in order of preference. The results are shown in the following table.

	Rankings				
Establish a Scholarship	1	2	3	3	4
Pay for Several Members to Travel to a Convention	2	1	2	1	5
Buy New Computers for the Club	3	3	1	4	1
Throw an End-of-Year Party	4	5	5	2	2
Donate to Charity	5	4	4	5	3
Number of votes:	8	5	12	9	7

a. Using the plurality voting system, how should the club spend the money?

b. Use the plurality with elimination method to determine how the money should be spent.

c. Using the Borda count method of voting, how should the money be spent?

d. In your opinion, which of the previous three methods seems most appropriate in this situation? Why?

15. **Recreation** A company is planning its annual summer retreat and has asked its employees to rank five different choices of recreation in order of preference. The results are given in the following table.

	Rankings				
Picnic in a Park	1	2	1	3	4
Water Skiing at a Lake	3	1	2	4	3
Amusement Park	2	5	5	1	2
Riding Horses at a Ranch	5	4	3	5	1
Dinner Cruise	4	3	4	2	5
Number of votes:	10	18	6	28	16

a. Using the plurality voting system, what activity should be planned for the retreat?

b. Use the plurality with elimination method to determine which activity should be chosen.

c. Using the Borda count method of voting, which activity should be planned?

16. **Star Wars Movies** Fans of some of the original Star Wars movies have been debating on an online forum regarding which of the films is the best. To see what the overall opinion is, visitors to the site can rank four films in order of preference. The results are shown in the following preference schedule.

	Rankings			
A New Hope	1	2	1	3
The Empire Strikes Back	4	4	2	1
Return of the Jedi	2	1	3	2
The Phantom Menace	3	3	4	4
Number of votes:	429	1137	384	582

Using pairwise comparison, which film is the favorite of the visitors to the site who voted?

17. **Family Reunion** A family is trying to decide where to take a vacation. All of the family members are asked to rank four choices in order of preference. The results are shown in the following preference schedule.

	Rankings				
Grand Canyon	3	1	2	3	1
Yosemite	1	2	3	4	4
Bryce Canyon	4	4	1	2	2
The Maldives	2	3	4	1	3
Number of votes:	7	3	12	8	13

Use the pairwise comparison method to determine the best choice for the reunion.

18. School Mascot A new college needs to pick a mascot for its football team. The students were asked to rank four choices in order of preference; the results are tallied in the following table.

	Rankings				
Bulldog	3	4	4	I	4
Panther	2	I	2	4	3
Hornet	4	2	I	2	2
Bobcat	I	3	3	3	I
Number of votes:	638	924	525	390	673

Using the pairwise comparison method of voting, which mascot should be chosen?

19. Election Five candidates are running for president of a charity organization. Interested persons were asked to rank the candidates in order of preference. The results are given in the following table.

	Rankings				
P. Gibson	5	I	2	I	2
E. Yung	2	4	5	5	3
R. Allenbaugh	3	2	I	3	5
T. Meckley	4	3	4	4	I
G. DeWitte	I	5	3	2	4
Number of votes:	I6	9	I4	9	4

Use the pairwise comparison method to determine the president of the organization.

20. Baseball Uniforms Use the pairwise comparison method to choose the colors for the Little League uniforms in Exercise 10.

21. Radio Stations Use the pairwise comparison method to determine the favorite radio station in Exercise 11.

22. Does the winner in Exercise 2c satisfy the Condorcet criterion?

23. Does the winner in Exercise 3 satisfy the Condorcet criterion?

24. Does the winner in Exercise 14c satisfy the Condorcet criterion?

25. Does the winner in Exercise 15c satisfy the Condorcet criterion?

26. Does the winner in Exercise 8 satisfy the majority criterion?

27. Does the winner in Exercise 11 satisfy the majority criterion?

28. Election Three candidates are running for mayor. A vote was taken in which the candidates were ranked in order of preference. The results are shown in the following preference schedule.

	Rankings		
John Lorenz	I	3	3
Marcia Beasley	3	I	2
Stephen Hyde	2	2	I
Number of votes:	269I	24I6	237

a. Use the Borda count method to determine the winner of the election.

b. Verify that the majority criterion has been violated.

c. Identify the candidate who wins all head-to-head comparisons.

d. Explain why the Condorcet criterion has been violated.

e. If Marcia Beasley drops out of the race for mayor (and voter preferences remain the same), determine the winner of the election again, using the Borda count method.

f. Explain why the independence of irrelevant alternatives criterion has been violated.

29. Film Competition Three films have been selected as finalists in a national student film competition. Seventeen judges have viewed each of the films and have ranked them in order of preference. The results are given in the following preference schedule.

	Rankings			
Film A	I	3	2	I
Film B	2	I	3	3
Film C	3	2	I	2
Number of votes:	4	6	5	2

a. Using the plurality with elimination method, which film should win the competition?

b. Suppose the first vote is declared invalid and a revote is taken. All of the judges' preferences remain the same except for the votes represented by the last column of the table. The judges who cast these votes both decide to switch their first place vote to film C, so their preference now is film C first, then film A, and then film B. Which film now wins using the plurality with elimination method?

c. Has the monotonicity criterion been violated?

Investigations

Approval voting is a system in which voters may vote for more than one candidate. Each vote counts equally, and the candidate with the most total votes wins the election. Many feel that this is a better system for large elections than simple plurality because it considers a voter's second choices and is a stronger measure of overall voter support for each candidate. Some organizations use approval voting to elect their officers. The United Nations uses this method to elect the secretary-general.

1. Suppose a math class is going to show a film involving mathematics or mathematicians on the last day of class. The options are *Stand and Deliver, Good Will Hunting, A Beautiful Mind, Proof*, and *Contact*. The students vote using approval voting. The results are as follows.

8 students vote for all five films.
8 students vote for *Good Will Hunting, A Beautiful Mind,* and *Contact*.
8 students vote for *Stand and Deliver, Good Will Hunting,* and *Contact*.
8 students vote for *A Beautiful Mind* and *Proof*.
8 students vote for *Stand and Deliver* and *Proof*.

8 students vote for *Good Will Hunting* and *Contact*.
1 student votes for *Proof*.

Which film will be chosen for the last day of class screening?

2. Suppose you and three friends, David, Sara, and Cofa, are trying to decide on a pizza restaurant. You like Pizza Hut best, Round Table pizza is acceptable to you, and you

definitely do not want to get pizza from Domino's. Domino's is David's favorite, and he also likes Round Table, but he won't eat pizza from Pizza Hut. Sara says she will only eat Round Table pizza. Cofa prefers Domino's, but he will also eat Round Table pizza. He doesn't like Pizza Hut. Use approval voting to determine the pizza restaurant that the group of friends should choose.

Section 9.3
Weighted Voting Systems

Learning Objectives:

- Determine a winning coalition in a weighted voting system.
- Compute the Banzhaf power index.
- Use the Banzhaf power index to compute a voter's power.

Biased Voting Systems

A **weighted voting system** is one in which some voters have more influence (weight) on the outcome of an election. Examples of weighted voting systems are fairly common. A few examples are the stockholders of a company, the Electoral College, the United Nations Security Council, and the European Union.

Consider a small company with a total of 100 shares of stock and three stockholders, A, B, and C. Suppose that A owns 45 shares of the stock (which means A has 45 votes), B owns 45 shares, and C owns 10 shares. If a vote of 51 or greater is required to approve any measure before the owners, then a measure cannot be passed without two of the three owners voting for the measure. Even though C has only 10 shares, C has the same voting power as A and B.

Now suppose that a new stockholder is brought into the company and the shares of the company are redistributed so that A has 27 shares, B has 26 shares, C has 25 shares, and D has 22 shares. Note, in this case, that any two of A, B, or C can pass a measure, but D paired with any of the other shareholders cannot pass a measure. D has virtually no power even though D has only three shares less than C.

The number of votes that are required to pass a measure is called a **quota**. For the two stockholder examples above, the quota was 51. The **weight of a voter** is the number of votes controlled by the voter. In the case of the company whose stock was split A–27 shares, B–26 shares, C–25 shares, and D–22 shares, the weight of A is 27, the weight of B is 26, the weight of C is 25, and the weight of D is 22. Rather than write out in sentence form the quota and weight of each voter, we use the notation

Quota ⎤ ⎡ Weights

$$\{51: 27, 26, 25, 22\}$$

• The four numbers after the colon indicate that there are a total of four voters in this system.

This notation is very convenient. We state its more general form in the following definition.

Weighted Voting System

A weighted voting system of n voters is written $\{q: w_1, w_2, \ldots, w_n\}$, where q is the quota and w_1 through w_n represent the weights of each of the n voters.

Using this notation, we can describe various voting systems.

- **One person, one vote:** For instance, {5: 1, 1, 1, 1, 1, 1, 1, 1, 1}. In this system, each person has one vote and five votes, a majority, are required to pass a measure.
- **Dictatorship:** For instance, {20: 21, 6, 5, 4, 3}. In this system, the person with 21 votes can pass any measure. Even if the remaining four people get together, their votes do not total the quota of 20.
- **Null system:** For instance, {28: 6, 3, 5, 2}. If all the members of this system vote for a measure, the total number of votes is 16, which is less than the quota. Therefore, no measure can be passed.
- **Veto power system:** For instance, {21: 6, 5, 4, 3, 2, 1}. In this case, the sum of all the votes is 21, the quota. Therefore, if any one voter does not vote for the measure, it will fail. Each voter is said to have **veto power**. In this case, this means that even the voter with one vote can veto a measure (cause the measure not to pass). A voter has veto power whenever a measure cannot be passed without that voter's vote. If at least one voter in a voting system has veto power, the system is a veto power system.

In a weighted voting system, a **coalition** is a set of voters each of whom votes the same way, either for or against a resolution. A **winning coalition** is a set of voters the sum of whose votes is greater than or equal to the quota. A **losing coalition** is a set of voters the sum of whose votes is less than the quota. A voter who leaves a winning coalition and thereby turns it into a losing coalition is called a **critical voter**.

As shown in the next theorem, for large numbers of voters, there are many possible coalitions.

Take Note: The number of coalitions of n voters is the number of subsets that can be formed from n voters; this is 2^n. Because a coalition must contain at least one voter, the empty set is not a possible coalition. Therefore, the number of coalitions is $2^n - 1$.

Number of Possible Coalitions of n Voters

The number of possible coalitions of n voters is $2^n - 1$.

As an example, if all electors of each state to the Electoral College cast their ballots for one candidate, then there are $2^{51} - 1 \approx 2.25 \times 10^{15}$ possible coalitions (the District of Columbia is included). The number of *winning* coalitions is far less. For instance, any coalition of 10 or fewer states cannot be a winning coalition because the largest 10 states do not have enough electoral votes to elect the president. As we proceed through this section, we will not attempt to list all the coalitions, only the winning coalitions.

Example 1 **Determine Winning Coalitions in a Weighted Voting System**

Suppose that the four owners of a company, Ang, Bonhomme, Carmel, and Diaz, own, respectively, 500 shares, 375 shares, 225 shares, and 400 shares. There are a total of 1500 votes; half of this is 750, so the quota is 751. The weighted voting system for this company is {751: 500, 375, 225, 400}.

a. Determine the winning coalitions.

b. For each winning coalition, determine the critical voters.

Solution

a. A winning coalition must represent at least 751 votes. We will list these coalitions in Table 9.36, in which we use A for Ang, B for Bonhomme, C for Carmel, and D for Diaz.

Take Note: The coalition {A, C} is not a winning coalition because the total number of votes for that coalition is 725, which is less than 751.

Table 9.36

Winning Coalition	Number of Votes
{A, B}	875
{A, D}	900
{B, D}	775
{A, B, C}	1100
{A, B, D}	1275
{A, C, D}	1125
{B, C, D}	1000
{A, B, C, D}	1500

b. A voter who leaves a winning coalition and thereby creates a losing coalition is a critical voter. For instance, for the winning coalition {A, B, C}, if A leaves, the number of remaining votes is 600, which is not enough to pass a resolution. If B leaves, the number of remaining votes is 725—again, not enough to pass a resolution. If C leaves, the number of remaining votes is 875, which is greater than the quota. Therefore, A and B are critical voters for the coalition {A, B, C} and C is not a critical voter. Table 9.37 shows the critical voters for each winning coalition.

Table 9.37

Winning Coalition	Number of Votes	Critical Voters
{A, B}	875	A, B
{A, D}	900	A, D
{B, D}	775	B, D
{A, B, C}	1100	A, B
{A, B, D}	1275	None
{A, C, D}	1125	A, D
{B, C, D}	1000	B, D
{A, B, C, D}	1500	None

Banzhaf Power Index

There are a number of measures of the *power* of a voter. For instance, as we saw from the Electoral College example, some electors represent fewer people and therefore their votes may have more power. As an extreme case, suppose that two electors, A and B, each represent 10 people and that a third elector, C, represents 1000 people. If a measure passes when two of the three electors vote for the measure, then A and B voting together could pass a resolution even though they represent only 20 people.

Another measure of power, called the *Banzhaf power index*, was derived by John F. Banzhaf III in 1965. The purpose of this index is to determine the power of a voter in a weighted voting system.

Banzhaf Power Index

The **Banzhaf power index** of a voter v, symbolized by $BPI(v)$, is given by

$$BPI(v) = \frac{\text{number of times voter } v \text{ is a critical voter}}{\text{number of times any voter is a critical voter}}$$

Consider four people, A, B, C, and D, and the one-person, one-vote system given by {3: 1, 1, 1, 1}, as in Table 9.38.

Table 9.38

Winning Coalition	Number of Votes	Critical Voters
{A, B, C}	3	A, B, C
{A, B, D}	3	A, B, D
{A, C, D}	3	A, C, D
{B, C, D}	3	B, C, D
{A, B, C, D}	4	None

To find $BPI(A)$, we look under the critical voters column and find that A is a critical voter three times. The number of times any voter is a critical voter, the denominator of the Banzhaf power index, is 12. (A is a critical voter three times, B is a critical voter three times, C is a critical voter three times, and D is a critical voter three times. The sum is $3 + 3 + 3 + 3 = 12$.) Thus

Take Note: The Banzhaf power index is a number between 0 and 1. If the index for voter A is less than the index for voter B, then A has less power than B. This means that A has fewer opportunities to form a winning coalition than does B.

$$BPI(A) = \frac{3}{12} = 0.25$$

Similarly, we can calculate the Banzhaf power index for each of the other voters.

$$BPI(B) = \frac{3}{12} = 0.25 \qquad BPI(C) = \frac{3}{12} = 0.25 \qquad BPI(D) = \frac{3}{12} = 0.25$$

In this case, each voter has the same power. This is expected in a voting system in which each voter has one vote.

Now suppose that three people, A, B, and C, belong to a dictatorship given by {3: 3, 1, 1}, as in Table 9.39.

Table 9.39

Winning Coalition	Number of Votes	Critical Voters
{A}	3	A
{A, B}	4	A
{A, C}	4	A
{A, B, C}	5	A

The sum of the numbers of critical voters in all winning coalitions is 4. To find $BPI(A)$, we look under the critical voters column and find that A is a critical voter four times. Thus

$$BPI(A) = \frac{4}{4} = 1 \qquad BPI(B) = \frac{0}{4} = 0 \qquad BPI(C) = \frac{0}{4} = 0$$

Thus A has all the power. This is expected in a dictatorship.

Example 2 Compute the BPI for a Weighted Voting System

Suppose the stock in a company is held by five people, A, B, C, D, and E. The voting system for this company is {626: 350, 300, 250, 200, 150}. Determine the Banzhaf power index for A and E.

Solution

Determine all of the winning coalitions and the critical voters in each coalition.

Winning Coalition	Critical Voters	Winning Coalition	Critical Voters
{A, B}	A, B	{B, C, E}	B, C, E
{A, B, C}	A, B	{B, D, E}	B, D, E
{A, B, D}	A, B	{A, B, C, D}	None
{A, B, E}	A, B	{A, B, C, E}	None
{A, C, D}	A, C, D	{A, B, D, E}	None
{A, C, E}	A, C, E	{A, C, D, E}	A
{A, D, E}	A, D, E	{B, C, D, E}	B
{B, C, D}	B, C, D	{A, B, C, D, E}	None

The number of times any voter is critical is 28. To find $BPI(A)$, we look under the critical voters columns and find that A is a critical voter eight times. Thus

$$BPI(A) = \frac{8}{28} \approx 0.29$$

To find $BPI(E)$, we look under the critical voters columns and find that E is a critical voter four times. Thus

$$BPI(E) = \frac{4}{28} \approx 0.14$$

In many cities, the only time the mayor votes on a resolution is when there is a tie vote by the members of the city council. This is also true of the U.S. Senate. The vice president only votes when there is a tie vote by the senators.

In Example 3, we will calculate the Banzhaf power index for a voting system in which one voter votes only to break a tie.

Example 3 **Use the BPI to Determine a Voter's Power**

Suppose a city council consists of four members, A, B, C, and D, and a mayor, M, each with an equal vote. The mayor votes only when there is a tie vote among the members of the council. In all cases, a resolution receiving three or more votes passes. Show that the Banzhaf power index for the mayor is the same as the Banzhaf power index for each city council member.

Solution

We first list all of the winning coalitions that do not include the mayor. To this list, we add the winning coalitions in which the mayor votes to break a tie.

Winning Coalition (without mayor)	Critical Voters	Winning Coalition (mayor voting)	Critical Voters
{A, B, C}	A, B, C	{A, B, M}	A, B, M
{A, B, D}	A, B, D	{A, C, M}	A, C, M
{A, C, D}	A, C, D	{A, D, M}	A, D, M
{B, C, D}	B, C, D	{B, C, M}	B, C, M
{A, B, C, D}	None	{B, D, M}	B, D, M
		{C, D, M}	C, D, M

By examining the table, we see that A, B, C, D, and M each occur in exactly six winning coalitions. The total number of critical voters in all winning coalitions is 30. Therefore, each member of the council and the mayor have the same Banzhaf power index, which is $\frac{6}{30} = 0.2$.

9.3   Exercise Set

▶ **Think About It**

1. What is a weighted voting system?

2. What is a quota in a weighted voting system?

3. The weighted voting system for the United Nations Security Council is $\{39: 7, 7, 7, 7, 7, 1, 1, 1, 1, 1, 1, 1, 1, 1, 1\}$, where the 5 permanent members of the council, United States, France, United Kingdom, China, and Russia, each have 7 votes and the other 10 countries elected for a two-year term have 1 vote. Explain why this is a veto power system.

4. There have been proposals to change the United Nations Security Council voting system to $\{30: 5, 5, 5, 5, 5, 1, 1, 1, 1, 1, 1, 1, 1, 1, 1\}$. Is this voting system a veto power system?

▶ **Preliminary Exercises**

For Exercises P1 to P3, if necessary, use the referenced example following the exercise for assistance.

P1. Many countries must govern by forming coalitions from among many political parties. Suppose a country has five political parties named A, B, C, D, and E. The numbers of votes, respectively, for the five parties are 22, 18, 17, 10, and 5.

    a. Determine the winning coalitions if 37 votes are required to pass a resolution.

    b. For each winning coalition, determine the critical voters. **Example 1**

P2. Suppose that a government is composed of four political parties, A, B, C, and D. The voting system for this government is $\{26: 18, 16, 10, 6\}$. Determine the Banzhaf power index for A and D. **Example 2**

P3. The European Economic Community (EEC) was founded in 1958 and originally consisted of Belgium, France, Germany, Italy, Luxembourg, and the Netherlands. The weighted voting system was $\{12: 2, 4, 4, 4, 1, 2\}$. Find the Banzhaf power index for each country. **Example 3**

▶ **Section Exercises**

In the following exercises that involve weighted voting systems for voters A, B, C, . . . , the systems are given in the form $\{q: w_1, w_2, w_3, w_4, \ldots, w_n\}$. The weight of voter A is w_1, the weight of voter B is w_2, the weight of voter C is w_3, and so on.

1. A weighted voting system is given by $\{6: 4, 3, 2, 1\}$.

    a. What is the quota?

    b. How many voters are in this system?

    c. What is the weight of voter B?

    d. What is the weight of the coalition {A, C}?

    e. Is {A, D} a winning coalition?

    f. Which voters are critical voters in the coalition {A, C, D}?

    g. How many coalitions can be formed?

    h. How many coalitions consist of exactly two voters?

2. A weighted voting system is given by $\{16: 8, 7, 4, 2, 1\}$.

    a. What is the quota?

    b. How many voters are in this system?

    c. What is the weight of voter C?

    d. What is the weight of the coalition {B, C}?

    e. Is {B, C, D, E} a winning coalition?

    f. Which voters are critical voters in the coalition {A, B, D}?

    g. How many coalitions can be formed?

    h. How many coalitions consist of exactly three voters?

In Exercises 3 to 12, calculate, if possible, the Banzhaf power index for each voter. Round to the nearest hundredth.

3. $\{6: 4, 3, 2\}$

4. $\{10: 7, 6, 4\}$

5. $\{10: 7, 3, 2, 1\}$

6. $\{14: 7, 5, 1, 1\}$

7. $\{19: 14, 12, 4, 3, 1\}$

8. $\{3: 1, 1, 1, 1\}$

9. $\{18: 18, 7, 3, 3, 1, 1\}$

10. $\{14: 6, 6, 4, 3, 1\}$

11. $\{80: 50, 40, 30, 25, 5\}$

12. $\{85: 55, 40, 25, 5\}$

13. Which, if any, of the voting systems in Exercises 3 to 12 is

    a. a dictatorship?

    b. a veto power system? *Note:* A voting system is a veto power system if any of the voters has veto power.

    c. a null system?

    d. a one-person, one-vote system?

14. Explain why it is impossible to calculate the Banzhaf power index for any voter in the null system $\{8: 3, 2, 1, 1\}$.

15. **Music Education**   A music department consists of a band director and a music teacher. Decisions on motions are made by voting. If both members vote in favor of a motion, it passes. If both members vote against a motion, it fails. In the event of a tie vote, the principal of the school votes to break the tie. For this voting scheme, determine the Banzhaf power index for each department member and for the principal. *Hint:* See Example 3, page 445.

16. **Voting**   Four voters, A, B, C, and D, make decisions by using the voting scheme $\{4: 3, 1, 1, 1\}$, except when there is a tie. In the event of a tie, a fifth voter, E, casts a vote to break the tie. For this voting scheme, determine the Banzhaf power index for each voter, including voter E. *Hint:* See Example 3, page 445.

17. **Criminal Justice**   In a criminal trial, each of the 12 jurors has one vote and all of the jurors must agree to reach a verdict. Otherwise the judge will declare a mistrial.

    a. Write the weighted voting system, in the form $\{q: w_1, w_2, w_3, w_4, \ldots, w_{12}\}$, used by these jurors.

    b. Is this weighted voting system a one-person, one-vote system?

c. Is this weighted voting system a veto power system?

d. Explain an easy way to determine the Banzhaf power index for each voter.

18. **Criminal Justice**   In California civil court cases, each of the 12 jurors has one vote and at least 9 of the jury members must agree on the verdict.

 a. Write the weighted voting system, in the form $\{q: w_1, w_2, w_3, w_4, \ldots, w_{12}\}$, used by these jurors.

 b. Is this weighted voting system a one-person, one-vote system?

 c. Is this weighted voting system a veto power system?

 d. Explain an easy way to determine the Banzhaf power index for each voter.

A voter who has a weight that is greater than or equal to the quota is called a **dictator**. In a weighted voting system, the dictator has all the power. A voter who is never a critical voter has no power and is referred to as a **dummy**. This term is not meant to be a comment on the voter's intellectual powers; it just indicates that the voter has no ability to influence an election.

In Exercises 19 to 22, identify any dictator and all dummies for each weighted voting system.

19. $\{16: 16, 5, 4, 2, 1\}$ 20. $\{15: 7, 5, 3, 2\}$

21. $\{19: 12, 6, 3, 1\}$ 22. $\{45: 40, 6, 2, 1\}$

23. **Football**   At the beginning of each football season, the coaching staff at Vista High School must vote to decide which players to select for the team. They use the weighted voting system $\{4: 3, 2, 1\}$. In this voting system, the head coach, A, has a weight of 3, the assistant coach, B, has a weight of 2, and the junior varsity coach, C, has a weight of 1.

 a. Compute the Banzhaf power index for each of the coaches.

 b. Explain why it seems reasonable that the assistant coach and the junior varsity coach have the same Banzhaf power index in this voting system.

24. **Football**   The head coach in Exercise 23 has decided that next year the coaching staff should use the weighted voting system $\{5: 4, 3, 1\}$. The head coach is still voter A, the assistant coach is still voter B, and the junior varsity coach is still voter C.

 a. Compute the Banzhaf power index for each coach under this new system.

 b. How do the Banzhaf power indices for this new voting system compare with the Banzhaf power indices in Exercise 23? Did the head coach gain any power, according to the Banzhaf power indices, with this new voting system?

25. Consider the weighted voting system $\{60: 4, 56, 58\}$.

 a. Compute the Banzhaf power index for each voter in this system.

 b. Voter B has a weight of 56 compared with only 4 for voter A, yet the results of part a show that voter A and voter B both have the same Banzhaf power index. Explain why it seems reasonable, in this voting system, to assign voters A and B the same Banzhaf power index.

▶ **Investigation**

Most voting systems are susceptible to fraudulent practices. In the following exercise, the Banzhaf power index is used to examine the power shift that occurs when a fraudulent practice is used.

Voter Intimidation and Tampering   Consider the voting system $\{3: 1, 1, 1, 1, 1\}$ with voters A, B, C, D, and E. The Banzhaf power index of each voter is 0.2.

1. Intimidation is used to force B to vote exactly as A. Thus the original $\{3: 1, 1, 1, 1, 1\}$ voting system becomes

$$\{3: 2, 0, 1, 1, 1\}$$

 The weight of A's vote is now 2 instead of 1. Does this mean that $BPI(A)$ has doubled from 0.2 to 0.4? Explain.

2. Voter C tampers with the voting software so that the original $\{3: 1, 1, 1, 1, 1\}$ voting system becomes

$$\{3: 1, 1, 2, 1, 1\}$$

 The weight of C's vote is now 2 instead of 1. Does this mean that $BPI(C)$ has doubled from 0.2 to 0.4? Explain.

Test Prep ⑨ Voting and Apportionment

The Hamilton Plan To apply the Hamilton plan, first compute the standard divisor for the total population. The standard divisor is determined by dividing the total population of all the states by the number of representatives to apportion.

$$\text{standard divisor} = \frac{\text{total population}}{\text{number of representatives to apportion}}$$

Then compute the standard quota for each state. The standard quota for a given state is the whole number part of the quotient of the state's population divided by the standard divisor.

Initially each state is apportioned the number of representatives given by its standard quota. However, if the sum of all the standard quotas is less than the total number of representatives to apportion, then an additional representative is assigned to the state that has the largest decimal remainder of the quotients formed by the population of each state and the standard divisor. This process is continued until the total number of representatives equals the number of representatives to apportion.

The Hamilton plan is susceptible to three paradoxes.

- *The Alabama Paradox* The Alabama paradox occurs when an increase in the number of representatives to be apportioned results in a loss of a representative for a state.
- *The Population Paradox* The population paradox occurs when a state loses a representative to another state, even though its population is increasing at a faster rate than that of the other state.
- *The New States Paradox* The new states paradox occurs when the addition of a new state results in a reduction in the number of representatives of another state.

The Hamilton plan does satisfy the quota rule.

- *The Quota Rule* The number of representatives apportioned to a state is the standard quota or more than the standard quota.

See **Example 1a** on page 413, and then try Exercises 1a, 2a, and 5 to 8 on pages 450 and 451.

The Jefferson Plan The Jefferson plan uses a modified standard divisor that is determined by trial and error, so that the sum of all the standard quotas equals the total number of representatives.

The Jefferson plan is not susceptible to the three paradoxes that the Hamilton plan may produce, but it may violate the quota rule.

See **Example 1b** on page 414, and then try Exercises 1b, 2b, and 10 on pages 450 and 451.

The Webster Method The Webster method of apportionment is similar to the Jefferson method except that quotas are rounded up when the decimal remainder is 0.5 or greater and quotas are rounded down when the decimal remainder is less than 0.5. The Webster method can violate the quota rule but rarely. The Webster method does not violate the Alabama, population, or new-states paradox.

See **Example 1c** on page 415, and then try Exercises 1c and 2c on page 450.

Relative Unfairness of an Apportionment The average constituency of a state is defined as the quotient of the population of the state and the number of representatives from that state, rounded to the nearest whole number. One method that is used to gauge the fairness of an apportionment between two states is to examine the average constituency of the states. The state with the smaller average constituency has the more favorable representation.

The absolute value of the difference between the average constituency of state A and the average constituency of state B is called the absolute unfairness of an apportionment.

See **Example 2** on page 420, and then try Exercises 3a and 3b on page 450.

The relative unfairness of an apportionment is calculated when a new representative is assigned to one of two states. The relative unfairness of an apportionment that gives a new representative to state A rather than state B is the quotient of the absolute unfairness of apportionment between the two states and the average constituency of state A (the state receiving the new representative).

Apportionment Principle When adding a new representative, the representative is assigned to the state in such a way as to give the smallest relative unfairness of apportionment.	See **Example 3** on page 420, and then try Exercises 3c and 4 on pages 450 and 451.
The Huntington-Hill Apportionment Method The Huntington-Hill method is implemented by calculating the Huntington-Hill number and applying the Huntington-Hill apportionment principle, which states that when there is a choice of adding one representative to a state, the representative should be added to the state with the greatest Huntington-Hill number. The Huntington-Hill number for state A is $$\frac{(P_A)^2}{a(a+1)}$$ where P_A is the population of state A and a is the current number of representatives from state A. The Huntington-Hill method is not susceptible to the three paradoxes that the Hamilton plan may produce, but it may violate the quota rule. • *Balinski-Young Impossibility Theorem* Any apportionment method either will violate the quota rule or will produce paradoxes such as the Alabama paradox.	See **Example 4** on page 422, and then try Exercise 11 on page 451.

Section 9.2   Introduction to Voting

Plurality Method of Voting Each voter votes for one candidate, and the candidate with the most votes wins. The winning candidate does not have to have a majority of the votes.	See **Example 1** on page 428, and then try Exercises 12a and 13a on pages 451 and 452.
Borda Count Method of Voting With n candidates in an election, each voter ranks the candidates by giving n points to the voter's first choice, $n - 1$ points to the voter's second choice, and so on, with the voter's least favorite choice receiving 1 point. The candidate that receives the most total points is the winner.	See **Example 2** on page 430, and then try Exercises 12b and 13b on pages 451 and 452.
Plurality with Elimination Method of Voting Eliminate the candidate with the smallest number of first-place votes. Retake a vote, keeping the same ranking preferences, and eliminate the candidate with the smallest number of first-place votes. Continue until only one candidate remains.	See **Example 3** on pages 432 and 433, and then try Exercises 14 and 15 on page 452.
Pairwise Comparison Method of Voting Compare each candidate head-to-head with every other candidate. Award 1 point for a win, 0.5 point for a tie, and 0 points for a loss. The candidate with the greatest number of points wins the election.	See **Example 4** on pages 433 and 434, and then try Exercises 16 and 17 on page 452.
Fairness Criteria 1. *Majority Criterion* The candidate who receives a majority of the first-place votes is the winner. 2. *Monotonicity Criterion* If candidate A wins an election, then candidate A will also win that election if the only change in the voters' preferences is that supporters of a different candidate change their votes to support candidate A. 3. *Condorcet Criterion* A candidate who wins all possible head-to-head matchups should win an election when all candidates appear on the ballot. 4. *Independence of Irrelevant Alternatives Criterion* If a candidate wins an election, the winner should remain the winner in any recount in which losing candidates withdraw from the race. • *Arrow's Impossibility Theorem* There is no voting method involving three or more choices that satisfies all four fairness criteria.	See **Examples 5** and **6** on pages 435 to 437, and then try Exercises 18 to 20 on page 452.

Section 9.3 Weighted Voting Systems

Weighted Voting System A weighted voting system is one in which some voters have more weight on the outcome of an election. The number of votes required to pass a measure is called the quota. The weight of a voter is the number of votes controlled by the voter. A weighted voting system of n voters is written $\{q: w_1, w_2, \ldots, w_n\}$, where q is the quota and w_1 through w_n represent the weights of each of the n voters.

In a weighted voting system, a coalition is a set of voters each of whom votes the same way, either for or against a resolution. A winning coalition is a set of voters the sum of whose votes is greater than or equal to the quota. A losing coalition is a set of voters the sum of whose votes is less than the quota. A voter who leaves a winning coalition and thereby turns it into a losing coalition is called a critical voter. The number of possible coalitions of n voters is $2^n - 1$.

A voter who has a weight that is greater than or equal to the quota is called a dictator. A voter who is never a critical voter is referred to as a dummy.

See **Example 1** on pages 442 to 443, and then try Exercises 21, 22, 27, and 28 on pages 452 and 453.

Banzhaf Power Index The Banzhaf power index of a voter v, symbolized by $BPI(v)$, is given by

$$BPI(v) = \frac{\text{number of times voter } v \text{ is a critical voter}}{\text{number of times any voter is a critical voter}}$$

See **Examples 2** and **3** on pages 444 and 445, and then try Exercises 23 to 26 and 29 on page 453.

Review Exercises

1. **Education** The following table shows the enrollments for the four divisions of a college. There are 50 new overhead projectors that are to be apportioned among the divisions based on the enrollments.

Division	Enrollment
Health	1280
Business	3425
Engineering	1968
Science	2936
Total	9609

   a. Use the Hamilton method to determine the number of projectors to be apportioned to each division.
   b. Use the Jefferson method to determine the number of projectors to be apportioned to each division.
   c. Use the Webster method to determine the number of projectors to be apportioned to each division.

2. **Airline Industry** The following table shows the numbers of ticket agents at five airports for a small airline company. The company has hired 35 new security employees who are to be apportioned among the airports based on the number of ticket agents at each airport.

Airport	Number of Ticket Agents
Newark	28
Cleveland	19
Chicago	34
Philadelphia	13
Detroit	16
Total	110

   a. Use the Hamilton method to apportion the new security employees among the airports.
   b. Use the Jefferson method to apportion the new security employees among the airports.
   c. Use the Webster method to apportion the new security employees among the airports.

3. **Airline Industry** The following table shows how the average constituency changes when two different airports, High Desert Airport and Eastlake Airport, add a new air traffic controller.

	High Desert Airport Average Constituency	Eastlake Airport Average Constituency
High Desert Airport Receives New Controller	297	326
Eastlake Airport Receives New Controller	302	253

   a. Determine the relative unfairness of an apportionment that gives a new air traffic controller to High Desert Airport rather than to Eastlake Airport. Round to the nearest thousandth.
   b. Determine the relative unfairness of an apportionment that gives a new air traffic controller to Eastlake Airport rather than to High Desert Airport. Round to the nearest thousandth.
   c. Using the apportionment principle, determine which airport should receive the new air traffic controller.

4. **Education** The following table shows the number of English professors and the number of students taking English

at two campuses of a state university. The university is planning to add a new English professor to one of the campuses. Use the apportionment principle to determine which campus should receive the new professor.

University Campus	Number of English Professors	Number of Students Taking English
Morena Valley	38	1437
West Keyes	46	1504

5. **Technology** A company has four offices. The president of the company uses the Hamilton method to apportion 66 new computer printers among the offices based on the number of employees at each office.

Office	A	B	C	D
Number of Employees	19	195	308	402
Apportionment of 66 Printers	1	14	22	29

a. If the number of printers to be apportioned by the Hamilton method increases from 66 to 67, will the Alabama paradox occur? Explain.

b. If the number of printers to be apportioned by the Hamilton method increases from 67 to 68, will the Alabama paradox occur? Explain.

6. **Automobile Sales** Consider the apportionment of 27 automobiles to the sales departments of a business with five regional centers labeled A, B, C, D, and E. The following table shows the Hamilton apportionment of the automobiles based on the number of sales personnel at each center.

Center	A	B	C	D	E
Number of Sales Personnel	31	108	70	329	49
Apportionment of 27 Automobiles	2	5	3	15	2

a. If the number of automobiles to be apportioned increases from 27 to 28, what will be the apportionment if the Hamilton method is used? Will the Alabama paradox occur? Explain.

b. If the number of automobiles to be apportioned using the Hamilton method increases from 28 to 29, will the Alabama paradox occur? Explain.

7. **Music Company** A music streaming company has offices in Los Angeles and Newark. The number of employees at each office is shown in the following table. There are 11 new computer file servers to be apportioned between the offices according to their numbers of employees.

Office	Los Angeles	Newark
Employees	1430	235

a. Use the Hamilton method to find each office's apportionment of file servers.

b. The corporation opens an additional office in Kansas City with 111 employees and decides to have a total of 12 file servers.

If the file servers are reapportioned using the Hamilton method, will the new states paradox occur? Explain.

8. **Building Inspectors** A city apportions 34 building inspectors among three regions according to their populations. The following table shows the present population of each region.

Region	A	B	C
Population	14,566	3321	29,988

a. Use the Hamilton method to apportion the inspectors.

b. After a year the regions have the following populations.

Region	A	B	C
Population	15,008	3424	30,109

Region A has an increase in population of 442, which is an increase of 3.03%. Region B has an increase in population of 103, which is an increase of 3.10%. Region C has an increase in population of 121, which is an increase of 0.40%. If the inspectors are reapportioned using the Hamilton method, will the population paradox occur? Explain.

9. Is the Hamilton apportionment method susceptible to the population paradox?

10. Is the Jefferson apportionment method susceptible to the new states paradox?

11. **Corporate Security** The Huntington-Hill apportionment method has been used to apportion 86 security guards among three corporate office buildings according to the number of employees at each building. See the following table.

Building	Number of Security Guards	Number of Employees
A	25	414
B	43	705
C	18	293

The corporation has decided to hire a new security guard.

a. Use the Huntington-Hill apportionment principle to determine to which building the new security guard should be assigned.

b. If another security guard is hired, bringing the total number of guards to 88, to which building should this guard be assigned?

12. **Essay Contest** Four finalists are competing in an essay contest. Judges have read and ranked each essay in order of preference. The results are shown in the following preference schedule.

	Rankings			
Crystal Kelley	3	2	2	1
Manuel Ortega	1	3	4	3
Peter Nisbet	2	4	1	2
Sue Toyama	4	1	3	4
Number of votes:	8	5	4	6

a. Using the plurality voting system, who is the winner of the essay contest?

b. Does this winner have a majority?

c. Use the Borda count method of voting to determine the winner of the essay contest.

13. Ski Club A campus ski club is trying to decide where to hold its winter break ski trip. The members of the club were surveyed and asked to rank five choices in order of preference. Their responses are tallied in the following table.

	Rankings				
Aspen	1	1	3	2	3
Copper Mountain	5	4	2	4	4
Powderhorn	3	2	5	1	5
Telluride	4	5	4	5	2
Vail	2	3	1	3	1
Number of votes:	14	8	11	18	12

a. Use the plurality method of voting to determine which resort the club should choose.

b. Use the Borda count method to choose the ski resort the club should visit.

14. Campus Election Four students are running for the activities director position on campus. Students were asked to rank the four candidates in order of preference. The results are shown in the following table.

	Rankings			
G. Reynolds	2	3	1	3
L. Hernandez	1	4	4	2
A. Kim	3	1	2	1
J. Schneider	4	2	3	4
Number of votes:	132	214	93	119

Use the plurality with elimination method to determine the winner of the election.

15. Consumer Preferences A group of consumers were surveyed about their favorite candy bars. Each participant was asked to rank four candy bars in order of preference. The results are given in the following table.

	Rankings				
Nestle Crunch	1	4	4	2	3
Snickers	2	1	2	4	1
Milky Way	3	2	1	1	4
Twix	4	3	3	3	2
Number of votes:	15	38	27	16	22

Use the plurality with elimination method to determine the group's favorite candy bar.

16. Use the pairwise comparison method of voting to choose the winner of the election in Exercise 14.

17. Use the pairwise comparison method of voting to choose the group's favorite candy bar in Exercise 15.

18. Student Elections Three students are running for student-body president. Students at the school were allowed to rank the candidates in order of preference. The results are shown in the following preference schedule.

	Rankings		
Logan Moro	3	2	1
Sohail Hassip	1	3	3
Jules Abreu	2	1	2
Number of votes:	246	213	24

a. Use the Borda count method to find the winner.

b. Find a candidate who wins all head-to-head comparisons.

c. Explain why the Condorcet criterion has been violated.

d. Who wins using the plurality voting system?

e. Explain why the majority criterion has been violated.

19. Student-body President In Exercise 18, suppose Logan Moro withdraws from the student-body president election.

a. Assuming voter preferences between the remaining two candidates remain the same, who will be elected student-body president using the Borda count method?

b. Explain why the independence of irrelevant alternatives criterion has been violated.

20. Scholarship Awards A scholarship committee must choose a winner from three finalists, Jean, Margaret, and Terry. Each member of the committee ranked the three finalists, and Margaret was selected using the plurality with elimination method. This vote was later declared invalid and a new vote was taken. All members voted using the same rankings except one, who changed her first choice from Terry to Margaret. This time Jean won the scholarship using the same voting method. Which fairness criterion was violated and why?

21. A weighted voting system for voters A, B, C, and D is given by {18: 12, 7, 6, 1}. The weight of voter A is 12, the weight of voter B is 7, the weight of voter C is 6, and the weight of voter D is 1.

a. What is the quota?

b. What is the weight of the coalition {A, C}?

c. Is {A, C} a winning coalition?

d. Which voters are critical voters in the coalition {A, C, D}?

e. How many coalitions can be formed?

f. How many coalitions consist of exactly two voters?

22. A weighted voting system for voters A, B, C, D, and E is given by {35: 29, 11, 8, 4, 2}. The weight of voter A is 29, the weight of voter B is 11, the weight of voter C is 8, the weight of voter D is 4, and the weight of voter E is 2.

a. What is the quota?

b. What is the weight of the coalition {A, D, E}?

c. Is {A, D, E} a winning coalition?

d. Which voters are critical voters in the coalition {A, C, D, E}?

e. How many coalitions can be formed?

f. How many coalitions consist of exactly two voters?

23. Calculate the Banzhaf power indices for voters A, B, and C in the weighted voting system {9: 6, 5, 3}.

24. Calculate the Banzhaf power indices for voters A, B, C, D, and E in the one-person, one-vote system {3: 1, 1, 1, 1, 1}.

25. Calculate the Banzhaf power indices for voters A, B, C, and D in the weighted voting system {31: 19, 15, 12, 10}. Round to the nearest hundredth.

26. Calculate the Banzhaf power indices for voters A, B, C, D, and E in the weighted voting system {35: 29, 11, 8, 4, 2}. Round to the nearest hundredth.

In Exercises 27 and 28, identify any dictator and all dummies for each weighted voting system.

27. Voters A, B, C, D, and E: {15: 15, 10, 2, 1, 1}

28. Voters A, B, C, and D: {28: 19, 6, 4, 2}

29. Four voters, A, B, C, and D, make decisions by using the weighted voting system {5: 4, 2, 1, 1}. In the event of a tie, a fifth voter, E, casts a vote to break the tie. For this voting scheme, determine the Banzhaf power index for each voter, including voter E.

Project 1

Apportioning the 1790 House of Representatives

The first apportionment of the House of Representatives, using the 1790 census, is given in the following table. This apportionment was calculated by using the Jefferson method. (See our companion site at Cengage.com for an Excel spreadsheet that will help with the computations.)

Apportionment Using the Jefferson Method, 1790

State	Population	Number of Representatives
Connecticut	236,841	7
Delaware	55,540	1
Georgia	70,835	2
Maryland	278,514	8
Massachusetts	475,327	14
Kentucky	68,705	2
New Hampshire	141,822	4
Vermont	85,533	2
New York	331,591	10
New Jersey	179,570	5
Pennsylvania	432,879	13
North Carolina	353,523	10
South Carolina	206,236	6
Virginia	630,560	19
Rhode Island	68,446	2

Source: U.S. Census Bureau

Project Exercises

1. Verify this apportionment using the Jefferson method. You will have to experiment with various modified standard divisors until you reach the given representation. See the Calculator Note on page 410.

2. Find the apportionment that would have resulted if the Hamilton method had been used.

3. Give each state one representative. Use the Huntington-Hill method with $a = 1$ to determine the state that receives the next representative. The Calculator Note on page 410 will help. With the populations stored in L1, enter 2nd L1 x² ÷ 2 STO 2nd L2 ENTER. Now scroll through L2 to find the largest number.

4. Find the apportionment that would have resulted if the Huntington-Hill method (the one used for the 2020 census) had been used in 1790. See our companion site at Cengage.com for a spreadsheet that will help with the calculations.

Project 2

Blocking Coalitions and the Banzhaf Power Index

The four members, A, B, C, and D, of an organization adopted the weighted voting system {6: 4, 3, 2, 1}. The following table shows the winning coalitions.

Winning Coalition	Number of Votes	Critical Voters
{A, B}	7	A, B
{A, C}	6	A, C
{A, B, C}	9	A
{A, B, D}	8	A, B
{A, C, D}	7	A, C
{B, C, D}	6	B, C, D
{A, B, C, D}	10	None

Using the Banzhaf power index, we have $BPI(A) = \frac{5}{12}$.

A **blocking coalition** is a group of voters who can prevent passage of a resolution. In this case, a critical voter is one who leaves a blocking coalition, thereby producing a coalition that is no longer capable of preventing the passage of a resolution. For the voting system given, we have

Blocking Coalition	Number of Votes	Number of Remaining Votes	Critical Voters
{A, B}	7	3	A, B
{A, C}	6	4	A, C
{A, D}	5	5	A, D
{B, C}	5	5	B, C
{A, B, C}	9	1	None
{A, B, D}	8	2	A
{A, C, D}	7	3	A
{B, C, D}	6	4	B, C

If we count the number of times A is a critical voter in a winning or blocking coalition, we find what is called the *Banzhaf index*. In this case, the Banzhaf index is 10 and we write $BI(A) = 10$. Using both the winning coalition and the blocking coalition tables, we find that $BI(B) = 6$, $BI(C) = 6$, and $BI(D) = 2$. This information can be used to create an alternative definition of the Banzhaf power index.

Take Note: The Banzhaf index is always a *whole number*, whereas the Banzhaf power index is often a fraction between 0 and 1.

Banzhaf Power Index—Alternative Definition

$$BPI(A) = \frac{BI(A)}{\text{sum of all Banzhaf indices for the voting system}}$$

Applying this definition to the voting system given, we have

$$BPI(A) = \frac{BI(A)}{BI(A) + BI(B) + BI(C) + BI(D)}$$
$$= \frac{10}{10 + 6 + 6 + 2} = \frac{10}{24} = \frac{5}{12}$$

Project Exercises

1. Using the data in Example 1 on page 438, list all blocking coalitions.

2. For the data in Example 1 on page 438, calculate the Banzhaf power indices for A, B, C, and D using the alternative definition.

3. Create a voting system with three members that is a dictatorship. Calculate the Banzhaf power index for each voter for this system using the alternative definition.

4. Create a voting system with four members in which one member has veto power. Calculate the Banzhaf power index for this system using the alternative definition.

5. Create a voting system with five members that satisfies the one-person, one-vote rule. Calculate the Banzhaf power index for this system using the alternative definition.

Math in Practice

Are there enough coffee shops? That is the question your boss at Waited Coffee is asking about Riverside. According to the latest census information, this city has a population of 98,545. From the business registrations from the city, there are 52 coffee shops.

Part 1

Your company, Waited Coffee, owns coffee shops in Uptown, Midtown, and Downtown. The number of employees in each neighborhood is shown in the following table. There are 8 managers to be apportioned among the neighborhoods.

Shops	Uptown	Midtown	Downtown
Employees	37	54	109

Use the Hamilton method to find each neighborhood's apportionment of managers.

Part 2

Your coffee company, Waited Coffee, wants to expand its coffee shop business by opening another shop in Riverside. You research information about Riverside's population and registered coffee shops. Use the Huntington-Hill apportionment principle to determine which neighborhood a new coffee shop should be opened.

Neighborhood	Number of Coffee Shops	Population
Uptown	8	19,644
Midtown	13	33,891
Downtown	31	45,010

Chapter

Circuits and Networks

10

Section	How This Is Used ...
10.1 Graphs and Euler Circuits	A graph in the context of this chapter is a set of points and lines connecting some or all of the points. This very general concept of graphs is used, for instance, to determine the most efficient ways to connect utility customers or to design routes between cities for airlines.
10.2 Weighted Graphs	Besides knowing how cities are connected, it may be important to know the distance between two cities. By adding a weight to a graph, the distance between the cities can be taken into consideration.
10.3 Planarity and Euler's Formula	Trying to draw a graph as a planar drawing has many practical applications. For instance, the design of circuit boards used in computers and other electronic components depends on wires connecting different components without touching each other elsewhere. Circuit boards can be very complex, but in effect they require that collections of wires be arranged in a planar drawing of a planar graph. If this is impossible, special connections can be installed that allow one wire to "jump" over another without touching.
10.4 Graph Coloring	Instead of placing a bunch of black dots for a graph, it may be useful to place colored dots. For instance, suppose you must schedule meetings for clubs. If some people belong to different clubs, say, A and B, scheduling may be an issue if both club meetings are scheduled at 9:00 A.M. Using graphs with colored dots can help resolve this problem.

Math in Practice:
Delivery Routes, p. 506

Applications of Circuits and Networks

There are many applications of circuits and networks, including finding optimal routes for sweeping streets, connecting computers on a network, analyzing the social interactions among people, designing computer circuits, and many others.

In the Investigation on page 468, you can work through an example of how "close" (called degrees of separation) a person is to you on a social network.

457

Section

10.1

Graphs and Euler Circuits

Learning Objectives:

*   Draw a graph.
*   Find an Euler circuit.
*   Find an Euler path.

Be Prepared: Prep Test 10 reviews concepts used in this chapter. This can be found in your Companion Site at Cengage.com.

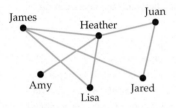

Figure 10.1

Take Note: Vertices are always clearly indicated with a "dot." Edges that intersect with no marked vertex are considered to cross over each other without touching.

Introduction to Graphs

Think of all the various connections we experience in our lives—friends are connected on social media, cities are connected by roads, computers are connected across the Internet. A branch of mathematics called **graph theory** illustrates and analyzes connections such as these.

For example, the diagram in Figure 10.1 could represent friends who are connected on social media. Each dot represents a person, and a line segment connecting two dots means that those two people are friends on social media. This type of diagram is called a *graph*.

> **Graph**
>
> A **graph** is a set of points called **vertices** and line segments or curves called **edges** that connect vertices.

Graphs can be used to represent many different scenarios. For instance, the two graphs in Figure 10.2 are the same graph as in Figure 10.1 but used in different contexts. In part (a), each vertex represents a baseball team, and an edge connecting two vertices might mean that the two teams played against each other during the current season. The graph in part (b) could be used to represent the flights available on a particular airline between a selection of cities; each vertex represents a city, and an edge connecting two cities means that there is a direct flight between the two cities.

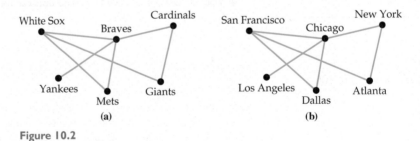

Figure 10.2

Example 1 **Constructing a Graph**

The following table lists five students at a college. An "X" indicates that the two students participate in the same study group this semester.

	Miao	Amber	Oscar	Leiko	Kayla
Miao	—	X		X	
Amber	X	—	X	X	
Oscar		X	—		X
Leiko	X	X		—	
Kayla			X		—

a. Draw a graph that represents this information where each vertex represents a student and an edge connects two vertices if the corresponding students study together.

b. Use your graph to answer the following questions: Which student is involved in the most study groups with the others? Which student has only one study group in common with the others? How many study groups does Leiko have in common with the others?

Solution

a. We draw five vertices (in any configuration we wish) to represent the five students, and connect vertices with edges according to the table.

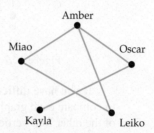

b. The vertex corresponding to Amber is connected to more edges than the others, so she is involved with more study groups (three) than the others. Kayla is the only student with one study group in common, as her vertex is the only one connected to just one edge. Leiko's vertex is connected to two edges, so she shares two study groups with the others.

•

In general, a graph can include vertices that are not joined to any edges, but all edges must begin and end at vertices. If two or more edges connect the same vertices, they are called **multiple edges**. If an edge begins and ends at the same vertex, it is called a **loop**.

A graph is called **connected** if any vertex can be reached from any other vertex by tracing along edges. (Essentially, the graph consists of one "piece.") A connected graph in which every possible edge is drawn between vertices (without any multiple edges) is called a **complete graph**. Several examples of graphs are shown in Figures 10.3, 10.4, 10.5, and 10.6.

Figure 10.3 This graph has five vertices but no edges. It is not connected.

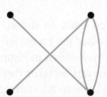

Figure 10.4 This is a connected graph that has a pair of multiple edges. Note that two edges cross in the center, but there is no vertex there. Unless a dot is drawn, the edges are considered to pass over each other without touching.

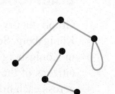

Figure 10.5 This graph is not connected; it consists of two different sections. It also contains a loop.

Figure 10.6 This is a complete graph with five vertices.

Note that it does not matter whether the edges are drawn straight or curved, and their lengths are not important. Nor is the placement of the vertices important. All that matters is which vertices are connected by edges.

Consequently, the three graphs shown in Figure 10.7 are considered **equivalent graphs** because the edges form the same connections of vertices in each graph.

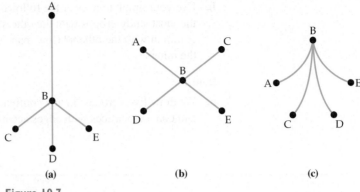

Figure 10.7

If you have difficulty seeing that these graphs are equivalent, use the labeled vertices to compare each graph. Note that in each case, vertex B has an edge connecting it to each of the other four vertices, and no other edges exist.

Example 2 **Equivalent Graphs**

Determine whether the following two graphs are equivalent.

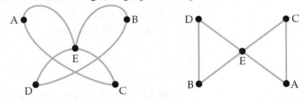

Solution

Despite the fact that the two graphs have different arrangements of vertices and edges, they are equivalent. To illustrate, we examine the edges of each graph. The first graph contains six edges; we can list them by indicating which two vertices they connect. The edges are AC, AE, BD, BE, CE, and DE. If we do the same for the second graph, we get the same six edges. Because the two graphs represent the same connections among the vertices, they are equivalent.

Take Note: The order in which the vertices of an edge are given is not important; AC and CA represent the same edge.

Euler Circuits

Graph theory is often traced back to the early eighteenth century to a city called Königsberg. Seven bridges crossed the Pregel River and connected four different land areas in Königsberg similar to the map drawn in Figure 10.8. A question that occurred to some citizens of the city at the time was whether they could take a stroll that would lead them across each bridge and return them to the starting point without traversing the same bridge twice. To solve the Königsberg bridges problem, we can represent the arrangement of land

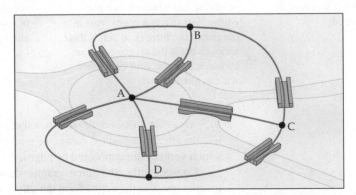

Figure 10.8

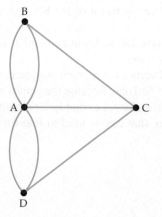

Figure 10.9

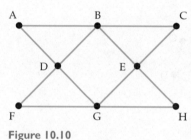

Figure 10.10

areas and bridges with a graph. Let each land area be represented by a vertex, and connect two vertices if there is a bridge spanning the corresponding land areas. Then the geographical configuration shown in Figure 10.8 becomes the graph shown in Figure 10.9.

In terms of a graph, the original problem can be stated as follows: Can we start at any vertex, move through each edge once (but not more than once), and return to the starting vertex? Again, try it with pencil and paper. Every attempt seems to end in failure.

Before we can examine how Euler proved this task impossible, we need to establish some terminology. A **path** in a graph can be thought of as a movement from one vertex to another by traversing edges. We can refer to our movement by vertex letters. For example, in the graph in Figure 10.9, one path would be A–B–A–C.

If a path ends at the same vertex at which it started, it is considered a **closed path**, or **circuit**. For the graph in Figure 10.10, the path A–D–F–G–E–B–A is a circuit because it begins and ends at the same vertex. The path A–D–F–G–E–H is not a circuit, as the path ends at a different vertex than the one it started at.

A circuit that uses every edge, but never uses the same edge twice, is called an **Euler circuit**. (The path may cross through vertices more than once.) The path B–D–F–G–H–E–C–B–A–D–G–E–B in Figure 10.10 is an Euler circuit. It begins and ends at the same vertex and uses each edge exactly once. (Trace the path with your pencil to verify!) The path A–B–C–E–H–G–E–B–D–A is not an Euler circuit: The path begins and ends at the same vertex but it does not use edges DF, DG, or FG. The path A–B–C–E–H–G–F–D–A–B–E–G–D–A begins and ends at A but uses edges AB and AD twice, so it is not an Euler circuit.

All of this relates to the Königsberg bridges problem in the following way: Finding a path that crosses each bridge exactly once and returns to the starting point is equivalent to finding an Euler circuit for the graph in Figure 10.9.

Euler essentially proved that the graph in Figure 10.9 could not have an Euler circuit. He accomplished this by examining the number of edges that met at each vertex. The number of edges that meet at a vertex is called the **degree** of a vertex. He made the observation that in order to complete the desired path, every time you approached a vertex you would then need to leave that vertex. If you traveled through that vertex again, you would again need an approaching edge and a departing edge. Thus for an Euler circuit to exist, the degree of every vertex would have to be an even number. Furthermore, he was able to show that any graph that has an even degree at every vertex must have an Euler circuit. Consequently, such graphs are called **Eulerian**.

Eulerian Graph Theorem

A connected graph is Eulerian if and only if every vertex of the graph is of even degree.

Example 3 **Identifying Eulerian Graphs**

Which of the following graphs has an Euler circuit?

a.

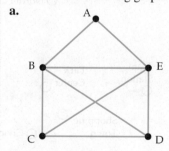

b.

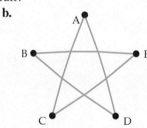

Solution

a. Vertices C and D are of odd degree. By the Eulerian graph theorem, the graph does not have an Euler circuit.

b. All vertices are of even degree. By the Eulerian graph theorem, the graph has an Euler circuit.

Take Note: For information on and examples of finding Eulerian circuits, search for "Fleury's algorithm" on the Internet.

Look again at Figure 10.9 on page 461, the graph representation of the Königsberg bridges. Because not every vertex is of even degree, we know by the Eulerian graph theorem that no Euler circuit exists. Consequently, it is not possible to begin and end at the same location near the river and cross each bridge exactly once.

The Eulerian graph theorem guarantees that when all vertices of a graph have an even degree, an Euler circuit exists, but it does not tell us how to find one. Because the graphs we will examine here are relatively small, we will rely on trial and error to find Euler circuits. There is a systematic method, called **Fleury's algorithm**, that can be used to find Euler circuits in graphs with large numbers of vertices.

| Example 4 | **Find an Euler Circuit** |

Determine whether the following graph is Eulerian. If it is, find an Euler circuit. If it is not, explain how you know.

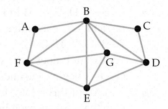

Solution

Each vertex is of even degree (2, 4, or 6), so by the Eulerian graph theorem, the graph is Eulerian. There are many possible Euler circuits in this graph. We do not have a formal method of locating one, but by trial and error, one Euler circuit is B–A–F–B–E–F–G–E–D–G–B–D–C–B.

| Example 5 | **An Application of Euler Circuits** |

The following subway map shows the tracks that subway trains traverse as well as the junctions where one can switch trains. Suppose an inspector needs to travel the full length of each track. Is it possible to plan a journey that traverses the tracks and returns to the starting point without traveling through any portion of a track more than once?

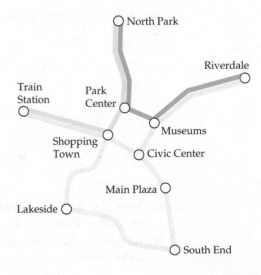

Solution

We can consider the subway map a graph, with a vertex at each junction. An edge represents a track that runs between two junctions. In order to find a travel route that does not traverse the same track twice, we need to find an Euler circuit in the graph. Note, however, that the vertex representing the Civic Center junction has a degree of 3. Because a vertex has an odd degree, the graph cannot be Eulerian, and it is impossible for the inspector not to travel at least one track twice.

●

Euler Paths

Perhaps the Königsberg bridges problem would have a solution if we did not need to return to the starting point. In this case, what we are looking for in Figure 10.4 on page 459 is a path (not necessarily a circuit) that uses every edge once and only once. We call such a path an **Euler path**. Euler showed that even with this relaxed condition, the bridge problem still was not solvable. The general result of his argument is given in the following theorem.

Take Note: Note that an Euler *path* does not require that we start and stop at the same vertex, whereas an Euler *circuit* does.

> **Euler Path Theorem**
>
> A connected graph contains an Euler path if and only if the graph has two vertices of odd degree with all other vertices of even degree. Furthermore, every Euler path must start at one of the vertices of odd degree and end at the other.

To see why this theorem is true, note that the only places at which an Euler path differs from an Euler circuit are the start and end vertices. If we never return to the starting vertex, only one edge meets there and the degree of the vertex is 1. If we do return, we cannot stop there. So we depart again, giving the vertex a degree of 3. Similarly, any return trip means that an additional two edges meet at the vertex. Thus the degree of the start vertex must be odd. By similar reasoning, the ending vertex must also be of odd degree. All other vertices, just as in the case of an Euler circuit, must be of even degree.

Example 6   **An Application of Euler Paths**

A photographer would like to travel across all of the roads shown on the following map. The photographer will rent a car that need not be returned to the same city, so the trip can begin in any city. Is it possible for the photographer to design a trip that traverses all of the roads exactly once?

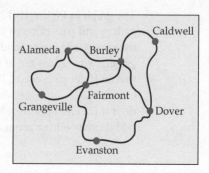

Solution

Looking at the map of roads as a graph, we see that a route that includes all of the roads but does not cover any road twice corresponds to an Euler path of the graph. Notice that only two vertices are of odd degree, the cities Alameda and Dover. Thus we know that an Euler path exists, and so it is possible for the photographer to plan a route that travels each road once. Because (abbreviating the cities) A and D are vertices of odd degree, the photographer must start at one of these cities. With a little experimentation, we find that one Euler path is A–B–C–D–B–F–A–G–F–E–D.

Example 7 An Application of Euler Paths

The following figure depicts the floor plan of an art gallery. Draw a graph that represents the floor plan, where vertices correspond to rooms and edges correspond to doorways. Is it possible to take a stroll that passes through every doorway without going through the same doorway twice? If so, does it matter whether we return to the starting point?

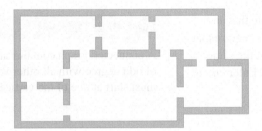

Solution

We can represent the floor plan by a graph if we let a vertex represent each room. Draw an edge between two vertices if there is a doorway between the two rooms, as shown in Figure 10.11.

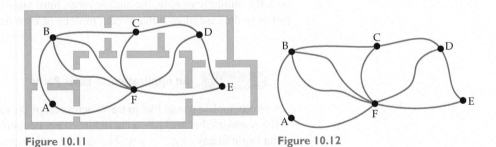

Figure 10.11   **Figure 10.12**

The graph in Figure 10.12 is equivalent to our floor plan. If we would like to tour the gallery and pass through every doorway once, we must find a path in our graph that uses every edge once (and no more). Thus we are looking for an Euler path. In the graph, two vertices are of odd degree and the others are of even degree. So we know that an Euler path exists, but not an Euler circuit. Therefore, we cannot pass through each doorway once and only once if we want to return to the starting point, but we can do it if we end up somewhere else. Furthermore, we know we must start at a vertex of odd degree—either room C or room D. By trial and error, one such path is C–B–F–B–A–F–E–D–C–F–D.

Exercise Set

Think About It

1. Is the following graph a complete graph?

2. Give an example of a connected graph and a disconnected graph.

3. What is the degree of a vertex?

4. What is a circuit in a graph?

5. What is an Euler circuit?

Preliminary Exercises

For Exercises P1 to P7, if necessary, use the referenced example following the exercise for assistance.

P1. The following table lists five mobile phone companies and indicates whether they have agreements to roam onto each other's networks. Draw a graph that represents this information, where each vertex represents a phone company and an edge connects two vertices if the corresponding companies have a roaming agreement. Then use the graph to answer the questions: Which phone company has roaming agreements with the most carriers? Which company can roam with only one other network? **Example 1**

	MobilePlus	TalkMore	SuperCell	Airwave	Lightning
MobilePlus	—	No	Yes	No	Yes
TalkMore	No	—	Yes	No	No
SuperCell	Yes	Yes	—	Yes	No
Airwave	No	No	Yes	—	Yes
Lightning	Yes	No	No	Yes	—

P2. Determine whether the following two graphs are equivalent. **Example 2**

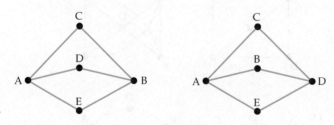

P3. Does the following graph have an Euler circuit? **Example 3**

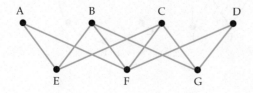

P4. Does the following graph have an Euler circuit? If so, find one. If not, explain why. **Example 4**

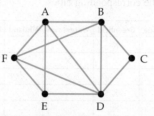

P5. Suppose the city of Königsberg had the arrangement of islands and bridges pictured here instead of the arrangement that we introduced previously. Would the citizens be able to complete a stroll across each bridge and return to their starting points without crossing the same bridge twice? **Example 5**

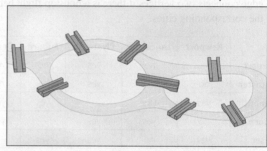

P6. A bicyclist wants to mountain bike through all the trails of a national park. A map of the park is shown. Because the bicyclist will be dropped off in the morning by friends and picked up in the evening, she does not have a preference where she begins and ends her ride. Is it possible for the cyclist to traverse all of the trails without repeating any portions of her trip? **Example 6**

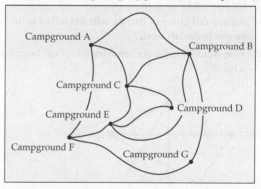

P7. Consider the following floor plan of a warehouse. Use a graph to represent the floor plan, and answer the following questions: Is it possible to walk through the warehouse so that you pass through every doorway once but not twice? Does it matter whether you return to the starting point? **Example 7**

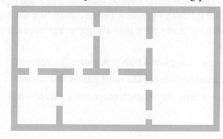

Section Exercises

1. Transportation An "X" in the following table indicates a direct train route between the corresponding cities. Draw a graph that represents this information, in which each vertex represents a city and an edge connects two vertices if there is a train route between the corresponding cities.

	Springfield	Riverside	Greenfield	Watertown	Midland	Newhope
Springfield	—		X		X	
Riverside		—		X	X	X
Greenfield	X		—	X	X	X
Watertown		X	X	—		
Midland	X	X	X		—	X
Newhope		X	X		X	—

2. Transportation The following table shows the nonstop flights offered by a small airline. Draw a graph that represents this information, where each vertex represents a city and an edge connects two vertices if there is a nonstop flight between the corresponding cities.

	Newport	Lancaster	Plymouth	Auburn	Dorset
Newport	—	no	yes	no	yes
Lancaster	no	—	yes	yes	no
Plymouth	yes	yes	—	yes	yes
Auburn	no	yes	yes	—	yes
Dorset	yes	no	yes	yes	—

3. Social Network A group of friends is represented by the following graph. An edge connecting two names means that the two friends have spoken to each other in the last week.

  a. Have John and Sigge talked to each other in the last week?

  b. How many of the friends in this group has Sen talked to in the last week?

  c. Among this group of friends, who has talked to the most people in the last week?

  d. Why would it not make sense for this graph to contain a loop?

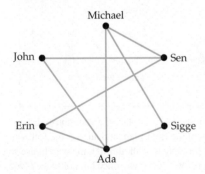

4. Baseball The local Little League baseball teams are represented by the following graph. An edge connecting two teams means that those teams have played a game against each other this season.

  a. Which team has played only one game this season?

  b. Which team has played the most games this season?

  c. Have any teams played each other twice this season?

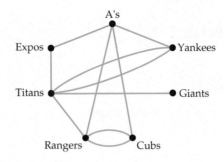

In Exercises 5 to 8, determine (a) the number of edges in the graph, (b) the number of vertices in the graph, (c) the number of vertices that are of odd degree, (d) whether the graph is connected, and (e) whether the graph is a complete graph.

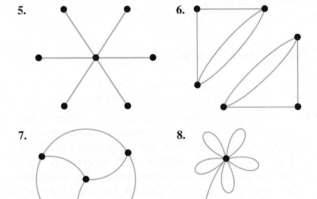

5.

6.

7.

8.

In Exercises 9 to 12, determine whether the two graphs are equivalent.

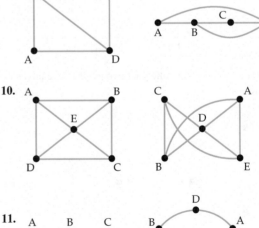

9.

10.

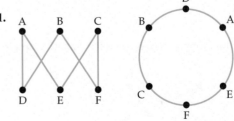

11.

12.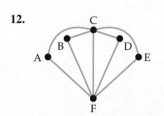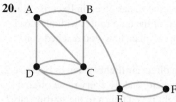

20.

13. Explain why the following two graphs cannot be equivalent.

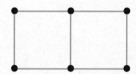

21.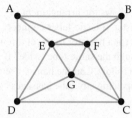

14. Label the vertices of the second graph so that it is equivalent to the first graph.

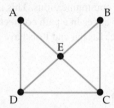

22.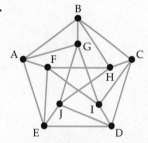

In Exercises 15 to 22, **(a)** determine whether the graph is Eulerian. If it is, find an Euler circuit. If it is not, explain why. **(b)** If the graph does not have an Euler circuit, does it have an Euler path? If so, find one. If not, explain why.

In Exercises 23 and 24, a map of a park is shown with bridges connecting islands in a river to the banks. **(a)** Represent the map as a graph. See Figures 10.8 and 10.9 on pages 460 and 461 for an example. **(b)** Is it possible to take a walk that crosses each bridge once and returns to the starting point without crossing any bridge twice? If not, can you do it if you do not end at the starting point? Explain how you know.

15. **16.**

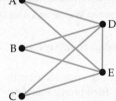 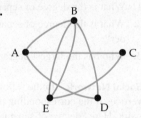

23.

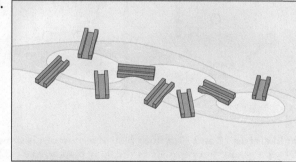

17. **18.**

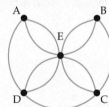 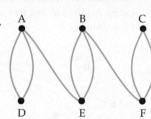

24.

19.

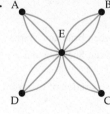

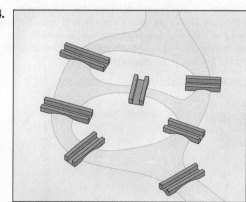

25. Transportation For the train routes given in Exercise 1, is it possible to travel along all of the train routes without traveling along any route twice? Explain how you reached your conclusion.

26. Transportation For the direct air flights given in Exercise 2, is it possible to start at one city and fly every route offered without repeating any flight if you return to the starting city? Explain how you reached your conclusion.

27. Pets The following diagram shows the arrangement of a Habitrail cage for a pet hamster. (Plastic tubes connect different cages.) Is it possible for a hamster to travel through every tube without going through the same tube twice? If so, find a route for the hamster to follow. Can the hamster return to its starting point without repeating any tube passages?

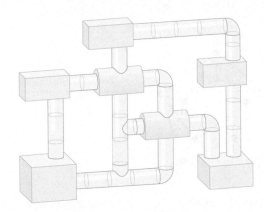

28. Transportation Consider the following subway map. Is it possible for a rider to travel the length of every subway route without repeating any segments? Justify your conclusion.

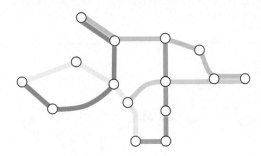

In Exercises 29 and 30, a floor plan of a museum is shown. Draw a graph that represents the floor plan, where each vertex represents a room and an edge connects two vertices if there is a doorway between the two rooms. Is it possible to walk through the museum and pass through each doorway without going through any doorway twice? Does it depend on whether you return to the room you started at? Justify your conclusion.

29.

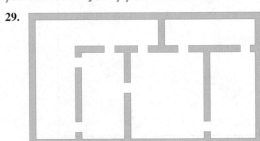

30.

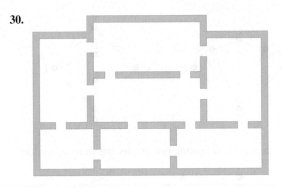

Investigations

Degrees of Separation In the following graph, an edge connects two vertices if the corresponding people have communicated by text message. Here we define the *degree of separation* between two individuals to be the minimum number of steps required to link one person to another through text message communications. This is equivalent to the (minimum) number of edges in a path connecting the corresponding vertices. For example, Karina and Lois have a degree of separation of 2.

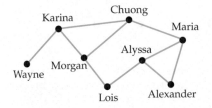

1. What is the degree of separation between Morgan and Maria?
2. What is the degree of separation between Wayne and Alexander?
3. Who has the largest degree of separation from Alyssa?

Social Network In the following graph, an edge connects two vertices if the corresponding people are friends on a social network. Here "degree of separation" refers to the minimum number of steps required to link two individuals through a social network of friendships.

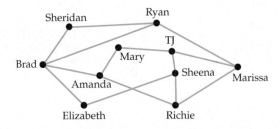

1. What is the degree of separation between Brad and Marissa?
2. What is the degree of separation between Sheridan and Sheena?
3. List the individuals with a degree of separation of two from Brad.
4. Do any two individuals have a degree of separation of more than three?

Weighted Graphs

Learning Objectives:

- Find a Hamiltonian circuit.
- Find a Hamiltonian circuit for a weighted graph.
- Apply the greedy algorithm to find a Hamiltonian circuit.
- Apply the edge-picking algorithm.
- Solve applications with weighted graphs.

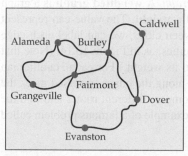

Figure 10.13

Hamiltonian Circuits

In Section 10.1 we looked at paths that use every edge of a graph exactly once. In some situations we may be more interested in paths that visit each vertex once, regardless of whether all edges are used or not.

For instance, Figure 10.13 shows the map of cities from Example 6 in Section 10.1. If our priority is to visit each city, we could travel along the route A–B–C–D–E–F–G–A (abbreviating the cities). This path visits each vertex once and returns to the starting vertex without visiting any vertex twice. This type of path is called a *Hamiltonian circuit*.

Hamiltonian Circuit

A **Hamiltonian circuit** is a path that begins and ends at the same vertex and passes through each vertex of a graph exactly once. A graph that contains a Hamiltonian circuit is called **Hamiltonian**.

Unfortunately we do not have a straightforward criterion to guarantee that a graph is Hamiltonian, but we do have the following helpful theorem.

Dirac's Theorem

Consider a connected graph with at least three vertices and no multiple edges. Let n be the number of vertices in the graph. If every vertex has a degree of at least $n/2$, then the graph must be Hamiltonian.

We must be careful, however; if our graph does not meet the requirements of this theorem, it still might be Hamiltonian. Dirac's theorem does not help us in this case.

Example 1   **Apply Dirac's Theorem**

The following graph shows the available flights of a small airline. An edge between two vertices in the graph means that the airline has direct flights between the two corresponding cities. Apply Dirac's theorem to verify that the following graph is Hamiltonian. Then find a Hamiltonian circuit. What does the Hamiltonian circuit represent in terms of flights?

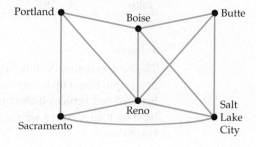

Solution

There are six vertices in the graph, so $n = 6$, and every vertex has a degree of at least $n/2 = 3$. By Dirac's theorem, the graph is Hamiltonian. This means that the graph contains a circuit that visits each vertex once and returns to the starting vertex without visiting any vertex twice. By trial and error, one Hamiltonian circuit is Portland–Boise–Butte–Salt Lake City–Reno–Sacramento–Portland, which represents a sequence of flights that visits each city and returns to the starting city without visiting any city twice.

Weighted Graphs

A Hamiltonian circuit can identify a route that visits all of the cities represented on a graph, as in Figure 10.13, but there are often a number of different paths we could use. If we are concerned with the distances we must travel between cities, chances are that some of the routes will involve a longer total distance than others. We might be interested in finding the route that minimizes the total number of miles traveled.

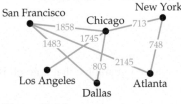

Figure 10.14

We can represent this situation with a *weighted graph*. A **weighted graph** is a graph in which each edge is associated with a value, called a **weight**. The value can represent any quantity we desire. In the case of distances between cities, we can label each edge with the number of miles between the corresponding cities, as in Figure 10.14. (Note that the length of an edge does not necessarily correlate to its weight.) For each Hamiltonian circuit in the weighted graph, the sum of the weights along the edges traversed gives the total distance traveled along that route. We can then compare different routes and find the one that requires the shortest total distance. This is an example of a famous problem called the *traveling salesman problem*.

Example 2   **Find Hamiltonian Circuits in a Weighted Graph**

The following table lists the distances in miles between six popular cities that a particular airline flies to. Suppose a traveler would like to start in Chicago, visit the other five cities this airline flies to, and return to Chicago. Find three different routes that the traveler could follow, and find the total distance flown for each route.

	Chicago	New York	Washington, D.C.	Philadelphia	Atlanta	Dallas
Chicago	—	713	597	665	585	803
New York	713	—	No flights	No flights	748	1374
Washington, D.C.	597	No flights	—	No flights	544	1185
Philadelphia	665	No flights	No flights	—	670	1299
Atlanta	585	748	544	670	—	No flights
Dallas	803	1374	1185	1299	No flights	—

Solution

The various options will be simpler to analyze if we first organize the information in a graph. Begin by letting each city be represented by a vertex. Draw an edge between two vertices if there is a flight between the corresponding cities, and label each edge with a weight that represents the number of miles between the two cities.

Take Note: Remember that the placement of the vertices is not important. There are many equivalent ways in which to draw the graph in Example 2.

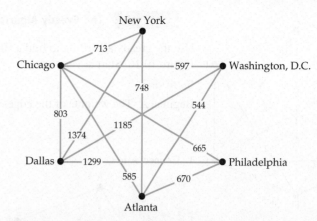

A route that visits each city just once corresponds to a Hamiltonian circuit. Beginning at Chicago, one such circuit is Chicago–New York–Dallas–Philadelphia–Atlanta–Washington, D.C.–Chicago. By adding the weights of each edge in the circuit, we see that the total number of miles traveled is

$$713 + 1374 + 1299 + 670 + 544 + 597 = 5197$$

By trial and error, we can identify two additional routes. One is Chicago–Philadelphia–Dallas–Washington, D.C.–Atlanta–New York–Chicago. The total weight of the circuit is

$$665 + 1299 + 1185 + 544 + 748 + 713 = 5154$$

A third route is Chicago–Washington, D.C.–Dallas–New York–Atlanta–Philadelphia–Chicago. The total mileage is

$$597 + 1185 + 1374 + 748 + 670 + 665 = 5239$$

Algorithms in Complete Graphs

In Example 2, the second route we found represented the smallest total distance out of the three options. Is there a way we can find the very best route to take? It turns out that this is no easy task. One method is to list every possible Hamiltonian circuit, compute the total weight of each one, and choose the smallest total weight. Unfortunately, the number of different possible circuits can be extremely large. For instance, the graph shown in Figure 10.15, with only 6 vertices, has 60 unique Hamiltonian circuits. If we have a graph with 12 vertices, and every vertex is connected to every other by an edge, there are almost 20 million different Hamiltonian circuits! Even by using computers, it can take too long to investigate each and every possibility.

Unfortunately, there is no known shortcut for finding the optimal Hamiltonian circuit in a weighted graph. There are, however, two *algorithms*, the greedy algorithm and the edge-picking algorithm, that can be used to find a pretty good solution. Both of these algorithms apply only to **complete graphs**—graphs in which every possible edge is drawn between vertices (without any multiple edges). For instance, the graph in Figure 10.15 is a complete graph with six vertices. The circuits found by the algorithms are not guaranteed to have the smallest total weight possible, but they are often better than you would find by trial and error.

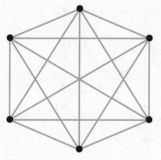

Figure 10.15

The Greedy Algorithm

1. Choose a vertex to start at, then travel along the connected edge that has the smallest weight. (If two or more edges have the same weight, pick any one.)
2. After arriving at the next vertex, travel along the edge of smallest weight that connects to a vertex not yet visited. Continue this process until you have visited all vertices.
3. Return to the starting vertex.

The **greedy algorithm** is so called because it has us choose the "cheapest" option at every chance we get.

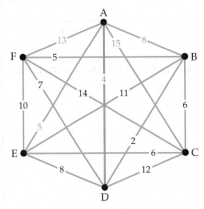

Figure 10.16

| Example 3 | **The Greedy Algorithm** |

Use the greedy algorithm to find a Hamiltonian circuit in the weighted graph shown in Figure 10.16. Start at vertex A.

Solution

Begin at A. The weights of the edges from A are 13, 5, 4, 15, and 8. The smallest is 4.

1. Connect A to D.

2. At D, the edge with the smallest weight is DB. Connect D to B.

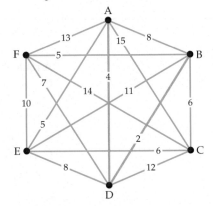

3. At B, the edge with the smallest weight is BF. Connect B to F.

4. At F, the edge with the smallest weight, 7, is FD. However, D has already been visited. Choose the next smallest weight, edge FE. Connect F to E.

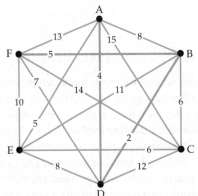

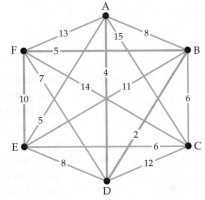

5. At E, the edge with the smallest weight whose vertex has not been visited is C. Connect E to C.

6. All vertices have been visited, so we are at step 3 of the algorithm. We return to the starting vertex by connecting C to A.

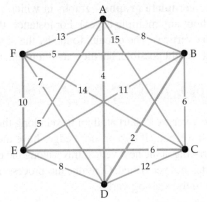

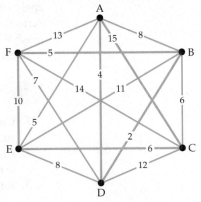

The Hamiltonian circuit is A–D–B–F–E–C–A. The weight of the circuit is

$$4 + 2 + 5 + 10 + 6 + 15 = 42$$

Take Note: A Hamiltonian circuit will always have exactly two edges at each vertex. In step 2 we are warned not to mark an edge that would allow three edges to meet at one vertex.

The Edge-Picking Algorithm

1. Mark the edge of smallest weight in the graph. (If two or more edges have the same weight, pick any one.)
2. Mark the edge of next smallest weight in the graph, as long as it does not complete a circuit and does not add a third marked edge to a single vertex.
3. Continue this process until you can no longer mark any edges. Then mark the final edge that completes the Hamiltonian circuit.

Example 4 **The Edge-Picking Algorithm**

Use the edge-picking algorithm to find a Hamiltonian circuit in Figure 10.16 (reprinted at the left).

Solution

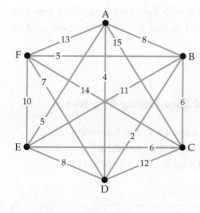

1. We first highlight the edge of smallest weight, namely BD with weight 2.

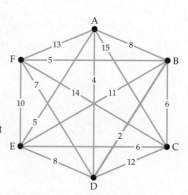

2. The edge of next smallest weight is AD with weight 4.

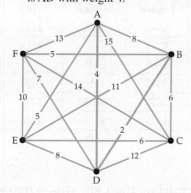

3. The next smallest weight is 5, which appears twice, with edges AE and FB. We can mark both of them.

4. There are two edges of weight 6 (the next smallest weight), BC and EC. We cannot use BC because it would add a third marked edge to vertex B. We mark edge EC.

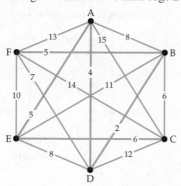

5. We are now at step 3 of the algorithm; any edge we mark will either complete a circuit or add a third edge to a vertex. So we mark the final edge to complete the Hamiltonian circuit, edge FC.

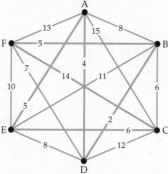

Beginning at vertex A, the Hamiltonian circuit is A–D–B–F–C–E–A. (In the reverse direction, an equivalent circuit is A–E–C–F–B–D–A.) The total weight of the circuit is

$$4 + 2 + 5 + 14 + 6 + 5 = 36$$

Take Note: A Hamiltonian circuit forms a complete loop, so we can follow along the circuit starting at any of the vertices. We can also reverse the direction in which we follow the circuit.

Note in Examples 3 and 4 that the two algorithms gave different Hamiltonian circuits, and in this case the edge-picking algorithm gave the more efficient route. Is this the best route? We mentioned before that the algorithms are helpful but that there is no known efficient method for finding the very best circuit. In fact, a third Hamiltonian circuit, A–D–F–B–C–E–A in Figure 10.16, has a total weight of 33, which is smaller than the weights of both routes given by the algorithms.

Applications of Weighted Graphs

In Example 2, we examined distances between cities. This is just one example of a weighted graph; the weight of an edge can be used to represent any quantity we like. For example, a traveler might be more interested in the cost of flights than the time or distance between cities. If we labeled each edge of the graph in Example 2 with the cost of traveling between the two cities, the total weight of a Hamiltonian circuit would be the total travel cost of the trip.

Example 5 **An Application of the Greedy and Edge-Picking Algorithms**

The cost of flying between various European cities is shown in the following table. Use both the greedy algorithm and the edge-picking algorithm to find a low-cost route that visits each city just once and starts and ends in London. Which route is more economical?

	London, England	Berlin, Germany	Paris, France	Rome, Italy	Madrid, Spain	Vienna, Austria
London, England	—	$325	$160	$280	$250	$425
Berlin, Germany	$325	—	$415	$550	$675	$375
Paris, France	$160	$415	—	$495	$215	$545
Rome, Italy	$280	$550	$495	—	$380	$480
Madrid, Spain	$250	$675	$215	$380	—	$730
Vienna, Austria	$425	$375	$545	$480	$730	—

Solution

First we draw a weighted graph with vertices representing the cities and each edge labeled with the price of the flight between the corresponding cities.

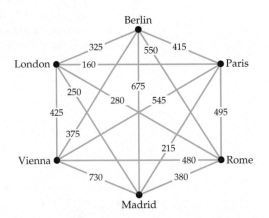

To use the greedy algorithm, start at London and travel along the edge with the smallest weight, 160, to Paris. The edge of smallest weight leaving Paris is the edge to Madrid. From Madrid, the edge of smallest weight (that we have not already traversed) is the edge to London, of weight 250. However, we cannot use this edge, because it would bring us to a city we have already seen. We can take the next-smallest-weight edge to Rome. We cannot yet return to London, so the next available edge is to Vienna, then to Berlin, and finally back to London.

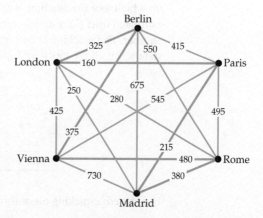

The total weight of the edges, and thus the total airfare for the trip, is

$$160 + 215 + 380 + 480 + 375 + 325 = \$1935$$

If we use the edge-picking algorithm, the edges with the smallest weights that we can highlight are London–Paris and Madrid–Paris. The edge of next smallest weight has a weight of 250, but we cannot use this edge because it would complete a circuit. We can take the edge of next smallest weight, 280, from London to Rome. We cannot take the edge of next smallest weight, 325, because it would add a third edge to the London vertex, but we can take the edge Vienna–Berlin of weight 375. We must skip the edges of weights 380, 415, and 425, but we can take the edge of weight 480, which is the Vienna–Rome edge. There are no more edges we can mark that will meet the requirements of the algorithm, so we mark the last edge to complete the circuit, Berlin–Madrid.

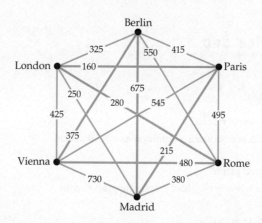

The resulting route is London–Paris–Madrid–Berlin–Vienna–Rome–London, for a total cost of

$$160 + 215 + 675 + 375 + 480 + 280 = \$2185$$

(We could also travel this route in the reverse order.)

A wide variety of problems are actually traveling salesman problems in disguise and can be analyzed using the algorithms from this section.

| Example 6 | **An Application of the Edge-Picking Algorithm** |

A toolmaker needs to use one machine to create four different tools. The toolmaker needs to make adjustments to the machine before starting each different tool. However, since the tools have parts in common, the amount of adjustment time required depends on which tool the machine was previously used to create. The following table lists the estimated time (in minutes) required to adjust the machine from making one tool to another. The machine is currently configured for tool A and should be returned to that state when all the tools are finished.

	Tool A	Tool B	Tool C	Tool D
Tool A	—	25	6	32
Tool B	25	—	18	9
Tool C	6	18	—	15
Tool D	32	9	15	—

Use the edge-picking algorithm to determine a sequence for creating the tools.

Solution

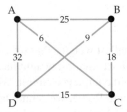

Draw a weighted graph in which each vertex represents a tool configuration of the machine and the weight of each edge is the number of minutes required to adjust the machine from one tool to another.

Using the edge-picking algorithm, we first choose edge AC, of weight 6. The next smallest weight is 9, or edge DB, followed by edge DC of weight 15. We end by completing the circuit with edge AB. The final circuit, of total weight 25 + 9 + 15 + 6 = 55, is A–B–D–C–A. So the machine starts with tool A, is reconfigured for tool B, then for tool D, then for tool C, and finally is returned to the settings used for tool A. (Note that we can equivalently follow this sequence in the reverse order: from tool A to C, to D, to B, and back to A.)

10.2 Exercise Set

Think About It

1. What is a Hamiltonian circuit?
2. What is the difference between a Hamiltonian circuit and an Euler circuit?
3. What is a weighted graph?
4. Give an instance where a weighted graph is appropriate for the situation.
5. What is a complete graph?

Preliminary Exercises

For Exercises P1 to P6, if necessary, use the referenced example following the exercise for assistance.

P1. A large law firm has offices in seven major cities. The firm has overnight document deliveries scheduled every day between certain offices. In the following graph, an edge between vertices indicates that there is delivery service between the corresponding offices. Use Dirac's theorem to answer the following question: Using the law firm's existing delivery service, is it possible to route a document to all the offices and return the document to its originating office without sending it through the same office twice? **Example 1**

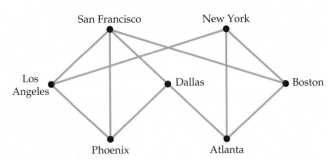

P2. A tourist visiting San Francisco is staying at a hotel near the Moscone Center. The tourist would like to visit five locations by bus tomorrow and then return to the hotel. The number of minutes spent traveling by bus between locations is given in the following table. (N/A in the table indicates that no convenient bus route is available.) Find two different routes for the tourist to follow and compare the total travel times. **Example 2**

	Moscone Center	Civic Center	Union Square	Embarcadero Plaza	Fisherman's Wharf	Coit Tower
Moscone Center	—	18	6	22	N/A	N/A
Civic Center	18	—	14	N/A	33	N/A
Union Square	6	14	—	24	28	36
Embarcadero Plaza	22	N/A	24	—	N/A	18
Fisherman's Wharf	N/A	33	28	N/A	—	14
Coit Tower	N/A	N/A	36	18	14	—

P3. Use the greedy algorithm to find a Hamiltonian circuit starting at vertex A in the following weighted graph. **Example 3**

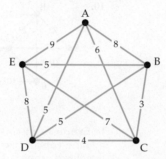

P4. Use the edge-picking algorithm to find a Hamiltonian circuit in the weighted graph in Preliminary Exercise P3. **Example 4**

P5. Susan needs to mail a package at the post office, pick up several items at the grocery store, drop off clothes at the dry cleaners, and make a deposit at her bank. The estimated driving time, in minutes, between each of these locations is given in the following table.

	Home	Post Office	Grocery Store	Dry Cleaners	Bank
Home	—	14	12	20	23
Post Office	14	—	8	12	21
Grocery Store	12	8	—	17	11
Dry Cleaners	20	12	17	—	18
Bank	23	21	11	18	—

Use both of the algorithms from this section to design routes for Susan to follow that will help minimize her total driving time. Assume she must start from home and return home when her errands are done. **Example 5**

P6. Businesses often network their various computers. One option is to run cables from a central hub to each computer individually; another is to connect one computer to the next, and that one to the next, and so on until you return to the first computer. Thus the computers are all connected in a large loop. Suppose a company wishes to use the latter method, and the lengths of cable (in feet) required between computers are given in the following table.

	Computer A	Computer B	Computer C	Computer D	Computer E	Computer F	Computer G
Computer A	—	43	25	6	28	30	45
Computer B	43	—	26	40	37	22	25
Computer C	25	26	—	20	52	8	50
Computer D	6	40	20	—	30	24	45
Computer E	28	37	52	30	—	49	20
Computer F	30	22	8	24	49	—	41
Computer G	45	25	50	45	20	41	—

Use the edge-picking algorithm to determine how the computers should be networked if the business wishes to use the smallest amount of cable possible. **Example 6**

Section Exercises

In Exercises 1 to 4, use Dirac's theorem to verify that the graph is Hamiltonian. Then find a Hamiltonian circuit.

1.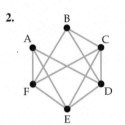

2.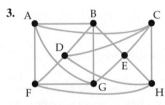

3.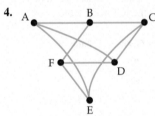

4.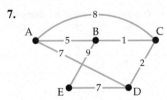

5. Transportation For the train routes given in Exercise 1 of Section 10.1, find a route that visits each city and returns to the starting city without visiting any city twice.

6. Transportation For the direct air flights given in Exercise 2 of Section 10.1, find a route that visits each city and returns to the starting city without visiting any city twice.

In Exercises 7 to 10, use trial and error to find two Hamiltonian circuits of different total weights, starting at vertex **A** in the weighted graph. Compute the total weight of each circuit.

7.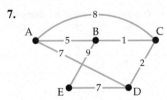

8.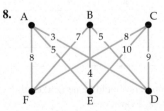

9.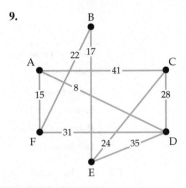

10.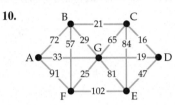

In Exercises 11 to 14, use the greedy algorithm to find a Hamiltonian circuit starting at vertex **A** in the weighted graph.

11.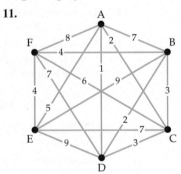

12.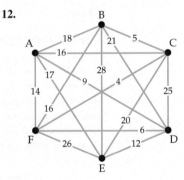

13.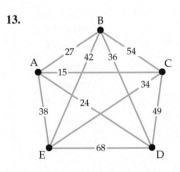

14.

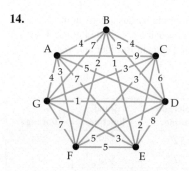

In Exercises 15 to 18, use the edge-picking algorithm to find a Hamiltonian circuit in the indicated graph.

15. Graph in Exercise 11

16. Graph in Exercise 12

17. Graph in Exercise 13

18. Graph in Exercise 14

19. Travel   A company representative lives in Louisville, Kentucky, and needs to visit offices in five different Indiana cities over the next few days. The representative wants to drive between cities and return to Louisville at the end of the trip. The estimated driving times, in hours, between cities are given in the following table. Represent the driving times by a weighted graph. Use the greedy algorithm to design an efficient route for the representative to follow.

	Louisville	Bloomington	Fort Wayne	Indianapolis	Lafayette	Evansville
Louisville	—	3.6	6.4	3.2	4.9	3.1
Bloomington	3.6	—	4.5	1.3	2.4	3.4
Fort Wayne	6.4	4.5	—	3.3	3.0	8.0
Indianapolis	3.2	1.3	3.3	—	1.5	4.6
Lafayette	4.9	2.4	3.0	1.5	—	5.0
Evansville	3.1	3.4	8.0	4.6	5.0	—

20. Travel   A resident of Toronto, Canada, and would like to visit four other Canadian cities by train. The visitor wants to go from one city to the next and return to Toronto while minimizing the total travel distance. The distances between cities, in kilometers, are given in the following table. Represent the distances between the cities using a weighted graph. Use the greedy algorithm to plan a route for this resident.

	Toronto	Kingston	Niagara Falls	Ottawa	Windsor
Toronto	—	259	142	423	381
Kingston	259	—	397	174	623
Niagara Falls	142	397	—	562	402
Ottawa	423	174	562	—	787
Windsor	381	623	402	787	—

21. Travel   Use the edge-picking algorithm to design a route for the company representative in Exercise 19.

22. Travel   Use the edge-picking algorithm to design a route for the tourist in Exercise 20.

23. Travel   Nicole wants to tour Asia. She will start and end her journey in Tokyo and visit Hong Kong, Bangkok, Seoul, and Beijing. The airfares available to her between cities are given in the following table. Draw a weighted graph that represents the travel costs between cities and use the greedy algorithm to find a low-cost route.

	Tokyo	Hong Kong	Bangkok	Seoul	Beijing
Tokyo	—	$845	$1275	$470	$880
Hong Kong	$845	—	$320	$515	$340
Bangkok	$1275	$320	—	$520	$365
Seoul	$470	$515	$520	—	$225
Beijing	$880	$340	$365	$225	—

24. **Travel** The prices for traveling between five cities in Colorado by bus are given in the following table. Represent the travel costs between cities using a weighted graph. Use the greedy algorithm to find a low-cost route that starts and ends in Boulder and visits each city.

	Boulder	Denver	Colorado Springs	Grand Junction	Durango
Boulder	—	$16	$25	$49	$74
Denver	$16	—	$22	$45	$72
Colorado Springs	$25	$22	—	$58	$59
Grand Junction	$49	$45	$58	—	$32
Durango	$74	$72	$59	$32	—

25. **Travel** Use the edge-picking algorithm to find a low-cost route for the traveler in Exercise 23.

26. **Travel** Use the edge-picking algorithm to find a low-cost bus route in Exercise 24.

27. **Route Planning** Brian needs to visit the pet store, the shopping mall, the local farmers market, and the pharmacy. His estimated driving times (in minutes) between the locations are given in the following table. Use the greedy algorithm and the edge-picking algorithm to find two possible routes, starting and ending at home, that will help Brian minimize his total travel time.

	Home	Pet Store	Shopping Mall	Farmers Market	Pharmacy
Home	—	18	27	15	8
Pet Store	18	—	24	22	10
Shopping Mall	27	24	—	20	32
Farmers Market	15	22	20	—	22
Pharmacy	8	10	32	22	—

28. **Route Planning** A bike messenger needs to deliver packages to five different buildings and return to the courier company. The estimated biking times (in minutes) between the buildings are given in the following table. Use the greedy algorithm and the edge-picking algorithm to find two possible routes for the messenger to follow that will help minimize the total travel time.

	Courier Company	Prudential Building	Bank of America Building	Imperial Bank Building	GE Tower	Design Center
Courier Company	—	10	8	15	12	17
Prudential Building	10	—	10	6	9	8
Bank of America Building	8	10	—	7	18	20
Imperial Bank Building	15	6	7	—	22	16
GE Tower	12	9	18	22	—	5
Design Center	17	8	20	16	5	—

29. **Scheduling** A research company has a large supercomputer that is used by different teams for a variety of computational tasks. In between each task, the software must be reconfigured. The time required depends on which tasks follow which, because some settings are shared by different tasks. The times (in minutes) required to reconfigure the machine from one task to another are given in the following table. Use the greedy algorithm and the edge-picking

algorithm to find time-efficient sequences in which to assign the tasks to the computer. The software configuration must start and end in the home state.

	Home State	Task A	Task B	Task C	Task D
Home State	—	35	15	40	27
Task A	35	—	30	18	25
Task B	15	30	—	35	16
Task C	40	18	35	—	32
Task D	27	25	16	32	—

30. **Computer Networks** A small office wishes to network its six computers in one large loop (see Example 6 on page 476). The lengths of cable, in meters, required between machines are given in the following table. Use the edge-picking algorithm to find an efficient cable configuration in which to network the computers.

	Computer A	Computer B	Computer C	Computer D	Computer E	Computer F
Computer A	—	10	22	9	15	8
Computer B	10	—	12	14	16	5
Computer C	22	12	—	14	9	16
Computer D	9	14	14	—	7	15
Computer E	15	16	9	7	—	13
Computer F	8	5	16	15	13	—

Investigations

1. **Route Planning** A security officer patrolling a city neighborhood needs to drive every street each night. The officer has drawn a graph to represent the neighborhood in which the edges represent the streets and the vertices correspond to street intersections. Would the most efficient way to drive the streets correspond to an Euler circuit, a Hamiltonian circuit, or neither? (The officer must return to the starting location when finished.) Explain your answer.

2. **Route Planning** A city engineer needs to inspect the traffic signs at each street intersection of a neighborhood. The engineer

has drawn a graph to represent the neighborhood, where the edges represent the streets and the vertices correspond to street intersections. Would the most efficient route to drive correspond to an Euler circuit, a Hamiltonian circuit, or neither? (The engineer must return to the starting location when finished.) Explain your answer.

3. **a.** Draw a connected graph with six vertices that has no Euler circuits and no Hamiltonian circuits.

   b. Draw a graph with six vertices that has a Hamiltonian circuit but no Euler circuits.

   c. Draw a graph with five vertices that has an Euler circuit but no Hamiltonian circuits.

Planarity and Euler's Formula

Learning Objectives:

* Draw a planar graph.
* Identify a nonplanar graph.
* Apply Euler's formula.

Planarity

A puzzle that was posed some time ago goes something like this: Three utility companies each need to run pipes to three houses. Can they do so without crossing over each other's pipes at any point? The puzzle is illustrated in Figure 10.17. Go ahead and try to draw pipes connecting each utility company to each house without letting any pipes cross over each other.

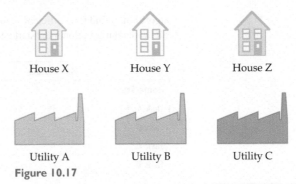

Figure 10.17

One way to approach the puzzle is to express the situation in terms of a graph. Each of the houses and utility companies will be represented by a vertex, and we will draw an edge between two vertices if a pipe needs to run from one building to the other. If we were not worried about pipes crossing, we could easily draw a solution, as in Figure 10.18.

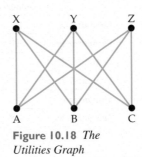

Figure 10.18 *The Utilities Graph*

To solve the puzzle, we need to draw an equivalent graph in which no edges cross over each other. If this is possible, the graph is called a *planar graph*.

Planar Graph

A **planar graph** is a graph that can be drawn so that no edges intersect each other (except at vertices).

If the graph is drawn in such a way that no edges cross, we say that we have a **planar drawing** of the graph.

Example 1 **Identify a Planar Graph**

Show that the following graph is planar.

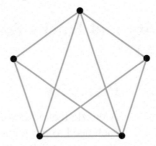

Solution

As given, the graph has several intersecting edges. However, we can redraw the graph in an equivalent form in which no edges touch except at vertices by redrawing the two orange edges, as shown in the following graphs. To verify that the second graph is equivalent to the first, we can label the vertices and check that the edges join the same vertices in each graph. Because the given graph is equivalent to a graph whose edges do not intersect, the graph is planar.

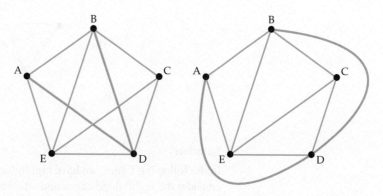

Take Note: The strategy we used here to show that the graph in Figure 10.18 is not planar can often be used to make planar drawings of graphs that *are* planar. The basic strategy is: Find a Hamiltonian circuit, redraw this circuit in a circular loop, and then draw in the remaining edges.

To see that the graph in Figure 10.18 representing the puzzle of connecting utilities is *not* planar, note that the graph is Hamiltonian, and one Hamiltonian circuit is A–X–B–Y–C–Z–A. If we redraw the graph so that this circuit is drawn in a loop (see Figure 10.19), we then need to add the edges AY, BZ, and CX. All three of these edges connect opposite vertices. We can draw only one of these edges inside the loop; otherwise two edges would cross. This means that the other two edges must be drawn outside the loop, but as you can see in Figure 10.20, those two edges would then have to cross. Thus the graph in Figure 10.18, which we will refer to as the *Utilities Graph,* is not planar, and so the utilities puzzle is not solvable.

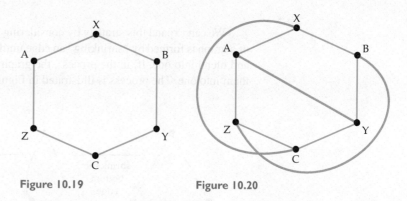

Figure 10.19   **Figure 10.20**

One strategy we can use to show that a graph is *not* planar is to find a **subgraph**, a graph whose edges and vertices come from the given graph, that is not planar. The Utilities Graph in Figure 10.18 is a common subgraph to watch for. Another graph that is not planar is the complete graph with five vertices, denoted K_5, shown in Figure 10.21.

Figure 10.21 The complete graph K_5

Subgraph Theorem

If a graph G has a subgraph that is not planar, then G is also not planar. In particular, if G contains the Utilities Graph or K_5 as a subgraph, G is not planar.

Example 2 **Identify a Nonplanar Graph**

Show that the following graph is not planar.

Solution

In the following figure, we have highlighted edges connecting the top six vertices. If we consider the highlighted edges and attached vertices as a subgraph, we can verify that the subgraph is the Utilities Graph. (The graph is slightly distorted compared with the version shown in Figure 10.18, but it is equivalent.) By the preceding theorem, we know that the graph is not planar.

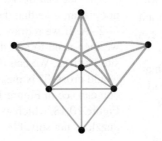

We can expand this strategy by considering *contractions* of a subgraph. A **contraction** of a graph is formed by "shrinking" an edge until the two vertices it connects come together and blend into one. If, in the process, the graph is left with any multiple edges, we merge them into one. The process is illustrated in Figure 10.22.

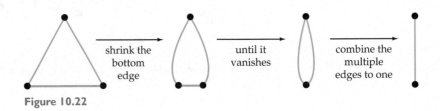

Figure 10.22

Example 3 **Contracting a Graph**

Show that the first graph below can be contracted to the second graph.

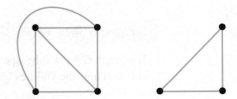

Solution

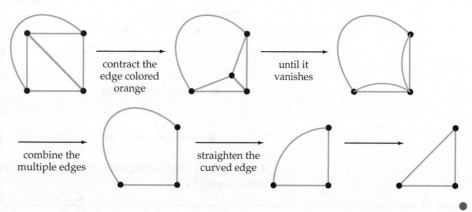

If we consider contractions, it turns out that the Utilities Graph and K_5 serve as building blocks for nonplanar graphs. In fact, it was proved in 1930 that any nonplanar graph will always have a subgraph that is the Utilities Graph or K_5, or a subgraph that can be contracted to the Utilities Graph or K_5. We can then expand our strategy, as given by the following theorem.

Nonplanar Graph Theorem

A graph is nonplanar if and only if it has the Utilities Graph or K_5 as a subgraph, or it has a subgraph that can be contracted to the Utilities Graph or K_5.

This gives us a definite test that will determine whether or not a graph can be drawn in such a way as to avoid crossing edges. (Note that the theorem includes the case in which the entire graph can be contracted to the Utilities Graph or K_5, as we can consider the graph a subgraph of itself.)

Example 4   **Identify a Nonplanar Graph**

Show that the following graph is not planar.

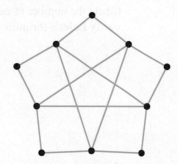

Solution

Note that the graph looks similar to K_5. In fact, we can contract some edges and make the graph look like K_5. Choose a pair of adjacent outside edges and contract one of them, as shown in the following figure.

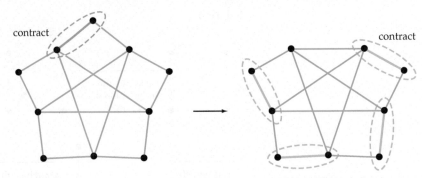

If we similarly contract the four edges colored green in the preceding figure, we arrive at the graph of K_5.

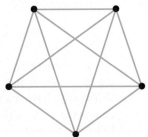

We were able to contract our graph to K_5, so by the nonplanar graph theorem, the given graph is not planar.

Euler's Formula

Euler noticed a connection between various features of planar graphs. In addition to edges and vertices, he looked at *faces* of a graph. In a planar drawing of a graph, the edges divide the graph into different regions called **faces**. The region surrounding the graph, or the exterior, is also considered a face, called the **infinite face**. (See Figure 10.23.) The following relationship, called Euler's formula, is always true.

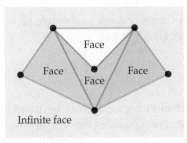

Figure 10.23

> ### Euler's Formula
>
> In a connected planar graph drawn with no intersecting edges, let v be the number of vertices, e the number of edges, and f the number of faces. Then $v + f = e + 2$.

Example 5 **Verify Euler's Formula in a Graph**

Count the number of edges, vertices, and faces in the following planar graph, and then verify Euler's formula.

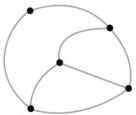

Solution

There are seven edges, five vertices, and four faces (counting the infinite face) in the graph. Thus $v + f = 5 + 4 = 9$ and $e + 2 = 7 + 2 = 9$, so $v + f = e + 2$, as Euler's formula predicts.

10.3 Exercise Set

Think About It

1. What is a planar graph?
2. Is this a planar drawing of a graph? Is the graph planar?

3. What is Euler's formula for a connected planar graph with no intersecting edges?

Preliminary Exercises

For Exercises P1 to P5, if necessary, use the referenced example following the exercise for assistance.

P1. Show that the following graph is planar. **Example 1**

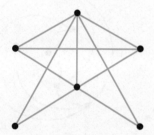

P2. Show that the following graph is not planar. **Example 2**

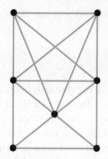

P3. Show that the first graph can be contracted to the second graph. **Example 3**

P4. Show that the following graph is not planar. **Example 4**

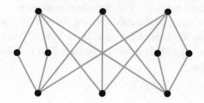

P5. Verify Euler's formula for the following planar graph. **Example 5**

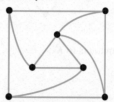

Section Exercises

In Exercises 1 to 8, show that the graph is planar by finding a planar drawing.

1.
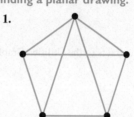

2.
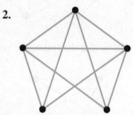

3.
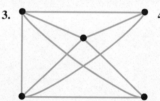

4.
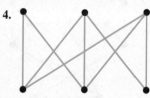

5.
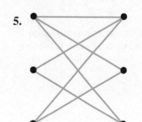

6.
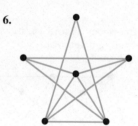

7.
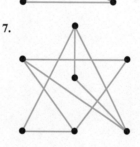

8.
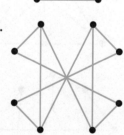

In Exercises 9 to 12, show that the graph is not planar.

9.
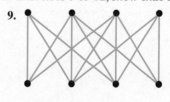

10.
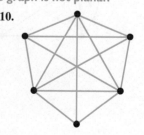

11.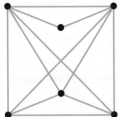

12.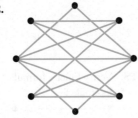

13. Show that the following graph contracts to K_5.

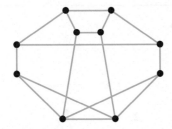

14. Show that the following graph contracts to the Utilities Graph.

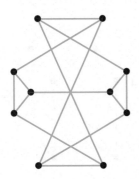

In Exercises 15 and 16, show that the graph is not planar by finding a subgraph whose contraction is the Utilities Graph or K_5.

15.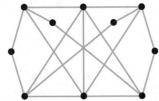

16.

In Exercises 17 to 22, count the number of vertices, edges, and faces, and then verify Euler's formula for the given graph.

17.    **18.**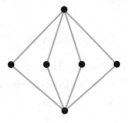

19. 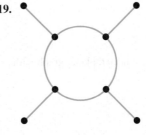   **20.**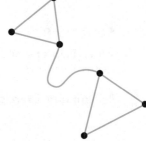

21. 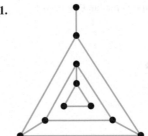   **22.**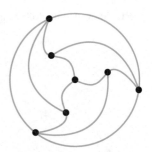

23. If a planar drawing of a graph has 15 edges and 8 vertices, how many faces does the graph have?

24. If a planar drawing of a graph has 100 vertices and 50 faces, how many edges are in the graph?

▶ **Investigation**

The Five Regular Convex Polyhedra   A **regular polyhedron** is a three-dimensional object in which all faces are identical. Specifically, each face is a regular polygon (meaning that each edge has the same length and each angle has the same measure). In addition, the same number of faces must meet at each vertex, or corner, of the object. Here we will discuss convex polyhedra, which means that any line joining one vertex to another is entirely contained within the object. (In other words, there are no indentations.)

It was proved long ago that only five objects fit this description—the tetrahedron, the cube, the octahedron, the dodecahedron, and the icosahedron. (See Figure 10.24.) One way mathematicians determined that there were only five such polyhedra was to visualize these three-dimensional objects in a two-dimensional way. We will use the cube to demonstrate.

Imagine a standard cube, but with the edges made of wire and the faces empty. Now suspend the cube above a flat surface and place a light above the top face of the cube. If the light is shining downward, a shadow is created on the flat surface. This is called a *projection* of the cube.

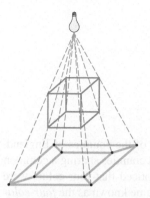

The shadow forms a graph. Note that each corner, or vertex, of the cube corresponds to a vertex of the graph, and each face of the cube corresponds to a face of the graph. The exception is the top face of the cube—its projection overlaps the entire graph. As a convention, we will identify the top face of the cube with the infinite face in the projected graph. Thus the three-dimensional cube is identified with the following graph.

Because the same number of faces must meet at each vertex of a polyhedron, the same number of faces, and so the same number of edges, must meet at each vertex of the graph. We also know that each face of a polyhedron has the same shape. Although the corresponding faces of the graph are distorted, we must have the same number of edges around each face (including the infinite face). In addition, Euler's formula must hold, because the projection is always a planar graph.

Thus we can investigate different polyhedra by looking at their features in two-dimensional graphs. It turns out that only five planar graphs satisfy these requirements, and the polyhedra that these graphs are the projections of are precisely the five pictured in Figure 10.24. The following exercises ask you to investigate some of the relationships between polyhedra and their projected graphs.

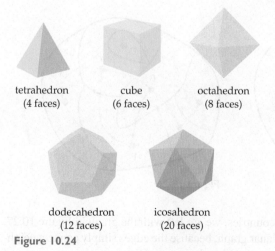

tetrahedron
(4 faces)

cube
(6 faces)

octahedron
(8 faces)

dodecahedron
(12 faces)

icosahedron
(20 faces)

Figure 10.24

Investigation Exercises

1. The tetrahedron in Figure 10.24 consists of four faces, each of which is an equilateral triangle. Draw the graph that results from a projection of the tetrahedron.

2. The following graph is the projection of one of the polyhedra in Figure 10.24. Identify the polyhedron it represents by comparing features of the graph to features of the polyhedron.

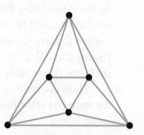

3. **a.** If we form a graph by a projection of the dodecahedron in Figure 10.24, what will the degree of each vertex be?

   b. The dodecahedron has 12 faces, each of which has 5 edges. Use this information to determine the number of edges in the projected graph.

   c. Use Euler's formula to determine the number of vertices in the projected graph, and hence in the dodecahedron.

4. **a.** Give a reason why the following graph cannot be the projection of a *regular* convex polyhedron.

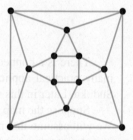

   b. If this graph is the projection of a three-dimensional convex polyhedron, how many faces does the polyhedron have?

   c. Describe, in as much detail as you can, additional features of the polyhedron corresponding to the graph. If you are feeling adventurous, sketch a polyhedron that this graph could be a projection of!

Graph Coloring

Learning Objectives:

- Use a graph to color a map.
- Find the chromatic number of a graph.
- Solve an application of graph coloring.

Coloring Maps

In the mid-1800s, Francis Guthrie was trying to color a map of the counties of England. So that it would be easy to distinguish the counties, he wanted counties sharing a common border to have different colors. After several attempts, he noticed that four colors were required to color the map, but not more. This observation became known as the *four-color problem*. It was not proved until over 100 years later.

Here is a map of the contiguous states of the United States colored similarly. Note that the map has only four colors and that no two states that share a common border have the same color.

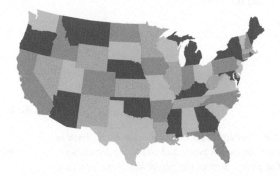

There is a connection between coloring maps and graph theory. This connection has many practical applications, from scheduling tasks, to designing computers, to playing Sudoku. Later in this section we will look more closely at some of these applications.

Suppose the map in Figure 10.25 shows the countries, labeled as letters, of a continent. We will assume that no country is split into more than one piece and countries that touch at just a corner point will not be considered neighbors. We can represent each country by a vertex, placed anywhere within the boundary of that country. We will then connect two vertices with an edge if the two corresponding countries are neighbors—that is, if they share a common boundary. The result is shown in Figure 10.26.

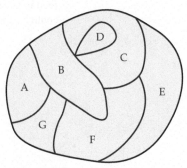

Figure 10.25

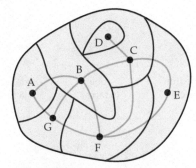

Figure 10.26

If we erase the boundaries of the countries, we are left with the graph in Figure 10.27. The resulting graph will always be a planar graph, because the edges simply connect neighboring countries. Our map-coloring question then becomes: Can we give each vertex of the graph a color such that no two vertices connected by an edge share the same color?

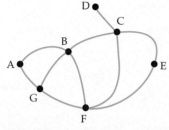

Figure 10.27

How many different colors will be required? If this can be accomplished using four colors, for instance, we will say that the graph is **4-colorable**. The graph in Figure 10.27 is actually *3-colorable;* only three colors are necessary. One possible coloring is given in Figure 10.28. A colored map for Figure 10.25 based on the colors of the vertices of the graph is shown in Figure 10.28. Figure 10.29 shows the coloring of the map in Figure 10.25 based on the graph in Figure 10.28.

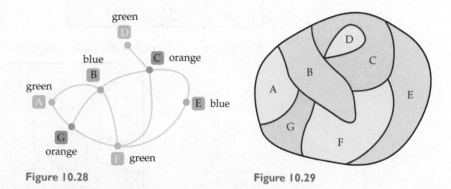

Figure 10.28   **Figure 10.29**

We can now formally state the four-color theorem.

Four-Color Theorem

Every planar graph is 4-colorable.

| Example I | **Using a Graph to Color a Map** |

The fictional map that follows shows the boundaries of countries on a rectangular continent. Represent the map as a graph, and find a coloring of the graph using the lowest possible number of colors. Then color the map according to the graph coloring.

Solution

First draw a vertex in each country and then connect two vertices with an edge if the corresponding countries are neighbors. (See the following figure on the left.) Now try to color the vertices of the resulting graph so that no edge connects two vertices of the same color. We know we will need at least two colors, so one strategy is simply to pick a starting vertex, give it a color, and then assign colors to the connected vertices one by one. Try to reuse the same colors, and use a new color only when there is no other option. For this graph we will need four colors. (The four-color theorem guarantees that we will not need more than that.) To see why we will need four colors, notice that the one vertex colored green in the following figure on the right connects to a ring of five vertices. Three different colors are required to color the five-vertex ring, and the green vertex connects to all these, so it requires a fourth color.

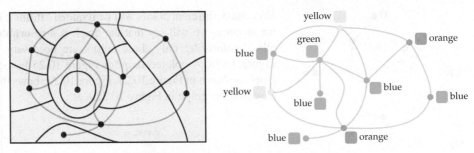

Now we color each country the same color as the corresponding vertex.

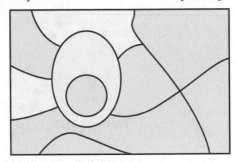

The Chromatic Number of a Graph

We mentioned previously that representing a map as a graph always results in a planar graph. The four-color theorem guarantees that we need only four colors to color a *planar* graph; however, if we wish to color a nonplanar graph, we may need more than four colors. The minimum number of colors needed to color a graph so that no edge connects vertices of the same color is called the **chromatic number** of the graph. In general, there is no efficient method of finding the chromatic number of a graph, but we do have a theorem that can tell us whether a graph is 2-colorable.

2-Colorable Graph Theorem

A graph is 2-colorable if and only if it has no circuits that consist of an odd number of vertices.

Example 2 **Determine the Chromatic Number of a Graph**

Find the chromatic number of the Utilities Graph from Section 10.3.

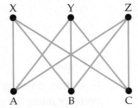

Solution

Note that the graph contains circuits such as A–Y–C–Z–B–X–A with six vertices and A–Y–B–X–A with four vertices. It seems that any circuit we find, in fact, involves an even number of vertices. It is difficult to determine whether we have looked at all possible circuits, but our observations suggest that the graph may be 2-colorable. A little trial and error confirms this if we simply color vertices A, B, and C one color and the remaining vertices another. Thus the Utilities Graph has a chromatic number of 2.

Applications of Graph Coloring

Determining the chromatic number of a graph and finding a corresponding coloring of the graph can solve a wide assortment of practical problems. One common application is in scheduling meetings or events. This is best shown by example.

Example 3 **A Scheduling Application of Graph Coloring**

Eight different school clubs want to schedule meetings on the last day of the semester. Some club members, however, belong to more than one of these clubs, so clubs that share members cannot meet at the same time. How many different time slots are required so that all members can attend all meetings? Clubs that have a member in common are indicated with an "X" in the following table.

	Ski Club	Student Government	Debate Club	Honor Society	Student Newspaper	Community Outreach	Campus Democrats	Campus Republicans
Ski Club	—	X		X			X	X
Student Government	X	—	X	X	X			
Debate Club		X	—	X		X		X
Honor Society	X	X	X	—	X	X		
Student Newspaper		X		X	—	X	X	
Community Outreach			X	X	X	—	X	X
Campus Democrats	X				X	X	—	
Campus Republicans	X		X			X		—

Solution

We can represent the given information by a graph. Each club is represented by a vertex, and an edge connects two vertices if the corresponding clubs have at least one common member.

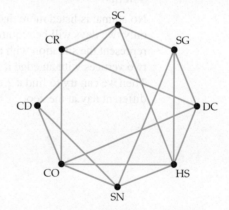

Two clubs that are connected by an edge cannot meet simultaneously. If we let a color correspond to a time slot, then we need to find a coloring of the graph that uses the lowest possible number of colors. The graph is not 2-colorable, because we can find circuits of odd length. However, by trial and error, we can find a 3-coloring. One example is shown in the following figure. Thus the chromatic number of the graph is 3, so we need three different time slots.

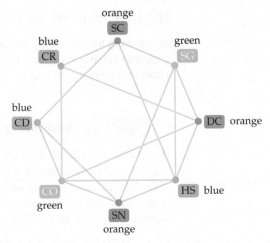

Each color corresponds to a time slot, so one scheduling is

> First time slot: ski club, debate club, student newspaper
>
> Second time slot: student government, community outreach
>
> Third time slot: honor society, campus Democrats, campus Republicans

---

Example 4   **A Scheduling Application of Graph Coloring**

Five classes at an elementary school have arranged a tour at a zoo where the students get to feed the animals.

> Class 1 wants to feed the elephants, giraffes, and hippos.
>
> Class 2 wants to feed the monkeys, rhinos, and elephants.
>
> Class 3 wants to feed the monkeys, deer, and sea lions.
>
> Class 4 wants to feed the parrots, giraffes, and polar bears.
>
> Class 5 wants to feed the sea lions, hippos, and polar bears.

If the zoo allows animals to be fed only once a day by one class of students, can the tour be accomplished in two days? (Assume that each class will visit the zoo only on one day.) If not, how many days will be required?

Solution

No animal is listed more than twice in the tour list, so you may be tempted to say that only two days will be required. However, to get a better picture of the problem, we can represent the situation with a graph. Use a vertex to represent each class, and connect two vertices with an edge if the corresponding classes want to feed the same animal. Then we can try to find a 2-coloring of the graph, where a different color represents a different day at the zoo.

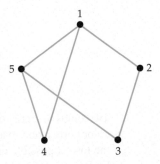

Note that the graph contains a circuit, 1–4–5–1, consisting of three vertices. This circuit will require three colors, and the remaining vertices will not require additional colors.

So the chromatic number of the graph is 3; one possible coloring is given in the following figure. Using this coloring, three days are required at the zoo. On the first day classes 2 and 5, represented by the blue vertices, will visit the zoo; on the second day classes 1 and 3, represented by the orange vertices, will visit the zoo; and on the third day class 4, represented by the green vertex, will visit the zoo.

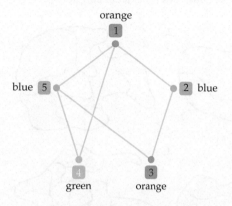

10.4 Exercise Set

Think About It

1. What is the four-color theorem?
2. What is the chromatic number of a graph?

Preliminary Exercises

For Exercises P1 to P4, if necessary, use the referenced example following the exercise for assistance.

P1. Represent the following fictional map of countries as a graph, and determine whether the graph is 2-colorable, 3-colorable, or 4-colorable by finding a suitable coloring of the graph. Then color the map according to the graph coloring. **Example 1**

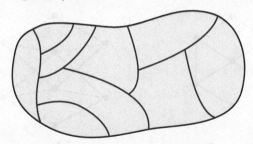

P2. Determine whether the following graph is 2-colorable. **Example 2**

P3. Six film students have collaborated on the creation of five films.

Film A was produced by Bako, Chris, and Damon.
Film B was produced by Allison and Fernando.
Film C was produced by Damon, Eavan, and Fernando.
Film D was produced by Bako and Eavan.
Film E was produced by Bako, Chris, and Eavan.

The college is scheduling a one-day film festival where each film will be shown once and the producers of each film will attend and participate in a discussion afterward. The college has several screening rooms available and two hours will be allotted for each film. If the showings begin at noon, create a screening schedule that allows the festival to end as early as possible while assuring that all of the producers of each film can attend that film's screening. **Example 3**

P4. Several delis in New York City have arranged deliveries to various buildings at lunchtime. The buildings' managements do not want more than one deli showing up at a particular building in one day, but the delis would like to deliver as often as possible. If they decide to agree on a delivery schedule, how many days will be required before each deli can return to the same building?

Deli A delivers to the Empire State Building, the Statue of Liberty, and Rockefeller Center.

Deli B delivers to the Chrysler Building, the Empire State Building, and the New York Stock Exchange.

Deli C delivers to the New York Stock Exchange, the American Stock Exchange, and the United Nations Building.

Deli D delivers to New York City Hall, the Chrysler Building, and Rockefeller Center.

Deli E delivers to Rockefeller Center, New York City Hall, and the United Nations Building. **Example 4**

Section Exercises

In Exercises 1 to 4, a fictional map of the countries of a continent is given. Represent the map by a graph and find a coloring of the graph that uses the lowest possible number of colors. Then color the map according to the graph coloring you found.

1.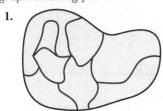

2.

3.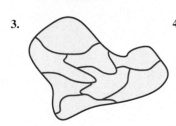

4.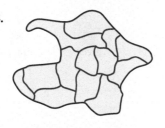

In Exercises 5 to 8, represent the map by a graph and find a coloring of the graph that uses the lowest possible number of colors.

5. Western portion of the United States

6. Counties of New Hampshire

7. Countries of South America

8. Provinces of South Africa

In Exercises 9 to 14, if possible, show that the graph is 2-colorable. If the graph is not 2-colorable, explain why.

9.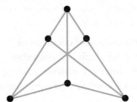

10.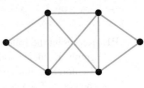

11.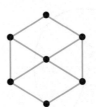

12.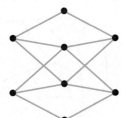

13.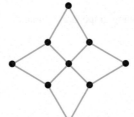

14.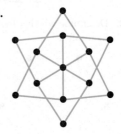

In Exercises 15 to 20, determine (by trial and error) the chromatic number of the graph.

15.

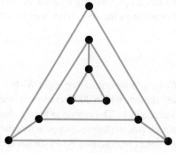

16.

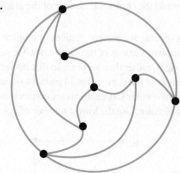

17. **18.**

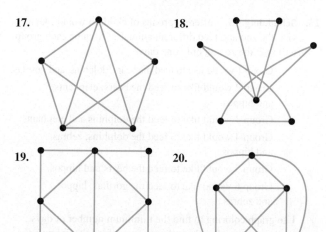

19. **20.**

21. Scheduling Six student clubs need to hold meetings on the same day, but some students belong to more than one club. In order to avoid members missing meetings, the meetings need to be scheduled during different time slots. An "X" in the following table indicates that the two corresponding clubs share at least one member. Use graph coloring to determine the minimum number of time slots necessary to ensure that all club members can attend all meetings.

	Student Newspaper	Honor Society	Biology Association	Gaming Club	Debate Team	Engineering Club
Student Newspaper	—		X		X	
Honor Society		—	X		X	X
Biology Association	X	X	—	X		
Gaming Club			X	—	X	X
Debate Team	X	X		X	—	
Engineering Club		X		X		—

22. Scheduling Eight political committees must meet on the same day, but some members are on more than one committee. Thus any committees that have members in common cannot meet at the same time. An "X" in the following table indicates that the two corresponding committees share a member. Use graph coloring to determine the minimum number of meeting times that will be necessary so that all members can attend the appropriate meetings.

	Appropriations	Budget	Finance	Judiciary	Education	Health	Foreign Affairs	Housing
Appropriations	—		X			X	X	
Budget		—		X		X		
Finance	X		—	X			X	X
Judiciary		X	X	—		X		X
Education					—		X	X
Health	X	X		X		—		
Foreign Affairs	X		X		X		—	
Housing			X	X	X			—

23. Scheduling Six different groups of children would like to visit the zoo and feed different animals. (Assume each group will visit the zoo on only one day.)

> Group 1 would like to feed the bears, dolphins, and gorillas.
>
> Group 2 would like to feed the bears, elephants, and hippos.
>
> Group 3 would like to feed the dolphins and elephants.
>
> Group 4 would like to feed the dolphins, zebras, and hippos.
>
> Group 5 would like to feed the bears and hippos.
>
> Group 6 would like to feed the gorillas, hippos, and zebras.

Use graph coloring to find the minimum number of days that are required so that all groups can feed the animals they would like to feed but no animals will be fed twice on the same day. Design a schedule to accomplish this goal.

24. Scheduling Five different charity organizations send trucks on various routes to pick up donations that residents leave on their doorsteps.

> Charity A covers Main St., First Ave., and State St.
>
> Charity B covers First Ave., Second Ave., and Third Ave.
>
> Charity C covers State St., City Dr., and Country Lane.
>
> Charity D covers City Dr., Second Ave., and Main St.
>
> Charity E covers Third Ave., Country Lane, and Fourth Ave.

Each charity has its truck travel down all three streets on its route on the same day, but no two charities wish to visit the same streets on the same day. Use graph coloring to design a schedule for the charities. Arrange their pickup routes so that no street is visited twice on the same day by different charities. The schedule should use the smallest possible number of days.

25. Scheduling Students in a film class have volunteered to form groups and create several short films. The class has three digital video cameras that may be checked out for one day only, and it is expected that each group will need the entire day to finish shooting. All members of each group must participate in the film they volunteered for, so a student cannot work on more than one film on any given day.

> Film 1 will be made by Brian, Angela, and Kotara.
>
> Film 2 will be made by Jessica, Vinh, and Brian.
>
> Film 3 will be made by Corey, Brian, and Vinh.
>
> Film 4 will be made by Ricardo, Sarah, and Lupe.
>
> Film 5 will be made by Sarah, Kotara, and Jessica.
>
> Film 6 will be made by Angela, Corey, and Lupe.

Use graph coloring to design a schedule for lending the cameras, using the smallest possible number of days, so that each group can shoot its film and all members can participate.

26. Animal Housing A researcher has discovered six new species of insects overseas and needs to transport them home. Some species will harm each other and so cannot be transported in the same container.

> Species A cannot be housed with species C or F.
>
> Species B cannot be housed with species D or F.
>
> Species C cannot be housed with species A, D, or E.
>
> Species D cannot be housed with species B, C, or F.
>
> Species E cannot be housed with species C or F.
>
> Species F cannot be housed with species A, B, D, or E.

Draw a graph where each vertex represents a species of insect and an edge connects two vertices if the species cannot be housed together. Then use graph coloring to determine the minimum number of containers the researcher will need to transport the insects.

Investigations

1. Wi-Fi Stations An office building is installing eight Wi-Fi transmitting stations. Any stations within 200 feet of each other must transmit on different channels. The engineers have a chart of the distance between each pair of stations. Suppose that they draw a graph where each vertex represents a Wi-Fi station and an edge connects two vertices if the distance between the stations is 200 feet or less. What would the chromatic number of the graph tell the engineers?

2. Edge Coloring In this section, we colored vertices of graphs so that no edge connected two vertices of the same color. We can also consider coloring edges, rather than vertices, so that no vertex connects two or more edges of the same color. In parts a to d, assign each edge in the graph a color so that no vertex connects two or more edges of the same color. Use the lowest number of colors possible.

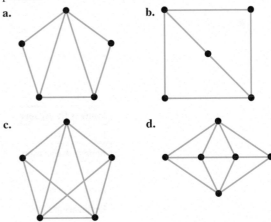

e. Explain why the number of colors required will always be at least the number of edges that meet at the vertex of highest degree in the graph.

3. Scheduling Edge colorings, as explained in Investigation 2, can be used to solve scheduling problems. For instance, suppose five players are competing in a tennis tournament. Each player needs to play every other player in a match (but not more than once). Each player will participate in no more than one match per day, and two matches can occur at the same time when possible. How many days will be required for the tournament? Represent the tournament as a graph, in which each vertex corresponds to a player and an edge joins two vertices if the corresponding players will compete against each other in a match. Next, color the edges, where each different color corresponds to a different day of the tournament. Because one player will not be in more than one match per day, no two edges of the same color can meet at the same vertex. If we can find an edge coloring of the graph that uses the lowest number of colors possible, it will correspond to the lowest number of days required for the tournament. Sketch a graph that represents the tournament, find an edge coloring using the lowest number of colors possible, and use your graph to design a schedule of matches for the tournament that minimizes the number of days required.

Test Prep ⑩ Circuits and Networks

Section 10.1 Graphs and Euler Circuits

Graphs A graph is a set of points called vertices and line segments or curves called edges that connect vertices. A graph is a connected graph provided any vertex can be reached from any other vertex by tracing along edges. Two graphs are equivalent graphs provided the edges form the same connections of vertices in each graph. The degree of a given vertex is equal to the number of edges that meet at the vertex.

See **Examples 1** and **2** on pages 458 and 459, and then try Exercises 1, 2, 5, and 6 on page 501.

Euler Circuits An Euler circuit is a path that uses every edge but does not use any edge more than once, and begins and ends at the same vertex.

Eulerian Graph Theorem A connected graph is Eulerian (has an Euler circuit) if and only if every vertex of the graph is of even degree.

See **Examples 3** to **5** on pages 461 and 463, and then try part b of Exercises 7 to 10 and Exercise 11 on pages 501 and 502.

Euler Paths An Euler path is a path that uses every edge but does not use any edge more than once.

Euler Path Theorem A connected graph contains an Euler path if and only if the graph has two vertices of odd degree with all other vertices of even degree. Furthermore, every Euler path must start at one of the vertices of odd degree and end at the other.

See **Examples 6** and **7** on pages 463 and 464, and then try part a of Exercises 7 to 10 on page 501.

Section 10.2 Weighted Graphs

Hamiltonian Circuits A Hamiltonian circuit is a path in a graph that uses each vertex exactly once and returns to the starting vertex. If a graph has a Hamiltonian circuit, the graph is said to be Hamiltonian.

Dirac's Theorem Dirac's theorem states that in a connected graph with at least three vertices and with no multiple edges, if n is the number of vertices in the graph and every vertex has degree of at least $n/2$, then the graph must be Hamiltonian. If it is not the case that every vertex has a degree of at least $n/2$, then the graph may or may not be Hamiltonian.

See **Example 1** on page 469, and then try Exercises 13 to 15 on page 502.

Weighted Graphs A weighted graph is a graph in which each edge is associated with a value, called a weight.

The Greedy Algorithm A method of finding a Hamiltonian circuit in a complete weighted graph is given by the following greedy algorithm.

1. Choose a vertex to start at, then travel along the connected edge that has the smallest weight. (If two or more edges have the same weight, pick any one.)
2. After arriving at the next vertex, travel along the edge of smallest weight that connects to a vertex not yet visited. Continue this process until you have visited all vertices.
3. Return to the starting vertex.

The greedy algorithm attempts to give a circuit of minimal total weight, although it does not always succeed.

See **Examples 2** and **3** on pages 470 and 472, and then try Exercises 17, 18, and 21 on pages 502 and 503.

The Edge-Picking Algorithm Another method of finding a Hamiltonian circuit in a complete weighted graph is given by the following edge-picking algorithm.

1. Mark the edge of smallest weight in the graph. (If two or more edges have the same weight, pick any one.)
2. Mark the edge of next smallest weight in the graph, as long as it does not complete a circuit and does not add a third marked edge to a single vertex.
3. Continue this process until you can no longer mark any edges. Then mark the final edge that completes the Hamiltonian circuit.

The edge-picking algorithm attempts to give a circuit of minimal total weight, although it does not always succeed.

See **Examples 4** to **6** on pages 473 to 476, and then try Exercises 19, 20, and 22 on pages 502 and 503.

Section 10.3 Planarity and Euler's Formula

Planar Graph A planar graph is a graph that can be drawn so that no edges intersect each other (except at vertices).

See **Example 1** on page 482, and then try Exercises 23 and 24 on page 503.

Subgraphs A subgraph of a graph is a graph whose edges and vertices come from the given graph.

The Nonplanar Graph Theorem A graph is nonplanar if and only if it has the Utilities Graph or K_5 as a subgraph, or if it has a subgraph that can be contracted to the Utilities Graph or K_5.

See **Examples 2** and **4** on pages 484 and 485, and then try Exercises 25 and 26 on page 503.

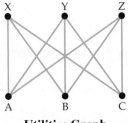
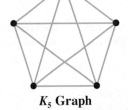

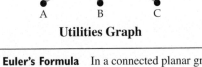

Utilities Graph K_5 **Graph**

Euler's Formula In a connected planar graph drawn with no intersecting edges, let v be the number of vertices, e the number of edges, and f the number of faces. Then $v + f = e + 2$.

See **Example 5** on page 486, and then try Exercises 27 and 28 on page 503.

Section 10.4 Graph Coloring

Representing Maps as Graphs Draw a vertex in each region (country, state, etc.) of any map. Connect two vertices if the corresponding regions share a common border.

The Four-Color Theorem Every planar graph is 4-colorable. (In some cases less than four colors may be required. Also, if the graph is not planar, more than four colors may be necessary.)

See **Example 1** on page 491, and then try Exercises 29 and 30 on page 503.

2-Colorable Graph Theorem A graph is 2-colorable if and only if it has no circuits that consist of an odd number of vertices.

See **Example 2** on page 492, and then try Exercises 31 and 32 on page 503.

Applications of Graph Coloring Determining the chromatic number of a graph and finding a corresponding coloring of the graph can solve some practical applications such as scheduling meetings or events.

See **Examples 3** and **4** on pages 493 and 494, and then try Exercises 33 to 35 on page 504.

Review Exercises

In Exercises 1 and 2, (a) determine the number of edges in the graph, (b) find the number of vertices in the graph, (c) list the degree of each vertex, and (d) determine whether the graph is connected.

1.

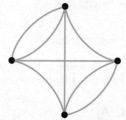

2.

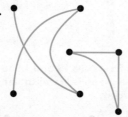

3. Soccer In the following table, an "X" indicates teams from a junior soccer league that have played each other in the current season. Draw a graph to represent the games by using a vertex to represent each team. Connect two vertices with an edge if the corresponding teams played a game against each other this season.

	Mariners	Scorpions	Pumas	Stingrays	Vipers
Mariners	—	X		X	X
Scorpions	X	—	X		
Pumas		X	—	X	X
Stingrays	X		X	—	
Vipers	X		X		—

4. Each vertex in the following graph represents a freeway in the Los Angeles area. An edge connects two vertices if the corresponding freeways have interchanges allowing drivers to transfer from one freeway to the other.

a. Can drivers transfer from the 105 freeway to the 10 freeway?

b. Among the freeways represented, how many have interchanges with the 5 freeway?

c. Which freeways have interchanges to all the other freeways in the graph?

d. Of the freeways represented in the graph, which has the fewest interchanges to the other freeways?

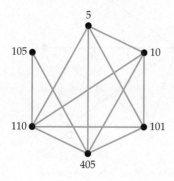

In Exercises 5 and 6, determine whether the two graphs are equivalent.

5.

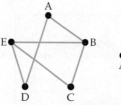

6.

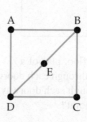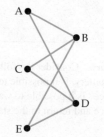

In Exercises 7 to 10, (a) find an Euler path if possible, and (b) find an Euler circuit if possible.

7.

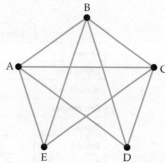

8.

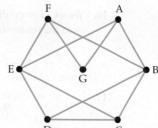

9.

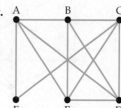

10.

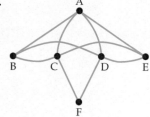

11. **Parks** The figure shows an arrangement of bridges connecting land areas in a park. Represent the map as a graph, and then determine whether it is possible to stroll across each bridge exactly once and return to the starting position.

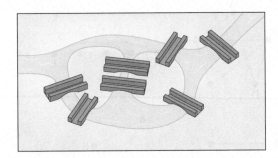

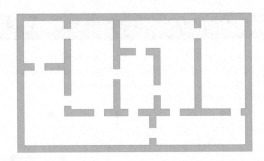

12. **Architecture** Consider the following floor plan of a sculpture gallery. Is it possible to walk through each doorway exactly once? Is it possible to walk through each doorway exactly once and return to the starting point?

In Exercises 13 and 14, use Dirac's theorem to verify that the graph is Hamiltonian, and then find a Hamiltonian circuit.

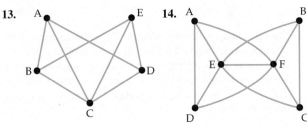

13.   14.

15. **Travel** The following table lists cities serviced by a small airline. An "X" in the table indicates a direct flight offered by the airline. Draw a graph that represents the direct flights, and use your graph to find a route that visits each city exactly once and returns to the starting city.

	Casper	Rapid City	Minneapolis	Des Moines	Topeka	Omaha	Boulder
Casper	—	X					X
Rapid City	X	—	X				
Minneapolis		X	—	X	X		X
Des Moines			X	—	X		
Topeka			X	X	—	X	X
Omaha					X	—	X
Boulder	X		X		X	X	—

16. **Travel** For the direct flights given in Exercise 15, find a route that travels each flight exactly once and returns to the starting city. (You may visit cities more than once.)

In Exercises 17 and 18, use the greedy algorithm to find a Hamiltonian circuit starting at vertex A in the weighted graph.

17.   18.

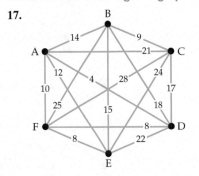

In Exercises 19 and 20, use the edge-picking algorithm to find a Hamiltonian circuit starting at vertex A in the weighted graph.

19. The graph in Exercise 17

20. The graph in Exercise 18

21. **Efficient Route** The distances, in miles, between five different cities are given in the following table. Sketch a weighted graph that represents the distances, and then use the greedy algorithm to design a route that starts in Memphis, visits each city, and returns to Memphis while attempting to minimize the total distance traveled.

	Memphis	Nashville	Atlanta	Birmingham	Jackson
Memphis	—	210	394	247	213
Nashville	210	—	244	189	418
Atlanta	394	244	—	148	383
Birmingham	247	189	148	—	239
Jackson	213	418	383	239	—

22. **Computer Networking** A small office needs to network five computers by connecting one computer to another and forming a large loop. The length of cable needed (in feet) between pairs of machines is given in the following table. Use the edge-picking algorithm to design a method to network the computers while attempting to use the smallest possible amount of cable.

	Computer A	Computer B	Computer C	Computer D	Computer E
Computer A	—	85	40	55	20
Computer B	85	—	35	40	18
Computer C	40	35	—	60	50
Computer D	55	40	60	—	30
Computer E	20	18	50	30	—

In Exercises 23 and 24, show that the graph is planar by finding a planar drawing.

23. 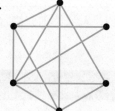   24.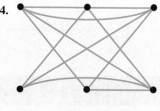

In Exercises 25 and 26, show that the graph is not planar.

25. 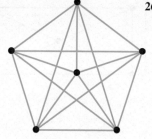   26.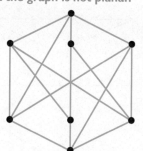

In Exercises 27 and 28, count the number of vertices, edges, and faces in the graph, and then verify Euler's formula.

27.   28.

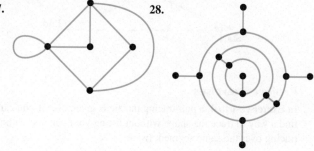

In Exercises 29 and 30, a fictional map is given showing the states of a country. Represent the map by a graph and find a coloring of the graph, using the minimum number of colors possible, such that no two connected vertices are the same color. Then color the map according to the graph coloring that you found.

29.

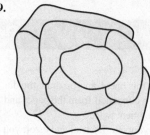

30.

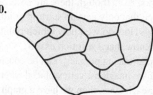

In Exercises 31 and 32, show that the graph is 2-colorable by finding a 2-coloring, or explain why the graph is not 2-colorable.

31.   32.

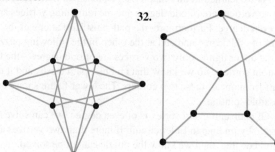

In Exercises 33 and 34, determine (by trial and error) the chromatic number of the graph.

33.

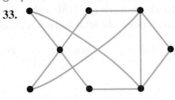

34.

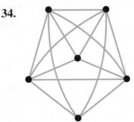

35. **Scheduling** A company has scheduled a retreat at a hotel resort. It needs to hold meetings the first day there, and several conference rooms are available. Some employees must attend more than one meeting, however, so the meetings cannot all be scheduled at the same time. An "X" in the following table indicates that at least one employee must attend both of the corresponding meetings, and so those two meetings must be held at different times. Draw a graph in which each vertex represents a meeting, and an edge joins two vertices if the corresponding meetings require the attendance of the same employee. Then use graph coloring to design a meeting schedule that uses the minimum number of time slots.

	Budget Meeting	Marketing Meeting	Executive Meeting	Sales Meeting	Research Meeting	Planning Meeting
Budget Meeting	—	X	X		X	
Marketing Meeting	X	—		X		
Executive Meeting	X		—		X	X
Sales Meeting		X		—		X
Research Meeting	X		X		—	X
Planning Meeting			X	X	X	—

Project 1

Pen-Tracing Puzzles

You may have seen puzzles like this one: Can you draw the diagram at the right without lifting your pencil from the paper and without tracing over the same segment twice?

Before reading on, try it for yourself. By trial and error, you may discover a tracing that works. Even though there are several possible tracings, you may notice that only certain starting points seem to allow a complete tracing. How do we know which point to start from? How do we even know that a solution exists?

Puzzles such as this, called "pen-tracing puzzles," can be considered problems in graph theory. If we imagine a vertex placed wherever two lines meet or cross over each other, then we have a graph. Our task is to start at a vertex and find a path that traverses every edge of the graph, without repeating any edges. In other words, we need an Euler path! (An Euler circuit would work as well.)

As we learned in this chapter, a graph has an Euler path only if two vertices are of odd degree and the remaining vertices are of even degree. Furthermore, the path must start at one of the vertices of odd degree and end at the other. In the following puzzle, in the upper figure, only two vertices are of odd degree—the two bottom corners. So we know that an Euler path exists, and it must start from one of these two corners. The lower figure shows one possible solution.

Of course, if every vertex is of even degree, we can solve the puzzle by finding an Euler circuit. If more than two vertices are of odd degree, then we know the puzzle cannot be solved.

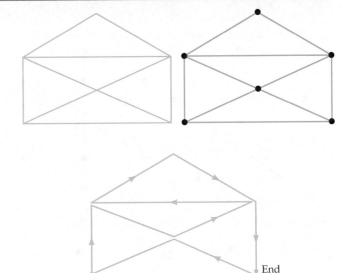

Project Exercises

In Exercises 1 to 4, a pen-tracing puzzle is given. See if you can find a way to trace the shape without lifting your pen and without tracing over the same segment twice.

1.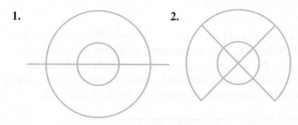

2.

3.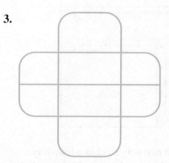

4.

5. Explain why the following pen-tracing puzzle is impossible to solve.

Project 2

Modeling Traffic Lights with Graphs

Have you ever watched the cycles that traffic lights go through while you were waiting for a red light to turn green? Some intersections have lights that go through several stages to allow all the different lanes of traffic to proceed safely.

Ideally, each stage of a traffic-light cycle should allow as many lanes of traffic to proceed through the intersection as possible. We can design a traffic-light cycle by modeling an intersection with a graph. Figure 10.30 shows a three-way intersection where two two-way roads meet. Each direction of traffic has turn lanes, with left-turn lights where possible. There are six different directions in which vehicles can travel, as indicated in the figure, and we have labeled each possibility with a letter.

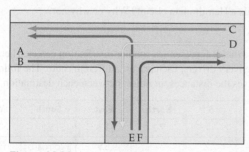

Figure 10.30

We can represent the traffic patterns with a graph; each vertex will represent one of the six possible traffic paths, and we will draw an edge between two vertices if the corresponding paths would allow vehicles to collide. The result is the graph shown in Figure 10.31. Because we do not want to allow vehicles to travel simultaneously along routes on which they could collide, any vertices connected by an edge can allow traffic to move only during different parts of the light cycle. We can represent each portion of the cycle by a color. Our job then is to color the graph using the fewest colors possible.

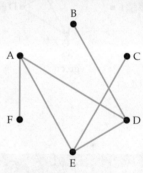

Figure 10.31

There is no 2-coloring of the graph because we have a circuit of length 3: A–D–E–A. We can, however, find a 3-coloring. One possibility is given in Figure 10.32.

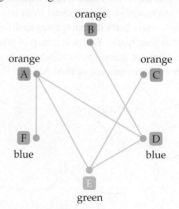

Figure 10.32

A 3-coloring of the graph means that the traffic lights at the intersection will have to go through a three-stage cycle. One stage will allow the traffic routes corresponding to the orange vertices A, B, and C to proceed, the next stage will let the paths corresponding to the blue vertices D and F to proceed, and finally, the third stage will let path E, colored green, proceed.

Although safety requires three stages for the lights, we can refine the design to allow more traffic to travel through the intersection. Note that at the third stage, only one route, path E, is scheduled to be moving. However, there is no harm in allowing path B to move at the same time, since it is a right turn that doesn't conflict with route E. We could also allow path F to proceed at the same time. Adding these additional paths corresponds to adding colors to the graph in Figure 10.32. We do not want to use more than three colors, but we can add a second color to some of the vertices while maintaining the requirement that no edge can connect two vertices of the same color. The result is shown in Figure 10.33. Notice that the vertices in the triangular circuit A–D–E–A can be assigned only a single color, but the remaining vertices can accommodate two colors.

In summary, our design allows traffic paths A, B, and C to proceed during one stage of the cycle, paths C, D, and F during another, and paths B, E, and F during the third stage.

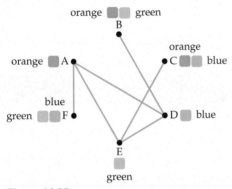

Figure 10.33

Project Exercises

1. A one-way road ends at a two-way street. The intersection and the different possible traffic routes are shown in the following figure. The one-way road has a left-turn light. Represent the traffic routes with a graph and use graph coloring to determine the minimum number of stages required for a light cycle.

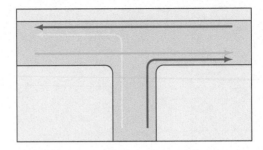

2. A one-way road intersects a two-way road in a four-way intersection. Each direction has turn lanes and left-turn lights. Represent the various traffic routes with a graph and use graph coloring to determine the minimum number of stages required for a light cycle. Then refine your design to allow as much traffic as possible to proceed at each stage of the cycle.

3. A two-way road intersects another two-way road in a four-way intersection. One road has left-turn lanes with left-turn lights, but on the other road cars are not allowed to make left turns. Represent the various traffic routes with a graph and use graph coloring to determine the minimum number of stages required for a light cycle. Then refine your design to allow as much traffic as possible to proceed at each stage of the cycle.

Math in Practice

There are many different industries that make weekly deliveries: flowers to venues for celebrations, milk deliveries from dairies to grocery stores and ice cream makers, eggs from farms to grocery stores and dessert making factories, metal from steel factories to vehicle factories, water bottled from springs to distributers, and even meal kits to your house. Below is a map of your company's delivery route. Your company's main warehouse is marked HOME and the delivery destinations are marked A, B, C, D, E, F, and G.

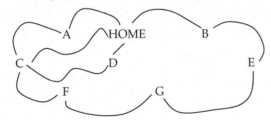

Part 1

Your company hires a new delivery driver who needs to learn all the roads on the delivery route.

a. Given the map, it is possible for the new delivery truck driver to travel all the roads exactly once starting and ending at HOME?

b. What are the benefits of the delivery truck traveling to HOME more than just at the beginning and end of the route?

c. Is this a Hamiltonian circuit? Why or why not?

Part 2

Suppose that B represents a smaller distribution center. There are three delivery drivers, each drives a different route. The following table gives the distances, in miles, between each destination.

	B	**North**	**East**	**South**	**West**
B	—	7.2	3.5	5.1	6.5
North	7.2	—	8.9	No roads	11.6
East	3.5	8.9	—	8.0	No roads
South	5.1	No roads	8.0	—	9.4
West	6.5	11.6	No roads	9.4	—

a. There are 8 different ways to start from B, visit each destination, and return to B. Name the different routes.

b. Suppose the first driver makes deliveries on Monday taking route B-West-North-East-South-B. A second driver makes deliveries on Wednesday taking route

B-North-East-South-West-B. A third driver makes deliveries on Friday taking route B-East-North-West-South-B. Find the total distance for each driver. Which driver has the shortest route?

c. The finance department is concerned about delivery costs. Taking the route from North to West or West to North requires a toll. If you eliminate all delivery routes that go either North-West or West-North, what delivery routes remain?

Comdas/Shutterstock.com

Geometry

Math in Practice: Feed Tanks, p. 587

Applications of Geometry

Geometry is one of the fundamental branches of mathematics. Most of our exposure is to Euclidean geometry, the geometry of flat surfaces. However, we can extend these concepts to *spherical* geometry, the geometry on the surface of a sphere. On a sphere, the concept of parallel lines is altered. In Euclidean geometry, parallel lines never meet. On Earth's surface, which can be approximated by a sphere, the longitude lines are parallel and all pass through the north and south poles.

There are other non-Euclidean geometries, geometries based on changing what it means to have parallel lines. These geometries have wide-ranging applications in physics. Albert Einstein's general relativity theory is based on such a geometry.

Although Earth's surface is not flat, we often consider it flat for "local" measurements and calculations. On page 555, Exercise 27 asks you to determine the distance across a river. The calculation is based on right triangles that have two parallel legs.

Section

11.1

Review of Measurement

Learning Objectives:

- Convert between units of length in the U.S. Customary System.
- Convert between units of weight in the U.S. Customary System.
- Convert between units of capacity in the U.S. Customary System.
- Convert between units of length in the Metric System.
- Convert between units of mass in the Metric System.
- Convert between units of capacity in the Metric System.
- Convert between units in the U.S. Customary System and the Metric System.

Be Prepared: Prep Test 11 reviews concepts used in this chapter. This can be found in your Companion Site at Cengage.com.

Please note that this section reviews material covered in Chapter 3. If you have already mastered this topic, you can begin with Section 11.2.

U.S. Customary Units of Measure

Units of measure are a way to give a magnitude or size to a number. For instance, there is quite a difference between 7 feet and 7 miles. The units *feet* and *miles* are units of measure in the U.S. Customary System. In this section we are going to focus on units of length, weight, and capacity.

Here are some of the U.S. Customary units and their equivalences.

Equivalences Between Units of Length in the U.S. Customary System

12 inches (in.) = 1 foot (ft)

3 ft = 1 yard (yd)

5280 ft = 1 mile (mi)

Equivalences Between Units of Weight in the U.S. Customary System

16 ounces (oz) = 1 pound (lb)

2000 lb = 1 ton

Equivalences Between Units of Capacity in the U.S. Customary System

8 fluid ounces (fl oz) = 1 cup (c)

2 c = 1 pint (pt)

2 pt = 1 quart (qt)

4 qt = 1 gallon (gal)

These equivalences can be used to change from one unit to another unit by using a **conversion rate**. For instance, because 12 in. = 1 ft, we can write two conversion rates,

$$\frac{12 \text{ in.}}{1 \text{ ft}} \text{ and } \frac{1 \text{ ft}}{12 \text{ in.}}$$

Dimensional analysis involves using conversion rates to change from one unit of measurement to another.

Example 1 **Convert Between Units of Length**

Convert 18 in. to feet.

Solution

Choose a conversion rate so that the unit in the numerator of the conversion rate is the same as the unit needed in the answer ("feet" for this problem). The unit in the denominator of the conversion rate is the same as the unit in the given measurement ("inches" for this problem). The conversion rate is $\frac{1 \text{ ft}}{12 \text{ in.}}$.

$$18 \text{ in.} = 18 \text{ in.} \times \frac{1 \text{ ft}}{12 \text{ in.}}$$

The unit in the numerator is the same as the unit needed in the answer.

The unit in the denominator is the same as the unit in the given measurement.

$$= \frac{18 \text{ ft}}{12} = 1.5 \text{ ft}$$

$$18 \text{ in.} = 1.5 \text{ ft}$$

Example 2   **Convert Between Units of Weight**

Convert $3\frac{1}{2}$ tons to pounds.

Solution

Write a conversion rate with pounds in the numerator and tons in the denominator. The conversion rate is $\frac{2000 \text{ lb}}{1 \text{ ton}}$.

$$3\frac{1}{2} \text{ tons} = \frac{7}{2} \text{ tons} \times \frac{2000 \text{ lb}}{1 \text{ ton}}$$

$$= \frac{14,000 \text{ lb}}{2} = 7000 \text{ lb}$$

$$3\frac{1}{2} \text{ tons} = 7000 \text{ lb}$$

Sometimes one unit cannot be converted directly to another. For instance, consider trying to convert quarts to cups using only the capacity equivalences given earlier. In this case, it is necessary to use more than one conversion factor, as shown next.

Convert 3 qt to cups.

$$3 \text{ qt} = 3 \text{ qt} \times \boxed{\frac{2 \text{ pt}}{1 \text{ qt}}} \times \boxed{\frac{2 \text{ c}}{1 \text{ pt}}}$$

• The direct equivalence between quarts and cups was not given earlier in the section. Use two conversion rates. First convert quarts to pints, and then convert pints to cups.

$$= \frac{3 \text{ qt}}{1} \times \frac{2 \text{ pt}}{1 \text{ qt}} \times \frac{2 \text{ c}}{1 \text{ pt}}$$

$$= \frac{12 \text{ c}}{1} = 12 \text{ c}$$

$$3 \text{ qt} = 12 \text{ c}$$

Example 3   **Convert Between Units of Capacity**

Convert 42 c to quarts.

Solution

First convert cups to pints, and then convert pints to quarts.

$$42 \text{ c} = 42 \text{ c} \times \frac{1 \text{ pt}}{2 \text{ c}} \times \frac{1 \text{ qt}}{2 \text{ pt}}$$

$$= \frac{42 \text{ qt}}{4} = 10\frac{1}{2} \text{ qt}$$

$$42 \text{ c} = 10\frac{1}{2} \text{ qt}$$

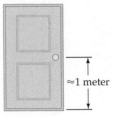

≈1 meter

Figure 11.1

The Metric System

In 1789, an attempt was made to standardize units of measurement internationally in order to simplify trade and commerce between nations. A commission in France developed a system of measurement known as the **metric system**.

The basic unit of length in the metric system is the **meter**. One meter is approximately the distance from a doorknob to the floor, as shown in Figure 11.1. All units of length in the metric system are derived from the meter. Prefixes to the basic unit denote the length of each unit. For example, the prefix "centi-" means one-hundredth, so 1 centimeter is 1 one-hundredth of a meter.

Prefixes and Units of Length in the Metric System

kilo- = 1000   1 kilometer (km) = 1000 meters (m)

hecto- = 100   1 hectometer (hm) = 100 m

deca- = 10   1 decameter (dam) = 10 m

           1 meter (m) = 1 m

deci- = 0.1   1 decimeter (dm) = 0.1 m

centi- = 0.01   1 centimeter (cm) = 0.01 m

milli- = 0.001   1 millimeter (mm) = 0.001 m

Conversion between units of length in the metric system involves moving the decimal point to the right or to the left. Listing the units in order from largest to smallest will indicate how many places to move the decimal point and in which direction.

To convert 4200 cm to meters, write the units in order from largest to smallest.

km   hm   dam   m   dm   cm   mm

2 positions

- Converting centimeters to meters requires moving 2 positions to the left.

4200 cm = 42.00 m

2 places

- Move the decimal point the same number of places and in the same direction.

A metric measurement that involves two units is customarily written in terms of one unit. Convert the smaller unit to the larger unit, and then add.

To convert 8 km 32 m to kilometers, first convert 32 m to kilometers.

km   hm   dam   m   dm   cm   mm

- Converting meters to kilometers requires moving 3 positions to the left.

32 m = 0.032 km

- Move the decimal point the same number of places and in the same direction.

8 km 32 m = 8 km + 0.032 km

= 8.032 km

- Add the result to 8 km.

Example 4   **Convert Between Units of Length**

Convert 0.38 m to millimeters.

Solution

0.38 m = 380 mm

Mass and weight are closely related. Weight is a measure of how strongly Earth is pulling on an object. Therefore, an object's weight is less in space than on Earth's surface. However, the amount of material in the object, its **mass**, remains the same. On the surface of Earth, *mass* and *weight* can be used interchangeably.

The gram is the unit of mass in the metric system to which prefixes are added. One gram is about the weight of a paper clip (Figure 11.2).

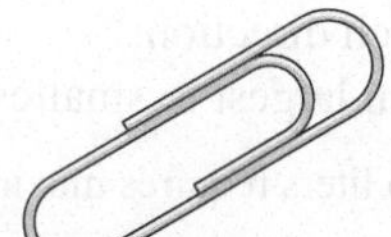

Weight ≈ 1 gram
Figure 11.2

Units of Mass in the Metric System

1 kilogram (kg) = 1000 grams (g)

1 hectogram (hg) = 100 g

1 decagram (dag) = 10 g

1 gram (g) = 1 g

1 decigram (dg) = 0.1 g

1 centigram (cg) = 0.01 g

1 milligram (mg) = 0.001 g

Conversion between units of mass in the metric system involves moving the decimal point to the right or to the left. Listing the units in order from largest to smallest will indicate how many places to move the decimal point and in which direction.

To convert 324 g to kilograms, write the units in order from largest to smallest.

kg   hg   dag   g   dg   cg   mg
        3 positions

$324 \text{ g} = 0.324 \text{ kg}$
        3 places

• Converting grams to kilograms requires moving 3 positions to the left.

• Move the decimal point the same number of places and in the same direction.

Example 5   **Convert Between Units of Mass**

Convert 4.23 g to milligrams.

Solution

$4.23 \text{ g} = 4230 \text{ mg}$

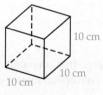

Figure 11.3

The basic unit of capacity in the metric system is the liter. One **liter** is defined as the capacity of a box that is 10 cm long on each side as shown in Figure 11.3.

The units of capacity in the metric system have the same prefixes as the units of length.

Units of Capacity in the Metric System

1 kiloliter (kl) = 1000 L

1 hectoliter (hl) = 100 L

1 decaliter (dal) = 10 L

1 liter (L) = 1 L

1 deciliter (dl) = 0.1 L

1 centiliter (cl) = 0.01 L

1 milliliter (ml) = 0.001 L

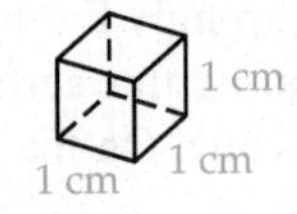

1 ml = 1 cm³

Figure 11.4

The milliliter is equal to 1 **cubic centimeter** (cm³). See Figure 11.4. In medicine, cubic centimeter is often abbreviated cc.

Conversion between units of capacity in the metric system involves moving the decimal point to the right or to the left. Listing the units in order from largest to smallest will indicate how many places to move the decimal point and in which direction.

To convert 824 ml to liters, first write the units in order from largest to smallest.

kl   hl   dal   L   dl   cl   ml   • Converting milliliters to liters requires moving
                                            3 positions to the left.

3 positions

824 ml = 0.824 L   • Move the decimal point the same number of places
                            and in the same direction.

3 places

Example 6   **Convert Between Units of Capacity**

Convert 4 L 32 ml to liters.

Solution

$$32 \text{ ml} = 0.032 \text{ L}$$

$$4 \text{ L } 32 \text{ ml} = 4 \text{ L} + 0.032 \text{ L}$$

$$= 4.032 \text{ L}$$

Convert Between U.S. Customary Units and Metric Units

More than 90% of the world's population uses the metric system of measurement. Therefore, converting U.S. Customary units to metric units is essential in trade and commerce—for example, in importing foreign goods and exporting domestic goods. Approximate equivalences between the two systems follow.

Units of Length	Units of Weight	Units of Capacity
1 in. = 2.54 cm	1 oz ≈ 28.35 g	1 L ≈ 1.06 qt
1 m ≈ 3.28 ft	1 lb ≈ 454 g	1 gal ≈ 3.79 L
1 m ≈ 1.09 yd	1 kg ≈ 2.2 lb	
1 mi ≈ 1.61 km		

These equivalences can be used to form conversion rates to change from one unit of measurement to another. For example, because 1 mi ≈ 1.61 km, the conversion rates $\frac{1 \text{ mi}}{1.61 \text{ km}}$ and $\frac{1.61 \text{ km}}{1 \text{ mi}}$ are both approximately equal to 1.

Convert 55 mi to kilometers.

$$55 \text{ mi} \approx 55 \text{ mi} \times \frac{1.61 \text{ km}}{1 \text{ mi}}$$

$$= \frac{55 \text{ mi}}{1} \times \frac{1.61 \text{ km}}{1 \text{ mi}}$$

$$= \frac{88.55 \text{ km}}{1}$$

$$55 \text{ mi} \approx 88.55 \text{ km}$$

• The conversion rate must contain kilometers in the numerator and miles in the denominator.

Example 7   **Convert U.S. Customary Units to Metric Units**

The price of gasoline is $3.89/gal. Find the cost per liter. Round to the nearest tenth of a cent.

Solution

$$\frac{\$3.89}{\text{gal}} \approx \frac{\$3.89}{\text{gal}} \times \frac{1 \text{ gal}}{3.79 \text{ L}} = \frac{\$3.89}{3.79 \text{ L}} \approx \frac{\$1.026}{1 \text{ L}}$$

$$\$3.89/\text{gal} \approx \$1.026/\text{L}$$

Metric units are used in the United States. Cereal is sold by the gram, 35-mm film is available, and soda is sold by the liter. The same conversion rates used to convert U.S. Customary units to metric units are used to convert metric units to U.S. Customary units.

Example 8   **Convert Metric Units to U.S. Customary Units**

Convert 200 m to feet.

Solution

$$200 \text{ m} \approx 200 \text{ m} \times \frac{3.28 \text{ ft}}{1 \text{ m}} = \frac{656 \text{ ft}}{1}$$

$$200 \text{ m} \approx 656 \text{ ft}$$

Example 9   **Convert Metric Units to U.S. Customary Units**

Convert 90 km/h to miles per hour. Round to the nearest hundredth.

Solution

$$\frac{90 \text{ km}}{\text{h}} \approx \frac{90 \text{ km}}{\text{h}} \times \frac{1 \text{ mi}}{1.61 \text{ km}} = \frac{90 \text{ mi}}{1.61 \text{ h}} \approx \frac{55.90 \text{ mi}}{1 \text{ h}}$$

$$90 \text{ km/h} \approx 55.90 \text{ mi/h}$$

11.1 Exercise Set

▸ Think About It

1. What is the conversion rate to convert inches to feet?
2. How many quarts in one gallon?
3. How many milliliters in one liter?
4. How many kilograms in one gram?
5. To the nearest liter, how many liters in one gallon?
6. Is the length of one yard more than or less than the length of one meter?
7. Is two pounds more than or less than one kilogram?

▸ Preliminary Exercises

For Exercises P1 to P9, if necessary, use the referenced example following the exercise for assistance.

P1. Convert 14 ft to yards. **Example 1**
P2. Convert 3 lb to ounces. **Example 2**
P3. Convert 18 pt to gallons. **Example 3**
P4. Convert 3.07 m to centimeters. **Example 4**
P5. Convert 42.3 mg to grams. **Example 5**
P6. Convert 2 kl 167 L to liters. **Example 6**

P7. The price of milk is \$3.69/gal. Find the cost per liter. Round to the nearest cent. **Example 7**

P8. Convert 45 cm to inches. Round to the nearest hundredth. **Example 8**

P9. Express 75 km/h in miles per hour. Round to the nearest hundredth. **Example 9**

▶ Section Exercises

For Exercises 1 to 24, convert between the two measurements.

1. 6 ft = _____ in.

2. 7920 ft = _____ mi

3. 5 yd = _____ in.

4. $\frac{1}{2}$ mi = _____ yd

5. 64 oz = _____ lb

6. 9000 lb = _____ tons

7. $1\frac{1}{2}$ lb = _____ oz

8. $\frac{4}{5}$ ton = _____ lb

9. $2\frac{1}{2}$ c = _____ fl oz

10. 10 qt = _____ gal

11. 7 gal = _____ pt

12. $1\frac{1}{2}$ qt = _____ c

13. 62 cm = _____ mm

14. 6804 m = _____ km

15. 3.21 m = _____ cm

16. 260 cm = _____ m

17. 7421 g = _____ kg

18. 43 mg = _____ g

19. 0.45 g = _____ dg

20. 0.0456 g = _____ mg

21. 7.5 ml = _____ L

22. 0.037 L = _____ ml

23. 0.435 L = _____ cm³

24. 897 L = _____ kl

For Exercises 25 to 34, solve. Round to the nearest hundredth if necessary.

25. Find the weight in kilograms of a 145-pound person.

26. Find the number of liters in 14.3 gal of gasoline.

27. Express 30 mi/h in kilometers per hour.

28. Seedless watermelon costs \$0.59/lb. Find the cost per kilogram.

29. Deck stain costs \$32.99/gal. Find the cost per liter.

30. Find the weight, in pounds, of an 86-kilogram person.

31. Find the width, in inches, of 35-mm film.

32. Express 30 m/s in feet per second.

33. A 5-kg ham costs \$10/kg. Find the cost per pound.

34. A 2.5-kg bag of grass seed costs \$10.99. Find the cost per pound.

▶ Investigation

As our scientific and technical knowledge has increased, so has our need for ever-smaller and ever-larger units of measure. The prefixes used to denote some of these units of measure are listed in Table 11.1.

Table 11.1

Prefixes for Large Units			Prefixes for Small Units		
Name	Symbol	10^n	Name	Symbol	10^n
yotta	Y	10^{24}	yocto	y	10^{-24}
zetta	Z	10^{21}	zepto	z	10^{-21}
exa	E	10^{18}	atto	a	10^{-18}
peta	P	10^{15}	femto	f	10^{-15}
tera	T	10^{12}	pico	p	10^{-12}
giga	G	10^9	nano	n	10^{-9}
mega	M	10^6	micro	μ	10^{-6}

These new units of measure are quickly working their way into our everyday lives. For example, it is quite easy to purchase a 1-terabyte hard drive. One terabyte (TB) is equal to 10^{12} bytes. As another example, many computers can do multiple operations in 1 nanosecond (ns). One nanosecond is equal to 10^{-9} s.

For Exercises 1 to 10, convert between the two units.

1. 2.3 T = _____ Y

2. 4.51 n = _____ p

3. 0.65 Z = _____ G

4. 9.46 a = _____ μ

5. 4.01 G = _____ E

6. 7.15 y = _____ f

7. The speed of light is approximately 3×10^8 m/s. What is the speed of light in Zm (zettameters) per second?

8. A light year is approximately 6,000,000,000,000,000 mi. Describe a light year using Ym (yottameters).

9. In the 1980s, it became possible to measure optical events in nanoseconds and picoseconds. Express 1 ps as a decimal.

10. A tau lepton, which is an extremely small elementary particle, has a lifetime of 3×10^{-13} s. What is the lifetime of a tau lepton in femtoseconds?

Section

11.2 Basic Concepts of Euclidean Geometry

Learning Objectives:

- Find the measures of complementary angles.
- Find the measures of supplementary angles.
- Find the measures of adjacent angles.
- Solve problems involving intersecting lines.
- Solve problems involving the angles of a triangle.

Lines, Rays, Line Segments, and Angles

The word *geometry* comes from the Greek words for "earth" and "measure." Geometry has applications in such disciplines as physics, medicine, and geology. Geometry is also used in applied fields such as mechanical drawing and astronomy. Geometric forms are used in art and design. We will begin our study by introducing two basic geometric concepts: point and line.

A **point** is symbolized by drawing a dot. A **line** is determined by two distinct points and extends indefinitely in both directions, as the arrows on the line in Figure 11.5 indicate. This line contains points A and B and is represented by $\overleftrightarrow{AB}$. A line can also be represented by a single letter, such as ℓ.

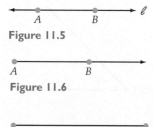

Figure 11.5

Figure 11.6

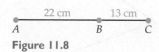

Figure 11.7

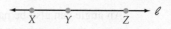

Figure 11.8

A **ray** starts at a point and extends indefinitely in *one* direction. The point at which a ray starts is called the **endpoint** of the ray. The ray shown in Figure 11.6 is denoted $\overrightarrow{AB}$. Point A is the endpoint of the ray.

A **line segment** is part of a line and has two endpoints. The line segment shown in Figure 11.7 is denoted by $\overline{AB}$. The distance between the endpoints of $\overline{AB}$ is denoted by AB.

Given $AB = 22$ cm and $BC = 13$ cm in Figure 11.8, then AC is the sum of the distances AB and BC.

$$AB + BC = AC$$
$$22 + 13 = AC$$
$$35 = AC$$
$$AC = 35 \text{ cm}$$

Example 1   **Use an Equation to Find a Distance on a Line Segment**

X, Y, and Z are all on line ℓ. Given $XY = 9$ m and YZ is twice XY, find XZ.

Solution

$$XZ = XY + YZ$$
$$XZ = XY + 2(XY) \qquad \bullet \; YZ \text{ is twice } XY.$$
$$XZ = 9 + 2(9) \qquad \bullet \text{ Replace } XY \text{ by } 9.$$
$$XZ = 9 + 18 \qquad \bullet \text{ Solve for } XZ.$$
$$XZ = 27$$

$$XZ = 27 \text{ m}$$

In this section, we are discussing figures that lie in a plane. A **plane** is a flat surface with no thickness and no boundaries. It can be pictured as a desktop or whiteboard that extends forever. Figures that lie in a plane are called **plane figures**.

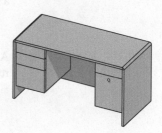

Lines in a plane can be intersecting or parallel. **Intersecting lines** cross at a point in the plane.

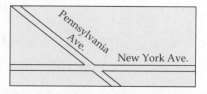

Parallel lines never intersect. The distance between them is always the same.

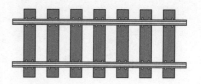

The symbol $\parallel$ means "is parallel to." In Figure 11.9, $j \parallel k$ and $\overleftrightarrow{AB} \parallel \overleftrightarrow{CD}$.

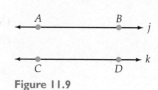

Figure 11.9

Angles

An **angle** is formed by two rays with the same endpoint. The **vertex** of the angle is the point at which the two rays meet. The rays are called the **sides** of the angle.

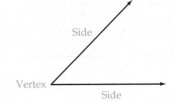

If *A* is a point on one ray of an angle, *C* is a point on the other ray, and *B* is the vertex, then the angle is called ∠*B* or ∠*ABC*, where ∠ is the symbol for angle. Note that an angle can be named by the vertex, or by giving three points, where the second point listed is the vertex. ∠*ABC* could also be called ∠*CBA*.

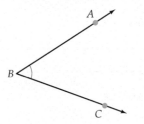

An angle can also be named by a variable written between the rays close to the vertex. In Figure 11.10, ∠*x* and ∠*QRS* are two different names for the same angle. ∠*y* and ∠*SRT* are two different names for the same angle. Note that in Figure 11.10, more than two rays meet at *R*. In this case, the vertex alone cannot be used to name ∠*QRT*.

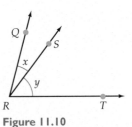

Figure 11.10

An angle is often measured in **degrees**. The symbol for degrees is a small raised circle, °. The angle formed by rotating a ray through a complete circle has a measure of 360°.

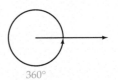

360°

A 90° angle is called a **right angle**. The symbol ⌐ represents a right angle.

90°

Perpendicular lines are intersecting lines that form right angles.

The symbol ⊥ means "is perpendicular to." In Figure 11.11, *p* ⊥ *q* and $\overline{AB}$ ⊥ $\overline{CD}$.

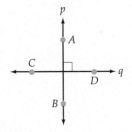

Figure 11.11

A **straight angle** is an angle whose measure is 180°. ∠*AOB* is a straight angle.

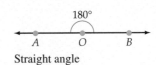

Straight angle

Complementary angles are two angles whose measures have the sum 90°.

∠*A* and ∠*B* in Figure 11.12 are complementary angles.

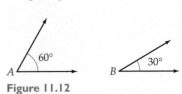

Figure 11.12

Supplementary angles are two angles whose measures have the sum 180°.

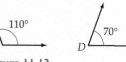

Figure 11.13

∠C and ∠D in Figure 11.13 are supplementary angles.

An **acute angle** is an angle whose measure is between 0° and 90°. ∠D is an acute angle. An **obtuse angle** is an angle whose measure is between 90° and 180°. ∠C is an obtuse angle.

The measure of ∠C is 110°. This is often written as $m \angle C = 110°$, where m is an abbreviation for "the measure of."

Example 2 **Find the Measure of the Complement of an Angle**

Find the measure of the complement of a 38° angle.

Solution

Complementary angles are two angles the sum of whose measures is 90°. To find the measure of the complement, let x represent the complement of a 38° angle. Write an equation and solve for x.

$$x + 38° = 90°$$
$$x = 52°$$

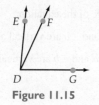

Figure 11.14

Adjacent angles are two co-planar (in the same plane), nonoverlapping angles that share a common vertex and a common side. In Figure 11.14, ∠DAC and ∠CAB are adjacent angles.

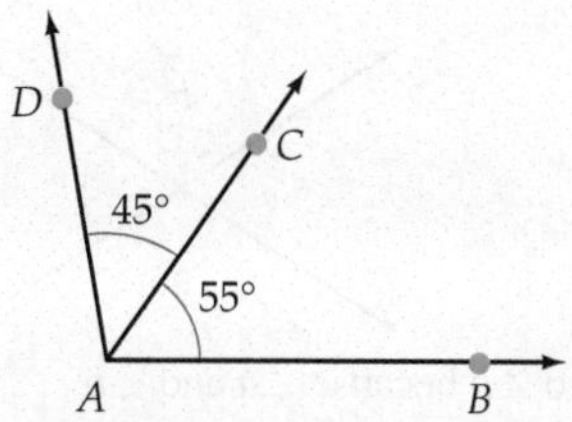

$$m \angle DAC = 45° \text{ and } m \angle CAB = 55°.$$
$$m \angle DAB = m \angle DAC + m \angle CAB$$
$$= 45° + 55° = 100°$$

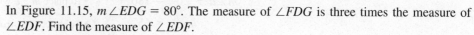

In Figure 11.15, $m \angle EDG = 80°$. The measure of ∠FDG is three times the measure of ∠EDF. Find the measure of ∠EDF.

Figure 11.15

Let x = the measure of ∠EDF.

Then $3x$ = the measure of ∠FDG.

Write an equation and solve for x.

$$m \angle EDF + m \angle FDG = m \angle EDG$$
$$x + 3x = 80°$$
$$4x = 80°$$
$$x = 20°$$

$$m \angle EDF = 20°$$

Example 3 **Find the Measure of an Adjacent Angle**

Given that $m \angle ABC$ is 84°, find the measure of ∠x.

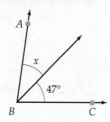

Solution

To find the measure of ∠x, write an equation using the fact that the sum of the measures of ∠x and 47° is 84°. Solve for $m \angle x$.

$$m \angle x + 47° = 84°$$
$$m \angle x = 37°$$

Angles Formed by Intersecting Lines

Four angles are formed by the intersection of two lines. If the two lines are not perpendicular, then two of the angles formed are acute angles and two of the angles formed are obtuse angles. The two acute angles are always opposite each other, and the two obtuse angles are always opposite each other.

The nonadjacent angles formed by two intersecting lines are called **vertical angles**. $\angle w$ and $\angle y$ in Figure 11.16 are vertical angles. $\angle x$ and $\angle z$ are vertical angles.

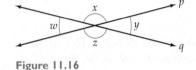

Figure 11.16

Vertical angles have the same measure.

$$m\angle w = m\angle y \qquad\qquad m\angle x = m\angle z$$

In Figure 11.16, $\angle x$ and $\angle y$ are adjacent angles, as are $\angle y$ and $\angle z$, $\angle z$ and $\angle w$, and $\angle w$ and $\angle x$.

Adjacent angles formed by intersecting lines are supplementary angles.

$$m\angle x + m\angle y = 180° \qquad\qquad m\angle z + m\angle w = 180°$$

$$m\angle y + m\angle z = 180° \qquad\qquad m\angle w + m\angle x = 180°$$

Example 4	Solve a Problem Involving Intersecting Lines

In the diagram at the right, $m\angle b = 115°$.
Find the measures of angles a, c, and d.

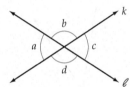

Solution

$$m\angle a + m\angle b = 180° \qquad \text{• } \angle a \text{ is supplementary to } \angle b \text{ because } \angle a \text{ and } \angle b$$
are adjacent angles of intersecting lines.

$$m\angle a + 115° = 180° \qquad \text{• Replace } m\angle b \text{ with } 115°.$$

$$m\angle a = 65° \qquad \text{• Subtract } 115° \text{ from each side of the equation.}$$

$$m\angle c = 65° \qquad \text{• } m\angle c = m\angle a \text{ because } \angle c \text{ and } \angle a \text{ are vertical angles.}$$

$$m\angle d = 115° \qquad \text{• } m\angle d = m\angle b \text{ because } \angle d \text{ and } \angle b \text{ are vertical angles.}$$

$m\angle a = 65°$, $m\angle c = 65°$, and $m\angle d = 115°$.

A line that intersects two other lines at different points is called a **transversal**. In Figure 11.17, ℓ_1 and ℓ_2 are parallel lines. The eight angles formed by the transversal t that intersects the parallel lines have certain properties.

$\angle c$ and $\angle w$ are **alternate interior angles**; $\angle d$ and $\angle x$ are also alternate interior angles.

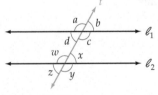

Figure 11.17

Alternate interior angles are equal.

$$m\angle c = m\angle w \qquad\qquad m\angle d = m\angle x$$

$\angle a$ and $\angle y$ are **alternate exterior angles**; $\angle b$ and $\angle z$ are also alternate exterior angles.

Alternate exterior angles are equal.

$$m\angle a = m\angle y \qquad\qquad m\angle b = m\angle z$$

$\angle a$ and $\angle w$, $\angle d$ and $\angle z$, $\angle b$ and $\angle x$, and $\angle c$ and $\angle y$ are **corresponding angles**.

Corresponding angles are equal.

$$m\angle a = m\angle w \qquad\qquad m\angle b = m\angle x$$

$$m\angle d = m\angle z \qquad\qquad m\angle c = m\angle y$$

Example 5 **Solve a Problem Involving Parallel Lines Cut by a Transversal**

In the diagram at the right, $\ell_1 \| \ell_2$ and $m \angle f = 58°$.
Find the measures of $\angle a$, $\angle c$, and $\angle d$.

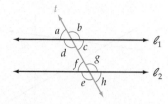

Solution

$$m \angle a = m \angle f = 58°$$ • $\angle a$ and $\angle f$ are corresponding angles.

$$m \angle c = m \angle f = 58°$$ • $\angle c$ and $\angle f$ are alternate interior angles.

$$m \angle d + m \angle a = 180°$$ • $\angle d$ is supplementary to $\angle a$.

$$m \angle d + 58° = 180°$$ • Replace $m \angle a$ with $58°$.

$$m \angle d = 122°$$ • Subtract $58°$ from each side of the equation.

$m \angle a = 58°$, $m \angle c = 58°$, and $m \angle d = 122°$.

Angles of a Triangle

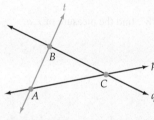

Figure 11.18

Figure 11.18 shows three intersecting lines. The plane figure formed by the line segments AB, BC, and AC is called a **triangle**.

The angles within the region enclosed by the triangle are called **interior angles**. In Figure 11.19, angles a, b, and c are interior angles. The sum of the measures of the interior angles of a triangle is $180°$.

$$m \angle a + m \angle b + m \angle c = 180°$$

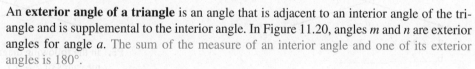

The Sum of the Measures of the Interior Angles of a Triangle

The sum of the measures of the interior angles of a triangle is $180°$.

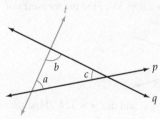

Figure 11.19

An **exterior angle of a triangle** is an angle that is adjacent to an interior angle of the triangle and is supplemental to the interior angle. In Figure 11.20, angles m and n are exterior angles for angle a. The sum of the measure of an interior angle and one of its exterior angles is $180°$.

$$m \angle a + m \angle m = 180°$$
$$m \angle a + m \angle n = 180°$$

Example 6 **Find the Measure of the Third Angle of a Triangle**

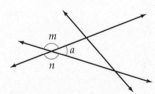

Figure 11.20

Two angles of a triangle measure $43°$ and $86°$. Find the measure of the third angle.

Solution

Use the fact that the sum of the measures of the interior angles of a triangle is $180°$. Write an equation using x to represent the measure of the third angle. Solve the equation for x.

$$x + 43° + 86° = 180°$$

$$x + 129° = 180°$$ • Add $43° + 86°$.

$$x = 51°$$ • Subtract $129°$ from each side of the equation.

The measure of the third angle is $51°$.

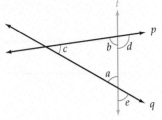

Figure II.21

| Example 7 | Solve a Problem Involving the Angles of a Triangle |

In Figure 11.21, $m\angle c = 40°$ and $m\angle e = 60°$. Find the measure of $\angle d$.

Solution

$$m\angle a = m\angle e = 60°$$ • $\angle a$ and $\angle e$ are vertical angles.

$$m\angle c + m\angle a + m\angle b = 180°$$ • The sum of the interior angles is $180°$.

$$40° + 60° + m\angle b = 180°$$ • Replace $m\angle c$ with $40°$ and $m\angle a$ with $60°$.

$$100° + m\angle b = 180°$$ • Add $40° + 60°$.

$$m\angle b = 80°$$ • Subtract $100°$ from each side of the equation.

$$m\angle b + m\angle d = 180°$$ • $\angle b$ and $\angle d$ are supplementary angles.

$$80° + m\angle d = 180°$$ • Replace $m\angle b$ with $80°$.

$$m\angle d = 100°$$ • Subtract $80°$ from each side of the equation.

$$m\angle d = 100°$$

11.2 Exercise Set

Think About It

1. Classify each diagram as a line, a ray, or a line segment.

   a.
   E   F

   b.
   C   D

   c.
   J   K

2. Use the following figure for parts a through d.

   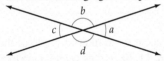

   a. $\angle a = \angle$ _____
   b. $\angle d = \angle$ _____
   c. $m\angle c + m\angle D =$ _____°
   d. $\angle a$ and $\angle c$ are called _____ angles.

3. What is the measure of a right angle?

4. The measure of an obtuse angle is less than or greater than a right angle?

5. An _____ angle has a measure of less than a right angle.

6. If a, b, and c are the measures of three angles of a triangle, then the sum of the measures is _____°.

Preliminary Exercises

For Exercises P1 to P7, if necessary, use the referenced example following the exercise for assistance.

P1. A, B, and C are all on line ℓ. Given $BC = 16$ ft and $AB = \frac{1}{4}(BC)$, find AC. **Example 1**

A   B   C

P2. Find the measure of the supplement of a 129° angle.
Example 2

P3. Given that the $m\angle DEF$ is 118°, find the measure of $\angle a$.
Example 3

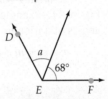

P4. In the following diagram, $m\angle a = 35°$. Find the measures of angles b, c, and d. **Example 4**

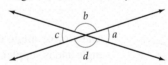

P5. In the following diagram, $\ell_1 \parallel \ell_2$ and $m\angle g = 124°$. Find the measures of $\angle b$, $\angle c$, and $\angle d$. **Example 5**

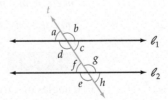

P6. One angle in a triangle is a right angle, and one angle measures 27°. Find the measure of the third angle. **Example 6**

P7. In the following diagram, $m\angle c = 35°$ and $m\angle d = 105°$. Find the measure of $\angle e$. **Example 7**

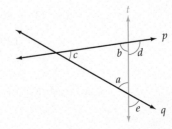

Section Exercises

1. Provide three names for this angle.

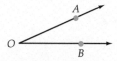

2. State the number of degrees in a full circle, a straight angle, and a right angle.

3. Find the complement of a 62° angle.

4. Find the complement of a 31° angle.

5. Find the supplement of a 162° angle.

6. Find the supplement of a 72° angle.

For Exercises 7 to 10, determine whether the described angle is an acute angle, is a right angle, is an obtuse angle, or does not exist.

7. The complement of an acute angle

8. The supplement of a right angle

9. The supplement of an acute angle

10. The supplement of an obtuse angle

11. Given $AB = 12$ cm, $CD = 9$ cm, and $AD = 35$ cm, find the length of $\overline{BC}$.

12. Given $AB = 21$ mm, $BC = 14$ mm, and $AD = 54$ mm, find the length of $\overline{CD}$.

13. Given $QR = 7$ ft and RS is three times the length of $\overline{QR}$, find the length of $\overline{QS}$.

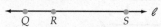

14. Given $QR = 15$ in. and RS is twice the length of $\overline{QR}$, find the length of $\overline{QS}$.

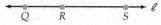

15. Given $m \angle LOM = 53°$ and $m \angle LON = 139°$, find the measure of $\angle MON$.

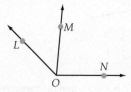

16. Given $m \angle MON = 38°$ and $m \angle LON = 85°$, find the measure of $\angle LOM$.

In Exercises 17 and 18, find the measure of $\angle x$.

17.   **18.**

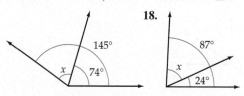

In Exercises 19 and 20, given that $\angle LON$ is a right angle, find the measure of $\angle x$.

19.   **20.**

In Exercises 21 to 24, find the measure of $\angle a$.

21.

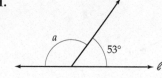

22.

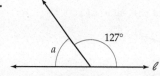

23.   **24.**

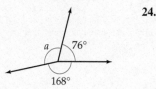   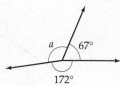

In Exercises 25 to 28, find the value of x.

25.

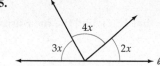

26.

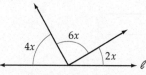

27.   **28.**

29. Given $m \angle a = 51°$, find the measure of $\angle b$.

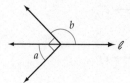

30. Given $m \angle a = 38°$, find the measure of $\angle b$.

38.

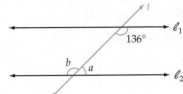

In Exercises 31 and 32, find the measure of $\angle x$.

31.

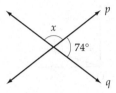

32.

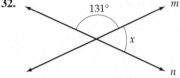

In Exercises 33 and 34, find the value of x.

33.

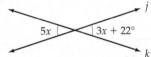

34.

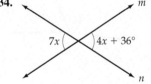

In Exercises 35 to 38, given that $\ell_1 \parallel \ell_2$, find the measures of angles a and b.

35.

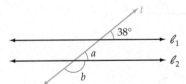

36.

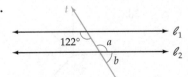

37.

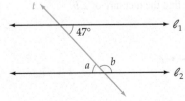

In Exercises 39 to 42, given that $\ell_1 \parallel \ell_2$, find x.

39.

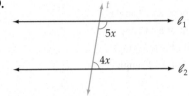

40.

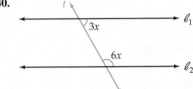

41.

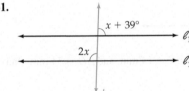

42.

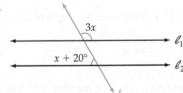

43. Given that $m \angle a = 95°$ and $m \angle b = 70°$, find the measures of angles x and y.

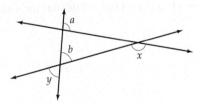

44. Given that $m \angle a = 35°$ and $m \angle b = 55°$, find the measures of angles x and y.

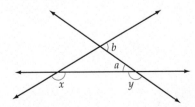

45. Given that $m \angle y = 45°$, find the measures of angles a and b.

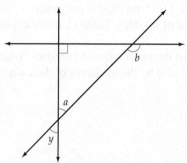

46. Given that $m \angle y = 130°$, find the measures of angles a and b.

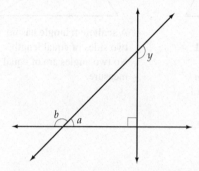

47. One angle in a triangle is a right angle, and one angle is equal to 30°. What is the measure of the third angle?

48. A triangle has a 45° angle and a right angle. Find the measure of the third angle.

49. Two angles of a triangle measure 42° and 103°. Find the measure of the third angle.

50. Two angles of a triangle measure 62° and 45°. Find the measure of the third angle.

For Exercises 51 to 53, determine whether the statement is true or false.

51. A triangle can have two obtuse angles.

52. The legs of a right triangle are perpendicular.

53. If the sum of two angles of a triangle is less than 90°, then the third angle is an obtuse angle.

▶ **Investigations**

1. Cut out a triangle and then tear off two of the angles, as shown in the following diagram. Position angle a so that it is to the left of angle b and is adjacent to angle b. Now position angle c so that it is to the right of angle b and is adjacent to angle b. Describe what you observe. What does this demonstrate?

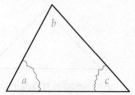

2. For the following figure, find the sum of the measures of angles x, y, and z.

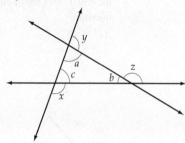

3. For the above figure, explain why $m \angle a + m \angle b = m \angle x$. Write a rule that describes the relationship between the measure of an exterior angle of a triangle and the sum of the measures of its two opposite interior angles (the interior angles that are non-adjacent to the exterior angle). Use the rule to write an equation involving angles a, c, and z.

4. If $\overline{AB}$ and $\overline{CD}$ intersect at point O, and $m \angle AOC = m \angle BOC$, explain why $\overline{AB} \perp \overline{CD}$.

Section **11.3**

Perimeter and Area of Plane Figures

Learning Objectives:

• Find the perimeter of plane geometric figures.

• Find the area of plane geometric figures.

Perimeter of Plane Geometric Figures

A **polygon** is a closed figure determined by three or more line segments that lie in a plane. The line segments that form the polygon are called its **sides**. Figure 11.22 shows examples of polygons.

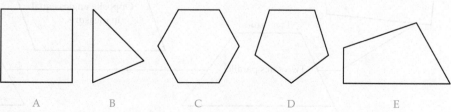

| A | B | C | D | E |

Figure 11.22

Table 11.2

Number of Sides	Name of Polygon
3	Triangle
4	Quadrilateral
5	Pentagon
6	Hexagon
7	Heptagon
8	Octagon
9	Nonagon
10	Decagon

A **regular polygon** is one in which each side has the same length and each angle has the same measure. The polygons A, C, and D in Figure 11.22 are regular polygons.

The name of a polygon is based on the number of its sides. Table 11.2 lists the names of polygons that have from 3 to 10 sides.

Triangles and quadrilaterals are two of the most common types of polygons. Triangles are distinguished by the number of equal sides and also by the measures of their angles.

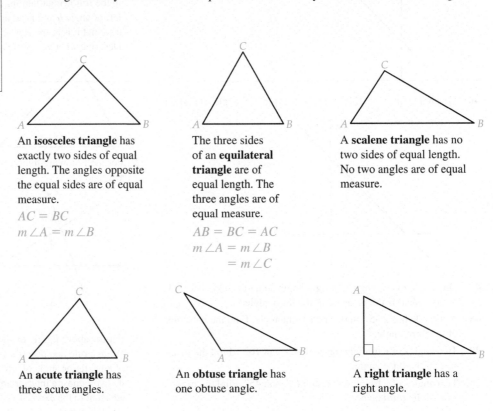

An **isosceles triangle** has exactly two sides of equal length. The angles opposite the equal sides are of equal measure.
$AC = BC$
$m \angle A = m \angle B$

The three sides of an **equilateral triangle** are of equal length. The three angles are of equal measure.
$AB = BC = AC$
$m \angle A = m \angle B$
$\qquad = m \angle C$

A **scalene triangle** has no two sides of equal length. No two angles are of equal measure.

An **acute triangle** has three acute angles.

An **obtuse triangle** has one obtuse angle.

A **right triangle** has a right angle.

A **quadrilateral** is a four-sided polygon. Quadrilaterals are also distinguished by their sides and angles, as shown in Figure 11.23. Note that a rectangle, a square, and a rhombus are different forms of a parallelogram.

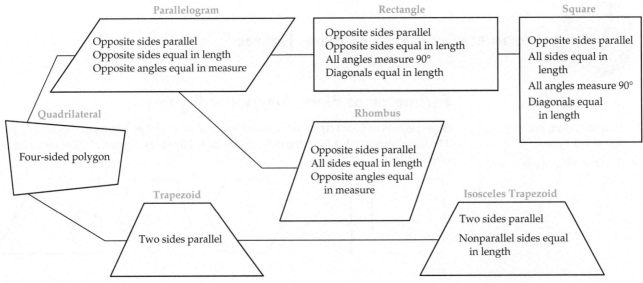

Figure 11.23 •

The **perimeter** of a plane geometric figure is a measure of the distance around the figure. Perimeter is used, for example, when buying fencing for a garden or determining how much baseboard is needed for a room.

The perimeter of a triangle is the sum of the lengths of the three sides.

Perimeter of a Triangle

Let a, b, and c be the lengths of the sides of a triangle. The perimeter, P, of the triangle is given by $P = a + b + c$.

$$P = a + b + c$$

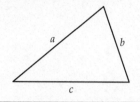

To find the perimeter of the triangle shown in Figure 11.24, add the lengths of the three sides.

$$P = 5 + 7 + 10 = 22$$

The perimeter is 22 ft.

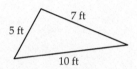

Figure 11.24

A **rectangle** is a quadrilateral with all right angles and opposite sides of equal length. Usually the length, L, of a rectangle refers to the length of one of the longer sides of the rectangle and the width, W, refers to the length of one of the shorter sides. The perimeter can then be represented as $P = L + W + L + W$.

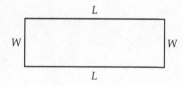

The formula for the perimeter of a rectangle is derived by combining like terms.

$$P = L + W + L + W$$
$$P = 2L + 2W$$

Perimeter of a Rectangle

Let L represent the length and W the width of a rectangle. The perimeter, P, of the rectangle is given by $P = 2L + 2W$.

A **square** is a rectangle in which each side has the same length. Letting s represent the length of each side of a square, the perimeter of the square can be represented by $P = s + s + s + s$.

The formula for the perimeter of a square is derived by combining like terms.

$$P = s + s + s + s$$
$$P = 4s$$

Perimeter of a Square

Let s represent the length of a side of a square. The perimeter, P, of the square is given by $P = 4s$.

Figure $ABCD$ is a parallelogram. $\overline{BC}$ is the **base** of the parallelogram. Opposite sides of a parallelogram are equal in length, so $\overline{AD}$ is the same length as $\overline{BC}$, and $\overline{AB}$ is the same length as $\overline{CD}$.

Let b represent the length of the base and s the length of an adjacent side. Then the perimeter of a parallelogram can be represented as $P = b + s + b + s$.

$$P = b + s + b + s$$

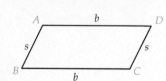

The formula for the perimeter of a parallelogram is derived by combining like terms.

$$P = 2b + 2s$$

Perimeter of a Parallelogram

Let b represent the length of the base of a parallelogram and s the length of a side adjacent to the base. The perimeter, P, of the parallelogram is given by $P = 2b + 2s$.

Example 1   **Find the Perimeter of a Rectangle**

You want to trim a rectangular frame with a metal strip. The frame measures 30 in. by 20 in. Find the length of metal strip you will need to trim the frame.

Solution

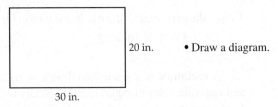

20 in.   • Draw a diagram.

30 in.

$P = 2L + 2W$ • Use the formula for the perimeter of a rectangle.

$P = 2(30) + 2(20)$ • The length is 30 in. Substitute 30 for L. The width is 20 in. Substitute 20 for W.

$P = 60 + 40$

$P = 100$

You will need 100 in. of the metal strip.

Example 2   **Find the Perimeter of a Square**

Find the length of fencing needed to surround a square corral that measures 60 ft on each side.

Solution

60 ft   • Draw a diagram.

$P = 4s$ • Use the formula for the perimeter of a square.

$P = 4(60)$ • The length of a side is 60 ft. Substitute 60 for s.

$P = 240$

240 ft of fencing are needed.

A **circle** is a plane figure in which all points are the same distance from point O, called the **center** of the circle.

A **diameter** of a circle is a line segment with endpoints on the circle and passing through the center. $\overline{AB}$ is a diameter of the circle in Figure 11.25. The variable d is used to designate the length of a diameter of a circle.

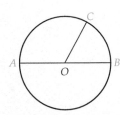

Figure 11.25

A **radius** of a circle is a line segment from the center of the circle to a point on the circle. $\overline{OC}$ is a radius of the circle in Figure 11.25. The variable r is used to designate the length of a radius of a circle.

The length of the diameter is twice the length of the radius.

$$d = 2r \text{ or } r = \frac{1}{2}d$$

The distance around a circle is called the **circumference**.

Circumference of a Circle

The circumference, C, of a circle with diameter d and radius r is given by $C = \pi d$ or $C = 2\pi r$.

The formula for circumference uses the number π (pi), which is an irrational number. The value of π can be approximated by a fraction or by a decimal.

$$\pi \approx 3\frac{1}{7} = \frac{22}{7} \text{ or } \pi \approx 3.14$$

The π key on a scientific calculator gives a closer approximation of π than 3.14. A scientific calculator is used in this section to find approximate values in calculations involving π.

Example 3 **Find the Circumference of a Circle**

Find the circumference of a circle with a radius of 15 cm. Round to the nearest hundredth of a centimeter.

Solution

$C = 2\pi r$ • The radius is given. Use the circumference formula that involves the radius.

$C = 2\pi(15)$ • Replace r with 15.

$C = 30\pi$ • Multiply 2 times 15.

$C \approx 94.25$ • An approximation is asked for. Use the π key on a calculator.

The circumference of the circle is approximately 94.25 cm.

Example 4 **Application of Finding the Circumference of a Circle**

A bicycle tire has a diameter of 24 in. How many feet does the bicycle travel when the wheel makes 8 revolutions? Round to the nearest hundredth of a foot.

24 in.

Solution

24 in. = 2 ft • The diameter is given in inches, but the answer must be expressed in feet. Convert the diameter (24 in.) to feet. There are 12 in. in 1 ft. Divide 24 by 12.

$C = \pi d$ • The diameter is given. Use the circumference formula that involves the diameter.

$C = \pi(2)$ • Replace d with 2.

$C = 2\pi$ • This is the distance traveled in 1 revolution.

$8C = 8(2\pi) = 16\pi \approx 50.27$ • Find the distance traveled in 8 revolutions.

The bicycle will travel about 50.27 ft when the wheel makes 8 revolutions.

1 in²

1 cm²

Area of Plane Geometric Figures

Area is the amount of surface in a region. Area can be used to describe, for example, the size of a rug, a parking lot, a farm, or a national park. Area is measured in square units.

A square that measures 1 in. on each side has an area of 1 square inch, written 1 in².

A square that measures 1 cm on each side has an area of 1 square centimeter, written 1 cm².

Larger areas are often measured in square feet (ft²), square meters (m²), square miles (mi²), acres (43,560 ft²), or any other square unit.

Area of a Rectangle

Let L represent the length and W the width of a rectangle. The area, A, of the rectangle is given by $A = LW$.

Example 5 Find the Area of a Rectangle

How many square feet of sod are needed to cover a football field? A football field measures 360 ft by 160 ft.

Solution

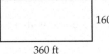

160 ft
360 ft

• Draw a diagram.

$A = LW$

• Use the formula for the area of a rectangle.

$A = 360(160)$

• The length is 360 ft. Substitute 360 for L.
The width is 160 ft. Substitute 160 for W.
Remember that LW means "L times W."

$A = 57{,}600$

57,600 ft² of sod is needed.

• Area is measured in square units.

---

s

A square is a rectangle in which all sides are the same length. Therefore, both the length and the width of a square can be represented by s, and $A = LW = s \cdot s = s^2$.

$A = s \cdot s$
$A = s^2$

Area of a Square

Let s represent the length of a side of a square. The area, A, of the square is given by $A = s^2$.

Example 6 Find the Area of a Square

A homeowner wants to carpet the family room. The floor is square and measures 6 m on each side. How much carpet should be purchased?

Solution

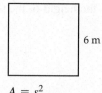

6 m   • Draw a diagram.

$A = s^2$ • Use the formula for the area of a square.

$A = 6^2$ • The length of a side is 6 m.
                                                   Substitute 6 for s.
$A = 36$

36 m² of carpet should be purchased.   • Area is measured in square units.

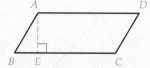

Figure $ABCD$ is a parallelogram. $\overline{BC}$ is the **base** of the parallelogram. $\overline{AE}$, perpendicular to the base, is the **height** of the parallelogram.

Any side of a parallelogram can be designated as the base. The corresponding height is found by drawing a line segment perpendicular to the base from the opposite side. In the figure to the right, $\overline{CD}$ is the base and $\overline{AE}$ is the height.

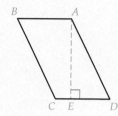

A rectangle can be formed from a parallelogram by cutting a right triangle from one end of the parallelogram and attaching it to the other end. The area of the resulting rectangle will equal the area of the original parallelogram.

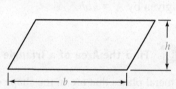   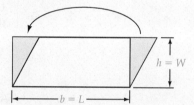

Area of a Parallelogram

Let b represent the length of the base and h the height of a parallelogram. The area, A, of the parallelogram is given by $A = bh$.

Example 7   **Find the Area of a Parallelogram**

A solar panel is in the shape of a parallelogram that has a base of 2 ft and a height of 3 ft. Find the area of the solar panel.

Solution

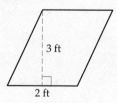

3 ft   • Draw a diagram.

2 ft

$A = bh$ • Use the formula for the area of a parallelogram.

$A = 2(3)$ • The base is 2 ft. Substitute 2 for b.
                                                   The height is 3 ft. Substitute 3 for h.
                                                   Remember that bh means "b times h."

$A = 6$

The area is 6 ft².   • Area is measured in square units.

Figure *ABC* is a triangle. $\overline{AB}$ is the **base** of the triangle. $\overline{CD}$, perpendicular to the base, is the **height** of the triangle.

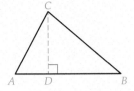

Any side of a triangle can be designated as the base. The corresponding height is found by drawing a line segment perpendicular to the base from the vertex opposite the base.

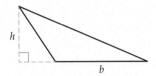

Consider triangle *ABC* with base *b* and height *h*, shown at the right. By extending a line segment from *C* parallel to the base $\overline{AB}$ and equal in length to the base, a parallelogram is formed. The area of the parallelogram is *bh* and is twice the area of the original triangle. Therefore, the area of the triangle is one half the area of the parallelogram, or $\frac{1}{2}bh$.

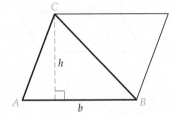

Area of a Triangle

Let *b* represent the length of the base and *h* the height of a triangle. The area, *A*, of the triangle is given by $A = \frac{1}{2}bh$.

Example 8   **Find the Area of a Triangle**

A riveter uses metal plates that are in the shape of a triangle with a base of 12 cm and a height of 6 cm. Find the area of one metal plate.

Solution

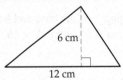

• Draw a diagram.

$A = \dfrac{1}{2}bh$ • Use the formula for the area of a triangle.

$A = \dfrac{1}{2}(12)(6)$ • The base is 12 cm. Substitute 12 for *b*. The height is 6 cm. Substitute 6 for *h*. Remember that *bh* means "*b* times *h*."

$A = 6(6)$

$A = 36$

The area is 36 cm².   • Area is measured in square units.

Take Note: The bases of a trapezoid are the parallel sides of the figure.

Figure *ABCD* is a trapezoid. $\overline{AB}$, with length b_1, is one **base** of the trapezoid and $\overline{CD}$, with length b_2, is the other base. $\overline{AE}$, perpendicular to the two bases, is the **height**.

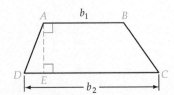

In the trapezoid in Figure 11.26 the line segment $\overline{BD}$ divides the trapezoid into two triangles, *ABD* and *BCD*. In triangle *ABD*, b_1 is the base and *h* is the height. In triangle *BCD*, b_2 is the base and *h* is the height. The area of the trapezoid is the sum of the areas of the two triangles.

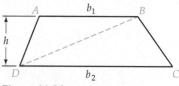

Figure 11.26

Area of trapezoid *ABCD* = Area of triangle *ABD* + area of triangle *BCD*

$$= \frac{1}{2}b_1 h + \frac{1}{2}b_2 h = \frac{1}{2}h(b_1 + b_2)$$

Area of a Trapezoid

Let b_1 and b_2 represent the lengths of the bases and *h* the height of a trapezoid. The area, *A*, of the trapezoid is given by $A = \frac{1}{2}h(b_1 + b_2)$.

Example 9   **Find the Area of a Trapezoid**

A boat dock is built in the shape of a trapezoid with bases measuring 14 ft and 6 ft and a height of 7 ft. Find the area of the dock.

Solution

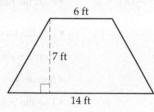

• Draw a diagram.

$A = \dfrac{1}{2}h(b_1 + b_2)$ • Use the formula for the area of a trapezoid.

$A = \dfrac{1}{2} \cdot 7(14 + 6)$ • The height is 7 ft. Substitute 7 for *h*. The bases measure 14 ft and 6 ft. Substitute 14 and 6 for b_1 and b_2.

$A = \dfrac{1}{2} \cdot 7(20)$

$A = 70$

The area is 70 ft^2.   • Area is measured in square units.

The area of the circle in Figure 11.27 is the product of π and the square of the radius.

$$A = \pi r^2$$

Figure 11.27

The Area of a Circle

The area, *A*, of a circle with radius of length *r* is given by $A = \pi r^2$.

| Example 10 | Find the Area of a Circle |

Find the area of a circle with a diameter of 10 m. Round to the nearest hundredth of a square meter.

Solution

$$r = \frac{1}{2}d = \frac{1}{2}(10) = 5 \qquad \text{• Find the radius of the circle.}$$

$$A = \pi r^2 \qquad \text{• Use the formula for the area of a circle.}$$

$$A = \pi(5)^2 \qquad \text{• Replace } r \text{ with } 5.$$

$$A = \pi(25) \qquad \text{• Square 5.}$$

$$A \approx 78.54 \qquad \text{• An approximation is asked for. Use the } \pi \text{ key on a calculator.}$$

The area of the circle is approximately 78.54 m².

| Example 11 | Application of Finding the Area of a Circle |

How large a cover is needed for a circular hot tub that is 8 ft in diameter? Round to the nearest tenth of a square foot.

Solution

$$r = \frac{1}{2}d = \frac{1}{2}(8) = 4 \qquad \text{• Find the radius of a circle with a diameter of 8 ft.}$$

$$A = \pi r^2 \qquad \text{• Use the formula for the area of a circle.}$$

$$A = \pi(4)^2 \qquad \text{• Replace } r \text{ with } 4.$$

$$A = \pi(16) \qquad \text{• Square 4.}$$

$$A \approx 50.3 \qquad \text{• Use the } \pi \text{ key on a calculator.}$$

The cover for the hot tub must be 50.3 ft².

11.3 Exercise Set

Think About It

1. What distinguishes a rectangle from other parallelograms?

2. What distinguishes a rhombus from a square?

3. What is the area of a square that measures 1 meter on each side?

4. How many squares, each 1 foot on each side, are required to cover a rectangle that measures 20 ft²?

5. True or false: $\pi = \frac{22}{7}$

6. Which of the following are possible measures of the area of a triangle?
   a. 27 in²
   b. 37 cm
   c. π m²
   d. $\sqrt{5}$ ft

7. Which of the following are possible measures of the circumference of a circle?
   a. 27 in²
   b. 37 cm
   c. π m²
   d. $\frac{2}{3}$ ft

Preliminary Exercises

For Exercises P1 to P11, if necessary, use the referenced example following the exercise for assistance.

P1. Find the length of decorative molding needed to edge the top of the walls in a rectangular room that is 12 ft long and 8 ft wide. **Example 1**

P2. A homeowner plans to fence in the area around the swimming pool in the backyard. The area to be fenced in is a square measuring 24 ft on each side. How many feet of fencing should the homeowner purchase? **Example 2**

P3. Find the circumference of a circle with a diameter of 9 km. Give the exact measure. **Example 3**

P4. A tricycle tire has a diameter of 12 in. How many feet does the tricycle travel when the wheel makes 12 revolutions? Round to the nearest hundredth of a foot. **Example 4**

P5. Find the amount of fabric needed to make a rectangular flag that measures 308 cm by 192 cm. **Example 5**

P6. Find the area of the floor of a two-car garage that is in the shape of a square that measures 24 ft on a side. **Example 6**

P7. A fieldstone patio is in the shape of a parallelogram that has a base measuring 14 m and a height measuring 8 m. What is the area of the patio? **Example 7**

P8. Find the amount of felt needed to make a banner that is in the shape of a triangle with a base of 18 in. and a height of 9 in. **Example 8**

P9. Find the area of a patio that has the shape of a trapezoid with a height of 9 ft and bases measuring 12 ft and 20 ft. **Example 9**

P10. Find the area of a circle with a diameter of 12 km. Give the exact measure. **Example 10**

P11. How much material is needed to make a circular tablecloth that is to have a diameter of 4 ft? Round to the nearest hundredth of a square foot. **Example 11**

▶ Section Exercises

1. What is wrong with each statement?
  a. The perimeter is 40 m².
  b. The area is 120 ft.

In Exercises 2 to 8, find (a) the perimeter and (b) the area of the figure.

2.
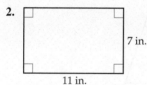
7 in.
11 in.

3.
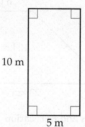
10 m
5 m

4.
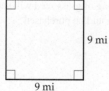
9 mi
9 mi

5.
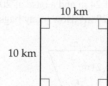
10 km
10 km

6.
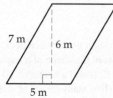
7 m
6 m
5 m

7.
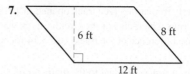
6 ft
8 ft
12 ft

8.
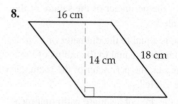
16 cm
14 cm
18 cm

In Exercises 9 to 14, find (a) the circumference and (b) the area of the figure. State an exact answer and a decimal approximation rounded to the nearest hundredth.

9.

4 cm

10.
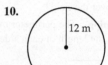
12 m

11.
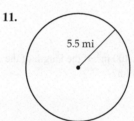
5.5 mi

12.
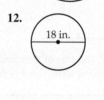
18 in.

13.

17 ft

14.
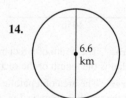
6.6 km

15. Perimeter Find the perimeter of a regular pentagon that measures 4 in. on each side.

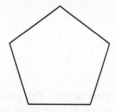

16. Interior Decorating Wall-to-wall carpeting is installed in a room that is 15 ft long and 10 ft wide. The edges of the carpet are held down by tack strips. How many feet of tack-strip material are needed?

17. Cross Country A cross-country course is in the shape of a parallelogram with a base of length 3 mi and a side of length 2 mi. What is the total length of the cross-country course?

18. Parks and Recreation A rectangular playground has a length of 160 ft and a width of 120 ft. Find the length of hedge that surrounds the playground.

19. Sewing Bias binding is to be sewn around the edge of a rectangular tablecloth measuring 68 in. by 42 in. If the bias binding comes in packages containing 15 ft of binding, how many packages of bias binding are needed for the tablecloth?

20. **Race Tracks** The first circular dog race track opened in 1919 in Emeryville, California. The radius of the circular track was 157.64 ft. Find the circumference of the track. Use 3.14 for π. Round to the nearest foot.

21. The length of a side of a square is equal to the diameter of a circle. Which is greater, the perimeter of the square or the circumference of the circle?

22. The length of a rectangle is equal to the diameter of a circle, and the width of the rectangle is equal to the radius of the same circle. Which is greater, the perimeter of the rectangle or the circumference of the circle?

23. **Construction** What is the area of a square patio that measures 12 m on each side?

24. **Athletic Fields** Artificial turf is being used to cover a playing field. If the field is rectangular with a length of 110 yd and a width of 80 yd, how much artificial turf must be purchased to cover the field?

25. **Framing** The perimeter of a square picture frame is 36 in. Find the length of each side of the frame.

26. **Area** The area of a rectangle is 400 in². If the length of the rectangle is 40 in., what is the width?

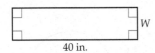

27. **Area** The width of a rectangle is 8 ft. If the area is 312 ft², what is the length of the rectangle?

28. **Area** The area of a parallelogram is 56 m². If the height of the parallelogram is 7 m, what is the length of the base?

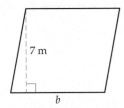

29. **Storage Units** You want to rent a storage unit. You estimate that you will need 175 ft² of floor space. You see the ad below on the Internet. You want to rent the smallest possible unit that will hold everything you want to store. Which of the six units pictured in the ad should you select?

30. **Sailing** A sail is in the shape of a triangle with a base of 12 m and a height of 16 m. How much canvas was needed to make the body of the sail?

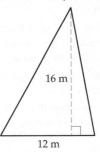

31. **Gardens** A vegetable garden is in the shape of a triangle with a base of 21 ft and a height of 13 ft. Find the area of the vegetable garden.

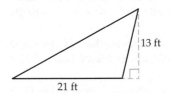

32. **Athletic Fields** How much artificial turf should be purchased to cover an athletic field that is in the shape of a trapezoid with a height of 15 m and bases that measure 45 m and 36 m?

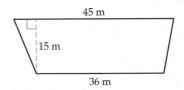

33. **Land Area** A township is in the shape of a trapezoid with a height of 10 km and bases measuring 9 km and 23 km. What is the land area of the township?

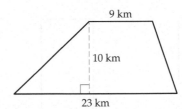

34. **Parks and Recreation** A city plans to plant grass seed in a public playground that has the shape of a triangle with a height of 24 m and a base of 20 m. Each bag of grass seed will seed 120 m². How many bags of seed should be purchased?

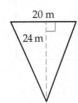

35. **Home Maintenance** You plan to stain the wooden deck at the back of your house. The deck is in the shape of a trapezoid with bases that measure 10 ft and 12 ft and a height of 10 ft. A quart of stain will cover 55 ft². How many quarts of stain should you purchase?

36. Interior Decorating A fabric wall hanging is in the shape of a triangle that has a base of 4 ft and a height of 3 ft. An additional 1 ft² of fabric is needed for hemming the material. How much fabric should be purchased to make the wall hanging?

37. Interior Decorating You are wallpapering two walls of a den, one measuring 10 ft by 8 ft and the other measuring 12 ft by 8 ft. The wallpaper costs $96 per roll, and each roll will cover 40 ft². What is the cost to wallpaper the two walls?

38. Gardens An urban renewal project involves reseeding a garden that is in the shape of a square, 80 ft on each side. Each bag of grass seed costs $12 and will seed 1500 ft². How much money should be budgeted for buying grass seed for the garden?

39. Carpeting You want to install wall-to-wall carpeting in the family room. The floor plan is shown. If the cost of the carpet you would like to purchase is $38 per square yard, what is the cost of carpeting your family room? Assume that there is no waste. *Hint:* 9 ft² = 1 yd².

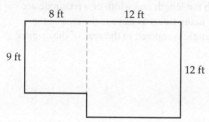

40. Interior Decorating You want to paint the rectangular walls of your bedroom. Two walls measure 16 ft by 8 ft, and the other two walls measure 12 ft by 8 ft. The paint you wish to purchase costs $28 per gallon, and each gallon will cover 400 ft² of wall. Find the minimum amount you will need to spend on paint.

41. Landscaping A walkway 2 m wide surrounds a rectangular plot of grass. The plot is 25 m long and 15 m wide. What is the area of the walkway?

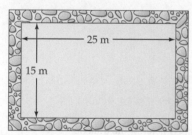

42. Draperies The material used to make pleated draperies for a window must be twice as wide as the width of the window. Draperies are being made for four windows, each 3 ft wide and 4 ft high. Because the drapes will fall slightly below the window sill and extra fabric is needed for hemming the drapes, 1 ft must be added to the height of the window. How much material must be purchased to make the drapes?

43. Carpentry Find the length of molding needed to put around a circular table that is 4.2 ft in diameter. Round to the nearest hundredth of a foot.

44. Sewing How much binding is needed to bind the edge of a circular rug that is 3 m in diameter? Round to the nearest hundredth of a meter.

45. Pulleys Consider the following diagram of a pulley system. If pulley B has a diameter of 16 in. and is rotating at 240 revolutions per minute, how far does a given point on the belt travel each minute that the pulley system is in operation? Assume the belt does not slip as the pulley rotates. Round to the nearest inch.

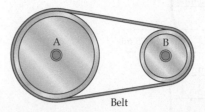

Belt

46. Bicycles A bicycle tire has a diameter of 18 in. How many feet does the bicycle travel when the wheel makes 20 revolutions? Round to the nearest hundredth of a foot.

47. Tricycles The front wheel of a tricycle has a diameter of 16 in. How many feet does the tricycle travel when the wheel makes 15 revolutions? Round to the nearest hundredth of a foot.

48. Telescopes The circular lens located on an astronomical telescope has a diameter of 24 in. Find the exact area of the lens.

49. Irrigation An irrigation system waters a circular field that has a 50-foot radius. Find the exact area watered by the irrigation system.

50. Pizza How much greater is the area of a pizza that has a radius of 10 in. than the area of a pizza that has a radius of 8 in.? Round to the nearest hundredth of a square inch.

51. Pizza A restaurant serves a small pizza that has a radius of 6 in. The restaurant's large pizza has a radius that is twice the radius of the small pizza. How much larger is the area of the large pizza? Round to the nearest hundredth of a square inch. Is the area of the large pizza more or less than twice the area of the small pizza?

52. Satellites A geostationary (GEO) satellite orbits Earth over the Equator. The orbit is circular and at a distance of 36,000 km above Earth. An orbit at this altitude allows the satellite to maintain a fixed position in relation to Earth. What is the distance traveled by a GEO satellite in one orbit around Earth? The radius of Earth is 6380 km. Round to the nearest kilometer.

Earth

53. Lake Tahoe One way to measure the area of an irregular figure, such as a lake, is to divide the area into trapezoids that have the same height. Then measure the length of each base, calculate the area of each trapezoid, and add the areas. The following figure gives approximate dimensions for Lake Tahoe, which straddles the California and Nevada borders.

Approximate the area of Lake Tahoe using the given trapezoids. Round to the nearest tenth of a square mile.

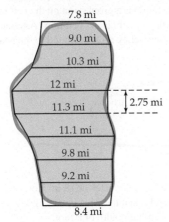

7.8 mi
9.0 mi
10.3 mi
12 mi
2.75 mi
11.3 mi
11.1 mi
9.8 mi
9.2 mi
8.4 mi

54. **Ball Fields** How much farther is it around the bases of a baseball diamond than around the bases of a softball diamond? *Hint:* Baseball and softball diamonds are squares.

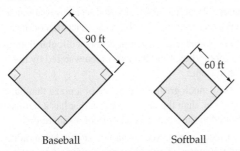

90 ft

60 ft

Baseball   Softball

55. **Area** Write an expression for the area of the shaded portion of the diagram. Leave the answer in terms of π and r.

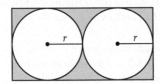

56. **Area** Write an expression for the area of the shaded portion of the diagram. Leave the answer in terms of π and r.

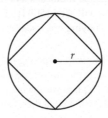

57. **Area** If both the length and width of a rectangle are doubled, how many times as large is the area of the resulting rectangle compared to the area of the original rectangle?

▶ Investigations

1. A circle with radius r and circumference C is sliced into 16 identical sectors that are then arranged as shown in the following diagram.

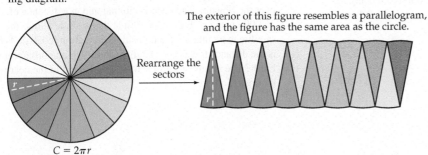

The exterior of this figure resembles a parallelogram, and the figure has the same area as the circle.

Rearrange the sectors

$C = 2\pi r$

The exterior of the figure shown by the rearranged sectors resembles a parallelogram.

   a. What dimension of the circle approximates the height of the parallelogram?
   b. What dimension of the circle approximates the base of the parallelogram?
   c. Explain how the formula for the area of a circle can be derived by using this slicing approach.

2. Heron's (or Hero's) formula is sometimes used to calculate the area of a triangle.

Heron's Formula
The area of a triangle with sides of lengths a, b, and c is given by

$$A = \sqrt{s(s-a)(s-b)(s-c)}$$

where s is the semiperimeter of the triangle:

$$s = \frac{a+b+c}{2}$$

   a. Use Heron's formula to find the area of a triangle with sides that measure 4.4 in., 5.7 in., and 6.2 in. Round to the nearest tenth of a square inch.
   b. Use Heron's formula to find the area of an equilateral triangle with sides that measure 8.3 cm. Round to the nearest tenth of a square centimeter.
   c. Find the lengths of the sides of a triangle that has a perimeter of 12 in., given that the length of each side, in inches, is a counting number and the area of the triangle, in square inches, is also a counting number. *Hint:* All three sides are different lengths.

Volume and Surface Area

Section
11.4

Learning Objectives:

- Find the volume of geometric solids.
- Find the surface area of geometric solids.

Volume

In Section 3 of this chapter, we developed the geometric concepts of perimeter and area. Perimeter and area refer to plane figures (figures that lie in a plane). We are now ready to introduce *volume* of geometric solids.

Geometric solids are three-dimensional shapes that are bounded by surfaces. Common geometric solids include the rectangular solid, sphere, cylinder, cone, and pyramid. Despite being called "solids," these figures are actually hollow; they do not include the points inside their surfaces.

Volume is a measure of the amount of space occupied by a geometric solid. Volume can be used to describe, for example, the amount of trash in a landfill, the amount of concrete poured for the foundation of a house, or the amount of water in a town's reservoir.

The **rectangular solid** in Figure 11.28 is one in which all six sides, called **faces**, are rectangles. The variable L is used to represent the length of a rectangular solid, W is used to represent its width, and H is used to represent its height. A shoebox is an example of a rectangular solid.

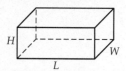

Figure 11.28

The **cube** in Figure 11.29 is a special type of rectangular solid. Each of the six faces of a cube is a square. The variable s is used to represent the length of one side of a cube. A baby's block is an example of a cube.

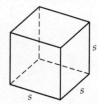

Figure 11.29

The cube in Figure 11.30 measures 1 cm on each side and has a volume of 1 cubic centimeter, written 1 cm^3.

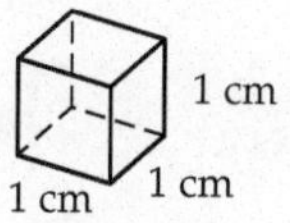

Figure 11.30

The volume of a solid is the number of cubes, each of volume 1 cubic unit, that are necessary to exactly fill the solid. The volume of the rectangular solid in Figure 11.31 is 24 cm^3 because it will hold exactly 24 cubes, each 1 cm on a side. Note that the volume can be found by multiplying the length times the width times the height.

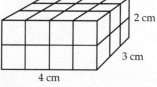

Figure 11.31

$$4 \cdot 3 \cdot 2 = 24$$

The volume of the solid is 24 cm^3.

Volume of a Rectangular Solid

The volume, V, of a rectangular solid with length L, width W, and height H is given by $V = LWH$.

Volume of a Cube

The volume, V, of a cube with side of length s is given by $V = s^3$.

A **sphere** is a solid in which all points are the same distance from a point O, called the **center** of the sphere. A **diameter** of a sphere is a line segment with endpoints on the sphere and passing through the center. A **radius** is a line segment from the center to a point on the sphere. $\overline{AB}$ is a diameter and $\overline{OC}$ is a radius of the sphere shown in Figure 11.32. A basketball is an example of a sphere.

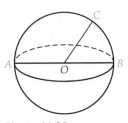

If we let d represent the length of a diameter and r represent the length of a radius, then $d = 2r$ or $r = \frac{1}{2}d$.

Figure 11.32

$$d = 2r \quad \text{or} \quad r = \frac{1}{2}d$$

Volume of a Sphere

The volume, V, of a sphere with radius of length r is given by $V = \frac{4}{3}\pi r^3$.

Find the volume of a rubber ball that has a diameter of 6 in.

First find the length of a radius of the sphere.

$$r = \frac{1}{2}d = \frac{1}{2}(6) = 3$$

Use the formula for the volume of a sphere.

$$V = \frac{4}{3}\pi r^3$$

Replace r with 3.

$$V = \frac{4}{3}\pi(3)^3$$

$$V = \frac{4}{3}\pi(27)$$

The exact volume of the rubber ball is 36π in^3.

$$V = 36\pi$$

An approximate measure can be found by using the π key on a calculator.

$$V \approx 113.10$$

The volume of the rubber ball is approximately 113.10 in^3.

The most common cylinder, called a **right circular cylinder** as shown in Figure 11.33, is one in which the bases are circles and are perpendicular to the height of the cylinder. The variable r is used to represent the length of the radius of a base of a cylinder, and h represents the height of the cylinder. In this text, only right circular cylinders are discussed.

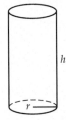

Figure 11.33

Volume of a Right Circular Cylinder

The volume, V, of a right circular cylinder is given by $V = \pi r^2 h$, where r is the radius of the base and h is the height of the cylinder.

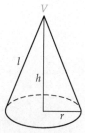

Figure 11.34

A **right circular cone**, as shown in Figure 11.34, is obtained when one base of a right circular cylinder is shrunk to a point, called the **vertex**, V. The variable r is used to represent the radius of the base of the cone, and h represents the height of the cone. The variable l is used to represent the **slant height**, which is the distance from a point on the circumference of the base to the vertex. In this text, only right circular cones are discussed. An ice cream cone is an example of a right circular cone.

Volume of a Right Circular Cone

The volume, V, of a right circular cone is given by $V = \frac{1}{3}\pi r^2 h$, where r is the length of a radius of the circular base and h is the height of the cone.

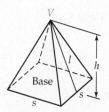

Figure 11.35

The base of a **regular pyramid** is a regular polygon, and the sides are isosceles triangles (two sides of the triangle are the same length). The height, h, is the distance from the vertex, V, to the base and is perpendicular to the base. The variable l is used to represent the **slant height**, which is the height of one of the isosceles triangles on the face of the pyramid. The regular square pyramid in Figure 11.35 has a square base. This is the only type of pyramid discussed in this text. Many Egyptian pyramids are regular square pyramids.

Volume of a Regular Square Pyramid

The volume, V, of a regular square pyramid is given by $V = \frac{1}{3}s^2 h$, where s is the length of a side of the base and h is the height of the pyramid.

Example 1 **Find the Volume of a Geometric Solid**

Find the volume of a cube that measures 1.5 m on a side.

Solution

$$V = s^3 \qquad \text{• Use the formula for the volume of a cube.}$$

$$V = 1.5^3 \qquad \text{• Replace } s \text{ with } 1.5.$$

$$V = 3.375$$

The volume of the cube is 3.375 m^3.

Example 2 **Find the Volume of a Geometric Solid**

The radius of the base of a cone is 8 cm. The height of the cone is 12 cm. Find the volume of the cone. Round to the nearest hundredth of a cubic centimeter.

Solution

$$V = \frac{1}{3}\pi r^2 h \qquad \text{• Use the formula for the volume of a cone.}$$

$$V = \frac{1}{3}\pi (8)^2(12) \qquad \text{• Replace } r \text{ with } 8 \text{ and } h \text{ with } 12.$$

$$V = \frac{1}{3}\pi (64)(12)$$

$$V = 256\pi \qquad \text{• Exact volume}$$

$$V \approx 804.25 \qquad \text{• Use the } \pi \text{ key on a calculator.}$$

The volume of the cone is approximately 804.25 cm^3.

Example 3   **Find the Volume of a Geometric Solid**

An oil storage tank in the shape of a cylinder is 4 m high and has a diameter of 6 m. The oil tank is two-thirds full. Find the number of cubic meters of oil in the tank. Round to the nearest hundredth of a cubic meter.

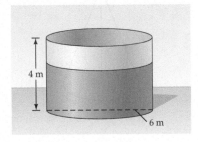

Solution

$$r = \frac{1}{2}d = \frac{1}{2}(6) = 3$$ • Find the radius of the base.

$$V = \pi r^2 h$$ • Use the formula for the volume of a cylinder.

$$V = \pi(3)^2(4)$$ • Replace r with 3 and h with 4.

$$V = \pi(9)(4)$$

$$V = 36\pi$$

$$\frac{2}{3}(36\pi) = 24\pi$$ • Multiply the volume by $\frac{2}{3}$.

$$\approx 75.40$$ • Use the π key on a calculator.

There are approximately 75.40 m³ of oil in the storage tank.

●

Surface Area

The **surface area** of a solid is the total area on the surface of the solid. Suppose you want to cover a geometric solid with wallpaper. The amount of wallpaper needed is equal to the surface area of the figure.

When the rectangular solid in Figure 11.36 is cut open and flattened out, each face is a rectangle. See Figure 11.37. The surface area, S, of the rectangular solid is the sum of the areas of the six rectangles:

$$S = LW + LH + WH + LW + WH + LH$$

which simplifies to

$$S = 2LW + 2LH + 2WH$$

If the rectangular solid is a cube, then all three sides L, W, and H are equal. Therefore, each side can be represented by s. The surface area of a cube is

$$S = 2LW + 2LH + 2WH$$

$$S = 2 \cdot s \cdot s + 2 \cdot s \cdot s + 2 \cdot s \cdot s = 2s^2 + 2s^2 + 2s^2$$

$$S = 6s^2$$

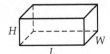

Figure 11.36

Figure 11.37

When a cylinder is cut open and flattened out, the top and bottom of the cylinder are circles. The side of the cylinder flattens out to a rectangle. See Figure 11.38. The length of the rectangle is the circumference of the base, which is $2\pi r$; the width is h, the height of the cylinder. Therefore, the area of the rectangle is $2\pi rh$. The surface area, S, of the cylinder is

$$S = \pi r^2 + 2\pi rh + \pi r^2$$

which simplifies to

$$S = 2\pi r^2 + 2\pi rh$$

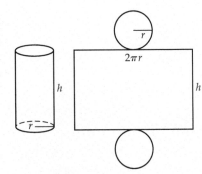

Figure 11.38

The surface area of the regular square pyramid in Figure 11.39 is the area of the base plus the area of the four isosceles triangles. The length of a side of the square base is s; therefore, the area of the base is s^2. The slant height, l, is the height of each triangle, and s is the length of the base of each triangle. The surface area, S, of a regular square pyramid is

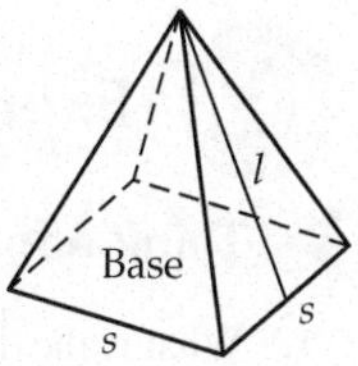

Figure 11.39

$$S = s^2 + 4\left(\frac{1}{2}sl\right)$$

which simplifies to

$$S = s^2 + 2sl$$

Formulas for the surface areas of geometric solids are given here.

Surface Areas of Geometric Solids

The surface area, S, of a **rectangular solid** with length L, width W, and height H is given by $S = 2LW + 2LH + 2WH$.

The surface area, S, of a **cube** with sides of length s is given by $S = 6s^2$.

The surface area, S, of a **sphere** with radius r is given by $S = 4\pi r^2$.

The surface area, S, of a **right circular cylinder** is given by $S = 2\pi r^2 + 2\pi rh$, where r is the radius of the base and h is the height.

The surface area, S, of a **right circular cone** is given by $S = \pi r^2 + \pi rl$, where r is the radius of the circular base and l is the slant height.

The surface area, S, of a **regular square pyramid** is given by $S = s^2 + 2sl$, where s is the length of a side of the base and l is the slant height.

Example 4 **Find the Surface Area of a Geometric Solid**

The diameter of the base of a cone is 5 m and the slant height is 4 m. Find the surface area of the cone. Round to the nearest hundredth of a square meter.

Solution

$$r = \frac{1}{2}d = \frac{1}{2}(5) = 2.5 \qquad \bullet \text{ Find the radius of the cone.}$$

$$S = \pi r^2 + \pi rl \qquad \bullet \text{ Use the formula for the surface area of a cone.}$$

$$S = \pi(2.5)^2 + \pi(2.5)(4) \qquad \bullet \text{ Replace } r \text{ with } 2.5 \text{ and } l \text{ with } 4.$$

$$S = \pi(6.25) + \pi(2.5)(4)$$

$$S = 6.25\pi + 10\pi$$

$$S = 16.25\pi$$

$$S \approx 51.05$$

The surface area of the cone is approximately 51.05 m².

The ratio of an animal's surface area to the volume of its body is a crucial factor in its survival. The more square units of skin for every cubic unit of volume, the more rapidly the animal loses body heat. Therefore, animals living in a warm climate benefit from a higher ratio of surface area to volume, whereas those living in a cool climate benefit from a lower ratio.

11.4 Exercise Set

Think About It

1. What is the difference between volume and surface area?
2. If a figure is measured in meters, what are the units of the volume of the figure?
3. If a figure is measured in inches, what are the units of the surface area of the figure?
4. What is the shape of the base of a right circular cylinder?

Preliminary Exercises

For Exercises P1 to P4, if necessary, use the referenced example following the exercise for assistance.

P1. The length of a rectangular solid is 5 m, the width is 3.2 m, and the height is 4 m. Find the volume of the solid. **Example 1**

P2. The length of a side of the base of a regular square pyramid is 15 m and the height of the pyramid is 25 m. Find the volume of the pyramid. **Example 2**

P3. A silo in the shape of a cylinder is 16 ft in diameter and has a height of 30 ft. The silo is three-fourths full. Find the volume of the portion of the silo that is not being used for storage. Round to the nearest hundredth of a cubic foot. **Example 3**

P4. The diameter of the base of a cylinder is 6 ft and the height is 8 ft. Find the surface area of the cylinder. Round to the nearest hundredth of a square foot. **Example 4**

Section Exercises

In Exercises 1 to 6, find the volume of the figure. For calculations involving π, give both the exact value and an approximation to the nearest hundredth of a unit.

1.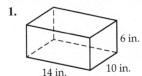

2.

3.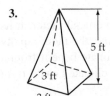

4.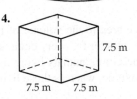

5.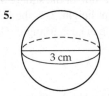

6.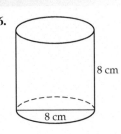

In Exercises 7 to 12, find the surface area of the figure. For calculations involving π, give both the exact value and an approximation to the nearest hundredth of a unit.

7.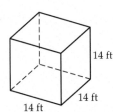

8.

9.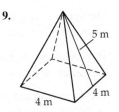

10.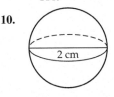

11.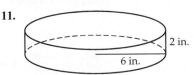

12.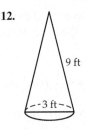

In Exercises 13 to 45, solve.

13. **Volume** A rectangular solid has a length of 6.8 m, a width of 2.5 m, and a height of 2 m. Find the volume of the solid.

14. **Volume** Find the volume of a rectangular solid that has a length of 4.5 ft, a width of 3 ft, and a height of 1.5 ft.

15. **Volume** Find the volume of a cube whose side measures 2.5 in.

16. **Volume** The length of a side of a cube is 7 cm. Find the volume of the cube.

17. **Volume** The diameter of a sphere is 6 ft. Find the exact volume of the sphere.

18. **Volume** Find the volume of a sphere that has a radius of 1.2 m. Round to the nearest hundredth of a cubic meter.

19. **Volume** The diameter of the base of a cylinder is 24 cm. The height of the cylinder is 18 cm. Find the volume of the cylinder. Round to the nearest hundredth of a cubic centimeter.

20. **Volume** The height of a cylinder is 7.2 m. The radius of the base is 4 m. Find the exact volume of the cylinder.

21. **Volume** The radius of the base of a cone is 5 in. The height of the cone is 9 in. Find the exact volume of the cone.

22. **Volume** The height of a cone is 15 cm. The diameter of the cone is 10 cm. Find the volume of the cone. Round to the nearest hundredth of a cubic centimeter.

23. **Volume** The length of a side of the base of a regular square pyramid is 6 in. and the height of the pyramid is 10 in. Find the volume of the pyramid.

24. **Volume** The height of a regular square pyramid is 8 m and the length of a side of the base is 9 m. What is the volume of the pyramid?

25. **Comparing Volumes** The length of a side of a cube is equal to the radius of a sphere. Which solid has the greater volume?

26. **Comparing Volumes** A sphere and a cylinder have the same radius. The height of the cylinder is equal to the radius of its base. Which solid has the greater volume?

27. **The Statue of Liberty** The index finger of the Statue of Liberty is 8 ft long. The circumference at the second joint is 3.5 ft. Use the formula for the volume of a cylinder to approximate the volume of the index finger on the Statue of Liberty. Round to the nearest hundredth of a cubic foot.

28. **Fish Hatchery** A rectangular tank at a fish hatchery is 9 m long, 3 m wide, and 1.5 m deep. Find the volume of the water in the tank when the tank is full.

29. **The Panama Canal** When the lock is full, the water in the Pedro Miguel Lock near the Pacific Ocean side of the Panama Canal fills a rectangular solid of dimensions 1000 ft long, 110 ft wide, and 43 ft deep. There are 7.48 gal of water in each cubic foot. How many gallons of water are in the lock?

30. **Surface Area** The width of a rectangular solid is 32 cm, the length is 60 cm, and the height is 14 cm. What is the surface area of the solid?

31. **Surface Area** The side of a cube measures 3.4 m. Find the surface area of the cube.

32. **Surface Area** Find the surface area of a cube with a side measuring 1.5 in.

33. **Surface Area** Find the exact surface area of a sphere with a diameter of 15 cm.

34. **Surface Area** The radius of a sphere is 2 in. Find the surface area of the sphere. Round to the nearest hundredth of a square inch.

35. **Surface Area** The radius of the base of a cylinder is 4 in. The height of the cylinder is 12 in. Find the surface area of the cylinder. Round to the nearest hundredth of a square inch.

36. **Surface Area** The diameter of the base of a cylinder is 1.8 m. The height of the cylinder is 0.7 m. Find the exact surface area of the cylinder.

37. **Surface Area** The slant height of a cone is 2.5 ft. The radius of the base is 1.5 ft. Find the exact surface area of the cone.

38. **Surface Area** The diameter of the base of a cone is 21 in. The slant height is 16 in. What is the surface area of the cone? Round to the nearest hundredth of a square inch.

39. **Surface Area** The length of a side of the base of a regular square pyramid is 9 in., and the pyramid's slant height is 12 in. Find the surface area of the pyramid.

40. **Surface Area** The slant height of a regular square pyramid is 18 m, and the length of a side of the base is 16 m. What is the surface area of the pyramid?

41. **Appliances** The volume of a freezer that is a rectangular solid with a length of 7 ft and a height of 3 ft is 52.5 ft^3. Find the width of the freezer.

42. **Aquariums** The length of a rectangular solid aquarium is 18 in. and the width is 12 in. If the volume of the aquarium is 1836 in^3, what is the height of the aquarium?

43. **Paint** A can of paint will cover 300 ft^2 of surface. How many cans of paint should be purchased to paint a cylinder that has a height of 30 ft and a radius of 12 ft?

44. **Ballooning** A hot air balloon is in the shape of a sphere. Approximately how much fabric was used to construct the balloon if its diameter is 32 ft? Round to the nearest square foot.

45. **Surface Area** The length of a side of the base of a regular square pyramid is 5 cm and the slant height of the pyramid is 8 cm. How much larger is the surface area of this pyramid than the surface area of a cone with a diameter of 5 cm and a slant height of 8 cm? Round to the nearest hundredth of a square centimeter.

In Exercises 46 to 51, find the volume of the figure. Round to the nearest hundredth of a unit.

46.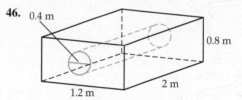

47.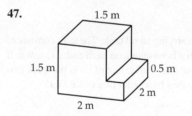

48.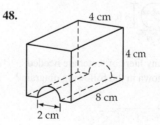

49.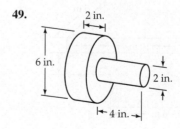

50.

51.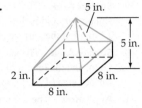

In Exercises 52 to 55, find the surface area of the figure. Round to the nearest hundredth of a unit.

52.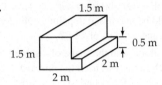

53.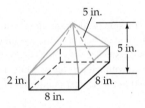

54.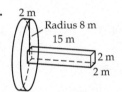

55.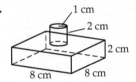

56. Oil Tanks A truck is carrying an oil tank. The tank consists of a circular cylinder with a hemisphere on each end, as shown. If the tank is half full, how many cubic feet of oil is the truck carrying? Round to the nearest hundredth of a cubic foot.

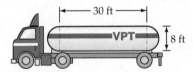

57. Swimming Pools How many liters of water are needed to fill the swimming pool shown in the following diagram? (1 m³ contains 1000 L.)

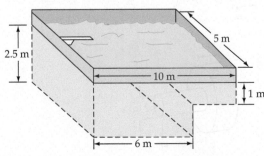

58. Metallurgy A piece of sheet metal is cut and formed into the shape shown in the following diagram. Given that there are 0.24 g in 1 cm² of the metal, find the total number of grams of metal used. Round to the nearest hundredth of a gram.

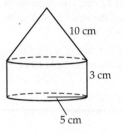

59. Gold A solid sphere of gold alloy with a radius of 0.5 cm has a value of $180. Find the value of a solid sphere of the same alloy with a radius of 1.5 cm.

60. Swimming Pools A swimming pool is built in the shape of a rectangular solid. It holds 32,000 gal of water. If the length, width, and height of the pool are each doubled, how many gallons of water will be needed to fill the pool?

▶ **Investigations**

1. a. Draw a two-dimensional figure that can be cut out and made into a right circular cone.
   b. Draw a two-dimensional figure that can be cut out and made into a regular square pyramid.

2. A sphere fits inside a cylinder as shown in the following figure. The height of the cylinder equals the diameter of the sphere. Show that the surface area of the sphere equals the surface area of the side of the cylinder.

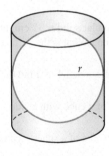

3. Determine whether the statement is always true, sometimes true, or never true.
   a. The slant height of a regular square pyramid is longer than the height.
   b. The slant height of a cone is shorter than the height.
   c. The four triangular faces of a regular square pyramid are equilateral triangles.

4. a. What is the effect on the surface area of a rectangular solid of doubling the width and height?
   b. What is the effect on the volume of a rectangular solid of doubling the length and width?
   c. What is the effect on the volume of a cube of doubling the length of each side of the cube?
   d. What is the effect on the surface area of a cylinder of doubling the radius and height?

Section 11.5

Properties of Triangles

Learning Objectives:

- Determine whether two triangles are similar.
- Solve applications using similar triangles.
- Determine whether two triangles are congruent.
- Solve problems using the Pythagorean Theorem.

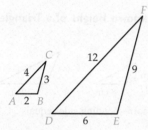

Figure 11.40

Similar Triangles

Similar objects have the same shape but not necessarily the same size. A tennis ball is similar to a basketball. A model ship is similar to an actual ship.

Similar objects have corresponding parts; for example, the rudder on the model ship corresponds to the rudder on the actual ship. The relationship between the sizes of the corresponding parts can be written as a ratio. All corresponding parts of two similar figures share the same ratio. If the rudder on the model ship is $\frac{1}{100}$ the size of the rudder on the actual ship, then the model mast is $\frac{1}{100}$ of the size of the actual mast, the width of the model is $\frac{1}{100}$ the width of the actual ship, and so on.

The two triangles ABC and DEF shown in Figure 11.40 are similar. Side $\overline{AB}$ corresponds to side $\overline{DE}$, side $\overline{BC}$ corresponds to side $\overline{EF}$, and side $\overline{AC}$ corresponds to side $\overline{DF}$. The ratios of the lengths of corresponding sides are equal.

$$\frac{AB}{DE} = \frac{2}{6} = \frac{1}{3}, \quad \frac{BC}{EF} = \frac{3}{9} = \frac{1}{3}, \quad \text{and}$$

$$\frac{AC}{DF} = \frac{4}{12} = \frac{1}{3}$$

Because the ratios of corresponding sides are equal, several proportions can be formed.

$$\frac{AB}{DE} = \frac{BC}{EF}, \quad \frac{AB}{DE} = \frac{AC}{DF}, \quad \text{and} \quad \frac{BC}{EF} = \frac{AC}{DF}$$

The measures of corresponding angles in similar triangles are equal. Therefore,

$$m \angle A = m \angle D, \quad m \angle B = m \angle E, \quad \text{and}$$

$$m \angle C = m \angle F$$

Triangles ABC and DEF in Figure 11.41 are similar triangles. AH and DK are the heights of the triangles. The ratio of the heights of similar triangles equals the ratio of the lengths of corresponding sides.

$$\text{Ratio of corresponding sides} = \frac{1.5}{6} = \frac{1}{4}$$

$$\text{Ratio of heights} = \frac{1}{4}$$

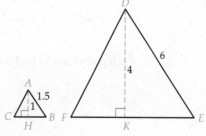

Figure 11.41

Properties of Similar Triangles

For similar triangles, the ratios of corresponding sides are equal. The ratio of corresponding heights is equal to the ratio of corresponding sides. The measures of corresponding angles are equal.

The two triangles in Figure 11.42 are similar triangles. Find the length of side $\overline{EF}$. Round to the nearest tenth of a meter.

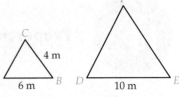

Figure 11.42

The triangles are similar, so the ratios of the lengths of corresponding sides are equal.

$$\frac{EF}{BC} = \frac{DE}{AB}$$

$$\frac{EF}{4} = \frac{10}{6}$$

$$6(EF) = 4(10)$$

$$6(EF) = 40$$

$$EF \approx 6.7$$

The length of side EF is approximately 6.7 m.

Example 1 **Use Similar Triangles to Find the Unknown Height of a Triangle**

Triangles ABC and DEF are similar. Find FG, the height of triangle DEF.

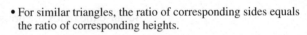

Solution

$$\frac{AB}{DE} = \frac{CH}{FG}$$ • For similar triangles, the ratio of corresponding sides equals the ratio of corresponding heights.

$$\frac{8}{12} = \frac{4}{FG}$$ • Replace AB, DE, and CH with their values.

$$8(FG) = 12(4)$$ • The cross products are equal.

$$8(FG) = 48$$

$$FG = 6$$ • Divide both sides of the equation by 8.

The height FG of triangle DEF is 6 cm.

●

Triangles ABC and DEF shown below are similar triangles. Find the area of triangle ABC.

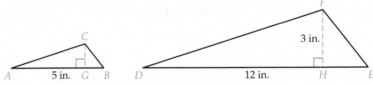

Solve a proportion to find the height of triangle ABC.

$$\frac{AB}{DE} = \frac{CG}{FH}$$

$$\frac{5}{12} = \frac{CG}{3}$$

$$12(CG) = 5(3)$$

$$12(CG) = 15$$

$$CG = 1.25$$

Use the formula for the area of a triangle.

$$A = \frac{1}{2}bh$$

The base is 5 in. The height is 1.25 in.

$$A = \frac{1}{2}(5)(1.25)$$

The area of triangle ABC is 3.125 in^2.

$$A = 3.125$$

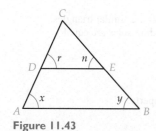

Figure 11.43

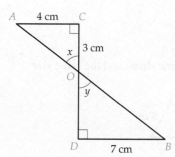

Figure 11.44

If the three angles of one triangle are equal in measure to the three angles of another triangle, then the triangles are similar.

In triangle *ABC* in Figure 11.43, line segment $\overline{DE}$ is drawn parallel to the base $\overline{AB}$. Because the measures of corresponding angles are equal, $m \angle x = m \angle r$ and $m \angle y = m \angle n$. We know that $m \angle C = m \angle C$. Thus the measures of the three angles of triangle *ABC* are equal, respectively, to the measures of the three angles of triangle *DEC*. Therefore, triangles *ABC* and *DEC* are similar triangles.

The sum of the measures of the three angles of a triangle is 180°. If two angles of one triangle are equal in measure to two angles of another triangle, then the third angles must be equal. Thus we can say that if two angles of one triangle are equal in measure to two angles of another triangle, then the two triangles are similar.

In Figure 11.44, $\overline{AB}$ intersects $\overline{CD}$ at point *O*. Angles *C* and *D* are right angles. Find the length of $\overline{DO}$.

First determine whether triangles *AOC* and *BOD* are similar.

$m \angle C = m \angle D$ because they are both right angles.

$m \angle x = m \angle y$ because vertical angles have the same measure.

Because two angles of triangle *AOC* are equal in measure to two angles of triangle *BOD*, triangles *AOC* and *BOD* are similar.

Use a proportion to find the length of the unknown side.

$$\frac{AC}{BD} = \frac{CO}{DO}$$

$$\frac{4}{7} = \frac{3}{DO}$$

$$4(DO) = 7(3)$$

$$4(DO) = 21$$

$$DO = 5.25$$

The length of $\overline{DO}$ is 5.25 cm.

Example 2   Solve a Problem Involving Similar Triangles

In the figure at the right, $\angle B$ and $\angle D$ are right angles, $AB = 12$ m, $DC = 4$ m, and $AC = 18$ m. Find the length of $\overline{CO}$.

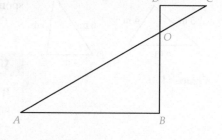

Solution

$\angle B$ and $\angle D$ are right angles. Therefore, $\angle B = \angle D$. $\angle AOB$ and $\angle COD$ are vertical angles. Therefore, $\angle AOB = \angle COD$.

Because two angles of triangle *AOB* are equal in measure to two angles of triangle *COD*, triangles *AOB* and *COD* are similar triangles.

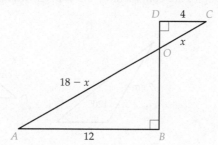

• Label the diagram using the given information. Let *x* represent *CO*. $AC = AO + CO$. Because $AC = 18$, $AO = 18 - x$.

$$\frac{DC}{BA} = \frac{CO}{AO}$$

$$\frac{4}{12} = \frac{x}{18 - x}$$

$$12x = 4(18 - x)$$

$$12x = 72 - 4x$$

$$16x = 72$$

$$x = 4.5$$

• Triangles *AOB* and *COD* are similar triangles. The ratios of corresponding sides are equal.

• Use the distributive property.

The length of $\overline{CO}$ is 4.5 m.

Congruent Triangles

The two triangles below are **congruent**. They have the same shape and the same size.

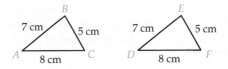

The notation $\triangle ABC \cong \triangle DEF$ is used to indicate that triangle *ABC* is congruent to triangle *DEF*.

The corresponding angles of congruent triangles have the same measure and the corresponding sides are equal in length. In contrast, for similar triangles, corresponding angles have the same measure but corresponding sides are not necessarily the same length.

Three major theorems are used to determine whether two triangles are congruent.

Side-Side-Side Theorem (SSS)

If the three sides of one triangle are equal in measure to the corresponding three sides of a second triangle, the two triangles are congruent.

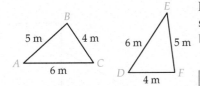

Figure 11.45

In the triangles shown in Figure 11.45, $AC = DE$, $AB = EF$, and $BC = DF$. The corresponding sides of triangles *ABC* and *DEF* are equal in measure. The triangles are congruent by the SSS theorem.

Side-Angle-Side Theorem (SAS)

If two sides and the included angle of one triangle are equal in measure to two sides and the included angle of a second triangle, the two triangles are congruent.

Take Note: $\angle A$ is *included* between sides $\overline{AB}$ and $\overline{AC}$. $\angle E$ is *included* between sides $\overline{DE}$ and $\overline{EF}$.

In the two triangles shown in Figure 11.46, $AB = EF$, $AC = DE$, and $m\angle BAC = m\angle DEF$. The triangles are congruent by the SAS theorem.

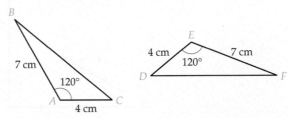

Figure 11.46

> ### Angle-Side-Angle Theorem (ASA)
>
> If two angles and the included side of one triangle are equal in measure to two angles and the included side of a second triangle, the two triangles are congruent.

Take Note: Side $\overline{AC}$ is *included* between $\angle A$ and $\angle C$. Side $\overline{EF}$ is *included* between $\angle E$ and $\angle F$.

For triangles *ABC* and *DEF* in Figure 11.47, $m\angle A = m\angle F$, $m\angle C = m\angle E$, and $AC = EF$. The triangles are congruent by the ASA theorem.

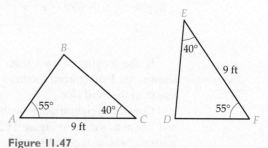

Figure 11.47

Given triangles *PQR* and *MNO*, do the conditions $m\angle P = m\angle O$, $m\angle Q = m\angle M$, and $PQ = MO$ guarantee that triangle *PQR* is congruent to triangle *MNO*?

Draw a sketch of the two triangles and determine whether one of the theorems for congruence is satisfied.

See Figure 11.48. Because two angles and the included side of one triangle are equal in measure to two angles and the included side of the second triangle, the triangles are congruent by the ASA theorem.

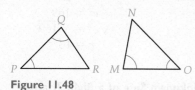

Figure 11.48

Example 3   **Determine Whether Two Triangles Are Congruent**

In the figure at the right, is triangle *ABC* congruent to triangle *DEF*?

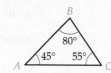
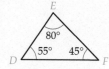

Solution

To determine whether the triangles are congruent, determine whether one of the theorems for congruence is satisfied.

The triangles do not satisfy the SSS theorem, the SAS theorem, or the ASA theorem. The triangles are not necessarily congruent.

The Pythagorean Theorem

Recall that a right triangle contains one right angle. The side opposite the right angle is called the **hypotenuse**. The other two sides are called **legs**. See Figure 11.49.

The angles in a right triangle are usually labeled with the capital letters *A*, *B*, and *C*, with *C* reserved for the right angle. The side opposite angle *A* is side *a*, the side opposite angle *B* is side *b*, and *c* is the hypotenuse.

Figure 11.50 shows a right triangle with legs measuring 3 units and 4 units and a hypotenuse measuring 5 units. Each side of the triangle is also the side of a square. The number of square units in the area of the largest square is equal to the sum of the numbers of square units in the areas of the smaller squares.

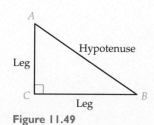

Figure 11.49

$$\begin{array}{c} \text{Square of the} \\ \text{hypotenuse} \end{array} = \begin{array}{c} \text{Sum of the squares} \\ \text{of the two legs} \end{array}$$

$$5^2 = 3^2 + 4^2$$
$$25 = 9 + 16$$
$$25 = 25$$

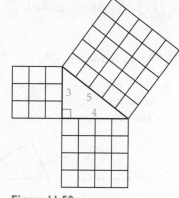

Figure 11.50

The Greek mathematician Pythagoras is generally credited with the discovery that the square of the hypotenuse of a right triangle is equal to the sum of the squares of the two legs. This is called the **Pythagorean theorem**.

> ### The Pythagorean Theorem
>
> If a and b are the lengths of the legs of a right triangle and c is the length of the hypotenuse, then $c^2 = a^2 + b^2$.

If the lengths of two sides of a right triangle are known, the Pythagorean theorem can be used to find the length of the third side.

Consider a right triangle with legs that measure 5 cm and 12 cm as shown in Figure 11.51. Use the Pythagorean theorem, with $a = 5$ and $b = 12$, to find the length of the hypotenuse. (If you let $a = 12$ and $b = 5$, the result will be the same.) Take the square root of each side of the equation. The length of the hypotenuse is 13 cm.

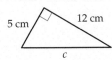

Figure 11.51

$$c^2 = a^2 + b^2$$
$$c^2 = 5^2 + 12^2$$
$$c^2 = 25 + 144$$
$$c^2 = 169$$
$$\sqrt{c^2} = \sqrt{169}$$
$$c = 13$$

Example 4   **Determine the Length of the Unknown Side of a Right Triangle**

The length of one leg of a right triangle is 8 in. The length of the hypotenuse is 12 in. Find the length of the other leg. Round to the nearest hundredth of an inch.

Solution

$a^2 + b^2 = c^2$	• Use the Pythagorean theorem.
$8^2 + b^2 = 12^2$	• $a = 8, c = 12$
$64 + b^2 = 144$	
$b^2 = 80$	• Solve for b^2. Subtract 64 from each side.
$\sqrt{b^2} = \sqrt{80}$	• Take the square root of each side of the equation.
$b \approx 8.94$	• Use a calculator to approximate $\sqrt{80}$.

The length of the other leg is approximately 8.94 in.

11.5 Exercise Set

Think About It

1. If two triangles are similar, what can be said about the angles of the two triangles?

2. The following two triangles are similar.

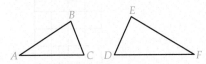

What is the side corresponding to AB?

3. Do similar triangles have the same area? Do congruent triangles have the same area?

4. If two triangles are similar, are they congruent? If two triangles are congruent, are they similar?

5. What is the side-angle-side theorem for congruent triangles?

6. If the measures of three sides of a triangle are 9 ft, 12 ft, and 15 ft, is the triangle a right triangle?

▶ **Preliminary Exercises**

For Exercises P1 to P4, if necessary, use the referenced example following the exercise for assistance.

P1. Triangles *ABC* and *DEF* are similar. Find *FG*, the height of triangle *DEF*. **Example 1**

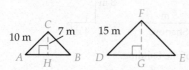

P2. In the following figure, ∠*A* and ∠*D* are right angles, *AB* = 10 cm, *CD* = 4 cm, and *DO* = 3 cm. Find the area of triangle *AOB*. **Example 2**

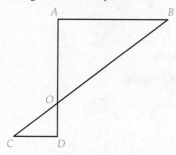

P3. In the following figure, is triangle *PQR* congruent to triangle *MNO*? **Example 3**

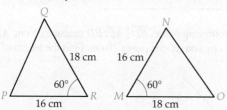

P4. The hypotenuse of a right triangle measures 6 m, and one leg measures 2 m. Find the measure of the other leg. Round to the nearest hundredth of a meter. **Example 4**

▶ **Section Exercises**

In Exercises 1 to 4, find the ratio of the lengths of corresponding sides for the similar triangles.

1.

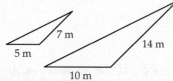

2.

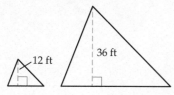

3.

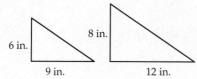

4.

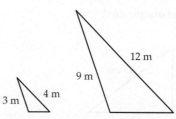

In Exercises 5 to 14, triangles *ABC* and *DEF* are similar triangles. Use this fact to solve each exercise. Round to the nearest tenth.

5. Find side *DE*.

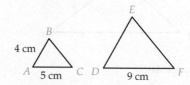

6. Find side *DE*.

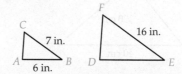

7. Find the height of triangle *DEF*.

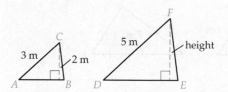

8. Find the height of triangle *ABC*.

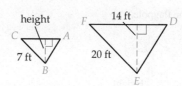

9. Find the perimeter of triangle *ABC*.

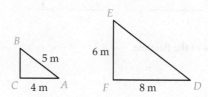

10. Find the perimeter of triangle *DEF*.

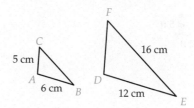

11. Find the perimeter of triangle *ABC*.

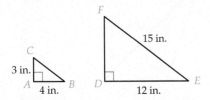

12. Find the area of triangle *DEF*.

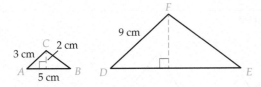

13. Find the area of triangle *ABC*.

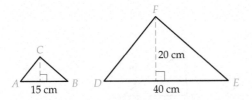

14. Find the area of triangle *DEF*.

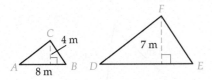

In Exercises 15 to 19, the given triangles are similar triangles. Use this fact to solve each exercise.

15. Find the height of the flagpole.

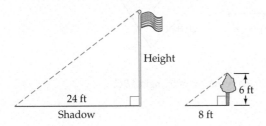

16. Find the height of the flagpole.

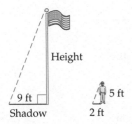

17. Find the height of the building.

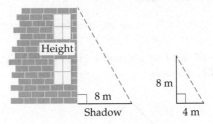

18. Find the height of the building.

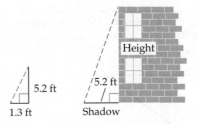

19. Find the height of the flagpole.

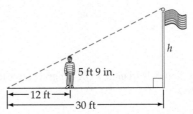

20. In the following figure, $\overline{BD} \parallel \overline{AE}$, *BD* measures 5 cm, *AE* measures 8 cm, and *AC* measures 10 cm. Find the length of $\overline{BC}$.

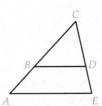

21. In the following figure, $\overline{AC} \parallel \overline{DE}$, *BD* measures 8 m, *AD* measures 12 m, and *BE* measures 6 m. Find the length of $\overline{BC}$.

22. In the following figure, $\overline{DE} \parallel \overline{AC}$, *DE* measures 6 in., *AC* measures 10 in., and *AB* measures 15 in. Find the length of $\overline{DA}$.

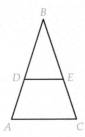

23. In the following figure, $\overline{AE} \parallel \overline{BD}$, $AB = 3$ ft, $ED = 4$ ft, and $BC = 3$ ft. Find the length of $\overline{CE}$.

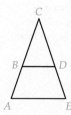

24. In the following figure, $\overline{MP}$ and $\overline{NQ}$ intersect at O, $NO = 25$ ft, $MO = 20$ ft, and $PO = 8$ ft. Find the length of $\overline{QO}$.

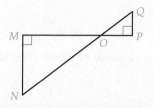

25. In the following figure, $\overline{MP}$ and $\overline{NQ}$ intersect at O, $NO = 24$ cm, $MN = 10$ cm, $MP = 39$ cm, and $QO = 12$ cm. Find the length of $\overline{OP}$.

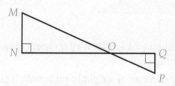

26. In the following figure, $\overline{MQ}$ and $\overline{NP}$ intersect at O, $NO = 12$ m, $MN = 9$ m, $PQ = 3$ m, and $MQ = 20$ m. Find the perimeter of triangle OPQ.

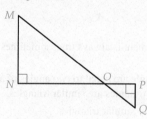

Surveyors use similar triangles to measure distances that cannot be measured directly. This is illustrated in Exercises 27 and 28.

27. The following diagram represents a river of width CD. Triangles AOB and DOC are similar. The distances AB, BO, and OC were measured and found to have the lengths given in the diagram. Find CD, the width of the river.

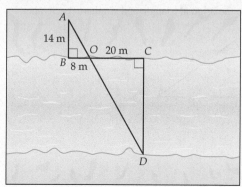

28. The following diagram shows how surveyors laid out similar triangles along the Winnepaugo River. Find the width, d, of the river.

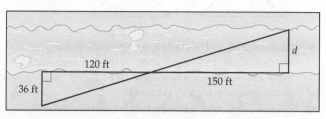

In Exercises 29 to 36, determine whether the two triangles are congruent. If they are congruent, state by what theorem (*SSS*, *SAS*, or *ASA*) they are congruent.

29.

30.

31.

32.

33.

34.

35.

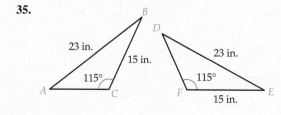

36.

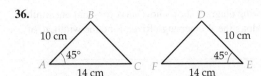

37. Given triangle *ABC* and triangle *DEF*, do the conditions $m\angle C = m\angle E$, $AC = EF$, and $BC = DE$ guarantee that triangle *ABC* is congruent to triangle *DEF*? If they are congruent, by what theorem are they congruent?

38. Given triangle *PQR* and triangle *MNO*, do the conditions $PR = NO$, $PQ = MO$, and $QR = MN$ guarantee that triangle *PQR* is congruent to triangle *MNO*? If they are congruent, by what theorem are they congruent?

39. Given triangle *LMN* and triangle *QRS*, do the conditions $m\angle M = m\angle S$, $m\angle N = m\angle Q$, and $m\angle L = m\angle R$ guarantee that triangle *LMN* is congruent to triangle *QRS*? If they are congruent, by what theorem are they congruent?

40. Given triangle *DEF* and triangle *JKL*, do the conditions $m\angle D = m\angle K$, $m\angle E = m\angle L$, and $DE = KL$ guarantee that triangle *DEF* is congruent to triangle *JKL*? If they are congruent, by what theorem are they congruent?

41. Given triangle *ABC* and triangle *PQR*, do the conditions $m\angle B = m\angle P$, $BC = PQ$, and $AC = QR$ guarantee that triangle *ABC* is congruent to triangle *PQR*? If they are congruent, by what theorem are they congruent?

42. True or false? If the ratio of the corresponding sides of two similar triangles is 1 to 1, then the two triangles are congruent.

In Exercises 43 to 51, find the length of the unknown side of the triangle. Round to the nearest tenth.

43.

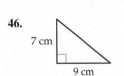

44.

45.

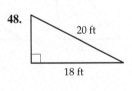

46.

47.

48.

49.

50.

51.

In Exercises 52 to 56, use the given information to solve each exercise. Round to the nearest tenth.

52. **Home Maintenance** A ladder 8 m long is leaning against a building. How high on the building will the ladder reach when the bottom of the ladder is 3 m from the building?

53. **Mechanics** Find the distance between the centers of the holes in the metal plate.

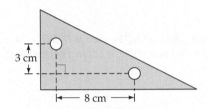

54. **Travel** If you travel 18 mi east and then 12 mi north, how far are you from your starting point?

55. **Perimeter** Find the perimeter of a right triangle with legs that measure 5 cm and 9 cm.

56. **Perimeter** Find the perimeter of a right triangle with legs that measure 6 in. and 8 in.

Investigations

1. Determine whether the statement is always true, sometimes true, or never true.

a. If two angles of one triangle are equal to two angles of a second triangle, then the triangles are similar triangles.

b. Two isosceles triangles are similar triangles.

c. Two equilateral triangles are similar triangles.

d. If an acute angle of a right triangle is equal to an acute angle of another right triangle, then the triangles are similar triangles.

2. In the following figure, the height of a right triangle is drawn from the right angle perpendicular to the hypotenuse. (Recall that the hypotenuse of a right triangle is the side opposite the right angle.) Verify that the two smaller triangles formed are similar to the original triangle and similar to each other.

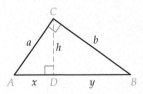

Section 11.6

Right Triangle Trigonometry

Learning Objectives:

- Find the values of trigonometric ratios of an acute angle.
- Find the length of a side of a right triangle.
- Evaluate the inverse sine, cosine, and tangent.
- Solve problems involving angles of elevation and depression.

Trigonometric Ratios of an Acute Angle

Consider the problem of engineers trying to determine the distance across a ravine in order to design a bridge. Look at Figure 11.52. It is fairly easy to measure the length of the side of the triangle that is on the land (100 ft), but the lengths of sides a and c cannot be measured easily because of the ravine.

The study of *trigonometry*, a term that comes from two Greek words meaning "triangle measurement," began about 2000 years ago, partially as a means of solving surveying problems such as the one described above. In this section, we will examine *right triangle* trigonometry—that is, trigonometry that applies only to right triangles.

When working with right triangles, it is convenient to refer to the side *opposite* an angle and the side *adjacent to* (next to) an angle. The hypotenuse of a right triangle is not adjacent to or opposite either of the acute angles of the triangle.

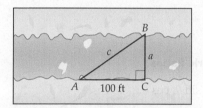

Figure 11.52

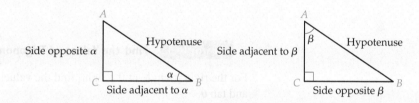

Take Note: In trigonometry, it is common practice to use Greek letters for the angles of a triangle. Here are some frequently used letters: α (alpha), β (beta), and θ (theta). The word *alphabet* is derived from the first two letters of the Greek alphabet, α and β.

Consider the right triangle shown in Figure 11.53. Six possible ratios can be formed using the lengths of the sides of the triangle.

$$\frac{\text{length of opposite side}}{\text{length of hypotenuse}} \qquad \frac{\text{length of hypotenuse}}{\text{length of opposite side}}$$

$$\frac{\text{length of adjacent side}}{\text{length of hypotenuse}} \qquad \frac{\text{length of hypotenuse}}{\text{length of adjacent side}}$$

$$\frac{\text{length of opposite side}}{\text{length of adjacent side}} \qquad \frac{\text{length of adjacent side}}{\text{length of opposite side}}$$

These ratios are called the **sine** (sin), **cosine** (cos), **tangent** (tan), **cosecant** (csc), **secant** (sec), and **cotangent** (cot) of the right triangle.

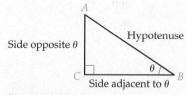

Figure 11.53

The Trigonometric Ratios of an Acute Angle of a Right Triangle

If θ is an acute angle of a right triangle ABC, then

$$\sin \theta = \frac{\text{length of opposite side}}{\text{length of hypotenuse}} \qquad \csc \theta = \frac{\text{length of hypotenuse}}{\text{length of opposite side}}$$

$$\cos \theta = \frac{\text{length of adjacent side}}{\text{length of hypotenuse}} \qquad \sec \theta = \frac{\text{length of hypotenuse}}{\text{length of adjacent side}}$$

$$\tan \theta = \frac{\text{length of opposite side}}{\text{length of adjacent side}} \qquad \cot \theta = \frac{\text{length of adjacent side}}{\text{length of opposite side}}$$

As a convenience, we will write opp, adj, and hyp as abbreviations for *the length of the opposite side, adjacent side,* and *hypotenuse,* respectively. Using this convention, the definitions of the trigonometric ratios are written as follows:

$$\sin \theta = \frac{\text{opp}}{\text{hyp}} \qquad \csc \theta = \frac{\text{hyp}}{\text{opp}}$$

$$\cos \theta = \frac{\text{adj}}{\text{hyp}} \qquad \sec \theta = \frac{\text{hyp}}{\text{adj}}$$

$$\tan \theta = \frac{\text{opp}}{\text{adj}} \qquad \cot \theta = \frac{\text{adj}}{\text{opp}}$$

For the remainder of this section, we will focus on the sine, cosine, and tangent ratios.

When working with trigonometric ratios, be sure to draw a diagram and label the adjacent and opposite sides of an angle. For instance, in the definition above, if we had placed θ at angle A, then the triangle would have been labeled as shown in Figure 11.54. The definitions of the ratios remain the same.

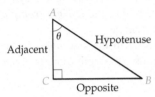

Figure 11.54

$$\sin \theta = \frac{\text{opp}}{\text{hyp}} \qquad \cos \theta = \frac{\text{adj}}{\text{hyp}} \qquad \tan \theta = \frac{\text{opp}}{\text{adj}}$$

Example 1 **Find the Values of Trigonometric Ratios**

For the right triangle at the right, find the values of $\sin \theta$, $\cos \theta$, and $\tan \theta$.

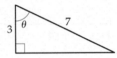

Solution

Use the Pythagorean theorem to find the length of the side opposite θ.

$a^2 + b^2 = c^2$	• See the figure at the right.
$3^2 + b^2 = 7^2$	• $a = 3, c = 7$
$9 + b^2 = 49$	
$b^2 = 40$	
$b = \sqrt{40} = 2\sqrt{10}$	

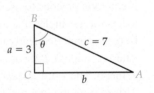

Using the definitions of the trigonometric ratios, we have

$$\sin \theta = \frac{\text{opp}}{\text{hyp}} = \frac{2\sqrt{10}}{7} \qquad \cos \theta = \frac{\text{adj}}{\text{hyp}} = \frac{3}{7} \qquad \tan \theta = \frac{\text{opp}}{\text{adj}} = \frac{2\sqrt{10}}{3}$$

In Example 1, we gave the exact answers. In many cases, approximate values of trigonometric ratios are given. The answers to Example 1, rounded to the nearest ten-thousandth, are

$$\sin \theta = \frac{2\sqrt{10}}{7} \approx 0.9035 \qquad \cos \theta = \frac{3}{7} \approx 0.4286 \qquad \tan \theta = \frac{2\sqrt{10}}{3} \approx 2.1082$$

We will sometimes want to know the value of a trigonometric ratio for a given angle. Triangle ABC in Figure 11.55 is an equilateral triangle with sides of length 2 units and angle bisector $\overline{BD}$. Because $\overline{BD}$ bisects $\angle ABC$, the measures of $\angle ABD$ and $\angle DBC$ are both 30°.

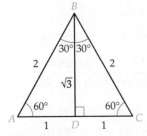

Figure 11.55

The angle bisector $\overline{BD}$ also bisects $\overline{AC}$. Therefore, $AD = 1$ and $DC = 1$. Using the Pythagorean theorem, we can find the measure of BD.

$$(DC)^2 + (BD)^2 = (BC)^2$$

$$1^2 + (BD)^2 = 2^2 \qquad \bullet\, DC = 1, BC = 2$$

$$1 + (BD)^2 = 4 \qquad \bullet\, \text{Solve for } BD.$$

$$(BD)^2 = 3$$

$$BD = \sqrt{3}$$

Using the definitions of the trigonometric ratios and triangle BCD, we can find the values of the sine, cosine, and tangent of 30° and 60°.

$$\sin 30° = \frac{\text{opp}}{\text{hyp}} = \frac{1}{2} = 0.5 \qquad\qquad \sin 60° = \frac{\text{opp}}{\text{hyp}} = \frac{\sqrt{3}}{2} \approx 0.8660$$

$$\cos 30° = \frac{\text{adj}}{\text{hyp}} = \frac{\sqrt{3}}{2} \approx 0.8660 \qquad \cos 60° = \frac{\text{adj}}{\text{hyp}} = \frac{1}{2} = 0.5$$

$$\tan 30° = \frac{\text{opp}}{\text{adj}} = \frac{1}{\sqrt{3}} \approx 0.5774 \qquad \tan 60° = \frac{\text{opp}}{\text{adj}} = \sqrt{3} \approx 1.7321$$

The properties of an equilateral triangle enabled us to calculate the values of the trigonometric ratios for 30° and 60°. Calculating values of the trigonometric ratios for most other angles, however, would be quite difficult. Fortunately, many calculators have been programmed to allow us to estimate these values.

To find tan 30° on a TI-83/84 calculator, first confirm that your calculator is in "degree mode." Press the **TAN** key and type in 30). Then press **ENTER**.

$$\tan 30° \approx 0.5774$$

On a scientific calculator, type in 30, and then press the **TAN** key.

Use a calculator to find sin 43.8° and tan 37.1° to the nearest ten-thousandth.

$$\sin 43.8° \approx 0.6921$$

$$\tan 37.1° \approx 0.7563$$

The engineers mentioned at the beginning of this section could use trigonometry to determine the distance across the ravine shown in Figure 11.56. Suppose the engineers measure angle A as 33.8°. Now they would ask, "Which trigonometric ratio, sine, cosine, or tangent, involves the side opposite angle A and the side adjacent to angle A?" Knowing that the tangent ratio is the required ratio, the engineers could write and solve the equation

$$\tan 33.8° = \frac{a}{100}$$

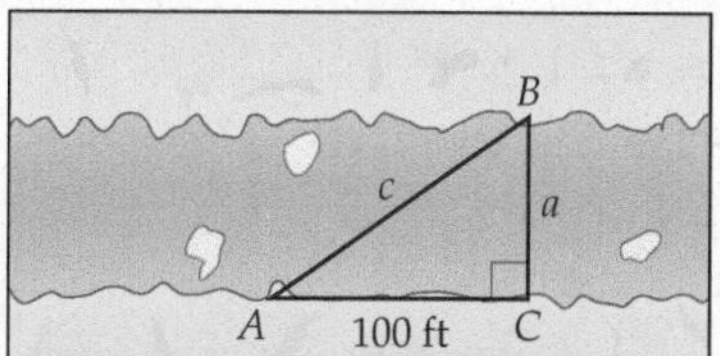

$$\tan 33.8° = \frac{a}{100}$$

$$100(\tan 33.8°) = a \qquad \bullet\, \text{Multiply each side of the equation by } 100.$$

$$66.9 \approx a \qquad \bullet\, \text{Use a calculator to find tan 33.8°. Multiply the result in the display by 100.}$$

Figure 11.56

The distance across the ravine is approximately 66.9 feet.

Example 2   **Find the Length of a Side of a Triangle**

For the right triangle, find the length of side a. Round to the nearest tenth of a meter.

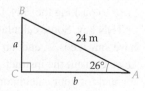

Solution

We are given the measure of $\angle A$ and the hypotenuse. We want to find the length of side a. Side a is opposite $\angle A$. The sine function involves the side opposite an angle and the hypotenuse.

$$\sin A = \frac{\text{opp}}{\text{hyp}}$$

$$\sin 26° = \frac{a}{24} \qquad \bullet \; A = 26°, \text{ hypotenuse} = 24 \text{ m}$$

$$24(\sin 26°) = a \qquad \bullet \; \text{Multiply each side by } 24.$$

$$10.5 \approx a \qquad \bullet \; \text{Use a calculator to find sin } 26°. \text{ Multiply the result in the}$$
$$\qquad\qquad\qquad\qquad \text{display by } 24.$$

The length of side a is approximately 10.5 m.

Inverse Sine, Inverse Cosine, and Inverse Tangent

Suppose it is necessary to find the measure of $\angle A$ in Figure 11.57. Because the length of the side adjacent to $\angle A$ is known and the length of the hypotenuse is known, we can write

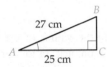

Figure 11.57

$$\cos A = \frac{\text{adj}}{\text{hyp}}$$

$$\cos A = \frac{25}{27}$$

The solution of this equation is the angle whose cosine is $\frac{25}{27}$. This angle can be found by using the $\cos^{-1}$ key on a calculator. The expression $\cos^{-1}$ is read "the inverse cosine of."

$$\cos^{-1}\left(\frac{25}{27}\right) \approx 22.19160657$$

To the nearest tenth of a degree, the measure of $\angle A$ is 22.2°.

Take Note: The expression $\sin^{-1}(x)$ is sometimes written $\arcsin(x)$. The two expressions are equivalent. The expressions $\cos^{-1}(x)$ and $\arccos(x)$ are equivalent, as are $\tan^{-1}(x)$ and $\arctan(x)$.

Inverse Sine, Inverse Cosine, and Inverse Tangent

$\sin^{-1}(x)$ is defined as the angle whose sine is x, $-1 \leq x \leq 1$.
$\cos^{-1}(x)$ is defined as the angle whose cosine is x, $-1 \leq x \leq 1$.
$\tan^{-1}(x)$ is defined as the angle whose tangent is x, $-\infty < x < \infty$.

Calculator Note: To find an inverse on a calculator, usually the INV or 2nd key is pressed prior to pushing the SIN, COS, or TAN key. Some calculators have $\sin^{-1}$, $\cos^{-1}$, and $\tan^{-1}$ keys. Consult the instruction manual for your calculator.

Example 3   **Evaluate an Inverse Sine Expression**

Use a calculator to find $\sin^{-1}(0.9171)$. Round to the nearest tenth of a degree.

Solution

$$\sin^{-1}(0.9171) \approx 66.5° \qquad \bullet \; \text{The calculator must be in degree mode. Press the keys for}$$
$$\qquad\qquad\qquad\qquad\qquad \text{inverse sine followed by 0.9171). Press ENTER.}$$

Example 4 Find the Measure of an Angle Using Inverse Sine

Given $\sin \theta = 0.7239$, find θ. Use a calculator. Round to the nearest tenth of a degree.

Solution

This is equivalent to finding $\sin^{-1}(0.7239)$. The calculator must be in degree mode.

$$\sin^{-1}(0.7239) \approx 46.4°$$

$$\theta \approx 46.4°$$

Example 5 Find the Measure of an Angle in a Right Triangle

For the right triangle, find the measure of $\angle B$. Round to the nearest tenth of a degree.

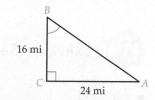

Solution

We want to find the measure of $\angle B$, and we are given the lengths of the sides opposite $\angle B$ and adjacent to $\angle B$. The tangent ratio involves the side opposite an angle and the side adjacent to that angle.

$$\tan B = \frac{\text{opp}}{\text{adj}}$$

$$\tan B = \frac{24}{16}$$

$$B = \tan^{-1}\left(\frac{24}{16}\right)$$

$$B \approx 56.3° \qquad \text{• Use the } \tan^{-1} \text{ key on a calculator.}$$

The measure of $\angle B$ is approximately 56.3°.

Angles of Elevation and Depression

One application of trigonometry, called **line-of-sight problems**, concerns an observer looking at an object.

Angles of elevation and depression are measured with respect to a horizontal line. If the object being sighted is above the observer, the acute angle formed by the line of sight and the horizontal line is an **angle of elevation**. If the object being sighted is below the observer, the acute angle formed by the line of sight and the horizontal line is an **angle of depression**.

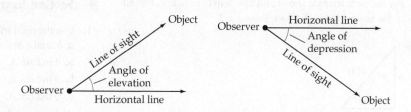

Example 6 Solve an Angle of Elevation Problem

The angle of elevation from a point 62 ft away from the base of a flagpole to the top of the flagpole is 34°. Find the height of the flagpole. Round to the nearest tenth of a foot.

Solution

Draw a diagram. To find the height, h, write a trigonometric ratio that relates the given information and the unknown side of the triangle.

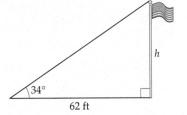

$$\tan 34° = \frac{h}{62}$$

$62(\tan 34°) = h$ • Solve for h.

$41.8 \approx h$ • Use a calculator to find $\tan 34°$. Multiply the result in the display by 62.

The height of the flagpole is approximately 41.8 ft.

11.6 Exercise Set

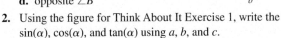

▶ Think About It

1. For the right triangle shown, indicate which side is
   a. adjacent to $\angle A$
   b. opposite θ
   c. adjacent to α
   d. opposite $\angle B$

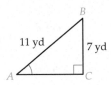

2. Using the figure for Think About It Exercise 1, write the $\sin(\alpha)$, $\cos(\alpha)$, and $\tan(\alpha)$ using a, b, and c.

3. $\tan(45°) = 1$. What is the value of $\tan^{-1}(1)$?

4. If $\sin^{-1}(\theta) = x$, what is $\sin(x)$?

▶ Preliminary Exercises

For Exercises P1 to P6, if necessary, use the referenced example following the exercise for assistance.

P1. For the right triangle shown, find $\sin(\theta)$, $\cos(\theta)$, and $\tan(\theta)$. **Example 1**

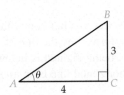

P2. For the right triangle shown, find the length of side a. Round to the nearest tenth of a foot. **Example 2**

P3. Use a calculator to find $\tan^{-1}(0.3165)$. Round to the nearest tenth of a degree. **Example 3**

P4. Given $\tan \theta = 0.5681$, find θ. Use a calculator. Round to the nearest tenth of a degree. **Example 4**

P5. For the right triangle shown, find the measure of $\angle A$. Round to the nearest tenth of a degree. **Example 5**

P6. The angle of depression from the top of a lighthouse that is 20 m high to a boat on the water is 25°. How far is the boat from the base of the lighthouse? Round to the nearest tenth of a meter. **Example 6**

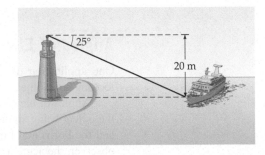

▶ Section Exercises

1. Use the right triangle shown in the following figure and sides a, b, and c to do the following:
   a. Find $\sin A$.
   b. Find $\sin B$.
   c. Find $\cos A$.
   d. Find $\cos B$.
   e. Find $\tan A$.
   f. Find $\tan B$.

2. Explain the meaning of the notation $\sin^{-1}(x)$, $\cos^{-1}(x)$, and $\tan^{-1}(x)$.

In Exercises 3 to 10, find the values of sin θ, cos θ, and tan θ for the given right triangle. Give the exact values.

3.
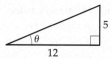

4.
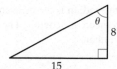

5.
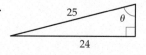

6.

7.

8.
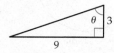

9.
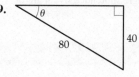

10.
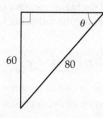

In Exercises 11 to 16, use a calculator to estimate the value of each of the following. Round to the nearest ten-thousandth.

11. cos 47°   **12.** cos 21.9°   **13.** tan 63.4°

14. tan 41.6°   **15.** sin 57.7°   **16.** sin 58.3°

In Exercises 17 to 28, use a calculator. Round to the nearest tenth of a degree.

17. Given sin θ = 0.6239, find θ.

18. Given sin β = 0.7349, find β.

19. Given cos β = 0.9516, find β.

20. Given cos θ = 0.3007, find θ.

21. Given tan α = 0.3899, find α.

22. Given tan α = 1.588, find α.

23. Find $\sin^{-1}(0.4478)$.

24. Find $\sin^{-1}(0.0105)$.

25. Find $\tan^{-1}(0.7815)$.

26. Find $\tan^{-1}(0.2438)$.

27. Find $\cos^{-1}(0.7536)$.

28. Find $\cos^{-1}(0.6032)$.

For Exercises 29 to 42, draw a picture and label it. Then set up an equation and solve it. Show all your work. Round the measure of each angle to the nearest tenth of a degree. Round the length of a side to the nearest tenth of a unit. Assume the ground is level unless indicated otherwise.

29. Ballooning   A balloon, tethered by a cable 997 ft long, was blown by a wind so that the cable made an angle of 57.6° with the ground. Find the height of the balloon off the ground.

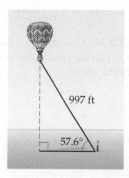

997 ft

57.6°

30. Roadways   A road is inclined at an angle of 9.8° with the horizontal. Find the distance that one must drive on this road in order to be elevated 14.8 ft above the horizontal.

31. Home Maintenance   A ladder 30.8 ft long leans against a building. If the foot of the ladder is 7.25 ft from the base of the building, find the angle the top of the ladder makes with the building.

32. Aviation   A plane takes off from a field and rises at an angle of 11.4° with the horizontal. Find the height of the plane after it has traveled a distance of 1250 ft.

33. Guy Wires   A guy wire whose grounded end is 16 ft from the telephone pole it supports makes an angle of 56.7° with the ground. How long is the wire?

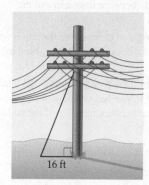

16 ft

34. Angle of Depression   A lighthouse built at sea level is 169 ft tall. From its top, the angle of depression to a boat below measures 25.1°. Find the distance from the boat to the foot of the lighthouse.

35. Angle of Elevation   At a point 39.3 ft from the base of a tree, the angle of elevation of its top measures 53.4°. Find the height of the tree.

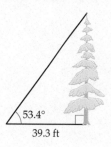

53.4°

39.3 ft

36. Angle of Depression An artillery spotter in a plane that is at an altitude of 978 ft measures the angle of depression of an enemy tank as 28.5°. How far is the enemy tank from the point on the ground directly below the spotter?

37. Home Maintenance A 15-foot ladder leans against a house. The ladder makes an angle of 65° with the ground. How far up the side of the house does the ladder reach?

38. Angle of Elevation Find the angle of elevation of the sun when a tree 40.5 ft high casts a shadow 28.3 ft long.

39. Guy Wires A television transmitter tower is 600 ft high. If the angle between the guy wire (attached at the top) and the tower is 55.4°, how long is the guy wire?

40. Ramps A ramp used to load a racing car onto a flatbed carrier is 5.25 m long, and its upper end is 1.74 m above the lower end. Find the angle between the ramp and the road.

41. Angle of Elevation The angle of elevation of the sun is 51.3° at a time when a tree casts a shadow 23.7 yd long. Find the height of the tree.

42. Angle of Depression From the top of a building 312 ft tall, the angle of depression to a flower bed on the ground below is 12.0°. What is the distance between the base of the building and the flower bed?

▶ **Investigations**

As we noted in this section, angles can also be measured in *radians*. For physicists, engineers, and other applied scientists who use calculus, radians are preferred over degrees because they simplify many calculations. To define a radian, first consider a circle of radius r and two radii $\overline{OA}$ and $\overline{OB}$. The angle θ formed by the two radii is a **central angle**. The portion of the circle between A and B is an **arc** of the circle and is written $\overset{\frown}{AB}$. We say that $\overset{\frown}{AB}$ *subtends* the angle θ. The length of the arc is s. (See Figure 1.)

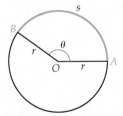

Figure 1

Radian

One **radian** is the measure of the central angle subtended by an arc of length r. The measure of θ in Figure 2 is 1 radian.

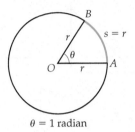

$\theta = 1$ radian

Figure 2

To find the radian measure of an angle subtended by an arc of length s, use the following formula.

Radian Measure

Given an arc of length s on a circle of radius r, the measure of the central angle subtended by the arc is

$$\theta = \frac{s}{r} \text{ radians.}$$

For example, to find the measure in radians of the central angle subtended by an arc of 9 in. in a circle of radius 12 in., divide the length of the arc ($s = 9$ in.) by the length of the radius ($r = 12$ in.). See Figure 3.

$$\theta = \frac{9 \text{ in.}}{12 \text{ in.}} \text{ radian}$$

$$= \frac{3}{4} \text{ radian}$$

Figure 3

1. Find the measure in radians of the central angle subtended by an arc of 12 cm in a circle of radius 3 cm.

2. Find the measure in radians of the central angle subtended by an arc of 4 cm in a circle of radius 8 cm.

3. Find the measure in radians of the central angle subtended by an arc of 6 in. in a circle of radius 9 in.

4. Find the measure in radians of the central angle subtended by an arc of 12 ft in a circle of radius 10 ft.

Recall that the circumference of a circle is given by $C = 2\pi r$. Therefore, the radian measure of the central angle subtended by the circumference is $\theta = \dfrac{2\pi r}{r} = 2\pi$. In degree measure, the central angle has a measure of 360°. Thus we have 2π radians = 360°. Dividing each side of the equation by 2 gives π radians = 180°. From the last equation, we can establish the conversion factors $\dfrac{\pi \text{ radians}}{180°}$ and $\dfrac{180°}{\pi \text{ radians}}$. These conversion factors are used to convert between radians and degrees.

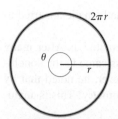

$2\pi r$

θ r

To convert 2 radians to degrees, multiply 2 by $\dfrac{180°}{\pi \text{ radians}}$.

$$2 \text{ radians} = 2\left(\dfrac{180°}{\pi \text{ radians}}\right)$$

$$= \left(\dfrac{360}{\pi}\right)° \quad \bullet \text{ Exact answer}$$

$$\approx 114.5916° \quad \bullet \text{ Approximate answer}$$

5. What is the measure in degrees of 1 radian?

6. Is the measure of 1 radian larger or smaller than the measure of 1°?

Conversion Between Radians and Degrees

- To convert from degrees to radians, multiply by $\dfrac{\pi \text{ radians}}{180°}$.

- To convert from radians to degrees, multiply by $\dfrac{180°}{\pi \text{ radians}}$.

For instance, to convert 30° to radians, multiply 30° by $\dfrac{\pi \text{ radians}}{180°}$.

$$30° = 30°\left(\dfrac{\pi \text{ radians}}{180°}\right)$$

$$= \dfrac{\pi}{6} \text{ radian} \quad \bullet \text{ Exact answer}$$

$$\approx 0.5236 \text{ radian} \quad \bullet \text{ Approximate answer}$$

In Exercises 7 to 12, convert degree measure to radian measure. Find an exact answer and an answer rounded to the nearest ten-thousandth.

7. 45°   **8.** 180°   **9.** 315°

10. 90°   **11.** 210°   **12.** 18°

In Exercises 13 to 18, convert radian measure to degree measure. For Exercises 16 to 18, find an exact answer and an answer rounded to the nearest ten-thousandth.

13. $\dfrac{\pi}{3}$ radians   **14.** $\dfrac{11\pi}{6}$ radians

15. $\dfrac{4\pi}{3}$ radians   **16.** 1.2 radians

17. 3 radians   **18.** 2.4 radians

Section 11.7

Non-Euclidean Geometry

Learning Objectives:

- Find the area of spherical geometry.

- Find Euclidean distance and city distance between two points.

Take Note: In addition to postulates, Euclid's geometry, referred to as *Euclidean geometry*, involves definitions and some *undefined terms*. For instance, the term *point* is an undefined term in Euclidean geometry because any definition of the term *point* would require additional undefined terms. In a similar way, the term *line* is also an undefined term in Euclidean geometry.

Euclidean Geometry vs. Non-Euclidean Geometry

Some of the most popular games are based on a handful of rules that are easy to learn but still allow the game to develop into complex situations. The ancient Greek mathematician Euclid wanted to establish a geometry that was based on the *fewest* possible number of rules. He called these rules **postulates**. Euclid based his geometry on the following five postulates.

Euclid's Postulates

P1: A line segment can be drawn from any point to any other point.

P2: A line segment can be extended continuously in a straight line.

P3: A circle can be drawn with any center and any radius.

P4: All right angles have the same measure.

P5: *The Parallel Postulate*   Through a given point not on a given line, exactly one line can be drawn parallel to the given line.

For many centuries, the truth of these postulates was felt to be self-evident. However, a few mathematicians suspected that the fifth postulate, known as the parallel postulate, could be deduced from the other postulates. Over the years, many mathematicians tried to prove the parallel postulate, but none were successful.

Carl Friedrich Gauss (gaus′) (1777–1855) was one such mathematician. After many failed attempts to establish the parallel postulate as a theorem, Gauss came to the conclusion that the parallel postulate was an independent postulate. However, he noted that by changing this one postulate, he could create a whole new type of geometry! This is analogous to changing one of the rules of a game to create a new game.

Gauss's Alternative to the Parallel Postulate

Through a given point not on a given line, there are *at least two* lines parallel to the given line.

Another pioneer in **non-Euclidean geometry** (any geometry that does not include Euclid's parallel postulate) was the Russian mathematician Nikolai Lobachevsky. In a series of monthly articles that appeared in the academic journal of the University of Kazan in 1829, Lobachevsky provided a detailed investigation into the problem of the parallel postulate. He proposed that a consistent new geometry could be developed by replacing the parallel postulate with the alternative postulate, which assumes that *more than one* parallel line can be drawn through a point not on a given line. This new geometry, developed independently by both Gauss and Lobachevsky, is often called *hyperbolic geometry*.

The year 1826, in which Lobachevsky first lectured about a new non-Euclidean geometry, also marks the birth of the mathematician Bernhard Riemann. Although Riemann died of tuberculosis at age 39, he made major contributions in several areas of mathematics and physics. Riemann was the first person to consider a geometry in which the parallel postulate was replaced with the following postulate.

Riemann's Alternative to the Parallel Postulate

Through a given point not on a given line, there exist *no* lines parallel to the given line.

Unlike the geometry developed by Lobachevsky, which was not based on a physical model, the non-Euclidean geometry of Riemann was closely associated with a sphere and the remarkable idea that because a line is an undefined term, a line on the surface of a sphere can be different from a line on a plane. It seems reasonable to suspect that "spherical lines" should retain some of the properties of lines on a plane. For example, on a plane, the shortest distance between two points is measured along the line that connects the points. The line that connects the points is an example of what is called a *geodesic*.

Geodesic

A **geodesic** is a curve C on a surface S such that for any two points on C, the portion of C between these points is the shortest path on S that joins these points.

On a sphere, the geodesic between two points is a *great circle* that connects the points.

Great Circle

A **great circle** of a sphere is a circle on the surface of the sphere whose center is at the center of the sphere. Any two points on a great circle divide the circle into two arcs. The shorter arc is the **minor arc**, and the longer arc is the **major arc**.

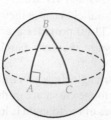

Center of
the sphere

Two great circles
of the sphere

Figure 11.58 A sphere and its great circles serve as a physical model for Riemannian geometry.

In *Riemannian geometry*, which is also called *spherical geometry* or *elliptical geometry*, great circles, which are the geodesics of a sphere, are thought of as lines. Figure 11.58 shows a sphere and two of its great circles. Because all great circles of a sphere intersect, a sphere provides us with a model of a geometry in which there are no parallel lines.

In Riemannian geometry, a triangle may have as many as three right angles. Figure 11.59 illustrates a spherical triangle with one right angle, a spherical triangle with two right angles, and a spherical triangle with three right angles.

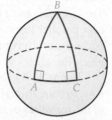

a. A spherical triangle
with one right angle

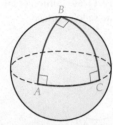

b. A spherical triangle
with two right angles

c. A spherical triangle
with three right angles

Figure 11.59

The Spherical Triangle Area Formula

The area S of the spherical triangle ABC on a sphere with radius r is given by

$$S = (m\angle A + m\angle B + m\angle C - 180°)\left(\frac{\pi}{180°}\right)r^2$$

where each angle is measured in degrees.

Example 1   Find the Area of a Spherical Triangle

Find the area of a spherical triangle with three right angles on a sphere with a radius of 1 ft. Find both the exact area and the approximate area rounded to the nearest hundredth of a square foot.

Solution

Apply the spherical triangle area formula.

$$S = (m\angle A + m\angle B + m\angle C - 180°)\left(\frac{\pi}{180°}\right)r^2$$

$$= (90° + 90° + 90° - 180°)\left(\frac{\pi}{180°}\right)(1)^2$$

$$= (90°)\left(\frac{\pi}{180°}\right)$$

$$= \frac{\pi}{2} \text{ ft}^2 \qquad \bullet \text{ Exact area}$$

$$\approx 1.57 \text{ ft}^2 \qquad \bullet \text{ Approximate area}$$

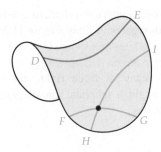

Figure 11.60 Portion of an infinite saddle surface

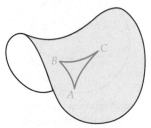

Figure 11.61 Hyperbolic triangle

Mathematicians have not been able to create a three-dimensional model that *perfectly* illustrates all aspects of hyperbolic geometry. However, an infinite saddle surface can be used to visualize some of the basic aspects of hyperbolic geometry. Figure 11.60 shows a portion of an infinite saddle surface.

A line (geodesic) can be drawn through any two points on the saddle surface. Most lines on the saddle surface have a concave curvature, as shown by $\overleftrightarrow{DE}$, $\overleftrightarrow{FG}$, and $\overleftrightarrow{HI}$. Keep in mind that the saddle surface is an infinite surface. Figure 11.60 shows only a portion of the surface.

Parallel lines on an infinite saddle surface are defined as two lines that do not intersect. In Figure 11.60, $\overleftrightarrow{FG}$ and $\overleftrightarrow{HI}$ are *not* parallel because they intersect at a point. The lines $\overleftrightarrow{DE}$ and $\overleftrightarrow{FG}$ are parallel because they do not intersect. The lines $\overleftrightarrow{DE}$ and $\overleftrightarrow{HI}$ are also parallel lines. Figure 11.60 provides a geometric model of a hyperbolic geometry because for a given line, *more than one* parallel line exists through a point not on the given line.

Figure 11.61 shows a triangle drawn on a saddle surface. The triangle is referred to as a *hyperbolic triangle*. Due to the curvature of the sides of the hyperbolic triangle, the sum of the measures of the angles of the triangle is less than 180°. This is true for all hyperbolic triangles.

Table 11.3 summarizes some of the properties of plane, hyperbolic, and spherical geometries.

Table 11.3

Euclidean Geometry	Non-Euclidean Geometries	
Euclidean or Plane Geometry (ca. 300 BC):	**Lobachevskian or Hyperbolic Geometry (1826):**	**Riemannian or Spherical Geometry (1855):**
Through a given point not on a given line, exactly one line can be drawn parallel to the given line.	***Through a given point not on a given line, there are at least two lines parallel to the given line.***	***Through a given point not on a given line, there exist no lines parallel to the given line.***
Geometry on a plane	Geometry on an infinite saddle surface	Geometry on a sphere
For any triangle *ABC*, $m\angle A + m\angle B + m\angle C = 180°$	For any triangle *ABC*, $m\angle A + m\angle B + m\angle C < 180°$	For any triangle *ABC*, $180° < m\angle A + m\angle B + m\angle C < 540°$
A triangle can have at most one right angle.	A triangle can have at most one right angle.	A triangle can have one, two, or three right angles.
The shortest path between two points is the line segment that connects the points.	The curves shown in the above figure illustrate some of the geodesics of an infinite saddle surface.	The shortest path between two points is the minor arc of a great circle that passes through the points.

Example 2   **Euclidean and Non-Euclidean Geometries**

Determine the type of geometry (Euclidean, Riemannian, or Lobachevskian) in which two lines can intersect at a point and both of the lines can be parallel to a third line that does not pass through the intersection point.

Solution

Euclid's parallel postulate states that through a given point not on a given line, exactly one line can be drawn parallel to the given line.

Riemann's alternative to the parallel postulate states that through a given point not on a given line, there exist no lines parallel to the given line.

Gauss's alternative to the parallel postulate, which is assumed in Lobachevskian geometry, states that through a given point not on a given line, there are at least two lines parallel to the given line.

Thus, if we consider only Euclidean, Riemannian, and Lobachevskian geometries, then the condition that "two lines can intersect at a point and both of the lines can be parallel to a third line that does not pass through the intersection point" can be true only in Lobachevskian geometry.

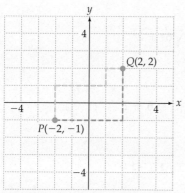

Figure 11.62 Two city paths from P to Q

City Geometry: A Contemporary Geometry

Consider the geometric model of a city shown in Figure 11.62. In this city, all of the streets run either straight north and south or straight east and west. The distance between adjacent north–south streets is 1 block, and the distance between adjacent east–west streets is 1 block. In a city it is generally not possible to travel from P to Q along a straight path. Instead, one must travel between P and Q by traveling along the streets. As you travel from P to Q, we assume that you always travel in a direction that gets you closer to point Q. Two such paths are shown by the purple and the green dashed line segments in Figure 11.62.

We will use the notation $d_C(P, Q)$ to represent the *city distance* between the points P and Q. For P and Q as shown in Figure 11.62, $d_C(P, Q) = 7$ blocks. This distance can be determined by counting the number of blocks needed to travel along the streets from P to Q or by using the following formula.

The City Distance Formula

If $P(x_1, y_1)$ and $Q(x_2, y_2)$ are two points in a city, then the **city distance** between P and Q is given by

$$d_C(P, Q) = |x_2 - x_1| + |y_2 - y_1|$$

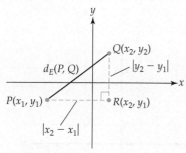

Figure 11.63

In Euclidean geometry, the distance between the points P and Q is defined as the length of $\overline{PQ}$. To determine a *Euclidean distance formula* for the distance between $P(x_1, y_1)$ and $Q(x_2, y_2)$, we first locate the point $R(x_2, y_1)$. See Figure 11.63.

Note that R has the same x-coordinate as Q and that R has the same y-coordinate as P. The horizontal distance between P and R is $|x_2 - x_1|$, and the vertical distance between R and Q is $|y_2 - y_1|$. Apply the Pythagorean theorem to the right triangle PRQ to produce

$$[d_E(P, Q)]^2 = |x_2 - x_1|^2 + |y_2 - y_1|^2$$

Because the square of a number cannot be negative, the absolute value signs are not necessary.

$$[d_E(P, Q)]^2 = (x_2 - x_1)^2 + (y_2 - y_1)^2$$

Take the square root of each side of the equation to produce

$$d_E(P, Q) = \sqrt{(x_2 - x_1)^2 + (y_2 - y_1)^2}$$

The Euclidean Distance Formula

If $P(x_1, y_1)$ and $Q(x_2, y_2)$ are two points in a plane, then the **Euclidean distance** between P and Q is given by

$$d_E(P, Q) = \sqrt{(x_2 - x_1)^2 + (y_2 - y_1)^2}$$

Example 3 Find the Euclidean Distance and the City Distance Between Two Points

For each of the following, find $d_E(P, Q)$ and $d_C(P, Q)$. Assume that both $d_E(P, Q)$ and $d_C(P, Q)$ are measured in blocks. Round approximate results to the nearest tenth of a block.

a. $P(-4, -3), Q(2, -1)$ **b.** $P(2, -3), Q(-5, 4)$

Solution

a. $d_E(P, Q) = \sqrt{(x_2 - x_1)^2 + (y_2 - y_1)^2}$

$\qquad\qquad = \sqrt{[2 - (-4)]^2 + [(-1) - (-3)]^2} = \sqrt{6^2 + 2^2}$

$\qquad\qquad = \sqrt{40} \approx 6.3$ blocks

$\quad d_C(P, Q) = |x_2 - x_1| + |y_2 - y_1|$

$\qquad\qquad = |2 - (-4)| + |(-1) - (-3)| = |6| + |2| = 6 + 2$

$\qquad\qquad = 8$ blocks

b. $d_E(P, Q) = \sqrt{(x_2 - x_1)^2 + (y_2 - y_1)^2}$

$\qquad\qquad = \sqrt{[(-5) - 2]^2 + [4 - (-3)]^2} = \sqrt{(-7)^2 + 7^2}$

$\qquad\qquad = \sqrt{98} \approx 9.9$ blocks

$\quad d_C(P, Q) = |x_2 - x_1| + |y_2 - y_1|$

$\qquad\qquad = |(-5) - 2| + |4 - (-3)| = |-7| + |7| = 7 + 7$

$\qquad\qquad = 14$ blocks

Recall that a circle is a plane figure in which all points are the same distance from a given center point and the length of the radius r of the circle is the distance from the center point to a point on the circle. Figure 11.64 shows a *Euclidean circle* centered at $(0, 0)$ with a radius of 3 blocks. Figure 11.65 shows all the points in a city that are 3 blocks from the center point $(0, 0)$. These points form a *city circle* with a radius of 3 blocks.

It is interesting to observe that the *city circle* shown in Figure 11.65 consists of just 12 points and that these points all lie on a square with vertices $(3, 0)$, $(0, 3)$, $(-3, 0)$, and $(0, -3)$.

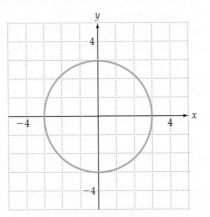

Figure 11.64 A Euclidean circle with center $(0, 0)$ and a radius of 3 blocks

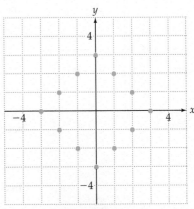

Figure 11.65 A city circle with center $(0, 0)$ and a radius of 3 blocks

11.7 Exercise Set

Think About It

1. State the parallel postulate for each of the following.
   a. Euclidean geometry
   b. Lobachevskian geometry
   c. Riemannian geometry

2. What is the maximum number of right angles a triangle can have in
   a. Euclidean geometry?
   b. Lobachevskian geometry?
   c. Riemannian geometry?

3. What is a geodesic?

4. In which geometry can two distinct lines be parallel to a third line but not parallel to each other?

5. In which geometry do all perpendiculars to a given line intersect each other?

Preliminary Exercises

For Exercises P1 to P3, if necessary, use the referenced example following the exercise for assistance.

P1. Find the area of the spherical triangle whose angles measure $200°$, $90°$, and $90°$ on a sphere with a radius of 6 in. Find both the exact area and the approximate area rounded to the nearest hundredth of a square inch. Example 1

P2. Determine the type of geometry in which there are no lines parallel to a given line. Example 2

P3. For each of the following, find $d_E(P, Q)$ and $d_C(P, Q)$. Assume that both $d_E(P, Q)$ and $d_C(P, Q)$ are measured in blocks. Round approximate results to the nearest tenth of a block. Example 3
   a. $P(-1, 4), Q(3, 2)$
   b. $P(3, -4), Q(-1, 5)$

Section Exercises

1. Find the exact area of a spherical triangle with angles of $150°$, $120°$, and $90°$ on a sphere with a radius of 1.

2. Find the area of a spherical triangle with three right angles on a sphere with a radius of 1980 mi. Round to the nearest ten thousand square miles.

In Exercises 3 to 10, find the Euclidean distance between the points and the city distance between the points. Assume that both $d_E(P, Q)$ and $d_C(P, Q)$ are measured in blocks. Round approximate results to the nearest tenth of a block.

3. $P(-3, 1), Q(4, 1)$ 4. $P(-2, 4), Q(3, -1)$
5. $P(2, -3), Q(-3, 5)$ 6. $P(-2, 0), Q(3, 7)$
7. $P(-1, 4), Q(5, -2)$ 8. $P(-5, 2), Q(3, -4)$
9. $P(2, 0), Q(3, -6)$ 10. $P(2, -2), Q(5, -2)$

A Distance Conversion Formula   The following formula can be used to convert the Euclidean distance between the points P and Q to the city distance between P and Q. In this formula, the variable m represents the slope of the line segment $\overline{PQ}$.

$$d_C(P, Q) = \frac{1 + |m|}{\sqrt{1 + m^2}} d_E(P, Q)$$

In Exercises 11 to 16, use the preceding formula to find the city distance between P and Q.

11. $d_E(P, Q) = 5$ blocks, slope of $\overline{PQ} = \dfrac{3}{4}$

12. $d_E(P, Q) = \sqrt{29}$ blocks, slope of $\overline{PQ} = \dfrac{2}{5}$

13. $d_E(P, Q) = \sqrt{13}$ blocks, slope of $\overline{PQ} = -\dfrac{2}{3}$

14. $d_E(P, Q) = 2\sqrt{10}$ blocks, slope of $\overline{PQ} = -3$

15. $d_E(P, Q) = \sqrt{17}$ blocks, slope of $\overline{PQ} = \dfrac{1}{4}$

16. $d_E(P, Q) = 4\sqrt{2}$ blocks, slope of $\overline{PQ} = -1$

17. Explain why there is no formula that can be used to convert $d_C(P, Q)$ to $d_E(P, Q)$. Assume that no additional information is given other than the value of $d_C(P, Q)$.

18. a. If $d_E(P, Q) = d_E(R, S)$, must $d_C(P, Q) = d_C(R, S)$? Explain.
    b. If $d_C(P, Q) = d_C(R, S)$, must $d_E(P, Q) = d_E(R, S)$? Explain.

19. Plot the points in the city circle with center $(-2, -1)$ and radius $r = 2$ blocks.

20. Plot the points in the city circle with center $(1, -1)$ and radius $r = 3$ blocks.

21. Plot the points in the city circle with center $(0, 0)$ and radius $r = 2.5$ blocks.

22. Plot the points in the city circle with center $(0, 0)$ and radius $r = 3.5$ blocks.

23. How many points are on the city circle with center $(0, 0)$ and radius $r = n$ blocks, where n is a natural number?

24. Which of the following city circles has the most points, a city circle with center $(0, 0)$ and radius 4.5 blocks or a city circle with center $(0, 0)$ and radius 5 blocks?

Investigation

Apartment Hunting in a City   Use the following information to answer each of the questions in Exercises 1 and 2.

Amy and her spouse Ryan are looking for an apartment located adjacent to a city street. Amy works at $P(-3, -1)$ and Ryan works at $Q(2, 3)$. Both Amy and Ryan plan to walk from their apartment to work along routes that follow the north–south and the east–west streets.

1. a. Plot the points where Amy and Ryan should look for an apartment if they wish the sum of the city distances they need to walk to work to be a minimum.
   b. Plot the points where Amy and Ryan should look for an apartment if they wish the sum of the city distances they

need to walk to work to be a minimum and they both will walk the same distance. *Hint:* Find the intersection of the city circle with center P and radius of 4.5 blocks, and the city circle with center Q and radius of 4.5 blocks.

2. a. Plot the points where Amy and Ryan should look for an apartment if they wish the sum of the city distances they need to walk to work to be less than or equal to 10 blocks.

b. Plot the points where Amy and Ryan should look for an apartment if they wish the sum of the city distances they need to walk to work to be less than or equal to 10 blocks and they both will walk the same distance. *Hint:* Find the intersection of the city circle with center P and radius of 5.5 blocks, and the city circle with center Q and radius of 5.5 blocks.

Section 11.8

Fractals

Learning Objectives:

- Draw stages of a fractal.
- Find the similarity dimension of a fractal.

Fractals—Endlessly Repeated Geometric Figures

Have you ever used a computer program to enlarge a portion of a photograph? Sometimes the result is a satisfactory enlargement; however, if the photograph is enlarged too much, the image becomes blurred. For example, the photograph in Figure 11.66 is shown at its original size. The image in Figure 11.67 is an enlarged portion of Figure 11.66, and the image in Figure 11.68 is an enlarged portion of Figure 11.67. If we continue to make enlargements of enlargements, we will produce extremely blurred images that provide little information about the original photograph.

Figure 11.66

Figure 11.67

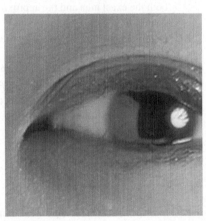

Figure 11.68

A computer monitor displays an image using small dots called *pixels*. If a computer image is enlarged using a software program, the program must determine the color of each pixel in the enlargement. If the image file for the photograph cannot supply the needed color information for each pixel, the color of some pixels is calculated by *averaging* the numerical color values of neighboring pixels for which the image file has the color information.

In the 1970s, the mathematician Benoit Mandelbrot discovered some remarkable methods that enable us to create geometric figures with a special property: if any portion of the figure is enlarged repeatedly, then additional details (not fewer details, as with the enlargement of a photograph) of the figure are displayed. Mandelbrot called these endlessly repeated geometric figures *fractals*. The fractals that we will study in this lesson can be defined as follows: a **fractal** is a geometric figure in which a self-similar motif repeats itself on an ever-diminishing scale.

Fractals are generally constructed by using **iterative processes** in which the fractal is more closely approximated as a repeated cycle of procedures is performed. For example, a fractal known as the *Koch curve* is constructed as follows.

Construction of the Koch Curve

Step 0: Start with a line segment. This initial segment is shown as stage 0 in Figure 11.69. Stage 0 of a fractal is called the **initiator** of the fractal.

Step 1: On the middle third of the line segment, draw an equilateral triangle and remove its base. The resulting curve is stage 1 in Figure 11.69. Stage 1 of a fractal is called the **generator** of the fractal.

Step 2: Replace each initiator shape (line segment, in this example) with a *scaled version* of the generator to produce the next stage of the Koch curve. The width of the scaled version of the generator is the same as the width of the line segment it replaces. Continue to repeat this step ad infinitum to create additional stages of the Koch curve.

Three applications of step 2 produce stage 2, stage 3, and stage 4 of the Koch curve, as shown in Figure 11.69.

None of the curves shown in Figure 11.69 is the Koch curve. The Koch curve is the curve that would be produced if step 2 in the above construction process were repeated

Take Note: The line segments at any stage of the Koch curve are $\frac{1}{3}$ the length of the line segments at the previous stage. To create the next stage of the Koch curve (after stage 1), replace each line segment with a scaled version of the generator.

At any stage after stage 2, the scaled version of the generator is $\frac{1}{3}$ the size of the preceding scaled generator.

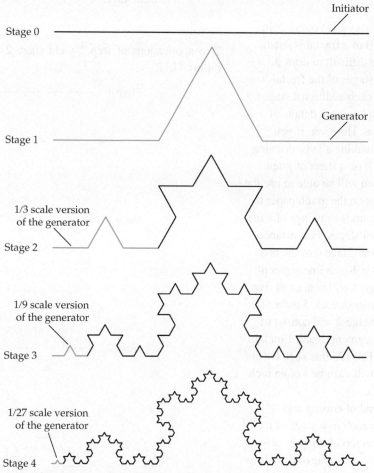

Figure 11.69 The first five stages of the Koch curve

ad infinitum. No one has ever seen the Koch curve, but we know that it is a very jagged curve in which the self-similar motif shown in Figure 11.69 repeats itself on an ever-diminishing scale.

The curves shown in Figure 11.70 are the first five stages of the *Koch snowflake.*

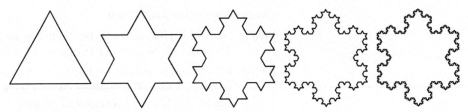

Figure 11.70 The first five stages of the Koch snowflake

Example 1 **Draw Stages of a Fractal**

Draw stage 2 and stage 3 of the *box curve*, which is defined by the following iterative process.

Step 0: Start with a line segment as the initiator. See stage 0 in Figure 11.71.

Step 1: On the middle third of the line segment, draw a square and remove its base. This produces the generator of the box curve. See stage 1 in Figure 11.71.

Step 2: Replace each initiator shape with a scaled version of the generator to produce the next stage.

Solution

Two applications of step 2 yield stage 2 and stage 3 of the box curve, as shown in Figure 11.71.

Stage 0

Stage 1

Stage 2

Stage 3

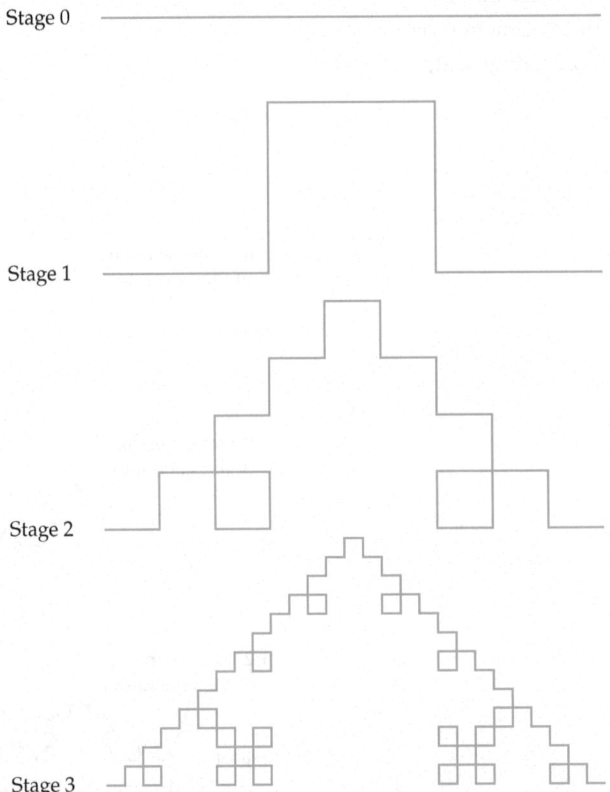

Figure 11.71 The first four stages of the box curve

Take Note: If your drawing of stage 0 of a fractal is small, it will be difficult to draw additional stages of the fractal, because each additional stage must display more details of the fractal. However, if you start by making a *large* drawing of stage 0 on a sheet of graph paper, you will be able to use the grid lines on the graph paper to make accurate drawings of a few additional stages. For instance, if you draw stage 0 of the box curve as a 9-inch line segment, then stage 1 will consist of five line segments, each 3 inches in length. Stage 2 will consist of 25 line segments, each 1 inch in length. The 125 line segments of stage 3 will each be $\frac{1}{3}$ of an inch in length.

Instead of erasing and drawing each new stage of the fractal on top of the previous stage, it is advantageous to draw each new stage directly below the previous stage, as shown in Figure 11.71.

In each of the previous fractals, the initiator was a line segment. In Example 2, we use a triangle and its interior as the initiator.

Stage 0 (the initiator)

Example 2 Draw Stages of a Fractal

Draw stage 2 and stage 3 of the *Sierpinski* triangle, which is defined by the following iterative process.

Step 0: Start with an equilateral triangle and its interior. This is stage 0 of the Sierpinski gasket. See Figure 11.72.

Step 1: Form a new triangle by connecting the midpoints of the sides of the triangle. Remove this center triangle. The result is the three green triangles shown in stage 1 in Figure 11.72.

Step 2: Replace each initiator (green triangle) with a scaled version of the generator.

Solution

Two applications of step 2 of the above process produce stage 2 and stage 3 of the Sierpinski gasket, as shown in Figure 11.73.

Stage 1 (the generator)

Figure 11.72 The initiator and generator of the Sierpinski gasket

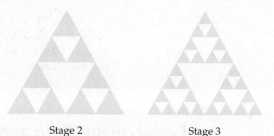

Stage 2   Stage 3

Figure 11.73 Stages 2 and 3 of the Sierpinski triangle

Strictly Self-Similar Fractals

All fractals show a self-similar motif on an ever-diminishing scale; however, some fractals are *strictly self-similar* fractals, according to the following definition.

> **Strictly Self-Similar Fractal**
>
> A fractal is said to be **strictly self-similar** if any arbitrary portion of the fractal contains a replica of the entire fractal.

Example 3 Determine Whether a Fractal Is Strictly Self-Similar

Determine whether the following fractals are strictly self-similar.

a. The Koch snowflake   **b.** The Koch curve

Solution

a. The Koch snowflake is a closed figure. Any portion of the Koch snowflake (like the portion circled in Figure 11.74) is not a closed figure. Thus the Koch snowflake is *not* a strictly self-similar fractal.

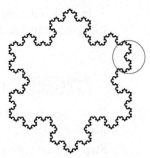

Figure 11.74 The portion of the Koch snowflake shown in the circle is not a replica of the entire snowflake.

b. Because any portion of the Koch curve replicates the entire fractal, the Koch curve is a strictly self-similar fractal. See Figure 11.75.

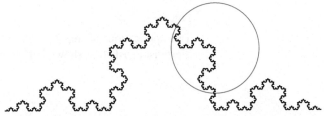

Figure 11.75 Any portion of the Koch curve is a replica of the entire Koch curve.

Replacement Ratio and Scaling Ratio

Mathematicians like to assign numbers to fractals so that they can objectively compare fractals. Two numbers that are associated with many fractals are the *replacement ratio* and the *scaling ratio*.

Replacement Ratio and Scaling Ratio of a Fractal

- If the generator of a fractal consists of N replicas of the initiator, then the **replacement ratio** of the fractal is N.

- If the initiator of a fractal has linear dimensions that are r times the corresponding linear dimensions of its replicas in the generator, then the **scaling ratio** of the fractal is r.

Example 4   **Find the Replacement Ratio and the Scaling Ratio of a Fractal**

Find the replacement ratio and the scaling ratio of the

a. box curve.   **b.** Sierpinski gasket.

Solution

a. Figure 11.71 on page 574 shows that the generator of the box curve consists of five line segments and that the initiator consists of only one line segment. Thus the replacement ratio of the box curve is $5:1$, or 5.

   The initiator of the box curve is a line segment that is 3 times as long as the replica line segments in the generator. Thus the scaling ratio of the box curve is $3:1$, or 3.

b. Figure 11.72 on page 575 shows that the generator of the Sierpinski gasket consists of three triangles and that the initiator consists of only one triangle. Thus the replacement ratio of the Sierpinski gasket is 3 : 1, or 3.

The initiator triangle of the Sierpinski gasket has a width that is 2 times the width of the replica triangles in the generator. Thus the scaling ratio of the Sierpinski gasket is 2 : 1, or 2.

Similarity Dimension

A number called the *similarity dimension* is used to quantify how densely a strictly self-similar fractal fills a region.

Similarity Dimension

The **similarity dimension** D of a strictly self-similar fractal is given by

$$D = \frac{\log N}{\log r}$$

where N is the replacement ratio of the fractal and r is the scaling ratio.

Example 5   **Find the Similarity Dimension of a Fractal**

Find the similarity dimension, to the nearest thousandth, of the

a. Koch curve.

b. Sierpinski gasket.

Solution

a. Because the Koch curve is a strictly self-similar fractal, we can find its similarity dimension. Figure 11.69 on page 573 shows that stage 1 of the Koch curve consists of four line segments and stage 0 consists of only one line segment. Hence the replacement ratio is 4 : 1, or 4. The line segment in stage 0 is 3 times as long as each of the replica line segments in stage 1, so the scaling ratio is 3. Thus the Koch curve has a similarity dimension of

$$D = \frac{\log 4}{\log 3} \approx 1.262$$

b. In Example 4, we found that the Sierpinski gasket has a replacement ratio of 3 and a scaling ratio of 2. Thus the Sierpinski gasket has a similarity dimension of

$$D = \frac{\log 3}{\log 2} \approx 1.585$$

The results of Example 5 show that the Sierpinski gasket has a larger similarity dimension than the Koch curve. This means that the Sierpinski gasket fills a flat two-dimensional surface more densely than does the Koch curve.

Computers are used to generate fractals such as those shown in Figure 11.76. These fractals were *not* rendered by using an initiator and a generator, but they were rendered using iterative procedures.

(a)   (b)   (c)

Figure 11.76 Computer-generated fractals

Fractals have other applications in addition to being used to produce intriguing images. For example, computer scientists have recently developed fractal image compression programs based on self-transformations of an image. An image compression program is a computer program that converts an image file to a smaller file that requires less computer memory. In some situations, these fractal compression programs outperform standardized image compression programs such as JPEG (*jay-peg*), which was developed by the Joint Photographic Experts Group.

11.8 Exercise Set

Think About It

1. What is a fractal?
2. What is a strictly self-similar fractal?

Preliminary Exercises

For Exercises P1 to P5, if necessary, use the referenced example following the exercise for assistance.

P1. Draw stage 2 of the *zig-zag curve*, which is defined by the following iterative process. **Example 1**

Step 0: Start with a line segment. See stage 0 of Figure 1.

Step 1: Remove the middle half of the line segment and draw a zig-zag, as shown in stage 1 of Figure 1. Each of the six line segments in the generator is a $\frac{1}{4}$-scale replica of the initiator.

Step 2: Replace each initiator shape with the scaled version of the generator to produce the next stage. Repeat this step to produce additional stages.

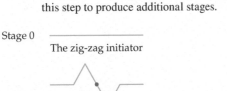

Stage 0

The zig-zag initiator

Stage 1

The zig-zag generator

Figure 1 The initiator and generator of the zig-zag curve

P2. Draw stage 2 of the *Sierpinski carpet*, which is defined by the following process. **Example 2**

Step 0: Start with a square and its interior. See stage 0 in Figure 2.

Step 1: Subdivide the square into nine smaller congruent squares and remove the center square. This yields stage 1 (the generator) shown in Figure 2.

Step 2: Replace each initiator (tan square) with a scaled version of the generator. Repeat this step to create additional stages of the Sierpinski carpet.

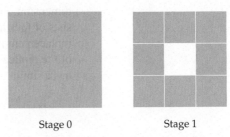

Stage 0   Stage 1

Figure 2 The initiator and generator of the Sierpinski carpet

P3. Determine whether the following fractals are strictly self-similar.

   a. The box curve (see Example 1)

   b. The Sierpinski gasket (see Example 2) **Example 3**

P4. Find the replacement ratio and the scaling ratio of the

   a. Koch curve (see Figure 11.75).

   b. zig-zag curve (see Figure 1 in Exercise P1). **Example 4**

P5. Compute the similarity dimension, to the nearest thousandth, of the

   a. box curve (see Example 1).

   b. Sierpinski carpet (see Exercise P2). **Example 5**

▶ Section Exercises

In Exercises 1 and 2, use an iterative process to draw stage 2 and stage 3 of the fractal with the given initiator (stage 0) and the given generator (stage 1).

1. The Cantor point set

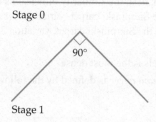

Stage 0

Stage 1

2. Lévy's curve

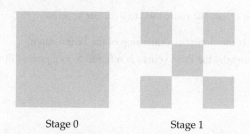

Stage 0

90°

Stage 1

In Exercises 3 to 8, use an iterative process to draw stage 2 of the fractal with the given initiator (stage 0) and the given generator (stage 1).

3. The Sierpinski carpet, variation 1

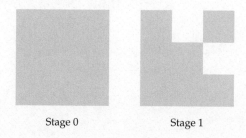

Stage 0   Stage 1

4. The Sierpinski carpet, variation 2

Stage 0   Stage 1

5. The river tree of Peano Cearo

Stage 0   Stage 1

6. Minkowski's fractal

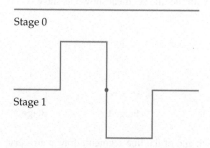

Stage 0

Stage 1

7. The square fractal

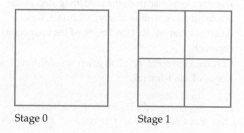

Stage 0   Stage 1

8. The cube fractal

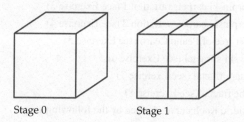

Stage 0   Stage 1

In Exercises 9 and 10, draw stage 3 and stage 4 of the fractal defined by the given iterative process.

9. The binary tree

Step 0: Start with a "⊤." This is stage 0 of the binary tree. The vertical line segment is the trunk of the tree, and the horizontal line segment is the branch of the tree. The branch is half the length of the trunk.

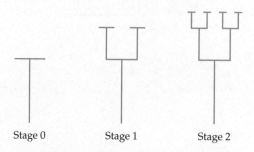

Stage 0   Stage 1   Stage 2

Step 1: At the ends of each branch, draw an upright ⊤ that is half the size of the ⊤ in the preceding stage.

Step 2: Continue to repeat step 1 to generate additional stages of the binary tree.

10. The I-fractal

Step 0: Start with the line segment shown as stage 0.

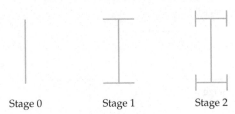

Stage 0   Stage 1   Stage 2

Step 1: At each end of the line segment, draw a crossbar that is half the length of the line segment it contacts. (This produces stage 1 of the I-fractal.)

Step 2: Use each crossbar from the preceding step as the connecting segment of a new "I." Attach new crossbars that are half the length of the connecting segment.

Step 3: Continue to repeat step 2 to generate additional stages of the I-fractal.

In Exercises 11 to 20, compute, if possible, the similarity dimension of the fractal. Round to the nearest thousandth.

11. The Cantor point set (see Exercise 1)

12. Lévy's curve (see Exercise 2)

13. The Sierpinski carpet, variation 1 (see Exercise 3)

14. The Sierpinski carpet, variation 2 (see Exercise 4)

15. The river tree of Peano Cearo (see Exercise 5)

16. Minkowski's fractal (see Exercise 6)

17. The square fractal (see Exercise 7)

18. The cube fractal (see Exercise 8)

19. The quadric Koch curve, defined by the following stages

Stage 0

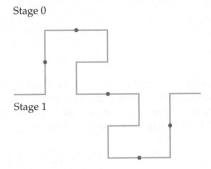

Stage 1

20. The Menger sponge, defined by the following stages

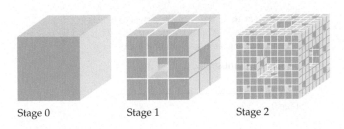

Stage 0   Stage 1   Stage 2

▶ **Investigations**

1. Compare Similarity Dimensions

a. Rank, from largest to smallest, the similarity dimensions of the Sierpinski carpet; the Sierpinski carpet, variation 1 (see Section Exercise 3); and the Sierpinski carpet, variation 2 (see Section Exercise 4).

b. Which of the three fractals is the most dense?

2. The Peano Curve The *Peano curve* is defined by the following stages.

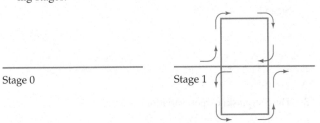

Stage 0   Stage 1

The arrows show the route used to trace the generator.

a. What is the similarity dimension of the Peano curve?

b. Explain why the Peano curve is referred to as a plane-filling curve.

Test Prep ⓫ Geometry

Section 11.1   Review of Measurement

U.S. Customary Units of Measure
- Measures of length: inches, feet, yards, miles
- Measures of weight: ounces, pounds, tons
- Measures of capacity: fluid ounces, cups, pints, quarts, gallons

See **Examples 1** to **3** on pages 510 and 511. Then try Exercises 1 and 2 on page 584.

The Metric System
The metric system uses a prefix of a base unit to denote the magnitude of a measurement.
- Base unit of length: meter
- Base unit of mass: gram
- Base unit of capacity: liter

Some of the prefixes used in the metric system are kilo (1000), hecto (100), deca (10), deci (0.1), centi (0.01), and milli (0.001).

See **Examples 4, 5,** and **6** on pages 512 to 514. Then try Exercises 3, 4, and 5 on page 584.

Convert Between U.S. Customary Units and Metric Units

See **Example 7** on page 515. Then try Exercise 6 on page 584.

Section 11.2   Basic Concepts of Euclidean Geometry

Vertical Angles   The nonadjacent angles formed by two intersecting lines are called vertical angles. Vertical angles have the same measure.

See **Example 4** on page 520, and then try Exercise 14 on page 584.

Lines and Angles   A line that intersects two other lines at different points is a transversal. If the lines cut by a transversal are parallel lines, several angles of equal measure are formed.
- Pairs of alternate interior angles have the same measure.
- Pairs of alternate exterior angles have the same measure.
- Pairs of corresponding angles have the same measure.

See **Example 5** on page 521, and then try Exercise 18 on page 584.

Triangles   The sum of the measures of the interior angles of a triangle is 180°.

See **Examples 6** and **7** on pages 521 and 522, and then try Exercises 7 and 21 on page 584.

Section 11.3   Perimeter and Area of Plane Figures

Perimeter   The perimeter of a plane geometric figure is a measure of the distance around the figure.

Triangle:	$P = a + b + c$
Rectangle:	$P = 2L + 2W$
Square:	$P = 4s$
Parallelogram:	$P = 2b + 2s$
Circle:	$C = \pi d$ or $C = 2\pi r$

See **Examples 1** to **3** on pages 528 and 529, and then try Exercises 17, 24, and 26 on page 584.

Area   Area is the amount of surface in a region. Area is measured in square units.

Triangle:	$A = \dfrac{1}{2}bh$
Rectangle:	$A = LW$
Square:	$A = s^2$
Parallelogram:	$A = bh$
Trapezoid:	$A = \dfrac{1}{2}h(b_1 + b_2)$
Circle:	$A = \pi r^2$

See **Examples 5** to **10** on pages 530 to 534, and then try Exercises 15, 22, 27, and 28 on page 584.

581

Section 11.4 Volume and Surface Area

Volume   Volume is a measure of the amount of space occupied by a geometric solid. Volume is measured in cubic units.

Rectangular solid:	$V = LWH$
Cube:	$V = s^3$
Sphere:	$V = \dfrac{4}{3}\pi r^3$
Right circular cylinder:	$V = \pi r^2 h$
Right circular cone:	$V = \dfrac{1}{3}\pi r^2 h$
Regular square pyramid:	$V = \dfrac{1}{3}s^2 h$

See **Examples 1, 2,** and **3** on pages 541 and 542, and then try Exercises 9, 16, 20, and 23 on page 584.

Surface Area   The surface area of a solid is the total area on the surface of the solid.

Rectangular solid:	$S = 2LW + 2LH + 2WH$
Cube:	$S = 6s^2$
Sphere:	$S = 4\pi r^2$
Right circular cylinder:	$S = 2\pi r^2 + 2\pi rh$
Right circular cone:	$S = \pi r^2 + \pi rl$
Regular square pyramid:	$S = s^2 + 2sl$

See **Example 4** on page 543, and then try Exercises 11, 12, and 25 on page 584.

Section 11.5 Properties of Triangles

Similar Triangles   The ratios of corresponding sides are equal. The ratio of corresponding heights is equal to the ratio of corresponding sides. The measures of corresponding angles are equal.

See **Examples 1** and **2** on pages 548 and 549, and then try Exercise 8 on page 584.

Congruent Triangles

Side-Side-Side Theorem (SSS)   If the three sides of one triangle are equal in measure to the three sides of a second triangle, the two triangles are congruent.

Side-Angle-Side Theorem (SAS)   If two sides and the included angle of one triangle are equal in measure to two sides and the included angle of a second triangle, the two triangles are congruent.

Angle-Side-Angle Theorem (ASA)   If two angles and the included side of one triangle are equal in measure to two angles and the included side of a second triangle, the two triangles are congruent.

See **Example 3** on page 551, and then try Exercise 29 on page 585.

The Pythagorean Theorem   If a and b are the lengths of the legs of a right triangle and c is the length of the hypotenuse, then

$$c^2 = a^2 + b^2.$$

See **Example 4** on page 552, and then try Exercise 30 on page 585.

Section 11.6 Right Triangle Trigonometry

The Trigonometric Ratios of an Acute Angle of a Right Triangle   If θ is an acute angle of a right triangle ABC, then

$$\sin \theta = \frac{\text{length of opposite side}}{\text{length of hypotenuse}}$$

$$\cos \theta = \frac{\text{length of adjacent side}}{\text{length of hypotenuse}}$$

$$\tan \theta = \frac{\text{length of opposite side}}{\text{length of adjacent side}}$$

See **Examples 1** and **2** on pages 558 and 559, and then try Exercises 31 and 32 on page 585.

$$\csc \theta = \frac{\text{length of hypotenuse}}{\text{length of opposite side}}$$

$$\sec \theta = \frac{\text{length of hypotenuse}}{\text{length of adjacent side}}$$

$$\cot \theta = \frac{\text{length of adjacent side}}{\text{length of opposite side}}$$

Inverse Sine, Inverse Cosine, and Inverse Tangent $\sin^{-1}(x)$ is defined as the angle whose sine is x, $-1 \le x \le 1$. $\cos^{-1}(x)$ is defined as the angle whose cosine is x, $-1 \le x \le 1$. $\tan^{-1}(x)$ is defined as the angle whose tangent is x, $-\infty < x < \infty$.	See **Examples 3, 4,** and **5** on pages 560 and 561, and then try Exercises 33 and 36 on page 585.
Applications of Trigonometry Trigonometry is often used to solve applications that involve right triangles.	See **Example 6** on page 561, and then try Exercises 37 and 38 on page 585.

Section 11.7   Non-Euclidean Geometry

Parallel Postulates **Euclidean Parallel Postulate** (*Euclidean or Plane Geometry*) Through a given point not on a given line, exactly one line can be drawn parallel to the given line. **Gauss's Alternate to the Parallel Postulate** (*Lobachevskian or Hyperbolic Geometry*) Through a given point not on a given line, there are at least two lines parallel to the given line. **Riemann's Alternative to the Parallel Postulate** (*Riemannian or Spherical Geometry*) Through a given point not on a given line, there exist no lines parallel to the given line.	See **Example 2** on page 568, and then try Exercises 42 and 43 on page 585.
The Spherical Triangle Area Formula The area S of the spherical triangle ABC on a sphere with radius r is $$S = (m\angle A + m\angle B + m\angle C - 180°)\left(\frac{\pi}{180°}\right)r^2$$	See **Example 1** on page 567, and then try Exercises 44 and 45 on page 585.

The Euclidean Distance Formula and the City Distance Formula
• The Euclidean distance between $P(x_1, y_1)$ and $Q(x_2, y_2)$ is
  $$d_E(P, Q) = \sqrt{(x_2 - x_1)^2 + (y_2 - y_1)^2}$$
• The city distance between $P(x_1, y_1)$ and $Q(x_2, y_2)$ is
  $$d_C(P, Q) = |x_2 - x_1| + |y_2 - y_1|$$
	See **Example 3** on page 570, and then try Exercises 46 and 49 on page 585.

Section 11.8   Fractals

Strictly Self-Similar Fractal A fractal is a strictly self-similar fractal if any arbitrary portion of the fractal contains a replica of the entire fractal.	See **Example 3** on page 575, and then try Exercise 51 on page 585.
Replacement Ratio and Scaling Ratio of a Fractal • If the generator of a fractal consists of N replicas of the initiator, then the replacement ratio of the fractal is N. • If the initiator of a fractal has linear dimensions that are r times the corresponding linear dimensions of its replicas in the generator, then the scaling ratio of the fractal is r.	See **Example 4** on page 576, and then try Exercise 53 on page 585.
Similarity Dimension of a Fractal The similarity dimension D of a strictly self-similar fractal is $$D = \frac{\log N}{\log r}$$ where N is the replacement ratio of the fractal and r is the scaling ratio.	See **Example 5** on page 577, and then try Exercise 54 on page 585.

Review Exercises

For Exercises 1 to 5, convert one measurement to another.

1. 27 in. = _____ ft
2. 15 cups = _____ pints
3. 37 mm = _____ cm
4. 0.678 g = _____ mg
5. 1273 ml = _____ L

6. The price of a beverage is $3.56 per liter. What is the price of this beverage in dollars per quart? Round to the nearest cent.

7. Given that $m \angle a = 74°$ and $m \angle b = 52°$, find the measures of angles x and y.

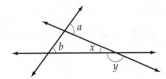

8. Triangles ABC and DEF are similar. Find the perimeter of triangle ABC.

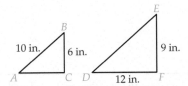

9. Find the volume of the geometric solid.

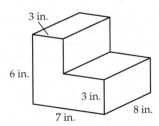

10. Find the measure of $\angle x$.

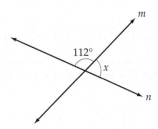

11. Find the surface area of the rectangular solid.

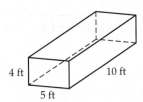

12. The length of a diameter of the base of a cylinder is 4 m, and the height of the cylinder is 8 m. Find the surface area of the cylinder. Give the exact value.

13. Given that $BC = 11$ cm and that AB is three times the length of BC, find the length of AC.

14. Given that $m \angle x = 150°$, find the measures of $\angle w$ and $\angle y$.

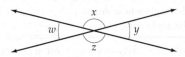

15. Find the area of a parallelogram that has a base of 6 in. and a height of 4.5 in.

16. Find the volume of the square pyramid.

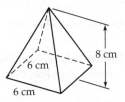

17. Find the circumference of a circle that has a diameter of 4.5 m. Round to the nearest tenth of a meter.

18. Given that $\ell_1 \| \ell_2$, find the measures of angles a and b.

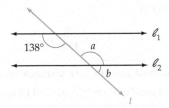

19. Find the supplement of a 32° angle.

20. Find the volume of a rectangular solid with a length of 6.5 ft, a width of 2 ft, and a height of 3 ft.

21. Two angles of a triangle measure 37° and 48°. Find the measure of the third angle.

22. The height of a triangle is 7 cm. The area of the triangle is 28 cm². Find the length of the base of the triangle.

23. Find the volume of a sphere that has a diameter of 12 mm. Give the exact value.

24. **Framing** The perimeter of a square picture frame is 86 cm. Find the length of each side of the frame.

25. **Paint** A can of paint will cover 200 ft² of surface. How many cans of paint should be purchased to paint a cylinder that has a height of 15 ft and a radius of 6 ft?

26. **Parks and Recreation** The length of a rectangular park is 56 yd. The width is 48 yd. How many yards of fencing are needed to surround the park?

27. **Patios** What is the area of a square patio that measures 9.5 m on each side?

28. **Landscaping** A walkway 2 m wide surrounds a rectangular plot of grass. The plot is 40 m long and 25 m wide. What is the area of the walkway?

29. Determine whether the two triangles are congruent. If they are congruent, state by what theorem they are congruent.

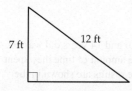

30. Find the unknown side of the triangle. Round to the nearest tenth of a foot.

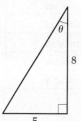

In Exercises 31 and 32, find the values of sin θ, cos θ, and tan θ for the given right triangle.

31.

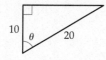

32.

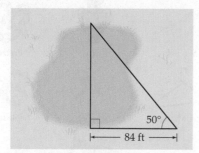

In Exercises 33 to 36, use a calculator. Round to the nearest tenth of a degree.

33. Find $\cos^{-1}(0.9013)$.

34. Find $\sin^{-1}(0.4871)$.

35. Given $\tan \beta = 1.364$, find β.

36. Given $\sin \theta = 0.0325$, find θ.

37. **Surveying** Find the distance across the marsh in the following figure. Round to the nearest tenth of a foot.

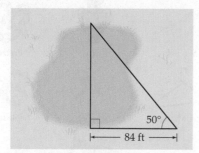

38. **Angle of Depression** The distance from a plane to a radar station is 200 mi, and the angle of depression is 40°. Find the number of ground miles from a point directly under

the plane to the radar station. Round to the nearest tenth of a mile.

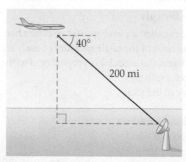

39. **Angle of Elevation** The angle of elevation from a point A on the ground to the top of a space shuttle is 27°. If point A is 110 ft from the base of the space shuttle, how tall is the space shuttle? Round to the nearest tenth of a foot.

40. What is another name for Riemannian geometry?

41. What is another name for Lobachevskian geometry?

42. Name a geometry in which the sum of the measures of the interior angles of a triangle is less than 180°.

43. Name a geometry in which there are no parallel lines.

In Exercises 44 and 45, determine the exact area of the spherical triangle.

44. Radius: 12 in.; angles: 90°, 150°, 90°

45. Radius: 5 ft; angles: 90°, 60°, 90°

Euclidean and City Distances   In Exercises 46 to 49, find the Euclidean distance and the city distance between the points. Assume that the distances are measured in blocks. Round approximate results to the nearest tenth of a block.

46. $P(-1, 1)$, $Q(3, 4)$ 47. $P(-5, -2)$, $Q(2, 6)$

48. $P(2, 8)$, $Q(3, 2)$ 49. $P(-3, 3)$, $Q(5, -2)$

50. Consider the points $P(1, 1)$, $Q(4, 5)$, and $R(-4, 2)$.

    a. Which two points are closest together if you only use the Euclidean distance formula to measure distance?

    b. Which two points are closest together if you only use the city distance formula to measure distance?

51. Draw stage 0, stage 1, and stage 2 of the Koch curve. Is the Koch curve a strictly self-similar fractal?

52. Draw stage 2 of the fractal with the following initiator and generator.

Stage 0   Stage 1
Initiator   Generator

53. For the fractal defined in Exercise 52, determine the

    a. replacement ratio.

    b. scaling ratio.

    c. similarity dimension.

54. Compute the similarity dimension of a strictly self-similar fractal with a replacement ratio of 5 and a scaling ratio of 4. Round to the nearest thousandth.

Project 1

Preparing a Circle Graph

A circle graph, sometimes called a pie chart, is a circular chart divided into sectors. The ratio of the angle measure of each sector to 360 (the number of degrees in a circle) is proportional to the ratio of the number of data values represented by the sector to the total number of data values.

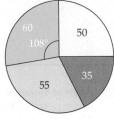

The circle graph shown represents the preferences of 200 people who were asked about the temperature range they considered most comfortable for walking. The data are given in the following table.

Temperature, °F	Number of People
60–64	50
65–69	55
70–74	60
75–79	35

The angle measure of the sector that represents people who chose a temperature between 70°F and 74°F is 108°. Note that $\frac{108}{360} = \frac{60}{200}$. That is, the ratio of the size of the angle, 108, to the total number of degrees in a circle, 360, is proportional to the magnitude of the data in that sector, 60, to the sum of all the data in the circle graph, 200.

Spreadsheet programs such as Excel and Numbers have built-in functions that will create a circle graph. Use such a program to prepare a circle graph for each of the following exercises.

Project Exercises

Prepare a circle graph for the data provided in each exercise.

1. A survey asked adults to name their favorite pizza topping. The results are shown in the following table.

Pepperoni	43%
Sausage	19%
Mushrooms	14%
Vegetables	13%
Other	7%
Onions	4%

2. A survey of children between 10 and 14 years old was conducted to determine the average amount of time they spent consuming media each day. The results are shown in the following table.

Watching TV	63 minutes
Listening to music	105 minutes
Nonschool computer use	42 minutes
Text messaging	84 minutes
Playing video games	76 minutes
Talking on cell phones	50 minutes

Project 2

Topology: A Brief Introduction

In this section, we discussed similar figures—that is, figures with the same shape. The branch of geometry called **topology** is the study of even more basic properties of geometric figures than their sizes and shapes. In topology, figures that can be stretched, shrunk, molded, or bent into the same shape without puncturing or cutting belong to the same family. They are said to be **topologically equivalent**. For instance, if a doughnut-shaped figure were made out of modeling clay, then it could be molded into a coffee cup, as shown in the following image.

A transformation of a doughnut into a coffee cup

In topology, figures are classified according to their **genus**, where the genus is given by the number of holes in the figure. An inlet in a figure is considered to be a hole if water poured into it passes through the figure. For example, a coffee cup has a hole that is created by its handle; however, the inlet at the top of

a coffee cup is not considered to be a hole, because water that is poured into this inlet does not pass through the coffee cup.

Consider the following drawings of several common geometric figures with genuses 0, 1, 2, and 3. Figures with the same genus are topologically equivalent.

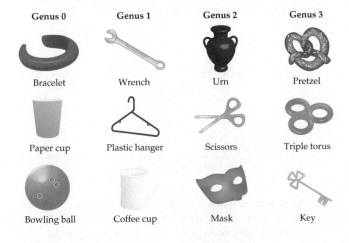

Project Exercises

1. Name the genus of each figure.

a. Funnel   **b.** Ship's wheel   **c.** Axe   **d.** Car steering wheel

2. Which one of the following figures is not topologically equivalent to the others?

Cleaver   Shovel   Nail   Class ring

3. Which one of the following figures is not topologically equivalent to the others?

Comb   Spatula   Block   Oar

4. In parts a and b, the letters of the alphabet are displayed using a particular font. List all the topologically equivalent letters according to their genus of 0, 1, or 2.

a. ABCDEFGHIJKLM NOPQRSTUVWXYZ

b. abcdefghijklm nopqrstuvwxyz

5. In parts a and b, the numerals 1 through 9 are displayed using a particular font. List all the topologically equivalent numerals according to their genus of 0, 1, or 2.

a. 1 2 3 4 5 6 7 8 9

b. 1 2 3 4 5 6 7 8 9

6. It has been said that a topologist doesn't know the difference between a doughnut and a coffee cup. In parts a through f, determine whether a topologist would classify the two items in the same genus.

a. A spoon and a fork

b. A screwdriver and a hammer

c. A salt shaker and a sugar bowl

d. A bolt and its nut

e. A slice of American cheese and a slice of Swiss cheese

f. A mixing bowl and a strainer

Math in Practice

Buying Feed Tanks for a Farm

You are a farmer who is interested in an additional feed tank for your livestock. The place you are thinking of building it allows for a silo that is 9 feet in diameter and at most 30 feet high.

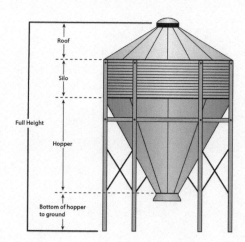

Part 1

Feed tanks are structured with several parts: the hopper, the silo which is made up of metal rings, a roof, footings, and a ladder that scales the entire structure.

You want a silo that will hold at least 1200 cubic feet of grain. The hopper is a cone with a 9-foot diameter and a 60° slant. The hopper comes initially with one silo ring that is 8.9 feet tall. Each additional ring is 2.6 feet tall. The roof is a cone with a 30° slant, but silos are not filled past the top ring so the roof section holds no grain. How many additional rings should be factored into the silo so that it can hold at least 1200 cubic feet of grain? Assume the hopper is an entire cone for this calculation.

Part 2

There is a 30-foot restriction in the location where you want your silo to go. You also want to ensure that there is enough clearance below the hopper to load feed carts or wheelbarrows. How much space below your hopper will there be if you need the silo's total height to not exceed 30 feet? Is this enough room for a wheelbarrow or feed cart to fit under and load? Explain your answer.

Part 3

The company you want to buy your silo from, Hanz Silos, estimates that silo construction costs roughly $4.00 per bushel of feed. Considering 1 bushel is about 1.25 ft^3, how much can you estimate this silo costing?

A Review of Integers, Rational Numbers, and Percents

Introduction to Integers

Objective A.IA Use Inequality Symbols with Integers

Mathematicians place objects with similar properties in groups called *sets*. A **set** is a collection of objects. The objects in a set are called the **elements of the set**.

The **roster method** of writing a set encloses a list of the elements in braces. Thus the set of sections within an orchestra is written {brass, percussion, string, woodwind}. When the elements of a set are listed, each element is listed only once. For instance, if the list of numbers 1, 2, 3, 2, 3 were placed in a set, the set would be {1, 2, 3}.

The symbol $\in$ means "is an element of." $2 \in B$ is read "2 is an element of set B."

Given $C = \{3, 5, 9\}$, then $3 \in C$, $5 \in C$, and $9 \in C$. $7 \notin C$ is read "7 is not an element of set C."

The numbers that we use to count objects, such as the students in a classroom or the horses on a ranch, are the *natural numbers*.

Natural numbers = $\{1, 2, 3, 4, 5, 6, 7, 8, 9, 10, \ldots\}$

The three dots mean that the list of natural numbers continues on and on, and that there is no largest natural number.

The natural numbers alone do not provide all the numbers that are useful in applications. For instance, a meteorologist also needs the number zero and numbers below zero.

Integers = $\{\ldots, -5, -4, -3, -2, -1, 0, 1, 2, 3, 4, 5, \ldots\}$

Each integer can be shown on a number line. The integers to the left of zero on the number line are called **negative integers**. The integers to the right of zero are called **positive integers**, or natural numbers. Zero is neither a positive nor a negative integer (see Figure A.1).

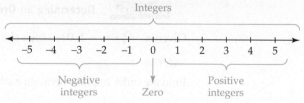

Figure A.1

589

The **graph** of an integer is shown by placing a heavy dot on the number line directly above the number. The graphs of -3 and 4 are shown on the number line in Figure A.2.

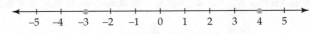

Figure A.2

Consider the following sentences:

The quarterback threw the football and the receiver caught *it*.

A student purchased a computer and used *it* to write history papers.

In the first sentence, *it* is used to mean the football; in the second sentence, *it* means the computer. In language, the word *it* can stand for many different objects. Similarly, in mathematics, a letter of the alphabet can be used to stand for a number. Such a letter is called a **variable**. Variables are used in the following definition of inequality symbols.

Inequality Symbols

If a and b are two numbers and a is to the left of b on the number line, then a **is less than** b. This is written $a < b$.

If a and b are two numbers and a is to the right of b on the number line, then a **is greater than** b. This is written $a > b$.

Examples

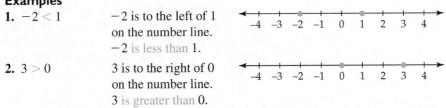

1. $-2 < 1$ -2 is to the left of 1 on the number line. -2 is less than 1.

2. $3 > 0$ 3 is to the right of 0 on the number line. 3 is greater than 0.

There are also inequality symbols for **is less than or equal to ($\leq$) and is greater than or equal to ($\geq$)**.

$7 \leq 15$ 7 is less than or equal to 15. $6 \leq 6$ 6 is less than or equal to 6.

This is true because $7 < 15$.   This is true because $6 = 6$.

Example 1   **Write a Set Using the Roster Method**

Use the roster method to write the set of negative integers greater than or equal to -4.

Solution

$A = \{-4, -3, -2, -1\}$ • A set is designated by a capital letter.

Example 2   **Determine an Order between Two Numbers**

Given $A = \{-6, -2, 0\}$, which elements of set A are less than or equal to -2?

Solution

Find the order relation between each element of set A and -2.

$$-6 < -2$$
$$-2 = -2$$
$$0 > -2$$

The elements -6 and -2 are less than or equal to -2.

Objective A.1A Practice

1. Place the correct symbol, $<$ or $>$, between the numbers below.
   $-14 \quad 16$
2. Place the correct symbol, $<$ or $>$, between the numbers below.
   $0 \quad -31$
3. Use the roster method to write the set of positive integers less than 4.
4. Given $C = \{-33, -24, -10, 0\}$, which elements of set C are less than -10?

Objective A.1B Simplify Expressions with Absolute Value

Two numbers that are the same distance from zero on the number line but are on opposite sides of zero are **opposite numbers**, or **opposites**. The opposite of a number is also called its **additive inverse**.

The opposite of 5 is -5.

The opposite of -5 is 5.

The negative sign can be read "the opposite of."

$$-(5) = -2 \qquad \text{The opposite of 5 is } -5.$$

$$-(-5) = 5 \qquad \text{The opposite of } -5 \text{ is 5.}$$

The absolute value of a number is a measure of the distance the number is from 0. Figure A.3 shows both 5 and -5 are 5 units from 0.

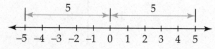

Figure A.3

Example 3 **Find an Additive Inverse**

Given $A = \{-12, 0, 4\}$, find the additive inverse of each element of set A.

Solution

$$-(-12) = 12$$

$$-0 = 0 \qquad \bullet \text{ Zero is neither positive nor negative.}$$

$$-(4) = -4$$

The **absolute value of a number** is its distance from zero on the number line. Therefore, the absolute value of a number is a positive number or zero. The symbol for absolute value is two vertical bars, $\|$.

The distance from 0 to 3 is 3. Therefore, the absolute value of 3 is 3 (see Figure A.4).

$$|3| = 3$$

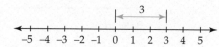

Figure A.4

The distance from 0 to -3 is 3. Therefore, the absolute value of -3 is 3 (see Figure A.5).

$$|-3| = 3$$

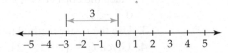

Figure A.5

Absolute Value

The absolute value of a positive number is the number itself.

The absolute value of a negative number is the opposite of the number.
The absolute value of zero is zero.

Examples

1. $|6| = 6$
2. $|-8| = 8$
3. $|0| = 0$

Using variables, the definition of absolute value is

$$|x| = \begin{cases} x, & x > 0 \\ 0, & x = 0 \\ -x, & x < 0 \end{cases}$$

Example 4 **Evaluate an Absolute Value Expression**

Evaluate $|-4|$ and $-|-10|$.

Solution

$$|-4| = 4$$
$$-|-10| = -10$$

Objective A.IB Practice

1. Find the additive inverse of 4.
2. Find the additive inverse of -9.
3. Find the additive inverse of -36.
4. Evaluate $-(-40)$.
5. Evaluate $|-74|$.
6. Evaluate $-|81|$.

Section

A.2

Operations on Integers

Objective A.2A Add Integers

Addition of Integers

To add two numbers with the same sign, add the absolute values of the numbers. Then attach the sign of the addends.

   To add two numbers with different signs, find the absolute value of each number. Subtract the smaller of the absolute values from the larger. Then attach the sign of the number with the larger absolute value.

Examples

1. Add: $-12 + (-26)$ — • The signs are the same.

   $-12 + (-26) = -38$ — • Add the absolute values of the numbers $(12 + 26)$. Attach the sign of the addends.

2. Add: $-19 + 8$ — • The signs are different.

   $|-19| = 19; |8| = 8$ — • Find the absolute value of each number.

   $19 - 8 = 11$ — • Subtract the smaller absolute value from the larger.

   $-19 + 8 = -11$ — • Attach the sign of the number with the larger absolute value.

Example 1 **Add Integers**

a. Add: $-52 + (-39)$

b. Add: $37 + (-52) + (-14)$

Solution

a. $-52 + (-39) = -91$

b. $37 + (-52) + (-14) = -15 + (-14)$

   $= -29$

All of the following phrases in Table A.1 are used to indicate addition.

Table A.1

added to	-6 added to 9	$9 + (-6) = 3$
more than	3 more than -8	$-8 + 3 = -5$
the sum of	the sum of -2 and -8	$-2 + (-8) = -10$
increased by	-7 increased by 5	$-7 + 5 = -2$
the total of	the total of 4 and -9	$4 + (-9) = -5$
plus	6 plus -10	$6 + (-10) = -4$

Example 2 **Translate an Expression**

a. Find 11 more than -23.

b. Find the sum of $-23, 47, -18,$ and -10.

Solution

a. $-23 + 11 = -12$

b. A sum is the answer to an addition problem.

$$-23 + 47 + (-18) + (-10)$$

$$= 24 + (-18) + (-10)$$

$$= 6 + (-10)$$

$$= -4$$

• To add more than two numbers, add the first two numbers. Then add the sum to the third number. Continue until all the numbers are added.

Objective A.2A Practice

1. Add: $-6 + (-9)$
2. Add: $-17 + 17$
3. Add: $-27 + (-42) + (-18)$
4. Add: $-6 + (-8) + 14 + (-4)$
5. What is 4 more than -8?
6. What is -8 added to -21?

Objective A.2B Subtract Integers

Subtraction of Integers

To subtract one number from another, add the opposite of the second number to the first number.

Examples
1. Subtract: $-21 - (-40)$

   $-21 - (-40) = -21 + 40$ • Rewrite the subtraction as addition of the opposite.

   $\qquad\qquad = 19$ • Add.

2. Subtract: $15 - 51$

   $15 - 51 = 15 + (-51)$ • Rewrite the subtraction as addition of the opposite.

   $\qquad\quad = -36$ • Add.

Example 3   **Subtract Integers**

Subtract: $-12 - (-21) - 15$

Solution

$$-12 - (-21) - 15 = -12 + 21 + (-15)$$

$$= 9 + (-15)$$

$$= -6$$

• Rewrite each subtraction as addition of the opposite. Then add.

All of the following phrases in Table A.2 are used to indicate subtraction.

Table A.2

minus	-5 minus 11	$-5 - 11 = -16$
less	-3 less 5	$-3 - 5 = -8$
less than	-8 less than -2	$-2 - (-8) = 6$
the difference between	the difference between -5 and 4	$-5 - 4 = -9$
decreased by	-4 decreased by 9	$-4 - 9 = -13$
subtract ... from	subtract 8 from -3	$-3 - 8 = -11$

Example 4 **Translate an Expression**

a. Find 9 less than -4.

b. Find the difference between -8 and 7.

Solution

a. $-4 - 9 = -4 + (-9) = -13$

b. $-8 - 7 = -8 + (-7)$ • Rewrite the subtraction as addition of the opposite.
$$= -15$$ Then add.

Objective A.2B Practice

1. Subtract: $16 - 8$
2. Subtract: $6 - (-12)$
3. Subtract: $-19 - (-19) - 18$
4. What is 9 less than -12?
5. Find -21 decreased by 19.

Objective A.2C Multiply Integers

Multiplication of Integers

To multiply two numbers with the same sign, multiply the absolute values of the numbers. The product is positive.

   To multiply two numbers with different signs, multiply the absolute values of the numbers. The product is negative.

Examples

1. $-4(-8) = 32$ • The signs are the same. The product is positive.
2. $5(-12) = -60$ • The signs are different. The product is negative.

Example I **Multiply More Than Two Integers**

Multiply: $-2(5)(-7)(-4)$

Solution

$-2(5)(-7)(-4) = -10(-7)(-4)$ • To multiply more than two numbers, multiply the first two numbers.
Then multiply the product by the third number.

$$= 70(-4)$$

$$= -280$$ Continue until all the numbers are multiplied.

Consider the products shown below. Note that when there is an even number of negative factors, the product is positive. When there is an odd number of negative factors, the product is negative.

$$(-3)(-5) = 15$$

$$(-2)(-5)(-6) = -60$$

$$(-4)(-3)(-5)(-7) = 420$$

$$(-3)(-3)(-5)(-4)(-5) = -900$$

$$(-6)(-3)(-4)(-2)(-10)(-5) = 7200$$

This idea can be summarized by the following useful rule: The product of an even number of negative factors is positive; the product of an odd number of negative factors is negative.

All of the following phrases in Table A.3 indicate multiplication.

Table A.3

times	−7 times −9	−7(−9) = 63
the product of	the product of 12 and −8	12(−8) = −96
multiplied by	−15 multiplied by 11	−15(11) = −165
twice	twice −14	2(−14) = −28

Example 2 **Translate an Expression**

Find the product of 15 and −9.

Solution

$$15(-9) = -135$$

Objective A.2C Practice

1. Multiply: $(-13)(-9)$
2. Multiply: $17(-13)$
3. Multiply: $-7(-6)(-5)$
4. Multiply: $-12(-4)7(-2)$
5. Find the product of -2, -3, -4, and -5.

Objective A.2D Divide Integers

For every division problem there is a related multiplication problem.

$$\underbrace{\frac{8}{2} = 4}_{\text{Division}} \quad \text{because} \quad \underbrace{4 \cdot 2 = 8.}_{\text{Related multiplication}}$$

Take Note: Think of the fraction bar as meaning "divided by." Thus $\frac{8}{2}$ means 8 divided by 2. The number 2 is the **divisor**. The number 8 is the **dividend**. The result of the division, 4, is called the **quotient**.

This fact and the rules for multiplying integers can be used to illustrate the rules for dividing integers.

Note in the following examples that the quotient of two numbers with the same sign is positive.

$$\frac{12}{3} = 4 \text{ because } 4 \cdot 3 = 12. \qquad \frac{-12}{-3} = 4 \text{ because } 4(-3) = -12.$$

The next two examples illustrate that the quotient of two numbers with different signs is negative.

$$\frac{12}{-3} = -4 \text{ because } (-4)(-3) = 12. \qquad \frac{-12}{3} = -4 \text{ because } (-4)\,3 = -12.$$

Division of Integers

To divide two numbers with the same sign, divide the absolute values of the numbers. The quotient is positive.

   To divide two numbers with different signs, divide the absolute values of the numbers. The quotient is negative.

Examples

1. $(-18) \div (-2) = 9$ • The signs are the same.
                               The quotient is positive.

2. $-36 \div 9 = -4$ • The signs are different.
                            The quotient is negative.

The fact that $\frac{-12}{3} = -4$, $\frac{12}{-3} = -4$, and $-\frac{12}{3} = -4$ suggests the following rule.

If a and b are integers, and $b \neq 0$, then $\frac{-a}{b} = \frac{a}{-b} = -\frac{a}{b}$.

Example 3   **Divide Integers**

a. Divide: $(-120) \div (-8)$

b. Divide: $\dfrac{95}{-5}$

Solution

a. $(-120) \div (-8) = 15$ • The signs are the same.
                                    The quotient is positive.

b. $\dfrac{95}{-5} = -19$ • The signs are different.
                                   The quotient is negative.

Example 4   **Simplify a Division Expression**

Simplify: $-\dfrac{-81}{3}$

Solution

$$-\frac{-81}{3} = -(-27) = 27$$

All of the following phrases in Table A.4 indicate division.

Table A.4

divided by	15 divided by −3	$15 \div (-3) = -5$
the quotient of	the quotient of −56 and −8	$(-56) \div (-8) = 7$
the ratio of	the ratio of 45 and −5	$45 \div (-5) = -9$
divide ... by	divide −100 by −20	$-100 \div (-20) = 5$

| Example 5 | **Translate an Expression** |

Find the quotient of 98 and -14.

Solution

$$98 \div (-14) = -7$$

The properties of division are stated below. In these statements, the symbol $\neq$ is read "is not equal to."

Properties of Zero and One in Division

If $a \neq 0$, $\dfrac{0}{a} = 0$.   Zero divided by any number other than zero is zero.

If $a \neq 0$, $\dfrac{a}{a} = 1$.   Any number other than zero divided by itself is 1.

$\dfrac{a}{1} = a$ A number divided by 1 is the number.

$\dfrac{a}{0}$ is undefined.   Division by zero is not defined.

Examples

1. $\dfrac{0}{-5} = 0$

2. $\dfrac{-4}{-4} = 1$

3. $\dfrac{-7}{1} = -7$

4. $\dfrac{-12}{0}$ is undefined.

Objective A.2D Practice

1. Divide: $18 \div (-3)$
2. Divide: $57 \div (-3)$
3. Divide: $\dfrac{85}{-5}$
4. Divide: $0 \div (-14)$
5. What is 15 divided by -15?

Section A.3

Exponents and the Order of Operations Agreement

Objective A.3A Simplify Expressions Containing Exponents

Repeated multiplication of the same factor can be written using an exponent.

$$2 \cdot 2 \cdot 2 \cdot 2 \cdot 2 = 2^5 \xleftarrow{} \text{Exponent} \qquad a \cdot a \cdot a \cdot a = a^4 \xleftarrow{} \text{Exponent}$$

$$\underset{\text{Base}}{\Big\uparrow} \qquad\qquad\qquad\qquad \underset{\text{Base}}{\Big\uparrow}$$

The **exponent** indicates how many times the factor, which is called the **base**, occurs in the multiplication. The multiplication $2 \cdot 2 \cdot 2 \cdot 2 \cdot 2$ is in **factored form**. The expression 2^5 is in **exponential form**.

2^1 is read "2 to the first power" or just "2." Usually the exponent 1 is not written.

2^2 is read "2 to the second power" or "2 squared."

2^3 is read "2 to the third power" or "2 cubed."

2^4 is read "2 to the fourth power."

a^4 is read "a to the fourth power."

The first three natural-number powers can be interpreted geometrically as length, area, and volume, respectively (see Figure A.6).

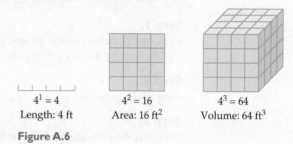

$4^1 = 4$ $4^2 = 16$ $4^3 = 64$
Length: 4 ft   Area: 16 ft^2 Volume: 64 ft^3

Figure A.6

To evaluate an exponential expression, write each factor as many times as indicated by the exponent. Then multiply. Be especially careful when negative numbers are used.

Take Note: Note the difference between $(-2)^4$ and -2^4. $(-2)^4$ is the fourth power of -2:

$$(-2)^4 = 16$$

-2^4 is the opposite of the fourth power of 2:

$$-2^4 = -16$$

$(-2)^4 = (-2)(-2)(-2)(-2)$ • Write -2 as a factor 4 times.

$\qquad\quad = 16$ • Multiply.

$-2^4 = -(2 \cdot 2 \cdot 2 \cdot 2)$ • Write 2 as a factor 4 times.

$\qquad = -16$ • Multiply.

Example 1   **Evaluate an Exponential Expression**

Evaluate.
a. $(-3)^2 \cdot 2^3$
b. $(-1)^6$
c. $-2 \cdot (-3)^2 \cdot (-1)^9$

Solution

a. $(-3)^2 \cdot 2^3 = (-3)(-3) \cdot (2)(2)(2)$

$\qquad\qquad\quad = 9 \cdot 8 = 72$

b. The product of an even number of negative factors is positive. Therefore, $(-1)^6 = 1$.

c. $-2 \cdot (-3)^2 \cdot (-1)^9 = -2 \cdot 9 \cdot (-1)$ • $(-3)^2 = 9; (-1)^9 = -1$

$\qquad\qquad\qquad\qquad = 18$

Objective A.3A Practice

1. Evaluate: $(-3)^2$
2. Evaluate: -4^4
3. Evaluate: $(-1)^9 \cdot 3^3$
4. Evaluate: $(-3) \cdot 2^2$
5. Evaluate: $(-7) \cdot 4^2 \cdot 3^2$
6. Evaluate: $(-2)^3 \cdot (-3)^2 \cdot (-1)^7$

Objective A.3B Use the Order of Operations Agreement to Simplify Expressions

The Order of Operations Agreement

Step 1.
Perform operations inside grouping symbols. Grouping symbols include parentheses ( ), brackets [ ], braces { }, the absolute value symbol ||, and the fraction bar.

Step 2.
Simplify exponential expressions.

Step 3.
Do multiplication and division as they occur from left to right.

Step 4.
Do addition and subtraction as they occur from left to right.

Example
Simplify: $(-4)^2 - 2(3 - 8)$

$$(-4)^2 - 2(3 - 8)$$

$= (-4)^2 - 2(-5)$ • Perform operations inside parentheses.

$= 16 - 2(-5)$ • Simplify exponential expressions.

$= 16 + 10$ • Do multiplication and division from left to right.

$= 26$ • Do addition and subtraction from left to right.

One or more of the steps listed above may not be needed to evaluate an expression. In that case, proceed to the next step in the Order of Operations Agreement.

Example 2 **Use the Order of Operations Agreement**

Evaluate $6 \div [4 - (6 - 8)] - 2^3$.

Solution

$6 \div [4 - (6 - 8)] - 2^3 = 6 \div [4 - (-2)] - 2^3$ • Perform operations inside grouping symbols.

$= 6 \div 6 - 2^3$

$= 6 \div 6 - 8$ • Simplify exponential expressions.

$= 1 - 8$ • Do multiplication and division from left to right.

$= -7$ • Do addition and subtraction from left to right.

Example 3 **Use the Order of Operations Agreement**

Evaluate $4 - 3[4 - 2(6 - 3)] \div 2$.

Solution

$4 - 3[4 - 2(6 - 3)] \div 2 = 4 - 3[4 - 2 \cdot 3] \div 2$ • Perform operations inside grouping symbols.

$= 4 - 3[4 - 6] \div 2$

$= 4 - 3[-2] \div 2$

$= 4 + 6 \div 2$ • Do multiplication and division from left to right.

$= 4 + 3$

$= 7$ • Do addition and subtraction from left to right.

Example 4 **Use the Order of Operations Agreement**

Evaluate $27 \div (5 - 2)^2 + (-3)^2 \cdot 4$.

Solution

$27 \div (5 - 2)^2 + (-3)^2 \cdot 4 = 27 \div 3^2 + (-3)^2 \cdot 4$ • Perform operations inside grouping symbols.

$= 27 \div 9 + 9 \cdot 4$ • Simplify exponential expressions.

$= 3 + 9 \cdot 4$ • Do multiplication and division from left to right.

$= 3 + 36$

$= 39$ • Do addition and subtraction from left to right.

Objective A.3B Practice

Evaluate using the Order of Operations Agreement.

1. $4 - 8 \div 2$

2. $24 - 18 \div 3 + 2$

3. $27 - 18 \div (-3^2)$

4. $16 + 15 \div (-5) - 2$

5. $6 + \dfrac{16 - 4}{2^2 + 2} - 2$

6. $18 \div 2 - 4^2 - (-3^2)$

Section
A.4   **Operations on Fractions**

Objective A.4A Multiply Fractions

To multiply two fractions, multiply the numerators and multiply the denominators.

Multiplication of Fractions

The product of two fractions is the product of the numerators over the product of the denominators.

$$\frac{a}{b} \cdot \frac{c}{d} = \frac{ac}{bd}, \quad \text{where} \quad b \neq 0 \quad \text{and} \quad d \neq 0$$

Take Note: Note that fractions do not need to have the same denominator in order to be multiplied.

Examples

1. $\frac{2}{5} \cdot \frac{1}{3} = \frac{2 \cdot 1}{5 \cdot 3} = \frac{2}{15}$ww

2. $\frac{3}{4} \cdot \frac{5}{8} = \frac{3 \cdot 5}{4 \cdot 8} = \frac{15}{32}$

The product $\frac{2}{5} \cdot \frac{1}{3}$ shown in example (1) can be read "$\frac{2}{5}$ times $\frac{1}{3}$" or "$\frac{2}{5}$ of $\frac{1}{3}$." Reading the times sign as "of" is useful in diagramming the product of two fractions.

$\frac{1}{3}$ of the bar in Figure A.7 is shaded.

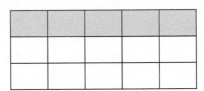

Figure A.7

We want to shade $\frac{2}{5}$ of the $\frac{1}{3}$ already shaded. $\frac{2}{15}$ of the bar is now shaded (see Figure A.8).

$$\frac{2}{5} \text{ of } \frac{1}{3} = \frac{2}{5} \cdot \frac{1}{3} = \frac{2}{15}$$

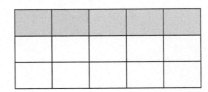

Figure A.8

After multiplying two fractions, write the product in simplest form.

Example 1 **Multiply Fractions**

a. $\dfrac{3}{8} \cdot \dfrac{4}{9}$

b. $\dfrac{3}{8}\left(\dfrac{2}{5}\right)\left(\dfrac{10}{21}\right)$

Solution

a. $\dfrac{3}{8} \cdot \dfrac{4}{9} = \dfrac{3 \cdot 4}{8 \cdot 9}$ • Multiply the numerators. Multiply the denominators.

$= \dfrac{3 \cdot 2 \cdot 2}{2 \cdot 2 \cdot 2 \cdot 3 \cdot 3}$ • Express the fraction in simplest form by first writing the prime factorization of each number.

$= \dfrac{1}{6}$ • Divide by the common factors and write the product in simplest form.

b. $\dfrac{3}{8}\left(\dfrac{2}{5}\right)\left(\dfrac{10}{21}\right) = \dfrac{3 \cdot 2 \cdot 10}{8 \cdot 5 \cdot 21}$ • Multiply the numerators. Multiply the denominators.

$= \dfrac{3 \cdot 2 \cdot 2 \cdot 5}{2 \cdot 2 \cdot 2 \cdot 5 \cdot 3 \cdot 7}$ • Write the product in simplest form.

$= \dfrac{1}{14}$

To multiply a whole number by a fraction or a mixed number, first write the whole number as a fraction with a denominator of 1.

$$3 \cdot \dfrac{5}{8}$$

$3 \cdot \dfrac{5}{8} = \dfrac{3}{1} \cdot \dfrac{5}{8}$ • Write the whole number 3 as the fraction $\dfrac{3}{1}$.

$= \dfrac{3 \cdot 5}{1 \cdot 8}$ • Multiply the fractions. There are no common factors in the numerator and denominator.

$= \dfrac{15}{8} = 1\dfrac{7}{8}$ • The answer can be written as an improper fraction or a mixed number.

Example 2 **Multiply a Fraction and a Whole Number**

What is the product of $\dfrac{7}{12}$ and 4?

Solution

$\dfrac{7}{12} \cdot 4 = \dfrac{7}{12} \cdot \dfrac{4}{1}$ • Write 4 as $\dfrac{4}{1}$.

$= \dfrac{7 \cdot 4}{12 \cdot 1}$ • Multiply numerators and multiply denominators.

$= \dfrac{7 \cdot 2 \cdot 2}{2 \cdot 2 \cdot 3 \cdot 1}$ • Write the product in simplest form.

$= \dfrac{7}{3}$ • The answer can be written as an improper fraction or a mixed number.

Objective A.4A Practice

1. Multiply: $\dfrac{3}{8} \cdot \dfrac{4}{5}$

2. Multiply: $\dfrac{15}{16} \cdot \dfrac{4}{9}$

3. Multiply: $\dfrac{2}{3} \cdot \dfrac{3}{8} \cdot \dfrac{4}{9}$

4. Multiply: $\dfrac{7}{9} \cdot \dfrac{15}{28}$

5. Find the product of $\dfrac{12}{25}$ and $\dfrac{5}{16}$.

Objective A.4B Divide Fractions

The **reciprocal** of a fraction is that fraction with the numerator and denominator interchanged. The reciprocal of a number is also called the *multiplicative inverse* of the number.

$\qquad$ The reciprocal of $\dfrac{3}{4}$ is $\dfrac{4}{3}$. $\qquad$ The reciprocal of $\dfrac{a}{b}$ is $\dfrac{b}{a}$.

The product of a number and its reciprocal is 1.

The process of interchanging the numerator and denominator of a fraction is called **inverting** the fraction.

To find the reciprocal of a whole number, first rewrite the whole number as a fraction with a denominator of 1. Then invert the fraction.

$$6 = \dfrac{6}{1}$$

The reciprocal of 6 is $\dfrac{1}{6}$.

Reciprocals are used to rewrite division problems as related multiplication problems. Look at the following two problems:

$$6 \div 2 = 3 \qquad\qquad 6 \cdot \dfrac{1}{2} = 3$$

6 divided by 2 equals 3. $\qquad$ 6 times the reciprocal of 2 equals 3.

Division is defined as multiplication by the reciprocal. Therefore, "divided by 2" is the same as "times $\dfrac{1}{2}$." Fractions are divided by making this substitution.

Division of Fractions

To divide two fractions, multiply by the reciprocal of the divisor.

$$\dfrac{a}{b} \div \dfrac{c}{d} = \dfrac{a}{b} \cdot \dfrac{d}{c}, \qquad \text{where } b \neq 0,\ c \neq 0, \text{ and } d \neq 0$$

Example

Divide: $\dfrac{2}{5} \div \dfrac{3}{4}$

$\dfrac{2}{5} \div \dfrac{3}{4} = \dfrac{2}{5} \cdot \dfrac{4}{3}$ $\qquad$ • Rewrite the division as multiplication by the reciprocal.

$\qquad\quad = \dfrac{2 \cdot 4}{5 \cdot 3} = \dfrac{8}{15}$ $\qquad$ • Multiply the fractions.

Example 3 **Divide Fractions**

a. $\dfrac{4}{5} \div \dfrac{8}{15}$

b. $\dfrac{7}{10} \div \dfrac{14}{15}$

Solution

a. $\dfrac{4}{5} \div \dfrac{8}{15} = \dfrac{4}{5} \cdot \dfrac{15}{8}$ • Multiply by the reciprocal.

$\phantom{\dfrac{4}{5} \div \dfrac{8}{15}} = \dfrac{4 \cdot 15}{5 \cdot 8}$

$\phantom{\dfrac{4}{5} \div \dfrac{8}{15}} = \dfrac{2 \cdot 2 \cdot 3 \cdot 5}{5 \cdot 2 \cdot 2 \cdot 2}$

$\phantom{\dfrac{4}{5} \div \dfrac{8}{15}} = \dfrac{3}{2}$

b. $\dfrac{7}{10} \div \dfrac{14}{15}$

$ = \dfrac{7}{10} \cdot \dfrac{15}{14}$ • Rewrite the division as multiplication by the reciprocal.

$ = \dfrac{7 \cdot 15}{10 \cdot 14}$ • Multiply the fractions.

$ = \dfrac{7 \cdot 3 \cdot 5}{2 \cdot 5 \cdot 2 \cdot 7}$ • Write the product in simplest form.

$ = \dfrac{3}{4}$

Objective A.4B Practice

1. Divide: $\dfrac{3}{8} \div \dfrac{2}{3}$

2. Divide: $\dfrac{3}{4} \div \dfrac{9}{10}$

3. Divide: $\dfrac{2}{5} \div \dfrac{4}{9}$

4. Divide: $\dfrac{7}{8} \div \dfrac{5}{16}$

5. Divide: $\dfrac{3}{5} \div \dfrac{9}{20}$

Objective A.4C Add Fractions

Addition of Fractions with the Same Denominator

To add fractions with the same denominator, add the numerators and place the sum over the common denominator.

Examples

1. $\dfrac{2}{7} + \dfrac{3}{7} = \dfrac{2+3}{7} = \dfrac{5}{7}$

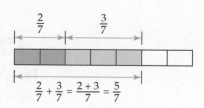

$$\dfrac{2}{7} + \dfrac{3}{7} = \dfrac{2+3}{7} = \dfrac{5}{7}$$

2. $\dfrac{1}{8} + \dfrac{5}{8} = \dfrac{1+5}{8} = \dfrac{6}{8} = \dfrac{3}{4}$

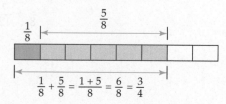

$$\dfrac{1}{8} + \dfrac{5}{8} = \dfrac{1+5}{8} = \dfrac{6}{8} = \dfrac{3}{4}$$

Addition of Fractions with Different Denominators

To add fractions with different denominators, first rewrite the fractions as equivalent fractions with a common denominator. Then add the numerators and place the sum over the common denominator. The least common multiple, LCM, of the denominators of the fractions is the **least common denominator (LCD)**.

Example

Add: $\frac{1}{2} + \frac{1}{3}$

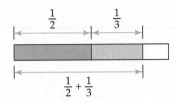

The LCM of the denominators 2 and 3 is 6.

$$\frac{1}{2} + \frac{1}{3} = \frac{3}{6} + \frac{2}{6} \qquad \text{• Write equivalent fractions with 6 as the denominator.}$$

$$= \frac{3 + 2}{6} = \frac{5}{6} \qquad \text{• Add the numerators.}$$

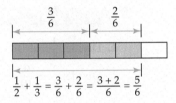

$$\frac{1}{2} + \frac{1}{3} = \frac{3}{6} + \frac{2}{6} = \frac{3+2}{6} = \frac{5}{6}$$

Example 4 Add Fractions

Add: $\frac{5}{8} + \frac{7}{9}$

Solution

$$\frac{5}{8} + \frac{7}{9} = \frac{45}{72} + \frac{56}{72} \qquad \text{• Write equivalent fractions using 72 (the LCM of the denominators) as the common denominator.}$$

$$= \frac{45 + 56}{72} \qquad \text{• Add the numerators.}$$

$$= \frac{101}{72}$$

Example 5 Add Fractions

Find $\frac{7}{12}$ more than $\frac{3}{8}$.

Solution

$$\frac{3}{8} + \frac{7}{12} = \frac{9}{24} + \frac{14}{24} \qquad \text{• Write equivalent fractions using 24 (the LCM of the denominators) as the common denominator.}$$

$$= \frac{9 + 14}{24} \qquad \text{• Add the numerators.}$$

$$= \frac{23}{24}$$

Example 6   Add Fractions

Add: $\dfrac{2}{3} + \dfrac{3}{5} + \dfrac{5}{6}$

Solution

$\dfrac{2}{3} + \dfrac{3}{5} + \dfrac{5}{6} = \dfrac{20}{30} + \dfrac{18}{30} + \dfrac{25}{30}$ • Write equivalent fractions using 30 (the LCM of the denominators) as the common denominator.

$= \dfrac{20 + 18 + 25}{30}$ • Add the numerators.

$= \dfrac{63}{30} = \dfrac{21}{10}$ • Write the answer in simplest form.

Objective A.4C Practice

1. Add: $\dfrac{2}{9} + \dfrac{4}{9}$

2. Add: $\dfrac{5}{12} + \dfrac{5}{16}$

3. Add: $\dfrac{2}{3} + \dfrac{1}{5} + \dfrac{7}{12}$

4. Find the sum of $\dfrac{3}{8}, \dfrac{5}{6}$, and $\dfrac{7}{12}$.

Objective A.4D Subtract Fractions

Subtraction of Fractions with the Same Denominator

To subtract fractions with the same denominator, subtract the numerators and place the difference over the common denominator.

Example

$\dfrac{6}{7} - \dfrac{2}{7} = \dfrac{6-2}{7} = \dfrac{4}{7}$

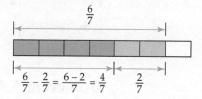

$\dfrac{6}{7} - \dfrac{2}{7} = \dfrac{6-2}{7} = \dfrac{4}{7}$ $\dfrac{2}{7}$

Subtraction of Fractions with Different Denominators

To subtract fractions with different denominators, first rewrite the fractions as equivalent fractions with a common denominator. Then subtract the numerators and place the difference over the common denominator.

Example

Subtract: $\dfrac{5}{6} - \dfrac{1}{4}$

The LCD of the fractions is 12.

$\dfrac{5}{6} - \dfrac{1}{4} = \dfrac{10}{12} - \dfrac{3}{12}$ • Rewrite each fraction as an equivalent fraction with 12 as the denominator.

$= \dfrac{10 - 3}{12} = \dfrac{7}{12}$ • Subtract the numerators and place the difference over the common denominator.

Example 7 Subtract Fractions

Subtract: $\dfrac{11}{16} - \dfrac{5}{12}$

Solution

$\dfrac{11}{16} - \dfrac{5}{12} = \dfrac{33}{48} - \dfrac{20}{48}$ • Write equivalent fractions using 48 (the LCM of the denominators) as the common denominator.

$= \dfrac{33 - 20}{48}$ • Subtract the numerators.

$= \dfrac{13}{48}$ • Write the answer in simplest form.

Example 8 Subtract Fractions

Find the difference between $\dfrac{7}{8}$ and $\dfrac{5}{12}$.

Solution

$\dfrac{7}{8} - \dfrac{5}{12} = \dfrac{21}{24} - \dfrac{10}{24}$ • Write equivalent fractions using 24 (the LCM of the denominators) as the common denominator.

$= \dfrac{21 - 10}{24}$ • Subtract the numerators.

$= \dfrac{11}{24}$ • Write the answer in simplest form.

Objective A.4D Practice

1. Subtract: $\dfrac{9}{20} - \dfrac{7}{20}$

2. Subtract: $\dfrac{7}{9} - \dfrac{1}{6}$

3. Subtract: $\dfrac{9}{14} - \dfrac{5}{42}$

4. Subtract: $\dfrac{7}{12} - \dfrac{5}{18}$

Objective A.4E Identify the Order Relation Between Two Fractions

Recall that whole numbers can be graphed as points on the number line. Fractions can also be graphed as points on the number line (see Figure A.9).

The graph of $\dfrac{3}{4}$ on the number line:

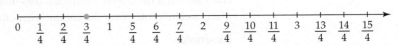

Figure A.9

The number line can be used to determine the order relation between two fractions. A fraction that appears to the left of a given fraction is less than the given fraction. A fraction that appears to the right of a given fraction is greater than the given fraction (see Figure A.10).

$\dfrac{1}{8} < \dfrac{3}{8}$ $\qquad$ $\dfrac{6}{8} > \dfrac{3}{8}$

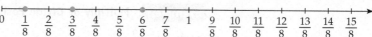

Figure A.10

To find the order relation between two fractions with the same denominator, compare the numerators. The fraction that has the smaller numerator is the smaller fraction. When the denominators are different, begin by writing equivalent fractions with a common denominator; then compare the numerators.

Example 9   **Find the Order Relation Between Two Fractions**

Find the order relation between $\frac{11}{18}$ and $\frac{5}{8}$.

Solution

The LCD of the fractions is 72.

$$\frac{11}{18} = \frac{44}{72} \longleftarrow \text{Smaller numerator}$$

$$\frac{5}{8} = \frac{45}{72} \longleftarrow \text{Larger numerator}$$

$$\frac{11}{18} < \frac{5}{8} \text{ or } \frac{5}{8} > \frac{11}{18}$$

Objective A.4E Practice

Place the correct symbol, $<$ or $>$, between the two numbers.

1. $\frac{11}{40}$ $\frac{19}{40}$

2. $\frac{2}{5}$ $\frac{3}{8}$

3. $\frac{13}{18}$ $\frac{7}{12}$

4. $\frac{5}{12}$ $\frac{7}{15}$

Section A.5 Operations on Positive and Negative Fractions

Objective A.5A Multiply and Divide Positive and Negative Fractions

The sign rules for multiplying positive and negative fractions are the same rules used to multiply integers.

The product or quotient of two numbers with the same sign is positive.

The product or quotient of two numbers with different signs is negative.

Example 1	Multiply or Divide Positive and Negative Fractions

a. Multiply: $-\dfrac{3}{4} \cdot \dfrac{8}{15}$

b. Multiply: $-\dfrac{3}{4}\left(\dfrac{1}{2}\right)\left(-\dfrac{8}{9}\right)$

c. Divide: $-\dfrac{7}{10} \div \left(-\dfrac{14}{15}\right)$

Solution

a. The signs are different. The product is negative.

$$-\frac{3}{4} \cdot \frac{8}{15} = -\frac{3 \cdot 8}{4 \cdot 15}$$
• Multiply the numerators. Multiply the denominators.

$$= -\frac{3 \cdot 2 \cdot 2 \cdot 2}{2 \cdot 2 \cdot 3 \cdot 5}$$
• Write the product in simplest form.

$$= -\frac{2}{5}$$

b. $-\dfrac{3}{4}\left(\dfrac{1}{2}\right)\left(-\dfrac{8}{9}\right) = \dfrac{3 \cdot 1 \cdot 8}{4 \cdot 2 \cdot 9}$
• The product is positive. Multiply the numerators. Multiply the denominators.

$$= \frac{3 \cdot 1 \cdot 2 \cdot 2 \cdot 2}{2 \cdot 2 \cdot 2 \cdot 3 \cdot 3}$$
• Write the product in simplest form.

$$= \frac{1}{3}$$

c. The signs are the same. The quotient is positive.

$$-\frac{7}{10} \div \left(-\frac{14}{15}\right) = \frac{7}{10} \div \frac{14}{15}$$

$$= \frac{7}{10} \cdot \frac{15}{14}$$
• Rewrite the division as multiplication by the reciprocal.

$$= \frac{7 \cdot 15}{10 \cdot 14}$$
• Multiply the fractions.

$$= \frac{7 \cdot 3 \cdot 5}{2 \cdot 5 \cdot 2 \cdot 7}$$
• Write the product in simplest form.

$$= \frac{3}{4}$$

Objective A.5A Practice

1. Multiply: $\dfrac{5}{6} \cdot \left(-\dfrac{2}{5}\right)$

2. Multiply: $\left(-\dfrac{3}{4}\right)\left(-\dfrac{2}{9}\right)$

3. Multiply: $-\dfrac{3}{4} \cdot \dfrac{1}{2}$

4. Multiply: $-1\dfrac{2}{3} \cdot \left(-\dfrac{3}{5}\right)$

5. Evaluate xy for $x = -\dfrac{5}{16}$ and $y = \dfrac{7}{15}$.

6. Divide: $-\dfrac{5}{7} \div \left(-\dfrac{5}{6}\right)$

7. Divide: $\dfrac{3}{8} \div \left(-\dfrac{5}{12}\right)$

8. Divide: $5\dfrac{3}{5} \div \left(-\dfrac{7}{10}\right)$

9. Find the quotient of $\dfrac{3}{5}$ and $\dfrac{12}{25}$.

10. Evaluate $x \div y$ for $x = -\dfrac{5}{8}$ and $y = -\dfrac{15}{2}$.

Objective A.5B Add and Subtract Positive and Negative Fractions

To add a fraction with a negative sign, rewrite the fraction with the negative sign in the numerator. Then add the numerators and place the sum over the common denominator.

Example 2 **Add Signed Fractions**

a. $-\dfrac{5}{6} + \dfrac{3}{4}$

b. $-\dfrac{2}{3} + \left(-\dfrac{4}{5}\right)$

c. $-\dfrac{3}{8} + \dfrac{3}{4} + \left(-\dfrac{5}{6}\right)$

Solution

a. $-\dfrac{5}{6} + \dfrac{3}{4} = \dfrac{-5}{6} + \dfrac{3}{4}$ • Rewrite the first fraction with the negative sign in the numerator.

$= \dfrac{-10}{12} + \dfrac{9}{12}$ • Rewrite each fraction as an equivalent fraction using the LCD (12) of the fractions.

$= \dfrac{-10 + 9}{12}$ • Add the fractions.

$= \dfrac{-1}{12} = -\dfrac{1}{12}$ • Simplify the numerator and write the negative sign in front of the fraction.

b. $-\dfrac{2}{3} + \left(-\dfrac{4}{5}\right) = \dfrac{-2}{3} + \dfrac{-4}{5}$ • Rewrite each negative fraction with the negative sign in the numerator.

$= \dfrac{-10}{15} + \dfrac{-12}{15}$ • Rewrite each fraction as an equivalent fraction using the LCD (15) of the fractions.

$= \dfrac{-10 + (-12)}{15}$ • Add the fractions.

$= -\dfrac{22}{15}$

c. $-\dfrac{3}{8} + \dfrac{3}{4} + \left(-\dfrac{5}{6}\right) = \dfrac{-3}{8} + \dfrac{3}{4} + \dfrac{-5}{6}$ • Rewrite each negative fraction with the negative sign in the numerator.

$= \dfrac{-9}{24} + \dfrac{18}{24} + \dfrac{-20}{24}$ • Rewrite each fraction as an equivalent fraction using the LCD (24) of the fractions.

$= \dfrac{-9 + 18 + (-20)}{24}$ • Add the fractions.

$= \dfrac{-11}{24} = -\dfrac{11}{24}$

To subtract fractions with negative signs, first rewrite the fractions with the negative signs in the numerators.

Example 3 **Subtract Signed Fractions**

a. $-\dfrac{2}{9} - \dfrac{5}{12}$

b. $\dfrac{2}{3} - \left(-\dfrac{4}{5}\right)$

c. $-\dfrac{5}{6} - \left(-\dfrac{3}{8}\right)$

Solution

a. $-\dfrac{2}{9} - \dfrac{5}{12} = \dfrac{-2}{9} - \dfrac{5}{12}$
• Rewrite the negative fraction with the negative sign in the numerator.

$= \dfrac{-8}{36} - \dfrac{15}{36}$
• Write the fractions as equivalent fractions with a common denominator.

$= \dfrac{-8 - 15}{36} = \dfrac{-23}{36}$
• Subtract the numerators and place the difference over the common denominator.

$= -\dfrac{23}{36}$
• Write the negative sign in front of the fraction.

b. $\dfrac{2}{3} - \left(-\dfrac{4}{5}\right) = \dfrac{2}{3} + \dfrac{4}{5}$
• Rewrite subtraction as addition of the opposite.

$= \dfrac{10}{15} + \dfrac{12}{15}$
• Write the fractions as equivalent fractions with a common denominator.

$= \dfrac{10 + 12}{15}$
• Add the fractions.

$= \dfrac{22}{15}$

c. $-\dfrac{5}{6} - \left(-\dfrac{3}{8}\right) = -\dfrac{5}{6} + \dfrac{3}{8}$
• Rewrite subtraction as addition of the opposite.

$= \dfrac{-20}{24} + \dfrac{9}{24}$
• Rewrite the fractions using the LCD (24).

$= \dfrac{-20 + 9}{24} = \dfrac{-11}{24}$

$= -\dfrac{11}{24}$

Objective A.5B Practice

1. Add: $-\dfrac{3}{4} + \dfrac{2}{3}$

2. Add: $-\dfrac{7}{12} + \dfrac{5}{8}$

3. Add: $\dfrac{2}{5} + \left(-\dfrac{11}{15}\right)$

4. Add: $\dfrac{3}{8} + \left(-\dfrac{1}{2}\right) + \dfrac{7}{12}$

5. Evaluate $x + y$ when $x = \dfrac{3}{10}$ and $y = -\dfrac{7}{15}$.

6. Subtract: $-\dfrac{1}{2} - \dfrac{3}{8}$

7. Subtract: $-\dfrac{5}{12} - \left(-\dfrac{2}{3}\right)$

8. Subtract: $-\dfrac{5}{9} - \left(-\dfrac{11}{12}\right)$

9. Subtract: $-\dfrac{5}{8} - \left(-\dfrac{7}{12}\right)$

10. Evaluate $x - y$ when $x = -\dfrac{5}{12}$ and $y = -\dfrac{5}{9}$.

Section A.6 Operations on Decimals

Objective A.6A Add and Subtract Decimals

To add decimals, write the numbers so that the decimal points are on a vertical line. Add as you would with whole numbers. Then write the decimal point in the sum directly below the decimal points in the addends.

Example 1 **Add Decimals**

Find the sum of 0.64, 8.731, 12, and 5.9.

Solution

Arrange the numbers vertically, placing the decimal points on a vertical line.

Add the numbers in each column.

Write the decimal point in the sum directly below the decimal points in the addends.

$$
\begin{array}{r}
\overset{1\ 2}{} \\
0.64 \\
8.731 \\
12. \\
+\ \ 5.9 \\
\hline
27.271
\end{array}
$$

To subtract decimals, write the numbers so that the decimal points are on a vertical line. Subtract as you would with whole numbers. Then write the decimal point in the difference directly below the decimal point in the subtrahend.

Example 2 **Subtract Decimals**

Subtract: $5.4 - 1.6832$

Insert zeros in the minuend so that it has the same number of decimal places as the subtrahend.

Subtract.

$$
\begin{array}{r}
5.\ 4\ 0\ 0\ 0 \\
-1.\ 6\ 8\ 3\ 2 \\
\end{array}
$$

$$
\begin{array}{r}
4\ \ 13\ \ 9\ \ 9\ \ 10 \\
\cancel{5}.\ \cancel{4}\ \cancel{0}\ \cancel{0}\ \cancel{0} \\
-1.\ 6\ 8\ 3\ 2 \\
\hline
3.\ 7\ 1\ 6\ 8
\end{array}
$$

The sign rules for adding and subtracting decimals are the same rules used to add and subtract integers.

| Example 3 | Add and Subtract Signed Decimal Numbers |

a. $-36.087 + 54.29$

b. $-2.86 - 10.3$

Take Note: Recall that the absolute value of a number is the distance from zero to the number on the number line. The absolute value of a number is a positive number or zero.

$|54.29| = 54.29$

$|-36.087| = 36.087$

Solution

a. The signs of the addends are different. Subtract the smaller absolute value from the larger absolute value.

$54.29 - 36.087 = 18.203$

Attach the sign of the number with the larger absolute value.

$|54.29| > |-36.087|$

The sum is positive. $-36.087 + 54.29 = 18.203$

b. Rewrite subtraction as addition of the opposite. $-2.86 - 10.3$
The opposite of 10.3 is -10.3. $= -2.86 + (-10.3)$
The signs of the addends are the same.

Add the absolute values of the numbers.
Attach the sign of the addends. $= -13.16$

| Example 4 | Evaluate a Variable Expression |

Evaluate $x + y + z$ for $x = -1.6$, $y = 7.9$, and $z = -4.8$.

Solution

$x + y + z$

$-1.6 + 7.9 + (-4.8) = 6.3 + (-4.8)$

$= 1.5$

Objective A.6A Practice

1. Subtract: $-42.1 - 8.6$
2. Add: $-9.37 + 3.465$
3. Find 382.9 more than -430.6.
4. What is 4.793 less than -6.82?
5. Evaluate $x + y$ for $x = -125.41$ and $y = 361.55$.
6. Evaluate $x - y$ for $x = -3.69$ and $y = -1.527$.

Objective A.6B Multiply Decimals

Decimals are multiplied as though they were whole numbers; then the decimal point is placed in the product. Writing the decimals as fractions shows where to write the decimal point in the product.

$$0.4 \cdot 2 = \frac{4}{10} \cdot \frac{2}{1} = \frac{8}{10} = 0.8$$

1 decimal place in 0.4   1 decimal place in 0.8

$$0.4 \cdot 0.2 = \frac{4}{10} \cdot \frac{2}{10} = \frac{8}{100} = 0.08$$

1 decimal place in 0.4   1 decimal place in 0.2
2 decimal places in 0.08

$$0.4 \cdot 0.02 = \frac{4}{10} \cdot \frac{2}{100} = \frac{8}{1000} = 0.008$$

1 decimal place in 0.4   2 decimal places in 0.02
3 decimal places in 0.008

To multiply decimals, multiply the numbers as you would whole numbers. Then write the decimal point in the product so that the number of decimal places in the product is the sum of the numbers of decimal places in the factors.

| Example 5 | **Multiply Decimal Numbers** |

a. (32.41)(7.6)

b. 0.061(0.08)

c. (−3.2)(−0.008)

Solution

a.
```
    32.41    2 decimal places
  ×   7.6    1 decimal place
   19446
   22687
  246.316    3 decimal places
```

b.
```
    0.061    3 decimal places
  × 0.08     2 decimal places
  0.00488    5 decimal places
```
• Insert two zeros between the 4 and the decimal point so that there are 5 decimal places in the product.

c. (−3.2)(−0.008) = 0.0256

• The signs are the same.
The product is positive.
Multiply the absolute values of the numbers.

To multiply a decimal by a power of 10 (10, 100, 1000, …), move the decimal point to the right the same number of places as there are zeros in the power of 10.

2.7935 · 10   = 27.935

1 zero   1 decimal place

2.7935 · 100   = 279.35

2 zeros   2 decimal places

2.7935 · 1000   = 2793.5

3 zeros   3 decimal places

2.7935 · 10,000   = 27,935.

4 zeros   4 decimal places

2.7935 · 100,000   = 279,350.

5 zeros   5 decimal places

• A zero must be inserted before the decimal point.

Note that if the power of 10 is written in exponential notation, the exponent indicates how many places to move the decimal point.

$$2.7935 \cdot 10^1 = 27.935$$
1 decimal place

$$2.7935 \cdot 10^2 = 279.35$$
2 decimal places

$$2.7935 \cdot 10^3 = 2793.5$$
3 decimal places

$$2.7935 \cdot 10^4 = 27,935.$$
4 decimal places

$$2.7935 \cdot 10^5 = 279,350.$$
5 decimal places

Example 6 **Multiply by a Power of 10**

a. What is 835.294 multiplied by 1000?
b. Find the product of 64.18 and 10^3.

Solution
a. Move the decimal point 3 places to the right.

$$835.294 \cdot 1000 = 835{,}294$$

b. $64.18 \cdot 10^3 = 64{,}180$ • The exponent on 10 is 3. Move the decimal point in 64.18 three places to the right.

Example 7 **Evaluate an Expression**

Evaluate $50ab$ for $a = -0.9$ and $b = -0.2$.

Solution
$$50ab$$
$$50(-0.9)(-0.2) = -45(-0.2)$$
$$= 9$$

Objective A.6B Practice

1. Multiply: $(3.4)(0.5)$
2. Multiply: $(-6.3)(-2.4)$
3. Multiply: $1.31(-0.006)$
4. Find the product of 6.71 and 10^4.
5. Find the product of 2.7, -16, and 3.04.
6. Evaluate ab for $a = 452$ and $b = -0.86$.

Objective A.6C Divide Decimals

To divide decimals, move the decimal point in the divisor to the right so that the divisor is a whole number. Move the decimal point in the dividend the same number of places to the right. Place the decimal point in the quotient directly above the decimal point in the dividend. Then divide as you would with whole numbers.

Example 8   **Divide Decimal Numbers**

Divide: $29.585 \div 4.85$

Solution

$$
4.85\overline{)29.58.5} \qquad\qquad 485\overline{)\begin{array}{r} 6.1 \\ 2958.5 \\ -2910 \\ \hline 485 \\ -485 \\ \hline 0 \end{array}}
$$

Move the decimal point 2 places to the right in the divisor. Move the decimal point 2 places to the right in the dividend. Place the decimal point in the quotient. Then divide as shown at the right.

Moving the decimal point the same number of places in the divisor and the dividend does not change the quotient, because the process is the same as multiplying the numerator and denominator of a fraction by the same number. For the previous example,

$$
4.85\overline{)29.585} = \frac{29.585}{4.85} = \frac{29.585 \cdot 100}{4.85 \cdot 100} = \frac{2958.5}{485} = 485\overline{)2958.5}
$$

In division of decimals, rather than writing the quotient with a remainder, we usually round the quotient to a specified place value. The symbol $\approx$ is read "is approximately equal to"; it is used to indicate that the quotient is an approximate value after being rounded.

Example 9   **Divide Decimal Numbers**

a. Divide and round to the nearest tenth: $0.86 \div 0.7$

b. Divide and round to the nearest hundredth: $448.2 \div 53$

c. Divide and round to the nearest tenth: $-6.94 \div (-1.5)$

Solution

a.
$$
0.7\overline{)\begin{array}{r} 1.22 \approx 1.2 \\ 0.8.60 \\ -7 \\ \hline 1\,6 \\ -1\,4 \\ \hline 2\,0 \\ -1\,4 \\ \hline 6 \end{array}}
$$

To round the quotient to the nearest tenth, the division must be carried to the hundredths place. Therefore, zeros must be inserted in the dividend so that the quotient has a digit in the hundredths place.

b.
$$
\begin{array}{r}
8.45\,6 \approx 8.46 \\
53\,)\overline{448.20\,0} \\
-4\,24 \\
\hline
2\,42 \\
-2\,12 \\
\hline
3\,00 \\
-2\,65 \\
\hline
3\,50 \\
-3\,18 \\
\hline
3\,2
\end{array}
$$

c. The quotient is positive.

$-6.94 \div (-1.5) \approx 4.6$

●

To divide a decimal by a power of 10 (10, 100, 1000, 10,000, …), move the decimal point to the left the same number of places as there are zeros in the power of 10.

$462.81 \div 1\underline{0} \qquad = 46.281$

         1 zero   1 decimal place

$462.81 \div 1\underline{00} \qquad = 4.6281$

         2 zeros   2 decimal places

$462.81 \div 1\underline{000} \qquad = 0.46281$

         3 zeros   3 decimal places

$462.81 \div 1\underline{0,000} \quad = 0.046281$

         4 zeros   4 decimal places

- A zero must be inserted between the decimal point and the 4.

$462.81 \div 1\underline{00,000} = 0.0046281$

         5 zeros   5 decimal places

- Two zeros must be inserted between the decimal point and the 4.

If the power of 10 is written in exponential notation, the exponent indicates how many places to move the decimal point.

$462.81 \div 10^1 = 46.281$

         1 decimal place

$462.81 \div 10^2 = 4.6281$

         2 decimal places

$462.81 \div 10^3 = 0.46281$

         3 decimal places

$462.81 \div 10^4 = 0.046281$

         4 decimal places

$462.81 \div 10^5 = 0.0046281$

         5 decimal places

Example 10   **Divide by Power of 10**

a. Find the quotient of 592.4 and 10^4.
b. Find the quotient of 3.59 and 100.

Solution

a. Move the decimal point 4 places to the left.

$$592.4 \div 10^4 = 0.05924$$

b. $3.59 \div 100 = 0.0359$ • There are two zeros in 100. Move the decimal point in 3.59 two places to the left.

Example 11   **Evaluate a Variable Expression**

Evaluate $\frac{x}{y}$ for $x = -76.8$ and $y = 0.8$.

Solution

$$\frac{x}{y} = \frac{-76.8}{0.8} = -96$$

Objective A.6C Practice

1. Divide: $7.02 \div 3.6$
2. Divide: $(-3.312) \div (-0.8)$
3. Divide: $84.66 \div (-1.7)$
4. Find 9.407 divided by 10^3.
5. Evaluate $\frac{x}{y}$ for $x = 26.22$ and $y = -6.9$.

Section **A.7**

Comparing and Converting Fractions and Decimals

Objective A.7A Convert Fractions to Decimals

To convert a fraction to a decimal, divide the numerator by the denominator.

For instance, to convert $\frac{5}{8}$ to a decimal, divide the numerator by the denominator. In this case, the remainder is 0. This fraction can be written as a **terminating decimal**, which is a decimal whose remainder is eventually 0.

$$
\begin{array}{r}
0.625 \\
8\overline{)5.000} \\
-48 \\
\hline
20 \\
-16 \\
\hline
40 \\
-40 \\
\hline
0
\end{array}
$$

$$\frac{5}{8} = 0.625$$

If the remainder is never 0, the fraction can be written as a **repeating decimal**. When a remainder begins to repeat (as 4 does in the following example), the digits in the quotient will begin to repeat.

For instance, to convert $\frac{7}{11}$ to a decimal, divide the numerator by the denominator.

$$
\begin{array}{r}
0.6363 \\
11\overline{)7.0000} \\
-6.6 \\
\hline
40 \\
-33 \\
\hline
70 \\
-66 \\
\hline
40 \\
-33 \\
\hline
7
\end{array}
$$

Take Note: The decimal representation of every fraction either terminates or repeats. However, the repeating cycle may be quite long. For instance, there are 22 digits in the repeating cycle of the decimal representation of $\frac{9}{23}$.

$\frac{9}{23} = 0.\overline{3913043478260869565217}$

$\frac{7}{11} = 0.6363\ldots = 0.\overline{63}$

It is common practice to write a bar over the repeating digits of a decimal.

Example 1	Convert a Fraction to a Decimal

a. Convert $\frac{5}{12}$ to a decimal.

b. Convert $\frac{5}{16}$ to a decimal.

c. Convert $\frac{32}{15}$ to a decimal.

Solution

a.
$$
\begin{array}{r}
0.4166 \\
12\overline{)5.0000} \\
-48 \\
\hline
20 \\
-12 \\
\hline
80 \\
-72 \\
\hline
80
\end{array}
$$

• Divide the numerator by the denominator. Continue to divide until the remainder is 0 or the remainder repeats.

$\frac{5}{12} = 0.41\overline{6}$

b.
$$
\begin{array}{r}
0.3125 \\
16\overline{)5.0000} \\
-48 \\
\hline
20 \\
-16 \\
\hline
40 \\
-32 \\
\hline
80 \\
-80 \\
\hline
0
\end{array}
$$

• Divide the numerator by the denominator. Continue to divide until the remainder is 0 or the remainder repeats.

$\frac{5}{16} = 0.3125$

c.

$$
\begin{array}{r}
2.1\,3\,3 \\
15\overline{)\,3\,2.000} \\
-3\,0 \\
\hline
2\,0 \\
-1\,5 \\
\hline
5\,0 \\
-4\,5 \\
\hline
5\,0
\end{array}
$$

• Divide the numerator by the denominator. Continue to divide until the remainder is 0 or the remainder repeats.

$$\frac{32}{15} = 2.1\overline{3}$$

Objective A.7A Practice

1. Convert $\frac{5}{9}$ to a terminating or repeating decimal. Place a bar over any repeating digits.

2. Convert $\frac{8}{11}$ to a terminating or repeating decimal. Place a bar over any repeating digits.

3. Convert $\frac{71}{111}$ to a terminating or repeating decimal. Place a bar over any repeating digits.

4. Convert $\frac{17}{22}$ to a terminating or repeating decimal. Place a bar over any repeating digits.

5. Convert $\frac{3}{40}$ to a terminating or repeating decimal. Place a bar over any repeating digits.

6. Convert $\frac{29}{4}$ to a terminating or repeating decimal. Place a bar over any repeating digits.

Objective A.7B Convert Decimals to Fractions

To convert a decimal to a fraction, remove the decimal point and place the decimal part over a denominator equal to the place value of the last digit in the decimal. Then write the fraction in simplest form.

Example 2 **Convert a Decimal to a Fraction**

a. Convert 0.275 to a fraction.

b. Convert 0.33 to a fraction.

Solution

a. $0.275 = \overset{\text{thousandths}}{\dfrac{275}{1000}} = \dfrac{\overset{1}{\cancel{5}} \cdot \overset{1}{\cancel{5}} \cdot 11}{2 \cdot 2 \cdot 2 \cdot \underset{1}{\cancel{5}} \cdot \underset{1}{\cancel{5}} \cdot 5} = \dfrac{11}{40}$

b. $0.33 = \overset{\text{hundredths}}{\dfrac{33}{100}} = \dfrac{3 \cdot 11}{2 \cdot 2 \cdot 5 \cdot 5} = \dfrac{33}{100}$

Objective A.7B Practice

1. Convert 0.4 to a fraction.
2. Convert 0.48 to a fraction.
3. Convert 0.485 to a fraction.
4. Convert 3.75 to a mixed number.
5. Convert 0.052 to a fraction.
6. Convert 0.00015 to a fraction.

Objective A.7C Compare a Fraction and a Decimal

One way to determine the order relation between a decimal and a fraction is to write the decimal as a fraction and then compare the fractions.

Example 3 **Compare a Decimal and a Fraction**

Place the correct symbol, $<$ or $>$, between the two numbers.

$0.7 \quad \dfrac{3}{4}$

Solution

$0.7 \quad \dfrac{3}{4}$

$\dfrac{7}{10} \quad \dfrac{3}{4}$ • Write the decimal as a fraction.

$\dfrac{14}{20} \quad \dfrac{15}{20}$ • Write the fractions with a common denominator.

$\dfrac{14}{20} < \dfrac{15}{20}$ • Compare the fractions.

$0.7 < \dfrac{3}{4}$

Objective A.7C Practice

Place the correct symbol, $<$ or $>$, between the numbers.

1. $0.7 \quad \dfrac{3}{4}$

2. $\dfrac{7}{8} \quad 0.9$

3. $0.13 \quad \dfrac{5}{40}$

4. $\dfrac{12}{55} \quad 0.22$

5. $0.55 \quad \dfrac{5}{9}$

6. $\dfrac{22}{7} \quad 3.14$

Objective A.7D Write Ratios and Rates

In previous work, we have used quantities with units, such as 12 ft, 3 hr, 2¢, and 15 acres. In these examples, the units are feet, hours, cents, and acres.

    A **ratio** is the quotient or comparison of two quantities with the *same* unit. We can compare the measure of 3 ft to the measure of 8 ft by writing a quotient.

$\dfrac{3\text{ ft}}{8\text{ ft}} = \dfrac{3}{8} \qquad$ 3 ft is $\dfrac{3}{8}$ of 8 ft.

A ratio can be written in three ways:

1. As a fraction $\frac{3}{8}$
2. As two numbers separated by a colon $3:8$
3. As two numbers separated by the word *to* $3 \text{ to } 8$

The ratio of 15 mi to 45 mi is written as

$$\frac{15 \text{ mi}}{45 \text{ mi}} = \frac{15}{45} = \frac{1}{3} \text{ or } 1:3 \text{ or } 1 \text{ to } 3$$

A ratio is in **simplest form** when the two numbers do not have a common factor. The units are not written in a ratio.

Example 4 **Write a Ratio**

a. Write the comparison \$6 to \$8 as a ratio in simplest form using a fraction, a colon, and the word *to*.

b. Write the comparison 18 qt to 6 qt as a ratio in simplest form using a fraction, a colon, and the word *to*.

Solution

a. $\dfrac{\$6}{\$8} = \dfrac{6}{8} = \dfrac{3}{4}$

   $\$6 : \$8 = 6 : 8 = 3 : 4$

   $\$6 \text{ to } \$8 = 6 \text{ to } 8 = 3 \text{ to } 4$

b. $\dfrac{18 \text{ qt}}{6 \text{ qt}} = \dfrac{18}{6} = \dfrac{3}{1}$

   $18 \text{ qt} : 6 \text{ qt} = 18 : 6 = 3 : 1$

   $18 \text{ qt to } 6 \text{ qt} = 18 \text{ to } 6$

   $\qquad\qquad\qquad\quad = 3 \text{ to } 1$

A **rate** is the comparison of two quantities with *different* units.

A catering company prepares 9 gal of coffee for every 50 people at a reception. This rate is written

$$\frac{9 \text{ gal}}{50 \text{ people}}$$

You traveled 200 mi in 6 hr. The rate is written

$$\frac{200 \text{ mi}}{6 \text{ hr}} = \frac{100 \text{ mi}}{3 \text{ hr}}$$

A rate is in **simplest form** when the numbers have no common factors. The units are written as part of the rate.

Many rates are written as unit rates. A **unit rate** is a rate in which the number in the denominator is 1. The word *per* generally indicates a unit rate. It means "for each" or "for every." For example,

23 mi per gallon   • The unit rate is $\dfrac{23 \text{ mi}}{1 \text{ gal}}$.

65 mi per hour   • The unit rate is $\dfrac{65 \text{ mi}}{1 \text{ hr}}$.

\$4.78 per pound   • The unit rate is $\dfrac{\$4.78}{1 \text{ lb}}$.

Take Note: Unit rates make comparisons easier. For example, if you travel 37 mph and I travel 43 mph, we know that I am traveling faster than you are. It is more difficult to compare speeds if we are told that you are traveling $\dfrac{111 \text{ mi}}{3 \text{ hr}}$ and I am traveling $\dfrac{172 \text{ mi}}{4 \text{ hr}}$.

| Example 5 | **Write a Unit Rate** |

a. Write "285 mi in 5 hr" as a unit rate.

b. Write "300 ft in 8 s" as a unit rate.

Solution

a. $\dfrac{285 \text{ mi}}{5 \text{ hr}}$ • Write the rate as a fraction.

$285 \div 5 = 57$ • Divide the numerator by the denominator.

The unit rate is 57 mph.

b. $\dfrac{300 \text{ ft}}{8 \text{ s}}$

$$8\overline{)3\,0\,0.0}^{\,37.5}$$

The unit rate is 37.5 ft/s.

Objective A.7D Practice

1. Write the ratio of 28 in. to 36 in. as a ratio in simplest form using a fraction, a colon, and the word *to*.
2. Write the ratio of 32 oz to 16 oz as a ratio in simplest form using a fraction, a colon, and the word *to*.
3. Write 10 ft in 4 s as a unit rate.
4. Write $51,000 earned in 12 months as a unit rate.
5. Write $11.05 for 3.4 lb as a unit rate.

Section A.8

Introduction to Percents

Objective A.8A Write a Percent as a Decimal or a Fraction

Percent means "parts of 100." In Figure A.11, there are 100 parts. Because 13 of the 100 parts are shaded, 13% of the figure is shaded. The symbol % is the **percent sign**.

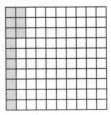

Figure A.11

In most applied problems involving percents, it is necessary either to rewrite a percent as a decimal or a fraction, or to rewrite a fraction or a decimal as a percent.

Take Note: Recall that division is defined as multiplication by the reciprocal. Therefore, multiplying by $\frac{1}{100}$ is equivalent to dividing by 100.

To write a percent as a decimal, remove the percent sign and multiply by 0.01.

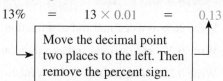

$$13\% \quad = \quad 13 \times 0.01 \quad = \quad 0.13$$

Move the decimal point two places to the left. Then remove the percent sign.

To write a percent as a fraction, remove the percent sign and multiply by $\frac{1}{100}$.

$$13\% = 13 \times \frac{1}{100} = \frac{13}{100}$$

Example 1   **Write a Percent as a Decimal and as a Fraction**

Write each percent as a decimal and as a fraction.

a. 120%
b. 4.3%
c. 0.45%

Solution

a. $120\% = 120 \times 0.01 = 1.2$

$$120\% = 120 \times \frac{1}{100} = \frac{120}{100} = 1\frac{1}{5}$$

b. $4.3\% = 4.3 \times 0.01 = 0.043$

$$4.3\% = 4.3 \times \frac{1}{100}$$

$$= 4\frac{3}{10} \times \frac{1}{100} \qquad \bullet\ 4.3 = 4\frac{3}{10}$$

$$= \frac{43}{10} \times \frac{1}{100} = \frac{43}{1000} \qquad \bullet \text{ Multiply the fractions.}$$

c. $0.45\% = 0.45 \times 0.01 = 0.0045$

$$0.45\% = 0.45 \times \frac{1}{100}$$

$$= \frac{9}{20} \times \frac{1}{100} \qquad \bullet\ 0.45 = \frac{45}{100} = \frac{9}{20}$$

$$= \frac{9}{2000} \qquad \bullet \text{ Multiply the fractions.}$$

Example 2   **Write a Percent as a Fraction**

Write $16\frac{2}{3}\%$ as a fraction.

Solution

$$16\frac{2}{3}\% = 16\frac{2}{3} \times \frac{1}{100}$$

$$= \frac{50}{3} \times \frac{1}{100} = \frac{50}{300} = \frac{1}{6}$$

Objective A.8A Practice

1. Write 36% as a decimal and as a fraction.

2. Write 6.2% as a decimal and as a fraction.

3. Write 0.25% as a decimal and as a fraction.

4. Write $12\frac{1}{2}$% as a fraction.

5. Write $4\frac{2}{7}$% as a fraction.

Objective A.8B Write a Decimal or a Fraction as a Percent

A decimal or a fraction can be written as a percent by multiplying by 100%.

Example 3 **Write a Decimal as a Percent**

Write 0.37 as a percent.

Solution

$$0.37 \quad = \quad 0.37 \times 100\% \quad = \quad 37\%$$

Move the decimal point two places to the right. Then write the percent sign.

When changing a fraction to a percent, if the fraction can be written as a terminating decimal, the percent is written in decimal form. If the decimal representation of the fraction is a repeating decimal, the answer is written with a fraction.

Take Note: The decimal form of $\frac{3}{8}$ terminates.

$$
\begin{array}{r}
0.375 \\
8\overline{)3.000} \\
-24 \\
\hline
60 \\
-56 \\
\hline
40 \\
-40 \\
\hline
0
\end{array}
$$

Example 4 **Write a Fraction as a Percent**

Write $\frac{3}{8}$ as a percent.

Solution

$$\frac{3}{8} = \frac{3}{8} \times \frac{100\%}{1} \qquad \bullet \frac{3}{8} = 0.375 \text{ is a terminating decimal.}$$

$$= \frac{300\%}{8}$$

$$= 37.5\% \qquad \bullet \text{ The answer is written in decimal form.}$$

Take Note: The decimal form of $\frac{1}{6}$ repeats.

$$
\begin{array}{r}
0.16\overline{6} \\
6\overline{)1.000} \\
-6 \\
\hline
40 \\
-36 \\
\hline
40 \\
-36 \\
\hline
4
\end{array}
$$

Example 5 **Write a Fraction as a Percent**

Write $\frac{1}{6}$ as a percent.

Solution

$$\frac{1}{6} = \frac{1}{6} \times \frac{100\%}{1} \qquad \bullet \frac{1}{6} = 0.1\overline{6} \text{ is a repeating decimal.}$$

$$= \frac{100\%}{6}$$

$$= 16\frac{2}{3}\% \qquad \bullet \text{ The answer is written with a fraction.}$$

Objective A.8B Practice

1. Write 0.73 as a percent.

2. Write 1.012 as a percent.

3. Write $\dfrac{17}{20}$ as a percent.

4. Write $\dfrac{9}{4}$ as a percent.

5. Write $\dfrac{7}{12}$ as a percent.

Radical Expressions and Real Numbers

Objective A.9A Find the Square Root of a Perfect Square

Recall that the square of a number is equal to the number multiplied by itself.

$3^2 = 3 \cdot 3 = 9$

The square of an integer is called a **perfect square**.

9 is a perfect square because 9 is the square of 3: $3^2 = 9$.

The numbers 1, 4, 9, 16, 25, 36, 49, 64, 81, and 100 are perfect squares.

$1^2 = 1$	$6^2 = 36$
$2^2 = 4$	$7^2 = 49$
$3^2 = 9$	$8^2 = 64$
$4^2 = 16$	$9^2 = 81$
$5^2 = 25$	$10^2 = 100$

Larger perfect squares can be found by squaring 11, squaring 12, squaring 13, and so on.

Note that squaring the negative integers results in the same list of numbers.

$(-1)^2 = 1 \quad (-3)^2 = 9$

$(-2)^2 = 4 \quad (-4)^2 = 16$, and so on.

Perfect squares are used in simplifying square roots. The symbol for square root is $\sqrt{\ }$.

Square Root

A **square root** of a positive number x is a number whose square is x.
  If $a^2 = x$, then $\sqrt{x} = a$.

Example
The expression $\sqrt{9}$, read "the square root of 9," is equal to the number that, when squared, is equal to 9.
  Since $3^2 = 9$, $\sqrt{9} = 3$.

Every positive number has two square roots, one a positive number and one a negative number. The symbol $\sqrt{\ }$ is used to indicate the positive square root of a number. When the negative square root of a number is to be found, a negative sign is placed in front of the square root symbol. For example,

$$\sqrt{9} = 3 \quad \text{and} \quad -\sqrt{9} = -3$$

The square root symbol, $\sqrt{\ }$, is also called a **radical**. The number under the radical is called the **radicand**. In the radical expression $\sqrt{9}$, 9 is the radicand.

Example 1 **Simplify Perfect Square Radical Expressions**

Simplify:

a. $\sqrt{121}$

b. $\sqrt{\dfrac{4}{25}}$

c. $\sqrt{36} - 9\sqrt{4}$

Solution

a. Since $11^2 = 121$, $\sqrt{121} = 11$.

b. Since $\left(\dfrac{2}{5}\right)^2 = \dfrac{4}{25}$, $\sqrt{\dfrac{4}{25}} = \dfrac{2}{5}$.

c. $\sqrt{36} - 9\sqrt{4} = 6 - 9 \cdot 2$

$\phantom{\sqrt{36} - 9\sqrt{4}} = 6 - 18$

$\phantom{\sqrt{36} - 9\sqrt{4}} = 6 + (-18)$

$\phantom{\sqrt{36} - 9\sqrt{4}} = -12$

Example 2 **Evaluate an Expression**

Evaluate $6\sqrt{ab}$ for $a = 2$ and $b = 8$.

Solution

$6\sqrt{ab}$

$6\sqrt{2 \cdot 8} = 6\sqrt{16}$

$\phantom{6\sqrt{2 \cdot 8}} = 6(4)$

$\phantom{6\sqrt{2 \cdot 8}} = 24$

Objective A.9A Practice

1. Simplify: $-\sqrt{1}$

2. Simplify: $\sqrt{81}$

3. Simplify: $\sqrt{144} - \sqrt{25}$

4. Simplify: $\sqrt{144} + 3\sqrt{9}$

5. Evaluate $7\sqrt{x + y}$ for $x = 34$ and $y = 15$.

6. For each **a** through **e**, indicate whether the number is a natural number, integer, rational number, irrational number, or real number.

a. 101

b. -17

c. $\dfrac{17}{3}$

d. $-5.34\overline{5}$

e. $2.34334333433334\ldots$

Objective A.9B Approximate the Square Root of a Real Number

If the radicand is not a perfect square, the square root can only be approximated. For example, the radicand in the radical expression $\sqrt{2}$ is 2, and 2 is not a perfect square. The square root of 2 can be approximated to any desired place value.

To the nearest tenth:	$\sqrt{2} \approx 1.4$	$(1.4)^2 = 1.96$
To the nearest hundredth:	$\sqrt{2} \approx 1.41$	$(1.41)^2 = 1.9881$
To the nearest thousandth:	$\sqrt{2} \approx 1.414$	$(1.414)^2 = 1.999396$
To the nearest ten-thousandth:	$\sqrt{2} \approx 1.4142$	$(1.4142)^2 = 1.99996164$

The square of each decimal approximation gets closer and closer to 2 as the number of place values in the approximation increases. But no matter how many place values are used to approximate $\sqrt{2}$, the digits never terminate or repeat. In general, the square root of any number that is not a perfect square can only be approximated.

Recall that a rational number has a decimal representation that terminates or repeats. A number such as $\sqrt{2}$ has a non-terminating, non-repeating decimal representation. Such a number is called an **irrational number**.

Irrational Numbers

An **irrational number** is a number whose decimal representation never terminates or repeats.

Examples of Irrational Numbers
1. π
2. $\sqrt{3}$
3. $0.23233233323333\ldots$

The rational numbers and the irrational numbers taken together are called the **real numbers**.

Real Numbers

The **real numbers** are all the rational numbers together with all the irrational numbers.

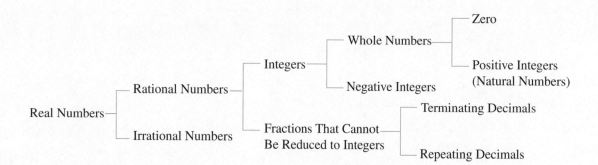

Example 3   **Approximate a Square Root**

Approximate $\sqrt{11}$ to the nearest ten-thousandth.

Solution

11 is not a perfect square.

Use a calculator to approximate $\sqrt{11}$.

$\sqrt{11} \approx 3.3166$

•

Example 4   **Find a Natural Number Inequality for a Square Root**

Between what two whole numbers is the value of $\sqrt{41}$?

Solution

Since the number 41 is between the perfect squares 36 and 49, the value of $\sqrt{41}$ is between $\sqrt{36}$ and $\sqrt{49}$.

Because $\sqrt{36} = 6$ and $\sqrt{49} = 7$, the value of $\sqrt{41}$ is between the whole numbers 6 and 7.

This can be written using inequality symbols as $6 < \sqrt{41} < 7$, which is read "the square root of 41 is greater than 6 and less than 7."

Use a calculator to verify that $\sqrt{41} \approx 6.4$, which is between 6 and 7.

•

Objective A.9B Practice

For Exercises 1 to 4, approximate the number to the nearest ten-thousandths.
1. $\sqrt{10}$
2. $6\sqrt{15}$
3. $10\sqrt{21}$
4. $-12\sqrt{53}$

Between what two smallest whole numbers is the value of the square root.
5. $\sqrt{29}$
6. $\sqrt{130}$

Variable Expressions and Equations

Evaluating Variable Expressions

Objective B.1A Evaluate a Variable Expression

Often we discuss a quantity without knowing its exact value—for example, the price of gold next month, the cost of a new automobile next year, or the tuition cost for next semester. Recall that a letter of the alphabet, called a **variable**, is used to stand for a quantity that is unknown or that can change, or *vary*. An expression that contains one or more variables is called a **variable expression**.

A variable expression is shown below. The expression can be rewritten by writing subtraction as the addition of the opposite.

$$3x^2 - 5y + 2xy - x - 7$$

$$3x^2 + (-5y) + 2xy + (-x) + (-7)$$

Note that the expression has five addends. The **terms** of a variable expression are the addends of the expression. The expression has five terms.

Five terms

$$\underbrace{3x^2 \quad - \quad 5y \quad + \quad 2xy \quad - \quad x}_{\text{Variable terms}} \quad \underbrace{- \quad 7}_{\substack{\text{Constant} \\ \text{term}}}$$

The terms $3x^2$, $-5y$, $2xy$, and $-x$ are **variable terms**.

The term -7 is a **constant term**, or simply a **constant**.

Each variable term is composed of a **numerical coefficient** and a **variable part** (the variable or variables and their exponents).

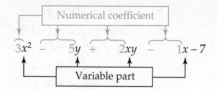

When the numerical coefficient is 1 or -1, the 1 is usually not written ($x = 1x$ and $-x = -1x$).

Variable expressions can be used to model scientific phenomena. In a physics lab, a student may discover that a weight of 1 pound will stretch a spring $\frac{1}{2}$ inch. Two pounds will stretch the spring 1 inch. By experimenting, the student can discover that the distance the spring will stretch is found by multiplying the weight by $\frac{1}{2}$. By letting W represent the weight attached to the spring, the student can represent the distance the spring stretches by the variable expression $\frac{1}{2}W$.

With a weight of W pounds attached, the spring will stretch $\frac{1}{2} \cdot W = \frac{1}{2}W$ inches.

With a weight of 10 pounds attached, the spring will stretch $\frac{1}{2} \cdot 10 = 5$ inches. The number 10 is called the **value of the variable W**.

With a weight of 3 pounds attached, the spring will stretch $\frac{1}{2} \cdot 3 = 1\frac{1}{2}$ inches.

Replacing each variable by its value and then simplifying the resulting numerical expression is called **evaluating a variable expression**.

Example 1 Name the Terms of an Expression

Name the variable terms of the expression $2a^2 - 5a + 7$.

Solution

$2a^2$ and $-5a$

Example 2 Evaluate a Variable Expression

a. Evaluate $x^2 - 3xy$ when $x = 3$ and $y = -4$.

b. Evaluate $\dfrac{a^2 - b^2}{a - b}$ when $a = 3$ and $b = -4$.

c. Evaluate $x^2 - 3(x - y) - z^2$ when $x = 2$, $y = -1$, and $z = 3$.

Solution

a. $x^2 - 3xy$

$3^2 - 3(3)(-4) = 9 - 3(3)(-4)$ • $x = 3, y = -4$

$= 9 - 9(-4)$

$= 9 - (-36)$

$= 9 + 36 = 45$

b. $\dfrac{a^2 - b^2}{a - b}$

$\dfrac{3^2 - (-4)^2}{3 - (-4)} = \dfrac{9 - 16}{3 - (-4)}$ • $a = 3, b = -4$

$= \dfrac{-7}{7} = -1$

c. $x^2 - 3(x - y) - z^2$

$2^2 - 3[2 - (-1)] - 3^2$ • $x = 2, y = -1, z = 3$

$= 2^2 - 3(3) - 3^2$

$= 4 - 3(3) - 9$

$= 4 - 9 - 9$

$= -5 - 9 = -14$

Objective B.1A Practice

1. Evaluate $3a + 2b$ for $a = 2$ and $b = 3$.
2. Evaluate $bc \div (2a)$ for $a = 2$, $b = 3$, and $c = -4$.
3. Evaluate $(b - 2a)^2 + bc$ for $a = 2$, $b = 3$, and $c = -4$.
4. Evaluate $(d - a)^2 - 3c$ for $a = -2$, $c = -1$, and $d = 3$.
5. Evaluate $-\dfrac{3}{4}b + \dfrac{1}{2}(ac + bd)$ for $a = -2$, $b = 4$, $c = -1$, and $d = 3$.

Section B.2

Simplifying Variable Expressions

Objective B.2A Simplify a Variable Expression Using the Properties of Addition

Like terms of a variable expression are terms with the same variable part. (Because $x^2 = x \cdot x$, x^2 and x are not like terms.)

Constant terms are like terms. 4 and 9 are like terms.

To simplify a variable expression, we use the Distributive Property to add the numerical coefficients of like variable terms. The variable part remains unchanged. This is called **combining like terms**.

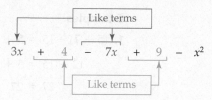

$$3x \;+\; 4 \;-\; 7x \;+\; 9 \;-\; x^2$$

Take Note: Here is an example of the Distributive Property using just numbers.

$$2(5 + 9) = 2(5) + 2(9)$$
$$= 10 + 18 = 28$$

This is the same result we would obtain using the Order of Operations Agreement.

$$2(5 + 9) = 2(14) = 28$$

The usefulness of the Distributive Property will become more apparent as we explore variable expressions.

The Distributive Property

If a, b, and c are real numbers, then $a(b + c) = ab + ac$ or $(b + c)a = ba + ca$.

By the Distributive Property, the term outside the parentheses is multiplied by each term inside the parentheses.

Examples

1. $2(3 + 4) = 2 \cdot 3 + 2 \cdot 4$
$$2(7) = 6 + 8$$
$$14 = 14$$

2. $(4 + 5)2 = 4 \cdot 2 + 5 \cdot 2$
$$(9)2 = 8 + 10$$
$$18 = 18$$

The Distributive Property in the form $(b + c)a = ba + ca$ is used to simplify a variable expression.

To simplify $2x + 3x$, use the Distributive Property to add the numerical coefficients of the like variable terms.

$$2x + 3x = (2 + 3)x$$
$$= 5x$$

Example I   **Simplify an Expression Using the Distributive Property**

Simplify: $5y - 11y$

Solution

$$5y - 11y = (5 - 11)y \quad \text{• Use the Distributive Property.}$$
$$= -6y$$

An expression such as $5 + 7p$ is in simplest form. The terms 5 and $7p$ are not like terms.

●

The following Properties of Addition are used to simplify variable expressions.

The Associative Property of Addition

If a, b, and c are real numbers, then $(a + b) + c = a + (b + c)$.

When three or more terms are added, the terms can be grouped (with parentheses, for example) in any order; the sum is the same.

Examples

1. $(5 + 7) + 15 = 5 + (7 + 15)$

$$12 + 15 = 5 + 22$$
$$27 = 27$$

2. $(3x + 5x) + 9x = 3x + (5x + 9x)$

$$8x + 9x = 3x + 14x$$
$$17x = 17x$$

The Commutative Property of Addition

If a and b are real numbers, then $a + b = b + a$.

When two terms are added, the terms can be added in either order; the sum is the same.

Examples

1. $15 + (-28) = (-28) + 15$

$$-13 = -13$$

2. $2x + (-4x) = -4x + 2x$

$$-2x = -2x$$

The Addition Property of Zero

If a is a real number, then $a + 0 = a$ and $0 + a = a$.
   The sum of a term and zero is the term.

Examples

1. $-9 + 0 = -9$ and $0 + (-9) = -9$

2. $0 + 5x = 5x$ and $5x + 0 = 5x$

The Inverse Property of Addition

If a is a real number, then $a + (-a) = 0$ and $(-a) + a = 0$.
   The sum of a term and its additive inverse (or opposite) is zero.

Examples

1. $8 + (-8) = 0$ and $-8 + 8 = 0$

2. $-7x + 7x = 0$ and $7x + (-7x) = 0$

Example 2 **Use the Properties of Addition to Simplify a Variable Expression**

a. Simplify: $3x + 4y - 10x + 7y$
b. Simplify: $x^2 - 7 + 4x^2 - 16$
c. Simplify: $8x + 4y - 8x + y$

Solution

a. $3x + 4y - 10x + 7y = (3x - 10x) + (4y + 7y)$ • Use the Commutative and Associative Properties of Addition to rearrange and group like terms.

$\qquad = -7x + 11y$ • Combine like terms.

b. $x^2 - 7 + 4x^2 - 16 = (x^2 + 4x^2) + (-7 - 16)$ • Use the Commutative and Associative Properties of Addition to rearrange and group like terms.

$\qquad = 5x^2 - 23$ • Combine like terms.

c. $8x + 4y - 8x + y = (8x - 8x) + (4y + y)$ • Use the Commutative and Associative Properties of Addition to rearrange and group like terms.

$\qquad = 0 + 5y = 5y$ • Combine like terms.

Objective B.2A Practice

1. Simplify: $6x + 8x$
2. Simplify: $7 - 3b$
3. Simplify: $-3ab + 3ab$
4. Simplify: $5a - 3a + 5a$
5. Simplify: $\frac{3}{4}x - \frac{1}{3}x - \frac{7}{8}x$
6. Simplify: $3x + (-8y) - 10x + 4x$

Objective B.2B Simplify a Variable Expression Using the Properties of Multiplication

In simplifying variable expressions, the following Properties of Multiplication are used.

The Associative Property of Multiplication

If a, b, and c are real numbers, then $(ab)c = a(bc)$.

When three or more factors are multiplied, the factors can be grouped in any order; the product is the same.

Examples

1. $3(5 \cdot 6) = (3 \cdot 5)6$

   $3(30) = (15)6$

   $90 = 90$

2. $2(3x) = (2 \cdot 3)x$

   $= 6x$

Take Note: The Associative Property of Multiplication allows us to multiply a coefficient by a number. Without this property, the expression $2(3x)$ could not be simplified.

The Commutative Property of Multiplication

If a and b are real numbers, then $ab = ba$.

Two factors can be multiplied in either order; the product is the same.

Examples

1. $5(-7) = -7(5)$

   $-35 = -35$

2. $(5x) \cdot 3 = 3 \cdot (5x)$ • Commutative Property of Multiplication

   $= (3 \cdot 5)x$ • Associative Property of Multiplication

   $= 15x$

Take Note: The Commutative Property of Multiplication allows us to rearrange factors. This property, along with the Associative Property of Multiplication, enables us to simplify some variable expressions.

The Multiplication Property of One

If a is a real number, then $a \cdot 1 = a$ and $1 \cdot a = a$.

The product of a term and 1 is the term.

Examples

1. $9 \cdot 1 = 9$

2. $(8x) \cdot 1 = 8x$

The Inverse Property of Multiplication

If a is a real number and a is not equal to zero, then $a \cdot \frac{1}{a} = 1$ and $\frac{1}{a} \cdot a = 1$.

$\frac{1}{a}$ is called the **reciprocal** of a. $\frac{1}{a}$ is also called the **multiplicative inverse** of a.

The product of a number and its reciprocal is 1.

Examples

1. $7 \cdot \frac{1}{7} = 1$ and $\frac{1}{7} \cdot 7 = 1$

2. $x \cdot \frac{1}{x} = 1$ and $\frac{1}{x} \cdot x = 1, x \neq 0$

Take Note: We must state that $x \neq 0$ because division by zero is undefined.

The multiplication properties are used to simplify variable expressions.

Example 3 **Use the Properties of Multiplication to Simplify a Variable Expression**

a. Simplify: $-2(3x^2)$

b. Simplify: $-5(-10x)$

c. Simplify: $-\frac{3}{4}\left(\frac{2}{3}x\right)$

d. Simplify: $2(-x)$

e. Simplify: $\frac{3}{2}\left(\frac{2x}{3}\right)$

Solution

a. $-2(3x^2) = (-2 \cdot 3)x^2$ • Use the Associative Property of Multiplication to group factors.
$= -6x^2$

b. $-5(-10x) = [(-5)(-10)]x$ • Use the Associative Property of Multiplication to group factors.
$= 50x$

c. $\left(-\frac{3}{4}\right)\left(\frac{2}{3}x\right) = \left(-\frac{3}{4} \cdot \frac{2}{3}\right)x$ • Use the Associative Property of Multiplication to group factors.
$= -\frac{1}{2}x$

d. $2(-x) = 2(-1 \cdot x)$ • Use the Associative Property of Multiplication to group factors.
$= [2(-1)]x$
$= -2x$

e. $\frac{3}{2}\left(\frac{2x}{3}\right) = \frac{3}{2}\left(\frac{2}{3}x\right)$ • Note that $\frac{2x}{3} = \frac{2}{3}x$
$= \left(\frac{3}{2} \cdot \frac{2}{3}\right)x$ • Use the Associative Property of Multiplication to group factors.
$= 1 \cdot x$
$= x$

Objective B.2B Practice

1. Simplify: $4(3x)$

2. Simplify: $(4x)2$

3. Simplify: $-\frac{1}{2}(-2x)$

4. Simplify: $-0.2(10x)$

5. Simplify: $-0.5(-16y)$

6. Simplify: $(-8a)\left(-\frac{3}{4}\right)$

Objective B.2C Simplify a Variable Expression Using the Distributive Property

Recall that the Distributive Property states that if a, b, and c are real numbers, then

$$a(b + c) = ab + ac$$

The Distributive Property is used to remove parentheses from a variable expression. For instance, to simplify $3(2x + 7)$, use the Distributive Property.

$3(2x + 7) = 3(2x) + 3(7)$ • Use the Distributive Property. Multiply each term inside
$\qquad\qquad\;\; = 6x + 21$ the parentheses by 3.

Example 4   Use the Distributive Property to Simplify a Variable Expression

a. Simplify: $-3(-5a + 7b)$
b. Simplify: $3(x^2 - x - 5)$
c. Simplify: $-2(x^2 + 5x - 4)$
d. Simplify: $-(2x - 4)$
e. Simplify: $3(4x - 2y - z)$

Solution

a. Use the Distributive Property.
$$-3(-5a + 7b) = 15a - 21b$$

b. Use the Distributive Property.
$$3(x^2 - x - 5) = 3x^2 - 3x - 15$$

c. Use the Distributive Property.
$$-2(x^2 + 5x - 4) = -2x^2 - 10x + 8$$

d. $-(2x - 4) = -1(2x - 4)$ • Use the Distributive Property.
$\qquad\qquad\quad = -1(2x) - (-1)(4)$
$\qquad\qquad\quad = -2x + 4$

e. $3(4x - 2y - z) = 3(4x) - 3(2y) - 3(z)$ • Use the Distributive Property.
$\qquad\qquad\qquad\quad = 12x - 6y - 3z$

Objective B.2C Practice

1. Simplify: $(5 - 3b)7$
2. Simplify: $3(5x^2 + 2x)$
3. Simplify: $(-3x - 6)5$
4. Simplify: $4(x^2 - 3x + 5)$
5. Simplify: $\frac{3}{4}(2x - 6y + 8)$
6. Simplify: $-(8b^2 - 6b + 9)$

Objective B.2D   Simplify General Variable Expressions

When simplifying variable expressions, use the Distributive Property to remove parentheses and brackets used as grouping symbols.

Example 5   **Simplify a General Variable Expression**

a. Simplify: $4(x - y) - 2(-3x + 6y)$
b. Simplify: $2x - 3(2x - 7y)$
c. Simplify: $7(x - 2y) - (-x - 2y)$
d. Simplify: $2x - 3[2x - 3(x + 7)]$

Solution

a. $4(x - y) - 2(-3x + 6y) = 4x - 4y + 6x - 12y$ • Use the Distributive Property.
$$= 10x - 16y$$ • Combine like terms.

b. $2x - 3(2x - 7y) = 2x - 6x + 21y$ • Use the Distributive Property.
$$= -4x + 21y$$ • Combine like terms.

c. $7(x - 2y) - (-x - 2y) = 7x - 14y + x + 2y$ • Use the Distributive Property.
$$= 8x - 12y$$ • Combine like terms.

d. $2x - 3[2x - 3(x + 7)] = 2x - 3[2x - 3x - 21]$ • Use the Distributive Property.
$$= 2x - 3[-x - 21]$$ • Combine like terms.
$$= 2x + 3x + 63$$ • Use the Distributive Property.
$$= 5x + 63$$ • Combine like terms.

Objective B.2D Practice

1. Simplify: $4x - 2(3x + 8)$
2. Simplify: $5n - (7 - 2n)$
3. Simplify: $12(y - 2) + 3(7 - 3y)$
4. Simplify: $-2[3x + 2(4 - x)]$
5. Simplify: $-3[2x - (x + 7)]$
6. Simplify: $0.05x + 0.02(4 - x)$

Section B.3

Introduction to Equations

Objective B.3A Determine Whether a Given Number Is a Solution of an Equation

An **equation** expresses the equality of two mathematical expressions. The expressions can be either numerical or variable expressions.

$$\left.\begin{array}{l} 9 + 3 = 12 \\ 3x - 2 = 10 \\ y^2 + 4 = 2y - 1 \\ z = 2 \end{array}\right\} \text{Equations}$$

$$x + 8 = 13$$

The equation at the right is true if the variable is replaced by 5. $5 + 8 = 13$ A true equation

The equation is false if the variable is replaced by 7. $7 + 8 = 13$ A false equation

A **solution of an equation** is a number that, when substituted for the variable, results in a true equation. 5 is a solution of the equation $x + 8 = 13$. 7 is not a solution of the equation $x + 8 = 13$.

> **Example 1**   Determine Whether a Number Is a Solution of an Equation

Is -2 a solution of $2x + 5 = x^2 - 3$?

Take Note: The Order of Operations Agreement applies when evaluating $2(-2) + 5$ and $(-2)^2 - 3$.

Solution

$$
\begin{array}{c|c}
2x + 5 & = x^2 - 3 \\
\hline
2(-2) + 5 & (-2)^2 - 3 \\
-4 + 5 & 4 - 3 \\
1 & = 1
\end{array}
$$

• Replace x by -2.
• Evaluate the numerical expressions.
• If the results are equal, -2 is a solution of the equation. If the results are not equal, -2 is not a solution of the equation.

Yes, -2 is a solution of the equation.

Objective B.3A Practice

1. Is 3 a solution of $y + 4 = 7$?
2. Is 2 a solution of $7 - 3n = 2$?
3. Is 4 a solution of $3y - 4 = 2y$?
4. Is 3 a solution of $z^2 + 1 = 4 + 3z$?
5. Is $\dfrac{3}{4}$ a solution of $8x - 1 = 12x + 3$?

Objective B.3B Solve an Equation of the Form $x + a = b$

Tips for Success: To learn mathematics, you must be an active participant. Listening and watching your professor do mathematics is not enough. Take notes in class, mentally think through every question your instructor asks, and try to answer it even if you are not called on to do so. Ask questions when you have them.

To **solve an equation** means to find a solution of the equation. The simplest equation to solve is an equation of the form *variable = constant*, because the constant is the solution.
The solution of the equation $x = 5$ is 5 because $5 = 5$ is a true equation.
The solution of the equation below is 7 because $7 + 2 = 9$ is a true equation.

$$x + 2 = 9 \qquad 7 + 2 = 9$$

Note that if 4 is added to each side of the equation $x + 2 = 9$, the solution is still 7.

$$
\begin{aligned}
x + 2 &= 9 \\
x + 2 + 4 &= 9 + 4 \\
x + 6 &= 13 \qquad 7 + 6 = 13
\end{aligned}
$$

If -5 is added to each side of the equation $x + 2 = 9$, the solution is still 7.

$$
\begin{aligned}
x + 2 &= 9 \\
x + 2 + (-5) &= 9 + (-5) \\
x - 3 &= 4 \qquad 7 - 3 = 4
\end{aligned}
$$

Equations that have the same solution are called **equivalent equations**. The equations $x + 2 = 9$, $x + 6 = 13$, and $x - 3 = 4$ are equivalent equations; each equation has 7 as its solution. These examples suggest that adding the same number to each side of an equation produces an equivalent equation. This is called the *Addition Property of Equations*.

Addition Property of Equations

The same number can be added to each side of an equation without changing its solution. In symbols, the equation $a = b$ has the same solution as the equation $a + c = b + c$.

Example

The equation $x - 3 = 7$ has the same solution as the equation $x - 3 + 3 = 7 + 3$.

In solving an equation, the goal is to rewrite the given equation in the form *variable* = *constant*. The Addition Property of Equations is used to remove a *term* from one side of the equation by adding the opposite of that term to each side of the equation.

Example 2   **Solve an Equation**

Solve: $x - 4 = 2$

Solution

$$x - 4 = 2$$ • The goal is to rewrite the equation in the form *variable* = *constant*.

$$x - 4 + 4 = 2 + 4$$ • Add 4 to each side of the equation.

$$x + 0 = 6$$ • Simplify.

$$x = 6$$ • The equation is in the form *variable* = *constant*.

Check: $\dfrac{x - 4 = 2}{6 - 4 \mid 2}$

$$2 = 2$$ • A true equation

The solution is 6.

Take Note: An equation has some properties that are similar to those of a balance scale. For instance, if a balance scale is in balance and equal weights are added to each side of the scale, then the balance scale remains in balance. If an equation is true, then adding the same number to each side of the equation produces another true equation (see Figure B.1).

Figure B.1

Because subtraction is defined in terms of addition, the Addition Property of Equations also makes it possible to subtract the same number from each side of an equation without changing the solution of the equation.

Example 3   **Solve an Equation with Fractions**

Solve: $y + \dfrac{3}{4} = \dfrac{1}{2}$

Solution

$$y + \frac{3}{4} = \frac{1}{2}$$ • The goal is to rewrite the equation in the form *variable* = *constant*.

$$y + \frac{3}{4} - \frac{3}{4} = \frac{1}{2} - \frac{3}{4}$$ • Subtract $\dfrac{3}{4}$ from each side of the equation.

$$y + 0 = \frac{2}{4} - \frac{3}{4} \qquad \text{• Simplify.}$$

$$y = -\frac{1}{4} \qquad \text{• The equation is in the form } variable = constant.$$

The solution is $-\frac{1}{4}$. You should check this solution.

●

Objective B.3B Practice

1. Solve and check: $y - 5 = -5$

2. Solve and check: $-8 = n + 1$

3. Solve and check: $-9 = 5 + x$

4. Solve and check: $x - \frac{2}{5} = \frac{3}{5}$

5. Solve and check: $w + 2.932 = 4.801$

Objective B.3C Solve an Equation of the Form $ax = b$

The solution of the equation below is 3 because $2 \cdot 3 = 6$ is a true equation.

$$2x = 6 \qquad 2 \cdot 3 = 6$$

Note that if each side of $2x = 6$ is multiplied by 5, the solution is still 3.

$$2x = 6$$
$$5(2x) = 5 \cdot 6$$
$$10x = 30 \qquad 10 \cdot 3 = 30$$

If each side of $2x = 6$ is multiplied by -4, the solution is still 3.

$$2x = 6$$
$$(-4)(2x) = (-4)6$$
$$-8x = -24 \qquad -8 \cdot 3 = -24$$

The equations $2x = 6$, $10x = 30$, and $-8x = -24$ are equivalent equations; each equation has 3 as its solution. These examples suggest that multiplying each side of an equation by the same nonzero number produces an equivalent equation.

Multiplication Property of Equations

Each side of an equation can be multiplied by the same nonzero number without changing the solution of the equation. In symbols, if $c \neq 0$, then the equation $a = b$ has the same solutions as the equation $ac = bc$.

Example
The equation $3x = 21$ has the same solution as the equation $\frac{1}{3} \cdot 3x = \frac{1}{3} \cdot 21$.

The Multiplication Property of Equations is used to remove a coefficient by multiplying each side of the equation by the reciprocal of the coefficient.

Because division is defined in terms of multiplication, each side of an equation can be divided by the same nonzero number without changing the solution of the equation.

When using the Multiplication Property of Equations, multiply each side of the equation by the reciprocal of the coefficient when the coefficient is a fraction. Divide each side of the equation by the coefficient when the coefficient is an integer or a decimal.

Example 4 **Use the Multiplication Property of Equations to Solve an Equation**

a. Solve: $\frac{3}{4}z = 9$

b. Solve: $6x = 14$

c. Solve: $-9 = \frac{3x}{4}$

d. Solve: $5x - 9x = 12$

Solution

a.

$$\frac{3}{4}z = 9$$

• The goal is to rewrite the equation in the form *variable = constant*.

$$\frac{4}{3} \cdot \frac{3}{4}z = \frac{4}{3} \cdot 9$$

• Multiply each side of the equation by $\frac{4}{3}$.

$$1 \cdot z = 12$$

• Simplify.

$$z = 12$$

• The equation is in the form *variable = constant*.

The solution is 12.

b.

$$6x = 14$$

• The goal is to rewrite the equation in the form *variable = constant*.

$$\frac{6x}{6} = \frac{14}{6}$$

• Divide each side of the equation by 6.

$$x = \frac{7}{3}$$

• Simplify. The equation is in the form *variable = constant*.

The solution is $\frac{7}{3}$.

c.

$$-9 = \frac{3x}{4}$$

• $\frac{3x}{4} = \frac{3}{4}x$

$$\frac{4}{3}(-9) = \frac{4}{3} \cdot \frac{3}{4}x$$

• Multiply each side by $\frac{4}{3}$.

$$-12 = x$$

The solution is -12.

d. $5x - 9x = 12$

$$-4x = 12$$

• Combine like terms.

$$\frac{-4x}{-4} = \frac{12}{-4}$$

• Divide each side by -4.

$$x = -3$$

The solution is -3.

Objective B.3C Practice

1. Solve and check: $2a = 0$

2. Solve and check: $\frac{x}{4} = 3$

3. Solve and check: $\frac{2}{5}x = 6$

4. Solve and check: $\frac{z}{2.95} = -7.88$

5. Solve and check: $7d - 4d = 9$

Objective B.3D   Solve the Basic Percent Equation

What percent of the region shown in Figure B.2 is shaded?

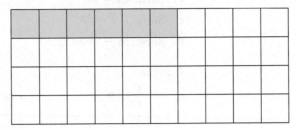

Figure B.2

To answer this question, first determine what fraction of the region is shaded.
There are a total of 40 squares in the region.
6 of the squares are shaded.
$\frac{6}{40}$ of the region is shaded.

Now write the fraction $\frac{6}{40}$ as a percent.

$$\frac{6}{40} = \frac{6}{40}(100\%) = \frac{600}{40}\% = 15\%$$

15% of the region is shaded.

Now consider the question, "How many squares should be shaded if we want to shade 7.5% of the region shown in Figure B.3?"

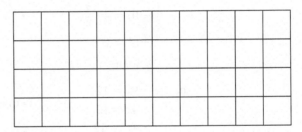

Figure B.3

Determining the number of squares to be shaded requires answering the question "7.5% of 40 is what number?"
This question can be translated into an equation and solved for the unknown number.

of	translates to	·	(times)
is	translates to	=	(equals)
what	translates to	*n*	(the unknown number)

Here is the translation of "7.5% of 40 is what number?" Note that the percent is written as a decimal.

7.5%   of   40   is   what number?

0.075   ·   40   =   *n*

$$0.075 \cdot 40 = n \qquad \text{• Solve this equation for } n.$$

$$3 = n$$

If we shade 3 squares, 7.5% of the region will be shaded. See Figure B.4.

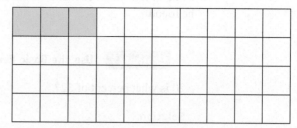

Figure B.4 7.5% of the region is shaded.

We can check this result.

$$\frac{3}{40} = \frac{3}{40}(100\%) = \frac{300}{40}\% = 7.5\%$$

3 of the 40 squares are shaded, and $\frac{3}{40} = 7.5\%$. The solution checks.

The Basic Percent Equation

Percent · base = amount

Example 5 **Use the Basic Percent Equation**

What is 25% of 800?

Solution

Percent · base = amount   • Use the basic percent equation.

$$0.25 \cdot 800 = n$$ • Percent = 25% = 0.25, base = 800, amount = n

$$200 = n$$

25% of 800 is 200.

Look for the number or phrase that follows the word *of* when determining the base in the basic percent equation.

In most cases, the percent is written as a decimal before the basic percent equation is solved. However, some percents are more easily written as a fraction than as a decimal. For example,

$$33\frac{1}{3}\% = \frac{1}{3} \qquad 66\frac{2}{3}\% = \frac{2}{3} \qquad 16\frac{2}{3}\% = \frac{1}{6} \qquad 83\frac{1}{3}\% = \frac{5}{6}$$

Example 6 **Use the Basic Percent Equation**

What is $33\frac{1}{3}\%$ of 90?

Solution

Percent · base = amount   • Use the basic percent equation.

$$\frac{1}{3} \cdot 90 = n$$ • Percent = $33\frac{1}{3}\% = \frac{1}{3}$, base = 90, amount = n

$$30 = n$$

$33\frac{1}{3}\%$ of 90 is 30.

The three elements of the basic percent equation are the percent, the base, and the amount. If any two elements of the basic percent equation are given, the third element can be found.

Example 7   Use the Basic Percent Equation

20 is what percent of 32?

Solution

$$\text{Percent} \cdot \text{base} = \text{amount}$$ • Use the basic percent equation.

$$n \cdot 32 = 20$$ • Percent $= n$, base $= 32$, amount $= 20$

$$\frac{32n}{32} = \frac{20}{32}$$ • Solve for n by dividing each side of the equation by 32.

$$n = 0.625$$

$$n = 62.5\%$$ • Write the decimal as a percent.

20 is 62.5% of 32.

Example 8   Use the Basic Percent Equation

60% of what number is 300?

Solution

$$\text{Percent} \cdot \text{base} = \text{amount}$$ • Use the basic percent equation.

$$0.60 \cdot n = 300$$ • Percent $= 60\% = 0.60$, base $= n$, amount $= 300$

$$\frac{0.60n}{0.60} = \frac{300}{0.60}$$ • Solve for n by dividing each side of the equation by 0.60.

$$n = 500$$

60% of 500 is 300.

Example 9   Use the Basic Percent Equation

a. Find 9.4% of 240.
b. What percent of 30 is 12?
c. 60 is 2.5% of what?

Solution

a. Use the basic percent equation.

Percent $= 9.4\% = 0.094$, base $= 240$, amount $= n$

$$\text{Percent} \cdot \text{base} = \text{amount}$$

$$0.094 \cdot 240 = n$$

$$22.56 = n$$

9.4% of 240 is 22.56.

b. Use the basic percent equation.

Percent $= n$, base $= 30$, amount $= 12$

$$\text{Percent} \cdot \text{base} = \text{amount}$$
$$n \cdot 30 = 12$$
$$\frac{30n}{30} = \frac{12}{30}$$
$$n = 0.4 = 40\%$$

12 is 40% of 30.

c. Use the basic percent equation.

Percent = 2.5% = 0.025, base = n, amount = 60

$$\text{Percent} \cdot \text{base} = \text{amount}$$
$$0.025 \cdot n = 60$$
$$\frac{0.025n}{0.025} = \frac{60}{0.025}$$
$$n = 2400$$

60 is 2.5% of 2400.

Objective B.3D Practice

1. Using the basic percent equation, find 0.05% of 150.
2. Using the basic percent equation, what percent of 16 is 6?
3. Using the basic percent equation, find 10.7% of 485.
4. Using the basic percent equation, 54 is what percent of 2000?
5. Using the basic percent equation, 16.4 is what percent of 4.1?

Objective B.3E Solve Percent Increase and Decrease Problems

Percent increase is used to show how much a quantity has increased over its original value. The statements "health care costs increased 4.8% last year" and "employees received an 8% pay increase" are illustrations of the use of percent increase.

Example 10   **Solve a Percent Increase Application**

A sales associate earned $11.60 per hour before an 8% increase in pay. What is the new hourly wage? Round to the nearest cent.

Strategy

To find the new hourly wage:
- Use the basic percent equation to find the increase in pay.

  Percent = 8% = 0.08, base = 11.60, amount = n
- Add the amount of increase to the original wage.

Take Note: In solving percent increase and percent decrease problems, remember that the base in the basic percent equation is always the quantity *before* the increase or decrease. That is, it is the *original value*. For this example, the base is 11.60—the quantity before the increase.

Solution

$$\text{Percent} \cdot \text{base} = \text{amount}$$
$$0.08 \cdot 11.60 = n$$
$$0.93 \approx n \qquad \text{• Increase in pay}$$
$$\$11.60 + \$0.93 = \$12.53 \qquad \text{• Add the increase to the original wage.}$$

The new hourly wage is $12.53.

Percent decrease is used to show how much a quantity has decreased from its original value. The statements "the number of family farms decreased 2% last year" and "the president's approval rating has decreased 9% over the last month" are illustrations of the use of percent decrease.

| Example 11 | **Solve a Percent Decrease Application**

During a recent year, violent crime in a small city decreased from 27 crimes per 1000 people to 24 crimes per 1000 people. Find the percent decrease in violent crime. Round to the nearest tenth of a percent.

Strategy
To find the percent decrease in crime:
- Find the decrease in the number of crimes.
- Use the basic percent equation to find the percent decrease in crime.

Solution
$$27 - 24 = 3$$ • Decrease in the number of crimes.

Percent · base = amount

$$n \cdot 27 = 3$$ • Amount is the decrease in the number of crimes.

$$n = \frac{3}{27} \approx 0.111$$

Violent crime decreased by approximately 11.1% during the year.

Take Note: Remember that the base in the basic percent equation is the *original value*—that is, the quantity *before* the decrease. For this example, the base is 27.

Objective B.3E Practice

1. In 1924, the number of events in the Winter Olympics was 14. The 2014 Winter Olympics included 98 medal events. Find the percent increase in the number of events in the Winter Olympics from 1924 to 2014.
2. Over a 20-year period, the number of millionaire households in the United States jumped from about 3.5 million to 10.8 million. Find the percent increase in the number of millionaire households from 1975 to 2010. Round to the nearest percent.
3. The average cost for tuition, fees, and room and board at a public college has increased from $7170 per year to $9270 over an 18-year period. Find the percent increase in the average cost of attending a public college.
4. Due to an increased number of service lines at a grocery store, the average amount of time a customer waits in line has decreased from 3.8 minutes to 2.5 minutes. Find the percent decrease.
5. It is estimated that the value of a new car is reduced by 30% after one year of ownership. Find the value after one year of a car that cost $21,900 new.

Objective B.4A Solve an Equation of the Form
$$ax + b = c$$

In solving an equation of the form $ax + b = c$, the goal is to rewrite the equation in the form *variable = constant*. This requires the application of both the Addition and Multiplication Properties of Equations.

Example 1 Solve an Equation

Solve: $\frac{3}{4}x - 2 = -11$

Solution

The goal is to write the equation in the form *variable = constant*.

$$\frac{3}{4}x - 2 = -11$$

$$\frac{3}{4}x - 2 + 2 = -11 + 2 \qquad \text{• Add 2 to each side of the equation.}$$

$$\frac{3}{4}x = -9 \qquad \text{• Simplify.}$$

$$\frac{4}{3} \cdot \frac{3}{4}x = \frac{4}{3}(-9) \qquad \text{• Multiply each side of the equation by } \frac{4}{3}.$$

$$x = -12 \qquad \text{• The equation is in the form } \textit{variable = constant.}$$

The solution is -12.

Take Note:

Check:

$$\frac{3}{4}x - 2 = -11$$

$\frac{3}{4}(-12) - 2$	-11
$-9 - 2$	-11
$-11 = -11$	

A true equation.

It may be easier to solve an equation containing two or more fractions by multiplying each side of the equation by the least common multiple (LCM) of the denominators. The result is an equation that does not contain any fractions. Multiplying each side of an equation that contains fractions by the LCM of the denominators is called **clearing denominators**. It is an alternative method, as we show in the next example, of solving an equation that contains fractions.

Example 2 Solve an Equation By Clearing Denominators

Solve: $\frac{2}{3}x + \frac{1}{2} = \frac{3}{4}$

Solution

$$\frac{2}{3}x + \frac{1}{2} = \frac{3}{4}$$

$$12\left(\frac{2}{3}x + \frac{1}{2}\right) = 12\left(\frac{3}{4}\right) \qquad \text{• Multiply each side of the equation by 12, the LCM of 3, 2, and 4.}$$

$$12\left(\frac{2}{3}x\right) + 12\left(\frac{1}{2}\right) = 12\left(\frac{3}{4}\right) \qquad \text{• Use the Distributive Property.}$$

$$8x + 6 = 9 \qquad \text{• Simplify.}$$

$$8x + 6 - 6 = 9 - 6 \qquad \text{• Subtract 6 from each side of the equation.}$$

$$8x = 3$$

Take Note: Observe that after we multiply both sides of the equation by the LCM of the denominators and then simplify, the equation no longer contains fractions.

Clearing denominators is a method of solving equations. The process applies only to equations, never to expressions.

$$\frac{8x}{8} = \frac{3}{8}$$

• Divide each side of the equation by 8.

$$x = \frac{3}{8}$$

The solution is $\frac{3}{8}$.

Example 3 Solve an Equation

Solve: $3x - 7 = -5$

Solution

$$3x - 7 = -5$$

$$3x - 7 + 7 = -5 + 7 \qquad \text{• Add 7 to each side.}$$

$$3x = 2$$

$$\frac{3x}{3} = \frac{2}{3} \qquad \text{• Divide each side by 3.}$$

$$x = \frac{2}{3}$$

The solution is $\frac{2}{3}$.

Example 4 Solve an Equation

Solve: $5 = 9 - 2x$

Solution

$$5 = 9 - 2x$$

$$5 - 9 = 9 - 9 - 2x \qquad \text{• Subtract 9 from each side.}$$

$$-4 = -2x$$

$$\frac{-4}{-2} = \frac{-2x}{-2} \qquad \text{• Divide each side by } -2.$$

$$2 = x$$

The solution is 2.

Objective B.4A Practice

1. Solve and check: $2x - 5 = -11$
2. Solve and check: $-4y + 15 = 15$
3. Solve and check: $7 - 9a = 4$
4. Solve and check: $\frac{2}{3}x - \frac{5}{6} = -\frac{1}{3}$
5. Solve and check: $5y + 9 + 2y = 23$

Objective B.4B Solve an Equation of the Form
$ax + b = cx + d$

Tips for Success: Have you considered joining a study group? Getting together regularly with other students in the class to go over material and quiz each other can be very beneficial.

In solving an equation of the form $ax + b = cx + d$, the goal is to rewrite the equation in the form *variable = constant*. Begin by rewriting the equation so that there is only one variable term in the equation. Then rewrite the equation so that there is only one constant term.

Example 5   **Solve an Equation with the Variable on Both Sides**

Solve: $2x + 3 = 5x - 9$

Solution

$$2x + 3 = 5x - 9$$
$$2x - 5x + 3 = 5x - 5x - 9$$ • Subtract $5x$ from each side of the equation.
$$-3x + 3 = -9$$ • Simplify. There is only one variable term.
$$-3x + 3 - 3 = -9 - 3$$ • Subtract 3 from each side of the equation.
$$-3x = -12$$ • Simplify. There is only one constant term.
$$\frac{-3x}{-3} = \frac{-12}{-3}$$ • Divide each side of the equation by -3.
$$x = 4$$ • The equation is in the form *variable = constant*.

The solution is 4. You should verify this by checking this solution.

Example 6   **Solve an Equation with the Variable on Both Sides**

Solve: $3x + 4 - 5x = 2 - 4x$

Solution

$$3x + 4 - 5x = 2 - 4x$$
$$-2x + 4 = 2 - 4x$$ • Combine like terms.
$$-2x + 4x + 4 = 2 - 4x + 4x$$ • Add $4x$ to each side.
$$2x + 4 = 2$$
$$2x + 4 - 4 = 2 - 4$$ • Subtract 4 from each side.
$$2x = -2$$
$$\frac{2x}{2} = \frac{-2}{2}$$ • Divide each side by 2.
$$x = -1$$

The solution is -1.

Objective B.4B Practice

1. Solve and check: $15x - 2 = 4x - 13$

2. Solve and check: $0.2b + 3 = 0.5b + 12$

3. Solve and check: $5 + 7x = 11 + 9x$

4. Solve and check: $7x - 8 = x - 3$

5. If $7x + 3 = 5x - 7$, evaluate $3x - 2$.

Objective B.4C Solve an Equation Containing Parentheses

When an equation contains parentheses, one of the steps in solving the equation is to use the Distributive Property. The Distributive Property is used to remove parentheses from a variable expression.

> **Example 7** Solve an Equation Containing Parentheses

Solve: $4 + 5(2x - 3) = 3(4x - 1)$

Solution

$4 + 5(2x - 3) = 3(4x - 1)$	
$4 + 10x - 15 = 12x - 3$	• Use the Distributive Property. Then simplify.
$10x - 11 = 12x - 3$	
$10x - 12x - 11 = 12x - 12x - 3$	• Subtract $12x$ from each side of the equation.
$-2x - 11 = -3$	• Simplify.
$-2x - 11 + 11 = -3 + 11$	• Add 11 to each side of the equation.
$-2x = 8$	• Simplify.
$\dfrac{-2x}{-2} = \dfrac{8}{-2}$	• Divide each side of the equation by -2.
$x = -4$	• The equation is in the form *variable = constant*.

The solution is -4. You should verify this by checking this solution.

> **Example 8** Solve an Equation Containing Parentheses

Solve: $3[2 - 4(2x - 1)] = 4x - 10$

Solution

$3[2 - 4(2x - 1)] = 4x - 10$	
$3[2 - 8x + 4] = 4x - 10$	• Distributive Property
$3[6 - 8x] = 4x - 10$	• Simplify.
$18 - 24x = 4x - 10$	• Distributive Property
$18 - 24x - 4x = 4x - 4x - 10$	• Subtract $4x$.
$18 - 28x = -10$	• Simplify.
$18 - 18 - 28x = -10 - 18$	• Subtract 18.
$-28x = -28$	
$\dfrac{-28x}{-28} = \dfrac{-28}{-28}$	• Divide by -28.
$x = 1$	

The solution is 1.

Objective B.4C Practice

1. Solve and check: $9n - 3(2n - 1) = 15$
2. Solve and check: $5(3 - 2y) + 4y = 3$
3. Solve and check: $0.05(4 - x) + 0.1x = 0.32$
4. Solve and check: $7 - (5 - 8x) = 4x + 3$
5. Solve and check: $5 + 3[1 + 2(2x - 3)] = 6(x + 5)$

Objective B.4D Solve a Literal Equation for One of the Variables

A **literal equation** is an equation that contains more than one variable. Examples of literal equations are shown below.

$$2x + 3y = 6$$

$$4w - 2x + z = 0$$

Formulas are used to express relationships among physical quantities. A **formula** is a literal equation that states a rule about measurements. Examples of formulas are shown below.

$$\frac{1}{R_1} + \frac{1}{R_2} = \frac{1}{R} \qquad \text{(Physics)}$$

$$s = a + (n - 1)d \qquad \text{(Mathematics)}$$

$$A = P + Prt \qquad \text{(Business)}$$

The Addition and Multiplication Properties can be used to solve a literal equation for one of the variables. The goal is to rewrite the equation so that the variable being solved for is alone on one side of the equation, and all the other numbers and variables are on the other side.

Example 9 Solve a Literal Equation

Solve $3x - 4y = 12$ for y.

Solution

$$3x - 4y = 12$$

$$3x - 3x - 4y = -3x + 12 \qquad \text{• Subtract } 3x.$$

$$-4y = -3x + 12$$

$$\frac{-4y}{-4} = \frac{-3x + 12}{-4} \qquad \text{• Divide by } -4.$$

$$y = \frac{-3x}{-4} + \frac{12}{-4} \qquad \text{• Recall that a fraction bar acts as a grouping symbol.}$$
$$\text{Divide each term in the numerator by } -4.$$

$$y = \frac{3}{4}x - 3$$

Example 10 Solve a Literal Equation by Removing Parentheses

Solve $A = P(1 + i)$ for i.

Solution

The goal is to rewrite the equation so that i is on one side of the equation and all other variables are on the other side.

$$A = P(1 + i)$$

$$A = P + Pi \qquad \text{• Use the Distributive Property to remove parentheses.}$$

$$A - P = P - P + Pi \qquad \text{• Subtract } P \text{ from each side of the equation.}$$

$$A - P = Pi$$

$$\frac{A - P}{P} = \frac{Pi}{P} \qquad \text{• Divide each side of the equation by } P.$$

$$\frac{A - P}{P} = i$$

In the previous example, the Distributive Property was used to remove parentheses, as can be symbolized by

$$a(x + y) = ax + ay$$

In a similar way, we can use the Distributive Property to rewrite two or more terms containing a common factor with parentheses.

$$ax + ay = a(x + y)$$

In this case, we say that we have *factored* a from the expression. This concept is used in solving some literal equations.

Example 11　**Solve a Literal Equation by Factoring**

Solve $S = C - rC$ for C.

Solution

$$S = C - rC \qquad \text{• } C \text{ is a common factor.}$$

$$S = C(1 - r) \qquad \begin{array}{l} \text{• Use the Distributive Property to factor } C \text{ from the two} \\ \text{terms. Recall that } C = C \cdot 1. \end{array}$$

$$\frac{S}{1 - r} = \frac{C(1 - r)}{1 - r} \qquad \text{• Divide by } (1 - r).$$

$$\frac{S}{1 - r} = C$$

●

Objective B.4D Practice

1. Solve $4x + 3y = 12$ for x.
2. Solve $E = IR$ for R.
3. Solve $F = \dfrac{9}{5}C + 32$ for C.
4. Solve $s = a(x - vt)$ for t.
5. Solve $a = S - Sr$ for S.

Objective B.4E Solve an Absolute Value Equation

Recall that the *absolute value* of a number is its distance from zero on the number line. Distance is always a positive number or zero. Therefore, the absolute value of a number is always a positive number or zero.

The distance from 0 to 3 or from 0 to -3 is 3 units.

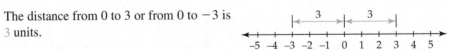

$$|3| = 3 \qquad |-3| = 3$$

Absolute value can be used to represent the distance between any two points on the number line. The **distance between two points** on the number line is the absolute value of the difference between the coordinates of the two points.

The distance between point a and point b is given by $|b - a|$.

The distance between 4 and -3 on the number line is 7 units. Note that the order in which the coordinates are subtracted does not affect the distance.

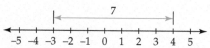

$$\text{Distance} = |-3 - 4| \qquad \text{Distance} = |4 - (-3)|$$
$$= |-7| \qquad\qquad\qquad = |7|$$
$$= 7 \qquad\qquad\qquad\quad = 7$$

For any two numbers a and b, $|b - a| = |a - b|$.

An equation containing a variable within an absolute value symbol is called an **absolute value equation**. Here are three examples.

$$|x| = 3 \qquad |x + 2| = 8 \qquad |3x - 4| = 5x - 9$$

Take Note: You should always check your answers. Here is the check for examples (1), (2), and (3).

$$\frac{|x| = 3}{\;|-3|\;\big|\;3\;}$$
$$3 = 3$$

$$\frac{|x| = 3}{\;|3|\;\big|\;3\;}$$
$$3 = 3$$

$$\frac{|-x| = 8}{\;|-(-8)|\;\big|\;8\;}$$
$$|8|\;\big|\;8$$
$$8 = 8$$

$$\frac{|-x| = 8}{\;|-8|\;\big|\;8\;}$$
$$8 = 8$$

$$\frac{|x + 3| = 4}{\;|1 + 3|\;\big|\;4\;}$$
$$|4|\;\big|\;4$$
$$4 = 4$$

$$\frac{|x + 3| = 4}{\;|-7 + 3|\;\big|\;4\;}$$
$$|-4|\;\big|\;4$$
$$4 = 4$$

Solutions of an Absolute Value Equation

If $a > 0$ and $|x| = a$, then $x = -a$ or $x = a$. If $|x| = 0$, then $x = 0$. If $a < 0$, then $|x| = a$ has no solution.

Examples

1. If $|x| = 3$, then $x = -3$ or $x = 3$.
2. If $|-x| = 8$, then $x = -8$ or $x = 8$.
3. If $|x + 3| = 4$, then $x + 3 = 4$ or $x + 3 = -4$. The solution of $x + 3 = 4$ is 1. The solution of $x + 3 = -4$ is -7.
4. If $|z| = 0$, then $z = 0$.
5. If $|y| = -2$, then the equation has no solution. The absolute value of any number is greater than or equal to zero.

Example 12 **Solve an Absolute Value Equation**

a. Solve: $|x + 2| = 8$

b. Solve: $|2 - x| = 12$

c. Solve: $3 - |2x - 4| = -5$

Solution

a.
$$|x + 2| = 8$$
$$x + 2 = 8 \qquad x + 2 = -8$$
$$x = 6 \qquad\quad x = -10$$

• Remove the absolute value sign and rewrite as two equations.
• Solve each equation.

The solutions are 6 and -10.

b.
$$|2 - x| = 12$$
$$2 - x = 12 \qquad 2 - x = -12$$
$$-x = 10 \qquad\quad -x = -14$$
$$x = -10 \qquad\quad x = 14$$

• Subtract 2.
• Multiply by -1.

The solutions are -10 and 14.

c.
$$3 - |2x - 4| = -5$$
$$-|2x - 4| = -8 \qquad \bullet \text{ Subtract 3.}$$
$$|2x - 4| = 8 \qquad \bullet \text{ Multiply by } -1.$$

$$2x - 4 = 8 \qquad\qquad 2x - 4 = -8$$
$$2x = 12 \qquad\qquad\quad 2x = -4$$
$$x = 6 \qquad\qquad\qquad x = -2$$

The solutions are 6 and -2.

Objective B.4E Practice

1. Solve: $|a - 2| = 0$
2. Solve: $|2x - 3| + 4 = -4$
3. Solve: $|3t + 2| + 3 = 4$
4. Solve: $5 - |2x + 1| = 5$
5. Solve: $8 - |1 - 3x| = -1$

Section B.5

Quadratic Equations

Objective B.5A Solve a Quadratic Equation by Using the Quadratic Formula

A **quadratic equation** is one that can be written in the form $ax^2 + bx + c = 0$, where $a \neq 0$. Examples of quadratic equations are:

$$2x^2 - 3x + 4 = 0 \qquad\qquad a = 3, b = -3, c = 4$$
$$x^2 + 6x = 0 \qquad\qquad\quad a = 1, b = 6, c = 0$$
$$-x^2 + 4 = 0 \qquad\qquad\quad a = -1, b = 0, c = 4$$

The Quadratic Formula

If $ax^2 + bx + c = 0$, $a \neq 0$, then

$$x = \frac{-b + \sqrt{b^2 - 4ac}}{2a} \quad \text{or} \quad x = \frac{-b - \sqrt{b^2 - 4ac}}{2a}$$

The quadratic formula is frequently written in the form

$$x = \frac{-b \pm \sqrt{b^2 - 4ac}}{2a}$$

Example 1 **Focus on Solving a Quadratic Equation by Using the Quadratic Formula**

Solve by using the quadratic formula.

a. $2x^2 - 3x + 1 = 0$

b. $2x^2 = 8x - 5$

c. $2x^2 = 4x - 1$

Solution

a. $2x^2 - 3x + 1 = 0$ • This is a quadratic equation in standard form. $a = 2$, $b = -3$, $c = 1$

$$x = \frac{-(-3) \pm \sqrt{(-3)^2 - 4(2)(1)}}{2 \cdot 2}$$

• Replace a, b, and c in the quadratic formula by their values.

$$= \frac{3 \pm \sqrt{9 - 8}}{4} = \frac{3 \pm \sqrt{1}}{4} = \frac{3 \pm 1}{4}$$

• Simplify.

$$x = \frac{3 + 1}{4} \qquad x = \frac{3 - 1}{4}$$

$$= \frac{4}{4} = 1 \qquad = \frac{2}{4} = \frac{1}{2}$$

The solutions are 1 and $\frac{1}{2}$.

b. $2x^2 = 8x - 5$ • This is a quadratic equation.

$2x^2 - 8x + 5 = 0$ • Write the equation in standard form.

$$x = \frac{-(-8) \pm \sqrt{(-8)^2 - 4(2)(5)}}{2 \cdot 2}$$

• Replace a, b, and c in the quadratic formula by their values.

$$= \frac{8 \pm \sqrt{64 - 40}}{4}$$

• Simplify.

$$= \frac{8 \pm \sqrt{24}}{4}$$

$$= \frac{8 \pm 2\sqrt{6}}{4}$$

$$= \frac{2(4 \pm \sqrt{6})}{2 \cdot 2} = \frac{4 \pm \sqrt{6}}{2}$$

The solutions are $\frac{4 - \sqrt{6}}{2}$ and $\frac{4 + \sqrt{6}}{2}$. The approximate solutions are 0.7753 and 3.2247.

c. $2x^2 = 4x - 1$

$2x^2 - 4x + 1 = 0$ • Write the equation in standard form. $a = 2$, $b = 24$, and $c = 1$.

$$x = \frac{-b \pm \sqrt{b^2 - 4ac}}{2a}$$

$$= \frac{-(-4) \pm \sqrt{(-4)^2 - 4 \cdot 2 \cdot 1}}{2 \cdot 2}$$

• Replace a, b, and c in the quadratic formula by their values.

$$= \frac{4 \pm \sqrt{16 - 8}}{4}$$

$$= \frac{4 \pm \sqrt{8}}{4}$$

• Simplify.

$$= \frac{4 \pm 2\sqrt{2}}{4} = \frac{2(2 \pm \sqrt{2})}{2 \cdot 2} = \frac{2 \pm \sqrt{2}}{2}$$

The solutions are $\frac{2 - \sqrt{2}}{2}$ and $\frac{2 + \sqrt{2}}{2}$.

The solutions can be approximated by using $\sqrt{2} \approx 1.414$.

$$\frac{2 - \sqrt{2}}{2} \approx \frac{2 - 1.414}{2} = 0.293$$

$$\frac{2 + \sqrt{2}}{2} \approx \frac{2 + 1.414}{2} = 1.707$$

The approximate solutions are 0.293 and 1.707.

Objective B.5A Practice

For Exercises 1, 2, 3, 4, and 5, solve by using the quadratic formula.

1. $z^2 + 6z - 7 = 0$

2. $x^2 - 3x - 6 = 0$

3. $2y^2 + 3 = 8y$

4. $6s^2 - s - 2 = 0$

5. $9v^2 = -30v - 23$

Review of the Rectangular Coordinates and Graphing

The Rectangular Coordinate System

Objective C.1A Graph an Equation in Two Variables

A **rectangular coordinate system** is formed by two number lines, one horizontal and one vertical, that intersect at the zero point of each line. The point of intersection is called the **origin**. The two lines are called **coordinate axes**, or simply **axes**.

The axes determine a **plane**, which can be thought of as a large, flat sheet of paper. The two axes divide the plane into four regions called **quadrants**, numbered counterclockwise from I to IV (see Figure C.1).

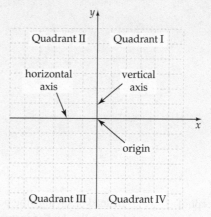

Figure C.1

Each point in the plane can be identified by a pair of numbers called an **ordered pair**. The first number of the pair measures a horizontal distance and is called the **abscissa**. The second number of the pair measures a vertical distance and is called the **ordinate**. The **coordinates of a point** are the numbers in the ordered pair associated with the point. The abscissa is also called the **first coordinate** of the ordered pair, and the ordinate is also called the **second coordinate** of the ordered pair.

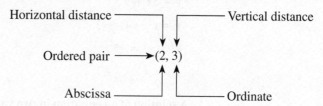

659

Graphing, or plotting, an ordered pair in the plane means placing a dot at the location given by the ordered pair. The **graph of an ordered pair** is the dot drawn at the coordinates of the point in the plane. The points whose coordinates are $(3, 4)$ and $(-2.5, -3)$ are graphed in Figure C.2.

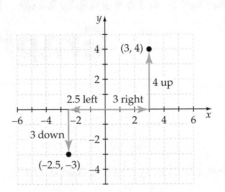

Figure C.2

The points whose coordinates are $(3, -1)$ and $(-1, 3)$ are graphed in Figure C.3. Note that the graphs are in different locations. The *order* of the coordinates of an ordered pair is important.

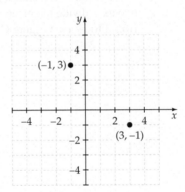

Figure C.3

Take Note: The concept of *ordered pair* is an important concept. Remember: There are two numbers (a *pair*), and the *order* in which they are listed is important.

Instead of saying "the point whose coordinates are $(3, -1)$," we will frequently use the notation $P(3, -1)$. For instance, $P(4, 3)$ is read "the point P whose coordinates are $(4, 3)$."

When drawing a rectangular coordinate system, we often label the horizontal axis x and the vertical axis y. In this case, the coordinate system is called an **xy-coordinate system**. The coordinates of the points are given by ordered pairs (x, y), where the abscissa is called the **x-coordinate** and the ordinate is called the **y-coordinate**.

The xy-coordinate system is used to graph equations in *two variables*. Examples of equations in two variables are shown below.

$$y = 3x + 7$$

$$y = x^2 - 4x + 3$$

$$x^2 + y^2 = 25$$

$$x = \frac{y}{y^2 + 4}$$

A **solution of an equation in two variables** is an ordered pair (x, y) whose coordinates make the equation a true statement.

Example 1 **Determine Whether an Ordered Pair Is a Solution of an Equation**

Is the ordered pair $(-3, 7)$ a solution of the equation $y = -2x + 1$?

Solution

$$y = -2x + 1$$

7	$-2(-3) + 1$	• Replace x by -3 and y by 7.
7	$6 + 1$	• Simplify.
$7 = 7$		• Compare the results. If the resulting equation is a true statement, the ordered pair is a solution of the equation. If the resulting equation is not a true statement, the ordered pair is not a solution of the equation.

Yes, the ordered pair $(-3, 7)$ is a solution of the equation.

In addition to the ordered pair $(-3, 7)$, there are many other ordered-pair solutions of the equation $y = -2x + 1$. For example, $(-5, 11)$, $(0, 1)$, $\left(-\frac{3}{2}, 4\right)$, and $(4, -7)$ are also solutions of the equation.

In general, an equation in two variables has an infinite number of solutions. By choosing any value of x and substituting that value into the equation, we can calculate a corresponding value for y. The resulting ordered-pair solution (x, y) of the equation can be graphed in a rectangular coordinate system.

Example 2 **Graph the Ordered Pair Solutions of an Equation**

Graph the solutions (x, y) of $y = x^2 - 1$ when x equals $-2, -1, 0, 1,$ and 2.

Solution

Substitute each value of x into the equation and solve for y. Then graph the resulting ordered pairs by placing a dot at the coordinates of each point. This is sometimes referred to as **plotting the points**. It is convenient to record the ordered-pair solutions in a table, similar to Table C.1. The graph of the ordered pairs is shown in Figure C.4.

Table C.1

x	$y = x^2 - 1$	y	(x, y)
-2	$y = (-2)^2 - 1$	3	$(-2, 3)$
-1	$y = (-1)^2 - 1$	0	$(-1, 0)$
0	$y = 0^2 - 1$	-1	$(0, -1)$
1	$y = 1^2 - 1$	0	$(1, 0)$
2	$y = 2^2 - 1$	3	$(2, 3)$

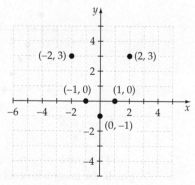

Figure C.4

Generally, when we graph an equation in two variables, we include *all* the solutions, not just some selected ones as we did above. The graph of an equation is the graph of all of the ordered-pair solutions of the equation.

Consider $y = -2x + 1$. We can find ordered-pair solutions when $x = -2, -1, 0, 1, 2$, and 3. The results are shown in Table C.2. The graph of these solutions is shown in Figure C.5.

Table C.2

x	$y = -2x + 1$	y	(x, y)
-2	$y = -2(-2) + 1$	5	$(-2, 5)$
-1	$y = -2(-1) + 1$	3	$(-1, 3)$
0	$y = -2(0) + 1$	1	$(0, 1)$
1	$y = -2(1) + 1$	-1	$(1, -1)$
2	$y = -2(2) + 1$	-3	$(2, -3)$
3	$y = -2(3) + 1$	-5	$(3, -5)$

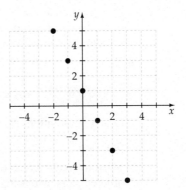

Figure C.5

If we find additional solutions, such as when $x = -1.5, -0.5, 0.5, 1.5$, and 2.5, we get more points, as shown in Figure C.6. If we continue to add more and more points, there would be so many dots that the graph would look like the straight line in Figure C.7, which is the graph of $y = -2x + 1$.

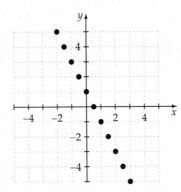

Figure C.6

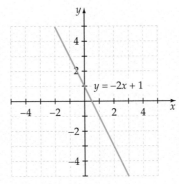

Figure C.7

The graph of $y = -2x + 1$ is shown again in Figure C.8. As can be seen from the graph, the point with coordinates $(1, 4)$ is not on the graph, because, as shown, $(1, 4)$ is not a solution of $y = -2x + 1$. The point with coordinates $(2, -3)$ is both a point on the graph and a solution of the equation.

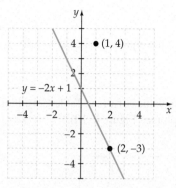

Figure C.8

$$y = -2x + 1$$

4	$-2(1) + 1$
4	$-2 + 1$

$$4 \neq -1$$

(1, 4) does *not* represent a point on the graph and is *not* a solution of the equation.

$$y = -2x + 1$$

-3	$-2(2) + 1$
-3	$-4 + 1$

$$-3 = -3$$

(2, −3) *does* represent a point on the graph and *is* a solution of the equation.

Every ordered pair on the graph of an equation is a solution of the equation, and every ordered-pair solution of an equation represents a point on the graph of the equation.

It may be necessary to plot a number of ordered pairs in order to create an accurate graph of an equation in two variables.

Example 3 **Graph an Equation in Two Variables**

Graph $y = \frac{1}{2}x + 1$ by first plotting the solutions of the equation when $x = -4, -2, 0, 2,$ and 4.

Solution

Determine the ordered-pair solutions (x, y) for the given values of x (see Table C.3). Plot the points, and then connect the points to form the graph (see Figure C.9).

Table C.3

x	$y = \frac{1}{2}x + 1$	y
-4	$y = \frac{1}{2}(-4) + 1$	-1
-2	$y = \frac{1}{2}(-2) + 1$	0
0	$y = \frac{1}{2}(0) + 1$	1
2	$y = \frac{1}{2}(2) + 1$	2
4	$y = \frac{1}{2}(4) + 1$	3

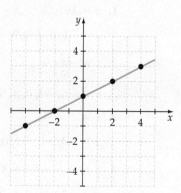

Figure C.9

Example 4   **Graph an Equation in Two Variables**

Graph $y = x^2 - 2x - 3$ by first plotting the solutions of the equation when $x = -2$, $-1, 0, 1, 2, 3,$ and 4.

Solution

Determine the ordered-pair solutions (x, y) for the given values of x (see Table C.4). Plot the points, and then connect the points to form the graph (see Figure C.10).

Table C.4

x	$y = x^2 - 2x - 3$	y
−2	$y = (-2)^2 - 2(-2) - 3$	5
−1	$y = (-1)^2 - 2(-1) - 3$	0
0	$y = (0)^2 - 2(0) - 3$	−3
1	$y = (1)^2 - 2(1) - 3$	−4
2	$y = (2)^2 - 2(2) - 3$	−3
3	$y = (3)^2 - 2(3) - 3$	0
4	$y = (4)^2 - 2(4) - 3$	5

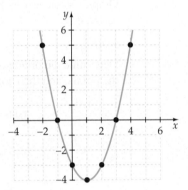

Figure C.10

Objective C.1A Practice

For Exercises 1, 2, 3, 4, and 5, graph the equation. First plot the solutions of the equation for the given values of x, and then connect the points with a smooth graph.

1. $y = 2x - 3$

   $x = -1, 0, 1, 2, 3, 4$

2. $y = -\frac{2}{3}x + 1$

   $x = -6, -3, 0, 3, 6$

3. $y = x^2 - 4$

   $x = -3, -2, -1, 0, 1, 2, 3$

4. $y = -x^2 + 2x + 3$

   $x = -2, -1, 0, 1, 2, 3, 4$

5. $y = |x + 2|$

   $x = -5, -4, -3, -2, -1, 0, 1$

Section C.2 Introduction to Functions

Objective C.2A Evaluate a Function

In mathematics and its applications, there are many times when it is necessary to investigate a relationship between two quantities. Here is a financial application: Consider a person who is planning to finance the purchase of a car. If the current interest rate for a 5-year loan is 5%, the equation that describes the relationship between the amount that is borrowed B and the monthly payment P is $P = 0.018871B$.

For each amount the purchaser may borrow (B), there is a certain monthly payment (P). The relationship between the amount borrowed and the payment can be recorded as a set of ordered pairs, where the first coordinate of each pair is the amount borrowed and the second coordinate is the monthly payment. Some of these ordered pairs are shown at the right below.

$$0.018871B = P$$

(6000, 113.23)
(7000, 132.10)
(8000, 150.97)
(9000, 169.84)

A relationship between two quantities is not always given by an equation. Table C.5 describes a grading scale that defines a relationship between a score on a test and a letter grade. For each score, the table assigns only one letter grade. The ordered pair (84, B) indicates that a score of 84 receives a letter grade of B.

Table C.5

Score	Grade
90–100	A
80–89	B
70–79	C
60–69	D
0–59	F

The bar graph in Figure C.11 shows the number of people who watched the Super Bowl for the years 2007 to 2012. The jagged line between 0 and 90 on the vertical axis indicates that a portion of the vertical axis has been omitted. The data in the graph can be written as a set of ordered pairs:

{(2007, 93.2), (2008, 97.4), (2009, 98.7), (2010, 106.5), (2011, 111.01),(2012, 111.35)}

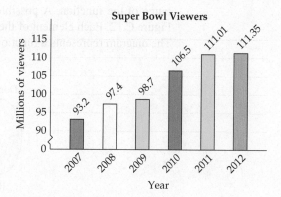

Figure C.11

This set is a function. There are no two ordered pairs with the same first coordinate. The ordered pair (2010, 106.5) means that in 2010, the number of people who watched the Super Bowl was 106.5 million.

In each of the above examples, there is a rule (an equation, a table, or a graph) that determines a certain set of ordered pairs.

Definition of a Function

A **function** is a set of ordered pairs in which no two ordered pairs have the same first coordinate. The **domain** of a function is the set of first coordinates of the ordered pairs; the **range** of a function is the set of second coordinates of the ordered pairs.

Examples

a. $\{(1, 2), (2, 4), (3, 6), (4, 8)\}$

    Domain $= \{1, 2, 3, 4\}$ Range $= \{2, 4, 6, 8\}$

b. $\{(-1, 0), (0, 0), (1, 0), (2, 0), (3, 0)\}$

    Domain $= \{-1, 0, 1, 2, 3\}$ Range $= \{0\}$

Now consider the set of ordered pairs $\{(1, 2), (4, 5), (7, 8), (4, 6)\}$. This set of ordered pairs is *not* a function. There are two ordered pairs, (4, 5) and (4, 6), with the same first coordinate. This set of ordered pairs is called a *relation*. A **relation** is any set of ordered pairs. A function is a special type of relation. The concepts of domain and range apply to relations as well as to functions.

Example 1   **Determine Whether a Set of Ordered Pairs Is a Function**

Determine whether each set of ordered pairs is a function.

a. $\{(2, 3), (4, 6), (6, 8), (10, 6)\}$
b. $\{(2, 2), (1, 1), (0, 0), (2, -2), (1, -1)\}$

Solution

a. No two ordered pairs have the same first element. The set of ordered pairs is a function.

b. The ordered pairs (2, 2) and (2, -2) have the same first coordinate. The set of ordered pairs is not a function.

For each element of the domain of a function there is a corresponding element in the range of the function. A possible diagram for the function in part a of Example 1 is in Figure C.12. Each element of the domain is paired with exactly one element in the range. The diagram represents a function.

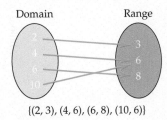

$\{(2, 3), (4, 6), (6, 8), (10, 6)\}$

Figure C.12

A diagram for part b of Example 1 is shown in Figure C.13. There are some elements in the domain that are paired with more than one element in the range. The diagram does not represent a function.

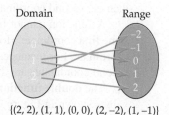

$$\{(2, 2), (1, 1), (0, 0), (2, -2), (1, -1)\}$$

Figure C.13

Take Note: For a set, the order in which the elements are listed is not important. For instance,

$$\{a, b, c\} = \{b, a, c\}.$$

Note that the elements of the domain of the grading-scale function were listed from smallest to largest. It is common practice to list both domain and range elements in order from smallest to largest.

Consider again the three examples of functions given at the beginning of this objective. For the equation $0.018871B = P$, the domain is the possible amounts a consumer might borrow to purchase a car. Let's assume that the most a person would borrow is $50,000. Then the domain is $\{B | 0 \le B \le 50,000\}$. The range is all possible monthly payments. The largest monthly payment is $P = 0.018871(50,000) \approx 943.55$, so the range is $\{P | 0 \le P \le 943.55\}$.

For the grading-scale function, the domain is all possible test scores. The domain is $\{0, 1, 2, 3, \dots, 97, 98, 99, 100\}$. The range is all possible grades. The range is $\{A, B, C, D, F\}$.

For a graph, the domain is represented on the horizontal axis and the range is represented on the vertical axis. For the graph of the Super Bowl data, the domain is the set of years. The domain is $\{2007, 2008, 2009, 2010, 2011, 2012\}$. The range is the number of people watching each year. The range is $\{93.2, 97.4, 98.7, 106.5, 111.01, 111.35\}$.

Example 2 **Determine the Domain and Range of a Function**

Find the domain and range of the function $\{(5, 3), (9, 7), (13, 7), (17, 3)\}$.

Solution

The domain is the set of first coordinates. The range is the set of second coordinates.

The domain is $\{5, 9, 13, 17\}$.

The range is $\{3, 7\}$.

Take Note: Pictorial representations of the square function and the double function follow in Figures C.14 and C.15, respectively. The function acts as a machine that changes a number from the domain into the square or double of the number.

The **square function**, which pairs each real number with its square, can be defined by the equation

$$y = x^2$$

This equation states that for a given value of x in the domain, the corresponding value of y in the range is the square of x. For instance, if $x = 6$, then $y = 36$, and if $x = -7$, then $y = 49$. Because the value of y *depends* on the value of x, y is called the **dependent variable** and x is called the **independent variable**.

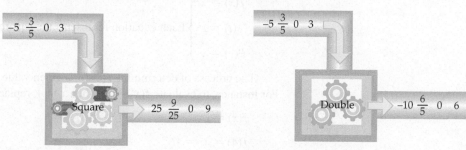

Figure C.14   **Figure C.15**

A function can be thought of as a rule that pairs one number with another number. For instance, the square function pairs a number with its square. The ordered pairs for the values shown in Figure C.14 are $(-5, 25)$, $\left(\frac{3}{5}, \frac{9}{25}\right)$, $(0, 0)$, and $(3, 9)$. For this function, the second coordinate is the square of the first coordinate. If x represents the first coordinate, then the second coordinate is x^2 and the ordered pair is (x, x^2).

A function cannot have two ordered pairs with *different* second coordinates and the same first coordinate. However, a function may contain ordered pairs with the *same* second coordinate. For instance, the square function has the ordered pairs $(-3, 9)$ and $(3, 9)$; the second coordinates are the same but the first coordinates are different.

The **double function** pairs a number with twice that number. The ordered pairs for the values shown in Figure C.15 are $(-5, -10)$, $\left(\frac{3}{5}, \frac{6}{5}\right)$, $(0, 0)$, and $(3, 6)$. For this function, the second coordinate is twice the first coordinate. If x represents the first coordinate, then the second coordinate is $2x$ and the ordered pair is $(x, 2x)$.

Not every equation in two variables defines a function. For instance, consider the equation

$$y^2 = x^2 + 9$$

Because

$$5^2 = 4^2 + 9 \qquad \text{and} \qquad (-5)^2 = 4^2 + 9$$

the ordered pairs $(4, 5)$ and $(4, -5)$ are both solutions of the equation. Consequently, there are two ordered pairs that have the same first coordinate (4) but *different* second coordinates (5 and -5). Therefore, the equation does not define a function.

The phrase "y is a function of x," or the same phrase with different variables, is used to describe an equation in two variables that defines a function. To emphasize that the equation represents a function, **function notation** is used.

Just as the variable x is commonly used to represent a number, the letter f is commonly used to name a function. The square function is written in function notation as follows:

This is the value of the function. It is the number that is paired with x.

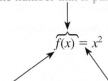

$$f(x) = x^2$$

The name of the function is f.   This is an algebraic expression that defines the relationship between the dependent and independent variables.

The symbol $f(x)$ is read "the value of f at x" or "f of x."

It is important to note that $f(x)$ does *not* mean f times x. The symbol $f(x)$ is the **value of the function** and represents the value of the dependent variable for a given value of the independent variable. We often write $y = f(x)$ to emphasize the relationship between the independent variable x and the dependent variable y. Remember that y and $f(x)$ are different symbols for the same number.

The letters used to represent a function are somewhat arbitrary. All of the following equations represent the same function.

$$\left.\begin{array}{l} f(x) = x^2 \\ s(t) = t^2 \\ P(v) = v^2 \end{array}\right\} \text{Each equation represents the square function.}$$

The process of determining $f(x)$ for a given value of x is called **evaluating a function**. For instance, to evaluate $f(x) = x^2$ when $x = 4$, replace x by 4 and simplify.

$$f(x) = x^2$$

$$f(4) = 4^2 = 16$$

The *value* of the function is 16 when $x = 4$. An ordered pair of the function is $(4, 16)$.

Example 3   **Evaluate a Function**

Evaluate $g(t) = 3t^2 - 5t + 1$ when $t = -2$.

Solution

$$g(t) = 3t^2 - 5t + 1$$

$$g(-2) = 3(-2)^2 - 5(-2) + 1 \quad \bullet \text{ Replace } t \text{ by } -2 \text{ and then simplify.}$$

$$= 3(4) - 5(-2) + 1$$

$$= 12 + 10 + 1 = 23$$

When t is -2, the value of the function is 23.

Take Note: Because $g(-2) = 23$, $(-2, 23)$ is an ordered pair of the function.

When a function is represented by an equation, the domain of the function is all real numbers for which the value of the function is a real number. For instance:

- The domain of $f(x) = x^2$ is all real numbers, because the square of every real number is a real number. In set-builder notation, the domain is $\{x \mid -\infty < x < \infty\}$.
- The domain of $g(x) = \dfrac{1}{x - 2}$ is all real numbers except 2, because when $x = 2$, $g(2) = \dfrac{1}{2 - 2} = \dfrac{1}{0}$, which is not a real number. The domain is $\{x \mid x \neq 2\}$.

Example 4   **Find the Domain of a Function Given by an Equation**

a. Find the domain of $f(x) = 2x^2 - 7x + 1$.
b. Find the domain of $f(x) = \dfrac{2}{x - 5}$.

Solution

a. Because the value of $2x^2 - 7x + 1$ is a real number for any value of x, no values are excluded from the domain of $f(x) = 2x^2 - 7x + 1$. The domain of the function is all real numbers, or $\{x \mid -\infty < x < \infty\}$.

b. For $x = 5$, $f(x) = f(5) = \dfrac{2}{5 - 5} = \dfrac{2}{0}$, which is undefined. So, 5 is excluded from the domain of f. The domain is $\{x \mid x \neq 5\}$.

Objective C.2A Practice

1. Determine whether $\{(0, 0), (1, 0), (2, 0), (3, 0), (4, 0)\}$ is a function. State the domain and range.
2. Given $f(x) = 5x - 4$, evaluate $f(-2)$.
3. Given $q(r) = r^2 - 4$, evaluate $q(-5)$.
4. Given $s(t) = t^3 - 3t + 4$, evaluate $s(2)$.

Objective C.2B   Graph a Function

The **graph of a function** is a graph of all of the ordered pairs of the function. Graphing a function is similar to graphing an equation in two variables. Evaluate the function for selected values of x. Plot the corresponding ordered pairs and then connect the points to form the graph.

Example 5   Graph a Function

Graph $f(x) = x^2 + 2x - 3$. Begin by creating a table of the ordered pairs (x, y) of the function when x equals $-4, -3, -2, -1, 0, 1,$ and 2.

Solution

Remember that the dependent variable is y, and $y = f(x)$ (see Table C.6).

Table C.6

x	$f(x) = x^2 + 2x - 3$	y	(x, y)
-4	$f(-4) = (-4)^2 + 2(-4) - 3$	5	$(-4, 5)$
-3	$f(-3) = (-3)^2 + 2(-3) - 3$	0	$(-3, 0)$
-2	$f(-2) = (-2)^2 + 2(-2) - 3$	-3	$(-2, -3)$
-1	$f(-1) = (-1)^2 + 2(-1) - 3$	-4	$(-1, -4)$
0	$f(0) = (0)^2 + 2(0) - 3$	-3	$(0, -3)$
1	$f(1) = (1)^2 + 2(1) - 3$	0	$(1, 0)$
2	$f(2) = (2)^2 + 2(2) - 3$	5	$(2, 5)$

Plot the points and then draw a smooth curve through the points as shown in Figure C.16.

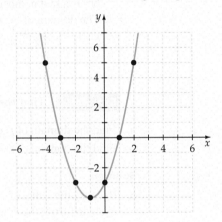

Figure C.16

The graph of $f(x) = x^2 + 2x - 3$ is shown again in Figure C.17. As we can see from the graph, the point with coordinates $(-1, 3)$ is not on the graph. Evaluating the function at -1, we have

$$f(x) = x^2 + 2x - 3$$

$$f(-1) = (-1)^2 + 2(-1) - 3 = -4 \neq 3$$

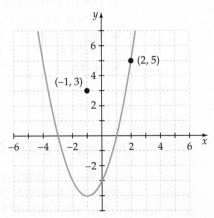

Figure C.17

Because $f(-1) \neq 3$, the ordered pair $(-1, 3)$ does not belong to the function, and the point with coordinates $(-1, 3)$ is not on the graph of f.

Every ordered pair (x, y) on the graph of a function f satisfies $y = f(x)$, and every ordered pair that satisfies $y = f(x)$ is a point on the graph of the function.

Example 6   Graph a Function

Graph $s = g(t) = -\frac{3}{2}t + 1$.

Solution

The variables in a function can be different from x and y. The key point is that the independent variable, t, is represented on the horizontal axis. We can choose any other variable for the dependent variable. In this case, we chose s. The dependent variable is represented on the vertical axis.

Begin by plotting the ordered pairs (t, s) of the function when $t = -2, 0$, and 2 (see Table C.7).

Table C.7

t	$g(t) = -\frac{3}{2}t + 1$	s	(t, s)
-2	$g(-2) = -\frac{3}{2}(-2) + 1$	4	$(-2, 4)$
0	$g(0) = -\frac{3}{2}(0) + 1$	1	$(0, 1)$
2	$g(2) = -\frac{3}{2}(2) + 1$	-2	$(2, -2)$

Take Note: We chose values of t that are multiples of 2 to make it easier to plot the points. For instance, if we had selected $t = 3$, then

$$s = g(3) = -\frac{3}{2}(3) + 1$$

$$= -\frac{9}{2} + 1 = -\frac{7}{2}$$

Now we would need to estimate $-\frac{7}{2}$ on the coordinate grid—not impossible, but not as easy as plotting an integer value.

Plot the points and then draw a smooth curve through the points as shown in Figure C.18.

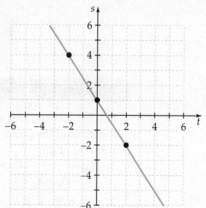

Figure C.18

Example 7   Graph an Absolute Value Function

Graph $f(x) = |2x + 4| - 1$.

Solution

The independent variable is x. The dependent variable is y.

Begin by plotting the ordered pairs (x, y) of the function when $x = -5, -4, -3, -2, -1, 0$, and 1 (see Table C.8).

Table C.8

x	$f(x) = \lvert 2x + 4 \rvert - 1$	y
−5	$f(-5) = \lvert 2(-5) + 4 \rvert - 1$	5
−4	$f(-4) = \lvert 2(-4) + 4 \rvert - 1$	3
−3	$f(-3) = \lvert 2(-3) + 4 \rvert - 1$	1
−2	$f(-2) = \lvert 2(-2) + 4 \rvert - 1$	−1
−1	$f(-1) = \lvert 2(-1) + 4 \rvert - 1$	1
0	$f(0) = \lvert 2(0) + 4 \rvert - 1$	3
1	$f(1) = \lvert 2(1) + 4 \rvert - 1$	5

Plot the points and then draw a smooth curve through the points as shown in Figure C.19.

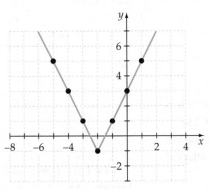

Figure C.19

Objective C.2B Practice

For Exercises 1, 2, 3, and 4, graph the function. First plot the ordered pairs of the function for the given values of the independent variable. Then connect the points to form the graph.

1. $f(x) = -x^2 - 4x + 1$
  $x = -5, -4, -3, -2, -1, 0, 1$

2. $f(x) = 2\lvert x + 1 \rvert - 1$
  $x = -4, -3, -2, -1, 0, 1, 2$

3. $h(u) = \lvert 2u \rvert - 3$
  $u = -4, -3, -2, -1, 0, 1, 2, 3, 4$

4. $f(t) = -\lvert 4 - 2t \rvert + 3$
  $t = -2, -1, 0, 1, 2, 3, 4, 5$

Selected Answers

Chapter 1

Section 1.1

Think About It

1. Inductive **2.** Specific **3.** One example is 5, an odd number.

For solutions to the Preliminary Exercises, see page 695.

Section Exercises

1. 64 **3.** $\frac{13}{15}$ **5.** -13

7.   **9.**

11.   **13.** $5000

15. More than **17.** 2100 feet **19.** Less than **21.** 64 feet

23. Answers will vary. For instance, $\frac{1}{4} \times \frac{1}{2} = \frac{1}{8}$.

25. A diamond shape can have four equal sides and is not a square.

27. Whales are mammals and do not have legs.

29. (a) Deductive (b) Inductive

31. See the solutions manual.

33. See the solutions manual.

35. Maria: the utility stock; Jose: the automotive stock; Anita: the technology stock; Tony: the oil stock

37. Atlanta: stamps; Chicago: baseball cards; Philadelphia: coins; San Diego: comic books

Section 1.2

Think About It

1. (a) Estimate (b) Estimate (c) Exact (d) Estimate
(e) Exact (f) Estimate

2. b **3.** True **4.** False **5.** Yes

For solutions to the Preliminary Exercises, see page 695.

Section Exercises

1. 500 miles **3.** 80°F **5.** 8000 square feet

7. Answers will vary depending on the grid. Our estimate is 360 cars.

9. Answers will vary depending on the grid. Our estimate is 160 cadets.

11. 1.6 million people **13.** 40 ft **15.** 8.9×10^{-12} meters

17. 3.5×10^{13} **19.** Yes. **21.** Yes

23. (a) 2018 through 2020 (b) Less than

Section 1.3

Think About It

1. Understand the problem, devise a plan, carry out the plan, review the solution

2. No. We would need to know the speed at which she ran.

For solutions to the Preliminary Exercises, see page 696.

Section Exercises

1. 195 girls **3.** 91 squares

5. $40 **7.** 18 direct routes **9.** $2^{12} = 4096$ ways

11. 28 handshakes **13.** 21 ducks, 14 pigs **15.** 12 ways

17. 6 **19.** 1 **21.** 1.5 inches

23. 2601 tiles **25.** Four more sisters than brothers

27. 11th day **29.** 91

31. (a) Place four coins each on the left and right balance pans. The pan that is the higher contains the fake coin. Take the four coins from the higher pan and use the balance scale to compare the weight of two of these coins to the weight of the other two. The pan that is the higher contains the fake coin. Take the two coins from the higher pan and use the balance scale to compare the weight of one of these coins to the weight of the other. The pan that is higher contains the fake coin. This procedure enables you to determine the fake coin in three weighings. (b) Place three of the coins on one of the balance pans and another three coins on the other. If the pans balance, then the fake coin is one of the two remaining coins. You can use the balance scale to determine which of the remaining coins is the fake coin because it will be lighter than the other coin. If the three coins on the left pan do not balance with the three coins on the right pan, then the fake coin must be one of the three coins on the higher pan. Pick any two coins from these three and place one on each balance pan. If these two coins do not balance, then the one that is the higher is the fake. If the coins balance, then the third coin (the one that you did not place on the balance pan) is the fake. In any case, this procedure enables you to determine the fake coin in two weighings.

33. (a) 1600. Sally likes perfect squares.

35. (d) 64. Each number is the cube of a term in the sequence 1, 2, 3, 4, 5, 6.

Section 1.4

Think About It

1. 25

2. Assuming Jeremiah divides the land as required by the rules of fair division, this is a fair division of the land. The size of a piece does not matter: the value of the piece is the important fact.

3. Craig. The divider must divide an asset so that each piece has the same value.

4. P_1: S_1, S_3; P_2: S_1, S_3; P_3: S_1, S_2, S_3

For solutions to the Preliminary Exercises, see page 697.

Section Exercises

1. (a) $18 (b) No.

3. Answers may vary. The easiest solution is to cut the cake so that each piece is one-half strawberry and one-half chocolate.

5. Yes.

7. Answers may vary. Here is one possibility.

	S_1	S_2	S_3	Bids
Caroline	45%	25%	30%	S_1
Jasmine	40%	30%	30%	S_1
Jean	$33\frac{1}{3}\%$	$33\frac{1}{3}\%$	$33\frac{1}{3}\%$	S_1, S_2, S_3

Because S_1 is contested among the choosers, set it aside. Now randomly give Jean S_2 or S_3. Assume she receives S_2. Now combine the remaining shares and use the divider–chooser method to distribute the remaining two shares. Randomly choose Caroline or Jasmine to be the divider. The other partner is the chooser. We will assume Caroline is the divider and divides the shares as T_1 and T_2. Assume Jasmine chooses T_1. Then Caroline gets T_2, Jasmine gets T_1, and Jean gets S_2.

9. Answers may vary. Here is one possibility.

	S_1	S_2	S_3	S_4	Bids
A	25%	25%	25%	25%	S_1, S_2, S_3, S_4
B	30%	10%	30%	30%	S_1, S_3, S_4
C	20%	35%	25%	20%	S_2, S_3
D	30%	30%	20%	20%	S_1, S_2

S_4 is uncontested among the choosers. Award S_4 to B. Combine S_1, S_2, and S_3. Then choose one of the remaining players to be the divider. We choose A. A divides the shares into three shares: T_1, T_2, and T_3. Everyone now submits bids for those shares. *This is where solutions may vary. It will depend on how you assign the bids. This is just one possibility.*

	T_1	T_2	T_3	Bids
A	$33\frac{1}{3}\%$	$33\frac{1}{3}\%$	$33\frac{1}{3}\%$	T_1, T_2, T_3
C	30%	30%	40%	T_3
D	30%	35%	35%	T_2, T_3

T_1 is uncontested, so award that to A. C only wants T_3 so award T_3 to C. That leaves D with T_2. The final result: A gets T_1; B gets S_4; C gets T_3; D gets T_2.

11. Answer may vary. Here is one possibility.

	S_1	S_2	S_3	S_4	Bids
Arianna	25%	25%	25%	25%	S_1, S_2, S_3, S_4
Richard	15%	30%	20%	35%	S_2, S_4
Michel	20%	45%	15%	20%	S_2
Carlos	30%	35%	15%	20%	S_1, S_2

Michel only values S_2, so he receives S_2. Because S_2 went to Michel, award S_4 to Richard and S_1 to Carlos. Arianna gets S_3.

13. Answers may vary. This is one possibility.

	S_1	S_2	S_3	S_4	Bids
Jorge	25%	25%	25%	25%	S_1, S_2, S_3, S_4
Yolanda	35%	20%	10%	35%	S_1, S_4
Maria	15%	40%	35%	10%	S_2, S_3
Daniel	25%	35%	10%	30%	S_1, S_2, S_4

S_3 is uncontested among the choosers. Award S_3 to Jorge. Now choose one of the remaining investors to be the divider. We choose Yolanda. She divides the shares into three shares: T_1, T_2, and T_3. Everyone now submits bids for those shares. *This is where solutions may vary. It will depend on how you assign the bids. This is just one possibility.*

	T_1	T_2	T_3	Bids
Yolanda	$33\frac{1}{3}\%$	$33\frac{1}{3}\%$	$33\frac{1}{3}\%$	T_1, T_2, T_3
Maria	30%	30%	40%	T_3
Daniel	30%	35%	35%	T_2, T_3

T_2 is uncontested among the choosers. Award T_2 to Daniel. Now use the divider–chooser method for two people, Yolanda and Maria, to divide the remaining two shares; call them U_1 and U_2. We will assume Maria divides and Yolanda chooses U_2. The final result: Jorge gets S_3, Yolanda gets U_2, Maria gets U_1, and Daniel gets T_2.

15. Chong gets the boat, condo, and 9.1% from the sale of the house. Riana gets 90.9% of the sale of the house.

17. Francine gets the residence, airplane, and 60% from the sale of the summer home. Elliot gets the sailboat, desk, and 40% of the sale of the summer home.

19. Amelia: receives \$359,776; Justine: medical suites and pays \$206,222; Colby: real estate and pays \$153,556

21. Leonard: home and pays \$311,667; Taylor: boat and receives \$198,333; Madison: SUV and painting and receives \$113,333

23. Clara: bedroom 4, pays \$512.50; Manuel: bedroom 1, pays \$662.50; Charles: bedroom 3, pays \$762.50; Beatrice: bedroom 2, pays \$562.50

25. Raven: chalet and pays \$275,000; Lenore: receives \$275,000

27. Marshall: sailboat, RV, and pays \$58,667; Jordan: cabin and pays \$12,667; Jan: receives \$71,333

29. June: piano and receives \$100,750; Melissa: organ, studio, condo, and pays \$345,500; Lloyd: receives \$122,750; Diane: receives \$122,000

Chapter 1 Review Exercises

1. $\frac{1}{32}$ **2.** -64 **3.** -2 **4.** 64 **5.** 1000 **6.** $1\ 6\ 15\ 20\ 15\ 6\ 1$

7. 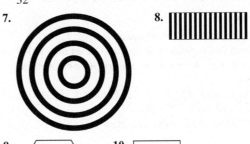 **8.**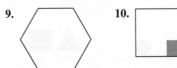

9. **10.**

11. More than **12.** 10 cases **13.** More than **14.** Less than **15.** Answers may vary. One possibility is 0.5.

16. 0 does not have reciprocal.

17. Answers may vary. One possibility is $-3(6) = -18$.

18. Answers may vary. One possibility is rectangle 1 has sides 10, 40, 10, and 40. The perimeter is 100 units; the area is 400 square units. Rectangle 2 has sides 20, 30, 20, and 30. The perimeter is 100 units; the area is 600 square units.

19. Michael, biology major; Clarissa, business major; Reggie, computer science major; and Ellen is the chemistry major.

20. Dodgers, drugstore; Pirates, supermarket; Tigers, bank; Giants, service station

21. First weighing: Place 5 coins at each side of the scale. Exclude all the 5 coins in the side that goes up (the heavier coin is on the pan that went down). Now you have only 5 coins. Second Weighing: Hold one coin in your hand and put 2 coins at each side of the scale. If the two sides balance, then the heavier one is the one you are holding in your hand. You are done. If one side is heavier than the other, one of the two coins is on that side is the heavier one. Now you have only 2 coins and a scale to find out which is the heavier by using the scale for a third time.

22. No **23.** More than **24.** 10 minutes

25. Answers may vary depending on how grids are placed. Our grid estimates 300 acai berries.

26. 9 inches **27.** 4.132×10^{13} **28.** 1.2×10^{-10} **29.** 2026 and 2027 **30.** physicians **31.** (a) 0 to 1 (b) Less than **32.** 48

33. Seth and Sarah cross the bridge, and Sarah returns. Time for round trip, 3 minutes. Kevin and Asher cross the bridge, and Seth returns. Time for round trip is 12 minutes. Total elapsed time is 15 minutes. Now Sarah and Seth cross the bridge. Total elapsed time is 17 minutes.

34. Red

35. No. A fair share is $6. If piece 1 is one-third pepperoni and one-sixth cheese, then the value of the piece is
$$\frac{1}{3}(9) + \frac{1}{6}(3) = 3 + 2 = 5 < 6.$$

36. *Answers may vary. Here is one possibility.* Jeffrey gets S_3, Clara gets U_1, Alyssa gets U_2, and Cameron gets T_3. See Solutions manual.

37. Alana: motorcycle, piano, and 80% of the proceeds from the sale of the painting; Taft: condo and 20% of the proceeds from the sale of the painting.

38. Miguel: house and pays $216,667; Cannon: receives $316,667; Ralph: condo and pays $100,000.

39. Thomas: no view, pays $733; Michael: garden view, pays $1033; Claudia: skyline view, pays $933

40. Jocelyn: receives $1,313,333; Harrison: vase, violin, pays $906,667; Colby: painting and pays $406,667

41. Santoro: Bentley and receives $652,437.50; Benjamin: sculpture and pays $1,028,812.50; Bianca: condo and pays $506,062.50; Vivian: harp and receives $882,437.50

Chapter 2

Section 2.1

Think About It

1. (a) True (b) True (c) True (d) False (e) False
(f) True (g) True (h) True

2. (a) No (b) Yes

3. A and B are disjoint—that is, they have no elements in common.

4. A **5.** False negative **6.** Correct outcome

For solutions to the Preliminary Exercises, see page 699.

Section Exercises

1. {penny, nickel, dime, quarter, 50-cent piece}

3. {Mercury, Mars} **5.** {0, 1, 2, 3, 4, 5, 6, 7}

7. {−5} **9.** True **11.** True **13.** True **15.** False

17. {0, 1, 3, 5, 8} **19.** U **21.** {0, 1, 4, 6, 8}

23. ∅, {a}, {z}, {a, z}

25. ∅, {I}, {II}, {III}, {IV}, {I, II}, {I, III}, {I, IV}, {II, III}, {II, IV}, {III, IV}, {I, II, III}, {I, II, IV}, {I, III, IV}, {II, III, IV}, {I, II, III, IV}

27. 32

29. The company can produce 4096 different versions of the truck.

31. {1, 2, 4, 5, 6, 8} **33.** {4, 6} **35.** {3, 7}

37. ∅ **39.** True **41.** True

43.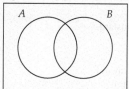

45.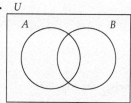

47. See solutions manual. **49.** See solutions manual.

51. (a) 180 (b) 200

53. (a) 3835 people (b) 245 people
(c) 1730 people (d) 190 people

55. (a) 101 people (b) 370 people (c) 380 people
(d) 373 people (e) 373 people (f) 530 people

57. (a)

	File Is Infected	File Is Not Infected
Found an Infected File	813	15
Did Not Find Infected File	27	42

(b) 15 (c) The software did not detect 27 files that were infected.

59. (a) 5 (b) There were 5 cases in which the test was positive but the child did not have the genetic marker. (c) 12 (d) There were 12 cases in which the test was negative but the child had the genetic marker. (e) 195 (f) There were 195 cases in which the test was positive and the child had the genetic marker. (g) 288 (h) There were 288 cases in which the test was negative and the child did not have the genetic marker.
(i)

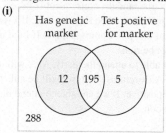

Section 2.2

Think About It

1. All humans are mammals. No cat is a dog.
2. Some pets are small. Some pets are not dogs.
3. No. The negative is "Some Ferraris are not red."
4. Yes. For instance, "If I like ice cream, all numbers are positive. I like ice cream. Therefore, all numbers are positive." is a valid argument but the conclusion is false.

For solutions to the Preliminary Exercises, see page 699.

Section Exercises

1. No lions are playful.
3. Some classic movies were not first produced in black and white.
5. Some even numbers are odd numbers.
7. Some cars do not run on gasoline.
9. Valid 11. Valid 13. Valid 15. Valid
17. Invalid 19. Invalid 21. Invalid 23. Valid
25. Valid 27. Invalid
29. All Reuben sandwiches need mustard.
31. 1001 ends with a 5.
33. Some horses are gray.
35. (a) Invalid (b) Invalid (c) Invalid
    (d) Invalid (e) Valid (f) Valid

Section 2.3

Think About It

1. Or 2. And 3. If–then
4. True 5. False 6. True

For solutions to the Preliminary Exercises, see page 700.

Section Exercises

1. Not a statement 3. Statement 5. Not a statement
7. The Giants did not lose the game.
9. The game did go into overtime.
11. $w \rightarrow t$ 13. $d \rightarrow f$ 15. $m \vee c$
17. I can play the piano and the violin.
19. If the painting is a watercolor, then James will not buy it.
21. If Carmella gets an A in mathematics, then she will major in physics or astronomy.
23. It did not rain and it did not snow.
25. Bao did not visit either France or Italy.
27. Antecedent: I had the money. Consequent: I would buy the painting.
29. Antecedent: Emene would go to the movies. Consequent: Ariel would go.
31. Antecedent: Cameron gets a raise. Consequent: Cameron will buy a new tablet computer.
33. False. The antecedent is true, and the consequent is false.
35. True: A false antecedent yields a true statement whether the consequent is true or false.
37. False. The antecedent is true, and the consequent is false.
39. Converse: If I quit this job, I would be rich. Inverse: If I were not rich, I would not quit this job. Contrapositive: If I did not quit this job, I would not be rich.
41. Converse: If we are not able to attend the party, Alex will not return soon. Inverse: If Alex returns soon, we will be able to attend the party. Contrapositive: If we are able to attend the party, Alex will return soon.

43. Converse: If we take the train, we'll be able to take the entire family. Inverse: If we do not take the entire family, we can't take the train. Contrapositive: If we do not take the train, we can't take the entire family.

45. $h \rightarrow r$
    $\sim h$
    ∴ $\sim r$
    Not valid

47. $e \rightarrow \sim s$
    $\sim s$
    ∴ e
    Invalid

49. Valid argument, *modus tollens*
51. Invalid argument, fallacy of the converse
53. Valid argument, transitive reasoning
55. Valid, transitive reasoning
57. Valid, transitive reasoning and rewriting some sentences using the contrapositive.
59. The moon is not made of green cheese.
61. Valid conclusion is not possible.
63. It is not a theropod.

Section 2.4

Think About It

1. Appeal to popularity fallacy
2. Appeal to common practice fallacy
3. Loaded question fallacy

For solutions to the Preliminary Exercises, see page 700.

Section Exercises

1. Ad hominem
   Premise: Dillon doesn't believe in equal rights.
   Conclusion: Don't listen to Dillon.
   This is an ad hominem fallacy because the premise is an attack on someone's character rather than the issue.
3. Black or white
   Premise: Either you buy me a car or I will lose my job.
   Conclusion: You must buy me a car so I don't lose my job.
   This is a black-or-white fallacy because the premise offers only two choices when in reality there are more.
5. Post hoc
   Premise: Immigration increased, and the economy improved.
   Conclusion: Immigration improved the economy.
   This is a post hoc fallacy; it reasons that because one event occurred before the other, it caused the second. Correlation does not always indicate causation.
7. Hasty generalization
   Premise: My cab driver was rude.
   Conclusion: Everyone in this country must be rude.
   This is a hasty generalization because it assumes that if something is true for one, then it is true for all.
9. Slippery Slope
   Premise: Without college you won't get a good job.
   Conclusion: You will be on the streets begging for change.
   This is a slippery slope because it assumes many factors in the premise are directly connected when the connection between each factor is actually weak. As the connection between factors weakens, so does the argument.

11. Appeal to emotion

Premise: He lost his partner a year ago.

Conclusion: Therefore, if he asks you on a date, you have to say yes.

This is an appeal to emotion because the argument attempts to connect an emotion, pity, to its conclusion.

13. Appeal to popularity

Premise: More than 100,000 people believe that astrology affects people's lives.

Conclusion: There must be some truth to it.

This is an appeal to popularity because the arguer is attempting to connect a popular idea to the reason for the conclusion.

15. Straw man

Premise: Going out of your way is an inconvenience.

Conclusion: There are more important things than convenience.

This is a straw man fallacy because a new premise was built to distract the audience from the original premise.

17. Red herring

Premise: Teachers don't get paid enough.

Conclusion: Teachers need to be paid extra for new programs.

This is a red herring fallacy because the premise is an attempt to distract the audience from the original statement.

19. Loaded question

No premise or conclusion.

This is a loaded question because a "yes" or a "no" answer both imply a false or deniable statement.

21. Circular reasoning

Premise: You have human parents.

Conclusion: You are a human being.

This is circular reasoning because the conclusion is the same as the premise, just reworded.

23. Appeal to common practice

Premise: Everyone drives 75 mph in a 65-mph zone.

Conclusion: Driving 75 mph in a 65-mph zone is acceptable.

This is an appeal to common practice because it relies on general acceptance in its premise rather than logic.

25. Appeal to authority

Premise: The actor stated that tight monetary policy is the best way to shorten a recession.

Conclusion: Tight monetary policy is the best way to shorten a recession.

This is an appeal to authority because it is drawing on expertise from an irrelevant authority.

27. Ad hominem

Premise: You are not a parent.

Conclusion: You don't know what's best for children.

This is an ad hominem fallacy because the premise is an attack on someone's character rather than the issue.

29. Slippery slope

Premise: I should make an exception for you.

Conclusion: I will have to make an exception for everyone.

This is a slippery slope because it assumes many factors in the premise are directly connected, when, actually, the connection between each factor is actually weak. As the connection between factors weakens, so does the argument.

31. Appeal to ignorance

Premise: The police could not prove that the strange activity was not from a ghost.

Conclusion: The strange activity was from a ghost.

This is an appeal to ignorance because it is drawing a conclusion based on lack of evidence for the converse of the premise.

33. Red herring

Premise: Students need a greater say in curriculum changes.

Conclusion: Teachers need a greater voice overall.

This is a red herring fallacy because the premise is an attempt to distract the audience from the original statement.

35. Black or white

Premise: Either you love soccer or you leave.

Conclusion: You must love soccer so that you don't have to leave.

This is a black-or-white fallacy because the premise offers only two choices when in reality there are more.

37. Appeal to emotion

Premise: My sister is beautiful and intelligent.

Conclusion: Therefore, she should give me a ride to school.

This is an appeal to emotion because the argument attempts to connect an emotion, flattery, to its conclusion.

39. Appeal to common practice

Premise: Most universities compensate athletes.

Conclusion: Compensating athletes is acceptable.

This is an appeal to common practice because it relies on general acceptance in its premise rather than logic.

41. Appeal to popularity

Premise: All of my friends have trucks, not cars.

Conclusion: Trucks are better than cars.

This is an appeal to popularity because the arguer is attempting to connect a popular opinion to the reason for the conclusion.

Chapter 2 Review Exercises

1. $\{-5, -4, -3, -2, -1\}$ **2.** $\{4, 6, 8, 9, 10, 12, 14, 15, 16\}$
3. $\{T, e, n, s\}$ **4.** $\{April, June, September, November\}$
5. $\{7\}$ **6.** $\{0, 1, 2, 3, 4\}$ **7.** $\{0, 1, 3, 5, 8\}$
8. $\{0, 1, 2, 4, 5, 7, 8\}$ **9.** U
10. $\varnothing$ **11.** True **12.** False **13.** True **14.** True
15. False **16.** False **17.** False **18.** False **19.** False
20. False **21.** $\{1, 2, 4, 5, 6, 8\}$ **22.** $\{2\}$ **23.** $\{4, 6\}$
24. $\{2, 5, 8\}$ **25.** $\{3, 7\}$ **26.** $\{2, 3, 4, 6, 7\}$
27. $\{1, 2, 4, 6\}$ **28.** $\{2\}$ **29.** U **30.** $\varnothing$
31.

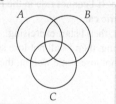

32. Not equal **33.** 32 **34.** 512
35. 1164 students **36.** 391 members
37. (a) 470 people **(b)** 20 people
  (c) 190 people **(d)** 70 people
38. (a) 342 people **(b)** 78 people
  (c) 581 people **(d)** 345 people
39. (a) 308 people **(b)** 319 people **(c)** 452 people
  (d) 138 people **(e)** 441 people
40. (a)

	Positive Test	Negative Test
Person Has Mumps Antigen	179	21
Person Does Not Have Mumps Antigen	83	717

(b) 21 **(c)** A person does not have the mumps antigen but tested positive for it. **(d)** 717

41. (a)

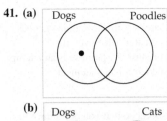

(b)

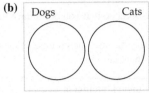

42. (a) All cars are fuel efficient.
   (b) Some parakeets are not birds.
43. Valid **44.** Invalid **45.** Invalid **46.** Valid
47. Invalid **48.** Invalid **49.** Valid **50.** Proposition
51. Not a proposition **52.** Not a proposition
53. The concert music was not too loud.
54. $p \wedge q$ **55.** $r \wedge \sim q$ **56.** $q \rightarrow \sim p$ **57.** $p \vee q \vee r$
58. The football game will be played on Friday and the football game will be played at home.
59. If the football game is played on Friday, then the visiting team is not from Boston.
60. The football game will be played at home, or the visiting team is from Boston, and the football game will be played on Friday.
61. This is not the time for action, and do not hold your peace.
62. I am not committed to this fine state or will seek the presidential nomination.
63. a. Antecedent: I practice my accent. Consequent: I will sound more like a native speaker **b.** Antecedent: Secure a loan on this house. Consequent: You make a 20% down payment.
64. True
65. False
66. True
67. True
68. Converse: If I can ride in the HOV lane, then I have an electric car. Inverse: If I do not have an electric car, then I cannot ride in the HOV lane. Contrapositive: If I cannot ride in the HOV lane, then I do not have an electric car.
69. Converse: If I lose some weight, then I start exercising. Inverse: If I do not start exercising, then I will not lose some weight. Contrapositive: If I do not lose some weight, then I did not start exercising.
70. Invalid
71. Valid
72. Valid
73. Valid
74. This is an ad hominem (personal attack) fallacy.
75. This is circular reasoning.
76. Slippery slope fallacy.
77. Appeal to emotion fallacy.

Chapter 3

Section 3.1

Think About It

1. 25 blue jelly beans **2.** Yes
3. 67 **4.** $22 per thousand

For solutions to the Preliminary Exercises, see page 701.

Section Exercises

1. Examples may vary. For instance, miles per gallon and cost per pound.
3. $\dfrac{1}{3}$ **5.** $\dfrac{3}{8}$
7. $\dfrac{2}{9}$ **9.** 48.5 mph
11. 68 words/min **13.** $26 per share
15. $ 19.50/hr **17. (a)** 336 ft/min **(b)** 89 s
19. 24 ounces for $3.89
21. **(a)** 9 to 1 **(b)** Explanations may vary. A lower ratio of at bats to home runs means the batter is more likely to hit a home run.
23. **(a)** 77.1, 76.4, 79.2 **(b)** 1.03 times
25. **(a)** 12 to 1; There is one faculty member for every 12 students. **(b)** Rice
27. 755 employed women per 1000 women
29. 60 dentists per 100,000 population
31. $1588.33 **33.** $276.59
35. $6,069,348.66
37. 2.378 Swedish krona per shekel

Section 3.2

Think About It

1. $\dfrac{1 \text{ foot}}{12 \text{ inches}}$ **2.** Kilogram **3.** 3, left **4.** Longer **5.** 1 liter

For solutions to the Preliminary Exercises, see page 701.

Section Exercises

1. 72 in. **3.** 180 in. **5.** 4 lb
7. 24 oz **9.** 20 fl oz **11.** 56 pt
13. 620 mm **15.** 321 cm **17.** 7.421 kg
19. 4.5 dg **21.** 0.0075 L **23.** 435 cm^3
25. 65.91 kg **27.** 48.3 km/h **29.** $8.70/L
31. 1.38 in. **33.** $4.55/lb

Section 3.3

Think About It

1. 3 and b are the extremes; 4 and a are the means
2. Equals **3.** Yes **4.** No
5. Corresponding units do not match.

For solutions to the Preliminary Exercises, see page 702.

Section Exercises

1. 4.5 **3.** 42.86 **5.** 216 **7.** 4.31 **9.** 4.35 **11.** 10.97
13. 29 lb **15.** 5.5 mg **17.** 24 ft; 15 ft
19. 64 in. **21.** 160,000 people
23. 11.25 g. Explanations will vary.
25. 39.0 pounds of flour; 9.8 pounds of sugar; 1.2 pounds of water.

Section 3.4

Think About It

1. Greater than **2.** Less than **3.** Less than
4. There may be people who favor both.
5. The five scores of 90% count more heavily than one score of 86%.

For solutions to the Preliminary Exercises, see page 702.

Section Exercises

1. $0.4, \dfrac{2}{5}$ 3. $0.73, \dfrac{73}{100}$ 5. $0.02, \dfrac{1}{50}$ 7. $0.005, \dfrac{1}{200}$
9. $0.015, \dfrac{3}{200}$ 11. $0.5625, \dfrac{9}{16}$ 13. 60%
15. 37.5% 17. 35% 19. 60% 21. 5%
23. 1.5% 25. 125 million returns were filed electronically.
27. $58.5 billion 29. 23.7%
31. 14 million oz 33. 3364 people 35. 51.4%
37. (a) 102.5% (b) 122.2% (c) 350% (d) It is 4.5 times as large. Convert the percent to a decimal and add 1.
39. 28.86 million
41. (a) 1.2% (b) 15.7 pounds
43. Less than

Section 3.5

Think About It

1. Increases 2. $y = kx^2$ 3. Increases 4. $y = \dfrac{k}{x^2}$
5. Directly 6. Inversely

For solutions to the Preliminary Exercises, see page 702.

Section Exercises

1. $y = kx, k = \dfrac{4}{3}$ 3. $r = kt^2, k = \dfrac{1}{81}$
5. $T = krs^2, k = \dfrac{7}{25}$ 7. $V = klwh, k = 1$
9. 1.02 liters 11. 62 semester hours
13. 11.7 fluid ounces 15. (a) 3.3 seconds (b) 3.7 feet
17. 40 revolutions per minute 19. 0.245 watt per square meter
21. (a) V is 9 times as large. (b) V is 3 times as large. (c) V is 27 times as large.
23. V is 6 times as large. 25. 3950 pounds
27. $d = 16$ inches

Chapter 3 Review Exercises

1. $\dfrac{1}{4}$ 2. 7.2 3. 12.3 to 1 4. Greater than
5. 28.4 mi/gal 6. 18-ounce box for $3.75
7. New York, Chicago, Los Angeles, Houston
8. 22.5 per 100,000 9. 343 people 10. $2797.50
11. $247,636.62 12. 2.25 ft 13. 40 oz 14. 4 quarts
15. 3.7 cm 16. 3,200,000 mg 17. 0.58 liter 18. $3.36/qt
19. $0.65/L 20. 72 km/hr 21. 1.7 22. 14 23. 9.7
24. 1.3 25. 7.5 tablespoons
26. 525 gallons, 50 minutes
27. 1236 elk 28. $80,000 29. $0.04, \dfrac{1}{25}$
30. $0.23, \dfrac{23}{100}$ 31. $0.125, \dfrac{1}{8}$ 32. $1.5, \dfrac{3}{2}$
33. 8% 34. 120% 35. 1.5% 36. 62.5%
37. 22,750 strikes 38. 200 students 39. 37.7%
40. 18.2% 41. 7.4% 42. $7095
43. (a) Less than (b) About 410,400,000 people
44. 3.0% 45. 3 feet per second squared
46. 23.85 feet 47. 7500 video games
48. 10 pounds 49. 75 pounds
50. 160 cubic centimeters

Chapter 4

Section 4.1

Think About It

1. Yes 2. Present value, $500; Maturity value, $525
3. $20 4. Raul

For solutions to the Preliminary Exercises, see page 703.

Section Exercises

1. 10% 3. $960.96 5. $132.99 7. $1222 9. $18
11. $145 13. $10,350 15. $42.50 17. 350% 19. 8.2%
21. 2.7% 23. $2.09 25. $43.50 27. 1.2%

Section 4.2

Think About It

1. Compounded daily
2. Present value, $5000; Future value, $5100
3. Nominal $100; real $94
4. Nominal, 5%; Annual percentage yield, 5.13%

For solutions to the Preliminary Exercises, see page 704.

Section Exercises

1. $2139.78 3. $22,325.47 5. $7068.25
7. $71,000; $72,062; $70,530 9. 3.04% 11. $8594.38
13. 13.4% 15. $5414.84 17. $5728.18 19. $391.24
21. $9612.75 23. $67,227.88 25. $10,253.43 27. 4.93%

Section 4.3

Think About It

1. Payment for an ordinary annuity is made at the end of the payment period; payment for a due annuity is made at the beginning of the payment period.
2. Answers may vary. A car loan payment is based on an ordinary annuity. The first payment is due at the end of the month. Funding a retirement account is usually a due annuity. When the account is opened, a payment is made, the beginning of the payment period.
3. The due annuity. It has one more interest period than an ordinary annuity.

For solutions to the Preliminary Exercises, see page 705.

Section Exercises

1. $197,473.60 3. (a) $179.03 (b) $36.27
5. $508.75 7. $121,791.14 9. $91,912.73 11. $1408.81
13. $267,623.57 15. $210,192.59 17. $385.46

Section 4.4

Think About It

1. 1.5% 2. $210

For solutions to the Preliminary Exercises, see page 706.

Section Exercises

1. $830.53 3. $586.55 5. $9.86 7. 19 months
9. $20.60 11. $179.52 13. (a) $3610.12 (b) $2000 (c) $1610.12 15. $1951.01
17. (a) $2501.04 (b) By making the minimum payment, your debt is increasing even though you are not using the card.

Section 4.5

Think About It

1. For a subsidized student loan, the government pays the interest on the loan while the student is in school. For a nonsubsidized loan, the student must pay the accrued interest on the loan.
2. The student will pay more interest on the nonsubsidized loan.
3. The discount fee is an amount subtracted from the loan amount. It is usually a percent of the loan amount.

For solutions to the Preliminary Exercises, see page 708.

Section Exercises

1. $7176.78　**3.** $205.07　**5.** $213.86
7. Approximately 120 months　**9.** $200.12
11. Katrina pays less interest for the subsidized loan. The interest paid on the nonsubsidized loan is $6392.96; the interest paid on the subsidized loan is $5393.60.
13. $676.66　**15.** 11.76%
17. Yes. The total interest on the two loans is $14,763.20; the interest on the 10-year $40,000 loan is $14,502.80.

Section 4.6

Think About It

1. No. The car depreciates by 20% of its current value, not 20% of its original value.
2. 4 gallons per 100 miles　**3.** 45%

For solutions to the Preliminary Exercises, see page 709.

Section Exercises

1. $1021.08　**3.** $133.04　**5.** $500.57　**7.** $25,493.08
9. $26,599　**11.** $655.63　**13.** (a) 20 mpg　(b) $2
15. $85.31　**17.** $528.36　**19.** $18,626.04
21.

License and registration	0.75% of purchase price	224.25
Sales tax	6.25% of purchase price	1868.75
Depreciation	21% of purchase price	6279.00
Fuel cost driving 14,000 miles	Average cost of gasoline: $2.74 per gallon	1370.00
Year 1 interest payment	Amount paid in interest on the car loan for first year of ownership	960.63
Insurance	$712	712.00
Maintenance	$356	356.00
	Total	11,770.63

23. The person going from 20 mpg to 30 mpg.
25. $639.50　**27.** $209.17　**29.** (a) $118.41　(b) $643.80
31. 4%　**33.** $1013.32

Section 4.7

Think About It

1. No. The DTI ratio must be less than 35.
2. The interest rate for a fixed-rate mortgage remains the same throughout the term of the loan. The interest rate for an adjustable-rate mortgage changes on a periodic basis.
3. More than　**4.** 1.5%

For solutions to the Preliminary Exercises, see page 710.

Section Exercises

1. 64　**3.** (a) $453,845.00　(b) 7.4%
5. $4025　**7.** $3280.00　**9.** $1451.22　**11.** $1255.48
13. $2771.47　**15.** 21 months　**17.** 3.789%
19. (a) $1844.77, $2031.52　(b) $54,661.20
21. (a) $1969.33, $3059.32　(b) $158,281.20
　　(c) No, it is about 64% of the 15-year-loan payment.
　　(d) More
23. (a) 176th month　(b) $412,028.22　(c) No.

Section 4.8

Think About It

1. Down
2. Decrease
3. Greater than
4. Discount
5. Real rate of return considers the effects of inflation on the value of an asset.

For solutions to the Preliminary Exercises, see page 711.

Section Exercises

1. 2.17%　**3.** 3.1%　**5.** $11.39　**7.** $1925　**9.** $2.16
11. 12.0　**13.** $1143　**15.** $983　**17.** 222.790　**19.** $15.03
21. 24.3　**23.** $297　**25.** $5.46　**27.** 2.58%

Chapter 4 Review Exercises

1. (a) $556.88　(b) $74.25
2. $510　**3.** 130.4%　**4.** $7215.41　**5.** 7.5%　**6.** 4.0%
7. $3654.90　**8.** $11,609.72　**9.** $7859.53　**10.** $200.23
11. $10,683.29　**12.** $19,225.50　**13.** $184,227.03　**14.** 1.0%
15. $285,722.89　**16.** No. It is shy by about $1852.
17. 8.3%　**18.** 5.4% compounded semiannually
19. $97,394.96　**20.** $59,281.08　**21.** $211.71
22. $411,513.73　**23.** $237,066.65　**24.** $100,065.72
25. $25,000 per month for 20 years
26. $231.00　**27.** $408.26, $5442.96　**28.** $14,864.60
29. $45.41　**30.** $632.93　**31.** $9.62
32. $16.16　**33.** $30.01　**34.** 11 months
35. $14,835　**36.** $266.74　**37.** $172.85
38. 103 months　**39.** $777.87　**40.** 11.8%　**41.** $664.40
42. (a) $3763.13　(b) $6007.11　(c) $1241.39
43. $11,290.74　**44.** $23,077.10　**45.** 3.0 gallons per 100 miles
46. $32,680　**47.** $17,062.18　**48.** $362.78　**49.** 47%
50. (a) $1659.11　(b) $597,279.60　(c) $341,479.60
51. (a) $1745.86　(b) $298,609.60
52. $276.67　**53.** $8145　**54.** 52 months　**55.** $3022.71
56. (a) $1396.69　(b) $150,665.74
57. 3.600%　**58.** $2658.53　**59.** $896　**60.** 3.1%
61. $4.40　**62.** 20.7　**63.** $1125　**64.** $90.32
65. 186.560　**66.** 4.8%

Chapter 5

Section 5.1

Think About It

1. $z = 7$
2. $x = -2, y = 5$
3. No. That would mean that (2, 3) and (2, 4) belong to the function, which means that the same value of the independent

variable, 2, is paired with two different values for the dependent variable. This contradicts the *one and only one* value of the dependent variable for a value of the independent variable.

4. Yes. Different values of the independent variable can be paired with the same value of the dependent variable. This cannot be reversed; that is, the same value of the independent variable cannot be paired with two different values of the dependent variable. See the preceding Exercise 3.

5. Independent variable, *x*; dependent variable, *y*; name of the function, *f*.

6. domain, range

For solutions to the Preliminary Exercises, see page 712.

Section Exercises

1. Independent variable, *h*, the height of the ball before it is dropped; dependent variable, *s*, the speed of the ball when it hits the ground. $s = Sp(h)$

3. Independent variable, *T*, the temperature; dependent variable, *s*, the speed of sound. $s = v(T)$

5. $45,000   **7.** 144 cm^2 **9. (a)** 100 ft   **(b)** 68 ft

11. **(a)** 1087 ft/s   **(b)** 1136 ft/s   **13. (a)** 1.92 s   **(b)** 1.0 s

15. **(a)** *p, H, A*   **(b)** *T*   **(c)** 147 beats per minute

17. **(a)** *w, h, a*   **(b)** BMR   **(c)** 1777 Calories

19. **(a)** (24, 60)   **(b)** It takes 32 seconds to download an 80-gigabyte file.

(c)

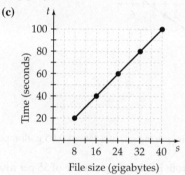

21. **(a)** (20, 78)   **(b)** If the price of the game is $25, the company expects to sell 66,000 games.

(c)

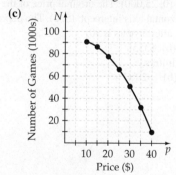

23. Approximately 24 feet

25. A little more than 200 yards

27. **(a)** −250 bacteria/hour   **(b)** The bacteria population is decreasing at an average rate of 250 bacteria per hour between 1 and 3 hrs after measurements begin.

29. **(a)** 1033 flu cases/week   **(b)** The number of flu cases is increasing at an average rate of 1033 flu cases per week between 2 and 5 weeks after measurements begin.

31. Decrease   **33.** *A*   **35. (a)** $1000/year   **(b)** $12,000

37. **(a)** −2 lb/week   **(b)** 170 pounds

Section 5.2

Think About It

1. **(a)** Not a linear function   **(b)** Linear function, $m = 1, b = 0$
   (c) Linear function, $m = -0.256, b = 4$
   (d) Not a linear function

2. A line with 0 slope is horizontal; a line with no slope is vertical.

3. $(0, -3)$ is the *y*-intercept; $(2, 0)$ is the *x*-intercept.

4. Yes, the slopes are the same. No, the *y*-intercepts are different.

5. *A*   **6.** *B*   **7.** b and d

For solutions to the Preliminary Exercises, see page 712.

Section Exercises

1.

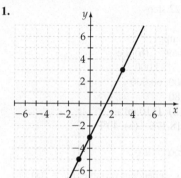

3.

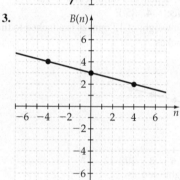

5. $-\dfrac{5}{6}$ **7.** 0

9. Yes. The slope between any two points is the same: 3 gallons/minute.

11. No. The slope between points 1 and 2 does not equal the slope between points 2 and 3.

13. $m = 40$ mi/hr. The motorist was traveling at 40 mph.

15. $m = -1000$ ft/min. The height of the plane is decreasing at a rate of 1000 feet per minute.

17. $m = -0.63$ million gallons/minute. The water in the lock is decreasing by 0.63 million gallons each minute.

19. $m = -15°F$/min. The temperature inside the oven is decreasing at a rate of 15°F per minute.

21. $m = -0.003°F$/foot. The temperature changes $-0.3°F$.

23. No. 1-to-12 ratio is approximately 0.083, and 0.09 > 0.083.

25. **(a)** 0.117 gallon/square foot   **(b)** approximately 36.7 gallons

27. **(a)** −0.01 dollar/drill sold   **(b)** For each 1-cent decrease in price, one additional drill can be sold.   **(c)** 3200 drills

29. **(a)** (0, 270)   **(b)** If the tuition were $0, 270 students would enroll.   **(c)** (900, 0)   **(d)** If the tuition were $900, no students would enroll.

31. **(a)** (0, −1000)   **(b)** If the business sells 0 smartwatches, it will lose $1000.   **(c)** (40, 0)   **(d)** The company must sell 40 smartwatches to break even.

Section 5.3

Think About It

1. **(a)** Decrease **(b)** Increase **(c)** Decrease **(d)** Increase
2. °F/min
3. Bucket B
4. In 4 hrs, bucket B will be empty ($y = 0$).

For solutions to the Preliminary Exercises, see page 713.

Section Exercises

1. $T = 0.02s + 4000$, $5600
3. $d = 2700 - 500t$, 1700 miles
5. $A = 1250t + 2000$, 27,000 feet
7. $S = -34h + 4562$, 2522 steps
9. $H = 3.5n + 3$, 38 feet
11. $mpg = -8w + 49$, 25 mpg
13. **(a)** $F(t) = 99 + 25t$; $699 **(b)** Less
15. $C(x) = 85x + 30,000$; $183,000
17. $C(x) = -10x + 230,000$; 60,000
19. **(a)** $y = 8x - 4$ **(b)** 80
21. **(a)** $y = 34x + 260$ **(b)** 1110 ml
23. **(a)** $y = -1.735x + 118.176$ **(b)** 40°F

Chapter 5 Review Exercises

1. Independent variable: s; dependent variable: t; $t = f(s)$
2. 5000 cameras **3.** 71 feet **4.** 85°F
5. **(a)** $(8, 5)$ **(b)** When a weight of 12 ounces is attached to the spring, its length is 6 inches.
   (c)

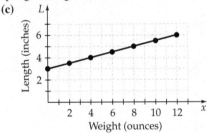

6. **(a)** $(200, 300)$ **(b)** When the company produces 250,000 parts, the expected profit is $275,000.
   (c)

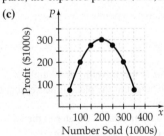

7. From the following graph, the amount borrowed is approximately $27,000.

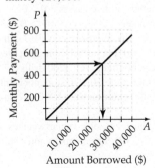

8. **(a)** -0.95 ft/s per foot **(b)** The velocity of the ball is decreasing at an average rate of 0.95 foot per second for each 1 foot of additional height.
9. B_1 and B_2
10. **(a)** 158 people/day **(b)** The number of people who have heard the rumor is increasing at an average rate of 158 people per day.
11. Greater than
12.

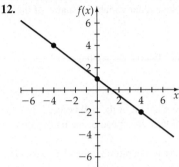

13.

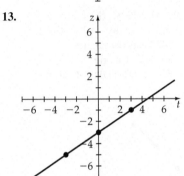

14. 5 **15.** $\dfrac{5}{2}$ **16.** 0
17. No slope
18. $m = -0.05$; The car is using gas at a rate of 0.05 gallon per mile driven.
19. **(a)** $m = 5$ **(b)** Profit is increasing at a rate of $5 per rose bush sold. **(c)** $550
20. -4 feet/second
21. Vertical axis intercept: $(0, 25,000)$. The original price of the van was $25,000. Horizontal axis intercept: $(5, 0)$. The depreciated value of the van after 5 years is $0.
22. **(a)** $C = 12x + 100$ **(b)** $3556
23. $y = 15t + 50$; 500 milliliters
24. **(a)** $y = -6x + 14$ **(b)** -28°C

Chapter 6

Section 6.1

Think About It

1. Growth, 3% **2.** Decay, 25%
3. False. It will be more than twice.
4. 50 seconds **5.** 5.5 hours

For solutions to the Preliminary Exercises, see page 714.

Section Exercises

1. $3128.62 **3.** 3,500,000 people **5.** $A = 2(3)^{t/8}$
7. $P = 100(0.81)^d$, 22.9% **9.** $C = 300,000(0.61)^{t/2}$, $4.43
11. **(a)** $N = 127(1.20)^{n/2}$. **(b)** Yes. There would be approximately 7.2 million viewers.

13. $A = 12.3(2^{t/4})$; 49.2 billion

15. $A = 5\left(\dfrac{1}{2}\right)^{t/110}$; 3.77 micrograms

17. $P = 100(0.75)^{x/4}$, 7.5% **19.** $9819.82

21. $2155.41 **23.** $N = 2(3)^n$

25. $P = 100(0.32)^{x/10}$. Yes. The percent of light passing through is approximately 6%.

27. 1.5 mg

29. (a) 100°F (b) No. The temperature of the oven cannot go below room temperature.

31. $A = 15(0.80)^{n/5280}$; 1.85 psi

33. $A = 33,000(1.02)^t$; 40,227 scientists

Section 6.2

Think About It

1. $x = \log(7)$ **2.** $10^y = 9$ **3.** No. **4.** Acidic

5. The time it takes for the investment to double in value.

For solutions to the Preliminary Exercises, see page 715.

Section Exercises

1. (a) 100,000 (b) 0.1 (c) 4 (d) −2

3. 25 dB **5.** 10^{10} times the threshold **7.** 8.8

9. 15.1 hours **11.** 20 minutes **13.** 174 years **15.** 75 dB

17. 341,526 gallons **19.** 3162 times

21. (a) 75.8% (b) 50,000 hours

23. 5.2 years **25.** 2.7 days **27.** $A = 26.3(2)^{t/3}$; 105.2 billion

29. $A = 1000(0.80)^{t/5}$; 67.1 hours

31. $A = 1000(2)^{t/84}$; 5209 ants

Section 6.3

Think About It

1. Down **2.** Yes

3. (a) quadratic (b) exponential

4. Vertex

For solutions to the Preliminary Exercises, see page 716.

Section Exercises

1. Up, (5, −25) **3.** Up, (0, −10) **5.** Down, (3, 10)

7. Up, $\left(\dfrac{3}{4}, \dfrac{47}{8}\right)$ **9.** Down, $\left(\dfrac{1}{8}, \dfrac{17}{16}\right)$ **11.** −16, minimum

13. 0, minimum **15.** 15, maximum **17.** −9, minimum

19. $\dfrac{33}{2}$, maximum **21.** 18.4 seconds **23.** 25 ft by 25 ft

25. (a) $3x + 2y = 2100$ (b) $y = 1050 - \dfrac{3}{2}x$

(c) $A = 1050x - \dfrac{3}{2}x^2$ (d) $x = 350$ ft, $y = 525$ ft

(e) Area = 183,750 square feet

27. (a) 1:00 pm (b) 91°F

29. (a) 4 seconds (b) 256 feet (c) 8 seconds

31. (a) $R = 6000x - 45x^2$ (b) $P = -45x^2 + 4500x - 3000$

(c) 50 televisions (d) $109,500

Chapter 6 Review Exercises

1. Exponential growth **2.** Exponential decay

3. Neither **4.** Exponential growth

5. (a) 262 (b) $\dfrac{27.5 \cdot 4^{51/24}}{27.5 \cdot 4^{39/24}} = 4^{51/24 - 39/24} = 4^{12/24} = 4^{1/2} = \sqrt{4} = 2$

6. 23.5 watts **7.** $A = 50,000 \cdot 1.2^t$, 94,646 bacteria

8. $H = 50 \cdot 1.05^t$, 90 centimeters

9. $W = 7.2 \cdot 2^{t/15}$, 38 pounds

10. $A = 5000 \cdot \left(1 + \dfrac{0.05}{12}\right)^{12t}$, $7090.18

11. $A = 10,000 \cdot 1.00010274^{365t}$, $13,498.39

12. (a) $A = 2000(1.12)^{t/4}$ (b) 6212

13. 141,000 *Streptococcus lactis*

14. 15 grams

15. $A = 25(0.7)^{t/30}$, 6.00 pounds

16. $A = 500,000(0.95)^{t/5}$, 270,000 pathogens

17. $H = 10(0.90)^n$ **18.** $A = 75(0.5)^{t/28.8}$, 22.5 g

19. $A = 50(0.5)^{t/15}$, 48.9 g

20. (a) 10,000 (b) 0.001 (c) 7 (d) −4

21. (a) 6.3096 (b) 0.0631 (c) 1.3617 (d) 1.4125

22. pH = 6.7, acid **23.** $H^+ = 0.0000063$ **24.** 5.9

25. 2512 times **26.** 188 db **27.** No. 31,594 times

28. $A = 5000(0.7)^{t/50}$, 548 seconds **29.** 6 years

30. 12.4 hours **31.** 33 years

32. (a) No (b) Yes, $a = \dfrac{1}{2}, b = 0, c = 3$ (c) No

(d) Yes, $a = -1, b = -1, c = 1$

33. (a) Up, (4, −22) (b) Down, $\left(\dfrac{3}{2}, \dfrac{9}{4}\right)$ (c) Down, (−1, −1)

(d) Up, $\left(\dfrac{1}{2}, \dfrac{11}{4}\right)$

34. (a) y-intercept: (0, −8); x-intercepts: (−2, 0), (4, 0)

(b) y-intercept: (0, 18); x-intercepts: (−3, 0), (6, 0)

35. 4.1 seconds **36.** 155 feet

37. (a) Minimum, −5; (b) Maximum, $\dfrac{15}{2}$

38. (a) 2 seconds (b) 70 feet

39. 18 ft by 36 ft **40.** 35 washing machines, $34,550

Chapter 7

Section 7.1

Think About It

1. An experiment is an activity with an observable outcome.

2. A sample space is the set of all possible outcomes from an experiment.

3. The sample space is the set of all possible outcomes of an experiment. An event contains one or more of the possible outcomes of an experiment, so an event is a subset of the sample space.

4. More. Each stage of an experiment (after the first) performed without replacement will have fewer possible outcomes than the preceding stage. Performed with replacement, each stage of the experiment has the same number of outcomes.

For solutions to the Preliminary Exercises, see page 717.

Section Exercises

1. {0, 2, 4, 6, 8}

3. {Monday, Tuesday, Wednesday, Thursday, Friday, Saturday, Sunday}

5. {HH, TT, HT, TH}

7. {1H, 2H, 3H, 4H, 5H, 6H, 1T, 2T, 3T, 4T, 5T, 6T}

9. {$S_1E_1D_1$, $S_1E_1D_2$, $S_1E_2D_1$, $S_1E_2D_2$, $S_1E_3D_1$, $S_1E_3D_2$, $S_2E_1D_1$, $S_2E_1D_2$, $S_2E_2D_1$, $S_2E_2D_2$, $S_2E_3D_1$, $S_2E_3D_2$}

11. {ABCD, ABDC, ACBD, ACDB, ADBC, ADCB}

13. 12 **15.** 4^{20} **17.** 7000 **19.** 90 **21.** 18

23. 62 **25.** 24 **27.** 15 **29.** 13^4 **31.** 1
33. Answers will vary.

Section 7.2

Think About It

1. $0! = 1$.
2. A permutation is an arrangement of objects in a definite order. A combination is a collection of objects in which the order of the objects is not important.
3. (a) permutation (b) permutation (c) combination
   (d) combination

For solutions to the Preliminary Exercises, see page 717.

Section Exercises

1. 40,320 **3.** 362,760 **5.** 120 **7.** 6720 **9.** 181,440
11. 1 **13.** 40,320 **15.** 3360 **17.** $\dfrac{1}{720}$ **19.** 36
21. 1 **23.** 525 **25.** $\dfrac{60}{143}$ **27.** $\dfrac{360}{1001}$ **29.** 2880
31. 21 **33.** 792 **35.** No **37.** 120 **39.** 43,680
41. (a) 362,880 (b) 8640 **43.** 10,080 **45.** 735,471
47. 35 **49.** 3136 **51.** 12 **53.** 9 **55.** 48,620
57. 24 **59.** 45,360 **61.** 252 **63.** 1728 **65.** 48
67. 4512 **69.** 103,776

Section 7.3

Think About It

1. No, all probabilities are between 0 and 1, inclusive.
2. Yes, each pencil has the same chance of being selected.
3. No, the probability of, for instance, a sum of 7 is not the same as the probability of a sum of 11.
4. $\dfrac{1}{2}$
5. Less than. There are fewer red balls.

For solutions to the Preliminary Exercises, see page 718.

Section Exercises

1. {HHH, HHT, HTH, HTT, THH, THT, TTH, TTT}
3. {Nov. 1, Nov. 2, Nov. 3, Nov. 4, Nov. 5, Nov. 6, Nov. 7, Nov. 8, Nov. 9, Nov. 10, Nov. 11, Nov. 12, Nov. 13, Nov. 14}
5. {Alaska, Alabama, Arizona, Arkansas}
7. {HHH, HHT, HTH, HTT, THH, THT, TTH, TTT}
9. {HTT, THT, TTH, TTT}
11. {HHT, HTH, HTT, THH, THT, TTH, TTT}
13. $\dfrac{1}{2}$ **15.** $\dfrac{7}{8}$ **17.** $\dfrac{1}{4}$ **19.** $\dfrac{1}{12}$ **21.** $\dfrac{1}{4}$ **23.** $\dfrac{5}{36}$
25. $\dfrac{1}{36}$ **27.** 0 **29.** $\dfrac{1}{6}$ **31.** $\dfrac{1}{2}$ **33.** $\dfrac{1}{6}$ **35.** $\dfrac{1}{13}$
37. $\dfrac{5}{13}$ **39.** $\dfrac{1267}{3228} \approx 0.39$ **41.** $\dfrac{804}{3228} \approx 0.25$
43. $\dfrac{150}{3228} \approx 0.05$ **45.** $\dfrac{26}{425}$ **47.** $\dfrac{18}{425}$ **49.** $\dfrac{58}{293}$
51. $\dfrac{36}{293}$ **53.** (a) $\dfrac{17}{334}$ (b) $\dfrac{44}{167}$ (c) $\dfrac{317}{334}$ (d) $\dfrac{24}{167}$
55. $\dfrac{1}{4}$ **57.** 0 **59.** $\dfrac{1}{3}$ **61.** $\dfrac{3}{10}$ **63.** $\dfrac{8}{13}$
65. 1 to 4 **67.** 3 to 5 **69.** 11 to 9 **71.** 1 to 14
73. 1 to 1 **75.** 15 to 1 **77.** $\dfrac{49}{97}$ **79.** $\dfrac{3}{11}$

Section 7.4

Think About It

1. Yes, a number cannot be both even and odd.
2. No, 3, 5, and 7 are both odd numbers and prime numbers.
3. $1 - 0.3 = 0.7$
4. $L = $ {HHHT, HHTH, HTHH, THHH, HHHH}
5. $M = $ {HHTT, HTHT, HTTH, TTHH, THHT, THTH, HTTT, THTT, TTHT, TTTH, TTTT}

For solutions to the Preliminary Exercises, see page 718.

Section Exercises

1. $\dfrac{2}{13}$ **3.** $\dfrac{1}{9}$ **5.** 0.6 **7.** 0.2 **9.** $\dfrac{7}{10}$ **11.** $\dfrac{4}{5}$
13. $\dfrac{5}{18}$ **15.** $\dfrac{1}{2}$ **17.** $\dfrac{11}{18}$ **19.** $\dfrac{4}{13}$ **21.** $\dfrac{3}{13}$
23. $\dfrac{3}{4}$ **25.** $\dfrac{1150}{3179}$ **27.** $\dfrac{1170}{3179}$ **29.** 0.96
31. $\dfrac{74}{75}$, or about 98.7% **33.** $\dfrac{5}{6}$ **35.** $\dfrac{11}{12}$ **37.** $\dfrac{12}{13}$ **39.** $\dfrac{15}{16}$
41. 42.1% **43.** 36.1% **45.** 54.5% **47.** 88.8% **49.** 20.4%

Section 7.5

Think About It

1. The probability that a number is odd given that it is a prime number.
2. Independent
3. Dependent
4. The product of the probability of A and the probability of B.

For solutions to the Preliminary Exercises, see page 719.

Section Exercises

1. $P(A|B) = 0.625$; $P(B|A) \approx 0.357$
3. $P(A|B) \approx 0.389$; $P(B|A) \approx 0.115$
5. $\dfrac{179}{864}$ **7.** $\dfrac{557}{921}$ **9.** 0.30 **11.** 0.15
13. $\dfrac{5}{18}$ **15.** $\dfrac{1}{5}$ **17.** 0.050 **19.** 0.127 **21.** $\dfrac{6}{1045}$
23. $\dfrac{3}{1045}$ **25.** $\dfrac{13}{102}$ **27.** $\dfrac{8}{5525}$ **29.** 0.000484
31. 0.001424 **33.** Independent **35.** Not independent
37. $\dfrac{25}{1296}$ **39.** $\dfrac{1}{72}$ **41.** $\dfrac{1}{4}$ **43.** $\dfrac{1}{16}$ **45.** $\dfrac{1}{216}$ **47.** $\dfrac{1}{169}$
49. $\dfrac{1}{16}$ **51.** $\dfrac{1}{16}$ **53.** (a) $\dfrac{1}{32}$ (b) $\dfrac{13}{425}$
55. (a) $\dfrac{100}{4913}$ (b) $\dfrac{1}{51}$ **57.** 0.46 **59.** 0.11

Section 7.6

Think About It

1. Yes. All casino games have a negative expectation for the player.
2. A person who plays this game many times can expect to lose an average of $2 each time the game is played.
3. $9.60, twice the expectation of the $5 wager.

For solutions to the Preliminary Exercises, see page 719.

Section Exercises

1. 49.5 **3.** -5 cents **5.** -24 cents **7.** -22 cents
9. $62.83 **11.** $-$209.71 **13.** More than $42.18
15. $39,100 **17.** $20,250 **19.** $2144.13

Chapter 7 Review Exercises

1. {11, 12, 13, 21, 22, 23, 31, 32, 33}
2. {26, 28, 62, 68, 82, 86}
3. {HHHH, HHHT, HHTH, HHTT, HTHH, HTHT, HTTH, HTTT, THHH, THHT, THTH, THTT, TTHH, TTHT, TTTH, TTTT}
4. {7A, 8A, 9A, 7B, 8B, 9B}   5. 72
6. 10,000   7. 2400   8. 64
9. 5040   10. 40,296
11. 1260   12. 151,200
13. 336   14. $\frac{60}{143}$ 15. 5040
16. 5040   17. 2520   18. 180
19. 495   20. 60   21. 3,268,760
22. 165   23. 660   24. 282,240
25. 624   26. $\frac{15}{64}$ 27. $\frac{3}{8}$
28. 0.56   29. 0.37   30. 0.85
31. $\frac{1}{9}$ 32. $\frac{17}{18}$ 33. $\frac{1}{6}$
34. $\frac{5}{9}$ 35. $\frac{2}{9}$ 36. $\frac{1}{6}$
37. $\frac{3}{4}$ 38. $\frac{4}{13}$ 39. $\frac{12}{13}$
40. $\frac{2}{3}$ 41. 5 to 31   42. 1 to 3
43. $\frac{5}{9}$ 44. $\frac{1}{2}$ 45. 0.036
46. $\frac{173}{1000}$ 47. $\frac{5}{8}$ 48. 1 to 5
49. $\frac{7}{16}$ 50. $\frac{646}{1771}$ 51. $\frac{7}{253}$
52. 0.37   53. 0.62   54. 0.07
55. 0.47   56. 0.29   57. $\frac{1}{216}$
58. 0.60   59. 0.16   60. $\frac{175}{256}$
61. 0.648   62. 0.029
63. 0.135   64. −50 cents
65. 25 cents   66. −37 cents
67. About 5.4   68. $296.20
69. $765.30   70. $11,000   71. $408,650

Chapter 8

Section 8.1

Think About It

1. Descriptive statistics involves the collection, organization, summarization, and presentation of data. Inferential statistics interprets and draws conclusions from the data.
2. The mean will change because 2700 replaces 2500 as part of a sum. The median will not change because 2700 does affect the number of scores above the median.
3. Yes. The mode is the most frequently occurring data point and therefore is in the data set.
4. The median is the data value in the middle of an ordered list of numbers, so the median of this data set is 3. The mean is the sum of the data values divided by the number of data values. Since the data value of 20 is quite large, the sum of the data values will be quite large. So, one can expect the mean to be greater than the median in this instance.
5. A weighted mean assigns more importance (weights) to some values over others. For instance, the calculation of GPA (grade point average) assigns different weights to different grades.

For solutions to the Preliminary Exercises, see page 720.

Section Exercises

1. 7; 7; 7   3. 22; 14; no mode
5. 22.9; 22; 21   7. $82,100; $83,000; no mode
9. 58.3; 59; 59   11. 59,560; 59,000; 62,000
13. 84.4; 89; 90
15. 3.22   17. 3.37   19. 82   21. 6.1 points   23. 7.2
25. 64°   27. −6°F

Section 8.2

Think About It

1. The range is the difference between the largest and smallest values in the data set.
2. Variance measures how spread out data are from the mean.
3. The standard deviation of a data set is the square root of the variance.
4. If the standard deviation of a data set is zero, all values of the data set are equal.
5. The data set 2, 4, 6, 8, 10 has the larger standard deviation because its values are more spread out from the mean.
6. Small. This would indicate that each box was getting very close to the same amount of cereal.

For solutions to the Preliminary Exercises, see page 721.

Section Exercises

1. 84°F   3. 23.67 mpg; 3.92 mpg
5. 493.6 cal; 20.30 cal   7. 4.27 hr; 0.69 hr
9. (a) 210 s, or 3 min 30 s; 26 s   (b) Yes; 2:27, 3:01, 4:02
11. 0.4 bps; 4.36 bps   13. Drop 2   15. Stock 1

Section 8.3

Think About It

1. The z-score for a given data value x is the number of standard deviations that x is above or below the mean of the data.
2. The value is one standard deviation below the mean.
3. It means that 58% of the data is smaller than 34.
4. Twenty-five percent of the data is less than 84.
5. Fifty percent of the data set lies between Q_1 and Q_3.

For solutions to the Preliminary Exercises, see page 721.

Section Exercises

1. (a) ≈0.87   (b) ≈1.74   (c) ≈−2.17   (d) 0.0
3. (a) ≈−0.32   (b) ≈0.21   (c) ≈1.16   (d) ≈−0.95
5. (a) ≈−0.67   (b) 147.78 mm Hg
7. (a) ≈0.72   (b) ≈112.16 mg/dl
9. The score in part a.
11. ≈59th percentile
13. 6396 students
15. (a) 50%   (b) 10%   (c) 40%
17. $Q_1 = 5, Q_2 = 10, Q_3 = 26$

19. Northeast

Midwest

South

West

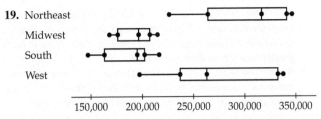

Answers will vary. Here are some possibilities. The region with the lowest median was the South; the region with the highest median was the Northeast. The range of prices was greatest for the West.

21.

CBX-21

PHT-34

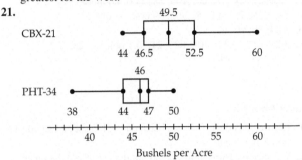

Bushels per Acre

Answers will vary. Here is one possibility: The maximum number of bushels cultivated per acre for PHT-34 is approximately equal to the median number of bushels cultivated per acre for CBX-21.

Section 8.4

Think About It

1. A normal distribution with mean 0 and standard deviation 1.
2. For data that are normally distributed, 50% of the data lie below the mean. For data that are normally distributed, 50% of the data lie above the median.
3. According to the Empirical Rule for normally distributed data, approximately 68% of the data lie between plus or minus 1 standard deviation from the mean.
4. If a data value is randomly selected from normally distributed data, the probability that it will be more than 3 standard deviations is approximately 1%.

For solutions to the Preliminary Exercises, see page 722.

Section Exercises

1. (a) 45% (b) 0.25
3. (a) 95% (b) 16% (c) 81.5%
5. (a) 81.5% (b) 0.15%
7. (a) 1280 vehicles (b) 12 vehicles
9. (a) 19.2% (b) 29.6%
11. (a) 80.6% (b) 7.1%
13. (a) 10.6% (b) 98.8%
15. (a) 0.106 (b) 0.459
17. (a) 0.747 (b) 0.022
19. (a) 0.0098 (b) 0.7475 (c) 0.7501 **21.** 0.0668

Section 8.5

Think About It

1. It is a line that approximates a bivariate data set.
2. No. They have perfect correlation, but the slope of the least-squares line is negative.

3. Decreases
4. Positive

For solutions to the Preliminary Exercises, see page 722.

Section Exercises

1. (a) a and b (b) c and d
3. (a) $\hat{y} = -0.1700621321x + 54545.58535$ (b) \$49,400
   (c) -0.9993480085 (d) The variables are negatively correlated, which means that as miles go up, retail value goes down.
5. (a) $\hat{y} = 0.03850038917x + 10.66206002$
   (b) 18.4 meters/second
7. (a) $\hat{y} = -0.3396749522x + 25.00745698$
   (b) approximately 15 hrs (c) -0.9987037488 (d) Yes
9. (a) $\hat{y} \approx -1.129773184x + 13.63881711$ (b) $\approx$ 87 mph
11. (a) $\hat{y} = 8.7333333x + 60.644444$
    (b) 104 bpm (c) Yes, $r = 0.9996409$

Chapter 8 Review Exercises

1. 14.7; 14; 12; 13; 17.6; 4.2 **2.** The mode
3. Answers will vary.
4. (a) Median (b) Mode (c) Mean
5. 1331.125 ft; 1223.5 ft; 1200 ft; 462 ft
6. 36 mph **7.** $\approx$3.10
8. (a) 1.25 (b) The 88th percentile
9. $\approx$3.16; $\approx$10.00 **10.** \$7.63; \$7.91; \$0.58
11. (a) <-1.73 (b) <0.58
12.

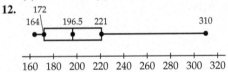

13. (a) 8 (b) 40
14. (a) 78% (b) 0.34
15. (a) 0.933 (b) 0.309
16. (a) 16% (b) 95%
17. (a) 6.7% (b) 64.5% (c) 78.8%
18. 93.15 million mi
19. (a) 0.9992 (b) $y = 0.0725547445x + 0.2883211679$
    (c) 10.1 inches

Chapter 9

Section 9.1

Think About It

1. It is the number of citizens represented by each representative.
2. Hamilton, Jefferson
3. Each representative represents 25,000 citizens living in that state.
4. The number of representatives apportioned to a state is the standard quota or one more than the standard quota.
5. An apportionment procedure in which an increase in the total number of representatives for a legislature would cause a district represented by the legislature to lose a seat.

For solutions to the Preliminary Exercises, see page 723.

Section Exercises

1. To calculate the standard divisor, divide the total population p by the number of items to apportion n.

3. The standard quota for a state is the whole number part of the quotient of the state's population divided by the standard divisor.

5. **(a)** 0.273 **(b)** 0.254 **(c)** Salinas

7. Seaside Mall

9. **(a)** 761,169; There is one representative for every 761,169 citizens in the United States **(b)** Underrepresented. The average constituency is greater than the standard divisor. **(c)** Overrepresented. The average constituency is less than the standard divisor.

11. **(a)** 37.10. There is one new nurse for every 37.10 beds. **(b)** Sharp: 7; Palomar: 10; Tri-City: 8; Del Raye: 5; Rancho Verde: 7; Bel Air: 11 **(c)** Sharp: 7; Palomar: 10; Tri-City: 8; Del Raye: 5; Rancho Verde: 7; Bel Air: 11 **(d)** They are identical.

13. The population paradox occurs when the population of one state is increasing faster than that of another state, yet the first state still loses a representative.

15. The Balinski-Young Impossibility Theorem states that any apportionment method either will violate the quota rule or will produce paradoxes such as the Alabama paradox.

17. **(a)** Yes **(b)** No **(c)** Yes

19. **(a)** Boston: 2; Chicago: 20 **(b)** Yes. Chicago lost a vice president while Boston gained one.

21. **(a)** Sixth grade **(b)** Sixth grade. Same

23. Valley

25. Texas's population is much larger than Utah's, so Texas has many more representatives, and a percentage change in its population adds or subtracts more representatives than for smaller states.

27. **(a)** They are the same. **(b)** Using the Jefferson method, the humanities division gets one less computer and the sciences division gets one more computer compared with using the Webster method.

29. The Jefferson and Webster methods

31. The Huntington-Hill method

Section 9.2

Think About It

1. An issue that passes by a majority of votes cast receives more than 50% of the votes cast. An issue that wins with a plurality of votes cast receives more votes than any other issue but not necessarily more than 50%.

2. There were five people who ranked the colors Red−3, Green−2, and Blue−1.

3. There is no voting method involving three or more choices that satisfies all four fairness criteria.

4. A candidate who wins all possible head-to-head matchups should win an election when all candidates appear on the ballot.

For solutions to the Preliminary Exercises, see page 725.

Section Exercises

1. There is a tie between Corn Flakes and Raisin Bran.

3. There is a tie between Italian and Mexican.

5. There is a tie between Corn Flakes and Raisin Bran.

7. Mexican **9.** Verizon

11. WKLS **13.** Verizon

15. **(a)** Amusement park **(b)** Amusement park **(c)** There is a tie between picnic in a park and amusement park.

17. Grand Canyon

19. There is a three-way tie among P. Gibson, R. Allenbaugh, and G. DeWitte.

21. There is a three-way tie among WKLS, WNNX, and WWVV.

23. No **25.** No **27.** No

29. **(a)** Film A **(b)** Film B **(c)** The monotonicity criterion has not been violated; Film A won the first vote.

Section 9.3

Think About It

1. A voting system in which some voters have more influence on the outcome of an election.

2. The number of votes required to pass a measure or elect a person to office.

3. If one of the permanent members votes against a resolution, there are only 38 votes remaining even if every other country voted for the resolution. This number is less than the quota, 39.

4. No. If one of the permanent members votes against a resolution, the four remaining members vote for it, and all the elected members vote for it, the measure would pass.

For solutions to the Preliminary Exercises, see page 726.

Section Exercises

1. **(a)** 6 **(b)** 4 **(c)** 3 **(d)** 6 **(e)** No **(f)** A and C **(g)** 15 **(h)** 6

3. 0.60, 0.20, 0.20 **5.** 0.50, 0.30, 0.10, 0.10

7. 0.36, 0.28, 0.20, 0.12, 0.04

9. 1.00, 0.00, 0.00, 0.00, 0.00, 0.00

11. 0.44, 0.20, 0.20, 0.12, 0.04

13. **(a)** Exercise 9 **(b)** Exercises 3, 5, 6, 9, and 12 **(c)** None **(d)** Exercise 8

15. 0.33, 0.33, 0.33

17. **(a)** {12: 1, 1, 1, 1, 1, 1, 1, 1, 1, 1, 1, 1} **(b)** Yes **(c)** Yes **(d)** Divide the voting power, 1, by the quota, 12.

19. Dictator: A; dummies: B, C, D, E

21. None **23.** **(a)** 0.60, 0.20, 0.20 **(b)** Answers will vary.

25. **(a)** 0.33, 0.33, 0.33 **(b)** This system has the same effect as a one-person, one-vote system.

Chapter 9 Review Exercises

1. **(a)** Health: 7; business: 18; engineering: 10; science: 15 **(b)** Health: 6; business: 18; engineering: 10; science: 16 **(c)** Health: 7; business: 18; engineering: 10; science: 15

2. **(a)** Newark: 9; Cleveland: 6; Chicago: 11; Philadelphia: 4; Detroit: 5 **(b)** Newark: 9; Cleveland: 6; Chicago: 11; Philadelphia: 4; Detroit: 5 **(c)** Newark: 9; Cleveland: 6; Chicago: 11; Philadelphia: 4; Detroit: 5

3. **(a)** 0.098 **(b)** 0.194 **(c)** High Desert

4. Morena Valley

5. **(a)** No. None of the offices loses a new printer. **(b)** Yes. Office A drops from two new printers to only one new printer.

6. **(a)** A: 2; B: 5; C: 3; D: 16; E: 2. No. None of the centers loses an automobile. **(b)** No. None of the centers loses an automobile.

7. **(a)** Los Angeles: 9; Newark: 2 **(b)** Yes. Newark loses a computer file server.

8. **(a)** A: 10; B: 3; C: 21 **(b)** Yes. The population of region B grew at a higher rate than the population of region A, yet region B lost an inspector to region A.

9. Yes   **10.** No

11. (a) A   (b) B

12. (a) Manuel Ortega   (b) No   (c) Crystal Kelley

13. (a) Vail   (b) Aspen

14. A. Kim   **15.** Snickers

16. A. Kim   **17.** Snickers

18. (a) Jules Abreu   (b) Sohail Hassip   (c) Sohail Hassip won all head-to-head comparisons but lost the overall election. (d) Sohail Hassip   (e) Sohail Hassip received a majority of the first-place votes but lost the overall election.

19. (a) Sohail Hassip   (b) Logan Moro, a losing candidate, withdrew from the race and caused a change in the overall winner of the election.

20. The monotonicity criterion was violated because the only change was that the supporter of a losing candidate changed his or her vote to support the original winner, but the original winner did not win the second vote.

21. (a) 18   (b) 18   (c) Yes   (d) A and C   (e) 15   (f) 6

22. (a) 35   (b) 35   (c) Yes   (d) A   (e) 31   (f) 10

23. 0.60, 0.20, 0.20

24. 0.20, 0.20, 0.20, 0.20, 0.20

25. 0.42, 0.25, 0.25, 0.08

26. 0.62, 0.14, 0.14, 0.05, 0.05

27. Dictator: A; dummies: B, C, D, E

28. Dummy: D

29. 0.50, 0.125, 0.125, 0.125, 0.125

Chapter 10

Section 10.1

Think About It

1. No. Not every possible edge is drawn. We can add two more edges:

2. Answers may vary. Here is a connected graph.

Here is a graph that is not connected.

3. The number of edges that meet at a vertex is called the degree of a vertex.

4. If a path ends at the same vertex at which it started, it is considered a circuit.

5. An Euler circuit is one that travels along each edge exactly once.

For solutions to the Preliminary Exercises, see page 727.

Section Exercises

1.

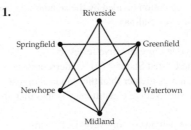

3. (a) No   (b) 3   (c) Ada   (d) A loop would correspond to a friend speaking to himself or herself.

5. (a) 6   (b) 7   (c) 6   (d) Yes   (e) No

7. (a) 6   (b) 4   (c) 4   (d) Yes   (e) Yes

9. Equivalent   **11.** Not equivalent

13. The graph on the right has a vertex of degree 4, and the graph on the left does not.

15. (a) D–A–E–B–D–C–E–D

17. (a) Not Eulerian   (b) A–E–A–D–E–D–C–E–C–B–E–B

19. (a) Not Eulerian   (b) No

21. (a) Not Eulerian   (b) E–A–D–E–G–D–C–G–F–C–B–F–A–B–E–F

23. (a)

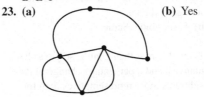

(b) Yes

25. Yes

27. Yes, but the hamster cannot return to its starting point.

29.

Yes. You can always return to the starting room.

Section 10.2

Think About It

1. A Hamiltonian path passes through each vertex (note not each edge) exactly once. If it ends at the initial vertex then it is a Hamiltonian circuit.

2. An Euler circuit is a path that crosses every edge exactly once without repeating and ends at the initial vertex. A Hamiltonian circuit passes through each vertex (note not each edge) exactly once and ends at the initial vertex.

3. A weighted graph is a graph in which each edge is associated with a value, called a weight.

4. Answers may vary. One possibility is designing the most efficient route to route packages for a package delivery company.

5. A complete graph is one in which every possible edge is drawn between vertices without any multiple edges.

For solutions to the Preliminary Exercises, see page 727.

Section Exercises

1. A–B–C–D–E–G–F–A   **3.** A–B–E–C–H–D–F–G–A

5. Springfield–Greenfield–Watertown–Riverside–Newhope–Midland–Springfield

7. A–B–E–D–C–A, total weight 31; A–D–E–B–C–A, total weight 32

9. A–D–C–E–B–F–A, total weight 114; A–C–D–E–B–F–A, total weight 158

11. A–D–B–C–F–E–A **13.** A–C–E–B–D–A

15. A–D–B–F–E–C–A **17.** A–C–E–B–D–A

19. Louisville–Evansville–Bloomington–Indianapolis–Lafayette–Fort Wayne–Louisville

21. Louisville–Evansville–Bloomington–Indianapolis–Lafayette–Fort Wayne–Louisville

23. Tokyo–Seoul–Beijing–Hong Kong–Bangkok–Tokyo

25. Tokyo–Seoul–Beijing–Hong Kong–Bangkok–Tokyo

27. Home–pharmacy–pet store–farmers market–shopping mall–home; home–pharmacy–pet store–shopping mall–farmers market–home

29. Home state–task B–task D–task A–task C–home state. Edge-picking algorithm gives the same sequence.

Section 10.3

Think About It

1. A planar graph is a graph that can be drawn so that no edges intersect each other (except at vertices).

2. The drawing is not planar because two edges cross. The graph is planar because we can make an equivalent planar drawing of it as shown at the right.

 redraw the edge colored orange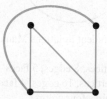

3. See page 486.

For solutions to the Preliminary Exercises, see page 728.

Section Exercises

1.  **3.**

5. 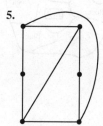 **7.**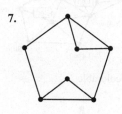

See the Student Solutions Manual for the answers to Exercises 9, 11, 13, and 15.

17. 5 faces, 5 vertices, 8 edges **19.** 2 faces, 8 vertices, 8 edges

21. 5 faces, 10 vertices, 13 edges **23.** 9

Section 10.4

Think About It

1. Every planar graph is 4-colorable.

2. The chromatic number of a graph is the minimum number of colors needed to color a graph so that no edge connects vertices of the same color.

For solutions to the Preliminary Exercises, see page 729.

Section Exercises

1. Requires three colors

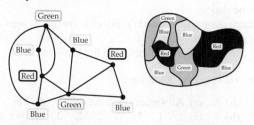

3. Requires three colors

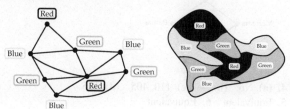

5.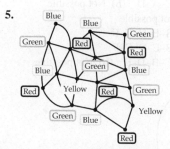

7.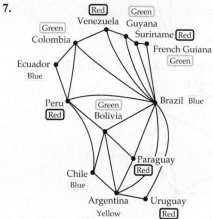

9.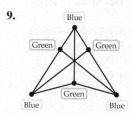

11. Not 2-colorable

13.

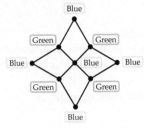

15. 3 **17.** 4 **19.** 3

21. Two time slots

23. Five days; one possible schedule: group 1, group 2, groups 3 and 5, group 4, group 6

25. Three days: films 1 and 4, films 2 and 6, films 3 and 5

Chapter 10 Review Exercises

1. (a) 8 **(b)** 4 **(c)** All vertices have degree 4. **(d)** Yes

2. (a) 6 **(b)** 7 **(c)** 1, 1, 2, 2, 2, 2, 2 **(d)** No

3.

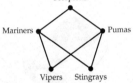

4. (a) No **(b)** 4 **(c)** 110, 405 **(d)** 105

5. Equivalent **6.** Equivalent

7. (a) E–A–B–C–D–B–E–C–A–D **(b)** Not possible

8. (a) Not possible **(b)** Not possible

9. (a) F–A–E–C–B–A–D–B–E–D–C–F
(b) F–A–E–C–B–A–D–B–E–D–C–F

10. (a) B–A–E–C–A–D–F–C–B–D–E **(b)** Not possible

11. Yes

12. Yes; no **13.** A–B–C–E–D–A **14.** A–D–F–B–C–E–A

15.

Casper–Rapid City–Minneapolis–Des Moines–Topeka–Omaha–Boulder–Casper

16. Casper–Boulder–Topeka–Minneapolis–Boulder–Omaha–Topeka–Des Moines–Minneapolis–Rapid City–Casper

17. A–D–F–E–B–C–A **18.** A–B–E–C–D–A

19. A–D–F–E–C–B–A **20.** A–B–E–D–C–A

21.

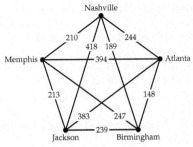

Memphis–Nashville–Birmingham–Atlanta–Jackson–Memphis

22. A–E–B–C–D–A

23.

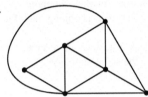

24.

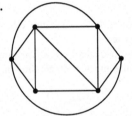

See the Student Solutions Manual for the answers to Exercises 25 and 26.

27. 5 vertices, 8 edges, 5 faces

28. 14 vertices, 16 edges, 4 faces

29. Requires four colors

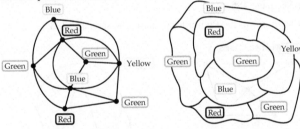

30. Requires four colors

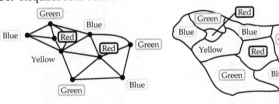

31. 2-colorable

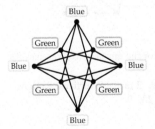

32. Not 2-colorable   **33.** 3   **34.** 5
35. Three time slots: budget and planning, marketing and executive, sales and research

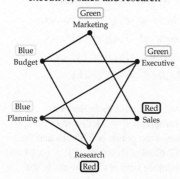

Chapter 11
Section 11.1

Think About It

1. $\dfrac{1 \text{ foot}}{12 \text{ inches}}$
2. 4 quarts in 1 gallon
3. 1000 ml = 1 L
4. 0.001 kg = 1 g
5. Approximately 4 liters in one gallon.
6. 1 yard is less than 1 meter.
7. Two pounds is less than 1 kilogram.

For solutions to the Preliminary Exercises, see page 730.

Section Exercises

1. 72 in.   **3.** 180 in.   **5.** 4 lb   **7.** 24 oz   **9.** 20 fl oz
11. 56 pt   **13.** 620 mm   **15.** 321 cm   **17.** 7.421 kg
19. 4.5 dg   **21.** 0.0075 L   **23.** 435 cm^3 **25.** 65.91 kg
27. 48.3 km/hr   **29.** $8.70/L   **31.** 1.38 in.   **33.** $4.55/lb

Section 11.2

Think About It

1. (a) ray   (b) line segment   (c) line
2. (a) $\angle a = \angle c$
   (b) $\angle d = \angle b$
   (c) $m\angle c + m\angle D = 180°$
   (d) $\angle a$ and $\angle c$ are called opposite angles
3. 90°
4. greater than
5. acute
6. 180°

For solutions to the Preliminary Exercises, see page 730.

Section Exercises

1. $\angle O$, $\angle AOB$, and $\angle BOA$ **3.** A 28° angle
5. An 18° angle   **7.** An acute angle   **9.** An obtuse angle
11. 14 cm   **13.** 28 ft   **15.** 86°   **17.** 71°   **19.** 30°
21. 127°   **23.** 116°   **25.** 20°   **27.** 20°   **29.** 141°
31. 106°   **33.** 11°   **35.** $m\angle a = 38°$, $m\angle b = 142°$
37. $m\angle a = 47°$, $m\angle b = 133°$ **39.** 20°
41. 47°   **43.** $m\angle x = 155°$, $m\angle y = 70°$
45. $m\angle a = 45°$, $m\angle b = 135°$ **47.** 60°
49. 35°   **51.** False   **53.** True

Section 11.3

Think About It

1. A rectangle has four 90° angles.
2. A square has four 90° angles and four equal sides. A rhombus has four equal sides.
3. 1 m^2
4. 20
5. False. $\pi \approx \dfrac{22}{7}$
6. a and c
7. b and d

For solutions to the Preliminary Exercises, see page 730.

Section Exercises

1. (a) Perimeter is not measured in square units.
   (b) Area is measured in square units.
3. (a) 30 m   (b) 50 m^2
5. (a) 40 km   (b) 100 km^2
7. (a) 40 ft   (b) 72 ft^2
9. (a) 8π cm; 25.13 cm   (b) 16π cm^2; 50.27 cm^2
11. (a) 11π mi; 34.56 mi   (b) 30.25π mi^2; 95.03 mi^2
13. (a) 17π ft; 53.41 ft   (b) 72.25π ft^2; 226.98 ft^2
15. 20 in.   **17.** 10 mi   **19.** 2 packages
21. Perimeter of the square   **23.** 144 m^2
25. 9 in.   **27.** 39 ft   **29.** 10×20 unit
31. 136.5 ft^2 **33.** 160 km^2 **35.** 2 qt
37. $480   **39.** $912   **41.** 176 m^2
43. 13.19 ft   **45.** 12,064 in.   **47.** 62.83 ft
49. 2500π ft^2 **51.** 339.29 in^2 larger; more than twice the size
53. 222.2 mi^2 **55.** $8r^2 - 2\pi r^2$ **57.** 4 times as large

Section 11.4

Think About It

1. Volume is a measure of capacity such as number of liters of milk in a container. Surface area is a measure of how much material is required to make the container.
2. m^3
3. in^2
4. circle

For solutions to the Preliminary Exercises, see page 731.

Section Exercises

1. 840 in^3 **3.** 15 ft^3 **5.** 4.5π cm^3; 14.14 cm^3
7. 94 m^2 **9.** 56 m^2 **11.** 96π in^2; 301.59 in^2
13. 34 m^3 **15.** 15.625 in^3 **17.** 36π ft^3
19. 8143.01 cm^3 **21.** 75π in^3 **23.** 120 in^3
25. Sphere   **27.** 7.80 ft^3 **29.** 35,380,400 gal
31. 69.36 m^2 **33.** 225π cm^2 **35.** 402.12 in^2
37. 6π ft^2 **39.** 297 in^2 **41.** 2.5 ft   **43.** 11 cans
45. 22.53 cm^2 **47.** 5 m^3 **49.** 69.12 in^3
51. 192 in^3 **53.** 208 in^2 **55.** 204.57 cm^2
57. 95,000 L   **59.** $4860

Section 11.5

Think About It

1. Corresponding angles are equal.
2. *DE*
3. Similar triangles may have the same area, but generally do not. Congruent triangles always have the same area.

4. Similar triangles are generally not congruent. Congruent triangles are always similar triangles.

5. If two sides and the included angle of one triangle are equal in measure to two sides and the included angle of a second triangle, the two triangles are congruent.

6. Yes. $9^2 + 12^2 = 15^2$

For solutions to the Preliminary Exercises, see page 731.

Section Exercises

1. $\frac{1}{2}$ **3.** $\frac{3}{4}$ **5.** 7.2 cm **7.** 3.3 m **9.** 12 m
11. 12 in. **13.** 56.3 cm^2 **15.** 18 ft **17.** 16 m
19. $14\frac{3}{8}$ ft **21.** 15 m **23.** 8 ft **25.** 13 cm **27.** 35 m
29. Yes, SAS theorem **31.** Yes, SSS theorem
33. Yes, ASA theorem **35.** No **37.** Yes, SAS theorem
39. No **41.** No **43.** 5 in. **45.** 8.6 cm **47.** 11.2 ft
49. 4.5 cm **51.** 12.7 yd **53.** 8.5 cm **55.** 24.3 cm

Section 11.6

Think About It

1. (a) b (b) a (c) a (d) b
2. $\sin(\alpha) = \dfrac{b}{c}$, $\cos(\alpha) = \dfrac{a}{c}$, $\tan(\alpha) = \dfrac{b}{a}$
3. 45°
4. θ

For solutions to the Preliminary Exercises, see page 731.

Section Exercises

1. (a) $\dfrac{a}{c}$ (b) $\dfrac{b}{c}$ (c) $\dfrac{b}{c}$ (d) $\dfrac{a}{c}$ (e) $\dfrac{a}{b}$ (f) $\dfrac{b}{a}$
3. $\sin\theta = \dfrac{5}{13}$, $\cos\theta = \dfrac{12}{13}$, $\tan\theta = \dfrac{5}{12}$
5. $\sin\theta = \dfrac{24}{25}$, $\cos\theta = \dfrac{7}{25}$, $\tan\theta = \dfrac{24}{7}$
7. $\sin\theta = \dfrac{8}{\sqrt{113}}$, $\cos\theta = \dfrac{7}{\sqrt{113}}$, $\tan\theta = \dfrac{8}{7}$
9. $\sin\theta = \dfrac{1}{2}$, $\cos\theta = \dfrac{\sqrt{3}}{2}$, $\tan\theta = \dfrac{1}{\sqrt{3}}$
11. 0.6820 **13.** 1.9970 **15.** 0.8453
17. 38.6° **19.** 17.9° **21.** 21.3°
23. 26.6° **25.** 38.0° **27.** 41.1°
29. 841.8 ft **31.** 13.6° **33.** 29.1 ft **35.** 52.9 ft
37. 13.6 ft **39.** 1056.6 ft **41.** 29.6 yd

Section 11.7

Think About It

1. (a) Through a given point not on a given line, exactly one line can be drawn parallel to the given line. (b) Through a given point not on a given line, there are at least two lines parallel to the given line. (c) Through a given point not on a given line, there exist no lines parallel to the given line.

2. (a) One (b) One (c) Three

3. A geodesic is a curve on a surface such that for any two points of the curve, the portion of the curve between the points is the shortest path on the surface that joins these points.

4. Lobachevskian or hyperbolic geometry

5. Riemannian geometry

For solutions to the Preliminary Exercises, see page 732.

Section Exercises

1. π square units
3. $d_E(P, Q) = \sqrt{49} = 7$ blocks, $d_C(P, Q) = 7$ blocks
5. $d_E(P, Q) = \sqrt{89} \approx 9.4$ blocks, $d_C(P, Q) = 13$ blocks
7. $d_E(P, Q) = \sqrt{72} \approx 8.5$ blocks, $d_C(P, Q) = 12$ blocks
9. $d_E(P, Q) = \sqrt{37} \approx 6.1$ blocks, $d_C(P, Q) = 7$ blocks
11. $d_C(P, Q) = 7$ blocks **13.** $d_C(P, Q) = 5$ blocks
15. $d_C(P, Q) = 5$ blocks
17. A city distance may be associated with more than one Euclidean distance. For example, if $P = (0, 0)$ and $Q = (2, 0)$, then the city distance between the points is 2 blocks and the Euclidean distance is also 2 blocks. However, if $P = (0, 0)$ and $Q = (1, 1)$, then the city distance between the points is still 2 blocks, but the Euclidean distance is $\sqrt{2}$ blocks.

19. 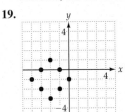 **21.**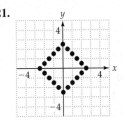

23. $4n$

Section 11.8

Think About It

1. A fractal is a geometric figure in which a self-similar motif repeats itself on an ever-diminishing scale.

2. A fractal is said to be strictly self-similar if any arbitrary portion of the fractal contains a replica of the entire fractal.

For solutions to the Preliminary Exercises, see page 732.

Section Exercises

1. Stage 2 —— ——   —— ——

Stage 3 – – – –   – – – –

3.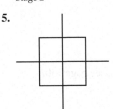

Stage 2

5.

Stage 2

7.

Stage 2

9.

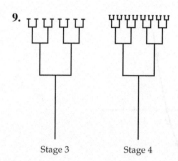

Stage 3   Stage 4

11. 0.631   **13.** 1.465   **15.** 2.000   **17.** 2.000   **19.** 1.613

Chapter 11 Review Exercises

1. $2\frac{1}{4}$ **2.** $7\frac{1}{2}$ **3.** 3.7   **4.** 678   **5.** 1.273

6. \$3.36 per qt   **7.** $m\angle x = 22°$; $m\angle y = 158°$

8. 24 in.   **9.** 240 in³   **10.** 68°

11. 220 ft²   **12.** 40π m²   **13.** 44 cm

14. $m\angle w = 30°$; $m\angle y = 30°$ **15.** 27 in²

16. 96 cm³   **17.** 14.1 m   **18.** $m\angle a = 138°$; $m\angle b = 42°$

19. A 148° angle   **20.** 39 ft³   **21.** 95°   **22.** 8 cm

23. 288π mm³   **24.** 21.5 cm   **25.** 4 cans

26. 208 yd   **27.** 90.25 m²   **28.** 276 m²

29. The triangles are congruent by the SAS theorem.

30. 9.7 ft   **31.** $\sin\theta = \dfrac{5\sqrt{89}}{89}$, $\cos\theta = \dfrac{8\sqrt{89}}{89}$, $\tan\theta = \dfrac{5}{8}$

32. $\sin\theta = \dfrac{\sqrt{3}}{2}$, $\cos\theta = \dfrac{1}{2}$, $\tan\theta = \sqrt{3}$

33. 25.7°   **34.** 29.2°   **35.** 53.8°   **36.** 1.9°

37. 100.1 ft   **38.** 153.2 mi   **39.** 56.0 ft

40. Spherical geometry or elliptical geometry

41. Hyperbolic geometry

42. Lobachevskian or hyperbolic geometry

43. Riemannian or spherical geometry

44. 120π in²   **45.** $\dfrac{25\pi}{3}$ ft²

46. $d_E(P, Q) = 5$ blocks, $d_C(P, Q) = 7$ blocks

47. $d_E(P, Q) = \sqrt{113} \approx 10.6$ blocks, $d_C(P, Q) = 15$ blocks

48. $d_E(P, Q) = \sqrt{37} \approx 6.1$ blocks, $d_C(P, Q) = 7$ blocks

49. $d_E(P, Q) = \sqrt{89} \approx 9.4$ blocks, $d_C(P, Q) = 13$ blocks

50. (a) P and Q (b) P and R

51.

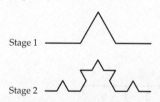

Stage 0 ——————————

Stage 1

Stage 2

Yes. The Koch curve is a strictly self-similar fractal.

52.

Stage 2

53. (a) 2   (b) 2   (c) $D = \dfrac{\log 2}{\log 2} = 1$ **54.** $\dfrac{\log 5}{\log 4} \approx 1.161$

Appendix A

Objective Practice

A.1A **1.** $-14 < 16$ **2.** $0 > -31$ **3.** $A = \{1, 2, 3\}$ **4.** $-33, -24$

A.1B **1.** -4 **2.** 9   **3.** 36   **4.** 40   **5.** 74   **6.** -81

A.2A **1.** -15 **2.** 0   **3.** -87 **4.** -4 **5.** -4 **6.** -29

A.2B **1.** 8   **2.** 18   **3.** -18 **4.** -21 **5.** -40

A.2C **1.** 117   **2.** -221 **3.** -210 **4.** -672 **5.** 120

A.2D **1.** -6 **2.** -19 **3.** -17 **4.** 0   **5.** -1

A.3A **1.** 9   **2.** -256 **3.** -27 **4.** -12 **5.** -1008 **6.** 72

A.3B **1.** 0   **2.** 20   **3.** 25   **4.** 11   **5.** 6   **6.** 2

A.4A **1.** $\dfrac{3}{10}$ **2.** $\dfrac{5}{12}$ **3.** $\dfrac{1}{9}$ **4.** $\dfrac{5}{12}$ **5.** $\dfrac{3}{20}$

A.4B **1.** $\dfrac{9}{16}$ **2.** $\dfrac{5}{6}$ **3.** $\dfrac{9}{10}$ **4.** $\dfrac{14}{5}$ **5.** $\dfrac{4}{3}$

A.4C **1.** $\dfrac{2}{3}$ **2.** $\dfrac{35}{48}$ **3.** $\dfrac{29}{20}$ **4.** $\dfrac{43}{24}$

A.4D **1.** $\dfrac{1}{10}$ **2.** $\dfrac{11}{18}$ **3.** $\dfrac{11}{21}$ **4.** $\dfrac{11}{36}$

A.4E **1.** $<$ **2.** $>$ **3.** $>$ **4.** $<$

A.5A **1.** $-\dfrac{1}{3}$ **2.** $\dfrac{1}{6}$ **3.** $-\dfrac{3}{8}$ **4.** 1   **5.** $-\dfrac{7}{48}$ **6.** $\dfrac{6}{7}$ **7.** $-\dfrac{9}{10}$

     8. -8 **9.** $\dfrac{5}{4}$ OR $1\dfrac{1}{4}$ **10.** $\dfrac{1}{12}$

A.5B **1.** $-\dfrac{1}{12}$ **2.** $\dfrac{1}{24}$ **3.** $-\dfrac{1}{3}$ **4.** $\dfrac{11}{24}$ **5.** $-\dfrac{1}{6}$ **6.** $-\dfrac{7}{8}$ **7.** $\dfrac{1}{4}$

     8. $\dfrac{13}{36}$ **9** $-\dfrac{1}{24}$ **10.** $\dfrac{5}{36}$

A.6A **1.** -50.7 **2.** -5.905 **3.** -47.7 **4.** -11.613

     5. 236.14   **6.** -2.163

A.6B **1.** 1.70   **2.** 15.12   **3.** -0.00786 **4.** 67,100

     5. -131.328 **6.** -388.72

A.6C **1.** 1.95   **2.** 4.14   **3.** -49.8 **4.** 0.009407   **5.** -3.8

A.7A **1.** $0.\overline{5}$ **2.** $0.\overline{72}$ **3.** $0.\overline{639}$ **4.** $0.7\overline{72}$

     5. 0.075   **6.** 7.25

A.7B **1.** $\dfrac{2}{5}$ **2.** $\dfrac{12}{25}$ **3.** $\dfrac{97}{200}$ **4.** $3\dfrac{3}{4}$ **5.** $\dfrac{13}{250}$ **6.** $\dfrac{3}{20,000}$

A.7C **1.** $<$ **2.** $<$ **3.** $>$ **4.** $<$ **5.** $<$ **6.** $>$

A.7D **1.** $\dfrac{7}{9}$, 7:9, 7 to 9   **2.** $\dfrac{2}{1}$, 2:1, 2 to 1   **3.** 2.5 ft/s

     4. \$4250/month   **5.** \$3.25/lb

A.8A **1.** $0.36, \dfrac{9}{25}$ **2.** $0.062, \dfrac{31}{500}$ **3.** $0.0025, \dfrac{1}{400}$ **4.** $\dfrac{1}{8}$

     5. $\dfrac{3}{70}$

A.8B **1.** 73%   **2.** 101.2%   **3.** 85%   **4.** 225%   **5.** $58\dfrac{1}{3}\%$

A.9A **1.** -1 **2.** 9   **3.** 7   **4.** 21   **5.** 49   **6a.** natural number, integer, rational number, real number   **6b.** integer, rational number, real number   **6c.** rational number, real number   **6d.** rational number, real number   **6e.** irrational number, real number

A.9B **1.** 3.1623   **2.** 23.2379   **3.** 45.8258   **4.** -87.3613

     5. 5, 6   **6.** 11, 12

Appendix B

Objective Practice

B.1A **1.** 12   **2.** -3 **3.** -11 **4.** 28   **5.** 4

B.2A **1.** $14x$ **2.** $7 - 3b$ **3.** 0   **4.** $7a$ **5.** $-\dfrac{11}{24}x$

     6. $-3x - 8y$

B.2B **1.** $12x$ **2.** $8x$ **3.** x **4.** $-2x$ **5.** $8y$ **6.** $6a$

B.2C **1.** $35 - 21b$ **2.** $15x^2 + 6x$ **3.** $-15x - 30$
4. $4x^2 - 12x + 20$ **5.** $\frac{3}{2}x - \frac{9}{2}y + 6$ **6.** $-8b^2 + 6b - 9$

B.2D **1.** $-2x - 16$ **2.** $7n - 7$ **3.** $3y - 3$ **4.** $-2x - 16$
5. $-3x + 21$ **6.** $0.03x + 0.08$

B.3A **1.** Yes **2.** No **3.** Yes **4.** No **5.** No

B.3B **1.** 0 **2.** -9 **3.** -14 **4.** 1 **5.** 1.869

B.3C **1.** 0 **2.** 12 **3.** 15 **4.** -23.246 **5.** 3

B.3D **1.** 0.075 **2.** 37.5% **3.** 51.895 **4.** 2.7% **5.** 400%

B.3E **1.** 600% **2.** 209% **3.** 29.3% **4.** 34.2% **5.** $15,330

B.4A **1.** -3 **2.** 0 **3.** $\frac{1}{3}$ **4.** $\frac{3}{4}$ **5.** 2

B.4B **1.** -1 **2.** -30 **3.** -3 **4.** $\frac{5}{6}$ **5.** -17

B.4C **1.** 4 **2.** 2 **3.** 2.4 **4.** $\frac{1}{4}$ **5.** $\frac{20}{3}$

B.4D **1.** $x = -\frac{3}{4}y + 3$ **2.** $R = \frac{E}{I}$ **3.** $C = \frac{5F - 160}{9}$
4. $t = -\frac{s - ax}{av}$ **5.** $S + \frac{a}{1 - r}$

B.4E **1.** 2 **2.** No solution **3.** $-\frac{1}{3}$ and -1 **4.** $-\frac{1}{2}$
5. $-\frac{8}{3}$ and $\frac{10}{3}$

B.5A **1.** $-7, 1$ **2.** $\frac{3 - \sqrt{33}}{2}, \frac{3 + \sqrt{33}}{2}$ **3.** $\frac{4 - \sqrt{10}}{2},$
$\frac{4 + \sqrt{10}}{2}$ **4.** $-\frac{1}{2}, \frac{2}{3}$ **5.** $\frac{-5 - \sqrt{2}}{3}, \frac{-5 + \sqrt{2}}{3}$

Appendix C

Objective Practice

C.1A **1.**

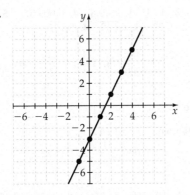

2.

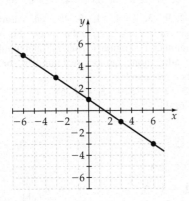

3.

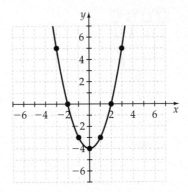

4.

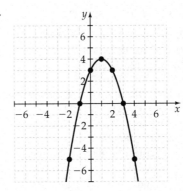

5.

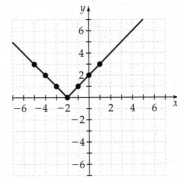

C.2A **1.** Yes; D = {0,1,2,3,4}; R = {0} **2.** -14 **3.** 21 **4.** 6

C.2B **1.**

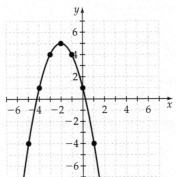

2.

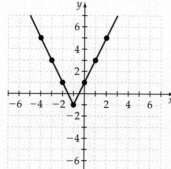

4.

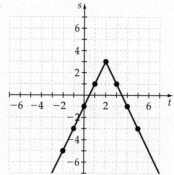

3.

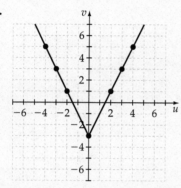

Solutions to Preliminary Exercises

Chapter 1

Section 1.1

P1. (a) The difference between the numbers is increasing as 3, 5, 7, 9. The next number is 11 more than 28, or 39.

(b) $1 = 1 \times 1$, $2 = 1 \times 2$, $6 = 1 \times 2 \times 3$, $24 = 1 \times 2 \times 3 \times 4$, $120 = 1 \times 2 \times 3 \times 4 \times 5$. The next number is $1 \times 2 \times 3 \times 4 \times 5 \times 6 = 720$.

P2. Horizontal followed by vertical lines dividing each region in half. The next image has horizontal line and looks like

P3. Each year, the increase is more than the previous year. Given that the increase from year 4 to year 5 is $303.88, we would expect that the increase between year 5 and year 6 will be more than $300. Thus, year 6 will be more than $6600.

P4. Each 10°C increase in temperature produces an increase in the amount of sugar that can be dissolved that is greater than the previous amount. Given that the increase from 50°C to 60°C is 1.72 g, we would expect that the increase between from 60°C to 70°C to be at least 1.72 more than 18.17. Thus, there will be more than 19 g of sugar that can be dissolved in 70°C water.

P5. $\frac{1}{2}$ is one instance. Any positive number less than 1 would work.

P6. Reasoning from premises known to be true is deductive reasoning.

P7. Label the coins 1, 2, 3, 4, 5, 6, 7, 8, 9. Place coins 1, 2, 3 on one side and coins 4, 5, 6 on the other scale. If they balance, one of 7, 8, or 9 is counterfeit. Place coin 7 on one side of the scale and coin 8 on the other side. If they balance, coin 9 is counterfeit. If coins 1, 2, 3 do not balance with coins 4, 5, 6 then one of them is counterfeit. Record whether the left side of the scale went down (heavier) or up (lighter).

Now place 1 and 5 on one pan with 2 and 4 on the other pan. If they have the same weight, then either 3 or 6 is bad. It can be shown which one is bad with one more weighing. If they do not balance and the heavier side this time was the same side that was heavier for the first weighing, then swapping 2 with 5 had no effect, which means that either 1 or 4 is counterfeit. It can be shown which one is counterfeit with one more weighing. If the side that is heavier this time was not the same side that was heavier the first time, then swapping 2 with 5 had an effect, and so it must be either 2 or 5 that is counterfeit. It can be shown which one is bad with one more weighing.

P8. From clue 1, we know that Ashley is not the president or the treasurer. In the following chart, write X1 (which stands for "ruled out by clue 1") in the president and treasurer columns of Ashley's row.

	Pres.	V. P.	Sec.	Treas.
Brianna				
Ryan				
Tyler				
Ashley	X1			X1

From clue 2, Brianna is not the secretary. We know from clue 1 that the president is not the youngest, and we know from clue 2 that Brianna and the secretary are the youngest members of the group. Thus, Brianna is not the president. In the chart, write X2 for these two conditions. Also, we know from clues 1 and 2 that Ashley is not the secretary because she is older than the treasurer. Write an X2 in the secretary column of Ashley's row.

	Pres.	V. P.	Sec.	Treas.
Brianna	X2		X2	
Ryan				
Tyler				
Ashley	X1		X2	X1

At this point, we see that Ashley must be the vice president and that none of the other members is the vice president. This also means Brianna is the treasurer. Thus, we can update the following chart.

	Pres.	V. P.	Sec.	Treas.
Brianna	X2	X2	X2	✓
Ryan		X2		X2
Tyler		X2		X2
Ashley	X1	✓	X2	X1

From clue 3, we know that Tyler is not the secretary. Thus, we can conclude that Tyler is the president and Ryan must be the secretary. See the following chart.

	Pres.	V. P.	Sec.	Treas.
Brianna	X2	X2	X2	✓
Ryan	X3	X2	✓	X2
Tyler	✓	X2	X3	X2
Ashley	X1	✓	X2	X1

Tyler is the president, Ashley is the vice president, Ryan is the secretary, and Brianna is the treasurer.

Section 1.2

P1. We are not looking for an exact answer but a mental approximation. $24,789 \approx \$25,000$ and $2529 \approx \$2500$. An approximate value is $\$25,000 + \$2500 = \$27,500$.

P2. 9 feet 8 inches is approximately 10 feet. $587 \approx 600$. $600 \div 10 = 60$. Approximately 60 cars can be parked in one row.

P3. Answers to this exercise may vary. Our solution is to divide the grid as shown in the following image and use the rectangle with the + sign as the representative sample.

Romariolen/Shutterstock.com

We counted 21 raspberries (you may get a different number depending on the pieces that were counted). There are 16 rectangles, so there are approximately $16 \times 21 = 336$ raspberries in the box.

P4. To estimate the height of the cell tower, use the fact that when the height of a known object is multiplied by a certain factor, the height of the shadow increases by the same factor. From the figure,

A 6-foot pole casts a shadow of 7.5 ft
A 12-ft (2×6) would cast a shadow of 15 ft (2×7.5)
A 18-ft (3×6) would cast a shadow of 22.5 ft (3×7.5)
A 24-ft (4×6) would cast a shadow of 30 ft (4×7.5)

......
......

A 144-ft (24×6) would cast a shadow of 180 ft (24×7.5)
A 150-ft (25×6) would cast a shadow of 187.5 ft (25×7.5)
Because a 150-foot tower would cast a shadow of 187.5 ft, the cell tower is approximately 150 feet tall.

P5. $0.000000495 = 4.95 \times 10^{-7}$

P6. (a) The smallest increase is between years 2015 and 2016.
(b) The largest increase is between 2016 and 2017.

P7. (a) The largest sector represents type O. Therefore, type O occurs most frequently. (b) The smallest sector represents type AB, so type AB occurs least frequently.

P8. From the graph, the item would cost approximately $16 in 2020.

Section 1.3

P1. *Understand the Problem* To go past Starbucks, Allison must walk along Third Avenue from Boardwalk to Park Avenue.

Devise a Plan Label each intersection that Allison can pass through with the number of routes to that intersection. If she can reach an intersection from two different routes, then the number of routes to that intersection is the sum of the numbers of routes to the two adjacent intersections.

Carry Out the Plan The following figure shows the number of routes to each intersection that Allison could pass through. Thus, there are a total of nine routes that Allison can take if she wishes to walk directly from point A to point B and pass by Starbucks.

Review the Solution The total of nine routes seems reasonable. We know from Example 1 that if Allison can take any route, the total number of routes is 35. Requiring Allison to go past Starbucks eliminates numerous routes.

P2. *Understand the Problem* There are several ways to answer the questions so that two answers are "false" and three answers are "true." One way is TTTFF and another is FFTTT.

Devise a Plan Make an organized list. Try the strategy of listing a T unless doing so will produce too many T's or a duplicate of one of the previous orders in your list.

Carry Out the Plan (Start with three T's in a row.)

TTTFF	(1)
TTFTF	(2)
TTFFT	(3)
TFTTF	(4)
TFTFT	(5)
TFFTT	(6)
FTTTF	(7)
FTTFT	(8)
FTFTT	(9)
FFTTT	(10)

Review the Solution Each entry in the list has two F's and three T's. Because the list is complete and has no duplications, we know there are 10 ways for a student to mark two questions with "false" and the other three with "true."

P3. *Understand the Problem* There are six people, and each person shakes hands with each of the other people.

Devise a Plan Each person will shake hands with five other people (a person will not shake his or her own hand; that would be silly). Because there are six people, we could multiply 6 times 5 to get the total number of handshakes. However, this procedure would count each handshake exactly twice, so we must divide this product by 2 for the actual answer.

Carry Out the Plan 6 times 5 is 30, and 30 divided by 2 is 15.

Review the Solution Denote the people by the letters A, B, C, D, E, and F. Make an organized list. Remember that AB and BA represent the same people shaking hands, so do not list both AB and BA.

AB AC AD AE AF

BC BD BE BF

CD CE CF

DE DF

EF

The method of making an organized list verifies that if six people shake hands with each other, there will be a total of 15 handshakes.

P4. ***Understand the Problem*** We need to find the ones digit of 4^{200}.

Devise a Plan Compute a few powers of 4 to see if there are any patterns. $4^1 = 4$, $4^2 = 16$, $4^3 = 64$, and $4^4 = 256$. It appears that the units digit (ones digit) of 4^{200} must be either a 4 or a 6.

Carry Out the Plan If the exponent n is an even counting number, then 4^n has a ones digit of 6. If the exponent n is an odd counting number, then 4^n has a ones digit of 4. Because 200 is an even counting number, we conjecture that 4^{200} has a ones digit of 6.

Review the Solution You could try to check the answer by using a calculator, but you would find that 4^{200} is too large to be displayed. Thus, we need to rely on the patterns we have observed to conclude that 6 is indeed the ones digit of 4^{200}.

P5. ***Understand the Problem*** We are asked to find the possible numbers that Melody could have started with.

Devise a Plan Work backward from 18 and do the inverse of each operation that Melody performed.

Carry Out the Plan To get 18, Melody subtracted 30 from a number, so that number was $18 + 30 = 48$. To get 48, she divided a number by 3, so that number was $48 \times 3 = 144$. To get 144, she squared a number. She could have squared either 12 or -12 to produce 144. If the number she squared was 12, then she must have doubled 6 to get 12. If the number she squared was -12, then the number she doubled was -6.

Review the Solution We can check by starting with 6 or -6. If we do exactly as Melody did, we end up with 18. The operation that prevents us from knowing with 100% certainty which number she started with is the squaring operation. We have no way of knowing whether the number she squared was positive or negative.

P6. ***Understand the Problem*** We need to find Diophantus's age when he died.

Devise a Plan Read the hint and then look for clues that will help you make an educated guess. You know from the given information that Diophantus's age must be divisible by 6, 12, 7, and 2. Find a number divisible by all of these numbers and check to see if it is a possible solution to the problem.

Carry Out the Plan All multiples of 12 are divisible by 6 and 2, but the smallest multiple of 12 that is divisible by 7 is $12 \times 7 = 84$. Thus we conjecture that Diophantus's age when he died was $x = 84$ years.
If $x = 84$, then $\frac{1}{6}x = 14$, $\frac{1}{12}x = 7$, $\frac{1}{7}x = 12$, and $\frac{1}{2}x = 42$. Then $\frac{1}{6}x + \frac{1}{12}x + \frac{1}{7}x + 5 + \frac{1}{2}x + 4 = 14 + 7 + 12 + 5 + 42 + 4 = 84$.
It seems that 84 years is a correct solution to the problem.

Review the Solution After 84, the next multiple of 12 that is divisible by 7 is 168. The number 168 also satisfies all the conditions of the problem, but it is unlikely that Diophantus

died at the age of 168 years or at any age older than 168 years. Hence, the only reasonable solution is 84 years.

P7. ***Understand the Problem*** We need to determine two U.S. coins that have a total value of 35¢, given that one of the coins is not a quarter.

Devise a Plan Experiment with different coins to try to produce 35¢. After a few attempts, you should conclude that one of the coins must be a quarter. Consider that the problem may be a *deceptive problem*.

Carry Out the Plan A total of 35¢ can be produced by using a dime and a quarter. One of the coins is a quarter, but it is also true that *one of the coins, the dime, is not a quarter.*

Review the Solution A dime and a quarter satisfy all the conditions of the problem. No other combination of coins satisfies the conditions of the problem. Thus, the only solution is a dime and a quarter.

Section 1.4

P1. (a) No. Chris could choose the all-strawberry part, leaving Melanie with the all chocolate. Her perceived share would be 0%. To be fair, she must have a piece that is at least worth 50% to her.
(b) Yes. Melanie would receive a piece that is 50% strawberry. Chris is 100% happy because he likes both flavors equally well.
(c) No. Because he likes each flavor equally well, he will be happy with whatever piece he receives.

P2. Because Macon values P_2 more than Jameson, give P_2 to Macon. Then P_1 to Jameson and P_3 to Kachina.

P3. The points awarded by Abril and Minh are shown in the following table.

Item	Abril	Minh	Winner
Home	48	46	Abril
Beachfront condo	25	27	Minh
Stamp collection	12	5	Abril
Sailboat	5	12	Minh
Pollock painting	10	10	Tie
Total of items won	60	39	

Because Minh has fewer points, he is awarded the painting. The table now looks like the following.

Item	Abril	Minh	Winner
Home	48	46	Abril
Beachfront condo	25	27	Minh
Stamp collection	12	5	Abril
Sailboat	5	12	Minh
Pollock painting	10	10	Minh
Total of items won	60	49	

Because Abril has the larger total points for items won (60 versus 49), Abril is the initial winner; Minh is the initial loser. For each item given to the *initial winner* (in this case, the home and condo), find the ratio of the points awarded by initial winner to that of the points awarded by the initial loser.

Item	Abril	Minh	Winner	Ratios
Home	48	46	Abril	$\frac{48}{46} \approx 1.043$
Beachfront condo	25	27	Minh	
Stamp collection	12	5	Abril	$\frac{12}{5} = 2.4$
Sailboat	5	12	Minh	
Pollock painting	10	10	Minh	
Total of items won	60	49		

Starting with *smallest* ratio (1.043 in this case), transfer the item from the initial winner to the initial loser *unless* that item would cause the initial winner to have fewer points than the initial loser. In that case, the item is shared between the two players. Transferring the home to Minh results in Minh having more points than Abril. Therefore, the home is shared between them.

Let x be the percent (as a decimal) that needs to be taken from Abril and given to Minh that will result in each of them having the same number of points. Then $1 - x$ is the percent Abril retains. After the transfer, the points should be equal. This is expressed by the equation

$$48(1 - x) + 12 = 46x + 27 + 12 + 10$$

Solve for x.

$$48(1 - x) + 12 = 46x + 27 + 12 + 10$$
$$60 - 48x = 46x + 49$$
$$11 = 94x$$
$$0.12 \approx x$$

Minh gets 12% of the proceeds from the sale of the home, the condo, the sailboat, and the painting. The home is sold and 88% of the sale price of the home goes to Abril along with the stamp collection.

P4. Here are the bids for each player. Because Carlos and Mina tie for the ring, toss a coin and award the ring to the winner. For this solution we will assume Mina gets the ring. The solution will be different if Carlos is awarded the ring. Such is the nature of some fair division problems.

	Carlos	Mina	Gordon	Winner
Ring	$ 5000	$ 5000	$ 4000	Mina
Car	$50,000	$ 45,000	$ 55,000	Gordon
Lamp	$25,000	$ 45,000	$ 40,000	Mina
Coin	$12,000	$ 9000	$ 9000	Carlos
Total Valuation	$92,000	$104,000	$108,000	
Fair Share	$30,667	$ 34,667	$ 36,000	
Value Items Won	$12,000	$ 50,000	$ 55,000	
Owed to Estate		$ 15,333	$ 19,000	
Estate Owes	$18,667			

Add the amounts owed to the estate and subtract the amount the estate owes.

$15,333 + $19,000 - $18,667 = $15,666

Divide this amount among the three players:

$$\frac{1}{3} \times 15,666 = 5222$$

	Carlos	Mina	Gordon	Winner
Ring	$ 5000	$ 5000	$ 4000	Mina
Car	$50,000	$ 45,000	$ 55,000	Gordon
Lamp	$25,000	$ 45,000	$ 40,000	Mina
Coin	$12,000	$ 9000	$ 9000	Carlos
Total Valuation	$92,000	$104,000	$108,000	
Fair Share	$30,667	$ 34,667	$ 36,000	
Value Items Won	$12,000	$ 50,000	$ 55,000	
Owed to Estate		-$ 15,333	-$ 19,000	
Estate Owes	$18,667			
Share of Surplus	$ 5222	$ 5222	$ 5222	
Disposition of Money	$23,889	-$ 10,111	-$ 13,778	
Disposition of Items	Coin	Ring and lamp	Car	

P5. Here are the bids by the players.

	Amy	Selma	Pieter	Winner
Cottage	$150,000	$225,000	$175,000	Selma
Condo	$200,000	$175,000	$150,000	Amy
Sailboat	$ 10,000	$ 5000	$ 15,000	Pieter

Selma wins the cottage and compensates the other players by paying $\frac{3-1}{3} = \frac{2}{3}$ of the value of the cottage as she valued it.

Selma owes: $\frac{2}{3} \times 225,000 = $150,000$

Amy receives: $\frac{1}{3} \times 150,000 = $50,000$

Pieter receives: $\frac{1}{3} \times 175,000 \approx $58,333$

Amount in joint account: $150,000 - $50,000 - $58,333 = $41,667

Distribute one-third of this amount to each player:
$$\frac{1}{3} \times 41,667 = 13,889$$

Final accounting for cottage:
Selma: cottage - $150,000 + $13,889 = cottage - $136,111
Amy: $50,000 + $13,889 = $63,889
Pieter: $58,333 + $13,889 = $72,222

Repeat the procedure for the condo.

Amy owes: $\frac{2}{3} \times 200,000 \approx $133,333$

Selma receives: $\frac{1}{3} \times 175,000 \approx $58,333$

Pieter receives: $\frac{1}{3} \times 150,000 = $50,000$

Amount in joint account: $133,333 - $58,333 - $50,000 = $25,000

Distribute one-third of this amount to each player:
$$\frac{1}{3} \times 25,000 \approx 8333$$

Final accounting for condo:
Amy: condo − $133,333 + $8333 = condo − $125,000
Selma: $58,333 + $8333 = $66,666
Pieter: $50,000 + $8333 = $58,333

Repeat the procedure for the sailboat.

Pieter owes: $\frac{2}{3} \times 15{,}000 = \$10{,}000$

Amy receives: $\frac{1}{3} \times 10{,}000 \approx \$3333$

Selma receives: $\frac{1}{3} \times 5000 \approx \$1667$

Amount in joint account: $10,000 − $3333 − $1667 = $5000
Distribute one-third of this amount to each player:
$\frac{1}{3} \times 5000 \approx 1667$

Final accounting for sailboat:
Pieter: sailboat − $10,000 + $1667 = sailboat − $8333
Amy: $3333 + $1667 = $5000
Selma: $1667 + $1667 = $3334

Final disposition of assets.
Selma: cottage − $136,111 + $66,666 + $3334 =
cottage −$66,111
Amy: condo − $125,000 + $63,889 + $5000 =
condo − $56,111
Pieter: sailboat − $8333 + $72,222 + $58,333 =
sailboat + $122,222

Chapter 2

Section 2.1

P1. {3, 6, 9, 12, 15, 18}
P2. {1, 4, 6, 8, 9, 10, 12, 14, 15, 16, 18}
P3. (a) True (b) False (c) True (d) True
P4. (a) ∅ (b) {−2, 0, 2}
P5. (a) {−6, −4, −3, −2, −1, 0, 1, 2, 3, 4, 6}
  (b) {−3, −2, −1, 0, 1, 2, 3, 4, 6}
P6.

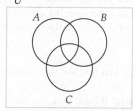

P7. 64 options
P8. 62 people
P9. (a) 261 people (b) 55 people (c) 143 people
  (d) 146 (e) 111 people
P10. (a) 524 people (b) 540 people (c) 517 people
P11.

	Positive Test	Negative Test
Athlete Has Banned Drug	20	5
Athlete Does Not Have Banned Drug	12	363

Section 2.2

P1. (a)

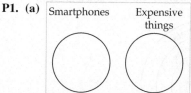

(b)

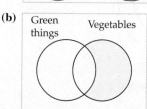

P2. (a) All fruit is good. (b) Some buildings are tall.

P3.

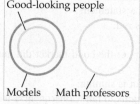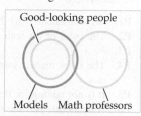

From the premises, it is possible to create two different Euler diagrams. Thus, the argument is invalid.

P4.

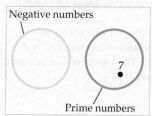

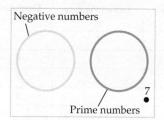

From the given premises we can conclude that 7 may or may not be a prime number. The argument is invalid.

P5.

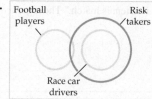

Because the circle that represents risk takers could intersect the circle that represents football players, the argument is invalid.

P6. The following Euler diagram illustrates that all squares are quadrilaterals, so the argument is a valid argument.

P7. No journalists are scientists.

Section 2.3

P1. (a) Statement   (b) Not a statement
P2. The color of the wall was red.
P3. $c \wedge s$
P4. The stock market went up, or the bond market did not go down.
P5. It is not raining, or the sun is shining.
P6. Antecedent: The snow begins to fall. Consequent: I will not go skiing.
P7. True
P8. Inverse: If it is not snowing, then I will go skiing.
Converse: If I do not go skiing, the snow will fall.
Contrapositive: If I go skiing, it will not snow.
P9. Let e represent "the gemstone is an emerald" and $\sim s$ represent "I will not sell it." The argument can be stated as

$$e \rightarrow \sim s \quad \text{First premise}$$
$$\underline{s \qquad\qquad} \quad \text{Second premise}$$
$$\therefore \sim e \quad \text{Conclusion}$$

e	s	$\sim s$	First Premise $e \rightarrow \sim s$	Second Premise s	Conclusion $\sim e$
T	T	F	T	T	F
T	F	T	F	F	F
F	T	F	T	T	T
F	F	T	T	F	T

The shaded row is the only row with both premises true and the conclusion true. The argument is valid.

P10. Let p represent "I go to the tennis match." and let q represent "My daughter is playing in a tennis match." The argument can be stated

$$\sim p \rightarrow \sim q$$
$$\underline{q \qquad\qquad}$$
$$\therefore p$$

p	q	$\sim p$	$\sim q$	First Premise $\sim p \rightarrow \sim q$	Second Premise q	Conclusion p
T	T	F	F	T	T	T
T	F	F	T	T	F	T
F	T	T	F	F	T	F
F	F	T	T	T	F	F

The shaded row is the only row with both premises true and the conclusion true. The argument is valid.

P11. Let p represent "Parker can run 100 yards in less than 12 seconds"; let q represent "Parker can qualify for the track team." Then the argument can be stated as

$$p \rightarrow q \quad \text{First premise}$$
$$\underline{q \qquad\qquad} \quad \text{Second premise}$$
$$\therefore p \quad \text{Conclusion}$$

This is the fallacy of the converse. The argument is not valid.

P12. Let p represent "I fall asleep"; let q represent "I read a math book"; let r represent "I drink a soda"; let c represent "eat a candy bar." Then the argument can be stated as

$$q \rightarrow p \quad \text{First premise}$$
$$p \rightarrow r \quad \text{Second premise}$$
$$\underline{r \rightarrow c} \quad \text{Third premise}$$
$$\therefore q \rightarrow c \quad \text{Conclusion}$$

Because q implies p, p implies r, and r implies c, it follows that q implies c by transitive reasoning. The argument is valid.

Section 2.4

P1. Red herring
Premise: I only made $80 per week when I was your age.
Conclusion: Therefore, it should not be hard to make a living on your salary.
This is a red herring fallacy because the premise is an attempt to distract the audience from the original statement.

P2. Straw man
Premise: Sometimes we use math without realizing it.
Conclusion: We are all mindless zombies.
This is a straw man fallacy because a new premise was built to distract the audience from the original premise.

P3. Ad hominem
Premise: The doctor gave me an exercise program to follow.
Conclusion: I'm not going to do it because the doctor is 60 years old.
This is an ad hominem fallacy because the doctor's age is being challenged and not the doctor's qualifications.

P4. Appeal to popularity
Premise: The best running back in the football league endorses these shoes.
Conclusion: They will make you run faster.
This is an appeal to popularity because the arguer is attempting to connect a popular figure to the reason for the conclusion.

P5. Appeal to ignorance
Premise: It has not been proven that she stole the jewelry from you.
Conclusion: She did not steal it from you.
This is an appeal to ignorance because it is drawing a conclusion based on lack of evidence for the converse of the premise.

P6. Appeal to authority
Premise: The policeman stated that eating too much red meat is unhealthy.
Conclusion: Eating too much red meat is unhealthy.
This is an appeal to authority because it is drawing on expertise from an irrelevant authority.

P7. Appeal to emotion
Premise: Buy Health Food Inc. nutrition bars.
Conclusion: If you don't, you will gain weight.
This is an appeal to emotion because the argument attempts to connect an emotion to its conclusion.

P8. Appeal to common practice
Premise: Everyone was making fun of another student.
Conclusion: Making fun of another student is acceptable.
This is an appeal to common practice because it relies on general acceptance in its premise rather than logic.

P9. Post hoc
Premise: The temperature dropped 10 degrees, and I got a headache.
Conclusion: The temperature drop gave me a headache.
This is a post hoc fallacy; it reasons that because one event occurred before the other, the first event caused the second. Correlation does not always indicate causation.

P10. Black-or-white (false dilemma)
Premise: Either you protest with us or you are against our cause.
Conclusion: You must protest with us so that you are not against our cause.
This is a black-or-white fallacy because the premise offers only two choices when in reality there are more.

P11. Slippery slope
Premise: We need more rain.
Conclusion: Without it, we will have to move to the city.
This is a slippery slope because it assumes many factors in the premise are directly connected when the connection between each factor is actually weak. As the connection between factors weakens, so does the argument.

P12. Circular reasoning
Premise: No other computer can compare with the quality of this computer.
Conclusion: This computer is the best.
This is circular reasoning because the conclusion is the same as the premise, just reworded.

P13. Hasty generalization
Premise: Not shaving during playoffs makes you play better.
Conclusion: We should all not shave during playoffs.
This is a hasty generalization because it assumes that if something is good for a few, then it is good for all.

P14. Loaded question
No premise or conclusion.
This is a loaded question because a "yes" or a "no" answer both imply a false or deniable statement.

Chapter 3

Section 3.1

P1. The veterinary technician's ratio is $\dfrac{900}{2700} = \dfrac{1}{3} > \dfrac{3}{10}$.
Therefore, the technician does not meet the guideline.

P2. Find the ratio of students to faculty:
$$\frac{41{,}309 + 11{,}075}{3133} = \frac{52{,}384}{3133} \approx 17$$
The student-to-faculty ratio is 17 to 1.

P3. The ratio 65 to 1 means there are 65 NCAA draft-eligible players to every one football player who gets drafted. To find the number of eligible football players, multiply the number who are drafted by 65.
$$65 \cdot 251 = 16{,}315$$
There were approximately 16,315 draft-eligible NCAA football players.

P4. To find the unit rate, divide the cost by the number of pounds:
$$\frac{\$6.75}{1.5 \text{ pounds}} = \$4.50/\text{pound}$$
The unit cost is $4.50 per pound.

P5. Find the unit rate for each purchase:
$$\frac{\$6.29}{32 \text{ oz}} \approx \$0.197/\text{oz} \qquad \frac{\$8.29}{48 \text{ oz}} \approx \$0.173/\text{oz}$$
The detergent for $8.29 for 48 ounces is more economical.

P6. The rate of physicians per 100,000 is
$$\frac{668 \text{ physicians}}{452{,}100 \text{ people}} \cdot 100{,}000 \approx 148$$
There are approximately 148 physicians per 100,000 people.

P7. The injury rate is 1405 per 100,000 drivers. Multiply 1405 per 100,000 by 231 million.
$$\frac{1405 \text{ injuries}}{100{,}000 \text{ drivers}} \cdot 231{,}000{,}000 = 3{,}245{,}550$$
There were approximately 3,245,550 injuries.

P8. Multiply the rate per thousand by the loan amount:
$$\frac{\$22.801}{\$1000} \cdot \$28{,}254 \approx \$644.22$$
The monthly car payment is $644.22.

P9. **(a)** Multiply the vacation package by the exchange rate:
$$\$4500 \cdot \frac{0.930789 \text{ franc}}{\$1} \approx 4188.55$$
The vacation package cost 4188.55 francs.

(b) Multiply the cost in francs by the reciprocal of the exchange rate:
$$254.67 \text{ franc} \cdot \frac{\$1}{0.930789} = \$273.61$$
The dinner cost $273.61.

Section 3.2

P1. $15 \text{ ft} \cdot \dfrac{1 \text{ yd}}{3 \text{ ft}} = 5 \text{ yd}$

P2. $3 \text{ lb} \cdot \dfrac{16 \text{ oz}}{1 \text{ lb}} = 48 \text{ oz}$

P3. $18 \text{ pt} \cdot \dfrac{1 \text{ qt}}{2 \text{ pt}} \cdot \dfrac{1 \text{ gal}}{4 \text{ qt}} = 2.25 \text{ gal}$

P4. Move the decimal point 3 places to the right.
0.38 m = 380 mm

P5. Move the decimal point 3 places to the left.
42.3 mg = 0.0423 g

P6. 2 kl + 167 L = 2000 L + 167 L = 2167 L

P7. Use the conversion from gallons to quarts and then the conversion from quarts to liters. (1 quart ≈ 0.946353 L)
$$\frac{\$3.69}{1 \text{ gal}} \cdot \frac{1 \text{ gal}}{4 \text{ qt}} \cdot \frac{1 \text{ qt}}{0.946353 \text{ L}} = \$0.97/\text{L}$$

P8. Use the conversion from centimeters to inches.
$$45 \text{ cm} \cdot \frac{1 \text{ in.}}{2.54 \text{ cm}} \approx 17.72 \text{ in.}$$
Use the conversion rates.

P9. $\dfrac{65 \text{ mi}}{1 \text{ hr}} \cdot \dfrac{1.61 \text{ km}}{1 \text{ mi}} \approx 104.65 \text{ km/hr}$

Section 3.3

P1. $\dfrac{42}{x} = \dfrac{5}{8}$

$42 \cdot 8 = 5x$

$336 = 5x$

$67.2 = x$

The solution is 67.2.

P2. $\dfrac{15 \text{ km}}{2 \text{ cm}} = \dfrac{x \text{ km}}{7 \text{ cm}}$

$\dfrac{15}{2} = \dfrac{x}{7}$

$15 \cdot 7 = 2 \cdot x$

$105 = 2x$

$\dfrac{105}{2} = \dfrac{2x}{2}$

$52.5 = x$

The distance between the two cities is 52.5 km.

P3. $\dfrac{7}{5} = \dfrac{\$84{,}000}{x \text{ dollars}}$

$\dfrac{7}{5} = \dfrac{84{,}000}{x}$

$7 \cdot x = 5 \cdot 84{,}000$

$7x = 420{,}000$

$\dfrac{7x}{7} = \dfrac{420{,}000}{7}$

$x = 60{,}000$

The other partner receives $60,000.

P4. Let x represent the number of people in the United States with the allergen. Write and solve a proportion.

$\dfrac{3}{10{,}000} = \dfrac{x}{325{,}000{,}000}$

$\dfrac{3 \cdot 325{,}000{,}000}{10{,}000} = x$

$98{,}000 = x$

P5. The ratio of $a : b : c$ is $3 : 4 : 5$, and b is given as 12 inches. Then

$\dfrac{a}{b} = \dfrac{3}{4}$ $\qquad \dfrac{c}{b} = \dfrac{5}{4}$

$\dfrac{a}{12} = \dfrac{3}{4}$ $\qquad \dfrac{c}{12} = \dfrac{5}{4}$

$a = 9$ $\qquad\qquad c = 15$

$a = 9$ inches, $c = 15$ inches

Section 3.4

P1. (a) $74\% = 0.74$ (b) $152\% = 1.52$
(c) $8.3\% = 0.083$ (d) $0.6\% = 0.006$

P2. (a) $0.3 = 30\%$ (b) $1.65 = 165\%$
(c) $0.072 = 7.2\%$ (d) $0.004 = 0.4\%$

P3. (a) $8\% = 8\left(\dfrac{1}{100}\right) = \dfrac{8}{100} = \dfrac{2}{25}$

(b) $180\% = 180\left(\dfrac{1}{100}\right) = \dfrac{180}{100} = 1\dfrac{80}{100} = 1\dfrac{4}{5}$

(c) $2.5\% = 2.5\left(\dfrac{1}{100}\right) = \dfrac{2.5}{100} = \dfrac{25}{1000} = \dfrac{1}{40}$

(d) $66\dfrac{2}{3}\% = \dfrac{200}{3}\% = \dfrac{200}{3}\left(\dfrac{1}{100}\right) = \dfrac{2}{3}$

P4. (a) $\dfrac{1}{4} = 0.25 = 25\%$

(b) $\dfrac{3}{8} = 0.375 = 37.5\%$

(c) $\dfrac{5}{6} = 0.83\overline{3} = 83.\overline{3}\%$

(d) $1\dfrac{2}{3} = 1.66\overline{6} = 166.\overline{6}\%$

P5. $P \cdot B = A$

$0.035 \cdot 32{,}500 = 1137.5$

The customer would receive $1137.50.

P6. $PB = A$

$0.03B = 14{,}370$

$\dfrac{0.03B}{0.03} = \dfrac{14{,}370}{0.03}$

$B = 479{,}000$

The selling price of the home was $479,000.

P7. Number of words not misspelled $= 329 - 8 = 321$.

$PB = A$

$P \cdot 329 = 321$

$\dfrac{329P}{329} = \dfrac{321}{329}$

$P \approx 0.976$

Approximately 97.6% of the words were not misspelled.

P8. $5.08 - 4.44 = 0.64$

$PB = A$

$P \cdot 4.44 = 0.64$

$\dfrac{4.44P}{4.44} = \dfrac{0.64}{4.44}$

$P \approx 0.144$

The percent increase in projected income tax revenue from 2022 to 2025 is 14.4%.

P9. $PB = A$

$P \cdot 55 = 2$

$\dfrac{55P}{55} = \dfrac{2}{55}$

$P \approx 0.036$

Approximately 3.6% of the athlete's body water will be lost in one hour.

Section 3.5

P1. Write a direct variation equation. Use the given information to find the proportionality constant.

$g = kA$

$1 = k(150)$

$\dfrac{1}{150} = k$

Then

$g = \dfrac{1}{150}A$

$= \dfrac{1}{150}(630)$

$= 4.2$

You must purchase 5 gallons.

P2. Write a direct variation equation. Use the given information to find the proportionality constant.

$$A = kr^2$$
$$113 = k(3)^2 = 9k$$
$$\frac{113}{9} = k$$

Then

$$A = \frac{113}{9}r^2$$
$$= \frac{113}{9}(4)^2$$
$$= \frac{1808}{9} \approx 200.9$$

The surface area is approximately 200.9 square inches.

P3. Write a direct variation equation. Use the given information to find the proportionality constant.

$$d = k\sqrt{h}$$
$$4 = k\sqrt{9}$$
$$4 = 3k$$
$$\frac{4}{3} = k$$

Then

$$d = \frac{4}{3}\sqrt{h}$$
$$= \frac{4}{3}\sqrt{10}$$
$$\approx 4.2$$

The lifeguard could see 4.2 miles.

P4. Write an inverse variation equation. Use the given information to find the proportionality constant.

$$i = \frac{k}{R}$$
$$40 = \frac{k}{25}$$
$$1000 = k$$

Then

$$i = \frac{1000}{R}$$
$$= \frac{1000}{30}$$
$$\approx 33.3$$

The current is approximately 33.3 amps.

P5. Write an inverse variation equation. Use the given information to find the proportionality constant.

$$p = \frac{k}{s^2}$$
$$75 = \frac{k}{125^2}$$
$$1,171,875 = k$$

Then

$$p = \frac{1,171,875}{s^2}$$
$$= \frac{1,171,875}{200^2}$$
$$\approx 29.3$$

Approximately 29.3% of the x-rays will pass through.

P6. Write a joint variation equation. Use the given information to find the proportionality constant.

$$P = kmh$$
$$490 = k \cdot 5 \cdot 10$$
$$490 = 50k$$
$$9.8 = k$$

Then

$$P = 9.8mh$$
$$500 = 9.8 \cdot m \cdot 5$$
$$500 = 49m$$
$$10.2 \approx m$$

The mass is approximately 10.2 kilograms.

P7. Write a joint variation equation. Use the given information to find the proportionality constant.

$$F = k\frac{mv^2}{r}$$
$$96 = k\frac{6 \cdot 20^2}{25}$$
$$96 = 96k$$
$$k = 1$$
$$F = k\frac{mv^2}{r}$$
$$4480 = 1 \cdot \frac{14 \cdot 40^2}{r}$$
$$4480 = \frac{22,400}{r}$$
$$4480r = 22,400$$
$$r = 5$$

The radius of the circle is 5 cm.

Chapter 4

Section 4.1

P1. $I = Prt$
$$= 1250(0.035)(4)$$
$$= 175$$

The interest on the loan is $175.

P2. $I = Prt$
$$= 2500(0.065)\left(\frac{3}{12}\right)$$
$$= 40.625$$

The interest is $40.63.

P3. Solve the formula $I = Prt$ for r.

$$I = Prt$$
$$7.75 = 1230(r)\left(\frac{65}{360}\right)$$
$$7.75 = \frac{2665}{12}r$$
$$\frac{12(7.75)}{2665} = r$$
$$0.0349 \approx r$$

The annual interest rate is 3.5%.

P4. Use the formula for maturity value, $A = P + Prt$

$$A = P + Prt$$
$$= 3300 + 3300(0.043)\left(\frac{18}{12}\right)$$
$$= 3512.85$$

The maturity value is $3512.85

P5. The interest on the loan is the difference between the maturity value and the loan amount.

$$I = 660 - 650 = 10$$

Solve the formula $I = Prt$ for r.

$$I = Prt$$
$$10 = 650(r)\left(\frac{50}{360}\right)$$
$$10 = \frac{1625}{18}r$$
$$\frac{18(10)}{1625} = r$$
$$0.111 \approx r$$

The annual simple interest rate is 11.1%.

P6. Solve the maturity value formula, $A = P + Prt$, for r.

$A = 2500, P = 2435, t = \dfrac{130}{360}$.

$$A = P + Prt$$
$$2500 = 2435 + 2435(r)\left(\frac{130}{360}\right)$$
$$65 = 2435(r)\left(\frac{130}{360}\right)$$
$$65 = \left(\frac{31655}{36}\right)r$$
$$\frac{36(65)}{31655} = r$$
$$0.074 \approx r$$

The annual simple interest rate is 7.4%.

Section 4.2

P1. Solution using a calculator. $PV = 1250, n = 365,$
$t = 3, r = 0.04$.

$$FV = PV\left(1 + \frac{r}{n}\right)^{nt}$$
$$FV = 1250\left(1 + \frac{0.04}{365}\right)^{365 \cdot 3}$$
$$\approx 1250(1.127489438)$$
$$\approx 1409.36$$

The future value is $1409.36.
Solution using a spreadsheet.
Use the future value formula.
=FV(B1/B2, B3*B2, B4, B5)
The payment is $0.

	A	B
1	Annual interest rate	4.00%
2	Compounding periods per year	365
3	Number of years	3
4	Payment	$0.00
5	Present value	-$1,250.00
6	Future value	$1,409.36

The future value is $1409.36.

P2. The amount of interest earned is the difference between the current present value, $50,000, and the future value.
=FV(B1/B2, B3*B2, B4, B5)

	A	B
1	Annual interest rate	5.10%
2	Compounding periods per year	12
3	Number of years	3
4	Payment	$0.00
5	Present value	-$50,000.00
6	Future value	$58,247.36

The future value is $58,247.36.
Interest earned = 58,247.36 − 50,000.00 = 8247.36.
The interest earned was $8247.36.

P3. Solution using a calculator.
$FV = 10,000, n = 4, t = 5, r = 0.07$

$$PV = \frac{FV}{\left(1 + \dfrac{r}{n}\right)^{nt}}$$
$$= \frac{10,000}{\left(1 + \dfrac{0.07}{4}\right)^{4 \cdot 5}}$$
$$\approx 7068.25$$

The amount deposited today must be $7068.25.
Solution using the Finance App from a TI-84 calculator.
$N = 4 \cdot 5 = 20$

```
N=20
I%=7
■PV=-7068.245772
PMT=0
FV=10000
P/Y=4
C/Y=4
PMT: END BEGIN
```

The amount deposited today must be $7068.25.

P4. The purchasing power is the present value of the investment.
$FV = 250,000, n = 1, t = 25, r = 0.04$
Solution using a spreadsheet.

	A	B
1	Annual interest rate	4.00%
2	Compounding periods per year	1
3	Number of years	25
4	Payment	$0.00
5	Future value	$250,000.00
6	Present value	-$93,779.20

The purchasing power will be $93,779.20.

P5. Find the percentage increase in the CPI between the two years.

$$\text{Inflation rate} = 100\left(\frac{\text{CPI in 2018} - \text{CPI in 2017}}{\text{CPI in 2017}}\right)$$

$$= 100\left(\frac{257.307 - 251.122}{251.122}\right)$$

$$\approx 2.463$$

The percentage increase was 2.463%.

P6. The CPI increased from 242.839 to 251.038 or $\dfrac{251.038}{242.839}$ times.

Therefore, an equivalent salary in 2018 is

$$54,500 \cdot \frac{251.038}{242.839} \approx 56,340.09$$

Because her new salary of $57,000 is greater than $56,340.09, her salary has kept up with inflation.

P7. Use the formula for APY, annual percentage yield.

$$\text{APY} = \left(1 + \frac{\text{APR}}{n}\right)^n - 1$$

$$= \left(1 + \frac{0.04}{12}\right)^{12} - 1$$

$$\approx 0.0407$$

The APY is 4.07%.

Section 4.3

P1. Payments are made at the beginning of the month. Use the future value of a due annuity formula with $PMT = 450$, $n = 20 \cdot 12 = 240$ (20 years is 240 months), and $i = \dfrac{0.035}{12}$.

$$FV = PMT\left[\frac{(1+i)^n - 1}{i}\right](1+i)$$

$$FV = 450\left[\frac{\left(1 + \dfrac{0.035}{12}\right)^{240} - 1}{\dfrac{0.035}{12}}\right]\left(1 + \frac{0.035}{12}\right)$$

$$\approx 156,546.44$$

The future value of the annuity is $156,546.44. For help with this calculation, you can access an Excel template from your digital resources.
= FV(B2/B3, B3*B4, B5, B6, B7)

	A	B
1	Calculate the Future Value of a Payment	
2	Annual interest rate	3.50%
3	Number of payments per year	12
4	Number of years	20
5	Payment per period	-$450.00
6	Present value	$0.00
7	Ordinary or due annuity	1
8	Future Value	$156,546.44

P2. Payments are made at the end of the month. Use the future value of an ordinary annuity formula with $PMT = 3000$, $n = 25 \cdot 12 = 300$, and $i = \dfrac{0.042}{12}$.

$$FV = PMT\left[\frac{(1+i)^n - 1}{i}\right]$$

$$FV = 3000\left[\frac{\left(1 + \dfrac{0.042}{12}\right)^{300} - 1}{\dfrac{0.042}{12}}\right]$$

$$\approx 1,587,786.17$$

The future value of the annuity is $1,587,786.17. For help with this calculation, you can access an Excel template from your digital resources.
= FV(B2/B3, B3*B4, B5, B6, B7)

	A	B
1	Calculate the Future Value of a Payment	
2	Annual interest rate	4.20%
3	Number of payments per year	12
4	Number of years	25
5	Payment per period	-$3,000.00
6	Present value	$0.00
7	Ordinary or due annuity	0
8	Future Value	$1,587,786.17

P3. Payments are made at the beginning of the week. Solve the future value of a due annuity formula for PMT. $FV = 20,000$, $n = 3 \cdot 52 = 156$, and $i = \dfrac{0.0625}{52}$.

$$FV = PMT\left[\frac{(1+i)^n - 1}{i}\right](1+i)$$

$$20,000 = PMT\left[\frac{\left(1 + \dfrac{0.0625}{52}\right)^{156} - 1}{\dfrac{0.0625}{52}}\right]\left(1 + \frac{0.0625}{52}\right)$$

$$20,000 \approx PMT(171.6766748)$$
$$116.50 \approx PMT$$

The manager must deposit $116.50 each week into the account.

For help with this calculation, you can access an Excel template from your digital resources.
= PMT(B11/B12, B12*B13, B14, B15, B16)

	A	B
10	Calculate a Payment for a Given Future Value	
11	Annual interest rate	6.25%
12	Number of payments per year	52
13	Number of years	3
14	Present value	$0.00
15	Future value	-$20,000.00
16	Ordinary or due annuity	1
17	Payment	$116.50

P4. For help with this calculation, you can access an Excel template from your digital resources.
= PMT(B11/B12, B12*B13, B14, B15, B16)

	A	B
10	Calculate a Payment for a Given Future Value	
11	Annual interest rate	4.00%
12	Number of payments per year	12
13	Number of years	25
14	Present value	$0.00
15	Future value	$300,000,000.00
16	Ordinary or due annuity	0
17	Payment	-$583,510.52

The monthly payment amount is $583,510.52.

P5. For help with this calculation, you can access an Excel template from your digital resources.
= PV(B3/B4, B5*B4, B6, B7, B8)

	A	B
2	Calculate Present Value Given payment Amount	
3	Annual interest rate	5.20%
4	Number of payments per year	1
5	Number of years	25
6	Payment	-$150,000.00
7	Future value	$0.00
8	Ordinary or due annuity	0
9	Present Value	$2,072,356.59

The present value is $2,072,356.59.

P6. For help with this calculation, you can access an Excel template from your digital resources.
= PV(B3/B4, B5*B4, B6, B7, B8)

	A	B
2	Calculate Present Value Given Payment Amount	
3	Annual interest rate	3.75%
4	Number of payments per year	12
5	Number of years	5
6	Payment	-$100,000.00
7	Future value	$0.00
8	Ordinary or due annuity	1
9	Present Value	$5,480,383.94

The present value is $5,480,383.94.

P7. For help with this calculation, you can access an Excel template from your digital resources.
= PMT(B12/B13, B13*B14, B15, B16, B17)

	A	B
11	Calculate a Payment for a Loan	
12	Annual interest rate	5.125%
13	Number of payments per year	12
14	Number of years	3
15	Present value	$16,500.00
16	Future value	$0.00
17	Ordinary or due annuity	0
18	Payment	-$495.45

The monthly payment is $495.45.

P8. For help with this calculation, you can access an Excel template from your digital resources.
= PMT(B12/B13, B13*B14, B15, B16, B17)

	A	B
11	Calculate a Payment for a Loan	
12	Annual interest rate	3.900%
13	Number of payments per year	12
14	Number of years	3
15	Present value	$6,100.00
16	Future value	$0.00
17	Ordinary or due annuity	0
18	Payment	-$179.83

The monthly payment is $179.83.

P9. Multiply the monthly payment, $179.83, by 36, the number of payments. Then subtract the amount borrowed, $6100.
179.83 × 36 − 6100 = 373.88
Tomasz paid $373.88 in interest.

P10. The amount of the payoff is the present value of the remaining payments. Because Aidan has made 15 payments, there are 21 remaining payments. Use a spreadsheet or calculator to find the present value. We will use a TI-84 for this example.

```
N=21
I%=6.5
■PV=7281.89906
PMT=-367.79
FV=0
P/Y=12
C/Y=12
PMT: END BEGIN
```

The loan payoff is $7281.90.

Section 4.4

P1. The results are recorded in the following table.

Date of Next Transaction	Purchase or Payment ($)	Balance ($)	Number of Days Until Balance Changes	Unpaid Balance Times Number of Days ($)
7/1/18		620.00		
7/7/18	315.00	935.00	6	6 · 620 + 315 = 4035.00
7/15/18	−400.00	535.00	8	8 · 935 − 400 = 7080.00
7/22/18	410.00	945.00	7	7 · 535 + 410 = 4155.00
			10	10 · 945 = 9450.00
			Total	$24,720.00

Average daily balance $\dfrac{24,720}{31} \approx \$797.42.$

P2. The results are recorded in the following table.

Date of Next Transaction	Purchase or Payment ($)	Balance ($)	Number of Days Until Balance Changes	Unpaid Balance Times Number of Days ($)
5/8/18		652.95		
5/15/18	325.00	977.95	7	$7 \cdot 652.95 + 325 = 4895.65$
5/17/18	114.78	1092.73	2	$2 \cdot 977.95 + 114.78 = 2070.68$
5/25/18	−350.00	742.73	8	$8 \cdot 1092.73 - 350 = 8391.84$
6/2/18	69.67	812.40	8	$8 \cdot 742.73 + 69.67 = 6011.51$
			6	$6 \cdot 812.40 = 4874.40$
			Total	26,244.08

Average daily balance $\dfrac{26{,}244.08}{31} \approx \$846.58.$

To calculate the finance charge, use the simple interest formula $I = Prt$.

$I = Prt$

$\quad = 846.58(0.018)(1)$ • $P = 846.58$, $r = 0.018$, $t = 1$ (1 month)

$\quad \approx 15.24$

The finance charge was $15.24.

P3. For help with this calculation, you can access an Excel template from your digital resources.

Date payment is due	9/3/18	• Enter the payment due date.
Date of next payment	10/3/18	• Enter the next payment due date.
Days in billing cycle	30	• Enter the annual interest rate.
Annual interest rate	20.80%	• Enter the beginning balance.
Beginning balance	915.22	• Enter the date of a transaction and its amount. • Do not enter values in any shaded cell.

Date of Next Transaction	Purchase or Payment ($)	Balance ($)	Number of Days Until Balance Changes	Unpaid Balance Times Number of Days ($)
9/3/18		915.22		
9/4/18	162.35	1077.57	1	1077.57
9/7/18	174.80	1252.37	3	3407.51
9/9/18	130.93	1383.30	2	2635.67
9/11/18	154.21	1537.51	2	2920.81
9/12/18	149.97	1687.48	1	1687.48
9/14/18	−425.00	1262.48	2	2949.96
9/17/18	156.97	1419.45	3	3944.41
9/19/18	170.52	1589.97	2	3009.42
9/20/18	28.15	1618.12	1	1618.12
9/21/18	16.54	1634.66	1	1634.66
9/23/18	48.77	1683.43	2	3318.09
9/24/18	137.19	1820.62	1	1820.62
9/25/18	155.49	1976.11	1	1976.11
9/26/18	86.22	2062.33	1	2062.33
9/27/18	183.97	2246.30	1	2246.30
9/28/18	70.77	2317.07	1	2317.07
9/29/18	70.52	2387.59	1	2387.59
9/30/18	40.50	2428.09	1	2428.09
10/1/18	53.15	2481.24	1	2481.24
10/2/18	163.66	2644.90	1	2644.90
			1	2644.90
			Total	$51,212.85
			Average daily balance	$ 1,707.10
			Finance charge	$ 27.28

The finance charge was $27.28.

P4. To calculate the minimum payment, use the simple interest formula $I = Prt$.

$$I = Prt$$
$$= 365.48(0.03)(1) \quad \bullet P = 365.48, r = 0.03, t = 1 \text{ (1 month)}$$
$$\approx 10.96$$

Emile must pay the greater of $10.96 or $30 so Emile must pay $30.

P5. For help with this calculation, you can access an Excel template from your digital resources.

$= \text{NPER(B30/B31, B32, B33, B34, B35)}$

	A	B
29	Calculate Number of Payments to Repay a Loan	
30	Annual interest rate	22.10%
31	Payments per year	12
32	Payment	-$175.00
33	Present value	$3,856.71
34	Future value	$0.00
35	Ordinary or due annuity	0
36	Number of Payments	28.53077163

It will take about 29 months to repay the credit card debt.

Section 4.5

P1. The loan fee is deducted from the loan amount.
Loan fee $= 5300 \cdot 0.0095 = 50.35$
Morgan receives $5300 - $50.35 = $5249.65.

P2. For help with this calculation, you can access an Excel template from your digital resources.
$= \text{PMT(B12/B13, B13*B14, B15, B16, B17)}$

	A	B
11	Calculate a Payment for a Loan	
12	Annual interest rate	4.100%
13	Number of payments per year	12
14	Number of years	10
15	Present value	$27,000.00
16	Future value	$0.00
17	Ordinary or due annuity	0
18	Payment	-$274.65

The monthly payment is $274.65.

P3. Because Mackenzie has a nonsubsidized loan, she must pay the interest on the loan during the period of time that payments on the loan are not being made, 4 years. The interest is

$$I = Prt$$
$$= 22,000(0.0475)(4)$$
$$= 4180$$

She owes $4180 in accrued interest. This amount is added to the loan amount, and the monthly payment determined the amount owed.

$22,000 + $4180 = $26,180

For help with this calculation, you can access an Excel template from your digital resources.
$= \text{PMT(B12/B13, B13*B14, B15, B16, B17)}$

	A	B
11	Calculate a Payment for a Loan	
12	Annual interest rate	4.750%
13	Number of payments per year	12
14	Number of years	15
15	Present value	26,180.00
16	Future value	$0.00
17	Ordinary or due annuity	0
18	Payment	-$203.64

The monthly payment is $203.64.

P4. For help with this calculation, you can access an Excel template from your digital resources.
$= \text{NPER(B30/B31,B32,B33,B34,B35)}$

	A	B
29	Calculate Number of Payments to Repay a Loan	
30	Annual interest rate	4.00%
31	Payments per year	12
32	Payment	-$325.00
33	Present value	$28,000.00
34	Future value	$0.00
35	Ordinary or due annuity	0
36	Number of payments	101.7268566

It will take approximately 102 months to repay the loan.

P5. Because interest on a PLUS Loan accrues from the time it is disbursed, Theodora owes interest on the $15,000 for 3 years (her last 2 years of law school plus the 1 year of deferred payments after graduation).

$$I = Prt$$
$$= 15,000(0.0568)(3)$$
$$= 2556$$

She owes $2556 in accrued interest. This amount is added to the loan amount, and the monthly payment determined by the amount owed.

Amount owed $= $15,000 + $2556 = 17,556$

For help with this calculation, you can access an Excel template from your digital resources. We will use a TI-84 calculator here.

```
N=96
I%=5.68
PV=17556
■PMT=-227.98519...
FV=0
P/Y=12
C/Y=12
PMT: END BEGIN
```

The monthly payment is $227.99.

P6. There are two steps to the calculation.
- Calculate the monthly payment based on the loan amount and the deferred interest.

$$I = Prt$$
$$= 5000(0.039)(8)$$
$$= 1560$$

He owes $1560 in deferred interest. This amount is added to the loan amount, and the monthly payment determined by the amount owed.

Amount owed $= $5000 + $1560 = 6560

For help with this calculation, you can access an Excel template from your digital resources. We will use a TI-84 calculator here.

```
N=96
I%=3.9
PV=6560
■PMT=-79.656843...
FV=0
P/Y=12
C/Y=12
PMT: END BEGIN
```

The monthly payment is $79.66.

- Because the student loan is discounted, the amount Cheng receives is

$$5000 - 5000 \cdot 0.0098 = 4951.$$

```
N=96
■I%=11.7022312...
PV=4951
PMT=-79.66
FV=0
P/Y=12
C/Y=12
PMT: END BEGIN
```

The APR of the loan is 11.70%.

Section 4.6

P1. For help with this calculation, you can access an Excel template from your digital resources. We will use the spreadsheet PMT function here.
= PMT(B12/B13, B13*B14, B15, B16, B17)

	A	B
11	Calculate a Payment for a Loan	
12	Annual interest rate	4.900%
13	Number of payments per year	12
14	Number of years	4
15	Present value	$39,895.66
16	Future value	$0.00
17	Ordinary or due annuity	0
18	Payment	-$916.96

The monthly payment is $916.96.

P2. (a) Sales tax = 26,500 · 0.0775 = 2053.75
The sales tax is $2053.75.

(b) Registration = 26,500 · 0.009 = 238.50
The registration fee is $238.50.

(c) Amount financed = 26,500 + 2053.75 + 238.50 − 5000 = $23,792.25

For help with this calculation, you can access an Excel template from your digital resources. We will use the spreadsheet PMT function here.
= PMT(B12/B13, B13*B14, B15, B16, B17)

	A	B
11	Calculate a Payment for a Loan	
12	Annual interest rate	3.500%
13	Number of payments per year	12
14	Number of years	5
15	Present value	$23,792.25
16	Future value	$0.00
17	Ordinary or due annuity	0
18	Payment	-$432.82

The monthly payment is $432.82.

P3. We will use a TI-84 calculator.
The firefighter has 24 payments remaining. Therefore, $n = 24$.

```
N=24
I%=3.75
■PV=17326.14339
PMT=-750.46
FV=0
P/Y=12
C/Y=12
PMT: END BEGIN
```

The firefighter owes $17,326.14 on the car.

P4. For help with this calculation, you can access an Excel template from your digital resources. We will use a spreadsheet here.
= PV(B3/B4, B5*B4, B6, B7, B8)

	A	B
2	Calculate Present Value Given Payment Amount	
3	Annual interest rate	3.900%
4	Number of payments per year	12
5	Number of years	4
6	Payment	-$475.00
7	Future value	$0.00
8	Ordinary or due annuity	0
9	Present value	$21,078.95

Kato can refinance approximately $21,000 for a car.

P5. Change 35 miles per gallon to gallons per 100 miles. Write and solve a proportion.

$$\frac{x \text{ gallons}}{100 \text{ miles}} = \frac{1 \text{ gallon}}{35 \text{ miles}}$$
$$35x = 100$$
$$x = \frac{100}{35} \approx 2.9$$

The car uses 2.9 gallons per 100 miles driven.

P6. Depreciation = $64,000 · 0.24 = $15,360

Cost per mile = $\dfrac{15,360}{14,200 \text{ miles}} \approx 1.08/\text{mile}$

The cost per mile is $1.08.

P7. License and registration = 48,000 · 0.008 = 384
Sales tax = 48,000 · 0.045 = 2160

Depreciation $= 48,000 \cdot 0.20 = 9600$

$$\text{Fuel cost} = \frac{\text{Miles in 1 year}}{\text{Fuel efficiency}} \cdot \text{Cost per gallon}$$

$$= \frac{12,000}{28} \cdot 2.97 \approx 1272.86$$

To find the year 1 interest payment, use the cumulative interest function (CUMINT) in a spreadsheet program. For help with this calculation, you can access an Excel template from your digital resources.

The cumulative interest for year 1 is $1478.12.

	A	B
	B9 ‹ ⊗ ⊘ ⌢ *fx* 1478.12	
1	Calculate Cumulative Interest Paid for a Loan	
2	Annual interest rate	4.200%
3	Payments per year	12
4	Number of payments	60
5	Present Value	$38,400.00
6	Start period	1
7	End period	12
8	Ordinary or due annuity	0
9	Cumulative Interest	1478.12

Enter the calculated values in the table and find the total.

License and registration	0.8% of purchase price	384.00
Sales tax	4.5% of purchase price	2160.00
Depreciation	20% of purchase price	9600.00
Fuel cost driving 14,000 miles	Average cost of gasoline: $2.56 per gallon	1493.33
Year 1 interest payment	Amount paid in interest on the car loan for first year of ownership	1478.12
Insurance	$789	789.00
Maintenance	$425	425.00
	Total	16,329.45

The total cost of ownership for the first year is $16,329.45.

P8. Follow the steps to Example 8.

Net capitalized cost $=$ Negotiated price $-$ Down payment $-$ Trade-in value

$= 48,000 - 4000 - 0$

$= 44,000$

Residual value $= 51,100 \cdot 0.48 = 24,528$

Finance charge $=$ (Net capitalized cost $+$ Residual value) $\cdot$ Money factor

$$= (44,000 + 24,528) \cdot \frac{7.8}{2400}$$

$$\approx 222.72$$

$$\text{Depreciation} = \frac{\text{Net capitalized cost} - \text{Residual value}}{\text{Number of months in lease}}$$

$$= \frac{44,000 - 24,528}{36}$$

$$\approx 540.89$$

Monthly payment $=$ Finance charge $+$ Monthly depreciation

$= 222.72 + 540.89$

$= 763.61$

The monthly lease payment is $763.61.

Section 4.7

P1. $\text{DTI} = \dfrac{\text{Monthly payments}}{\text{Gross monthly income}}$

$$= \frac{125 + 256.78 + 216 + 2031.79}{4875}$$

$$= \frac{2629.57}{4875} \approx 0.54$$

The debt-to-income ratio is 54.

P2. Real estate commission $= 365,000 \cdot 0.045 = 16,425$.

Title insurance $= 365,000 \cdot 0.01 = 3650$

Amount received $= 365,000 - 16,425 - 3650 - 900$

$= 344,025$

They receive $344,025.00.

To find the percentage of the selling price that are fees, subtract the fees from the selling price and then divide by the selling price.

$$\frac{365,000 - 344,025}{365,000} \approx 0.057$$

The fees were approximately 5.7% of the selling price.

P3. Use a calculator or PMT function of a spreadsheet program. We will use the TI-84 and the TMV solver app. Enter the values for each known quantity. Move the cursor to **PMT** and press **ALPHA ENTER**.

```
N=360
I%=5.25
PV=411000
■PMT=-2269.5572...
FV=0
P/Y=12
C/Y=12
PMT: END BEGIN
```

The monthly payment is $2269.56.

P4. PMI $= 410,000 \cdot 0.008 = 3280$

The PMI is $3280.

Monthly PMI $= \dfrac{3280}{12} \approx 273.33$

The monthly PMI payment is $273.33.

P5. Points $= 245,000 \cdot 0.02125 = 5206.25$

The payment for points is $5206.25.

P6. Difference in cost $= 485,000(0.0185) - 485,000(0.015)$

$= 8972.50 - 7275$

$= 1697.50$

The difference in points is $1697.50.

Now calculate the monthly payments based on the two interest rates.

```
N=360
I%=4.1
PV=485000
■PMT=-2343.5121
FV=0
P/Y=12
C/Y=12
PMT: END BEGIN
```

```
N=360
I%=3.8
PV=485000
■PMT=-2259.8931...
FV=0
P/Y=12
C/Y=12
PMT: END BEGIN
```

The difference in the two payments is $2343.51 - $2259.89 $=$ $83.62.

To find the number of months before the lesser annual interest rate makes sense, find $\dfrac{1697.50}{83.62} \approx 20.3$.

It will take about 20 months before the loan with more points is less expensive.

P7. We need to calculate the present value of her loan after 5 years. Use a calculator or PV function of a spreadsheet program. We will use the TI-84 and the TVM solver app. Enter the values for each known quantity. Because Zoey has made payments for 5 years (60 months), the value of N is 300, $360 - 60$. Move the cursor to PV and press ALPHA ENTER.

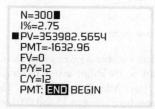

```
N=300█
I%=2.75
█PV=353982.5654
PMT=-1632.96
FV=0
P/Y=12
C/Y=12
PMT: END BEGIN
```

Use the current (353,982.57) present value and new interest rate (3.5%) to find the new monthly payment.

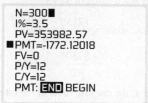

```
N=300█
I%=3.5
PV=353982.57
█PMT=-1772.12018
FV=0
P/Y=12
C/Y=12
PMT: END BEGIN
```

The new monthly payment is $1772.12.

P8. Use a calculator or PMT of a spreadsheet program. We will use the TI-84 and the TVM solver app. Enter the values for each known quantity. Move the cursor to PMT and press ALPHA ENTER. Because the pilot has made payments for 4 years (48 months), the number of remaining payments is $300 - 48 = 252$. Use this number for N, the number of payments.

```
N=252
I%=4.875
█PV=386549.6547
PMT=-2453.65
FV=0
P/Y=12
C/Y=12
PMT: END BEGIN
```

The loan payoff is $386,549.65.

P9. There are a few steps to this problem.
- Find the fee for points and the loan origination fee.

    Fee for points $= 512,000 \cdot 0.014 = 7168$
    Loan origination fee $= 512,000 \cdot 0.0075 = 3840$

- Add these amounts to the loan amount to give an adjusted loan balance.

    Adjusted loan balance $= 512,000 + 7168 + 3840$
    $= 523,008$

- Find the monthly payment based on the adjusted balance. We will use a TI-84 here.

```
N=360
I%=4.125
PV=523008
█PMT=-2534.7568...
FV=0
P/Y=12
C/Y=12
PMT: END BEGIN
```

The monthly payment is $2534.76.
- Replace the adjusted loan amount with the actual loan amount, $512,000, and then solve for I%. The result follows.

```
N=360
█I%=4.303386469
PV=512000
PMT=-2534.76
FV=0
P/Y=12
C/Y=12
PMT: END BEGIN
```

The loan APR is 4.303%.

Section 4.8

P1. Dividend $=$ Dividend per share $\cdot$ number of shares
$$= 2.03 \cdot 500 = 1015$$
The shareholder receives $1015 in dividends.

P2. Dividend yield $= \dfrac{2.035}{68.35} \approx 0.0298$
The dividend yield for the stock is 2.98%.

P3. EPS $= \dfrac{3,250,000,000}{850,000,000} \approx 3.82$
The earnings per share were $3.82.

P4. P/E $= \dfrac{\text{Stock price}}{\text{EPS}}$
$$= \dfrac{86.29}{5.06} \approx 17.1$$
The P/E for the stock is 17.1.

P5. Current value of bond $= \dfrac{\text{Coupon rate of bond}}{\text{Current interest rate}}$
$$\cdot \text{ Par value of the bond}$$
$$= \dfrac{0.0275}{0.0325} \cdot \$1000$$
$$\approx 846.15$$
The current value of the bond is $846.15.

P6. net asset value $= \dfrac{\text{Market value}}{\text{Number of shares}}$
$$= \dfrac{3,250,000,000}{120,000,000} \approx 27.08$$
The net asset value is $27.08.

P7. The amount available to invest is $7500 minus the transaction fee of 5% of $7500.

Amount to invest $= 7500 - 7500(0.05)$
$$= 7500 - 375 = 7125$$

The amount available to purchase shares is $7125.

Number of shares $= \dfrac{\text{Amount to invest}}{\text{Share price}}$
$$= \dfrac{7125}{38.26} \approx 186.226$$

Carley can purchase 186.226 shares.

P8. Real rate of return $= \dfrac{1 + \text{nominal rate}}{1 + \text{inflation rate}} - 1$
$$= \dfrac{1 + 0.056}{1 + 0.013} - 1$$
$$= \dfrac{1.056}{1.013} - 1$$
$$\approx 0.0424$$
The real rate of return is 4.24%.

Chapter 5

Section 5.1

P1. Evaluate $s(t) = -16t^2 + 64t + 4$ when $t = 3$.

$$s(t) = -16t^2 + 64t + 4$$
$$s(3) = -16(3)^2 + 64(3) + 4$$
$$= -144 + 192 + 4$$
$$= 52$$

The ball is 52 feet above the ground.

P2. (a) To complete the table, evaluate $C(t) = -50t^2 + 460t + 300$ when t is 3, 5, and 8.

$$C(t) = -50t^2 + 460t + 300$$
$$C(3) = -50(3)^2 + 460(3) + 300$$
$$= -450 + 1380 + 300$$
$$= 1230$$

$$C(t) = -50t^2 + 460t + 300$$
$$C(5) = -50(5)^2 + 460(5) + 300$$
$$= -1250 + 2300 + 300$$
$$= 1350$$

$$C(t) = -50t^2 + 460t + 300$$
$$C(8) = -50(8)^2 + 460(8) + 300$$
$$= -3200 + 3680 + 300$$
$$= 780$$

The completed table follows. Rather than use $C(t)$ in the table, we have used N to represent $C(t)$; that is, $N = C(t)$. See the remarks following Example 3.

Time (hours), t	Number, N	(t, N)
0	300	(0, 300)
1	710	(1, 710)
2	1020	(2, 1020)
3	1230	(3, 1230)
4	1340	(4, 1340)
5	1350	(5, 1350)
6	1260	(6, 1260)
7	1070	(7, 1070)
8	780	(8, 780)
9	390	(9, 390)

(b) Plot the points and then draw a smooth graph through the points. Note the use of N on the vertical axis for $C(t)$.

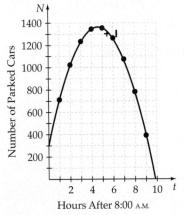

Hours After 8:00 A.M.

P3.

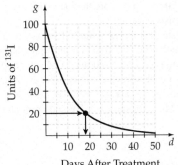

Days After Treatment

The value of the dependent variable is given as 20. Draw a line from the 20-unit mark on the iodine axis to the graph. Then draw a line to the "Days After Treatment" axis. The line touches the axis somewhere between 15 and 20 days and closer to 20 days than 15 days. An estimate is that after 18 days, there are 20 units of iodine in the bloodstream.

P4. (a) After 4 years, the car has a value of $8400.

(b) Average rate of depreciation $= \dfrac{17150 - 8400}{2 - 4}$

$$= \dfrac{8750}{-2} = -4375$$

Between year 2 and year 4, the car lost value at the rate of $4375 per year.

P5. (a) The speed is increasing as time increases.

(b) From the graph, the change between 1 and 2 seconds is greater than the change between 2 and 3 seconds.

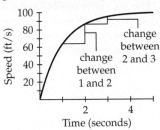

Time (seconds)

Section 5.2

P1. $f(v) = \frac{2}{3}v + 1$ is a linear function. Therefore, the graph is a straight line. The horizontal axis contains the values of the independent variable v. The vertical axis contains the values of the dependent variable, which we will call w. [We could have chosen any letter or even left it as $f(v)$.] Select three values of v and evaluate the function for those values. The results are recorded in the following table.

v	$w = \dfrac{2v}{3}$		w	(v, w)
−3	$w = \dfrac{2(-3)}{3} + 1$	=	−1	(−3, −1)
0	$w = \dfrac{2(0)}{3} + 1$	=	1	(0, 1)
3	$w = \dfrac{2(3)}{3} + 1$	=	3	(3, 3)

Plot the points and then draw a line through the points.

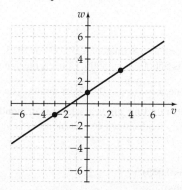

P2. Let $(x_1, y_1) = (3, 1)$ and $(x_2, y_2) = (-1, 3)$.

$$m = \frac{y_2 - y_1}{x_2 - x_1}$$

$$= \frac{3 - 1}{-1 - 3} = \frac{2}{-4}$$

$$= -\frac{1}{2}$$

P3. (a) $m = \dfrac{\text{Change in gallons of gas}}{\text{Change in distance}}$

$$= \frac{4 - 12}{504 - 216} = \frac{-8}{288}$$

$$= -\frac{1}{36} \approx -0.03 \text{ gal/mi}$$

(b) The number of gallons of gas remaining in the tank is decreasing (negative slope) at an approximate rate of 0.03 gallon per mile.

P4. (a) $m = \dfrac{\text{Change in distance}}{\text{Change in time}}$

$$= \frac{33 - 22}{15 - 10} = \frac{11}{5}$$

$$= 2.2 \text{ m/s}$$

(b) The number of meters the swimmer travels is increasing (positive slope) at a rate of 2.2 meters per second.
(c) From part b, the distance traveled by the swimmer is increasing at a rate of 2.2 meters per second. If the swimmer has traveled 44 meters in 20 seconds, then the swimmer will have traveled 44 m + 2.2 m = 46.2 meters in 21 seconds.

P5. (a) Average growth rate per day $= \dfrac{\text{Growth rate per month}}{30 \text{ days in 1 month}}$

$$= \frac{0.6 \text{ inch per month}}{30 \text{ days in 1 month}}$$

$$= 0.02 \text{ inches/day}$$

(b) To find the number of inches hair grows in 15 days, multiply the rate per day (slope) by 15 days:

$$\frac{0.02 \text{ inch}}{\text{day}} \cdot 15 \text{ days} = 0.3 \text{ inch}$$

P6. (a) To find the C-intercept, let $n = 0$ and solve for C.
$$C = 210n + 1000$$
$$C = 210(0) + 1000$$
$$C = 0 + 1000$$
$$C = 1000$$

The cost to produce 0 cell phones is $1000. This number is referred to as *fixed cost*. It is the cost for rent, equipment, and other items that are required to make a product.
(b) To find the n-intercept, let $C = 0$ and solve for n.
$$C = 210n + 1000$$
$$0 = 210n + 1000$$
$$-1000 = 210n$$
$$-4.76 \approx n$$

Because n is negative, the n-intercept does not make sense. A manufacturer cannot make a negative number of items.

Section 5.3

P1. The slope is 2 inches per hour, and the initial state was given as 20 inches of snow. Let y be the dependent variable (amount of snow on the ground) and t be the time the snow falls.

Linear model

$$\underset{\text{variable}}{\text{Dependent}} = \underset{\text{change}}{\text{Rate of}} \cdot \underset{\text{variable}}{\text{Independent}} + \underset{\text{state}}{\text{Initial}}$$

$$y = m \cdot t + b$$

$$y = 2 \cdot t + 20$$

A linear model is $y = 2t + 20$.
P2. The slope is 0.017 meter per second. The initial state is unknown. However, we know that at 1000 meters below sea level, the speed of sound is 1480 meters per second. Let y be the dependent variable (speed of sound) and x be the ocean depth.

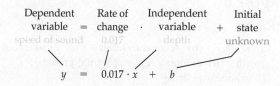

$$y = 0.017 \cdot x + b$$

Use the ordered pair (1000, 1480) to find b.
$$y = 0.017x + b$$
$$1480 = 0.17(1000) + b \qquad \bullet\, x = 1000,\ y = 1480$$
$$1480 = 170 + b$$
$$1310 = b$$

A linear model is $y = 0.017x + 1310$. To find the speed of sound at 2500 meters, replace x with 2500 and solve for y.
$$y = 0.017x + 1310$$
$$y = 0.017(2500) + 1310$$
$$= 1352.5$$

The speed of sound at a depth of 2500 meters is approximately 1353 meters/second.
P3. Use the two given data points to find the slope of the line between them.

$$m = \frac{\text{Change in dependent variable}}{\text{Change in independent variable}}$$

$$m = \frac{33{,}000 - 34{,}400}{50{,}000 - 25{,}000} = \frac{-1400}{25{,}000}$$

$$= -0.056$$

The slope of the line is -0.056 dollar/mile driven. Use the slope and one of the given points to find the initial state, b.

$$y = -0.056x + b$$
$$34,400 = -0.056(25,000) + b \quad \bullet x = 25,000, y = 34,400.$$
$$34,400 = -1400 + b$$
$$35,800 = b$$

A linear model is $y = -0.056x + 35,800$.

To find the value of the car that has been driven 45,000 miles, replace x by 45,000 and solve for y.

$$y = -0.056x + 35,800$$
$$y = -0.056(45,000) + 35,800$$
$$= 33,280$$

The value of a similarly equipped car with 45,000 miles on it is \$33,280.

P4. Use the two data points to find the slope of the line between them.

$$m = \frac{\text{Change in dependent variable}}{\text{Change in independent variable}}$$

$$m = \frac{15.50 - 13.50}{30 - 0} = \frac{2}{30}$$

$$\approx 0.067$$

The slope of the line is 0.067 centimeter per day. The initial state, b, is 13.50.

A linear model is $y = 0.067x + 13.5$.

To find the value of the circumference of a baby's head 18 days after birth, replace x with 18 and solve for y.

$$y = 0.067x + 13.5$$
$$y = 0.067(18) + 13.5$$
$$= 14.706$$

The circumference of a baby's head 18 days after birth is approximately 14.7 centimeters.

Chapter 6

Section 6.1

P1. **(a)** neither **(b)** exponential decay **(c)** exponential growth **(d)** neither **(e)** exponential growth **(f)** exponential decay

P2. The initial number of bacteria is 2500. $A_0 = 2500$. The growth rate is $r\% = 15$ or 0.15 as a decimal. The model is

$$A(t) = 2500(1 + 0.15)^t$$
$$A(t) = 2500(1.15)^t$$

To find the bacteria in the colony after 10 hours, evaluate the function for $t = 10$.

$$A(t) = 2500(1.15)^t$$
$$A(10) = 2500(1.15)^{10}$$
$$\approx 10,114$$

There are approximately 10,114 bacteria in the colony after 10 hours.

P3. The initial value of the coin is $A_0 = 50,000$. The value is doubling every 20 years. The value of r is 1 (100%) per 20 years. The model is to be written for time in years, but r is given in 20-year intervals (doubling every 20 years). Because 1 year is $\frac{1}{20}$ of a time interval, there are $\frac{t}{20}$ intervals in t years. $T = t/20$.

$$A = A_0(1 + r)^T$$
$$A = 50,000(1 + 1)^{t/20}$$

$$A = 50,000(2)^{t/20}$$
$$A = 50,000(2)^{45/20}$$
$$A \approx 238,000$$

The coin will be worth approximately \$238,000 in 2025.

P4. The initial deposit is \$5000. $A_0 = 5000$. Convert the annual rate to a monthly rate. $r = \frac{0.06}{12} = 0.005$ per month. The model is to be written for time t in years, but r is in months. Because 1 year is 12 months, there are $12t$ months in t years. $T = 12t$.

$$A = A_0(1 + r)^T$$
$$A = 5000(1 + 0.005)^{12t}$$
$$A = 5000(1.005)^{12t}$$
$$A = 5000(1.005)^{12 \cdot 5} = 5000(1.005)^{60}$$
$$A \approx 6744.25$$

The investment will be worth \$6744.25 in 5 years.

P5. The initial amount of *Listeria* is 1500. $A_0 = 1500$. The doubling time is 50 minutes. $d = 50$. A model for the amount of *Listeria* is

$$A = A_0(2)^{t/d}$$
$$A = 1500(2)^{t/50}$$

Use the model to find the number of *Listeria* after $t = 200$ minutes.

$$A = 1500(2)^{t/50}$$
$$A = 1500(2)^{200/50}$$
$$A = 24,000$$

There are 24,000 *Listeria* in the milk after 50 minutes.

P6. The initial amount of salt is 350 g. $A_0 = 350$. $r = 20\% = 0.20$ per 10 minutes. The model is to be written in terms of minutes, but r is given in 10-minute intervals. Because 1 minute is $\frac{1}{10}$ of a time interval, there are $\frac{t}{10}$ intervals in t minutes.

$$A = A_0(1 - r)^T$$
$$A = 350(1 - 0.20)^{t/10} \quad \bullet A_0 = 350, r = 0.20, T = t/10.$$
$$A = 350(0.8)^{t/10}$$
$$A = 350(0.8)^{45/10} \quad \bullet t = 45.$$
$$A \approx 128$$

Approximately 128 g of salt will remain after 45 minutes.

P7. $r = 50\% = 0.50$ per 5 hours. The initial amount of caffeine is 90 mg. $A_0 = 90$.

Find the number of 5-hour blocks in 1 hour.

$$\frac{1 \text{ hr}}{5 \text{ hrs}} = \frac{1}{5} \text{ block in 1 hour. In } t \text{ hours, there are } \frac{t}{5} \text{ blocks.}$$

$$A = A_0(1 - r)^T$$
$$A = 90(1 - 0.50)^{t/5} \quad \bullet A_0 = 90, r = 0.50, T = t/5.$$
$$A = 90(0.5)^{t/5}$$

An exponential function that gives the amount A of caffeine in the body t hours after drinking a cup of coffee is $A = 90(0.5)^{t/5}$.

P8. The half-life is 90 minutes. The initial amount of ibuprofen is 200 milligrams. $A_0 = 200$.

Change the time interval from 2:00 PM to 6:00 PM to hours. There are 4 hours which is 240 minutes.

$$A = A_0\left(\frac{1}{2}\right)^{t/d}$$

$$A = A_0\left(\frac{1}{2}\right)^{t/90}$$

$$A = 200\left(\frac{1}{2}\right)^{240/90}$$

$$A \approx 31.498$$

There are approximately 31 milligrams of ibuprofen in the bloodstream at 6:00 PM.

Section 6.2

P1. (a) 0.0001   (b) 0.5623   (c) 5   (d) 0.3979

P2. $pH = -\log[H^+]$
$$= -\log(0.000051)$$
$$\approx 4.3$$
The sauce is an acid.

P3. $M = \log(I)$
$$= \log(750{,}000)$$
$$\approx 5.9$$

P4. $\log(I_1) = 9.2 \rightarrow 10^{9.2} = I_1$
$\log(I_2) = 8.6 \rightarrow 10^{8.6} = I_2$
Find the ratio of the intensities.
$$\frac{I_1}{I_2} = \frac{10^{9.2}}{10^{8.6}} = 10^{9.2-8.6} = 10^{0.6} \approx 3.98$$
The Alaska earthquake was approximately four times the intensity of the Assam earthquake.

P5. $dB(I) = 10\log\left(\dfrac{I}{I_0}\right)$
$$= 10\log\left(\frac{9.2 \times 10^{12}I_0}{I_0}\right)$$
$$= 10\log(9.2 \times 10^{12})$$
$$\approx 130$$
The decibel level of a harp seal call is approximately 130 dB.

P6. Find the intensity of the first speaker.
$$dB = 10\log\left(\frac{I}{I_0}\right)$$
$$75 = 10\log\left(\frac{I}{I_0}\right)$$
$$7.5 = \log\left(\frac{I}{I_0}\right)$$
$$\frac{I}{I_0} = 10^{7.5}$$
$$I = 10^{7.5}I_0$$
The intensity is $10^{7.5}I_0$.
By bringing in a second speaker, the intensity is doubled to $2 \cdot 10^{7.5}I_0$.

Now find the decibel level of the new intensity.
$$dB = 10\log\left(\frac{2 \cdot 10^{7.5}I_0}{I_0}\right)$$
$$= 10\log(2 \cdot 10^{7.5})$$
$$\approx 78$$

The decibel level is approximately 78 decibels.

This is an important example. Decibels are logarithmic scales and therefore do not add. Think about it this way. Normal conversation is about 60 decibels. If two people are talking and we erroneously added the decibels, we would get 120 decibels, about the decibel level of a siren.

P7. Solve the equation $s = 5(6.2)^d$ for d when $s = 20$ tons. The equation is of the form $A = A_0(p)^t$.

$$s = 5(6.2)^d$$
$$20 = 5(6.2)^d$$
$$4 = 6.2^d$$ • Divide each side of the equation by 5.
$$\log(4) = \log(6.2^d)$$ • Take the logarithm of each side of the equation.
$$\log(4) = d\log(6.2)$$ • Use the power property of logarithms.
$$\frac{\log(4)}{\log(6.2)} = d$$ • Solve for d.
$$0.76 \approx d$$

The engineer needs a cable 0.76 inch in diameter.

P8. Use the equation for the half-life of a process.
$$h = \frac{t \cdot \log\left(\frac{1}{2}\right)}{\log\left(\frac{A}{A_0}\right)}$$
$$= \frac{2 \cdot \log\left(\frac{1}{2}\right)}{\log\left(\frac{3.9685}{5}\right)}$$ • $t = 2$, $A = 3.9685$, $A_0 = 5$.
$$\approx 6$$
The half-life is approximately 6 hours.

P9. Use the equation for the doubling time of a process.
$$d = \frac{t \cdot \log(2)}{\log\left(\frac{A}{A_0}\right)}$$
$$= \frac{1 \cdot \log(2)}{\log\left(\frac{1.05A_0}{A_0}\right)}$$ • $t = 1$, $A = 1.05A_0$.
$$= \frac{\log(2)}{\log(1.05)}$$
$$\approx 14.2$$
The doubling time is a little more than 14 years.

P10. Use the equation for the doubling time of a process.
$$d = \frac{t \cdot \log(2)}{\log\left(\frac{A}{A_0}\right)}$$
$$= \frac{10 \cdot \log(2)}{\log\left(\frac{122}{100}\right)}$$ • $t = 10$, $A = 122$, $A_0 = 100$.
$$\approx 35$$
The doubling time is approximately 35 years.

Section 6.3

P1. (a) No (b) Yes, $a = 1$, $b = -4$, $c = 1$
(c) Yes, $a = -16$, $b = 32$, $c = 1$

P2. $a = -1$, the graph opens down.

$$x = -\frac{b}{2a} = -\frac{3}{2(-1)} = \frac{3}{2}$$

$$y = -x^2 + 3x + 4$$

$$= -\left(\frac{3}{2}\right)^2 + 3\left(\frac{3}{2}\right) + 4$$

$$= \frac{25}{4}$$

The coordinates of the vertex are $\left(\frac{3}{2}, \frac{25}{4}\right)$.

P3. $y = x^2 - 3x - 2$
To find the y-intercept, let $x = 0$ and solve for y.

$$y = x^2 - 3x - 2$$
$$= 0^2 - 3(0) - 2$$
$$= -2$$

The y-intercept is $(0, -2)$.

To find the x-intercepts, let $y = 0$ and solve for x. Use the quadratic formula with $a = 1$, $b = -3$, and $c = -2$.

$$x = \frac{-b \pm \sqrt{b^2 - 4ac}}{2a}$$

$$= \frac{-(-3) \pm \sqrt{(-3)^2 - 4(1)(-2)}}{2(1)}$$

$$= \frac{3 \pm \sqrt{17}}{2} \approx \frac{3 \pm 4.12}{2}$$

$$x = \frac{3 + 4.12}{2} = 3.56; \quad x = \frac{3 - 4.12}{2} = -0.56$$

The coordinates of the x-intercepts are approximately $(3.56, 0)$ and $(-0.56, 0)$.

P4. The equation $y = -16t^2 + 96t + 5$ gives the height y of the ball t seconds after it is released. To find the time at which it is 20 feet above the ground, replace y by 20 and solve for t.

$$y = -16t^2 + 96t + 5$$
$$20 = -16t^2 + 96t + 5$$
$$0 = -16t^2 + 96t - 15$$

$$t = \frac{-b \pm \sqrt{b^2 - 4ac}}{2a}$$

$$t = \frac{-96 \pm \sqrt{96^2 - 4(-16)(-15)}}{2(-16)}$$

$\bullet\, a = -16, b = 96, c = -15.$

$$= \frac{-96 \pm \sqrt{8256}}{-32} \approx \frac{-96 \pm 90.86}{-32}$$

$$t = \frac{-96 + 90.86}{-32} \approx 0.16; \quad t = \frac{-96 - 90.86}{-32} \approx 5.84$$

The ball will be 20 feet above the ground 0.16 second (on its way up) and 5.84 seconds (on its way down) after it is released.

P5. Maximum. The maximum occurs at the vertex.

$$y = -\frac{x^2}{2} + 4x - 6$$

x-coordinate of the vertex

$$x = -\frac{b}{2a} = -\frac{4}{2\left(-\frac{1}{2}\right)} = 4$$

y-coordinate of the vertex

$$y = -\frac{x^2}{2} + 4x - 6$$

$$= -\frac{4^2}{2} + 4(4) - 6 = 2$$

The maximum value is 2.

P6. The minimum height is the vertex of
$h = 0.0088x^2 - 0.88x + 40$
x-coordinate of the vertex:

$$x = -\frac{b}{2a} = -\frac{-0.88}{2(0.0088)} = 50$$

h-coordinate of the vertex

$$h = 0.0088x^2 - 0.88x + 40$$
$$= 0.0088(50)^2 - 0.88(50) + 40$$
$$= 18$$

The minimum height of the cable is 18 feet above the road.

P7. Let x represent the width of the courtyard and y represent the length. We are given $x + y = 100$ and our task is to maximize the area, which means to maximize Area $= xy$.

Solve $x + y = 100$ for y (we could have chosen x just as well).

$$x + y = 100$$
$$y = 100 - x$$

Replace y in Area $= xy$.

$$\text{Area} = xy$$
$$= x(100 - x)$$
$$\text{Area} = -x^2 + 100x$$

This is a quadratic equation. To find the maximum area, find the vertex.

$$x = -\frac{b}{2a} = -\frac{100}{2(-1)} = 50$$

Replace 50 for x in $y = 100 - x$ to find y.

$$y = 100 - x$$
$$= 100 - 50$$
$$= 50$$

The dimensions that will maximize the area of the courtyard are 50 feet by 50 feet.

P8. (a) The demand function is $p = 137 - 4x$. The revenue function R is px.

$$R = px$$
$$= (137 - 4x)x$$
$$R = -4x^2 + 137x$$

(b) The cost function is given as $C = 25x + 200$.
Profit P is revenue minus cost.

$$P = R - C$$
$$= (-4x^2 + 137x) - (25x + 200)$$
$$P = -4x^2 + 112x - 200$$

(c) To find the maximum profit, find the coordinates of the vertex of $P = -4x^2 + 112x - 200$.

x-coordinate of the vertex:

$$x = -\frac{b}{2a} = -\frac{112}{2(-4)} = 14$$

P-coordinate of the vertex:

$$P = -4x^2 + 112x - 200$$
$$= -4(14^2) + 112(14) - 200$$
$$= 584$$

The maximum profit is $584.

Chapter 7

Section 7.1

P1. The possible outcomes are {M, i, s, p}. There are four possible outcomes.

P2. **(a)** {1, 3, 5, 7, 9} **(b)** {0, 3, 6, 9} **(c)** {8, 9}

P3.

	H	T
1	1H	1T
2	2H	2T
3	3H	3T
4	4H	4T
5	5H	5T
6	6H	6T

The sample space has 12 elements:
{1H, 1T, 2H, 2T, 3H, 3T, 4H, 4T, 5H, 5T, 6H, 6T}

P4.

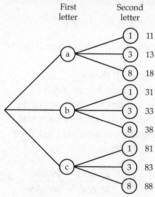

P5. Any of the 9 runners could win the gold medal, so $n_1 = 9$. That leaves $n_2 = 8$ runners that could win silver, and $n_3 = 7$ possibilities for bronze. By the counting principle, there are $9 \cdot 8 \cdot 7 = 504$ possible ways the medals can be awarded.

P6. **(a)** Because a student can receive more than 1 award and each award has 5 possible destinations, there are $5 \cdot 5 \cdot 5 = 125$ ways for the awards to be given.

(b) Once a student has an award, that student cannot receive another award, so there are $5 \cdot 4 \cdot 3 = 60$ different ways for the awards to be given.

Section 7.2

P1. **(a)** $7! + 4! = (7 \cdot 6 \cdot 5 \cdot 4 \cdot 3 \cdot 2 \cdot 1) + (4 \cdot 3 \cdot 2 \cdot 1)$
$$= 5040 + 24 = 5064$$

(b) $\dfrac{8!}{4!} = \dfrac{8 \cdot 7 \cdot 6 \cdot 5 \cdot 4!}{4!} = 8 \cdot 7 \cdot 6 \cdot 5 = 1680$

P2. Because the players are ranked, the number of different golf teams possible is the number of permutations of 8 players selected 5 at a time.

$$P(8, 5) = \frac{8!}{(8 - 5)!} = \frac{8!}{3!} = \frac{8 \cdot 7 \cdot 6 \cdot 5 \cdot 4 \cdot 3!}{3!}$$
$$= 8 \cdot 7 \cdot 6 \cdot 5 \cdot 4 = 6720$$

There are 6720 possible golf teams.

P3. The order in which the cars finish is important, so the number of ways to place first, second, and third is

$$P(43, 3) = \frac{43!}{(43 - 3)!} = \frac{43 \cdot 42 \cdot 41 \cdot 40!}{40!} = 74{,}046$$

There are 74,046 different ways to award the first, second, and third place prizes.

P4. **(a)** With no restrictions, there are 7 tutors available for 7 hr, so the number of schedules is

$$P(7, 7) = \frac{7!}{(7 - 7)!} = \frac{7!}{0!} = 7! = 5040$$

There are 5040 possible schedules.

(b) This is a multistage experiment; there are 3! ways to schedule the juniors and 4! ways to schedule the seniors. By the counting principle, the number of different tutoring schedules is $3! \cdot 4! = 6 \cdot 24 = 144$.
There are 144 tutor schedules.

P5. **(a)** With $n = 8$ coins and $k_1 = 3$ (number of pennies), $k_2 = 2$ (number of nickels), and $k_3 = 3$ (number of dimes), the number of different possible stacks is

$$\frac{8!}{3! \cdot 2! \cdot 3!} = \frac{8 \cdot 7 \cdot 6 \cdot 5 \cdot 4 \cdot 3!}{3! \cdot 2! \cdot 3!} = \frac{8 \cdot 7 \cdot 6 \cdot 5 \cdot 4}{6 \cdot 2} = 560$$

There are 560 possible stacks.

(b) Not including the dimes, there are $\dfrac{5!}{3! \cdot 2!} = 10$ ways to stack the pennies and nickels. The dimes are identical, so there is only one way to arrange the dimes together, but there are six different locations in the stack of pennies and nickels into which the dimes could be placed. By the counting principle, the total number of ways in which the stack of coins can be arranged if the dimes are together is $10 \cdot 6 = 60$.

P6. The order in which the waitstaff are chosen is not important, so the number of ways to choose 9 people from 16 is

$$C(16, 9) = \frac{16!}{9! \cdot (16 - 9)!} = \frac{16!}{9! \cdot 7!}$$

$$= \frac{16 \cdot 15 \cdot 14 \cdot 13 \cdot 12 \cdot 11 \cdot 10 \cdot 9!}{9! \cdot 7!}$$

$$= \frac{16 \cdot 15 \cdot 14 \cdot 13 \cdot 12 \cdot 11 \cdot 10}{7 \cdot 6 \cdot 5 \cdot 4 \cdot 3 \cdot 2 \cdot 1} = 11{,}440$$

There are 11,440 possible groups of 9 waitstaff.

P7. There are $C(4, 3)$ ways for the auditor to choose 3 corporate tax returns and $C(6, 2)$ ways to choose 2 individual tax returns. By the counting principle, the total number of ways in which the auditor can choose the returns is

$$C(4, 3) \cdot C(6, 2) = \frac{4!}{3! \cdot 1!} \cdot \frac{6!}{2! \cdot 4!} = 4 \cdot 15 = 60$$

There are 60 ways to choose the tax returns.

P8. For any single suit, there are $C(13, 4)$ ways of choosing 4 cards. That leaves $52 - 13 = 39$ cards of other suits from which to choose the fifth card. In addition, there are 4 different suits we could start with. By the counting principle, the number of 5-card combinations containing 4 cards of the same suit is

$$4 \cdot C(13, 4) \cdot 39 = 4 \cdot \frac{13!}{4! \cdot 9!} \cdot 39 = 4 \cdot 715 \cdot 39 = 111{,}540$$

There are 111,540 five-card combinations containing 4 cards of the same suit.

Section 7.3

P1. $S = \{HH, HT, TH, TT\}$

P2. The sample space for rolling a single die is $S = \{1, 2, 3, 4, 5, 6\}$. The elements in the event that an odd number is rolled are $E = \{1, 3, 5\}$. Then

$$P(E) = \frac{n(E)}{n(S)} = \frac{3}{6} = \frac{1}{2}$$

The probability that an odd number will be rolled is $\frac{1}{2}$.

P3. The sample space is shown in Figure 7.8 on page 319. Let E be the event that the sum of the pips on the upward faces is 7; the elements of this event are

$E = \{$ $\}$.

Then the probability of rolling a 7 is

$$P(E) = \frac{n(E)}{n(S)} = \frac{6}{36} = \frac{1}{6}$$

The probability that the sum is 7 is $\frac{1}{6}$.

P4. Let E be the event that a person between the ages of 39 and 49 is selected. Then

$$P(E) = \frac{773}{3228} \approx 0.24$$

The probability the selected person is between the ages of 39 and 49 is approximately 0.24.

P5. Make a Punnett square.

Parents	C	c
C	CC	Cc
c	Cc	cc

To be white, the offspring must be cc. From the table, only 1 of the 4 possible genotypes is cc, so the probability that an offspring will be white is $\frac{1}{4}$.

P6. Let E be the event of selecting a blue ball. Because there are 5 blue balls, there are 5 favorable outcomes, leaving 7 unfavorable outcomes.

$$\text{Odds against } E = \frac{\text{number of unfavorable outcomes}}{\text{number of favorable outcomes}} = \frac{7}{5}$$

The odds against selecting a blue ball from the box are 7 to 5.

P7. Let E represent the event of an earthquake of magnitude 6.7 or greater in the Bay Area in the next 30 years. Then the probability of E, $P(E)$, is 0.6. The odds in favor of this event are given by

$$\text{Odds in favor} = \frac{P(E)}{1 - P(E)}$$

$$= \frac{0.6}{1 - 0.6} = \frac{0.6}{0.4} = \frac{3}{2}$$

The odds in favor of this event are 3 to 2.

Section 7.4

P1. Let A be the event of rolling a 7, and let B be the event of rolling an 11. From the sample space on page 303, $P(A) = \frac{6}{36} = \frac{1}{6}$ and $P(B) = \frac{2}{36} = \frac{1}{18}$. Because A and B are mutually exclusive events,

$$P(A \text{ or } B) = P(A) + P(B) = \frac{1}{6} + \frac{1}{18} = \frac{4}{18} = \frac{2}{9}$$

The probability of rolling a 7 or an 11 is $\frac{2}{9}$.

P2. Let $A = \{\text{people with a degree in business}\}$ and $B = \{\text{people with a starting salary between \$20,000 and \$24,999}\}$. Then, from the table, $n(A) = 4 + 16 + 21 + 35 + 22 = 98$, $n(B) = 4 + 16 + 3 + 16 = 39$, and $n(A \text{ and } B) = 16$. The total number of people represented in the table is 206.

$$P(A \text{ or } B) = P(A) + P(B) - P(A \text{ and } B)$$

$$= \frac{98}{206} + \frac{39}{206} - \frac{16}{206} = \frac{121}{206} \approx 0.587$$

The probability of choosing a person who has a degree in business or a starting salary between \$20,000 and \$24,999 is about 58.7%.

P3. If E is the event that a person has type A blood, then E' is the event that the person does not have type A blood, and

$$P(E') = 1 - P(E) = 1 - 0.34 = 0.66$$

The probability that a person does not have type A blood is 66%.

P4. Let $E = \{\text{at least 1 roll of sum 7}\}$; then $E' = \{\text{no sum of 7 is rolled}\}$. Using Figure 7.8 on page 319, there are 36 possibilities for each toss of the dice. Thus $n(S) = 36 \cdot 36 \cdot 36 = 46{,}656$. For each roll of the dice, there are 30 numbers that do not total 7, so $n(E') = 30 \cdot 30 \cdot 30 = 27{,}000$. Then

$$P(E) = 1 - P(E') = 1 - \frac{27{,}000}{46{,}656} = \frac{19{,}656}{46{,}656} \approx 0.421$$

There is about a 42.1% chance of rolling a sum of 7 at least once.

P5. Let $E = \{\text{at least one \$100 bill}\}$; then $E' = \{\text{no \$100 bills}\}$. The number of elements in the sample space is the number of ways we can choose 4 bills from 35:

$$n(S) = C(35, 4) = \frac{35!}{4! (35 - 4)!} = \frac{35!}{4! \, 31!} = 52{,}360$$

To count the number of ways not to choose any $100 bills, we need to compute the number of ways we can choose 4 $1 bills from the 31 $1 bills available.

$$n(E') = C(31, 4) = \frac{31!}{4! \, (31 - 4)!} = \frac{31!}{4! \, 27!} = 31,465$$

$$P(E) = 1 - P(E') = 1 - \frac{n(E')}{n(S)}$$

$$= 1 - \frac{31,465}{52,360} = \frac{20,895}{52,360} \approx 0.399$$

The probability of pulling out at least one $100 bill is about 39.9%.

Section 7.5

P1. Let $B = \{$the sum is 6$\}$ and $A = \{$the first die is not a 3$\}$. From Figure 7.8 on page 319, there are 4 possible rolls of the dice for which the first die is not a 3 and the sum is 6. So $P(A \text{ and } B) = \frac{4}{36} = \frac{1}{9}$.

There are 30 possibilities for which the first die is not a 3, so $P(A) = \frac{30}{36} = \frac{5}{6}$. Then

$$P(B|A) = \frac{P(A \text{ and } B)}{P(A)} = \frac{\frac{1}{9}}{\frac{5}{6}} = \frac{2}{15}$$

The probability of rolling a 6 given that the first toss is not a 3 is $\frac{2}{15}$.

P2. Let $A = \{$a spade is dealt first$\}$, $B = \{$a heart is dealt second$\}$, and $C = \{$a spade is dealt third$\}$. Then

$$P(A \text{ and } B \text{ and } C) = P(A) \cdot P(B|A) \cdot P(C|A \text{ and } B)$$

$$= \frac{13}{52} \cdot \frac{13}{51} \cdot \frac{12}{50} = \frac{13}{850}$$

The probability is $\frac{13}{850}$, or about 0.015.

P3. Each coin toss is independent of the others because the probability of getting heads on any toss is not affected by the results of the other coin tosses. Let $E_1 = \{$heads on the first toss$\}$, $E_2 = \{$heads on the second toss$\}$, and $E_3 = \{$heads on the third toss$\}$. The events are independent, and the probability of flipping heads is $\frac{1}{2}$, so

$$P(E_1 \text{ and } E_2 \text{ and } E_3) = P(E_1) \cdot P(E_2) \cdot P(E_3) = \frac{1}{2} \cdot \frac{1}{2} \cdot \frac{1}{2} = \frac{1}{8}$$

P4. Let D be the event that a person has the genetic defect, and let T be the event that the test for the defect is positive. We are asked for $P(D|T)$, which can be calculated by

$$P(D|T) = \frac{P(D \text{ and } T)}{P(T)}$$

A tree diagram will help us compute the needed probabilities.

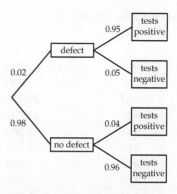

From the diagram, $P(D \text{ and } T) = P(D) \cdot P(T|D) = (0.02)(0.95)$. To compute $P(T)$, we need to combine two branches from the diagram, one corresponding to a correct positive test result when the person has the defect, and one corresponding to a false positive result when the person does not have the defect: $P(T) = (0.02)(0.95) + (0.98)(0.04)$. Then

$$P(D|T) = \frac{P(D \text{ and } T)}{P(T)} = \frac{(0.02)(0.95)}{(0.02)(0.95) + (0.98)(0.04)} \approx 0.326$$

There is only a 32.6% chance that a person who tests positive actually has the defect.

Section 7.6

P1. Let S_1 be the event that the roulette ball lands on a number from 1 to 12, in which case the player wins $10. There are 38 possible numbers, so $P(S_1) = \frac{12}{38}$.

Let S_2 be the event that a number from 1 to 12 does not come up, in which case the player loses $5. Then $P(S_2) = 1 - \frac{12}{38} = \frac{26}{38}$.

$$\text{Expectation} = P(S_1) \cdot S_1 + P(S_2) \cdot S_2$$

$$= \frac{12}{38}(10) + \frac{26}{38}(-5) = -\frac{5}{19} \approx -0.263$$

The player's expectation is about $-$0.263.

P2. Step 1 Enter the values in two columns

	A	B
	Annual Salary ($)	Probability
1	Annual Salary ($)	Probability
2	45,000	0.09
3	60,000	0.37
4	70,000	0.26
5	85,000	0.14
6	90,000	0.08
7	100,000	0.03
8	125,000	0.02
9	150,000	0.01

Step 2 Find the product of the number of goals and the probability of that number of goals for the second row.

	A	B	C
1	Annual Salary ($)	Probability	
2	45,000	0.09	=A2*B2

Step 3 Drag the formula in C2 through cell C9.

	A	B	C
1	Annual Salary ($)	Probability	Salary x Probability
2	45,000	0.09	4,050
3	60,000	0.37	22,200
4	70,000	0.26	18,200
5	85,000	0.14	11,900
6	90,000	0.08	7,200
7	100,000	0.03	3,000
8	125,000	0.02	2,500
9	150,000	0.01	1,500

Step 4 Sum the values from C2 to C9. Place the result in C10.

	A	B	C	
1	Annual Salary ($)	Probability	Salary x Probability	
2	45,000	0.09	4,050	
3	60,000	0.37	22,200	
4	70,000	0.26	18,200	
5	85,000	0.14	11,900	
6	90,000	0.08	7,200	
7	100,000	0.03	3,000	
8	125,000	0.02	2,500	
9	150,000	0.01	1,500	
10			70,550	SUM(C2:C9)

The entrepreneur's expected salary in the first year of the online business is $70,550. This is less than the current annual salary and suggests that the new business may not be a good idea.

P3. Let S_1 be the event that the person will die within 1 year. Then $P(S_1) = 0.000753$, and the company must pay out $10,000. Because the company received a premium of $45, the actual loss is $9955. Let S_2 be the event that the policy holder does not die during the year of the policy. Then $P(S_2) = 0.999247$ and the company keeps the premium. The expectation is

$$\text{Expectation} = P(S_1) \cdot S_1 + P(S_2) \cdot S_2$$
$$= 0.000753(-9955) + 0.999247(45)$$
$$= 37.47$$

The company's expectation is $37.47.

P4. Expectation $= 0.05(500,000) + 0.30(250,000) + 0.35(150,000) + 0.20(-100,000) + 0.10(-350,000)$
$= 97,500$

The company's profit expectation is $97,500.

Chapter 8

Section 8.1

P1. The four tests are a complete population. Use μ to represent the mean.

$$\mu = \frac{245 + 235 + 220 + 210}{4} = \frac{910}{4} = 227.5$$

The mean of the patient's blood cholesterol levels is 227.5.

P2. (a) The list 14, 27, 3, 82, 64, 34, 8, 51 contains 8 numbers. The median of a list of data with an even number of numbers is found by ranking the numbers and computing the mean of the two middle numbers. Ranking the numbers from smallest to largest gives 3, 8, 14, 27, 34, 51, 64, 82. The two middle numbers are 27 and 34. The mean of 27 and 34 is 30.5. Thus 30.5 is the median of the data.

(b) The list 21.3, 37.4, 11.6, 82.5, 17.2 contains 5 numbers. The median of a list of data with an odd number of numbers is found by ranking the numbers and finding the middle number. Ranking the numbers from smallest to largest gives 11.6, 17.2, 21.3, 37.4, 82.5. The middle number is 21.3. Thus 21.3 is the median.

P3. (a) In the list 3, 3, 3, 3, 3, 4, 4, 5, 5, 5, 8, the number 3 occurs more often than the other numbers. Thus 3 is the mode.

(b) In the list 12, 34, 12, 71, 48, 93, 71, the numbers 12 and 71 both occur twice and the other numbers occur only once. Thus 12 and 71 are both modes for the data.

P4. Using Excel, enter the data. Then use the AVERAGE, MEDIAN, and MODE.SNGL functions to calculate the arithmetic mean, median, and mode.

B2		f_x	=AVERAGE(A1:T1)

	A	B	C	D	E	F	G	H	I	J	K	L	M	N	O	P	Q	R	S	T
1	1		5	3	2	2	3	4	1	0	2	1	1	4	2	1	5	0	3	1	4
2	Mean	2.25																		

The arithmetic mean is 2.25 points.

B2		f_x	=MEDIAN(A1:T1)

	A	B	C	D	E	F	G	H	I	J	K	L	M	N	O	P	Q	R	S	T
1	1		5	3	2	2	3	4	1	0	2	1	1	4	2	1	5	0	3	1	4
2	Median	2																		

The median is 2 points.

B2		f_x	=MODE.SNGL(A1:T1)

	A	B	C	D	E	F	G	H	I	J	K	L	M	N	O	P	Q	R	S	T
1	1		5	3	2	2	3	4	1	0	2	1	1	4	2	1	5	0	3	1	4
2	Mode	1																		

The mode is 1 point.

P5. The A is worth 4 points, with a weight of 4; the B in Statistics is worth 3 points, with a weight of 3; the C is worth 2 points, with a weight of 3; the F is worth 0 points, with a weight of 2; and the B in CAD is worth 3 points, with a weight of 2. The sum of all the weights is $4 + 3 + 3 + 2 + 2$, or 14.

$$\text{Weighted mean} = \frac{(4 \times 4) + (3 \times 3) + (2 \times 3) + (0 \times 2) + (3 \times 2)}{14}$$
$$= \frac{37}{14} \approx 2.64$$

The student's GPA is approximately 2.64.

P6. The Excel function SUMPRODUCT can be used to calculate the weighted mean.

B7		f_x	=SUMPRODUCT(A2:A6,B2:B6)

	A	B	C	D	E	F
1	Score	Weight				
2	85	0.20				
3	98	0.25				
4	90	0.15				
5	85	0.15				
6	93	0.25				
7	Course grade	91				

Shasta's course grade is 91.

P7. weighted mean $= \dfrac{1589 \cdot 5 + 1705 \cdot 25 + 1847 \cdot 10 + 2164 \cdot 5}{5 + 25 + 10 + 5}$

$\qquad\qquad\quad = \dfrac{79{,}860}{45} \approx 1775$

Section 8.2

P1. The greatest number of ounces dispensed is 8.03, and the least is 7.95. The range of the number of ounces is $8.03 - 7.95 = 0.08$ oz.

P2. The mean for each sample of rope is 130 lb.
The rope from Trustworthy has a breaking point standard deviation of about 17.1 lb.

	B2			f_x	=STDEV.S(A1:G1)	
	A	B	C	D	E	F	G	H
1	122	141	151	114	108	149	125
2	s1=	17.08801					

The rope from Brand X has a breaking point standard deviation of about 22.6 lb.

	B2			f_x	=STDEV.S(A1:G1)	
	A	B	C	D	E	F	G	H
1	128	127	148	164	97	109	137
2	s2=	22.62742					

The rope from NeverSnap has a breaking point standard deviation of about 9.9 lb.

	B2			f_x	=STDEV.S(A1:G1)	
	A	B	C	D	E	F	G	H
1	112	121	138	131	134	139	135
2	s3=	9.93311					

The breaking point values of the rope from NeverSnap have the lowest standard deviation.

P3. The mean is approximately 71.9 lb.

	B3			f_x	=STDEV.P(A1:AD1)	
	A	B	C	D	E	F	G	H	I	J
1	71.7	72.8	70.9	75.3	71.5	76.7	67.4	74.1	71.2	75.3
2	mean	71.9								
3	σx=	2.45								

The population standard deviation is approximately 2.45.

P4. Find the mean.

$$\bar{x} = \frac{3.1 + 2.8 + 3.2 + 2.9 + 3.0}{5} = \frac{15}{5} = 3$$

Find the standard deviation.

$$\sigma = \sqrt{\frac{(3.1-3)^2 + (2.8-3)^2 + (3.2-3)^2 + (2.9-3)^2 + (3-3)^2}{5}}$$

$$= \sqrt{\frac{0.01 + 0.04 + 0.04 + 0.01 + 0}{5}} = \sqrt{\frac{0.10}{5}} = \sqrt{0.02}$$

Because the variance is the square of the standard deviation, the variance is $\left(\sqrt{0.02}\right)^2 = 0.02$

Section 8.3

P1. $z_{15} = \dfrac{15 - 12}{2.4} = 1.25 \qquad z_{14} = \dfrac{14 - 11}{2.0} = 1.5$

These z-scores indicate that in comparison to her classmates, Cameron did better on the second quiz than on the first quiz.

P2. $z_x = \dfrac{x - \mu}{s}$

$0.6 = \dfrac{70 - 65.5}{\sigma}$

$\sigma = \dfrac{4.5}{0.6} = 7.5$

The standard deviation for this set of test scores is 7.5.

P3. **(a)** By definition, the median is the 50th percentile. Therefore, 50% of the police dispatchers earned less than \$44,528 per year.
(b) Because \$32,761 is in the 25th percentile, $100\% - 25\% = 75\%$ of all police dispatchers made more than \$32,761.
(c) From parts a and b, $50\% - 25\% = 25\%$ of the police dispatchers earned between \$32,761 and \$44,528.

P4. Percentile $= \dfrac{\text{number of data values less than 405}}{\text{total number of data values}} \cdot 100$

$\qquad\quad = \dfrac{3952}{8600} \cdot 100$

$\qquad\quad \approx 46$

Hal's score of 405 places him at the 46th percentile.

P5. Using Excel, enter the data. Then use the PERCENTILE.INC function to calculate each percentile.

	A2			f_x	=PERCENTILE.INC(A1:AX1,0.9)	
	A	B	C	D	E	F	G	H	I	J	K	L
1	6.8	7.9	8.2	9.1	5.5	7.2	8.0	7.8	6.2	7.0	7.5	7.3
2	8.23											

The 90th percentile is approximately 8.2.

	A2			f_x	=PERCENTILE.INC(A1:AX1,0.5)	
	A	B	C	D	E	F	G	H	I	J	K	L
1	6.8	7.9	8.2	9.1	5.5	7.2	8.0	7.8	6.2	7.0	7.5	7.3
2	7.5											

The 50th percentile is 7.5.

	A2			f_x	=PERCENTILE.INC(A1:AX1,0.25)	
	A	B	C	D	E	F	G	H	I	J	K	L
1	6.8	7.9	8.2	9.1	5.5	7.2	8.0	7.8	6.2	7.0	7.5	7.3
2	7.03											

The 25th percentile is approximately 7.0.

P6. Rank the data.
7.5 9.8 10.2 10.8 11.4 11.4 12.2 12.4 12.6 12.8 13.1 14.2 14.5 15.6 16.4
The median of these 15 data values has a rank of 8. Thus the median is 12.4. The second quartile, Q_2, is the median of the data, so $Q_2 = 12.4$.
The first quartile is the median of the seven values less than Q_2. Thus Q_1 has a rank of 4, so $Q_1 = 10.8$.
The third quartile is the median of the values greater than Q_2. Thus Q_3 has a rank of 12, so $Q_3 = 14.2$.

P7.

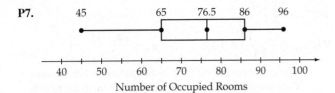

Number of Occupied Rooms

Section 8.4

P1. **(a)** The percent of data in all classes with an upper bound of 25 s or less is the sum of the percents for the first five classes in Table 8.7. Thus the percent of subscribers who required less than 25 s to download the file is 30.9%.
(b) The percent of data in all the classes with a lower bound of at least 10 s and an upper bound of 30 s or less is the sum of the percents in the third through sixth classes in Table 8.7. Thus the percent of subscribers who required from 10 s to 30 s to download the file is 47.8%. The probability that a subscriber chosen at random will require from 10 s to 30 s to download the file is 0.478.

P2. **(a)** 0.76 lb is 1 standard deviation above the mean of 0.61 lb. In a normal distribution, 34% of all data lie between the mean and 1 standard deviation above the mean, and 50% of all data lie below the mean. Thus 34% + 50% = 84% of the tomatoes weigh less than 0.76 lb.
(b) 0.31 lb is 2 standard deviations below the mean of 0.61 lb. In a normal distribution, 47.5% of all data lie between the mean and 2 standard deviations below the mean, and 50% of all data lie above the mean. This gives a total of 47.5% + 50% = 97.5% of the tomatoes that weigh more than 0.31 lb. Therefore

$$(97.5\%)(6000) = (0.975)(6000) = 5850$$

of the tomatoes can be expected to weigh more than 0.31 lb.
(c) 0.31 lb is 2 standard deviations below the mean of 0.61 lb and 0.91 lb is 2 standard deviations above the mean of 0.61 lb. In a normal distribution, 95% of all data lie within 2 standard deviations of the mean. Therefore

$$(95\%)(4500) = (0.95)(4500) = 4275$$

of the tomatoes can be expected to weigh from 0.31 lb to 0.91 lb.

P3. Using a TI-84
(a) Three standard deviations above the mean is 37.5. For an upper bound, we can use any number above 150 (the mean) + 37.5 = 187.5. We will use 200.

```
normalcdf(165,4⯈
   .1150697316
∎
```

The percent of students taking the GRE that had a score greater than 165 on the verbal reasoning portion was 11.5%.

(b) The upper bound is 155 and the lower bound is 140.

```
normalcdf(140,1⯈
   .4435663634
∎
```

The probability of a student who took the GRE and had a score greater than 140 and less than 155 is 0.4436.
(c) For a lower bound, we can use any number less than 150 − 37.5 = 112.5. We will use 110.

```
normalcdf(110,1⯈
   .7874574643
∎
```

The percent of students taking the GRE that had a score less than 160 on the verbal reasoning portion was 78.7%.

P4. **(a)** Use the Excel NORM.DIST function.

	B5		f_x	=1-NORM.DIST(B4,B2,B3,TRUE)		
	A	B	C	D	E	F	G
1	8.4 P4 a						
2	Mean	6.1					
3	Std Dev	1.8					
4	Lower Val	9					
5	Prob	0.053578					
6	Percent	5.4%					

Approximately 5.4% of professional football players have careers of more than 9 years.
(b) Use the Excel NORM.DIST function.

	B6		f_x	=NORM.DIST(B5,B2,B3,TRUE)-NORM.DIST(B4,B2,B3,TRUE)				
	A	B	C	D	E	F	G	H	I
1	8.4 P4 b								
2	Mean	6.1							
3	Std Dev	1.8							
4	Lower Val	3							
5	Upper Val	4							
6	Prob	0.079158							

The probability that a professional football player chosen at random will have a career of between 3 and 4 years is about 0.079.

Section 8.5

P1. Use a calculator or a spreadsheet program. Here is a result using Excel.

	A	B	C	D	E
	Stride length (meters)	Speed (meters/second)			
1					
2	2.5	2.3		Slope	3.164351852
3	3.0	3.9		Intercept	−5.704513889
4	3.2	4.1			
5	3.4	5.0			
6	3.5	5.5			
7	3.8	6.2			
8	4.0	7.1			
9	4.2	7.6			

The equation of the least-squares line is
$y = 3.164351852x − 5.704513889$.

P2. Use the FORECAST.LINEAR function from Excel.

	A	B	C	D	E
	Stride length (meters)	Speed (meters/second)			
1					
2	2.5	2.3		Slope	3.164351852
3	3.0	3.9		Intercept	−5.704513889
4	3.2	4.1		Forecast	5.687152778
5	3.4	5.0			
6	3.5	5.5			
7	3.8	6.2			
8	4.0	7.1			
9	4.2	7.6			

A camel with a stride length of 3.6 meters is predicted to run at 5.7 meters per second.

P3. Use the CORREL function from Excel.

	A	B	C	D	E
	Stride length (meters)	Speed (meters/second)			
1				Slope	3.164351852
2	2.5	2.3		Intercept	-5.704513889
3	3.0	3.9		Forecast	5.614473684
4	3.2	4.1		r	0.9958553395
5	3.4	5.0			
6	3.5	5.5			
7	3.8	6.2			
8	4.0	7.1			
9	4.2	7.6			

The linear correlation coefficient is 0.996.

Chapter 9

Section 9.1

P1. (a) The standard divisor is the sum of all the populations (268,730,000) divided by the number of representatives (20).

$$\text{Standard divisor} = \frac{268,730,000}{20} = 13,436,500$$

Country	Population	Quotient	Standard Quota	Number of Representatives
France	66,550,000	$\frac{66,550,000}{13,436,500} \approx 4.953$	4	5
Germany	80,850,000	$\frac{80,850,000}{13,436,500} \approx 6.017$	6	6
Italy	61,860,000	$\frac{61,860,000}{13,436,500} \approx 4.604$	4	5
Spain	48,150,000	$\frac{48,150,000}{13,436,500} \approx 3.584$	3	3
Belgium	11,320,000	$\frac{11,320,000}{13,436,500} \approx 0.842$	0	1
		Total	17	20

Because the sum of the standard quotas is 17 and not 20, we add one representative to each of the three countries with the largest decimal remainders. These are France, Belgium, and Italy. Thus the composition of the committee is France: 5, Germany: 6, Italy: 5, Spain: 3, and Belgium: 1.

(b) To use the Jefferson method, we must find a modified divisor such that the sum of the standard quotas is 20. This modified divisor is found by trial and error but is always less than or equal to the standard divisor. We are using 12,000,000 for the modified standard divisor.

Country	Population	Quotient	Number of Representatives
France	66,550,000	$\frac{66,550,000}{12,000,000} \approx 5.546$	5
Germany	80,850,000	$\frac{80,850,000}{12,000,000} \approx 6.738$	6
Italy	61,860,000	$\frac{61,860,000}{12,000,000} = 5.155$	5
Spain	48,150,000	$\frac{48,150,000}{12,000,000} \approx 4.013$	4
Belgium	11,320,000	$\frac{11,320,000}{12,000,000} \approx 0.943$	0
		Total	20

Thus the composition of the committee is France: 5, Germany: 6, Italy: 5, Spain: 4, and Belgium: 0.

(c) To use the Webster method, start with standard divisor. This does not work. Using various guesses, a modified divisor of 13,750,000 will work. Here are the calculations using an Excel spreadsheet.

Country	Population	Quota	Number of Representatives
France	66,550,000	4.84	5
Germany	80,850,000	5.88	6
Italy	61,860,000	4.498909091	4
Spain	48,150,000	3.501818182	4
Belgium	11,320,000	0.823272727	1

20

P2. $$\text{Relative unfairness of the apportionment} = \frac{\text{absolute unfairness of the apportionment}}{\text{average constituency of Shasta with a new representative}}$$

$$= \frac{210}{1390} \approx 0.151$$

The relative unfairness of the apportionment is approximately 0.151. Because the smaller relative unfairness results from adding the representative to Hampton, that state should receive the new representative.

P3. Calculate the relative unfairness of the apportionment that assigns the teacher to the first grade and the relative unfairness of the apportionment that assigns the teacher to the second grade. In this case, the average constituency is the number of students divided by the number of teachers.

	First Grade Number of Students per Teacher	Second Grade Number of Students per Teacher	Absolute Unfairness of Apportionment
First Grade Receives Teacher	$\frac{12{,}317}{512 + 1} \approx 24$	$\frac{15{,}439}{551} \approx 28$	$28 - 24 = 4$
Second Grade Receives Teacher	$\frac{12{,}317}{512} \approx 24$	$\frac{15{,}439}{551 + 1} \approx 28$	$28 - 24 = 4$

If the first grade receives the new teacher, then the relative unfairness of the apportionment is

$$\text{Relative unfairness of the apportionment} = \frac{\text{absolute unfairness of the apportionment}}{\text{first grade's average constituency with a new teacher}}$$

$$= \frac{4}{24} \approx 0.167$$

If the second grade receives the new teacher, then the relative unfairness of the apportionment is

$$\text{Relative unfairness of the apportionment} = \frac{\text{absolute unfairness of the apportionment}}{\text{second grade's average constituency with a new teacher}}$$

$$= \frac{4}{28} \approx 0.143$$

Because the smaller relative unfairness results from adding the teacher to the second grade, that class should receive the new teacher.

P4. Calculate the Huntington-Hill number for each of the classes. In this case, the population is the number of students.

First year:

$$\frac{2015^2}{12(12 + 1)} \approx 26{,}027$$

Second year:

$$\frac{1755^2}{10(10 + 1)} \approx 28{,}000$$

Third year:

$$\frac{1430^2}{9(9 + 1)} \approx 22{,}721$$

Fourth year:

$$\frac{1309^2}{8(8 + 1)} \approx 23{,}798$$

Because the second-year class has the greatest Huntington-Hill number, the new representative should represent the second-year class.

Section 9.2

P1. To answer the question, we will make a table showing the number of second-place votes for each candy.

	Second-Place Votes
Caramel Center	3
Vanilla Center	0
Almond Center	17 + 9 = 26
Toffee Center	2
Solid Chocolate	11 + 8 = 19

The largest number of second-place votes (26) was for almond centers. Almond centers would win second place using the plurality voting system.

P2. Using the Borda count method, each first-place vote receives 5 points, each second-place vote receives 4 points, each third-place vote receives 3 points, each fourth-place vote receives 2 points, and each last-place vote receives 1 point. The summaries for the five varieties are as follows.

Caramel:

0 first-place votes	$0 \cdot 5 = 0$
3 second-place votes	$3 \cdot 4 = 12$
0 third-place votes	$0 \cdot 3 = 0$
30 fourth-place votes	$30 \cdot 2 = 60$
17 fifth-place votes	$17 \cdot 1 = 17$
	Total 89

Vanilla:

17 first-place votes	$17 \cdot 5 = 85$
0 second-place votes	$0 \cdot 4 = 0$
0 third-place votes	$0 \cdot 3 = 0$
0 fourth-place votes	$0 \cdot 2 = 0$
33 fifth-place votes	$33 \cdot 1 = 33$
	Total 118

Almond:

8 first-place votes	$8 \cdot 5 = 40$
26 second-place votes	$26 \cdot 4 = 104$
16 third-place votes	$16 \cdot 3 = 48$
0 fourth-place votes	$0 \cdot 2 = 0$
0 fifth-place votes	$0 \cdot 1 = 0$
	Total 192

Toffee:

20 first-place votes	$20 \cdot 5 = 100$
2 second-place votes	$2 \cdot 4 = 8$
8 third-place votes	$8 \cdot 3 = 24$
20 fourth-place votes	$20 \cdot 2 = 40$
0 fifth-place votes	$0 \cdot 1 = 0$
	Total 172

Chocolate:

5 first-place votes	$5 \cdot 5 = 25$
19 second-place votes	$19 \cdot 4 = 76$
26 third-place votes	$26 \cdot 3 = 78$
0 fourth-place votes	$0 \cdot 2 = 0$
0 fifth-place votes	$0 \cdot 1 = 0$
	Total 179

Using the Borda count method, almond centers is the first choice.

P3.

	Rankings				
Italian	2	5	1	4	3
Mexican	1	4	5	2	1
Thai	3	1	4	5	2
Chinese	4	2	3	1	4
Indian	5	3	2	3	5
Number of ballots:	33	30	25	20	18

Indian food received no first place votes, so it is eliminated.

	Rankings				
Italian	2	4	1	3	3
Mexican	1	3	4	2	1
Thai	3	1	3	4	2
Chinese	4	2	2	1	4
Number of ballots:	33	30	25	20	18

In this ranking, Chinese food received the fewest first-place votes, so it is eliminated.

	Rankings				
Italian	2	3	1	2	3
Mexican	1	2	3	1	1
Thai	3	1	2	3	2
Number of ballots:	33	30	25	20	18

In this ranking, Italian food received the fewest first-place votes, so it is eliminated.

	Rankings				
Mexican	1	2	2	1	1
Thai	2	1	1	2	2
Number of ballots:	33	30	25	20	18

In this ranking, Thai food received the fewest first-place votes, so it is eliminated. The preference for the banquet food is Mexican.

P4. Do a head-to-head comparison for each of the restaurants and enter each winner in the following table. For instance, in the Sanborn's versus Apple Inn comparison, Sanborn's was favored by $31 + 25 + 11 = 67$ critics. In the Apple Inn versus Sanborn's comparison, Apple Inn was favored by $18 + 15 = 33$ critics. Therefore, Sanborn's wins this head-to-head match. The completed table is shown.

Versus	Sanborn's	Apple Inn	May's	Tory's
Sanborn's		Sanborn's	May's	Sanborn's
Apple Inn			May's	Tory's
May's				May's
Tory's				

From the table, May's has the most points, so it is the critics' choice.

P5. Do a head-to-head comparison for each of the candidates and enter each winner in the table below.

Versus	Alpha	Beta	Gamma
Alpha		Alpha	Alpha
Beta			Beta
Gamma			

From this table, Alpha is the winner. However, using the Borda count method (see Example 5), Beta is the winner. Thus the Borda count method violates the Condorcet criterion.

P6.

	Rankings		
Radiant Silver	1	3	3
Electric Red	2	2	1
Lightning Blue	3	1	2
Number of votes:	30	27	2

Using the Borda count method, we have

Silver:

30 first-place votes	$30 \cdot 3 =$	90
0 second-place votes	$0 \cdot 2 =$	0
29 third-place votes	$29 \cdot 1 =$	29
	Total	119

Red:

2 first-place votes	$2 \cdot 3 =$	6
57 second-place votes	$57 \cdot 2 =$	114
0 third-place votes	$0 \cdot 1 =$	0
	Total	120

Blue:

27 first-place votes	$27 \cdot 3 =$	81
2 second-place votes	$2 \cdot 2 =$	4
30 third-place votes	$30 \cdot 1 =$	30
	Total	115

Using this method, electric red is the preferred color. Now suppose we eliminate the third-place choice (lightning blue). This gives the following table.

	Rankings		
Radiant Silver	1	2	2
Electric Red	2	1	1
Number of votes:	30	27	2

Recalculating the results, we have

Silver:

30 first-place votes	$30 \cdot 2 = 60$
29 second-place votes	$29 \cdot 1 = 29$
	Total 89

Red:

29 first-place votes	$29 \cdot 2 = 58$
30 second-place votes	$30 \cdot 1 = 30$
	Total 88

Now radiant silver is the preferred color. By deleting an alternative, the result of the voting changed. This violates the irrelevant alternatives criterion.

Section 9.3

P1. (a) and (b)

Winning Coalition	Number of Votes	Critical Voters
{A, B}	40	A, B
{A, C}	39	A, C
{A, B, C}	57	A
{A, B, D}	50	A, B
{A, B, E}	45	A, B
{A, C, D}	49	A, C
{A, C, E}	44	A, C
{A, D, E}	37	A, D, E
{B, C, D}	45	B, C, D
{B, C, E}	40	B, C, E
{A, B, C, D}	67	None
{A, B, C, E}	62	None
{A, B, D, E}	55	A
{A, C, D, E}	54	A
{B, C, D, E}	50	B, C
{A, B, C, D, E}	72	None

P2.

Winning Coalition	Number of Votes	Critical Voters
{A, B}	34	A, B
{A, C}	28	A, C
{B, C}	26	B, C
{A, B, C}	44	None
{A, B, D}	40	A, B
{A, C, D}	34	A, C
{B, C, D}	32	B, C
{A, B, C, D}	50	None

The number of times any voter is critical is 12.

$$BPI(A) = \frac{4}{12} = \frac{1}{3}$$

$$BPI(D) = \frac{0}{12} = 0$$

P3. The countries are represented as follows: B, Belgium; F, France; G, Germany; I, Italy; L, Luxembourg; and N, Netherlands.

Winning Coalition	Number of Votes	Critical Voters
{F, G, I}	12	F, G, I
{B, F, G, I}	14	F, G, I
{B, F, G, I, L}	15	F, G, I
{B, F, G, I, N}	16	None
{B, F, G, I, L, N}	17	None
{B, F, G, N}	12	B, F, G, N
{B, F, I, N}	12	B, F, I, N
{B, G, I, N}	12	B, G, I, N
{B, G, I, N, L}	13	B, G, I, N
{B, F, I, N, L}	13	B, F, I, N
{B, F, G, N, L}	13	B, F, G, N
{F, G, I, L}	13	F, G, I
{F, G, I, L, N}	15	F, G, I
{F, G, I, N}	14	F, G, I

The number of times all votes are critical is 42. The BPIs of the nations are:

$$BPI(B) = \frac{6}{42} = \frac{1}{7}$$

$$BPI(F) = \frac{10}{42} = \frac{5}{21}$$

$$BPI(G) = \frac{10}{42} = \frac{5}{21}$$

$$BPI(I) = \frac{10}{42} = \frac{5}{21}$$

$$BPI(L) = \frac{0}{42} = 0$$

$$BPI(N) = \frac{6}{42} = \frac{1}{7}$$

Chapter 10

Section 10.1

P1.

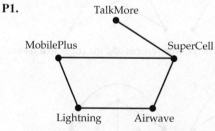

The vertex corresponding to SuperCell is connected to more edges than the others, so SuperCell has roaming agreements with the most carriers. TalkMore can roam with only one network because the corresponding vertex is connected to only one edge.

P2. Because the second graph has edge AB and the first graph does not, the two graphs are not equivalent.

P3. Vertices B, C, E, and G are of odd degree. By the Eulerian graph theorem, the graph does not have an Euler circuit.

P4. One vertex in the graph is of degree 3, and another is of degree 5. Because not all vertices are of even degree, the graph is not Eulerian.

P5. Represent the land areas and bridges with a graph, as we did for the Königsberg bridges earlier in the section. The vertices of the resulting graph, shown in the second figure, all have even degree. Thus we know that the graph has an Euler circuit. An Euler circuit corresponds to a stroll that crosses each bridge and returns to the starting point without crossing any bridge twice.

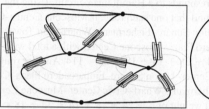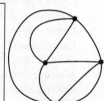

P6. Consider the campground map as a graph. A route through all the trails that does not repeat any trails corresponds to an Euler path. Because only two vertices (A and F) are of odd degree, we know that an Euler path exists. Furthermore, the path must begin at A and end at F or begin at F and end at A. By trial and error, one Euler path is A–B–C–D–E–B–G–F–E–C–A–F.

P7. Represent the floor plan with a graph, as in Example 7.

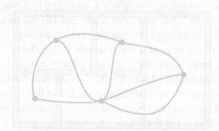

A stroll passing through each doorway just once corresponds to an Euler circuit or path. Because four vertices are of odd degree, no Euler circuit or path exists, so it is not possible to take such a stroll.

Section 10.2

P1. The graph has seven vertices, so $n = 7$ and $n/2 = 3.5$. Several vertices are of degree less than $n/2$, so Dirac's theorem does not apply. Still, a routing for the document may be possible. By trial and error, one such route is Los Angeles–New York–Boston–Atlanta–Dallas–Phoenix–San Francisco–Los Angeles.

P2. Draw a graph in which the vertices represent locations and the edges indicate available bus routes between locations. Each edge should be given a weight corresponding to the number of minutes for the bus ride.

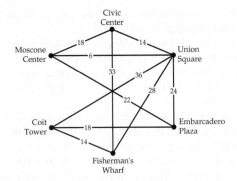

A route that visits each location and returns to the Moscone Center corresponds to a Hamiltonian circuit. Using the graph, we find that one such route is Moscone Center–Civic Center–Union Square–Fisherman's Wharf–Coit Tower–Embarcadero Plaza–Moscone Center, with a total weight of $18 + 14 + 28 + 14 + 18 + 22 = 114$. Another route is Moscone Center–Union Square–Embarcadero Plaza–Coit Tower–Fisherman's Wharf–Civic Center–Moscone Center, with a total weight of $6 + 24 + 18 + 14 + 33 + 18 = 113$. The travel time is one minute less for the second route.

P3. Starting at vertex A, the edge of smallest weight is the edge to D, with weight 5. From D, take the edge of weight 4 to C, and then the edge of weight 3 to B. From B, the edge of least weight to a vertex not yet visited is the edge to vertex E (with weight 5). This is the last vertex, so we return to A along the edge of weight 9. Thus the Hamiltonian circuit is A–D–C–B–E–A, with a total weight of 26.

P4. The smallest weight appearing in the graph is 3, so we mark edge BC. The next smallest weight is 4, on edge CD. Three edges have weight 5, but we cannot mark edge BD, because it would complete a circuit. We can, however, mark edge AD. The next valid edge of smallest weight is BE, also of weight 5. No more edges can be marked without completing a circuit or adding a third edge to a vertex, so we mark the final edge, AE, to complete the Hamiltonian circuit. In this case, the edge-picking algorithm generated the same circuit as the greedy algorithm did in Preliminary Exercise P3.

P5. Represent the time between locations with a weighted graph.

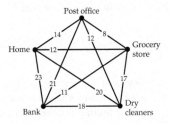

Starting at the home vertex and using the greedy algorithm, we first use the edge to the grocery store (of weight 12) followed by the edge of weight 8 to the post office and then the edge of weight 12 to the dry cleaners. The edge of next smallest weight is to the grocery store, but that vertex has already been visited, so we take the edge to the bank, with weight 18. All vertices have now been visited, so we select the last edge, of weight 23, to return home. The total weight is 73, corresponding to a total driving time of 73 minutes.

For the edge-picking algorithm, we first select the edge of weight 8, followed by the edge of weight 11. Two edges have weight 12, but one adds a third edge to the grocery store vertex, so we must choose the edge from the post office to the dry cleaners. The next smallest weight is 14, but that edge would add a third edge to a vertex, as would the edge of weight 17. The edge of weight 18 would complete a circuit too early, so the next edge we can select is that of weight 20, the edge from home to the dry cleaners. The final step is to select the edge from home to the bank to complete the circuit. The resulting route is home–dry cleaners–post office–grocery store–bank–home (we could travel the same route in the reverse order) with a total travel time of 74 minutes.

P6. Represent the computer network by a graph in which the weights of the edges indicate the distances between computers.

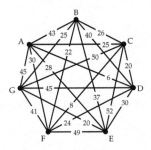

The edges with the smallest weights, which can all be chosen, are those of weights 6, 8, 20, 20, and 22. The edge of next smallest weight, 24, cannot be selected. There are two edges of weight 25; edge AC would add a third edge to vertex C, but edge BG can be chosen. All that remains is to complete the circuit with edge AE. The computers should be networked in this order: A, D, C, F, B, G, E, and back to A.

Section 10.3

P1. First redraw the highlighted edge as shown.

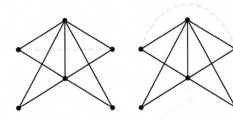

Now redraw the two lower vertices and the edges that meet there, as shown.

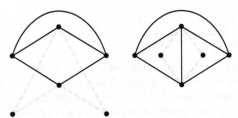

The result is a graph with no intersecting edges. Therefore, the graph is planar.

P2. The highlighted edges in the graph, considered as a subgraph, form the graph K_5. (It is upside down and slightly distorted compared with the version shown in Figure 10.21.)

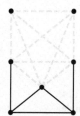

P3. First contract the highlighted edge and combine the multiple edges as shown.

Then contract the highlighted edge in the center of the graph.

P4. The graph looks similar to the Utilities Graph. Contract edges as shown, and combine the resulting multiple edges.

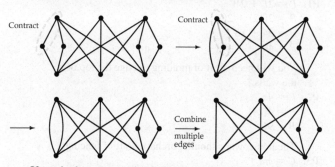

If we do the same on the right side of the graph, we are left with the Utilities Graph.

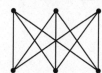

Therefore, the graph is not planar.

P5. There are 11 edges in the graph, seven vertices, and six faces (including the infinite face). Then $v + f = 7 + 6 = 13$ and $e + 2 = 11 + 2 = 13$, so $v + f = e + 2$.

Section 10.4

P1. Draw a graph on the map as in Example 1. More than two colors are required to color the resulting graph, but, by experimenting, the graph can be colored with three colors. Thus the graph is 3-colorable.

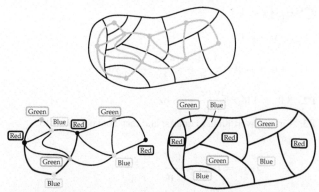

P2. There are several locations in the graph at which three edges form a triangle. Because a triangle is a circuit with an odd number of vertices, the graph is not 2-colorable.

P3. Draw a graph in which each vertex corresponds to a film and an edge joins two vertices if one person needs to attend both of the corresponding films. We can use colors to represent the different times at which the films can be viewed. No two vertices connected by an edge can share the same color, because that would mean one person would have to attend two films at the same time.

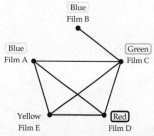

It is not possible to color the vertices with only three colors; one possible 4-coloring is shown. This means that four different time slots will be required to show the films, and the earliest that the festival can end is 8:00 P.M. A schedule can be set using the coloring in the graph. From 12 to 2, the films labeled blue, film A and film B, can be shown in two different rooms. The remaining films are represented by unique colors and so will require their own viewing times. Film C can be shown from 2 to 4, film D from 4 to 6, and film E from 6 to 8.

P4. Draw a graph in which each vertex represents a deli and an edge connects two vertices if the corresponding delis deliver to a common building. Try to color the vertices using the fewest number of colors possible; each color can correspond to a day of the week that the delis can deliver.

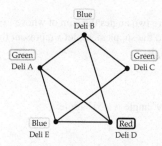

As shown, a 3-coloring is possible (but a 2-coloring is not). Therefore, three different delivery days will be necessary—delis A and C deliver on one day, delis B and E on another day, and deli D on a third day.

Chapter 11

Section 11.1

P1. $14 \text{ ft} = 14 \cancel{\text{ft}} \times \dfrac{1 \text{ yd}}{3 \cancel{\text{ft}}} = \dfrac{14 \text{ yd}}{3} = 4\dfrac{2}{3} \text{ yd}$

P2. $3 \text{ lb} = 3 \cancel{\text{lb}} \times \dfrac{16 \text{ oz}}{1 \cancel{\text{lb}}} = 48 \text{ oz}$

P3. First convert pints to quarts, and then convert quarts to gallons.

$$18 \text{ pt} = 18 \cancel{\text{pt}} \times \dfrac{1 \cancel{\text{qt}}}{2 \cancel{\text{pt}}} \times \dfrac{1 \text{ gal}}{4 \cancel{\text{qt}}}$$

$$= \dfrac{18 \text{ gal}}{8} = 2\dfrac{1}{4} \text{ gal}$$

P4. $3.07 \text{ m} = 307 \text{ cm}$

P5. $42.3 \text{ mg} = 0.0423 \text{ g}$

P6. $2 \text{ kl} = 2000 \text{ L}$

$2 \text{ kl } 167 \text{ L} = 2000 \text{ L} + 167 \text{ L}$
$\phantom{2 \text{ kl } 167 \text{ L}} = 2167 \text{ L}$

P7. $\dfrac{\$3.69}{\text{gal}} \approx \dfrac{\$3.69}{\cancel{\text{gal}}} \times \dfrac{1 \cancel{\text{gal}}}{3.79 \text{ L}}$

$\phantom{\dfrac{\$3.69}{\text{gal}}} = \dfrac{\$3.69}{3.79 \text{ L}} \approx \dfrac{\$.97}{\text{L}}$

$\$3.69/\text{gal} \approx \$0.97/\text{L}$

P8. $45 \text{ cm} = \dfrac{45 \cancel{\text{cm}}}{1} \times \dfrac{1 \text{ in.}}{2.54 \cancel{\text{cm}}}$

$\phantom{45 \text{ cm}} = \dfrac{45 \text{ in.}}{2.54} \approx 17.72 \text{ in.}$

$45 \text{ cm} \approx 17.72 \text{ in.}$

P9. $\dfrac{75 \text{ km}}{\text{hr}} \approx \dfrac{75 \cancel{\text{km}}}{\text{hr}} \times \dfrac{1 \text{ mi}}{1.61 \cancel{\text{km}}}$

$\phantom{\dfrac{75 \text{ km}}{\text{hr}}} = \dfrac{75 \text{ mi}}{1.61 \text{ hr}} \approx 46.58 \text{ mi/hr}$

$75 \text{ km/hr} \approx 46.58 \text{ mi/hr}$

Section 11.2

P1. $AB + BC = AC$

$\dfrac{1}{4}(BC) + BC = AC$

$\dfrac{1}{4}(16) + 16 = AC$

$4 + 16 = AC$

$20 = AC$

$AC = 20 \text{ ft}$

P2. Supplementary angles are two angles the sum of whose measures is $180°$. To find the supplement, let x represent the supplement of a $129°$ angle.

$x + 129 = 180$
$x = 51$

The supplement of a $129°$ angle is a $51°$ angle.

P3. $m\angle a + 68° = 118°$
$ m\angle a = 50°$

P4. $m\angle b + m\angle a = 180°$
$ m\angle b + 35° = 180°$
$ m\angle b = 145°$

$m\angle c = m\angle a = 35°$
$m\angle d = m\angle b = 145°$
$m\angle b = 145°, \ m\angle c = 35°, \text{ and } m\angle d = 145°.$

P5. $m\angle b = m\angle g = 124°$
$m\angle d = m\angle g = 124°$
$m\angle c + m\angle b = 180°$
$ m\angle c + 124° = 180°$
$ m\angle c = 56°$

$m\angle b = 124°, m\angle c = 56°, \text{ and } m\angle d = 124°.$

P6. Let x represent the measure of the third angle.

$x + 90° + 27° = 180°$
$ x + 117° = 180°$
$ x = 63°$

The measure of the third angle is $63°$.

P7. $m\angle b + m\angle d = 180°$
$m\angle b + 105° = 180°$
$ m\angle b = 75°$

$m\angle a + m\angle b + m\angle c = 180°$
$m\angle a + 75° + 35° = 180°$
$ m\angle a + 110° = 180°$
$ m\angle a = 70°$

$m\angle e = m\angle a = 70°$

Section 11.3

P1. $P = 2L + 2W$
$P = 2(12) + 2(8)$
$P = 24 + 16$
$P = 40$

You will need 40 ft of molding to edge the top of the walls.

P2. $P = 4s$
$P = 4(24)$
$P = 96$

The homeowner should purchase 96 ft of fencing.

P3. $C = \pi d$
$C = 9\pi$

The circumference of the circle is 9π km.

P4. $12 \text{ in.} = 1 \text{ ft}$

$C = \pi d$
$C = \pi(1)$
$C = \pi$

$12C = 12\pi \approx 37.70$

The tricycle travels approximately 37.70 ft when the wheel makes 12 revolutions.

P5. $A = LW$
$A = 308(192)$
$A = 59,136$

$59,136 \text{ cm}^2$ of fabric is needed.

P6. $A = s^2$
$A = 24^2$
$A = 576$

The area of the floor is 576 ft^2.

P7. $A = bh$
$A = 14(8)$
$A = 112$

The area of the patio is 112 m^2.

P8. $A = \dfrac{1}{2}bh$

$A = \dfrac{1}{2}(18)(9)$

$A = 9(9)$

$A = 81$

81 in^2 of felt is needed.

P9. $A = \dfrac{1}{2}h(b_1 + b_2)$

$A = \dfrac{1}{2} \cdot 9(12 + 20)$

$A = \dfrac{1}{2} \cdot 9(32)$

$A = \dfrac{9}{2} \cdot (32)$

$A = 144$

The area of the patio is 144 ft^2.

P10. $r = \dfrac{1}{2}d = \dfrac{1}{2}(12) = 6$

$A = \pi r^2$

$A = \pi(6)^2$

$A = 36\pi$

The area of the circle is 36π km^2.

P11. $r = \dfrac{1}{2}d = \dfrac{1}{2}(4) = 2$

$A = \pi r^2$

$A = \pi(2)^2$

$A = \pi(4)$

$A \approx 12.57$

Approximately 12.57 ft^2 of material is needed.

Section 11.4

P1. $V = LWH$
$V = 5(3.2)(4)$
$V = 64$

The volume of the solid is 64 m^3.

P2. $V = \dfrac{1}{3}s^2h$

$V = \dfrac{1}{3}(15)^2(25)$

$V = \dfrac{1}{3}(225)(25)$

$V = 1875$

The volume of the pyramid is 1875 m^3.

P3. $r = \dfrac{1}{2}d = \dfrac{1}{2}(16) = 8$

$V = \pi r^2 h$

$V = \pi(8)^2(30)$

$V = \pi(64)(30)$

$V = 1920\pi$

$\dfrac{1}{4}(1920\pi) = 480\pi$

≈ 1507.96

Approximately 1507.96 ft^3 are not being used for storage.

P4. $r = \dfrac{1}{2}d = \dfrac{1}{2}(6) = 3$

$S = 2\pi r^2 + 2\pi rh$
$S = 2\pi(3)^2 + 2\pi(3)(8)$
$S = 2\pi(9) + 2\pi(3)(8)$
$S = 18\pi + 48\pi$
$S = 66\pi$
$S \approx 207.35$

The surface area of the cylinder is approximately 207.35 ft^2.

Section 11.5

P1. $\dfrac{AC}{DF} = \dfrac{CH}{FG}$

$\dfrac{10}{15} = \dfrac{7}{FG}$

$10(FG) = (15)7$
$10(FG) = 105$
$FG = 10.5$

The height FG of triangle DEF is 10.5 m.

P2. $\angle A$ and $\angle D$ are right angles. Therefore, $\angle A = \angle D$. $\angle AOB$ and $\angle COD$ are vertical angles. Therefore, $\angle AOB = \angle COD$. Because two angles of triangle AOB are equal in measure to two angles of triangle DOC, triangles AOB and DOC are similar triangles.

$\dfrac{AO}{DO} = \dfrac{AB}{DC}$

$\dfrac{AO}{3} = \dfrac{10}{4}$

$4(AO) = 3(10)$
$4(AO) = 30$
$AO = 7.5$

$A = \dfrac{1}{2}bh$

$A = \dfrac{1}{2}(10)(7.5)$

$A = 5(7.5)$

$A = 37.5$

The area of triangle AOB is 37.5 cm^2.

P3. Because two sides and the included angle of one triangle are equal in measure to two sides and the included angle of the second triangle, the triangles are congruent by the SAS theorem.

P4. $a^2 + b^2 = c^2$ • Use the Pythagorean theorem.
$2^2 + b^2 = 6^2$ • $a = 2, c = 6$
$4 + b^2 = 36$
$b^2 = 32$ • Solve for b^2. Subtract 4 from each side.
$\sqrt{b^2} = \sqrt{32}$ • Take the square root of each side of the equation.
$b \approx 5.66$ • Use a calculator to approximate $\sqrt{32}$.

The length of the other leg is approximately 5.66 m.

Section 11.6

P1. Use the Pythagorean theorem to find the length of the hypotenuse.

$a^2 + b^2 = c^2$ $25 = c^2$
$3^2 + 4^2 = c^2$ $\sqrt{25} = \sqrt{c^2}$
$9 + 16 = c^2$ $5 = c$

$$\sin \theta = \frac{\text{opp}}{\text{hyp}} = \frac{3}{5},$$

$$\cos \theta = \frac{\text{adj}}{\text{hyp}} = \frac{4}{5},$$

$$\tan \theta = \frac{\text{opp}}{\text{adj}} = \frac{3}{4}$$

P2. We are given the measure of $\angle B$ and the hypotenuse. We want to find the length of side a. The cosine function involves the side adjacent and the hypotenuse.

$$\cos B = \frac{\text{adj}}{\text{hyp}}$$

$$\cos 48° = \frac{a}{12}$$

$$12(\cos 48°) = a$$

$$8.0 \approx a$$

The length of side a is approximately 8.0 ft.

P3. $\tan^{-1}(0.3165) \approx 17.6°$

P4. $\theta \approx \tan^{-1}(0.5681)$
$\theta \approx 29.6°$

P5. We want to find the measure of $\angle A$, and we are given the length of the side opposite $\angle A$ and the hypotenuse. The sine function involves the side opposite an angle and the hypotenuse.

$$\sin A = \frac{\text{opp}}{\text{hyp}}$$

$$\sin A = \frac{7}{11}$$

$$A = \sin^{-1}\frac{7}{11}$$

$$A \approx 39.5°$$

The measure of $\angle A$ is approximately 39.5°.

P6. Let d be the distance from the base of the lighthouse to the boat.

$$\tan 25° = \frac{20}{d}$$

$$d(\tan 25°) = 20$$

$$d = \frac{20}{\tan 25°}$$

$$d \approx 42.9$$

The boat is approximately 42.9 m from the base of the lighthouse.

Section 11.7

P1. $S = (m\angle A + m\angle B + m\angle C - 180°)\left(\dfrac{\pi}{180°}\right)r^2$

$$= (200° + 90° + 90° - 180°)\left(\frac{\pi}{180°}\right)(6)^2$$

$$= (200°)\left(\frac{\pi}{180°}\right)(36)$$

$$= 40\pi \text{ in}^2 \qquad \bullet \text{ Exact area}$$

$$\approx 125.66 \text{ in}^2 \qquad \bullet \text{ Approximate area}$$

P2. In Example 2, we observed that only Riemannian geometry has the property that there exist no lines parallel to a given line. Thus, in this exercise, the type of geometry must be Riemannian.

P3. (a) $d_E(P, Q) = \sqrt{(x_2 - x_1)^2 + (y_2 - y_1)^2}$
$$= \sqrt{[3 - (-1)]^2 + [2 - 4]^2}$$
$$= \sqrt{4^2 + (-2)^2}$$
$$= \sqrt{20} \approx 4.5 \text{ blocks}$$

$d_C(P, Q) = |x_2 - x_1| + |y_2 - y_1|$
$$= |3 - (-1)| + |2 - 4|$$
$$= |4| + |-2|$$
$$= 4 + 2$$
$$= 6 \text{ blocks}$$

(b) $d_E(P, Q) = \sqrt{(x_2 - x_1)^2 + (y_2 - y_1)^2}$
$$= \sqrt{[(-1) - 3]^2 + [5 - (-4)]^2}$$
$$= \sqrt{(-4)^2 + 9^2}$$
$$= \sqrt{97} \approx 9.8 \text{ blocks}$$

$d_C(P, Q) = |x_2 - x_1| + |y_2 - y_1|$
$$= |(-1) - 3| + |5 - (-4)|$$
$$= |-4| + 9$$
$$= 4 + 9$$
$$= 13 \text{ blocks}$$

Section 11.8

P1. Replace each line segment with a scaled version of the generator. As you move from left to right, your first zig should be to the left.

Stage 2 of the zig-zag curve

P2. Replace each square with a scaled version of the generator.

Stage 2 of the Sierpinski carpet

P3. (a) Any portion of the box curve replicates the entire fractal, so the box curve is a strictly self-similar fractal.

(b) Any portion of the Sierpinski gasket replicates the entire fractal, so the Sierpinski gasket is a strictly self-similar fractal.

P4. (a) The generator of the Koch curve consists of four line segments, and the initiator consists of only one line segment. Thus the replacement ratio of the Koch curve is $4:1$, or 4. The initiator of the Koch curve is a line segment that is 3 times as long as the replica line segments in the generator. Thus the scaling ratio of the Koch curve is $3:1$, or 3.

(b) The generator of the zig-zag curve consists of six line segments, and the initiator consists of only one line segment. Thus the replacement ratio of the zig-zag curve is $6:1$, or 6. The initiator of the zig-zag curve is a line segment that is 4 times as long as the replica line segments in the generator. Thus the scaling ratio of the zig-zag curve is $4:1$, or 4.

P5. (a) In Example 4, we determined that the replacement ratio of the box curve is 5 and the scaling ratio of the box curve is 3. Thus the similarity dimension of the box curve is
$$D = \frac{\log 5}{\log 3} \approx 1.465.$$

(b) The replacement ratio of the Sierpinski carpet is 8, and the scaling ratio of the Sierpinski carpet is 3. Thus the similarity dimension of the Sierpinski carpet is
$$D = \frac{\log 8}{\log 3} \approx 1.893.$$

Index of Applications

Index